PATTERN RECOGNITION IN COMPUTATIONAL MOLECULAR BIOLOGY

Wiley Series on

Bioinformatics: Computational Techniques and Engineering

A complete list of the titles in this series appears at the end of this volume.

PATTERN RECOGNITION IN COMPUTATIONAL MOLECULAR BIOLOGY

Techniques and Approaches

Edited by

Mourad Elloumi

Laboratory of Technologies of Information and Communication and Electrical Engineering (LaTICE), and University of Tunis-El Manar, Tunisia

Costas S. Iliopoulos

King's College London, UK

Jason T. L. Wang

New Jersey Institute of Technology, USA

Albert Y. Zomaya

The University of Sydney, Australia

WILEY

Published by John Wiley & Sons, Inc., Hoboken, New Jersey
Published simultaneously in Canada

Library of Congress Cataloging-in-Publication Data:

Pattern recognition in computational molecular biology : techniques and approaches / edited by Mourad Elloumi, Costas S. Iliopoulos, Jason T. L. Wang, Albert Y. Zomaya.
pages cm
Includes bibliographical references and index.
ISBN 978-1-118-89368-5 (cloth)
1. Molecular biology–Data processing. 2. Computational biology. 3. Pattern recognition systems. I. Elloumi, Mourad.
QH506.P38 2016
572′.8–dc23

2015029791

Cover image courtesy of iStock/Henrik5000

Typeset in 10/12pt, TimesLTStd by SPi Global, Chennai, India.

Printed in the United States of America

10 9 8 7 6 5 4 3 2 1

1 2016

CONTENTS

LIST OF CONTRIBUTORS

Andrej Aderhold, School of Biology, University of St Andrews, St Andrews, UK

Lefteris Angelis, Department of Informatics, Aristotle University of Thessaloniki, Thessaloniki, Greece

Miguel Arenas, Bioinformatics Unit, Centre for Molecular Biology "Severo Ochoa" (CSIC), Madrid, Spain; Institute of Molecular Pathology and Immunology, University of Porto (IPATIMUP), Porto, Portugal

Dunarel Badescu, Département d'informatique, Université du Québec à Montréal, Succ. Centre-Ville, Montréal, Québec, Canada

Carl Barton, The Blizard Institute, Barts and The London School of Medicine and Dentistry, Queen Mary University of London, London, UK

Sima Behpour, Department of Computer Science, University of Illinois at Chicago, Chicago, IL, USA

Paola Bertolazzi, Institute of Systems Analysis and Computer Science "A. Ruberti", National Research Council, Rome, Italy

Chengpeng Bi, Bioinformatics and Intelligent Computing Lab, Division of Clinical Pharmacology, Children's Mercy Hospitals, Kansas City, MO, USA

Kevin Byron, Department of Computer Science, New Jersey Institute of Technology, Newark, NJ, USA

Weiwen Cai, Institute of Applied Genomics, College of Biological Science and Engineering, Fuzhou University, Fuzhou, China

Virginio Cantoni, Department of Electrical, Computer and Biomedical Engineering, University of Pavia, Via Ferrata, Pavia, Italy

Kuo-Chen Chou, Gordon Life Science Institute, Belmont, MA, USA; King Abdulaziz University, Jeddah, Saudi Arabia

Matteo Comin, Department of Information Engineering, University of Padova, Padova, Italy

David Dao, Karlsruhe Institute of Technology, Institute for Theoretical Informatics, Postfach, Karlsruhe, Germany

Bhaskar DasGupta, Department of Computer Science, University of Illinois at Chicago, Chicago, IL, USA

Wajdi Dhifli, Clermont Université, Université Blaise Pascal, LIMOS, BP, Clermont-Ferrand, France; CNRS, UMR, LIMOS, Aubiére, France; Department of Computer Science, University of Quebec At Montreal, Downtown station, Montreal (Quebec) Canada

Carlotta Domeniconi, Department of Computer Science, George Mason University, Fairfax, VA, USA

Maryam Faridounnia, Bijvoet Center for Biomolecular Research, Utrecht University, Utrecht, The Netherlands

Giovanni Felici, Institute of Systems Analysis and Computer Science "A. Ruberti", National Research Council, Rome, Italy

Marco Ferretti, Department of Electrical, Computer and Biomedical Engineering, University of Pavia, Via Ferrata, Pavia, Italy

Giulia Fiscon, Institute of Systems Analysis and Computer Science "A. Ruberti", National Research Council, Rome, Italy; Department of Computer, Control and Management Engineering, Sapienza University, Rome, Italy

Tomáš Flouri, Heidelberg Institute for Theoretical Studies, Heidelberg, Germany

Adelaide Freitas, Department of Mathematics, and Center for Research & Development in Mathematics and Applications (CIDMA), University of Aveiro, Aveiro, Portugal

Valentina Fustaino, Cellular Biology and Neurobiology Institute, National Research Council, Rome, Italy; Institute of Systems Analysis and Computer Science "A. Ruberti", National Research Council, Rome, Italy

Yann Guermeur, LORIA, Université de Lorraine-CNRS, Nancy, France

Michael Hall, Department of Computing Science, University of Alberta, Edmonton, Canada

Robert Harrison, Department of Computer Science, Georgia State University, Atlanta, GA, USA

Dirk Husmeier, School of Mathematics and Statistics, University of Glasgow, Glasgow, UK

Costas S. Iliopoulos, Department of Informatics, King's College London, London, United Kingdom

Samad Jahandideh, Bioinformatics and Systems Biology Program, Sanford-Burnham Medical Research Institute, La Jolla, CA, USA

Giuseppe Lancia, Dipartimento di Matematica e Informatica, University of Udine, Udine, Italy

Fabien Lauer, LORIA, Université de Lorraine-CNRS, Nancy, France

Vladimir Makarenkov, Département d'informatique, Université du Québec à Montréal, Succ. Centre-Ville, Montréal, Québec, Canada

Christos Makris, Department of Computer Engineering and Informatics, University of Patras, Patras, Greece

Mina Maleki, School of Computer Science, University of Windsor, Windsor, Canada

Diego Mallo, Department of Biochemistry, Genetics and Immunology, University of Vigo, Vigo, Spain

David A. Morrison, Systematic Biology, Uppsala University, Norbyvägen, Uppsala, Sweden

Ahmed Moussa, LabTIC Laboratory, ENSA, Abdelmalek Essaadi University, Tangier, Morocco

Sami Muhaidat, ECE Department, Khalifa University, Abu Dhabi, UAE; EE Department, University of Surrey, Guildford, UK

Mirto Musci, Department of Electrical, Computer and Biomedical Engineering, University of Pavia, Via Ferrata, Pavia, Italy

Engelbert Mephu Nguifo, LIMOS–Blaise Pascal University–Clermont University, Clermont-Ferrand, France; LIMOS–CNRS UMR, Aubiére, France

Stavroula Ntoufa, Hematology Department and HCT Unit, G. Papanikolaou Hospital, Thessaloniki, Greece; Institute of Applied Biosciences, C.E.R.TH, Thessaloniki, Greece

Nahumi Nugrahaningsih, Department of Electrical, Computer and Biomedical Engineering, University of Pavia, Via Ferrata 3, 27100 Pavia, Italy

Hasan Oğul, Department of Computer Engineering, Başkent University, Ankara, Turkey

Nikos Papakonstantinou, Hematology Department and HCT Unit, G. Papanikolaou Hospital, Thessaloniki, Greece; Institute of Applied Biosciences, C.E.R.TH, Thessaloniki, Greece

Sheng-Lung Peng, Department of Computer Science and Information Engineering, National Dong Hwa University, Hualien, Taiwan

Solon P. Pissis, Department of Informatics, King's College London, London, United Kingdom

Huzefa Rangwala, Department of Computer Science, George Mason University, Fairfax, VA, USA

Sara Roque, Department of Mathematics, University of Aveiro, Aveiro, Portugal

Luis Rueda, School of Computer Science, University of Windsor, Windsor, Canada

Agustín Sánchez-Cobos, Bioinformatics Unit, Centre for Molecular Biology "Severo Ochoa" (CSIC), Madrid, Spain

Maad Shatnawi, Higher Colleges of Technology, Abu Dhabi, UAE

Soheila Shokrollahzade, Department of Medicinal Biotechnology, Iran University of Medical Sciences, Tehran, Iran

Carina Silva, Lisbon School of Health Technology, Lisbon, Portugal; Center of Statistics and Applications of Lisbon University (CEAUL), Lisbon, Portugal

Tiratha Raj Singh, Biotechnolgy and Bioinformatics Department, Jaypee University of Information and Technology (JUIT), Solan, Himachal Pradesh, India

V Anne Smith, School of Biology, University of St Andrews, St Andrews, UK

Jiangning Song, National Engineering Laboratory for Industrial Enzymes and Key Laboratory of Systems Microbial Biotechnology, Tianjin Institute of Industrial Biotechnology, Chinese Academy of Sciences, Tianjin, China; Department of Biochemistry and Molecular Biology, Faculty of Medicine, Monash University, Melbourne, VIC, Australia

Yang Song, Department of Computer Science, New Jersey Institute of Technology, Newark, NJ, USA

Lisete Sousa, Department of Statistics and Operations Research, Lisbon University and Center of Statistics and Applications of Lisbon University (CEAUL), Lisbon, Portugal, Lisbon, Portugal

Alexandros Stamatakis, Karlsruhe Institute of Technology, Institute for Theoretical Informatics, Postfach, Karlsruhe, Germany; Heidelberg Institute for Theoretical Studies, Heidelberg, Germany

Kostas Stamatopoulos, Hematology Department and HCT Unit, G. Papanikolaou Hospital, Thessaloniki, Greece; Institute of Applied Biosciences, C.E.R.TH, Thessaloniki, Greece

Kamal Taha, ECE Department, Khalifa University, Abu Dhabi, UAE

Nadia Tahiri, Département d'informatique, Université du Québec à Montréal, Succ. Centre-Ville, Montréal, Québec, Canada

Li Teng, Department of Internal Medicine, University of Iowa, Iowa City, IA, USA

Evangelos Theodoridis, Computer Technology Institute and Press "Diophantus," Patras, Greece

Athina Tsanousa, Department of Informatics, Aristotle University of Thessaloniki, Thessaloniki, Greece

Yu-Wei Tsay, Institute of Information Science, Academia Sinica, Taipei, Taiwan

Chinua Umoja, Department of Computer Science, Georgia State University, Atlanta, GA, USA

Brigitte Vannier, Receptors, Regulation and Tumor Cells (2RTC) Laboratory, University of Poitiers, Poitiers, France

Meghana Vasavada, Department of Computer Science, New Jersey Institute of Technology, Newark, NJ, USA

Akbar Vaseghi, Department of Genetics, Faculty of Biological Sciences, Tarbiat Modares University, Tehran, Iran

Davide Verzotto, Computational and Systems Biology, Genome Institute of Singapore, Singapore

Jason T.L. Wang, Department of Computer Science, New Jersey Institute of Technology, Newark, NJ, USA

Emanuel Weitschek, Department of Engineering, Uninettuno International University, Rome, Italy; Institute of Systems Analysis and Computer Science "A. Ruberti", National Research Council, Rome, Italy

Renxiang Yan, Institute of Applied Genomics, College of Biological Science and Engineering, Fuzhou University, Fuzhou, China

Paul D. Yoo, Department of Computing and Informatics, Bournemouth University, UK; Centre for Distributed and High Performance Computing, University of Sydney, Sydney, Australia

Guoxian Yu, College of Computer and Information Science, Southwest University, Chongqing, China

Xiaxia Yu, Department of Computer Science, Georgia State University, Atlanta, GA, USA

Ziding Zhang, State Key Laboratory of Agrobiotechnology, College of Biological Sciences, China Agricultural University, Beijing, China

Albert Y. Zomaya, Centre for Distributed and High Performance Computing, University of Sydney, Sydney, Australia

PREFACE

Pattern recognition is the automatic identification of *regularities*, that is, figures, characters, shapes, and forms present in data. Pattern recognition is the core process of many scientific discoveries, whereby researchers detect regularities in large amounts of data in fields as diverse as Geology, Physics, Astronomy, and Molecular Biology. Pattern recognition in biomolecular data is at the core of Molecular Biology research. Indeed, pattern recognition makes a very important contribution to the analysis of biomolecular data. In fact, it can reveal information about shared biological functions of biological macromolecules, that is, DNA, RNA, and proteins, originating from several different organisms, by the identification of patterns that are shared by structures related to these macromolecules. These patterns, which have been conserved during evolution, often play an important structural and/or functional role, and consequently, shed light on the mechanisms and the biological processes in which these macromolecules participate. Pattern recognition in biomolecular data is also used in evolutionary studies, in order to analyze relationships that exist between species and establish whether two, or several, biological macromolecules are *homologous*, that is, have a common biological ancestor, and to reconstruct the phylogenetic tree that links them to this ancestor. On the other hand, with the new sequencing technologies, the number of biological sequences in databases is increasing exponentially. In addition, the lengths of these sequences are large. Hence, the recognition of patterns in such databases requires the development of fast, low-memory requirements and high-performance techniques and approaches. This book provides an up-to-date forum of such techniques and approaches that deal with the most studied, the most important, and/or the newest topics in the field of pattern recognition. Some of these techniques and approaches represent improvements on old ones, while others are completely new. Most of current books on pattern recognition in biomolecular data either lack technical depth or focus on specific, narrow topics. This book is the first overview on techniques and approaches on pattern recognition in biomolecular data with both a broad coverage of this field and enough depth to be of practical use to working professionals. It surveys the most recent developments of techniques and approaches on pattern recognition in biomolecular data, offering enough fundamental and technical information on these techniques and approaches and the related problems, without overloading the reader. This book will thus be invaluable not only for practitioners and professional researchers in Computer Science, Life Science, and Mathematics but also for graduate students and young researchers looking for promising directions in their work. It will certainly point them to new techniques and approaches that may be the key to new and important discoveries in Molecular Biology.

This book is organized into seven parts: *Pattern Recognition in Sequences*, *Pattern Recognition in Secondary Structures*, *Pattern Recognition in Tertiary Structures*, *Pattern Recognition in Quaternary Structures*, *Pattern Recognition in Microarrays*, *Pattern Recognition in Phylogenetic Trees*, and *Pattern Recognition in Biological Networks*. The 29 chapters, which make up the seven parts of this book, were carefully selected to provide a wide scope with minimal overlap between the chapters so as to reduce duplications. Each contributor was asked to cover review material as well as current developments in his/her chapter. In addition, the choice of authors was made by selecting those who are leaders in their respective fields.

MOURAD ELLOUMI
TUNIS, TUNISIA

COSTAS S. ILIOPOULOS
LONDON, UK

JASON T. L. WANG
NEWARK, USA

ALBERT Y. ZOMAYA
SYDNEY, AUSTRALIA

NOVEMBER 1, 2015

PART I

PATTERN RECOGNITION IN SEQUENCES

CHAPTER 1

COMBINATORIAL HAPLOTYPING PROBLEMS

Giuseppe Lancia

Dipartimento di Matematica e Informatica, University of Udine, Udine,Italy,

1.1 INTRODUCTION

A few years back, the process of collection of a vast amount of genomic data culminated with the completion of the Human Genome Project [22]. The outcome of the project has brought the confirmation that the genetic makeup of humans (as well as other species) is remarkably well conserved. Generally speaking, the differences at DNA level between any two individuals amount to less than 5% of their genomic sequences. As a consequence, the differences at the phenotype level (i.e., in the way the organisms look) must be caused by small regions of differences in the genomes. The smallest possible region consists of a single nucleotide and is called the *Single Nucleotide Polymorphism* (SNP, pronounced "snip"). SNPs are the predominant form of human genetic variation and their importance can hardly be overestimated; they are used, for example, in medical, drug-design, diagnostic, and forensic applications.

Broadly speaking, a *polymorphism* is a trait common to everybody that can take different values, ranging in a limited set of possibilities, called *alleles* (for simple examples, think of blood group or eye color). In particular, a SNP is a nucleotide site, placed in the middle of a DNA region that is otherwise identical in everybody,

Pattern Recognition in Computational Molecular Biology: Techniques and Approaches,
First Edition. Edited by Mourad Elloumi, Costas S. Iliopoulos, Jason T. L. Wang, and Albert Y. Zomaya.

Individual 1, paternal:	`taggtccCtatttCccaggcgcCgtatacttcgacgggTctata`
Individual 1, maternal:	`taggtccGtatttAccaggcgcGgtatacttcgacgggTctata`
Individual 2, paternal:	`taggtccCtatttAccaggcgcGgtatacttcgacgggTctata`
Individual 2, maternal:	`taggtccGtatttCccaggcgcGgtatacttcgacgggCctata`
Individual 3, paternal:	`taggtccCtatttAccaggcgcGgtatacttcgacgggTctata`
Individual 3, maternal:	`taggtccGtatttAccaggcgcCgtatacttcgacgggCctata`

Figure 1.1 A chromosome in three individuals. There are four SNPs.

at which we observe a statistically significant variability. A SNP is almost always a polymorphism with only two alleles (out of the four possible). For a nucleotide site to be considered a SNP, it must be the case that the less frequent allele is found in the population with some nonnegligible frequency.

For many living species (e.g., mammals), there exist two sexes and each individual has two parents, one of each sex. The genome of these species is organized in pairs of chromosomes and a single copy of each chromosome pair is inherited from each of the two parents. Organisms in which the genome is organized in pairs of homologous chromosomes are called *diploid* organisms. For a diploid organism, at each SNP, an individual can either be *homozygous* (i.e., possess the same allele on both chromosomes) or *heterozygous* (i.e., possess two different alleles). The values of a set of SNPs on a particular chromosome copy define a *haplotype*.

In Figure 1.1, we illustrate a simplistic example, showing a specific chromosome in three individuals. For each individual, the pair of her chromosome copies is reported. There are four SNPs. The alleles for SNP 1, in this example, are **C** and **G**, while for SNP 4 they are **T** and **C**. Individual 1 is heterozygous for SNPs 1, 2, and 3, and homozygous for SNP 4. Her haplotypes are CCCT and GAGT. The haplotypes of individual 3 are CAGT and GACC.

Haplotyping an individual consists in determining her two haplotypes for a given chromosome. With the larger availability in SNP genomic data, recent years have seen the birth of many new computational problems related to haplotyping. These problems are motivated by the fact that it can be difficult and/or very expensive to determine the haplotypes experimentally, so *ad hoc* algorithms have been used to correct data errors or to infer missing data.

In the remainder of this chapter, we will address the haplotyping problem for both a single individual and a set of individuals (a population). In the former case, described in Section 1.2, the input is haplotype data inconsistent with the existence of exactly two parents for an individual. This inconsistency is due to experimental errors and/or missing data. In the latter case, described in Section 1.3, the input data specify for each SNP and each individual in a population whether her parents have contributed the same or different alleles, but do not specify which parent contributed which allele.

1.2 SINGLE INDIVIDUAL HAPLOTYPING

The process of passing from the sequence of nucleotides in a DNA molecule to a string over the DNA alphabet is called *sequencing*. A sequencer is a machine that is fed some DNA and whose output is a string of As, Ts, Cs, and Gs. To each letter, the sequencer attaches a value (confidence level) that essentially represents the probability that the letter has been correctly read.

The main problem with sequencing is that, owing to technological limitations, it is not possible to sequence a long DNA molecule at once. What we can do, however, is to sequence short DNA fragments (also called *reads*), of length of about 1000 nucleotides each, which provide small "windows" to look at the target molecule. To sequence a long DNA molecule, the molecule must first be replicated (*amplified*) by creating many copies of it. These copies are then broken, at random, into several smaller fragments, which are fed to a sequencer that will sequence those of the right size. The amplification phase is necessary so that the reads can have nonempty overlap. From the overlap of two reads, one may infer (through a process called *assembly*) a longer fragment, and so on, until the original DNA sequence has been reconstructed. This is, in essence, the principle of *shotgun sequencing*, a method used by Celera Genomics in the late 1990s to allow the completion of the sequencing of the human genome faster compared with other experimental techniques of the time [62]. In shotgun sequencing, the fragments were read and then assembled back into the original sequence by using sophisticated algorithms and powerful computers.

In Figure 1.2, we illustrate an example in which the two chromosome copies (a) and (b) have been amplified, and then some fragments (denoted by rectangular boxes) have been sequenced. The goal would then be to retrieve (a) and (b), given the set of sequenced fragments. The major difficulty obstructing this goal is that, during the amplification phase, both the paternal and the maternal chromosome copies are amplified together so that the random reads can belong to either the paternal or the maternal original copy. Of course, it is unknown which fragments are paternal and which are maternal, and one of the problems in reconstructing the haplotypes consists in segregating the paternal fragments from the maternal ones. In some cases, there may be pairs of reads, called *mate pairs*, for which it is known that they are either both paternal or both maternal. Mate pairs are due to a particular technique that allows to sequence the ends of a fragment several thousand nucleotides long. Again, one can only read about 1000 nucleotides at each end of the target, but the result is stronger than reading two individual fragments as now it is known that the two reads come from the same chromosome copy. Furthermore, the experiment returns a fairly precise estimate of the distance, expressed in bases, between the two reads of a mate pair.

Even with the best possible technology, sequencing errors are unavoidable. The main errors consist in bases that have been miscalled or skipped altogether. Furthermore, contaminants can be present, that is, DNA coming from organisms other than the one that had to be sequenced. Owing to the experimental errors, the reconstruction

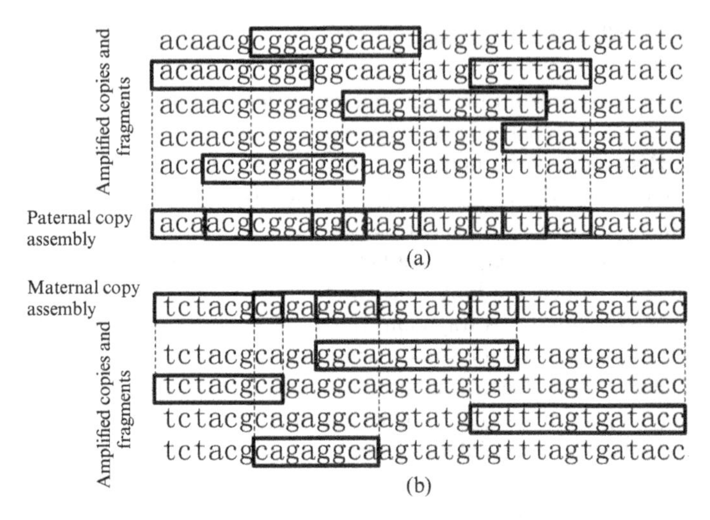

Figure 1.2 Sequence reads and assembly of the two chromosome copies.

of the haplotypes, given the reads coming from sequencing, is not always straightforward and may require correction of the input data. In a general way, the *haplotyping problem for an individual* can then be informally stated as follows:

> *Given inconsistent haplotype data coming from sequencing of an individual's chromosome, find and correct the errors in the data so as to retrieve a consistent pair of haplotypes.*

Depending on what type of errors one is after, there can be many versions of this problem (for surveys on individual haplotyping problems, see, e.g., Schwartz [59] or Zhang et al. [70]). Historically, the formalization of the first haplotyping problems for an individual was given by Lancia et al. [47]. In their work, the *Minimum Fragment Removal* (MFR) and *Minimum SNP Removal* (MSR) problems were introduced, which we briefly discuss here.

Given the fact that at each SNP only two alleles are possible, we can encode them by using a binary alphabet. Hence, in the sequel, the two values that a SNP can take are denoted by `0` and `1`. A haplotype is then simply a string over the alphabet $\{0, 1\}$.

The basic framework for a *single individual haplotyping problem* is as follows. There is a set $S = \{1, \ldots, n\}$ of SNPs and a set $\mathcal{F} = \{1, \ldots, m\}$ of fragments (i.e., reads coming from sequencing). Each SNP is covered by some of the fragments and can take the values `0` or `1`. Since there is a natural ordering of the SNPs, given by their physical location on the chromosome, the data can be represented by an $m \times n$ matrix M over the alphabet $\{0, 1, -\}$, called the *SNP matrix*. Each column corresponds to a SNP and each row corresponds to a fragment. If fragment i covers a SNP j, then $M[i, j]$ is the value of the allele for SNP j appearing in fragment i. The symbol "`-`" is used to represent a SNP not covered by a fragment (see Figure 1.3a and b for an example of a SNP matrix).

A *gapless* fragment is one covering a set of consecutive SNPs (i.e., the `0`s and `1`s appear consecutively in that row). We say that a fragment has k gaps if it covers $k+1$ blocks of consecutive SNPs (e.g., the fragment `00--101---01` has two gaps). Gaps are mainly due to two reasons: (i) thresholding of low-quality reads (when the sequencer cannot call a SNP `0` or `1` with enough confidence, the SNP is marked with a "`-`"); (ii) mate-pairing in shotgun sequencing (in this case $k = 1$, and the situation is equivalent to having two gapless fragments coming from the same chromosome copy).

Two fragments i and j are said to be *in conflict* if there exists a SNP k such that neither $M[i,k] = -$ nor $M[j,k] = -$ and it is $M[i,k] \neq M[j,k]$. The conflict of two fragments implies that either the fragments do not come from the same chromosome copy or there are errors in the data. Given a SNP matrix M, the *fragment conflict graph* is the graph $G_{\mathcal{F}}(M) = (\mathcal{F}, E_{\mathcal{F}})$ with an edge for each pair of fragments in conflict (see Figure 1.3c). Two SNPs i and j are said to be *in conflict* if there exist two fragments u and v such that the 2×2 submatrix defined by rows u and v and columns i and j contains three `0`s and one `1`, or three `1`s and one `0`. Given a SNP matrix M, the *SNP conflict graph* is the graph $G_{\mathcal{S}}(M) = (\mathcal{S}, E_{\mathcal{S}})$, with an edge for each pair of SNPs in conflict (see Figure 1.3d).

If $G_{\mathcal{F}}(M)$ is a bipartite graph, $\mathcal{F}$ can be segregated into two sets H_1 and H_2 of pair-wise compatible fragments. From each set, one can infer one haplotype by fragment overlap. Let h_1 and h_2 be the haplotypes thus obtained. Since h_1 and h_2 correspond to the assembly of the fragments, the single individual haplotyping problems are sometimes also referred to as *fragment assembly haplotyping* problems.

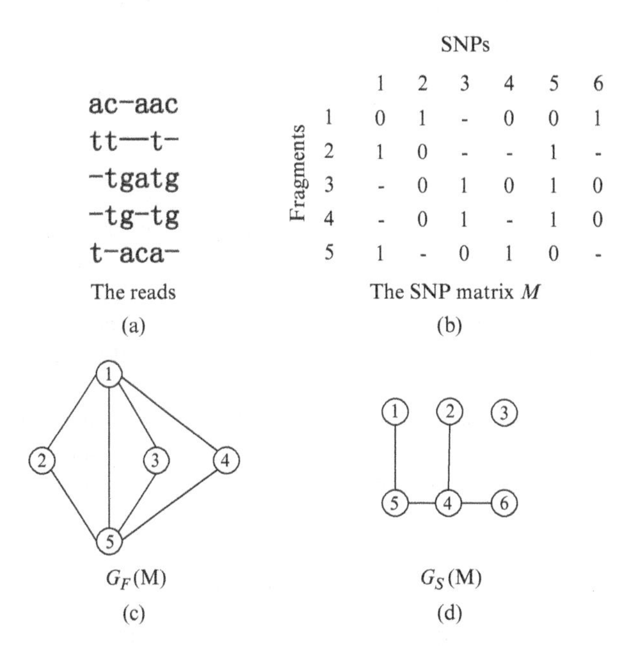

Figure 1.3 SNP matrix and conflict graphs.

We call a SNP matrix M *feasible* if $G_{\mathcal{F}}(M)$ is bipartite and *infeasible* otherwise. Note that a SNP matrix for error-free data must be feasible. Hence, the optimization problems to be defined aim at correcting an infeasible SNP matrix so that it becomes feasible.

The following are the first optimization problems defined with the above goal [47]. They arose at Celera Genomics in the context of sequencing the human genome.

- MFR: Given a SNP matrix, remove the minimum number of fragments (rows) so that the resulting matrix is feasible.
- MSR: Given a SNP matrix, remove the minimum number of SNPs (columns) so that the resulting matrix is feasible.

The first problem is mainly suited for a situation in which, more than sequencing errors, one is worried about the presence of contaminants. The second problem is more suited in the presence of sequencing errors only, when all the fragments are to be retained. These problems were shown to be NP-hard in general [47] so that exact algorithms for their solution are expected to be exponential branch-and-bound procedures. Lippert et al. [53] described a combinatorial branch-and-bound algorithm for MFR. They also described an *Integer Linear Programming* (ILP) formulation of the problem, based on the correspondence between MFR and the *maximum node-induced bipartite subgraph* problem.

While the general case for MFR and MSR is NP-hard, these problems were shown to be polynomial for gapless data [47]. Let us call M a *gapless matrix* if each row of M is a gapless fragment. The main connection between MFR and MSR is given by the following theorem.

Theorem 1.1 *[47] Let M be a gapless SNP matrix. Then, $G_{\mathcal{F}}(M)$ is a bipartite graph if and only if $G_{\mathcal{S}}(M)$ is a stable set.*

Later theoretical improvements extended these results to fragments with gaps of bounded length, giving $O(2^{2l}m^2n + 2^{3l}n^3)$ dynamic programming algorithms for MFR and $O(mn^{2l+2})$ for MSR for instances with gaps of total length l Rizzi et al. [57], Bafna et al. [58]. These algorithms are hardly practical, however, in instances for which the gaps can be rather large. To overcome this problem, Xie and Wang [67] (see also Xie et al. [68]) proposed an algorithm for MFR based on two new parameters: k, the maximum number of SNPs covered by any fragment and c, the maximum number of fragments covering any SNP site (also called *coverage*). Their solution has complexity $O(nc3^c + m\log\ m + mk)$. Since $k \leq n$ and c is generally at most 10, this method should be more suitable for mate-paired data, where l can be quite large.

1.2.1 The Minimum Error Correction Model

Owing to the nature of the sequencing process, most data errors are due to miscalling or missing a base in a read. As a consequence, a particular version of the single individual haplotyping problem was proposed in Reference [53] and has received great attention because of its practical relevance. This version is called *Minimum Error Correction* (MEC) problem (the problem is also known as *Minimum Letter Flip* (MLF), see Reference [32]):

> *MEC: Given a SNP matrix, change the minimum number of entries (`0` into `1`, or `1` into `0`) so that the resulting matrix is feasible.*

Particularly interesting is a weighted version of MEC in which each entry is associated with a nonnegative weight proportional to the confidence level with which the base corresponding to the entry has been called. The solution seeks to flip a set of entries with minimum total weight so as to make the matrix feasible.

The MEC problem was shown to be NP-hard, both for general [53] and for gapless matrices [20]. Many approaches were proposed in the literature for the solution of MEC. For a comparative survey of the various procedures, the reader is referred to Geraci [29] and to Geraci and Pellegrini [30].

One of the first heuristics for MEC was `FastHare` by Panconesi and Suozo [56]. `FastHare` starts by first sorting the fragments according to their leftmost ends and then it reconstructs the two final haplotypes by correcting the sorted fragments in a greedy manner. Being simple, this heuristic is very fast but nevertheless it provides quite accurate solutions in general, especially when the error rate in the data is not too high. For high error-rate data, Genovese et al. proposed `SpeedHap` [28], an effective heuristic procedure organized in phases. For each phase, they perform three tasks: (i) detect likely positions of errors, (ii) allocate fragments to the two partially built final haplotypes, and (iii) decide the final alleles in the two haplotypes via a majority rule on ambiguous SNPs.

`FAHR` (*Fast and Accurate Haplotype Reconstruction* by Wu and Liang [6]) is a somewhat similar greedy heuristic procedure, which builds the final haplotypes by partitioning the fragments at each SNP in a greedy manner. Yet another heuristic for MEC is `HMEC`, proposed by Bayzid et al. [7]. `HMEC` is a steepest–descent local search procedure in which each iteration takes time $O(m^3n)$. Because of its time complexity, `HMEC` may provide results slowly for instances with a large number of fragments.

Zhao et al. [72] considered the *weighted version of MLF* (called WMLF), for which they proposed the use of a dynamic clustering heuristic. Furthermore, they introduced an extended version of the problem to deal with the presence of contaminants. In this version, denoted as Complete Weighted Minimum Letter Flip (CWMLF), one is allowed not only to flip entries of the SNP matrix but also to remove some of the matrix rows. In addition, for CWMLF, they proposed the use of a dynamic clustering algorithm. Wu et al. [66] proposed a heuristic procedure for WMLF, with a performance similar to that of the best previous methods.

Among the various types of approaches to the solution of MEC, there is the use of evolutionary algorithms such as Genetic Algorithms (GA) and Particle Swarm Optimization (PSO). Wang et al. [63] proposed both a branch-and-bound and a GA for MEC. Being exponential, the branch-and-bound is only applicable to instances of very small size, but the GA can be used for large instances (e.g., more than

50 fragments over more than 50 SNPs). A PSO heuristic for MEC was proposed by Kargar et al. [46]. PSO turns out to be fast and suitable for instances with a low error rate in the data.

One of the most successful heuristics for MEC is HapCUT, proposed by Bansal and Bafna [3]. The algorithm makes use of a suitable graph obtained from the fragments, where each SNP corresponds to a node in the graph and two nodes are joined by an edge if there exists a fragment that covers both SNPs. The procedure then tries to minimize the MEC cost of the reconstructed haplotypes by iteratively finding max-cuts in the associated graph.

He et al. [41] proposed a dynamic programming algorithm for MEC with time complexity $O(2^L\ m\ n)$, where L is the maximum length of a fragment. The algorithm can be used for values of L up to about 15, but beyond that, it becomes impractical. For large values of L, the authors suggest to model the problem as a MaxSAT problem, to be solved by a MaxSAT solver. The quality of the solutions obtained through this approach is shown to be quite high, but the solving process is still slow.

Another dynamic programming algorithm, this time parameterized by the maximum coverage c of each SNP site, was proposed by Deng et al. [23]. The complexity of the dynamic program is $O(nc2^c)$, which can be quite high when c is large. For large values of c, the authors propose a heuristic procedure based on the dynamic programming algorithm. Experiments showed that the heuristic returns very accurate solutions on average.

Chen et al. [18] proposed an exact algorithm for MEC based on an ILP formulation of the problem. The solving process is considerably speeded up by a preprocessing phase, in which the input SNP matrix is decomposed into smaller, independent, blocks that are then individually solved by the ILP solver.

1.2.2 Probabilistic Approaches and Alternative Models

Some of the solutions to the assembly haplotyping problems have employed a probabilistic approach, either because of the type of algorithm used or because of the type of model and problem definition.

An example of probabilistic algorithm is HASH (Haplotype Assembly for Single Human), an MCMC (*Markov Chain Monte Carlo*) algorithm proposed by Bansal et al. [4] for assembling haplotype fragments under the MEC objective function. In HASH, the transitions of the Markov chain are generated using min-cut computations on graphs derived from the reads. The method was applied to infer the haplotypes from real data consisting of whole-genome shotgun sequence fragments of a human individual. The results showed the haplotypes inferred by using HASH to be significantly more accurate than the haplotypes estimated by using the best heuristics available at the time.

Another type of Markov chain approach was proposed by Wang et al. [64]. In this approach, the assignment of fragments to haplotypes is regarded as a Markov process in which successive allele pairs are generated on the basis of the value of a small number of preceding SNP sites in the sequence. In order to find the most probable haplotypes for the set of fragments, the authors propose a Viterbi-like dynamic programming procedure.

Chen et al. [19] described a probabilistic model for MEC in which they considered three error parameters, called α_1, α_2, and β. The parameter α_1 is the error rate with which entries of the SNP matrix have been miscalled (i.e., a `0` in place of `1` or a `1` in place of a `0`). The value α_2 is the error rate with which a "`-`" in the SNP matrix appears where, in fact, a SNP is covered by a fragment and therefore an allele `0` or `1` should be present. Finally, β is a measure of the expected dissimilarity between the haplotypes to be reconstructed. The authors gave a linear-time (in the size of M) probabilistic algorithm that reconstructs the correct haplotypes with high probability when the parameters are known. Even for the case when some of the parameters are unknown, they provided a probabilistic algorithm that outputs the correct haplotypes with probability depending on α_1, α_2, and β.

Li et al. [52] proposed a probabilistic model for the haplotype inference problem. In their model, they studied the conditional probability $P(h_1, h_2|M)$ of h_1 and h_2 being the correct haplotypes given that M was the SNP matrix measured. They then pursued the objective of determining the most likely pair of correct haplotypes, that is, a pair $\{h_1^*, h_2^*\}$ maximizing the above conditional probability. In order to solve this problem, they used Gibbs sampling and an *Expectation Maximization* (EM) algorithm.

Among the works proposed for the solution of the MEC model, there are some papers that introduced variants to the original framework in order to deal with some specific data problems. For instance, Zhang et al. [71] proposed a model called *Minimum Conflict Individual Haplotyping* (MCIH), suitable for data sets with particularly high error rate. The problem was shown to be NP-hard and a dynamic programming procedure for its solution was described.

Duitama et al. [25] proposed a model called *Maximum Fragments Cut* (MCF) whose objective is to identify a set of fragments (i.e., rows of the SNP matrix) maximizing a score proportional to their conflict with respect to the remaining rows. This set of fragments can be interpreted as the shore of a cut in a suitable graph, so that the problem can be reduced to a max-cut problem. The authors utilized a local optimization heuristic, called `ReFHap`, for the problem solution.

Xie et al. [69] introduced a model called *Balanced Optimal Partition* (BOP), which generalizes both MEC and MCF. The model is, in a way, a weighted combination of MEC and MCF, and by setting some model parameters to proper values, BOP degenerates into pure MEC or pure MCF. To solve the model, the authors proposed a dynamic programming algorithm called `H-BOP`.

Aguiar and Istrail [1] proposed `HapCompass`, an algorithm for haplotype assembly of densely sequenced human genome data. `HapCompass` operates on a graph where each SNP is a node and the edges are associated with the fragments. The edges are weighted, and the weight of an edge indicates what the best phasing of the alleles for the edge endpoints is, given the fragments in the data set that cover the corresponding SNPs. In this model, the reconstruction of the underlying haplotypes corresponds to a minimum-weight edge-removal problem until a special type of subgraph, called *happy graph* by the authors, is obtained. `HapCompass` is a heuristic with a local reoptimization step. Computational results showed the effectiveness of the procedure and the good quality of its solution also with respect to the MEC objective.

1.3 POPULATION HAPLOTYPING

Haplotype data are particularly sought after in the study of complex diseases (those affected by more than one gene) because they convey precious information about which set of gene alleles are inherited together. These types of polymorphism screenings are usually conducted on a *population*, that is, on sets of related individuals.

We have discussed at length the problem of haplotyping a single individual, which consists in determining her two haplotypes for a given chromosome. In a similar way, *haplotyping a population* consists in haplotyping each individual of the given population. Of course, one way to do this would be to solve a set of single individual haplotyping problems, one for each element of the population, but there might be better ways to proceed. In particular, with the larger availability in SNP data, recent years have seen the birth of a set of new computational problems related to population haplotyping. Most of these problems are motivated by the fact that while it is economically inconvenient to determine the haplotypes by sequencing and assembling, there is a cheap experiment that can determine a less informative and often ambiguous type of data, that is, the *genotypes*, from which the haplotypes can then be retrieved computationally.

A genotype of an individual contains the information about the two (possibly identical) alleles at each SNP, but without specifying their paternal or maternal origin. There may be many possible pairs of haplotypes that are consistent with a given genotype. For example, assume we only know that an individual is heterozygous for the alleles {`C`, `T`} at SNP 1 and for the alleles {`A`, `G`} at SNP 2. Then, either one of these alternatives may be true:

(i) One parent has contributed the alleles `C` and `A` and the other the alleles `T` and `G`.

(ii) One parent has contributed the alleles `C` and `G` and the other the alleles `T` and `A`.

Both possibilities are plausible. Associating the alleles with the parents is called *phasing* the alleles. For k heterozygous SNPs, there are 2^k possible phasings, which makes choosing the correct one a difficult problem. Once the alleles have been phased, the two haplotypes are inferred as phasing and haplotyping are in fact the same problem. The two haplotypes that are obtained by phasing the alleles are said to *resolve*, or *explain*, the genotype.

The most general *population haplotyping* problem can be stated as follows:

Given a set G of genotypes, corresponding to an existing, unknown, set H of haplotypes, retrieve H.

In other words, the goal is to compute a set H of haplotypes that contains, for each genotype $g \in G$, the two haplotypes h_1 and h_2 obtained by the correct phasing of g. As could be expected, on the basis of only the knowledge of Git, is not easy to describe constraints that define precisely which of the exponentially many phasings of a genotype is the correct one. Biologists have therefore described several sensible criteria for "good" phasings. For instance, under a widely accepted parsimony principle (inspired by the Occam's razor principle), a good solution may be one that minimizes the number of distinct haplotypes inferred.

Once it has been mathematically modeled, haplotyping gives rise to several nice and challenging combinatorial problems (for surveys on population haplotyping problems, see, e.g., References [11, 38, 39]). These problems have been extensively studied in the last few years. Some of them have been proven NP-hard and solved by exponential-time algorithms, while for others polynomial-time algorithms have been designed.

In this chapter, we address some of the most interesting combinatorial haplotyping models proposed in the literature. Each model and objective function has specific biological motivations, which are discussed in the following sections. While our focus will be on the combinatorial approach to haplotyping problems, it should be remarked that there is also a very important statistical approach to population haplotyping problems, which does not fall within the scope of this survey. The statistical approach has led to widely used software tools for haplotyping, such as the program `PHASE` [60].

Given a set of n SNPs, fix arbitrarily a binary encoding of the two alleles for each SNP (i.e., call one of the two alleles 0 and the other 1). Once the encoding has been fixed, each haplotype becomes a binary vector of length n.

A haplotype h is denoted by $h[i]$, the value of its ith component, with $i = 1, \dots, n$. Given two haplotypes h' and h'', we define a special sum whose result is a vector $g := h' \oplus h''$. The vector g has length n, and its components take values in $\{0, 1, 2\}$, according to the following rule:

$$g[i] := \begin{cases} 0 & \text{if } h'[i] = h''[i] = 0 \\ 1 & \text{if } h'[i] = h''[i] = 1 \\ 2 & \text{if } h'[i] \neq h''[i] \end{cases}$$

We call a vector g with entries in $\{0, 1, 2\}$ a *genotype*. Each position i such that $g[i] = 2$ is called an *ambiguous position* (or *ambiguous site*). We denote by $A(g) \subseteq \{1, \dots, n\}$ the set of ambiguous positions of g. Biologically, genotype entries of value 0 or 1 correspond to homozygous SNP sites, while entries of value 2 correspond to heterozygous sites. In Figure 1.4, we illustrate a case of three individuals, showing their haplotypes and genotypes.

A *resolution* of a genotype g is given by a pair of haplotypes $\{h', h''\}$ such that $g = h' \oplus h''$. The haplotypes h' and h'' are said to resolve g. A genotype is *ambiguous* if it has more than one possible resolution, that is, if it has at least two ambiguous positions. A haplotype h is said to be *compatible* with a genotype g if h can be used in a resolution of g. It can immediately be seen that h is compatible with g if and only if $g[i] = h[i]$ at each position where $g[i] \neq 2$. Two genotypes g and g' are compatible if there exists at least one haplotype compatible with both of them, otherwise, they are *incompatible*. It can immediately be seen that g and g' are compatible if and only if $g[i] = g'[i]$ at each position i where they are both nonambiguous.

As previously discussed, the experiment yielding each genotype is such that, at each SNP, it is known whether an individual is homozygous for allele `0` (in which case, we may set $g[i] = 0$), homozygous for allele `1` (in which case, we may set

Haplotype	Sequence	Genotype sequence	Genotype
Haplotype1,paternal:	0 1 0 1	2 2 2 1	Genotype1
Haplotype1,maternal:	1 0 1 1		
Haplotype2,paternal:	0 0 1 1	2 2 1 2	Genotype2
Haplotype2,maternal:	1 1 1 0		
Haplotype3,paternal:	0 0 1 1	2 0 2 2	Genotype3
Haplotype3,maternal:	1 0 0 0		

Figure 1.4 Haplotypes and corresponding genotypes.

$g[i] = 1$), or heterozygous (in which case, we may set $g[i] = 2$). Therefore, for the haplotyping problems described in this section, the input data consist in a set G of m genotypes $g_1, \ldots, g_m$, corresponding to m individuals in a population. The output is a set H of haplotypes such that, for each $g \in G$, there is at least one pair of haplotypes $\{h', h''\} \subseteq H$ with $g = h' \oplus h''$. Such a set H of haplotypes is said to explain G. In addition to explaining G, the set H is also required to satisfy some particular constraints. These constraints are different for different specific types of haplotyping problems. For each problem described in this survey, the particular constraints are given in the corresponding section.

1.3.1 Clark's Rule

The geneticist Clark proposed in Reference [21] a rule (today known as *Clark's rule*) to derive new haplotypes by inference from known ones as follows:

Clark's Inference Rule: Given a genotype g and a compatible haplotype h, obtain a new haplotype q by setting $q[j] := 1 - h[j]$ at all positions $j \in A(g)$ and $q[j] := h[j]$ at the remaining positions.

Note that q and h resolve g. In order to resolve all genotypes of G, Clark suggested the use of successive applications of his inference rule. The procedure requires a "bootstrap" set of haplotypes used to derive new haplotypes at the very beginning. These starting haplotypes are obtained by resolving, in the unique possible way, the unambiguous genotypes in G (of which it is assumed there is always at least one).

The following is the procedure that Clark proposed for haplotyping a population. He supported the validity of the approach by arguments from theoretical population genetics:

Clark's Algorithm: Let $G' \subset G$ be the set of nonambiguous genotypes and let H be the set of haplotypes obtained from G'. Reset $G := G - G'$. Then, repeat the following. If they exist, take a $g \in G$ and a compatible $h \in H$ and apply the inference rule, obtaining q. Set $G := G - \{g\}$, $H := H \cup \{q\}$, and iterate. When no such g and h exist, the algorithm succeeds if $G = \emptyset$ and fails otherwise.

Note that the procedure is nondeterministic as it does not specify how to choose g and h whenever there are more candidates to the application of the rule. For example, suppose $G = \{\texttt{2000, 2200,1122}\}$. The algorithm starts by setting $H = \{\texttt{0000,1000}\}$ and $G = \{\texttt{2200,1122}\}$. The inference rule can be used to resolve `2200` from `0000`, obtaining `1100`, which can, in turn, be used to resolve `1122`, obtaining `1111`. However, one could have started by using `1000` to resolve `2200` obtaining `0100`. At that point, there would be no way to resolve `1122`. The nondeterminism in the choice of the genotype g and haplotype h to which the inference rule is applied can be settled by fixing a deterministic rule based on the initial sorting of the data. In Reference [21], a large number of random sorting are used to run the algorithm and the best solution overall is reported. Tests on real and simulated data sets showed that although most of the time the algorithm could resolve all genotypes, often the algorithm failed.

The problem of finding an ordering of application of Clark's inference rule that leaves the fewest number of unresolved genotypes in the end was first defined and studied by Gusfield [33], who proved it is NP-hard and APX-hard (i.e., there is a value $\delta > 1$ for which it is NP-hard even to give a δ-approximation algorithm). As for practical algorithms, Gusfield [34, 35] proposed an integer programming approach for a graph-theoretic reformulation of the problem. The problem is first transformed, by an exponential-time reduction, into a problem on a digraph (N, E) defined as follows. Let $N = \bigcup_{g \in G} N(g)$, where $N(g) := \{(h, g) : h \quad \textit{is} \quad \textit{compatible} \quad \textit{with} \quad g\}$. Let $N' = \bigcup_{g \in G'} N(g)$ be the subset of haplotypes determined from the set G' of unambiguous genotypes. For each $v = (h, g')$ and $w = (q, g)$ in N, there is an arc $(v, w) \in E$ if g is ambiguous, $g' \neq g$, and $g = h \oplus q$ (i.e., q can be inferred from g via h). Then, any directed tree rooted at a node $v \in N'$ specifies a feasible history of successive applications of the inference rule starting at node $v \in N'$. The problem can then be stated as follows: find the largest number of nodes that can be reached by a set of node-disjoint-directed trees, where each tree is rooted at a node in N' and for every ambiguous genotype g, at most one node in $N(g)$ is reached.

The above graph problem was shown to be NP-hard [34]. For its solution, Gusfield proposed an ILP formulation and noticed that the solution of the LP relaxation was very often integer for the real-life instances tested. The model was applied to real data as well as random instances, with up to 80 genotypes, of which 60 were ambiguous, over 15 SNPs.

1.3.2 Pure Parsimony

Clark's algorithm tries to reuse existing haplotypes as much as possible and it introduces new haplotypes only when strictly needed. As a result, the procedure usually tends to produce solutions of small size. Note that the maximum size for a set H resolving G is $2|G|$, while the smallest possible size is $\Omega(\sqrt{|G|})$.

Pure Parsimony Haplotyping (PPH) has the explicit objective of minimizing the size of H. This objective has several biological motivations. For one, the number of distinct haplotypes observed in nature is vastly smaller than the number of possible haplotypes. Furthermore, as we all descend from a small number of ancestors, their haplotypes should be the same we possess today (if it were not for some recombination events and mutations). Finally, pure parsimony follows a general principle according to which, of many possible explanations of an observed phenomenon, one should always favor the simplest.

The PPH problem is NP-hard, as first shown by Hubbel [43]. Lancia et al. [48] showed that, in fact, the problem is APX-hard. This result holds also if each genotype is restricted to possess at most three ambiguous sites. Note that although the problem is APX-hard, there exist constant-ratio approximation algorithms under the restriction that each genotype has at most k ambiguous sites for each constant k [48].

Several algorithmic approaches, many of them employing mathematical programming techniques, have been used in PPH (for a survey of models and approaches for PPH, see Reference [17]). In particular, we recall the following works.

1.3.2.1 *Integer Programming Formulations of Exponential Size.*

Let us denote by $H(G)$ the set of all haplotypes compatible with at least one genotype of G and let $\chi(G) := |H(G)|$. The first ILP formulation for PPH, called *TIP*, was provided by Gusfield [37]. TIP has $\chi(G)$ variables and $O(m2^n)$ constraints. There is a binary variable x_h associated with every $h \in H(G)$ ($x_h = 1$ means that h is included in the solution, whereas $x_h = 0$ means that h is not included). After fixing a total ordering on $H(G)$, denote by $\mathcal{P}$ the set of those ordered pairs (h_1 and h_2) with h_1 and $h_2 \in H(G)$, $h_1 < h_2$. For every $g \in G$, let $\mathcal{P}_g := \{(h_1, h_2) \in \mathcal{P} \mid h_1 \oplus h_2 = g\}$. For every $g \in G$ and $(h_1, h_2) \in \mathcal{P}_g$, there is a binary variable y_{h_1,h_2} used to select an ordered pair (h_1, h_2) to resolve the genotype g. The following is then a valid ILP formulation of the PPH problem [37, 48]:

$$\min \sum_{h \in H(G)} x_h \tag{1.1}$$

subject to

$$\sum_{(h_1,h_2) \in \mathcal{P}_g} y_{h_1,h_2} \geq 1 \qquad \forall g \in G \tag{1.2}$$

$$y_{h_1,h_2} \leq x_{h_1} \qquad \forall (h_1, h_2) \in \bigcup_{g \in G} \mathcal{P}_g \tag{1.3}$$

$$y_{h_1,h_2} \leq x_{h_2} \qquad \forall (h_1, h_2) \in \bigcup_{g \in G} \mathcal{P}_g \tag{1.4}$$

$$x_h \in \{0, 1\}, y_{h_1,h_2} \in \{0, 1\} \qquad \forall h, h_1, h_2 \in H(G) \tag{1.5}$$

Constraints (1.2) impose that each genotype is resolved by at least one pair of haplotypes. Constraints (1.3) and (1.4) make sure that (h_1, h_2) can be used to resolve a genotype only if both h_1 and h_2 are included in the solution.

Gusfield managed to employ some preprocessing rules to get rid of variables that can be proved to be nonessential in the formulation. The resulting model is called RTIP (*Reduced TIP*). Although practically useful, the preprocessing still leaves an exponential model whose size grows quite quickly with respect to the instance size. The experimental results of Reference [37] showed that RTIP can be used to tackle problems with up to 50 genotypes over 30 SNPs, but there must be relatively small levels of heterozygosity in the input genotypes.

Lancia and Rizzi [50] showed that when each genotype has at most two ambiguous sites, the above ILP is totally unimodular. As a consequence, the ILP solution is always naturally integer and hence the problem can be solved in polynomial time.

Lancia and Serafini [51] proposed a new exponential ILP model for PPH. The model exploits an interpretation of PPH as a particular type of *Set Covering* (SC) based on the following observation: if *H is a set of haplotypes resolving G, then for each genotype g, ambiguous position $i \in A(g)$, and value $a \in \{0, 1\}$, there is a haplotype $h \in H \cap H(g)$ such that $h[i] = a$.*

This condition is a covering condition because it represents the covering constraint for a set cover problem in which the universe is the set of all triples (g, i, a), for $g \in G$, $i \in A(g)$, and $a \in \{0, 1\}$, and each haplotype h represents a subset of the universe, namely, $h \leftrightarrow \{(g, i, a) \mid h \in H \cap H(g), h_i = a\}$. The condition is only necessary, but not sufficient, for H to be a feasible solution of PPH. Consider, for example, the following "diagonal" instance $G = \{1222, 2122, 2212, 2221\}$. The set $\{0111, 1011, 1101, 1110\}$ satisfies the covering condition but does not resolve G.

The SC associated with PPH seeks to minimize the number of haplotypes needed to satisfy all covering conditions for the given set of genotypes G. As we have seen, this SC is in fact a relaxation of PPH. The formulation of the SC model has an exponential number of variables and constraints, and can be solved, as described in Reference [51], by branch-and-cut-and-price, that is, via the generation of variables and constraints at run-time. It is possible that the optimal solution of SC resolves G, in which case it is an optimal PPH solution as well. If not, one can try to obtain a good feasible PPH solution from the optimal cover by adding only a small number of haplotypes. This idea is exploited by an effective heuristic presented in Reference [51]. The computational results show that the SC approach can be orders of magnitude faster than RTIP and can be applied to instances with $n = m = 50$ and $\chi(G)$ up to 10^9. These are among the largest-size instances for which provably optimal solutions have been obtained in the literature.

1.3.2.2 Integer Programming Formulations of Polynomial Size and Hybrid Formulations.

Many authors Brown and Harrower [13], Bertolazzi et al. [8], Lancia et al. [48], Halldorsson et al. [40] have independently proposed polynomial-size integer programming formulations. More or less, all these formulations employ variables to represent the bits of the haplotypes in the solution (so there are, at most, $O(mn)$ such variables). The basic idea is that for each genotype $g^i \in G$ one must determine two haplotypes h_1^i and h_2^i such that $h_1^i \oplus h_2^i = g^i$. This implies that, for each position j such that $g^i[j] = 2$, one needs to decide a variable $x_{ij} \in \{0, 1\}$, and then set $h_1^i[j] = x_{ij}$ and $h_2^i[j] = 1 - x_{ij}$. The polynomial formulations for PPH consist in expressing the PPH objective function and constraints in terms of the x variables and possibly a (polynomial-size) set of additional variables.

The LP relaxation of these formulations is generally quite weak. The use of some valid cuts [8, 13] improves the quality of the bound, but the integrality gap between the integer optimum and the LP relaxation remains large. Brown and Harrower [14] also proposed a hybrid model in which variables for a fixed subset of haplotypes are explicitly present, while the rest of the haplotypes are implicitly represented by polynomially many variables and constraints. These polynomial/hybrid formulations were successfully used for the solution of problems of similar size as those solvable by using TIP. Furthermore, some tests were conducted on larger problems (30 genotypes over up to 75 SNPs), on which the exponential formulation could not be applied successfully owing to the IP size.

The last polynomial model for PPH was proposed by Catanzaro et al. [16] and is based on the class representatives with smallest index (a technique adopted by Campelo et al. for the vertex color problem [15]). The authors observed that a feasible solution to PPH can be represented by means of a bipartite graph whose two shores correspond to haplotypes and genotypes. Each vertex $g \in G$ has degree 2 and there are two vertices, say h' and h'' adjacent to g such that $g = h' \oplus h''$. The bipartite graph representation of a solution suggests that, in a feasible solution to PPH, the haplotypes induce a family of subsets of genotypes satisfying the following three properties: (i) each subset of genotypes shares one haplotype, (ii) each genotype belongs to exactly two subsets, and (iii) every pair of subsets intersects in at most one genotype. Catanzaro et al. [16] exploited the bipartite graph representation of a solution to PPH to provide a polynomial-size ILP model that turns out to be quite effective and to outperform other polynomial models for the PPH problem.

1.3.2.3 *Quadratic, Semidefinite Programming Approaches.* A quadratic formulation solved by semidefinite programming was proposed by Kalpakis and Namjoshi [45]. Similarly to the RTIP model, the formulation has a variable for each possible haplotype and hence it cannot be used to tackle instances for which $\chi(G)$ is too large. According to the computational results, the size of the problems solved is comparable to that of RTIP and of the best polynomial ILP models. On the basis of a similar formulation, an (exponential-time) approximation algorithm was presented in Reference [42].

1.3.2.4 *Combinatorial branch-and-bound Approaches.* Wang and Xu [65] proposed a simple combinatorial branch-and-bound approach. The solution is built by enumerating all possible resolutions for each of the genotypes in turn. The lower bound is the number of haplotypes used so far. Since the search space is exponential and the bound is weak, the method is not able to solve instances of size comparable to the other approaches. Even the solution for 20 genotypes over 20 SNPs can sometimes take an extremely long time to be found.

1.3.2.5 *Boolean Optimization.* One of the approaches applied to PPH is *Pseudo Boolean Optimization* (PBO). PBO is a technique by which a problem can be modeled via integer linear constraints over a set of boolean variables. The goal is to find an assignment of the variables that satisfies all constraints and minimizes a given objective function. The model is solved by a SAT solver (much in the same way as an ILP model is solved by an ILP solver), which employs solution-searching techniques specific for boolean models.

The PBO models for PPH are mostly based on the following feasibility question: given a tentative cardinality k, does there exist a set H of k haplotypes that resolves G? The feasibility question is expressed in terms of boolean variables similar to the binary variables of the polynomial ILP models. Starting at a lower bound, k is increased until the feasibility question has a positive answer. At that point, H represents the optimal solution.

The first PBO algorithm for PPH was presented by Lynce and Marques-Silva [54]. Later works aimed at breaking the symmetries in the original PBO model, and computing tight lower and upper bounds to be used for pruning the search space Graca et al. [31], Marques-Silva et al. [55]. The SAT approaches showed to be competitive with the best mathematical programming approaches for PPH.

1.3.2.6 Heuristics. In general, all the exact models mentioned so far run into troubles when trying to solve "large" instances (where the most critical parameter is the number of ambiguous positions per genotype). The exponential models imply the creation of too many variables and/or constraints for obtaining a solution within a reasonable time. The polynomial and combinatorial models, on the other hand, employ quite weak lower bounds so that closing the gap and terminating the branch-and-bound search is again impossible within a reasonable time. In order to solve large instances of PPH, one needs then to resort to the use of effective heuristics that can find near-optimal solutions to instances with hundreds of genotypes over hundreds of SNPs with high levels of heterozygosity.

One such heuristic procedure is `CollHaps`, proposed by Tininini et al. [61]. `CollHaps` is based on the representation of the solution as a matrix M' of $2m$ rows (the solving haplotypes, also called *symbolic haplotypes* because they may contain variables), in which some entries are fixed, while others have to be determined. The entries to be determined correspond to heterozygous positions in the input genotypes. To each such entry, `CollHaps` associates a 0–1 variable, which is then set to a fixed value in the course of the procedure. At each step, `CollHaps` tries to maintain as few distinct symbolic haplotypes as possible, which is done by trying to make identical some rows of M' via a greedy setting of the variables. The idea is to eventually end up with as few distinct actual haplotypes as possible. Experimental results have shown that `CollHaps` is a very effective and accurate heuristic for PPH, and ranks among the best available procedures for this problem.

In another heuristic approach for PPH, Di Gaspero and Roli [27] proposed the use of stochastic local search, which they considered to yield a reasonable compromise between the quality of the solutions and running time of the procedure. In their work, Di Gaspero and Roli utilized a family of local search strategies, such as *Best Improvement* (BI), *Stochastic First Improvement* (SFI), *Simulated Annealing* (SA), and *Tabu Search* (TS).

1.3.3 Perfect Phylogeny

One limitation of the PPH model is that it does not take into account the fact that haplotypes may evolve over time. Haplotypes evolve by mutation and recombination, and an individual can possess two haplotypes such that neither one is possessed by one of her parents. The haplotype regions in which no recombination and/or mutations have ever happened in a population over time are called *blocks*. For reconstructing haplotype blocks, starting from genotype data of a block in a population, the parsimony model is the most appropriate [37]. However, when genotypes span several blocks, different models of haplotyping have been considered. One of these is haplotyping for *perfect phylogeny*.

The perfect phylogeny model is used under the hypothesis that no recombination events happened, but there were only mutations. It is assumed that at the beginning there existed only one ancestral haplotype, and new haplotypes were derived over time from existing haplotypes as follows. If at some point there existed a haplotype h in the population and then a mutation of $h[i]$ happened, which was then passed on to the new generation, a new haplotype h' started to exist, with $h'[j] = h[j]$ for $j \neq i$, and $h'[i] = 1 - h[i]$. We say that h is the "father" of h' in the tree of haplotype evolution. In the *infinite-sites coalescent model* [44], once a site has mutated it cannot mutate back to its original state. Hence, the evolution of the haplotypes can be described by a rooted arborescence, in which the haplotypes are the vertices and each arc is directed from father to child.

A perfect phylogeny is such an arborescence. Given a set H of haplotypes and a haplotype h^* from which all other haplotypes have evolved, a perfect phylogeny for H is a rooted binary tree T such that the following holds:

1. The root of T corresponds to h^*.
2. The leaves of T are in one-to-one correspondence with H.
3. Each position $i \in 1, \ldots, n$ labels at most one edge in T.
4. For each leaf $h \in H$ and edge e along the path from h^* to h, if e is labeled with position i, then $h[i] \neq h^*[i]$.

Without loss of generality, it can be assumed that $h^* = \texttt{00} \ldots \texttt{0}$. It can be shown that a perfect phylogeny for H exists if and only if there are no four haplotypes $h^1, \ldots, h^4 \in H$ and two positions i, j such that

$$\{h^a[i]\ h^a[j]\ , 1 \leq a \leq 4\} = \{\texttt{00}, \texttt{01}, \texttt{10}, \texttt{11}\}.$$

The *haplotyping for perfect phylogeny* problem can then be stated as follows: given a set G of genotypes, find a set H of haplotypes such that H resolves G and there is a perfect phylogeny for H.

Haplotyping for perfect phylogeny was introduced by Gusfield [36], who first showed that the problem is polynomial and conjectured the existence of a linear-time algorithm for its solution. To prove that it is polynomial, Gusfield reduced the problem to a well-known graph theory problem, that is, the *graph realization*, with complexity $O(nm\ \alpha(n, m))$, where α is the slow-growing inverse Ackerman function. Although the complexity is nearly linear-time, implementing the algorithm for graph realization is very difficult. Much simpler to implement while still very effective algorithms were designed by Bafna et al. [2] and Eskin et al. [26]. These algorithms are based on combinatorial insights in the problem structure and on the analysis of the 2-bit patterns implied by the heterozygous sites. The complexity for both algorithms is $O(n^2 m)$. Eventually, Ding et al. [24] were able to obtain an algorithm for perfect phylogeny haplotyping of complexity $O(nm)$, that is, a linear-time algorithm. The main idea of the algorithm is to find the maximal subgraphs that are common to all feasible solutions. These subgraphs are represented via an original and effective data structure called a *shadow tree*, which is built and updated in linear time.

Almost at the same time as Ding et al. obtained their result, another linear-time algorithm for perfect phylogeny haplotyping was proposed by Bonizzoni [9], who showed that the problem solution can be obtained via the recognition of special posets of width 2.

The perfect phylogeny model does not take into account events such as back mutations. Bonizzoni et al. [10] considered a variant of the perfect phylogeny model called *Persistent Perfect Phylogeny* (P-PP). In the P-PP model, each SNP site can mutate to a new value and then back to its original value only once. Furthermore, they considered the case of incomplete data, that is, a SNP matrix in which some entries may be missing. They developed an exact algorithm for solving the P-PP problem that is exponential in the number of SNPs and polynomial in the number of individuals.

1.3.4 Disease Association

As discussed in the introduction, haplotypes are very useful for diagnostic and genetic studies as they are related to the presence/absence of genetic diseases.

In a very simplistic way, we can define a genetic disease as the malfunctioning of a specific gene. A gene does not function properly when its encoding sequence has been mutated with respect to one of its correct versions. Since each gene is present in two copies, it may be the case that either one of the two copies is malfunctioning. A genetic disease is said to be *recessive* if a person shows the symptoms of the disease only when *both* gene copies are malfunctioning (examples of recessive diseases are cystic fibrosis and sickle cell anemia). Note that, for a recessive disease, one can be a healthy carrier, that is, one copy is malfunctioning but the other is working properly. A genetic disease is called *dominant* if a person shows the symptoms of the disease when *at least one* gene copy is malfunctioning (examples of dominant diseases are Huntington's disease and Marfan's syndrome).

Let us consider the genotypes of a population consisting of healthy and diseased individuals. The haplotypes correspond to the gene sequences. Let us call "good" a haplotype corresponding to a sequence encoding a working gene, and "bad" a haplotype for which the encoded gene is malfunctioning. For a dominant disease, one individual is diseased if just one of her haplotypes is bad, while if the disease is recessive, both haplotypes must be bad for the individual to be diseased. In the context of haplotyping with respect to a disease, there should exist a coloring of the haplotypes into good and bad that is consistent with the genotypes of healthy and diseased individuals. Assume, for example, that the disease under study is recessive. Then we have the following problem, called *Haplotyping for Disease Association* (HDA): we are given a set G_D of *diseased* genotypes (i.e., belonging to diseased individuals) and a set G_H of *healthy* genotypes (i.e., belonging to healthy individuals). We want to find a set H of haplotypes, partitioned into H_G (good haplotypes) and H_B (bad haplotypes) such that

(i) H resolves $G_H \cup G_D$.
(ii) $\forall g \in G_H$, there is a resolution $g = h' \oplus h''$ in H, such that $|\{h', h''\} \cap H_G| \geq 1$.
(iii) $\forall g \in G_D$, there is a resolution $g = h' \oplus h''$ in H, such that $|\{h', h''\} \cap H_B| = 2$.

Note that by reversing the role of good versus bad and healthy versus diseased, the same formulation can be adopted to study a dominant disease.

The problem of HDA was introduced by Greenberg et al. [32] and was studied by Lancia et al. [49]. The problem is NP-hard, as shown in Reference [49]. However, real-life data are much simpler to solve than the artificial instances built in the NP-hardness reduction from satisfiability. In fact, in the same paper, it has been proved that, provided that the quite weak constraint of having at least two heterozygous sites per genotype holds, HDA is polynomially solvable.

1.3.5 Other Models

We have seen that if an individual is homozygous at a given SNP site, her genotype does not only specify that the individual is homozygous but it also specifies the allele at that particular site. This type of information carries a cost and there also exists a different type of genotype, called *XOR-genotype*, which is cheaper to obtain than a "normal" genotype. The XOR-genotype still distinguishes heterozygous sites from homozygous sites, but it does not identify the homozygous alleles. If we denote by the letter E the fact that an individual is heterozygous at a certain SNP site, and by the letter O the fact that she is homozygous, an X-OR genotype is then a vector over the alphabet $\{\mathtt{E}, \mathtt{O}\}$.

When the input data consist of XOR-genotypes rather than normal genotypes, similar population haplotyping problems as those described above can be defined. In particular, Bonizzoni et al. [12] considered the problem of PPH for XOR-genotype data. They gave exact polynomial-time solutions to some restricted cases of the problem, and both a fixed-parameter algorithm and a heuristic for the general case. These results are based on an insight into the combinatorial properties of a graph representation of the solutions.

The perfect phylogeny haplotyping problem for XOR-genotypes was considered by Barzuza et al. [5]. They showed how to resolve XOR-genotypes under the perfect phylogeny model and studied the degrees of freedom in such resolutions. In particular, they showed that when the input XOR-genotypes admit only one possible resolution, the full genotype of at most three individuals would have sufficed in order to determine all haplotypes of the phylogeny.

REFERENCES

1. Aguiar D, Istrail S. Hapcompass: a fast cycle basis algorithm for accurate haplotype assembly of sequence data. J Comput Biol 2012;19(6):577–590.
2. Bafna V, Gusfield D, Lancia G, Yooseph S. Haplotyping as perfect phylogeny: a direct approach. J Comput Biol 2003;10:323–340.
3. Bansal V, Bafna V. HapCUT: an efficient and accurate algorithm for the haplotype assembly problem. Bioinformatics 2008;24:i153–i159.
4. Bansal V, Halpern A, Axelrod N, Bafna V. An MCMC algorithm for haplotype assembly from whole-genome sequence data. Genome Res 2008;18(8):1336–1346.
5. Barzuza T, Beckmann JS, Shamir R, Pe'er I. Computational problems in perfect phylogeny haplotyping: XOR-genotypes and tag SNPs. In: *15th Annual Symposium on Combinatorial Pattern Matching (CPM)*. Volume 3109, Lecture Notes in Computer Science. Springer-Verlag; 2004. p 14–31.
6. Wu J, Liang B. A fast and accurate algorithm for diploid individual haplotype reconstruction. J Bioinform Comput Biol 2013;11(4).
7. Bayzid Sa, Alam M, Mueen A, Rahman S. Hmec: a heuristic algorithm for individual haplotyping with minimum error correction. ISRN Bioinformatics 2013;(Article ID.291741):1–10.
8. Bertolazzi P, Godi A, Labbè M, Tininini L. Solving haplotyping inference parsimony problem using a new basic polynomial formulation. Comput Math Appl 2008;55(5):900–911.
9. Bonizzoni P. A linear-time algorithm for the perfect phylogeny haplotype problem. Algorithmica 2007;48(3):267–285.
10. Bonizzoni P, Braghin C, Dondi R, Trucco G. The binary perfect phylogeny with persistent characters. Theor Comput Sci 2012;454:51–63.
11. Bonizzoni P, Della Vedova G, Dondi R, Li J. The haplotyping problem: an overview of computational models and solutions. J Comput Sci Technol 2004;19(1):1–23.
12. Bonizzoni P, Della Vedova G, Dondi R, Pirola Y, Rizzi R. Pure parsimony XOR haplotyping. IEEE/ACM Trans Comput Biol Bioinform 2010;7:598–610.
13. Brown DG, Harrower IM. A new integer programming formulation for the pure parsimony problem in haplotype analysis. In: *Proceedings of Annual Workshop on Algorithms in Bioinformatics (WABI)*, Lecture Notes in Computer Science. Springer-Verlag; 2004. p 254–265.
14. Brown DG, Harrower IM. A new formulation for haplotype inference by pure parsimony. Technical Report CS-2005-03. Waterloo: University of Waterloo, Department of Computer Science; 2005.
15. Campelo M, Campos V, Correa R. On the asymmetric representatives formulation for the vertex coloring problem. Discrete Appl Math 2008;156(7):1097–1111.
16. Catanzaro D, Godi A, Labbè M. A class representative model for pure parsimony haplotyping. INFORMS J Comput 2010;22:195–209.
17. Catanzaro D, Labbè M. The pure parsimony haplotyping problem: overview and computational advances. Int Trans Oper Res 2009;16:561–584.
18. Chen Z, Deng F, Wang L. Exact algorithms for haplotype assembly from whole-genome sequence data. Bioinformatics 2013;29(16):1938–1945.

19. Chen Z, Fu B, Schweller R, Yang B, Zhao Z, Zhu B. Linear time probabilistic algorithms for the singular haplotype reconstruction problem from SNP fragments. J Comput Biol 2008;15(5):535–546.
20. Cilibrasi R, Iersel LV, Kelk S, Tromp J. On the complexity of several haplotyping problems. In: *Proceedings of Annual Workshop on Algorithms in Bioinformatics (WABI)*. Volume 3692, Lecture Notes in Computer Science. Springer-Verlag; 2005. p 128–139.
21. Clark A. Inference of haplotypes from PCR amplified samples of diploid populations. Mol Biol Evol 1990;7:111–122.
22. Collins FS, Morgan M, Patrinos A. The human genome project: lessons from large-scale biology. Science 2003;300(5617):286–290.
23. Deng F, Cui W, Wang L. A highy accurate heuristic algorithm for the haplotype assembly problem. BMC Genomics 2013;14:1–10.
24. Ding Z, Filkov V, Gusfield D. A linear-time algorithm for the perfect phylogeny haplotyping problem. In: *Proceedings of the Annual International Conference on Computational Molecular Biology (RECOMB)*. New York: ACM Press; 2005.
25. Duitama J, Huebsch T, McEwen G, Suk EK, Hoehe M. Refhap: a reliable and fast algorithm for single individual haplotyping. Proceedings of the 1st ACM International conference on Bioinformatics and Computational Biology, DMTCS'03. New York: ACM; 2010. p 160–169.
26. Eskin E, Halperin E, Karp R. Efficient reconstruction of haplotype structure via perfect phylogeny. J Bioinform Comput Biol 2003;1(1):1–20.
27. Di Gaspero L, Roli A. Stochastic local search for large-scale instances of the haplotype inference problem by pure parsimony. J Algorithms 2008;63:55–69.
28. Genovese L, Geraci F, Pellegrini M. Speedhap: an accurate heuristic for the single individual SNP haplotyping problem with many gaps, high reading error rate and low coverage. IEEE/ACM Trans Comput Biol Bioinform 2008;5(4):492–502.
29. Geraci F. A comparison of several algorithms for the single individual SNP haplotyping reconstruction problem. Bioinformatics 2010;26(18):2217–2225.
30. Geraci F, Pellegrini M. Rehap: an integrated system for the haplotype assembly problem from shotgun sequencing data. In: Fred ALN, Filipe J, Gamboa H, editors. *BIOINFORMATICS 2010 - Proceedings of the 1st International Conference on Bioinformatics*. INSTICC Press: 2010. p 15–25.
31. Graca A, Marques-Silva J, Lynce I, Oliviera AL. Efficient haplotype inference with pseudo-boolean optimization. In: *2nd International Conference on Algebraic Biology (AB)*. Volume 4545, Lecture Notes in Computer Science. Springer-Verlag; 2007. p 125–139.
32. Greenberg H, Hart W, Lancia G. Opportunities for combinatorial optimization in computational biology. INFORMS J Comput 2004;16(3):1–22.
33. Gusfield D. Inference of haplotypes from PCR-amplified samples of diploid populations: complexity and algorithms. Technical Report cse–99–6. Davis (CA): University of California at Davis, Department of Computer Science; 1999.
34. Gusfield D. A practical algorithm for optimal inference of haplotypes from diploid populations. In: Altman R, Bailey TL, Bourne P, Gribskov M, Lengauer T, Shindyalov IN,

Ten Eyck LF, Weissig H, editors. *Proceedings of the Annual International Conference on Intelligent Systems for Molecular Biology (ISMB)*. Menlo Park (CA): AAAI Press; 2000. p 183–189.

35. Gusfield D. Inference of haplotypes from samples of diploid populations: complexity and algorithms. J Comput Biol 2001;8(3):305–324.
36. Gusfield D. Haplotyping as perfect phylogeny: conceptual framework and efficient solutions. In: Myers G, Hannenhalli S, Istrail S, Pevzner P, Waterman M, editors. *Proceedings of the Annual International Conference on Computational Molecular Biology (RECOMB)*. New York: ACM Press; 2002. p 166–175.
37. Gusfield D. Haplotype inference by pure parsimony. In: *Proceedings of the Annual Symposium on Combinatorial Pattern Matching (CPM)*. Volume 2676, Lecture Notes in Computer Science. Springer-Verlag; 2003. p 144–155.
38. Gusfield D, Orzack SH. Haplotype inference. In: Aluru S, editor. *Handbook of Computational Molecular Biology*. Boca Raton (FL): Champman and Hall/CRC-press; 2005. p 1–28.
39. Halldórsson B, Bafna V, Edwards N, Lippert R, Yooseph S, Istrail S. Combinatorial problems arising in SNP and haplotype analysis. In: *Proceedings of the 4th International Conference on Discrete Mathematics and Theoretical Computer Science, DMTCS'03*. Berlin Heidelberg: Springer-Verlag; 2003. p 26–47.
40. Halldorsson BV, Bafna V, Edwards N, Lippert R, Yooseph S, Istrail S. A survey of computational methods for determining haplotypes. In: *Computational Methods for SNP and Haplotype Inference: DIMACS/RECOMB Satellite Workshop*. Volume 2983, Lecture Notes in Computer Science. Springer-Verlag; 2004. p 26–47.
41. He D, Choi A, Pipatsrisawat K, Darwiche A, Eskin E. Optimal algorithms for haplotype assembly from whole-genome sequence data. Bioinformatics 2010;26(12):i83–i190.
42. Huang YT, Chao KM, Chen T. An approximation algorithm for haplotype inference by maximum parsimony. ACM Symposium on Applied Computing (SAC); 2005. p 146–150.
43. Hubbel E. Unpublished manuscript; 2002.
44. Hudson R. Gene genealogies and the coalescent process. Oxf Surv Evol Biol 1990;7:1–44.
45. Kalpakis K, Namjoshi P. Haplotype phasing using semidefinite programming. 5th IEEE Symposium on Bioinformatics and Bioengineering (BIBE). Minneapolis; 2005. p 145–152.
46. Kargar M, Poormohammadi H, Pirhaji L, Sadeghi M, Pezeshk H, Eslahchi C. Enhanced evolutionary and heuristic algorithms for haplotype reconstruction problem using minimum error correction model. MATCH Commun. Math. Comput. Chem. 2009;62:261–274.
47. Lancia G, Bafna V, Istrail S, Lippert R, Schwartz R. SNPs problems, complexity and algorithms. In: *Proceedings of the Annual European Symposium on Algorithms (ESA)*. Volume 2161, Lecture Notes in Computer Science. Springer-Verlag; 2001. p 182–193.
48. Lancia G, Pinotti C, Rizzi R. Haplotyping populations by pure parsimony: complexity, exact and approximation algorithms. INFORMS J Comput 2004;16(4):17–29.
49. Lancia G, Ravi R, Rizzi R. Haplotyping for disease association: a combinatorial approach. IEEE/ACM Trans Comput Biol Bioinform 2008;5(2):245–251.
50. Lancia G, Rizzi R. A polynomial solution to a special case of the parsimony haplotyping problem. Oper Res Lett 2006;34(3):289–295.

51. Lancia G, Serafini P. A set covering approach with column generation for parsimony haplotyping. INFORMS J Comput 2009;21(1):151–166.
52. Li L, Kim J, Waterman M. Haplotype reconstruction from SNP alignment. J Comput Biol 2004;11:507–518.
53. Lippert R, Schwartz R, Lancia G, Istrail S. Algorithmic strategies for the SNPs haplotype assembly problem. Brief Bioinform 2002;3(1):23–31.
54. Lynce I, Marques-Silva J. SAT in bioinformatics: making the case with haplotype inference. In: *Computational Methods for SNP and Haplotype Inference: DIMACS/RECOMB Satellite Workshop*. Volume 4121, Lecture Notes in Computer Science. Springer-Verlag; 2006. p 136–141.
55. Marques-Silva J, Lynce I, Oliviera AL. *Efficient and Tight Upper Bounds for Haplotype Inference by Pure Parsimony Using Delayed Haplotype Selection*. Volume 4874, Lecture Notes in Artificial Intelligence. Springer-Verlag; 2007. p 621–632.
56. Panconesi A, Sozio M. Fast hare: a fast heuristic for single individual SNP haplotype reconstruction. In: *Proceedings of Annual Workshop on Algorithms in Bioinformatics (WABI)*, Volume 3240, Algorithms in Bioinformatics. Springer-Verlag; 2004. p 266–277.
57. Rizzi R, Bafna V, Istrail S, Lancia G. Practical algorithms and fixed-parameter tractability for the single individual SNP haplotyping problem. In: Guigo R, Gusfield D, editors. *Proceedings of Annual Workshop on Algorithms in Bioinformatics (WABI)*. Volume 2452, Lecture Notes in Computer Science. Springer-Verlag; 2002. p 29–43.
58. Rizzi R, Bafna V, Istrail S, Lancia G. Polynomial and APX-hard cases of the individual haplotyping problem. Theor Comput Sci 2005;335:109–125.
59. Schwartz R. Theory and algorithms for the haplotype assembly problem. Commun Inf Syst 2010;10(1):23–38.
60. Stephens M, Smith N, Donnelly P. A new statistical method for haplotype reconstruction from population data. Am J Hum Genet 2001;68:978–989.
61. Tininini L, Bertolazzi P, Godi A, Lancia G. Collhaps: a heuristic approach to haplotype inference by parsimony. IEEE/ACM Trans Comput Biol Bioinform 2010;7(3):511–523.
62. Venter JC et al. The sequence of the human genome. Science 2001;291:1304–1351.
63. Wang R, Wu L, Li Z, Zhang X. Haplotype reconstruction from SNP fragments by minimum error correction. Bioinformatics 2005;21(10):2456–2462.
64. Wang R, Wu L, Zhang X, Chen L. A markov chain model for haplotype assembly from SNP fragments. Genome Inform 2006;17(2):162–171.
65. Wang L, Xu Y. Haplotype inference by maximum parsimony. Bioinformatics 2003; 19(14):1773–1780.
66. Wu J, Wang J, Chen J. A heuristic algorithm for haplotype reconstruction from aligned weighted SNP fragments. Int J Bioinform Res Appl 2013;9(1):13–24.
67. Xie M, Wang J. An improved (and practical) parametrized algorithm for the individual haplotyping problem MFR with mate pairs. Algorithmica 2008;52:250–266.
68. Xie M, Wang J, Chen J. A model of higher accuracy for the individual haplotyping problem based on weighted SNP fragments and genotype with errors. Bioinformatics 2008;24(13):i105–i113.
69. Xie M, Wang J, Jiang T. A fast and accurate algorithm for single individual haplotyping. BMC Syst Biol 2012;6 Suppl 2:1–10.

70. Zhang XS, Wang RS, Wu LY, Chen L. Models and algorithms for haplotyping problem. Curr Bioinform 2006;1:105–114.
71. Zhang X, Wang R, Wu A, Zhang W. Minimum conflict individual haplotyping from SNP fragments and related genotype. Evol Bioinform Online 2006;2:271–280.
72. Zhao Y, Wu L, Zhang J, Wang R, Zhang X. Haplotype assembly from aligned weighted SNP fragments. Comput Biol Chem 2005;29(4):281–287.

CHAPTER 2

ALGORITHMIC PERSPECTIVES OF THE STRING BARCODING PROBLEMS

Sima Behpour and Bhaskar DasGupta

Department of Computer Science, University of Illinois at Chicago, Chicago, USA

2.1 INTRODUCTION

Let Σ be a finite alphabet. A string is a concatenation of elements of Σ. The length of a string x, denoted by $|x|$, is the number of the characters that constitute this string. Let $\mathcal{S}$ be a set of strings over Σ. The simplest "binary-valued version" of the string barcoding problem discussed in this chapter is defined as follows [3, 17]:

Problem name: String barcoding problem ($\text{SB}^{\Sigma}(1)$).

Definition of a barcode: For a string s and a set of strings $\mathcal{T} = \{t_0, t_1, \ldots, t_{m-1}\}$, barcode $(s, \mathcal{T})$ is the boolean vector (b_0, b_1, b_{m-1}),

where $b_i = \begin{cases} 1, & \text{if } t_i \text{ is a } \textit{substring of } s \\ 0, & \text{otherwise} \end{cases}$.

Pattern Recognition in Computational Molecular Biology: Techniques and Approaches,
First Edition. Edited by Mourad Elloumi, Costas S. Iliopoulos, Jason T. L. Wang, and Albert Y. Zomaya.

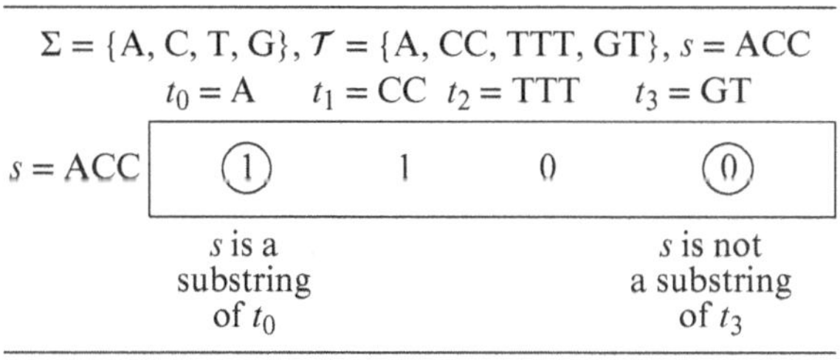

	$\Sigma = \{A, C, T, G\}, \mathcal{T} = \{A, CC, TTT, GT\}, s = ACC$			
	$t_0 = A$	$t_1 = CC$	$t_2 = TTT$	$t_3 = GT$
$s = ACC$	(1)	1	0	(0)
	s is a substring of t_0			s is not a substring of t_3

Input: A set of strings $\mathcal{S}$ over Σ.

Valid solutions: A set of strings $\mathcal{T}$ such that (see Figure 2.1):

$$\forall\, s, s' \in \mathcal{S} : s \neq s' \iff \text{barcode}(s, \mathcal{T}) \neq \text{barcode}(s', \mathcal{T})$$

Objective: *minimize* the length of the barcode $|\mathcal{T}|$.

The basic string barcoding problem $\text{SB}^{\Sigma}(1)$ was generalized in Reference [3] to a "grouped" string barcoding problem $\text{SB}^{\Sigma}(\kappa)$ in the following manner:

Problem name: Grouped string barcoding problem ($\text{SB}^{\Sigma}(\kappa)$).

Definition of a κ-string: A κ-string is a collection of at most κ strings.

Definition of a barcode: For a string s and a set of κ-strings $\mathcal{T} = \{t_0, t_1, \ldots, t_{m-1}\}$, barcode $(s, \mathcal{T})$ is the boolean vector $(b_0, b_1, \ldots, b_{m-1})$, where

$$b_i = \begin{cases} 1, & \text{if there exists a } t \in t_i \text{ for some } i \text{ such that } t \text{ is a } \textit{substring of } s \\ 0, & \text{otherwise} \end{cases}$$

Input: A set $\mathcal{S}$ of strings over Σ.

Valid solutions: A set of κ-strings $\mathcal{T}$ such that

$$\forall\, s, s' \in \mathcal{S} : s \neq s' \iff \text{barcode}(s, \mathcal{T}) \neq \text{barcode}(s', \mathcal{T})$$

Objective: *Minimize* the length of the barcode $|\mathcal{T}|$.

$\Sigma = \{A, C, T, G\}, \mathcal{T} = \{A, CC, TTT, GT\}, \mathcal{S} = \{S_1, S_2, S_3, S_4, S_5\}$

	$t_0 = A$	$t_1 = CC$	$t_2 = TTT$	$t_3 = GT$
$S_1 = AAC$	1	0	0	0
$S_2 = ACC$	1	1	0	0
$S_3 = GGGG$	0	0	0	0
$S_4 = GTGTGG$	0	0	0	1
$S_5 = TTTT$	0	0	1	0

Figure 2.1 An example of a valid barcode.

Finally, the binary-valued basic version of the string barcoding problem $SB^{\Sigma}(1)$ is actually a special case of the more general "integral-valued" version defined as follows [4]:

Problem name: Minimum cost probe set with threshold r ($MCP^{\Sigma}(r)$).

Definition of a r-barcode: For a string s and a set of strings $\mathcal{T} = \{t_0, t_1, \ldots, t_{m-1}\}$, r-barcode $(s, \mathcal{T})$ is the integer vector $(b_0, b_1, \ldots, b_{m-1})$, where

$$b_i = \min\{r, \text{number of occurrences of } t_i \text{ in } s\}$$

$\Sigma = \{A, C, T, G\}$, $\mathcal{T} = \{A, CC, AC, G\}$, $s = ACCCCA$, $r = 2$

	$t_0 = A$	$t_1 = CC$	$t_2 = AC$	$t_3 = G$
$s = ACCCCA$	(2)	(2)	1	0
	$\min\{r, 2\}$	$\min\{r, 3\}$		

Input: Sets $\mathcal{S}$ and $\mathcal{P}$ of strings over Σ and an integer $r > 0$.

Valid solutions: A set of strings $\mathcal{T} \subseteq \mathcal{P}$ such that

$$\forall\, s, s' \in \mathcal{S} : s \neq s' \iff r - \text{barcode}(s, \mathcal{T}) \neq r - \text{barcode}(s', \mathcal{T})$$

Objective: *Minimize* the "length" of the barcode $|\mathcal{T}|$.

Note that if $\mathcal{P}$ is the set of *all* substrings of all strings in $\mathcal{S}$, then $MCP^{\Sigma}(1)$ is precisely $SB^{\Sigma}(1)$. Inclusion relationships among the various barcoding problems defined above is shown in Figure 2.2.

In the rest of this chapter, $\mathcal{OPT}(I)$ (or simply $\mathcal{OPT}$ when I is clear from the context) will denote the optimum value of the objective function for the maximization or minimization problem under consideration. A ϵ-approximate solution (or simply a ϵ-approximation) of a minimization (respectively, maximization) problem is a solution with an objective value no larger than (respectively, no smaller than) ϵ times (respectively, $1/\epsilon$ times) the value of the optimum; an algorithm of *performance* or *approximation ratio* ϵ produces an ϵ-approximate solution. A problem is ϵ-inapproximable under a certain complexity-theoretic assumption means that the problem does not admit a polynomial-time ϵ-approximation algorithm assuming that the complexity-theoretic assumption is true. We assume that the reader is familiar with basic data structures and algorithmic methods found in graduate-level algorithms textbooks such as Reference [5].

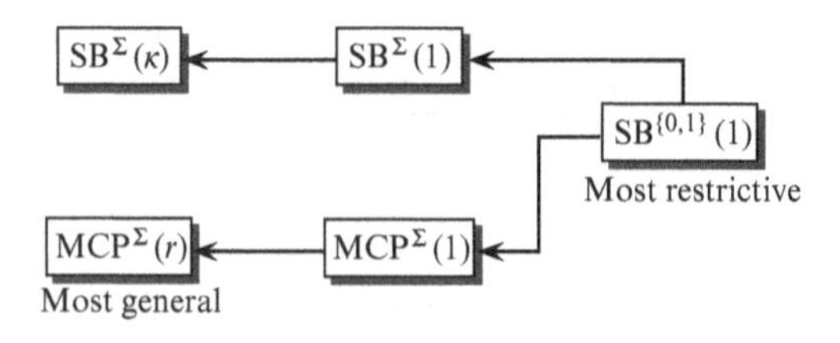

Figure 2.2 Inclusion relationships among the various barcoding problems.

- Motivating biological applications

Applications of barcoding techniques range over a diverse range of applications such as *rapid pathogen identification in epidemic outbreaks*, *database compression*, *point-of-care medical diagnosis*, and *monitoring of microbial communities in environmental studies*. A generic high-level description of most of these applications involving identification of microorganisms or similar entities is as follows. The identification is performed by synthesizing the Watson–Crick complements of the barcodes on a DNA microarray and then hybridizing the fluorescent-labeled DNA extracted from the unknown microorganism to the microarray. Assuming perfect hybridization, the hybridization pattern can be viewed as a string of zeros and ones, which in our terminology is the barcode of the microorganism. By definition, the barcodes corresponding to a set of microorganisms are distinct and thus the barcodes uniquely identify the organisms. Two specific applications of this nature are discussed below.

Pathogen identification in epidemic outbreaks: In the outbreak of an epidemic, possibly as a result of biological warfare, there is an urgent need to identify the pathogen and the family it belongs to *as early as possible*. Such *First Responder Pathogen Detection Systems* (FRPDS) must be able to recognize pathogens from *minute* amounts of genetic material. To enable reliable detection, one usually first amplifies the collected genetic material using high-efficiency techniques such as the *Multiplex Polymerase Chain Reaction*. Classical approaches to pathogen detection, based on sequencing and direct microarray hybridization [14, 21], are practically applicable only when the number of candidate pathogens is small. In a primer-based FRPDS, once the amplimers have been extracted, barcoding techniques can be used to efficiently generate *short* robust signatures (barcodes) via substrings (distinguishers) that can be detected by DNA or RNA hybridization chips such as DNA tag arrays. The compact size of the barcodes optimizes the cost of designing the hybridization array, reduces database size, and allows one to perform extremely rapid comparisons against large databases using a significantly small amount of memory. Moreover, robust barcoding can be error tolerant and may work with minute traces of the unknown sample. This was a main motivation for investigating various versions of the barcoding problems in publications such as References [3, 6, 7, 17].

Monitoring microbial communities: To minimize the number of oligonucleotide probes needed for analyzing populations of ribosomal RNA gene clones by hybridization experiments on DNA microarrays, the $\text{MCP}^{\Sigma}(r)$ problem was formulated and used in Reference [4]; the probes were selected in Reference [4] from a prespecified set ($\mathcal{P}$ in our notation).

In real applications, the string barcoding problems are further complicated by factors such as the occurrence of a substring that may be approximate due to several reasons such as noise and experimental errors. To address these issues, the *robustness* of designed barcodes is improved by using the grouped string barcoding problem $\text{SB}^{\Sigma}(k)$ that integrates the basic barcoding problem with group testing approach for designing probes [18] by allowing groups of distinguishers to differentiate between strings.

2.2 SUMMARY OF ALGORITHMIC COMPLEXITY RESULTS FOR BARCODING PROBLEMS

For the case when the alphabet Σ is allowed to have arbitrarily many symbols, the NP-hardness of $\text{MCP}^{\Sigma}(1)$ follows from a result by Garey and Johnson [11, p. 71] via a reduction from the three-dimensional matching problem, and heuristics algorithms for $\text{MCP}^{\Sigma}(1)$ were discussed by Moret and Shapiro [15].

The (unweighted) *Minimum Set Cover* (MSC) problem is a well-known combinatorial problem that is defined as follows:

Input: A universe of n elements $U = \{u_1, u_2, \ldots, u_n\}$ and a collection of m sets $\Delta_U = \{ S_1, S_2, \ldots, S_m \}$ with $\bigcup_{j=1}^{m} S_j \supseteq U$.

Valid solutions: A subset of indices $I \subseteq \{1, 2, \ldots, m\}$ of selected sets such that

$$\forall\, u_i \in U : \mid j \in I : \; u_i \in S_j \mid \geq 1$$

Objective: *Minimize* the number of selected sets $\mid I \mid$.

Let α denote the maximum number of elements in any set in Δ, that is, let $\alpha = \max_{i \in \{1,2,\ldots,m\}} \{ \mid S_i \mid \}$. A well-known greedy approach for solving MSC, shown in Figure 2.3, repeatedly selects a new set that covers a maximum number of "not yet covered" elements. This algorithm is known to have an approximation ratio of $(1 + \ln \alpha)$ [13, 20] and can be easily implemented to run in $O(n + m \log m)$ time.

For three strings x, y, and z, we say x "distinguishes" y from z if and only if x is a substring of exactly one of the two strings y and z. It is not very difficult to translate an instance of either $\text{SB}^{\Sigma}(1)$ or $\text{MCP}^{\Sigma}(r)$ to an instance of MSC as follows:

- For a given instance $S = \{ s_1, s_2, \ldots, s_n \}$ of $\text{SB}^{\Sigma}(1)$, the corresponding instance of MSC is defined as

$$U = \left\{ u_{i,j} \;\middle|\; i, j \in \{1, 2, \ldots, n\} \text{ and } i < j \right\}$$

$$\Delta_U = \left\{ S_x = \bigcup_{i,j} \{u_{i,j}\} \;\middle|\; \exists \ell : (x \text{ is a substring of } s_\ell) \text{ and } (x \text{ distinguishes } s_i \text{ from } s_j) \right\}$$

```
I = ∅, uncovered = U
while uncovered ≠ ∅ do
    select an index j ∈ {1, 2, ..., m} \ I that maximizes |uncovered ∩ S_j|
    uncovered = uncovered \ S_j ; I = I ∪ {j}
endwhile
```

Figure 2.3 A greedy algorithm for solving MSC [13].

- For a given instance $\mathcal{S} = \{ s_1, s_2, \ldots, s_n \}$ and $\mathcal{P}$ of $\text{MCP}^{\Sigma}(r)$, the corresponding instance of MSC is defined as

$$U = \left\{ u_{i,j} \mid i,j \in \{1,2,\ldots,n\} \text{ and } i < j \right\}$$

$$\Delta_U = \left\{ S_x = \bigcup_{i,j} \{u_{i,j}\} \;\middle|\; (x \in \mathcal{P}) \text{ and } (x \text{ distinguishes } s_i \text{ from } s_j) \right\}$$

Thus, we can use the greedy algorithm for MSC to approximate both $\text{SB}^{\Sigma}(1)$ and $\text{MCP}^{\Sigma}(r)$. For $\text{SB}^{\Sigma}(1)$, $|U| = \binom{n}{2} = O(n^2)$, $|\Delta_U| \leq \sum_{\ell=1}^{n} \binom{|s_\ell|}{2} = O\left(\sum_{\ell=1}^{n} |s_\ell|^2\right)$, and $\max_{S_x \in \Delta_U} \{|S_x|\} \leq \frac{|U|}{2} = \frac{n(n-1)}{4}$, giving an approximation algorithm that runs in $O\left(\left(\sum_{\ell=1}^{n} |s_\ell|^2\right) \log \left(\sum_{\ell=1}^{n} |s_\ell|^2\right) + n^2\right)$ time and has an approximation ratio of $1 + \ln \frac{n(n-1)}{4} \approx 2 \ln n - 0.4$. For $\text{MCP}^{\Sigma}(r)$, $|U| = \binom{n}{2} = O(n^2)$, $|\Delta_U| \leq |\mathcal{P}|$, and $\max_{S_x \in \Delta_U} |S_x| \leq \frac{|U|}{2} = \frac{n(n-1)}{4}$, giving an approximation algorithm that runs in $O\left(|\mathcal{P}| \log |\mathcal{P}| + n^2\right)$ time and has an approximation ratio of $1 + \ln \frac{n(n-1)}{4} \approx 2 \ln n - 0.4$.

The above-mentioned NP-hardness results and approximation algorithms were further improved by the authors of Reference [3] and shown in Table 2.1. In the next two sections, we will discuss some of the methodologies used to prove these improved results.

2.2.1 Average Length of Optimal Barcodes

Via simple information-theoretic argument it follows that for $\text{SB}^{\Sigma}(1)$ we must have $\mathcal{OPT}(I) \geq \log_2 |\mathcal{S}|$ for any instance I of the problem. On the other hand, it is trivial to see that $\mathcal{OPT}(I) \leq |\mathcal{S}| - 1$. Thus, it behooves to investigate the average value of $\mathcal{OPT}$ when the input strings are generated based on some probability distribution. Unfortunately, tight bounds for the average value of $\mathcal{OPT}$ is not currently known. However, the authors of Reference [7] provide a partial answer via the following theorem. The proof of the theorem uses some asymptotic bounds by Odlyzko on average occurrences of substrings in random strings via generating function [[16], Examples 6.4, 6.7, 6.8, 9.3, and 10.11].

Theorem 2.1 *[7] Consider a randomly generated instance of the* $\text{SB}^{\Sigma}(1)$ *of* n *strings over a fixed finite alphabet* Σ *in which each string in* $s_i = s_{i,0}, s_{i,1}, \ldots, s_{i,\ell-1} \in \mathcal{S}$ *is of length exactly* ℓ *and is generated independently randomly with* $\Pr[s_{i,j} = a] = \frac{1}{|\Sigma|}$ *for any* $j \in \{1, 2, \ldots, \ell\}$ *and any* $a \in \Sigma$. *Also assume that* ℓ *is sufficiently large compared to* n. *Then, for a random string* x *over* Σ *of length* $O(\log \ell)$, *the expected number of the strings in* $\mathcal{S}$ *that contain* x *as a substring is* $p\,n$ *for some constant* $0 < p < 1$.

Table 2.1 List of of a subset of approximability results proved in Reference [3].

$L = \max_{\ell \in \{1,2,\cdots,n\}} \{\lvert s_\ell \rvert\}$, $\mathcal{L} = \sum_{\ell=1}^{n} \lvert s_\ell \rvert$, ϵ and δ are constants				
Approximation Hardness			Approximation Algorithm	
Problem name	ρ-Inappro-ximable for $\rho =$	Minimal Assumptions Necessary	Running time	ρ-Approximation for $\rho =$
$SB^{\Sigma}(1)$	$(1-\epsilon)\ln n$	NP $\not\subset$DTIME($n^{\log\log n}$) $\lvert\Sigma\rvert > 1, 0 < \epsilon < 1$	$O(n^3 L^2)$	$1 + \ln n$
$MCP^{\Sigma}(r)$	$(1-\epsilon)\ln n$	NP $\not\subset$ DTIME($n^{\log\log n}$) $\lvert\Sigma\rvert > 1, 0 < \epsilon < 1$	$O((n^2\lvert + \mathcal{L})\lvert\mathcal{P}\rvert)$	$1 + \ln n + \ln(\log_2(\min\{r, n\} + 1))$
$SB^{\Sigma}(\kappa)$	$\Omega(n^{\epsilon})$	NP $\neq$ co-RP $\lvert\Sigma\rvert > 1, \kappa = n^{\delta}$ $0 < \epsilon < \delta < 1/2$		

DTIME($n^{\log\log n}$) denotes the class of problems that can be solved in deterministic quasi-polynomial time and co-RP denotes the class of decision problems that admits a randomized polynomial-time algorithm $\mathcal{A}$ with the property that if the answer to the problem is YES then $\mathcal{A}$ always outputs YES but if the answer to the problem is NO then $\mathcal{A}$ outputs NO with probability at least 1/2 (see Reference [2] for further details)

2.3 ENTROPY-BASED INFORMATION CONTENT TECHNIQUE FOR DESIGNING APPROXIMATION ALGORITHMS FOR STRING BARCODING PROBLEMS

This technique, introduced in Reference [3], is a *greedy* technique based on *information content* (entropy) of a partial solution; the notion of information content is directly related to the Shannon information complexity [1, 19]. In this approach, we seek to select an augmenting step for a partial solution of our optimization problem that optimizes the information content of the augmented partial solution as compared to the original partial solution. A key nontrivial step for applicability of this technique is to define a suitable efficiently computable measure of the information content of a partial solution such that the monotonicity of this measure is ensured with respect to any subset of an optimal solution. For the case of $SB^{\Sigma}(1)$, a high-level overview of the approach is shown below:

Input: A set of strings $\mathcal{S}$ over Σ.

Output: A set of strings $\mathcal{T}$ such that $\forall\ s, s' \in \mathcal{S} : s \neq s' \iff$ barcode $(s, \mathcal{T}) \neq$ barcode $(s', \mathcal{T})$.

Notation for the information content (entropy) of a partial solution:

$\mathcal{H}_{\mathcal{T}}$ for an arbitrary set $\mathcal{T}$ of strings (partial solution) over Σ

Algorithm:

compute $\Gamma(\mathcal{S}) = \{\ s | s$ is a substring of some string in $\mathcal{S}\ \}$
$\mathcal{T} = \emptyset$
while $\mathcal{H}_{\mathcal{T}} \neq 0$ **do**
select $x \in \Gamma(\mathcal{S}) \setminus \mathcal{T}$ that *maximizes* $IC(x, \mathcal{T}) = \mathcal{H}_{\mathcal{T}} - \mathcal{H}_{\mathcal{T} \cup \{x\}}$
$\mathcal{T} = \mathcal{T} \cup \{\ x\ \}$
endwhile

Of course, a key nontrivial step is to figure out a suitable value of the entropy $\mathcal{H}_{\mathcal{T}}$ such that an execution of the above algorithm produces a valid solution with the desired approximation ratio. The authors of Reference [3] define it in the following manner. For an arbitrary set $\mathcal{D}$ of strings over Σ:

- Define an *equivalence relation* $\overset{\mathcal{D}}{\equiv}$ on $\mathcal{S}$ as (for any two $s, s' \in \mathcal{S}$):

$$s \overset{\mathcal{D}}{\equiv} s' \text{ if and only if } \forall\, x \in \mathcal{D}: x \text{ is a substring of } s \equiv x \text{ is a substring of } s'$$

- If the equivalence relation $\overset{\mathcal{D}}{\equiv}$ has ℓ *equivalence classes* of size $p_1, p_2, \ldots, p_\ell > 0$, then $\mathcal{H}_{\mathcal{D}} = \log_2(\Pi_{i=1}^{\ell}(\ p_i!\ )\ )$.

The above definition of entropy is somewhat similar (but not the same) to the one suggested In Reference [15], namely $\frac{1}{|\mathcal{S}|}\log_2(\Pi_{i=1}^{\ell} p_i^{p_i})$, for empirical evaluation purposes.

Note that $\mathcal{H}_{\mathcal{D}} = 0$ implies the equivalence classes of $\overset{\mathcal{D}}{\equiv}$ are $|\mathcal{S}|$ singleton sets each containing one distinct string from $\mathcal{S}$, and if $\mathcal{H}_{\mathcal{D}} \neq 0$, then there exists a $x \in \mathcal{S} - \mathcal{D}$ with $IC(x, \mathcal{D}) > 0$; thus, the algorithm terminates in polynomial time with a valid solution. To prove the desired approximation ratio via an amortized analysis, the authors of Reference [3] first proved the following combinatorial properties of the function $IC(x, \mathcal{D}) > 0$:

- $\forall\, x: \mathcal{D} \subset \mathcal{D}' \ \Rightarrow\ IC(x, \mathcal{D}) \geq IC(x, \mathcal{D}')$, and
- $\forall\, x: IC(x, \emptyset) = h < |\mathcal{S}|$.

Thus, if the algorithm selected strings $x_1, x_2, \ldots, x_q$ in this order in $\mathcal{T}$, then $\sum_{i=1}^{q} IC(x_i, \{x_1, x_2, \ldots, x_{i-1}\}) = \mathcal{H}_{\emptyset} = h < |\mathcal{S}|$. The proof shows how to carefully distribute the cost 1 of adding each extra set in $\mathcal{T}$ to the strings in an optimal solution of $\text{SB}^{\Sigma}(1)$ using the $IC(x_i, \{x_1, x_2, \ldots, x_{i-1}\})$ quantities such that each element in this optimal solution receives a total cost of at most $1 + \int_1^{\sum_{i=1}^{q} IC(x_i, \{x_1, x_2, \ldots, x_{i-1}\})} \mathrm{d}x/x < 1 + \ln h < 1 + \ln|\mathcal{S}|$.

A very similar proof with appropriate modifications work for $\text{MCP}^{\Sigma}(r)$ as well. In this case, $IC(x, \emptyset) = h < |\mathcal{S}| \log_2(\min\{\ r+1, |\mathcal{S}|\}\ )$ and thus each element in this optimal solution receives a total cost of at most $1 + \ln h < 1 + \ln|\mathcal{S}| + \ln\log_2(\min\{\ r+1, |\mathcal{S}|\}\ )$.

2.4 TECHNIQUES FOR PROVING INAPPROXIMABILITY RESULTS FOR STRING BARCODING PROBLEMS

In this section, we review a few techniques from structural complexity theory that were used to prove inapproximability results for various string barcoding problems.

2.4.1 Reductions from Set Covering Problem

An usual starting point for this technique is the following well-known inapproximability result for MSC.

Theorem 2.2 *[9] Assuming* NP $\not\subset$*DTIME* $(n^{\log\log n})$*, instances of the MSC problem whose solution requires at least* $(\log_2 n)^2$ *sets cannot be approximated to within an approximation ratio of* $(1-\epsilon)\ln n$ *for any constant* $\epsilon > 0$ *in polynomial time.*

It seems difficult to transform the above inapproximability result for MSC to an inapproximability bound for $\text{SB}^{\{0,1\}}(1)$ of a similar quality because of the special restrictive nature of $\text{SB}^{\{0,1\}}$, and thus the techniques used by the authors of References [8, 12] does not seem to apply. To overcome this issue, the authors of Reference [3] introduced an intermediate problem, called *test set with order with parameter m problem* and denoted by TSO^m, which could be transformed to $\text{SB}^{\{0,1\}}$. TSO^m is a nontrivial generalization of the well-known NP-hard *minimum test collection problem* in diagnostic testing [[11], p. 71] and is defined as follows:

Problem name: Test set with order with parameter m (TSO^m).

Input: A universe $\mathcal{U} = \{u_1, u_2, \ldots, u_n\}$ of n elements, a collection $\mathcal{S}$ of subsets of $\mathcal{U}$ (tests) that includes the $2n-1$ special sets $S_1 = \{u_1\}, S_2 = \{u_1, u_2\}, S_3 = \{u_1, u_2, u_3\}, \ldots, S_n = \{u_1, u_2, u_3, \ldots, u_n\}, S_{n+1} = \{u_2\}, S_{n+2} = \{u_3\}, \ldots, S_{2n-1} = \{u_n\}$, and a positive integer m.

Valid solutions: A collection $\mathcal{T} \subseteq \mathcal{S}$ of subsets from $\mathcal{S}$ such that

$$\forall\, u_i, u_j \in \mathcal{U} : i \neq j \;\Rightarrow\; \exists T \in \mathcal{T} \text{ such that } |\{u_i, u_j\} \cap T| = 1$$

Objective: *Minimize* cost $(\mathcal{T}) = |\mathcal{T} \setminus \{S_1, S_2, \ldots, S_{2n-1}\}\,| + \frac{1}{m}|\mathcal{T} \cap \{S_1, S_2, \ldots, S_{2n-1}\}\,|$.

The proof is completed by having a transformation between these problems such that the corresponding optimal solutions are closely related as shown schematically below, where $\mathcal{OPT}_{\text{MSC}}$ and $\mathcal{OPT}_{\text{SB}^{\{0,1\}}}$ denote the objective values of an optimal solution of the generated instances of MSC and $\text{SB}^{\{0,1\}}$, respectively. This provides an $(1-\epsilon)$-inapproximability result for $\text{SB}^{\{0,1\}}$.

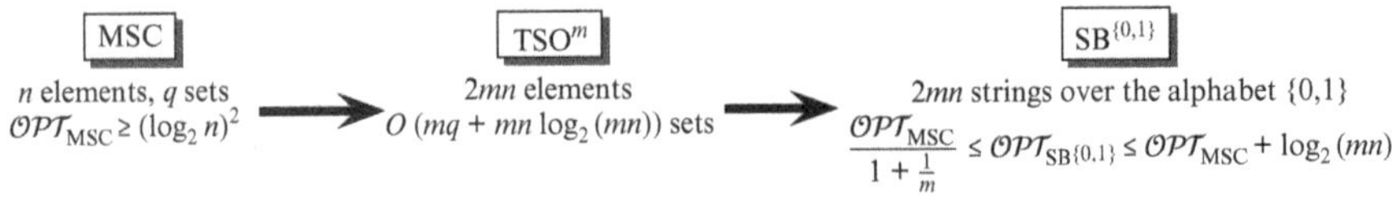

MSC		TSO^m		$\text{SB}^{\{0,1\}}$
n elements, q sets $\mathcal{OPT}_{\text{MSC}} \geq (\log_2 n)^2$	→	$2mn$ elements $O(mq + mn\log_2(mn))$ sets	→	$2mn$ strings over the alphabet $\{0,1\}$ $\frac{\mathcal{OPT}_{\text{MSC}}}{1+\frac{1}{m}} \leq \mathcal{OPT}_{\text{SB}\{0,1\}} \leq \mathcal{OPT}_{\text{MSC}} + \log_2(mn)$

A formal description of the transformations of the input instances among the problems is complicated, so here we just illustrate the transformation with the following simple example for $m = 1$:

Input instance of MSC:

$$U = \{u_1, u_2, u_3, u_4\}, \Delta_U = \{S_1, S_2, S_3\},$$

$$S_1 = \{u_1, u_2, u_4\}, S_2 = \{u_2, u_3\}, S_1 = \{u_2, u_3, u_4\}$$

Tranformed input instance of TSO^1 from input instance of MSC:

$\mathcal{U} = \{u_1, u_2, u_3, u_4, u_5, u_6, u_7, u_8\}$

$\mathcal{S}$ contains the following sets:

$\boldsymbol{B}_1 = \{u_1, u_3, u_7\}, \boldsymbol{B}_2 = \{u_3, u_5\}, \boldsymbol{B}_3 = \{u_3, u_5, u_7\}$ (corresponding to S_1, S_2, S_3 in MSC)

$\boldsymbol{B}_4 = \{u_3, u_4, u_7, u_8\}, \boldsymbol{B}_5 = \{u_5, u_6, u_7, u_8\}$ (additional sets)

(special sets):

$S_1 = \{u_1\}$, $S_2 = \{u_1, u_2\}$, $S_3 = \{u_1, u_2, u_3\}$, $S_4 = \{u_1, u_2, u_3, u_4\}$

$S_5 = \{u_1, u_2, u_3, u_4, u_5\}$, $S_6 = \{u_1, u_2, u_3, u_4, u_5, u_6\}$

$S_7 = \{u_1, u_2, u_3, u_4, u_5, u_6, u_7\}$, $S_8 = \boldsymbol{B}_6 = \{u_1, u_2, u_3, u_4, u_5, u_6, u_7, u_8\}$

$S_9 = \{u_2\}$, $S_{10} = \{u_3\}$, $S_{11} = \{u_4\}$, $S_{12} = \{u_5\}$, $S_{13} = \{u_6\}$, $S_{14} = \{u_7\}$, $S_{15} = \{u_8\}$

Tranformed input instance of SB$^{\{0,1\}}$ from input instance of TSO^1: The *eight* strings over $\Sigma = \{0, 1\}$ are as follows, where 0^i and 1^i indicate a string of i zeros or ones, respectively (e.g., $0^3 = 000$):

$S_1 = \mathbf{0\ 1^1\ 0\ 1^6\ 0 = 0\ 1\ 0\ 111111\ 0}$ (since $u_1 \in B_1, B_6$)

$S_2 = \mathbf{0^2\ 1^6\ 0^2 = 00\ 111111\ 00}$ (since $u_2 \in B_6$)

$S_3 = \mathbf{0^3\ 1^1\ 0^3\ 1^2\ 0^3\ 1^3\ 0^3\ 1^4\ 0^3\ 1^6\ 0^3}$

$\mathbf{= 000\ 1\ 000\ 11\ 000\ 111\ 000}$

$\mathbf{1111\ 000\ 111111\ 000}$ (since $u_3 \in B_1, B_2, B_3, B_4, B_6$)

$S_4 = \mathbf{0^4\ 1^4\ 0^4\ 1^6\ 0^4 = 0000\ 1111\ 0000}$

$\mathbf{111111\ 0000}$ (since $u_4 \in B_4, B_6$)

$S_5 = \mathbf{0^5\ 1^2\ 0^5\ 1^3\ 0^5\ 1^5\ 0^5\ 1^6\ 0^5}$

$\mathbf{= 00000\ 11\ 00000\ 111\ 00000\ 11111}$

$\mathbf{00000\ 111111\ 000000}$ (since $u_5 \in B_2, B_3, B_5, B_6$)

$S_6 = \mathbf{0^6\ 1^5\ 0^6\ 1^6\ 0^6 = 000000\ 11111\ 000000}$

$\mathbf{111111\ 000000}$ (since $u_6 \in B_5, B_6$)

$S_7 = \mathbf{0^7\ 1^1\ 0^7\ 1^3\ 0^7\ 1^4\ 0^7\ 1^5\ 0^7\ 1^6\ 0^7}$

$\mathbf{= 0000000\ 1\ 0000000\ 111\ 0000000}$

$\mathbf{1111\ 0000000\ 11111\ 0000000}$

$\mathbf{1111111\ 0000000}$ (since $u_7 \in B_1, B_3, B_4, B_5, B_6$)

$$S_8 = \mathbf{0^8\ 1^4\ 0^8\ 1^5\ 0^8\ 1^6\ 0^8}$$

$$= \mathbf{00000000\ 1111\ 00000000\ 11111}$$

$$\mathbf{00000000\ 111111\ 00000000} \qquad \text{(since } u_8 \in B_4, B_5, B_6\text{)}$$

For further details, see Reference [3].

2.4.2 Reduction from Graph-Coloring Problem

The $\Omega(n^{\epsilon})$-inapproximability result for $\text{SB}^{\{0,1\}}(n^{\delta})$ under the assumption of $0 < \epsilon < \delta < 1/2$ and NP $\neq$ co-RP is proved in Reference [3] by providing an approximation-preserving reduction from a strong inapproximability result for the graph-coloring problem. The graph-coloring problem is a well-known combinatorial optimization problem defined as follows [11]:

Problem name: Graph coloring.

Input: An undirected unweighted graph $G = (V, E)$.

Valid solutions: An assignment of colors to nodes such that no two adjacent nodes have the same color.

Objective: *Minimize* the number of colors used.

Let $\mathcal{OPT}_{\text{color}}(G)$ denotes the minimum number of colors used in a valid coloring of G. A set of nodes in G is said to be *independent* if no two of them are connected by an edge. Let $\mathcal{OPT}_{\text{ind}}(G)$ denotes the *maximum* number of independent nodes in G. The following strong inapproximability result for computing $\mathcal{OPT}_{\text{color}}(G)$ can be found in Reference [10].

Theorem 2.3 *[10] Assuming* NP $\neq$ *co-RP, there is no polynomial-time algorithm that computes* $\mathcal{OPT}_{\text{color}}(G)$ *with an approximation ratio of* $|V|^{\rho}$ *even if* $\mathcal{OPT}_{\text{ind}}(G) \leq |V|^{\delta}$ *for any two constants* $0 < \rho < \delta < 1$.

As in the case of reduction from the set covering problem in the previous section, the authors of [3] used an intermediate problem, namely the grouped test set (TS^{κ}) problem, that helps in the reduction from graph coloring to $\text{SB}^{\{0,1\}}$. The TS^{κ} problem can be thought as a generalization of the TSO^{m} problem defined in the previous section without the order property and the parameter m; formally, the problem is as follows.

Problem name: Grouped test set (TS^{κ}).

Input: A universe $\mathcal{U} = \{u_1, u_2, \ldots, u_n\}$ of n elements, a collection $\mathcal{S}$ of subsets of $\mathcal{U}$ (tests).

Definition of a κ-test: A κ-test is a union of at most k sets from $\mathcal{S}$.

Valid solutions: A collection $\mathcal{T} \subseteq \mathcal{S}$ of κ-tests such that

$$\forall\, u_i, u_j \in \mathcal{U} : i \neq j \;\Rightarrow\; \exists T \in \mathcal{T} \text{ such that } |\{u_i, u_j\} \cap T| = 1$$

Objective: *Minimize* $\text{cost}(\mathcal{T}) = |\mathcal{T}|$.

As before, one can define a version of TS^{κ} "with order" in the following manner:

Problem name: Grouped test set with order (TS^{κ} with order).

Input: A universe $\mathcal{U} = \{u_1, u_2, \ldots, u_n\}$ of n elements, a collection $\mathcal{S}$ of subsets of $\mathcal{U}$ (tests) such that

$$\{\{u_1\}, \{u_1, u_2\}, \{u_1, u_2, u_3\}, \ldots, \{u_1, u_2, u_3, \ldots, u_n\}, \{u_2\}, \{u_3\}, \ldots, \{u_n\}\} \subseteq \mathcal{S}$$

Valid solutions: A collection $\mathcal{T} \subseteq \mathcal{S}$ of κ-tests such that

$$\forall\, u_i, u_j \in \mathcal{U} : i \neq j \;\Rightarrow\; \exists T \in \mathcal{T} \text{ such that } |\{u_i, u_j\} \cap T| = 1$$

Objective: *Minimize* $\mathrm{cost}(\mathcal{T}) = |\mathcal{T}|$.

The proof is completed by having a transformation between these problems such that the corresponding optimal solutions are closely related as shown schematically below where $\mathcal{OPT}_{\mathcal{P}}$, for a problem $\mathcal{P}$, denotes the objective values of an optimal solution of the generated instances of the problem $\mathcal{P}$. This provides an $\Omega(n^{\epsilon})$-inapproximability result for $\mathrm{SB}^{\{0,1\}}(n^{\delta})$; for further details, see Reference [3].

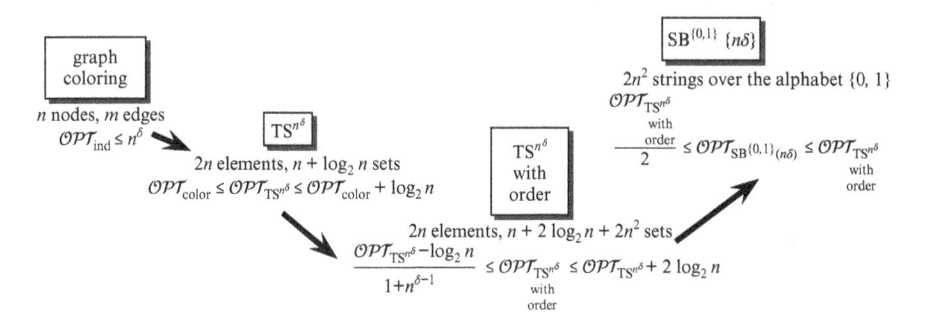

2.5 HEURISTIC ALGORITHMS FOR STRING BARCODING PROBLEMS

In addition to designing efficient algorithms with provable bounds on approximation ratio, one can also consider designing heuristic algorithms for barcoding problems that may not admit a proof of their approximation bounds but nonetheless work well in practice. For the basic binary-valued string barcoding problem SB^{Σ} (1), we outline a few possible heuristic approaches below.

2.5.1 Entropy-Based Method with a Different Measure for Information Content

The greedy approach described in Section 2.3 used a very specific definition of the measure of information content (entropy), namely $\mathcal{H}_{\mathcal{D}} = \log_2(\Pi_{i=1}^{\ell}(\,p_i!\,))$. In principle, the approach can be used with other entropy measures that decrease

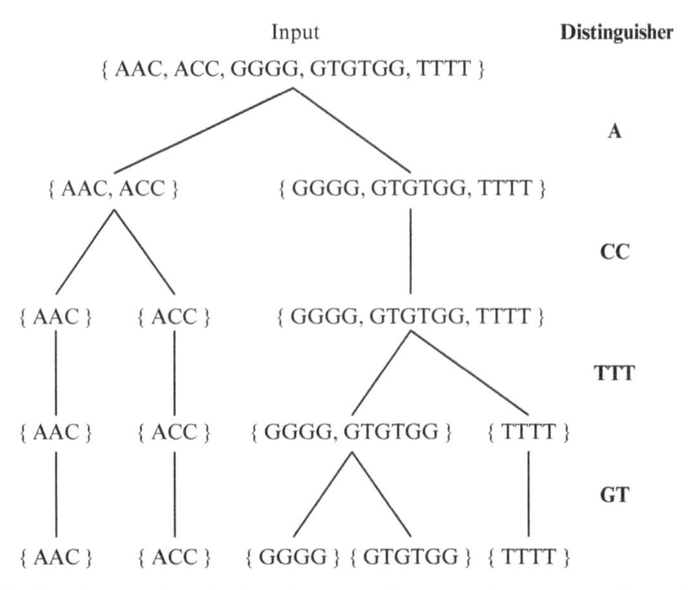

Figure 2.4 Greedy selection of strings in our solution generates a tree partition of the set of input sequences such that each leaf node has exactly one sequence.

monotonically as partial solutions progress toward a complete solution. An appealing candidate is $\mathcal{H}_D = \frac{1}{|S|}\log_2(\Pi_{i=1}^{\ell} p_i^{p_i})$ suggested in Reference [15] as this version of the measure follows the standard entropy definition more closely.

2.5.2 Balanced Partitioning Approach

The $\overset{\mathcal{T}}{\equiv}$ equivalence relation used in Section 2.3 suggests an alternate way of looking at greedy selection of strings to form a barcoding. At every step, the selected string (distinguisher) in the solution affects each equivalence set either by keeping it same or by partitioning the set into two parts. Equivalently, one can view the successive selection of strings in the solution as generating a tree partitioning the given set of input sequences S (see Figure 2.4). Note that the height of the tree is precisely the number of strings in our solution. Thus, one possible greedy strategy is to select a distinguisher greedily at each step that increases the height of the current partition tree by the *least* amount.

2.6 CONCLUSION

In this chapter, we have described a few versions of the string barcoding problems and have reviewed some algorithmic and inapproximability reduction tools to analyze algorithmic complexity questions about these problems. There are other aspects of these problems, such as robustness of barcodes against noises, that arise in practical applications of barcoding in pathogen detections, which we have not discussed here; the reader can find further information about them in the cited references. A software incorporating some of the algorithmic methodologies discussed in this review was reported in Reference [6] and can be found at the website http://dna.engr.uconn.edu/?page_id=23.

From an algorithmic and computational complexity perspective, the following research questions may be worthy of further investigation:

- If the set of strings is generated via a biologically relevant (nonuniform) distribution over the alphabet, what is the expected length of an optimal barcode and what is the computational complexity of finding such a barcode?
- Is there an efficient approximation algorithm for $\text{SB}^{\Sigma}(\kappa)$ when $\kappa > 1$ grows slowly with n (e.g., $\kappa = O(\log n)$) ?

ACKNOWLEDGMENTS

The authors were supported in part by NSF grants IIS-1160995 and DBI-1062328. The authors also thank all their collaborators in their joint barcoding research articles.

REFERENCES

1. Abu-Mostafa YS, editor. *Complexity in Information Theory*. New York: Springer-Verlag; 1986.
2. Balcázar JL, Díaz J, Gabarro J. *Structural Complexity I*, EATCS Monographs on Theoretical Computer Science. Berlin and New York: Springer-Verlag; 1988.
3. Berman P, DasGupta B, Kao M-Y. Tight approximability results for test set problems in bioinformatics. J Comput Syst Sci 2005;71(2):145–162.
4. Borneman J, Chrobak M, Vedova GD, Figueora A, Jiang T. Probe selection algorithms with applications in the analysis of microbial communities. Bioinformatics 2001;1:1–9.
5. Cormen TH, Leiserson CE, Rivest RL, Stein C. *Introduction to Algorithms*. Cambridge: The MIT Press; 2001.
6. DasGupta B, Konwar K, Mandoiu I, Shvartsman A. DNA-BAR: distinguisher selection for DNA barcoding. Bioinformatics 2005;21(16):3424–3426.
7. DasGupta B, Konwar K, Mandoiu I, Shvartsman A. Highly scalable algorithms for robust string barcoding. Int J Bioinform Res Appl 2005;1(2):145–161.
8. De Bontridder KMJ, Halldórsson BV, Halldórsson MM, Hurkens CAJ, Lenstra JK, Ravi R, Stougie L. Approximation algorithms for the test cover problem. Math Program, Ser B 2003;98(1–3):477–491.
9. Feige U. A threshold for approximating set cover. J ACM 1998;45:634–652.
10. Feige U, Kilian J. Zero knowledge and the chromatic number. J Comput Syst Sci 1998; 57(2):187–199.
11. Garey MR, Johnson DS. *Computers and Intractability: A Guide to the Theory of NP-Completeness*. New York: W. H. Freeman and Co.; 1979.

12. Halldórsson BV, Halldórsson MM, Ravi R. On the approximability of the minimum test collection problem. In: *Proceedings of the 9th Annual European Symposium on Algorithms*. Volume 2161, Lecture Notes in Computer Science. Berlin Heidelberg: Springer-Verlag; 2001. p 158–169.
13. Johnson DS. Approximation algorithms for combinatorial problems. J Comput Syst Sci 1974;9:256–278.
14. Ksiazek TG, Erdman D, Goldsmith CS, Zaki SR, Peret T, Emery S, Tong S, Urbani C, Comer JA, Lim W, Rollin PE, Dowell SF, Ling A-E, Humphrey CD, Shieh W-J, Guarner J, Paddock CD, Rota P, Fields B, DeRisi J, Yang J-Y, Cox N, Hughes JM, LeDuc JW, Bellini WJ, Anderson LJ. A novel coronavirus associated with severe acute respiratory syndrome. N Engl J Med 2003;348(20):1953–1966.
15. Moret BME, Shapiro HD. On minimizing a set of tests. SIAM J Sci Stat Comput 1985;6:983–1003.
16. Odlyzko AM. Asymptotic enumeration methods. In: Graham RL, Grötschel M, Lovász L, editors. *Handbook of Combinatorics*. Volume II. Cambridge: The MIT Press; 1995. p 1063–1230.
17. Rash S, Gusfield D. String barcoding: uncovering optimal virus signatures. 6th Annual International Conference on Computational Biology. Washington, DC; 2002. p 54–261.
18. Schliep A, Torney DC, Rahmann S. Group testing with DNA chips: generating designs and decoding experiments. Proc IEEE Comput Syst Conf 2003;2:84–91.
19. Shannon CE. Mathematical theory of communication. Bell Syst Tech J 1948;27:379–423, 623–658.
20. Vazirani V. *Approximation Algorithms*. Berlin Heidelberg: Springer-Verlag; 2001.
21. Wang D, Coscoy L, Zylberberg M, Avila PC, Boushey HA, Ganem D, DeRisi JL. Microarray-based detection and genotyping of viral pathogens. Proc Natl Acad Sci U S A 2002;99(24):15687–15692.

CHAPTER 3

ALIGNMENT-FREE MEASURES FOR WHOLE-GENOME COMPARISON

Matteo Comin[1] and Davide Verzotto[2]

[1]*Department of Information Engineering, University of Padova, Padova, Italy*
[2]*Computational and Systems Biology, Genome Institute of Singapore, Singapore*

3.1 INTRODUCTION

With the progress of modern sequencing technologies, a large number of complete genomes are now available. Traditionally, the comparison of two related genomes is carried out by sequence alignment. There are cases where these techniques cannot be applied, for example, if two genomes do not share the same set of genes, or if they are not alignable to each other due to low sequence similarity, rearrangements, and inversions, or more specifically due to their lengths when the organisms belong to different species. For these cases, the comparison of complete genomes can be carried out only with ad hoc methods that are usually called *alignment-free methods*.

The understanding of the whole human genome and of other species, and the mechanisms behind replication and evolution of genomes are some of the major

Pattern Recognition in Computational Molecular Biology: Techniques and Approaches,
First Edition. Edited by Mourad Elloumi, Costas S. Iliopoulos, Jason T. L. Wang, and Albert Y. Zomaya.

problems in genomics. Although most of the current methods in genome sequence analysis are based only on genetic and annotated regions, this could saturate the problem because of their limited size of information. In fact, recent evidence suggests that the evolutionary information is also carried by the nongenic regions [35], and in some cases we cannot even estimate a complete phylogeny by analyzing the genes shared by a clade of species [10]. Accordingly, this chapter addresses the phylogeny reconstruction problem for different organisms, namely viruses, prokaryotes, and unicellular eukaryotes, using whole-genome pairwise sequence comparison.

The organization of this chapter is as follows: in Section 3.2, we review the problem of whole-genome comparison and the most important methods, and in Section 3.3, we present the *Underlying Approach* (UA). Some results are discussed in Section 3.4, and the conclusions are presented in Section 3.5.

3.2 WHOLE-GENOME SEQUENCE ANALYSIS

The global spread of low-cost high-throughput sequencing technologies has made a large number of complete genomes publicly available, and this number is still growing rapidly. In contrast, only few computational methods can really handle as input entire chromosomes or entire genomes. In this section, we will discuss the use of computational tools for the comparison of whole-genome.

3.2.1 Background on Whole-Genome Comparison

Traditionally, the comparison of related genomes is carried out by sequence alignment. Popular methods extract gene-specific sequences from all species under examination and build a multiple sequence alignment for each gene [42]. Then all multiple sequence alignments are merged to form the final phylogeny. Other methods [24] use genes as a dictionary, counting the presence or absence of a gene. This gene profile is then used to derive a similarity score. However, if the genomes in question do not share a common set of genes, or if they cannot be aligned to each other, for example, due to substantially different lengths, these traditional techniques cannot be applied. As a general example, in a pairwise comparison of genomes, popular alignment tools rely on a specific order of elements for each genome sequence, and on a set of sparse shared seeds that should then be extended to obtain a global alignment. Therefore, low sequence similarity, rearrangements, and inversions can cause major problems in identifying a possible alignment and thus the actual sequence similarity. Furthermore, when considering whole genomes, the global alignment of large sequences has become a prohibitive task even for supercomputers, hence simply infeasible. To overcome these obstacles, in the last 10 years, a variety of alignment-free methods have been proposed. In principle, they are all based on counting procedures that characterize a sequence based on its constituents, for example, k-mers [11, 34].

An important aspect in phylogeny reconstruction is the fact that gene-based methods strictly focus on comparing the coding regions of genomes, which can account for as little as 1% of the genomic sequence in humans [40]. However, it is known that the use of whole genomes may provide more robust information when comparing different organisms [19]. Also most alignment-free methods in the literature use only a portion of complete genomes [39]. For instance, there are approaches that use only genic regions [11] or mitochondria; other methods filter out regions that are highly repetitive or with low complexity [34]. Recently, it has been shown that the evolutionary information is also carried by nongenic regions [35]. For certain viruses, we are not able to estimate a complete phylogeny by analyzing their genes, since these organisms may share a very limited genetic material [39].

3.2.2 Alignment-Free Methods

Sims et al. recently applied the *Feature Frequency Profiles* (FFP) method presented in Reference [34] Average Common Subwordhole-genome phylogeny of mammals [35] —that is, large eukaryotic genomes including the human genome—and of bacteria. This method needs to estimate the parameter k in order to compute a feature vector for each sequence, where the vector represents the frequency of each possible k-mer. Each feature vector is then normalized by the total number of k-mers found (i.e., by the sequence length minus $k - 1$), obtaining a probability distribution vector, or FFP, for each genome. FFP finally computes the distance matrix between all pairs of genomes by applying the Jensen–Shannon [31] divergence to their frequency profiles.

For completeness, we note that, for large eukaryotes, they filter out high-frequency and low-complexity features among all the k-mers found. Furthermore, in case the genomes have large differences in length, they use first the block method, similarly to Reference [43], by dividing the sequences into blocks having the size of the smallest genome (with possible overlaps between the blocks).

This general characterization of strings based on their subsequence composition closely resembles some of the information theory problems and is tightly related to the compression of strings. In fact, compositional methods can be viewed as the reinterpretation of data compression methods, well known in the literature [3, 6]. For a comprehensive survey on the importance and implications of data compression methods in computational biology, we refer the reader to Reference [22].

When comparing entire genomes, we want to avoid that large noncoding regions, which by nature tend to be highly repetitive, may contribute to our scoring function a multiple number of times, thus misleading the final similarity score. In fact, while analyzing massive genomes, the number of repeated patterns is very high, particularly in the nongenic regions. Furthermore, if we allow mismatches, the number of patterns can grow exponentially [4, 7, 25]. In this chapter, we will address this problem by controlling the distribution of subwords over the sequences under consideration so that their contribution will not be overcounted.

Moreover, when comparing genomes, it is well known that different evolutionary mechanisms can take place. In this framework, two closely related species are expected to share larger portions of DNA than two distant ones, whereby also other large complements and reverse complements, or inversions, may occur [26]. In this work, we will take into account all these symmetries in order to define a measure of similarity between whole genomes.

Table 3.1 Example of counters $l[i]$ for the ACS approach

$s_1[i]$	A	C	A	C	G	T	A	C
$l_1[i]$	2	1	4	3	3	3	2	1

$s_2[j]$	T	A	C	G	T	G	T	A
$l_2[j]$	3	4	3	2	1	3	2	1

Counters $l_1[i]$ and $l_2[j]$ provide the computation of ACS(s_1, s_2) and ACS(s_2, s_1), respectively, where s_1 = ACACGTAC, s_2 = TACGTGTA, and $i, j = 1, \dots, 8$.

3.2.3 Average Common Subword

Among the many distance measures proposed in the literature, which in most cases are dealing with k-mers, an effective and particularly elegant method is the *Average Common Subword* (ACS) approach, introduced by Ulitsky et al. [9]. They use a popular concept in the field of string algorithms, known as matching statistics [23]. In short, given two sequences s_1 and s_2, where s_1 is the reference sequence, it counts the length $l[i]$ of the longest subword starting at position i of s_1 that is also a subword of s_2, for every possible position i of s_1 (see Table 3.1). This counting is then averaged over the length of s_1. The general form of ACS follows:

$$ACS(s_1, s_2) = \frac{\sum_{i=1}^{|s_1|} l[i]}{|s_1|}.$$

The ACS measure is intrinsically asymmetric, but with simple operations can be reduced to a distance-like measure. In Section 3.3.5, we will give further detail on how to adjust this formula; for the moment, we note the similarity with the cross entropy of two probability distributions P and Q:

$$H(P, Q) = -\sum_x p(x) \log q(x),$$

where $p(x) \log q(x)$ measures the number of bits needed to code an event x from P if a different coding scheme based on Q is used, averaged over all possible events x.

The beauty of the ACS measure is that it is not based on fixed-length subwords, but it can capture also variable length matches, in contrast to most methods that are based on fixed sets of k-mers. In fact, with the latter the choice of the parameter k is critical, and every method needs to estimate k from the data under examination, typically using empirical measurements [34]. This may, however, overfit the problem and lead to loss of valuable information. In this spirit, ACS performs a massive genome sequence analysis without limiting the resolution of motifs that can be naturally captured from sequences. Moreover, it does not filter out any region of the genomes under consideration.

3.2.4 Kullback–Leibler Information Divergence

As a matter of fact, Burstein et al. [9] formally showed that the ACS approach indeed mimics the cross entropy estimated between two large sequences, supposed to have been generated by a finite-state Markov process. In practice, this is closely related to the Kullback–Leibler information divergence and represents the minimum number of bits needed to describe one string, given the other:

$$D_{KL}(P\|Q) = H(P,Q) - H(P)$$

For example, if we consider the Lempel–Ziv mutual compression of two strings [30, 44], it parses a string using a dictionary based on the other string, and this mechanism is exploited by ACS in a similar way. Since we are analyzing genome-wide sequences, this, asymptotically, can be seen as a natural distance measure between Markovian distributions.

The Kullback–Leibler divergence was introduced by Kullback and Leibler [29]. This is also known as relative entropy, information divergence, or information gain. Studied in detail by many authors, it is perhaps the most frequently used information-theoretic similarity measure [28]. The relative entropy is used to capture mutual information and differentiation between data, in the same way that the absolute entropy is employed in data compression frameworks. Given a source set of information, for example, s_2, the relative entropy is the quantity of data required to reconstruct the target, in this case s_1. Thus, this is a strong theoretical basis for the ACS measure. A drawback is that the Kullback–Leibler measure does not obey some of the fundamental axioms a distance measure must satisfy. In particular, Kullback–Leibler is not symmetric and does not satisfy the triangle inequality.

Despite these difficulties, ACS proved to be useful for reconstructing whole-genome phylogenies of viruses, bacteria, and eukaryotes, outperforming in most cases the state-of-the-art methods [10, 39]. Moreover, it is computationally less demanding than other notable phylogenomic inferences such as maximum parsimony and maximum likelihood, or other Bayesian estimations of divergence/correlation between entire genomes, where the correct estimation and use of the probability are often infeasible in practical problems—even when merely relegated to the analysis of genes and annotated regions, for example, [20]. Therefore, here we aim to characterize and improve the ACS method, filtering out motifs that might not be useful for a whole-genome phylogeny of different organisms. In particular, we want to discard common motifs occurring in regions covered by other more significant motifs, for example, according to the motif priority rule introduced in Reference [41].

3.3 UNDERLYING APPROACH

In this section, we discuss the use of alignment-free distance function based on subword compositions. We prove that the matching statistics, a popular concept in the field of string algorithms able to capture the statistics of common words between two sequences, can be derived from a small set of "independent" subwords, namely the irredundant common subwords. We define a distance-like measure based on these subwords such that each region of genomes contributes only once, thus avoiding to count shared subwords a multiple number of times. In a nutshell, this filter discards subwords occurring in regions covered by other more significant subwords.

The UA builds a scoring function based on this set of patterns, called *underlying* (see section 3.3.2). We prove that this set is by construction linear in the size of input, without overlaps, and can be efficiently constructed. Results show the validity of our method in the reconstruction of phylogenetic trees, where the UA outperforms the current state-of-the-art methods. Moreover, we show that the accuracy of the UA is achieved with a very small number of subwords, which in some cases carry meaningful biological information.

3.3.1 Irredundant Common Subwords

Here we propose a distance measure between entire genomes based on the notion of underlying subwords. In order to build a sound similarity measure between genomes, we need first to study the properties of the matching statistics. Our first contribution is the characterization of the subwords that are needed to compute the matching statistics. A second contribution is the selection of these subwords so that the resulting similarity measure does not contain overcounts. Our main idea is to avoid overlaps between selected subwords, more precisely by discarding common subwords occurring in regions covered by other more significant subwords.

In the literature, the values $l[i]$ used by the ACS approach are called the *matching statistics*, as described in detail by Gusfield [23]. Our first contribution is to characterize the matching statistics in order to identify which subwords are essentials.

It is well known that the total number of distinct subwords of any length found in a sequence of length n can be at most $\Theta(n^2)$. Remarkably, a notable family of fewer than $2n$ subwords exist that is maximal in the host sequence, in the sense that it is impossible to extend a word in this class by appending one or more characters to it without losing some of its occurrences [1]. It has been shown that the matching statistics can be derived from this set of maximal subwords [2]. Here we will further tighten this bound by showing that to compute the matching statistics it is enough to consider a subset of the maximal subwords, called *irredundant common subwords*.

The notion of irredundancy was introduced in Reference [8] and later modified for the problem of protein comparison [14, 15]. It proved useful in different contexts from data compression [5] to the identification of transcription factors [13]. In this section, we study the concept of *irredundant common subwords* (i.e., without mismatches/wildcards). This ensures that there exists a close correspondence between the irredundant common subwords and the matching statistics.

Definition 3.1 (Irredundant/redundant common subword} *A common subword w is* irredundant *if and only if at least an occurrence of w in s_1 or s_2 is not covered by other common subwords. A common subword that does not satisfy this condition is called a* redundant common subword.

We observe that the number of irredundant common subwords $\mathcal{I}_{s_1,s_2}$ is bounded by $m+n$, where $|s_1| = n$ and $|s_2| = m$, since it is a subset of the set of *maximal common subwords* (see References [2, 38] for a more complete treatment of this topic).

Proposition 3.1 *The matching statistics $l_{s_1}(i)$ can be computed by combining together all and only the irredundant common subwords of s_1 and s_2.*

Proof. To show that the vector $l_{s_1}(i)$ can be derived from the irredundant common subwords, we define a new vector of scores l_w for each subword w, where $l_w[j] = |w| - j + 1$ represents the length of each suffix j of w, with $j = 1, \ldots, |w|$. Then, for each subword w in $\mathcal{I}_{s_1,s_2}$, we superimpose the vector l_w on all the occurrences of w in s_1. For each position i, in s_1, $l_{s_1}(i)$ is the maximum value of the scores $\max_w(l_w[j])$ such that $k + j = i$ and k is an occurrence of w.

To complete the proof, we have to show that every occurrence of a common subword of s_1 and s_2 is covered by some occurrence of a subword in $\mathcal{I}_{s_1,s_2}$. By definition of irredundant common subword, any occurrence of a subword corresponds to an irredundant common subword or is covered by some subword in $\mathcal{I}_{s_1,s_2}$. Moreover, every irredundant common subword w has at least an occurrence i that is not covered by other subwords. Thus, $l_{s_1}(i)$ corresponds exactly to $|w|$ and the subword w is necessary to compute the matching statistics. In conclusion, by using the method described above for $l_{s_1}(i)$, we can compute for each position the length of the maximum common subword starting in that location, which corresponds to the matching statistics. ■

In summary, the notion of irredundant common subwords is useful to decompose the information provided by the matching statistics into several patterns. Unfortunately, these subwords can still overlap in some position. This might lead to an overcount in the matching statistics, in which the same region of the string contributes more than once. Our aim is to remove the possibility of overcount by filtering the most representative common subwords for each region of the sequences s_1 and s_2, which will also remove all overlaps.

3.3.2 Underlying Subwords

When comparing entire genomes, we want to avoid that large noncoding regions, which by nature tend to be highly repetitive, may contribute to the similarity score a multiple number of times, thus misleading the final score. In fact, while analyzing massive genomes, the number of repeated patterns is very high, particularly in the nongenic regions. Therefore, we need to filter out part of this information and select the "best" common subword, by some measure, for each region of the sequences.

In this regard, we must recall the definition of pattern priority and of underlying pattern, adapted from Reference [16], to the case of pairwise sequence comparison. We will take as input the irredundant common subwords and the underlying quorum $u = 2$, that is, they must appear at least twice. Let now w and w' be two distinct subwords. We say that w has priority over w', or $w \rightarrow w'$, if and only if either $|w| \geq |w'|$, or $|w| = |w'|$ and the first occurrence of w appears before the first occurrence of

w'. In this case, every subword can be defined just by its length and one of its starting positions in the sequences, meaning that any set of subwords is totally ordered with respect to the priority rule. We say that an occurrence l of w is *tied* to an occurrence l' of a subword w', if the two occurrences overlap, that is, $([l, l+|w|-1] \cap [l', l'+|w'|-1]) \neq \emptyset$, and $w' \rightarrow w$. Otherwise, we say that l is *untied* from l'.

Definition 3.2 (Underlying subword) *A set of subwords* $\mathcal{U}_{s_1,s_2} \subseteq \mathcal{I}_{s_1,s_2}$ *is said to be* underlying *if and only if*

(i) every subword w *in* $\mathcal{U}_{s_1,s_2}$, *called an* underlying subword, *has at least two occurrences, one in* s_1 *and the other in* s_2, *that are untied from all the untied occurrences of other subwords in* $\mathcal{U}_{s_1,s_2} \setminus w$, *and*

(ii) there does not exist a subword $w \in \mathcal{I}_{s_1,s_2} \setminus \mathcal{U}_{s_1,s_2}$ *such that* w *has at least two untied occurrences, one per sequence, from all the untied occurrences of subwords in* $\mathcal{U}_{s_1,s_2}$.

This subset of $\mathcal{I}_{s_1,s_2}$ is composed only of those subwords that rank higher with our priority rule with respect to s_1. In fact, if there are overlaps between subwords that are in $\mathcal{I}_{s_1,s_2}$, we will select only the subwords with the highest priority. Similarly to the score $\mathrm{ACS}(s_1, s_2)$, the set $\mathcal{U}_{s_1,s_2}$ is asymmetric and depends on the order of the two sequences; we will address this issue in Section 3.3.5. As for the underlying patterns [16], one can show that the set of underlying subwords exists, and is unique. As a corollary, we know that the untied occurrences of the underlying subwords can be mapped into the sequences s_1 and s_2 without overlaps. Moreover, by definition, the total length of the untied occurrences cannot exceed the length of the sequences. These two properties are crucial when building a similarity measure because any similarity that is based on these subwords will count the contribution of a region of the sequence only once.

3.3.3 Efficient Computation of Underlying Subwords

To summarize, we select the irredundant common subwords that best fit each region of s_1 and s_2, employing the UA. This approach is based on a simple pipeline. We first select the irredundant common subwords and subsequently filter out the subwords that are not underlying. From a different perspective, we start from the smallest set of subwords that captures the matching statistics and remove the overlaps by applying our priority rule. In the following, we show how to compute the irredundant common subwords and the matching statistics, and then we present an approach for the selection of the underlying subwords among these subwords. The general structure of the UA is the following:

(1) Compute the set of the irredundant common subwords $\mathcal{I}_{s_1,s_2}$.

(2) Rank all subwords in $\mathcal{I}_{s_1,s_2}$ according to the priority and initialize $\mathcal{U}$ to an empty set.

(3) Iteratively select a subwords p from $\mathcal{I}_{s_1,s_2}$ following the ranking.

(4a) If p has at least two untied occurrences: add p to $\mathcal{U}$ and update the corresponding regions of Γ (see section 3.3.3.2) in which p occurs;

(4b) otherwise, discard p and return to (3).

3.3.3.1 *Discovery of the Irredundant Common Subwords.* In step 1, we construct the generalized suffix tree T_{s_1,s_2} of s_1 and s_2. We recall that an occurrence of a subword is (left)right-maximal if it cannot be covered from the (left)right by some other common subword. The first step consists in making a depth-first traversal of all nodes of T_{s_1,s_2} and coloring each internal node with the colors of its leaves (each color corresponds to an input sequence). In this traversal, for each leaf i of T_{s_1,s_2}, we capture the lowest ancestor of i having both the colors c_1 and c_2, say the node w. Then, w is a common subword, and i is one of its right-maximal occurrences (in s_1 or in s_2); we select all subwords having at least one right-maximal occurrence. The resulting set will be linear in the size of the sequences, that is, $O(m+n)$. This is only a superset of the irredundant common subwords, since the occurrences of these subwords could not be left-maximal.

In a second phase, we map the length of each right-maximal occurrence i into $l_{s_1}(i)$, and using Proposition 3.1, we check which occurrences i have length greater than or equal to the length stored in the location $i-1$ (for locations $i \geq 2$). These occurrences are also left-maximal, since they cannot be covered by a subword appearing at position $i-1$. Finally, we can retain all subwords that have at least an occurrence that is both right- and left-maximal, that is, the set of irredundant common subwords $\mathcal{I}_{s_1,s_2}$. Note that by employing the above technique, we are able to directly discover the irredundant common subwords and the matching statistics $l_{s_1}(i)$.

The construction of the generalized suffix tree T_{s_1,s_2} and the subsequent extraction of the irredundant common subwords $\mathcal{I}_{s_1,s_2}$ can be completed in time and space that is linear in the size of sequences.

3.3.3.2 *Selection of the Underlying Subwords.* In this section, we describe, given the set of the irredundant common subwords $\mathcal{I}_{s_1,s_2}$, how to filter out the subwords that are not underlying, obtaining the set of underlying subwords $\mathcal{U}_{s_1,s_2}$.

The extraction of underlying subwords takes as input the set $\mathcal{I}_{s_1,s_2}$ and the tree T_{s_1,s_2} from the previous section. First, we need to sort all subwords in $\mathcal{I}_{s_1,s_2}$ according to the priority rule (step 2). Then, starting from the top subword, we analyze iteratively all subwords by checking their untied occurrences (step 3). If the subword passes a validity test, we select it as underlying (step 4a), otherwise we move on with the next subword (step 4b). The two key steps of this algorithm are sorting the subwords (step 2) and checking for their untied occurrences (step 4a).

Step 2 is implemented as follows. For all subwords, we retrieve their lengths and first occurrences in s_1 from the tree T_{s_1,s_2}. Then each subword is characterized by its length and the first occurrence. Since these are integers in the range $[0, n]$, we can apply radix sort [17], first by length and then by occurrence. This step can be done in linear time.

In order to implement step 4a, we need to define the vector Γ of n booleans, representing the locations of s_1. If $\Gamma[i]$ is TRUE, then the location i is covered by some untied occurrence. We also preprocess the input tree and add a link for all nodes v to the closest irredundant ancestor, say $prec(v)$. This can be done by traversing the tree in preorder. During the visit of a the node v, if it is not irredundant we transmit to the children $prec(v)$, and if v is irredundant we transmit v. This preprocess can be implemented in linear time and space.

For each subword w in $\mathcal{I}_{s_1,s_2}$, we consider the list $\mathcal{L}_w$ of occurrences to be checked. All $\mathcal{L}_w$ are initialized in the following way. Every leaf v, which represent a position i, send its value i to the location list of the closest irredundant ancestor using the link $prec(v)$. Again this preprocess takes linear time and space since all positions appear in exactly one location list. We will update these lists $\mathcal{L}_w$ only with the occurrences to be checked, that is, that are not covered by some underlying subword already discovered. We start analyzing the top subword w and, for this case, $\mathcal{L}_w$ is composed of all the occurrences of w.

For each occurrence i of w, we need to check only its first and last location in the vector Γ; that is, we need to check the locations $\Gamma[i]$ and $\Gamma[i + |w| - 1]$. If one of these two values is set to TRUE, then i is tied by some subword w'. Otherwise, if both the values are set to FALSE, then i must be untied from all other subwords. Since all subwords already evaluated are not shorter than w, then they cannot cover some locations in $\Gamma[i, i + |w| - 1]$ without also covering $\Gamma[i]$ or $\Gamma[i + |w| - 1]$. Thus, if $\Gamma[i]$ and $\Gamma[i + |w| - 1]$ are both set to FALSE, we mark this occurrence i as untied for the subword w and update the vector Γ accordingly.

If $\Gamma[i]$ is TRUE, we can completely discard the occurrence i, for the subword w and also for all its prefixes, which are represented by the ancestors of w in the tree T_{s_1,s_2}. Thus, the occurrence i will no longer be evaluated for any other subword.

If $\Gamma[i]$ is FALSE and $\Gamma[i + |w| - 1]$ is TRUE, we need to further evaluate this occurrence for some ancestors of w. In this case, one can compute the longest prefix, w', of w such that $\Gamma[i + |w'| - 1]$ is set to FALSE and w' is an irredundant common subword. Then the occurrence i is inserted into the list $\mathcal{L}_{w'}$.

This step is performed by first computing the length $d < |w|$ such that $\Gamma[i + d - 1]$ is FALSE and $\Gamma[i + d]$ is TRUE, and then retrieving the corresponding prefix w' of w in the tree that spells an irredundant common subword with length equal to or shorter than d. We can compute d by means of a *length table* χ in support (or in place) of the boolean vector Γ. For each untied occurrence i of w, χ stores the values $[1, 2, \dots , |w|]$ in the locations $[i, i + 1, \dots , i + |w| - 1]$, similarly to the proof of Proposition 3.1. Using this auxiliary table, we can compute the value of d for the location under study i as $d = |w| - \chi[i + |w| - 1]$.

Now, to select w', the longest prefix of w with $|w'| \leq d$, we employ an algorithm proposed by Kopelowitz and Lewenstein [27] for solving the *weighted ancestor problem*, where weights correspond to the length of words spelled in the path from the root to each node, in the case of a suffix tree. In the weighted ancestor problem, one preprocesses a weighted tree to support fast predecessor queries on the path from a query node to the root. That is, with a linear preprocessing on a tree of height n, using the above algorithm it is possible to locate any ancestor node w' that has a weight less than

d in time $O(\log\log n)$. In our case, the maximum length for an irredundant subword is $\min\{m,n\}$, thus we can find a suitable ancestor w' of w in time $O(\log\log\min\{m,n\})$, with $O(m+n)$ preprocessing of the tree T_{s_1,s_2}.

At the end of the process, if the subword w has at least one untied occurrence per sequence, then we mark w as underlying subword. Otherwise, all the occurrences of w that are not covered are sent to its ancestors, using the previous procedure.

To analyze the overall complexity, we need to compute how many times the same location i is evaluated. Suppose, for example, that i belongs to $\mathcal{L}_w$ of the subword w. The location i is evaluated again for some $\overline{w}$, and inserted into the list $\mathcal{L}_{\overline{w}}$, only if $\Gamma[i]$ is FALSE and $\Gamma[i+|w|-1]$ is TRUE. Note that the locations not already covered are in the range $[i, i+|w|-d-1]$, with $d>0$. Then, the subword $\overline{w}$ is the longest prefix of w that is an irredundant common subword and that lives completely in the locations $[i, i+|w|-d-1]$; however, $\overline{w}$ may not cover the entire interval. Now, the occurrence i will be evaluated again only if there exists another subword w' that overlaps with $\overline{w}$, and that has a higher priority with respect to $\overline{w}$. The worst case is when w' ends exactly at position $i+|w|-d-1$ and overlaps with $\overline{w}$ by only one location. Since w' must be evaluated before $\overline{w}$, then $|w'| \geq |\overline{w}|$. Thus, the worst case is when the two subwords have about the same length. In this settings, the length of the subword $\overline{w}$ can be at most $(|w|-d)/2$. We can iterate this argument at most $O(\log|w|)$ times for the same position i. Therefore, any location can be evaluated at most $O(\log\min\{m,n\})$ times. In conclusion, our approach requires $O((m+n)\log\min\{m,n\}\log\log\min\{m,n\})$ time and $O(m+n)$ space to discover the set of all underlying subwords $\mathcal{U}_{s_1,s_2}$.

3.3.4 Extension to Inversions and Complements

In this section, we discuss the extension of the algorithmic structure discussed above to accommodate also inversion and complement matches.

A simple idea is to concatenate each sequence with its inverse and its complement, while keeping separate the occurrences coming from direct matches, inversions, and complements. In brief, we first define $\hat{x}$ as the concatenation of a string x with its inverse, followed by its complement, in this exact order. Then, we compute the irredundant common subwords, $\mathcal{I}_{s_1,\hat{s}_2}$, on the sequences s_1 and $\hat{s}_2$. We subsequently select the underlying subwords by ranking all the irredundant common subwords in the set $\mathcal{I}_{s_1,\hat{s}_2}$. Using the same algorithm described above, we compute the set $\mathcal{U}_{s_1,\hat{s}_2}$, and then we map each subword occurrence to the reference sequences s_1. This will include also inversions and complements of s_2 that are shared by s_1. In this way, we can store all the untied occurrences and consider all possible matches for each region of s_1.

In this framework, we choose to take into account all these symmetries, and thus the experiments presented will use this extended approach. We will also measure the contribution of inversions and complements to our similarity measure.

3.3.5 A Distance-Like Measure Based on Underlying Subwords

In the following, we report the basic steps of our distance-like measure. Let us assume that we have computed $\mathcal{U}_{s_1,s_2}$, and the other specular set $\mathcal{U}_{s_2,s_1}$. For every subword

$w \in \mathcal{U}_{s_1,s_2}$, we sum up the score $h_w^{s_1} \sum_{i=1}^{|w|} i = h_w^{s_1}|w|(|w|+1)/2$ in $Score_{UA}(s_1, s_2)$, where $h_w^{s_1}$ is the number of its untied occurrences in s_1, similarly to ACS [39]. Then, we average $Score_{UA}(s_1, s_2)$ over the length of the first sequence, s_1, yielding

$$Score_{UA}(s_1, s_2) = \frac{\sum_{w \in \mathcal{U}_{s_1,s_2}} h_w^{s_1}|w|(|w|+1)}{2n}$$

This is a score that is large when two sequences are similar, therefore we take its inverse.

Moreover, for a fixed sequence s_1, this score can also grow with the length of s_2, since the probability of having a match in s_1 increases with the length of s_2. For this reason, we consider the measure $\log_4(|s_2|)/Score_{UA}(s_1, s_2)$; we use a base-4 logarithm since DNA sequences have four bases. Another issue with the above formula is the fact that it is not equal to zero for $s_1 = s_2$; thus, we subtract the correction term $\log_4(|s_1|)/Score_{UA}(s_1, s_1)$, which ensures that this condition is always satisfied. Since $\mathcal{U}_{s_1,s_1}$ contains only one subword, the sequence s_1 itself, which trivially has only one untied occurrence in s_1, it yields $Score_{UA}(s_1, s_1) = |s_1|(|s_1|+1)/(2|s_1|) = (|s_1|+1)/2$. The following formulas accommodate all of these observations in a symmetrical distance-like measure $UA(s_1, s_2)$ between the sequences s_1 and s_2:

$$Sim_{UA}(s_1, s_2) = \frac{\log_4(|s_2|)}{Score_{UA}(s_1, s_2)} - \frac{2\log_4(|s_1|)}{(|s_1|+1)}$$

$$UA(s_1, s_2) = \frac{Sim_{UA}(s_1, s_2) + Sim_{UA}(s_2, s_1)}{2}$$

We can easily see that the correction term rapidly converges to zero as $|s_1|$ increases. Moreover, we note that $UA(s_1, s_2)$ grows as the two sequences s_1 and s_2 diverge.

3.4 EXPERIMENTAL RESULTS

In this section, we assess the effectiveness of the UA on the estimation of whole-genome phylogenies of different organisms. We tested our distance function on three types of data sets: viruses, prokaryotes, and unicellular eukaryotes.

3.4.1 Genome Data sets and Reference Taxonomies

In the first data set, we selected 54 virus isolates of the 2009 human pandemic *Influenza A-subtype H1N1*, also called the *Swine Flu*. The Influenza A virion has eight segments of viral RNA with different functions. These RNAs are directly extracted from infected host cells, and synthesized into complementary DNA by reverse transcription reaction, where a specific gene amplification is performed for each segment [36]. We concatenate these segments according to their conventional order given by the literature [33]; this step, in general, does not affect the final

Table 3.2 Benchmark for prokaryotes—Archaea and Bacteria domains

Accession No.	Domain	Organism	Size
BA000002	Archaea	*Aeropyrum pernix* str. K1	1.7 Mbp
AE000782	Archaea	*Archaeoglobus fulgidus* str. DSM 4304	2.2 Mbp
AE009439	Archaea	*Methanopyrus kandleri* str. AV19	1.7 Mbp
AE010299	Archaea	*Methanosarcina acetivorans* str. C2A	5.8 Mbp
AE009441	Archaea	*Pyrobaculum aerophilum* str. IM2	2.3 Mbp
AL096836	Archaea	*Pyrococcus abyssi*	1.8 Mbp
AE009950	Archaea	*Pyrococcus furiosus* str. DSM 3638	1.9 Mbp
AE000520	Archaea	*Treponema pallidum* sp. pall. str. Nichols	1.2 Mbp
AE017225	Bacteria	*Bacillus anthracis* str. Sterne	5.3 Mbp
AL009126	Bacteria	*Bacillus subtilis* subsp. *subtilis* str. 168	4.3 Mbp
AE013218	Bacteria	*Buchnera aphidicola* str. Sg	651 kbp
AL111168	Bacteria	*Campylobacter jejuni* sp. jej. str. NCTC 11168	1.7 Mbp
AE002160	Bacteria	*Chlamydia muridarum* str. MoPn/Wiess-Nigg	1.1 Mbp
AM884176	Bacteria	*Chlamydia trachomatis* str. L2/434/Bu	1.1 Mbp
AE016828	Bacteria	*Coxiella burnetii* str. RSA 493	2.0 Mbp
AE017285	Bacteria	*Desulfovibrio vulgaris* sp. vulg. str. Hildenb.	3.6 Mbp
L42023	Bacteria	*Haemophilus influenzae* str. Rd KW20	1.9 Mbp
CP001037	Bacteria	*Nostoc punctiforme* str. PCC 73102	8.4 Mbp

Prokaryotic taxa used in our experiments, divided by domain. For each entity, we list the accession number in the NCBI genome database, the complete name and strain, and the genome size.

phylogeny computed by our algorithm and is used to sort subwords by location. The resulting sequences are very similar to each other and have lengths in the order of 13,200 nucleotides each, accounting for a total of 714,402 bases. To compute a reference taxonomic tree, we perform multiple sequence alignment using the ClustalW2 [37] tool [1] as suggested by many scientific articles on the 2009 Swine Flu [33, 36]. Then, we compute the tree using the *Dnaml* tool from the PHYLIP [21] software package, [2] which implements the maximum likelihood method for aligned DNA sequences. In *Dnaml*, we used the parameters suggested in References [33, 36], which consider empirical base frequencies, constant rate variation among sites (with no weights), a transition ratio of 2.0, and best tree search based on proper searching heuristics.

In the second data set, we selected 18 prokaryotic organisms among the species used in Reference [39] for a DNA phylogenomic inference. We chose the species whose phylogenomic tree can be inferred by well-established methods in the literature (see Table 3.2). The organisms come from both the major prokaryotic domains: Archaea, 8 organisms in total, and Bacteria, 10 organisms in total. The sequences in question have lengths ranging from 0.6 to 8 Mbp, accounting for a total of 48 Mbp. We compute their tree-of-life by using genes that code for the 16S RNA, the main

[1] ClustalW2 is available at http://www.ebi.ac.uk/Tools/msa/clustalw2.

[2] PHYLIP (phylogenetic inference package) is a free computational phylogenetics software package available at http://evolution.genetics.washington.edu/phylip.

Table 3.3 Benchmark for unicellular eukaryotes—genus *Plasmodium*

Parasite	Host	Region	Size (Mbp)
P. Berghei	Rodent	Africa	18.5
P. Chabaudi	Rodent	Africa	18.8
P. Falciparum	Human	Africa, Asia, and South/Central America	23.3
P. Knowlesi	Macaque	Southeast Asia	23.7
P. Vivax	Human	Africa, Asia, and South/Central America	22.6

Eukaryotic organisms of the unicellular genus *Plasmodium* whose genome has been completely sequenced. *Plasmodium* are parasites known as causative agents of malaria in different hosts and geographic regions. The right-most column lists the size of each complete DNA genome.

RNA molecule inside the small ribosomal subunit characterizing prokaryotes and widely used to reconstruct their phylogeny; the considered sequences are called 16S rDNA. We can extract a multiple alignment of 16S rDNA sequences of the selected organisms from the Ribosomal Database Project [12] [3]; our experiments are based on the release 8.1. Next, we perform a maximum likelihood estimation on the aligned set of sequences, employing *Dnaml* from PHYLIP with standard parameters, in order to compute a reference tree based on the resulting estimation.

In the third data set, we selected five eukaryotic organisms of the protozoan genus *Plasmodium* whose genomes have been completely sequenced (Table 3.3). *Plasmodium* are unicellular eukaryotic parasites best known as the etiological agents of malaria infectious disease. The sequences have lengths ranging from 18 to 24 Mbp, accounting for a total of 106 Mbp. We used as reference tree the taxonomy computed by Martinsen et al. [32], as suggested by the Tree of Life Project.

3.4.2 Whole-Genome Phylogeny Reconstruction

We exploited the above data sets to compare our approach, the UA, with other efficient state-of-the-art approaches in the whole-genome phylogeny reconstruction challenge: ACS [39], FFP [34], [4] and FFP_{RY}. The FFP_{RY} method, in contrast to FFP, employs the purine–pyrimidine reduced alphabet (RY) that is composed of two character classes: `[A,G]` (both purine bases, denoted by `R`) and `[C,T]` (both pyrimidines, denoted by `Y`). We implemented our ACS method, while for FFP and FFP_{RY} we used the FFP package release 3.14 available online.

We reconstruct the phylogenomic trees from the distance matrices using the Neighbor-joining method as implemented in the PHYLIP package. We compare the resulting topologies with the respective reference trees using the symmetric difference of *Robinson and Foulds* (R–F) and the triplet distance. For two unrooted binary trees with $n \geq 3$ leaves, the R–F score is in the range $[0, 2n-6]$. A score equal to 0 means that the two trees are isomorphic, while $2n-6$ means that all

[3] The Ribosomal Database Project is available at http://rdp.cme.msu.edu.

[4] The FFP software package release 3.14 is available at http://ffp-phylogeny.sourceforge.net.

Table 3.4 Comparison of whole-genome phylogeny reconstructions

Species	Group	UA	ACS	FFP	FFP_{RY}
Influenza A	Viruses	**80**/102	84/102	100/102	96/102
Archaea	Prokaryotes	**4**/10	**4**/10	6/10	6/10
Bacteria	Prokaryotes	**6**/14	10/14	**6**/14	10/14
Archaea and Bacteria	Prokaryotes	**20**/30	22/30	**20**/30	22/30
Plasmodium	Eukaryotes	**0**/4	**0**/4	4/4	**0**/4

Normalized Robinson–Foulds scores with the corresponding reference tree. For each data set, the best results are shown in bold.

nontrivial bipartitions are different. The R–F difference between two or more trees can be computed using the *TreeDist* tool from the PHYLIP package.

We ran FFP and FFP_{RY} for different values of k (the fixed subword length) as suggested in Reference [34], retaining the best results in agreement with the reference trees. Table 3.4 compares our method with the other state-of-the-art approaches, by showing the R–F difference with respect to the reference taxonomy tree.

Our approach, the UA, achieves good performance in every test considering the R–F difference with the reference taxonomy tree, and very good performance if we further analyze the resulting phylogenies, as in Figures 3.1–3.3. For every data set, the best results are shown in bold. We can observe that the UA is constantly the best performing method and that this advantage becomes more evident for large data set, where sequences share large parts, such as the Influenza A (H1N1) viruses.

The R–F distance is a standard method to evaluate topological discordance between trees. However, when dealing with large trees, it is known that small variations can generate very large R–F scores (typically, already for $n \sim 10$). For this reason, we conducted a second series of experiments using the triplet distance [18]. The triplet distance is a more refined measure that does not have this problem. Moreover, to better compare all taxonomies, we report the triplet distance between all trees. Tables 3.5–3.7 show the triplet distance between all trees for all data sets. This more refined measure confirms the applicability of the UA with respect to FFP and ACS.

In more detail, Figure 3.1 shows that the UA can distinguish the two main clades of the 2009 Influenza A-H1N1 (in different shades of gray), which have been outlined in Reference [33]. The origin of the flu could reside in the Mexican isolate (Mexico/4108, in dark gray), from which all other isolates may have ensued. Two subclades for the U.S. states of California and Texas are highlighted in light gray, most probably corresponding to the first major evolutions of the viral disease.

Similar results are obtained for the second data set, as shown in Figure 3.2. UA can easily distinguish the Archaea domain, top half of the tree, from the Bacteria domain, bottom half, and also other subclades with respect to the reference tree (these subclades are highlighted in the figure with different shades of gray). The organisms in black do not form a clade with other organisms in the reference tree. For the third data

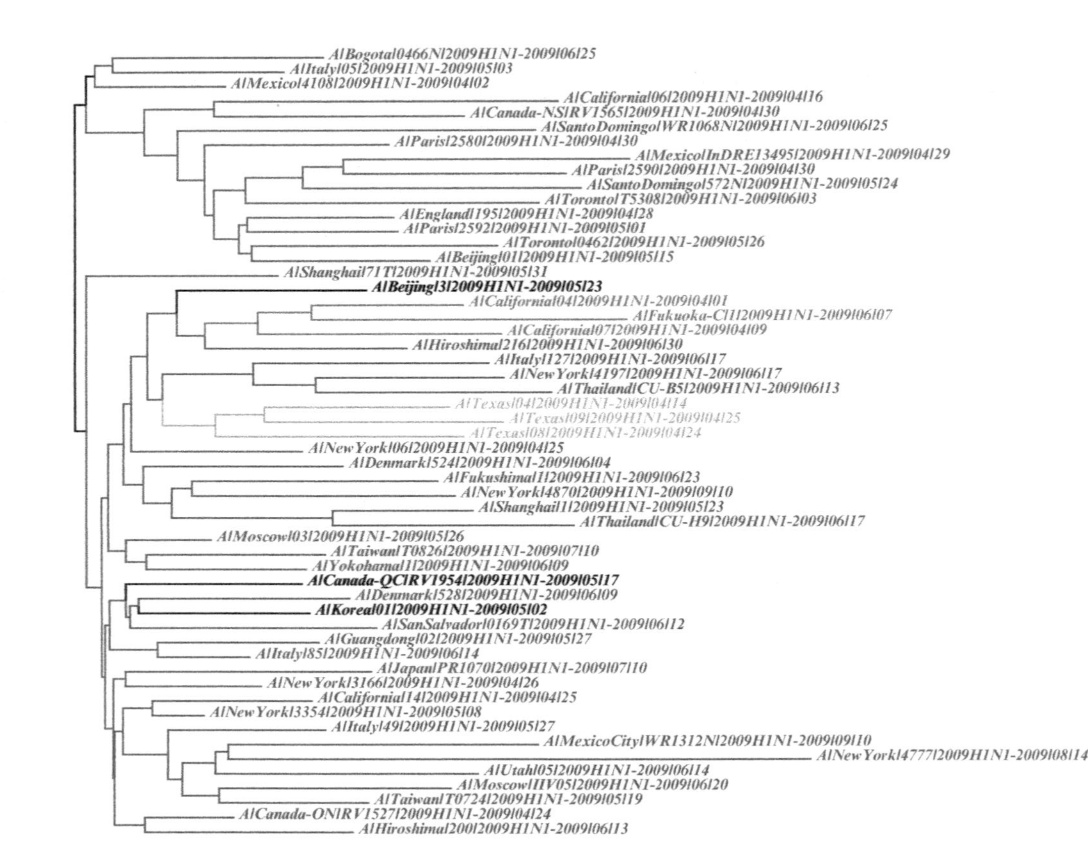

Figure 3.1 Whole-genome phylogeny of the 2009 world pandemic Influenza A (H1N1) generated by the UA. The clades are highlighted with different shades of gray. The node Mexico/4108 is probably the closest isolate to the origin of the influenza. The organisms that do not fall into one of the two main clades according to the literature are shown in bold.

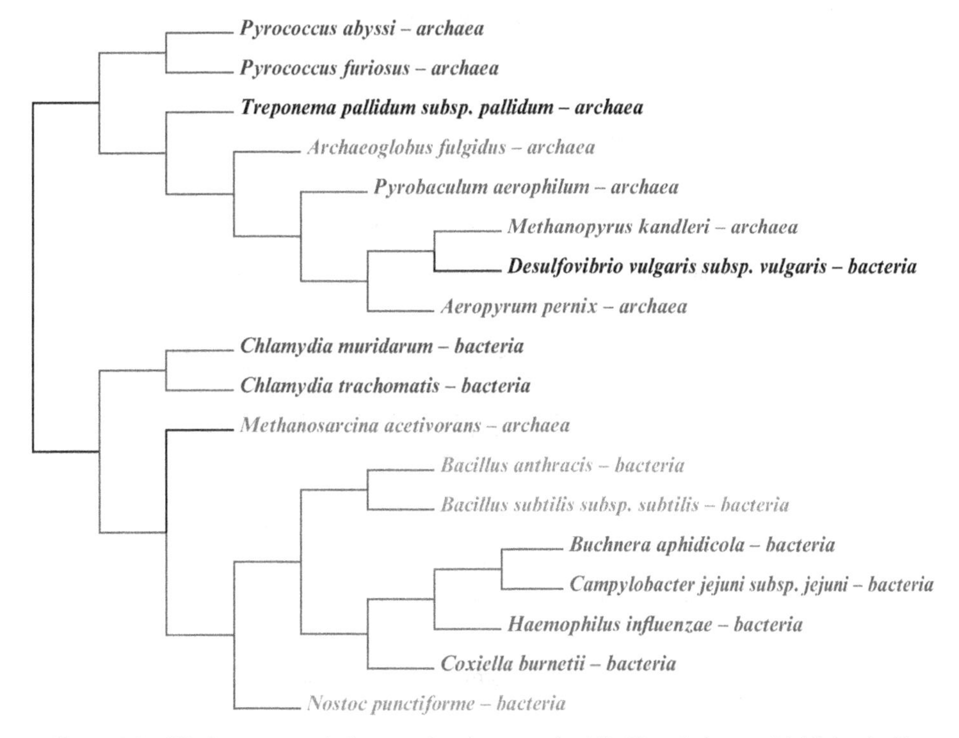

Figure 3.2 Whole-genome phylogeny of prokaryotes by UA. The clades are highlighted with different shades of gray. Only two organisms do not fall into the correct clade: *Methanosarcina acetivorans* (Archaea) and *Desulfovibrio vulgaris* subsp. *vulgaris* (Bacteria).

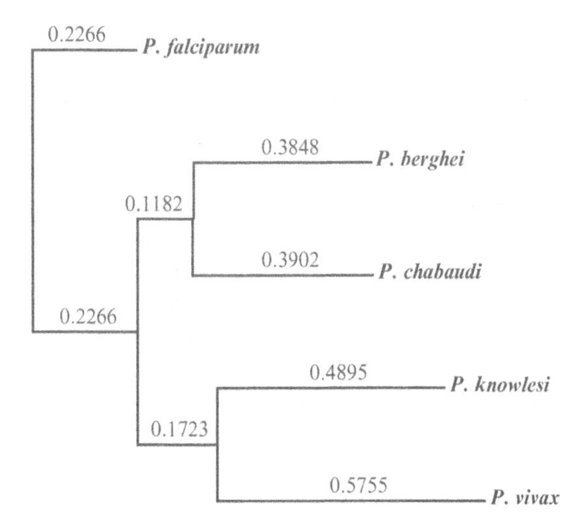

Figure 3.3 Whole-genome phylogeny of the genus *Plasmodium* by UA, with our whole-genome distance highlighted on the branches.

Table 3.5 Comparison of whole-genome phylogeny of Influenza virus

Viruses	Reference	UA	ACS	FFP	FFP_{RY}
Reference	0.0	0.60	0.63	0.86	0.88
UA	**0.60**	0.0	0.30	0.81	0.74
ACS	0.63	0.30	0.0	0.83	0.81
FFP	0.86	0.81	0.83	0.0	0.73
FFP_{RY}	0.88	0.74	0.81	0.73	0.0

Normalized triplet distance between all trees. The best results are shown in bold.

Table 3.6 Comparison of whole-genome phylogeny of prokaryotes

Prokaryotes	Reference	UA	ACS	FFP	FFP_{RY}
Reference	0.0	0.24	0.37	0.62	0.39
UA	**0.24**	0.0	0.37	0.55	0.47
ACS	0.37	0.37	0.0	0.59	0.48
FFP	0.62	0.55	0.59	0.0	0.57
FFP_{RY}	0.39	0.47	0.48	0.57	0.0

Normalized triplet distance between all trees. The best results are shown in bold.

Table 3.7 Comparison of whole-genome phylogeny of *Plasmodium*

Plasmodium	Reference	UA	ACS	FFP	FFP_{RY}
Reference	0.0	0.0	0.0	0.4	0.0
UA	**0.0**	0.0	0.0	0.3	0.0
ACS	**0.0**	0.0	0.0	0.3	0.0
FFP	0.4	0.3	0.0	0.0	0.3
FFP_{RY}	**0.0**	0.0	0.0	0.3	0.0

Normalized triplet distance between all trees. The best results are shown in bold.

set (Figure 3.3), the whole-genome phylogeny of the genus *Plasmodium* generated by the UA corresponds exactly to the taxonomy found in the literature.

The accuracy results are promising, but we believe that of equal interest are the patterns used for the classification. The UA, by construction, uses only a very small number of patterns. For this reason, we report in Table 3.8 some statistics for the underlying subwords selected, averaged over all experiments. We can note that the number of irredundant patterns is in general smaller than the length of the genomes, and this is a first form of information filtering. Moreover, we can observe that only a few underlying subwords are selected on average among the irredundant common subwords. This number is always very small when compared with all possible irredundant subwords, and much smaller than the length of the sequences.

Similar considerations can be drawn for the underlying subwords length. On average they can be very long, especially with respect to FFP that uses only k-mers with k in the range $[5, 10]$. Furthermore, each underlying subword occurs only a few

Table 3.8 Main statistics for the underlying approach averaged over all experiments

Counting	Influenza A	Archaea and Bacteria	*Plasmodium*
Minimum genome size	12,976 b	650 kbp	18,524 kbp
Maximum genome size	13,611 b	8350 kbp	23,730 kbp
Average genome size	13,230 b	2700 kbp	21,380 kbp
Irredundants $\lvert\mathcal{I}_{s_1,s_2}\rvert$	3722	3167 k	16,354 k
Underlying subwords $\lvert\mathcal{U}_{s_1,s_2}\rvert$	60	112 k	706 k
Minimum $\lvert w\rvert$ in $\mathcal{U}_{s_1,s_2}$	6	10	12
Maximum $\lvert w\rvert$ in $\mathcal{U}_{s_1,s_2}$	1615	25	266
Average $\lvert w\rvert$ in $\mathcal{U}_{s_1,s_2}$	264	14	20
Untied inversions	28%	31%	33%
Untied complements	22%	20%	19%

times per sequence, and in general about one occurrence per sequence. Removing the high-frequency subwords, we can notice that the underlying subwords typically have length $\geq \log_4 \min\{m, n\}$, and in the case of viruses they can be very large, capturing more information than FFP. The longest underlying subwords appear in the virus data set, and they are on the order of a thousand bases. We checked whether these subwords may have some biological meaning and found that, in some cases, they correspond to whole viral segments that are shared between two genomes. This confirms that, in some cases, the underlying subwords used for classification can capture some biological insight.

Another interesting aspect is the contribution of inversions and complements in our similarity measure, with respect to the classical notion of match. We compute the average number of occurrences used in our scoring function that is due to inversions and complements. The contribution of inversions and complements is about 28–33% and 19–20%, respectively. This fact may be due to the nature of the sequences considered, but we believe that this topic deserves more attention.

3.5 CONCLUSION

In conclusion, we have shown that the underlying subwords can be used for the reconstruction of phylogenetic trees. Preliminary experiments have shown very good performance in the identification of major clusters for viruses, prokaryotes, and unicellular eukaryotes. An important observation that distinguishes this method from the others is that only a small number of underlying subwords is used for the distance, nevertheless the results are promising. From this fact, we can speculate that only a very limited number of subwords are needed to establish the phylogeny of genomic sequences. Thus, an interesting problem that can be addressed using the underlying subwords is the selection of probes for DNA chips. In the future, a possible extension of this method is the comparison of whole genomes based on short reads coming from next-generation sequencing, instead of using assembled genomes.

AUTHOR'S CONTRIBUTIONS

All authors contributed equally to this study.

ACKNOWLEDGMENTS

Matteo Comin was partially supported by the Ateneo Project CPDA110239 and by the P.R.I.N. Project 20122F87B2. The authors thank Prof. Benny Chor for providing the data and for helpful discussions.

REFERENCES

1. Apostolico A. The myriad virtues of subword trees. In: Apostolico A, Galil Z, editors. *Combinatorial Algorithms on Words*. Volume 12. Berlin Heidelberg: Springer-Verlag; 1985. p 85–96.
2. Apostolico A. Maximal words in sequence comparisons based on subword composition. In: Elomaa T, Mannila H, Orponen P, editors. *Algorithms and Applications*. Volume 6060, Lecture Notes in Computer Science. Berlin Heidelberg: Springer-Verlag; 2010. p 34–44.
3. Apostolico A, Comin M, Parida L. Motifs in ziv-lempel-welch clef. Proceedings of IEEE DCC Data Compression Conference. Snowbird, UT. Computer Society Press; 2004. p 72–81.
4. Apostolico A, Comin M, Parida L. Conservative extraction of over-represented extensible motifs. Bioinformatics 2005;21 Suppl 1:i9–i18.
5. Apostolico A, Comin M, Parida L. Bridging Lossy and Lossless Compression by Motif Pattern Discovery. In: *General theory of information transfer and combinatorics*. Volume 4123, Lecture Notes in Computer Science. Berlin Heidelberg: Springer-Verlag; 2006. p 793–813.
6. Apostolico A, Comin M, Parida L. Mining, compressing and classifying with extensible motifs. Algorithms Mol Biol 2006;1:4.
7. Apostolico A, Comin M, Parida L. VARUN: discovering extensible motifs under saturation constraints. IEEE/ACM Trans Comput Biol Bioinform 2010;7(4):752–762.
8. Apostolico A, Parida L. Incremental paradigms of motif discovery. J Comput Biol 2004;11:15–25.
9. Burstein D, Ulitsky I, Tuller T, Chor B. Information theoretic approaches to whole genome phylogenies. In: Miyano S, Mesirov J, Kasif S, Istrail S, Pevzner P, Waterman M, editors. *Proceedings of the 9th Annual International Conference on Research in Computational Molecular Biology (RECOMB 2005)*. Berlin Heidelberg: Springer-Verlag; 2005. p 283–295.
10. Chor B, Cohen E. Detecting phylogenetic signals in eukaryotic whole genome sequences. J Comput Biol 2012;19(8):945–956.
11. Chor B, Horn D, Goldman N, Levy Y, Massingham T. Genomic DNA *k*-mer spectra: models and modalities. Genome Biol 2009;10(10):R108.

12. Cole JR, Wang Q, Cardenas E, Fish J, Chai B, Farris RJ, Kulam-Syed-Mohideen AS, McGarrell DM, Marsh T, Garrity GM, Tiedje JM. The Ribosomal Database Project: improved alignments and new tools for RRNA analysis. Nucleic Acids Res 2009;37:D141–145.
13. Comin M, Parida L. Detection of subtle variations as consensus motifs. Theor Comput Sci 2008;395(2-3):158–170.
14. Comin M, Verzotto D. Classification of protein sequences by means of irredundant patterns. BMC Bioinformatics 2010;11 Suppl 1:S16.
15. Comin M, Verzotto D. The Irredundant Class method for remote homology detection of protein sequences. J Comput Biol 2011;18(12):1819–1829.
16. Comin M, Verzotto D. Comparing, ranking and filtering motifs with character classes: application to biological sequences analysis. In: Elloumi M, Zomaya A, editors. *Biological Knowledge Discovery Handbook: Preprocessing, Mining and Postprocessing of Biological Data*. Chapter 13. John Wiley & Sons, Inc.; 2014. p. 307–332.
17. Cormen TH, Leiserson CE, Rivest RL. *Introduction to Algorithms*. Chapter 9. Cambridge, MA: MIT Press; 1990. p 178–180.
18. Critchlow DE, Pearl DK, Qian C. The triples distance for rooted bifurcating phylogenetic trees. Syst Biol 1996;45(3):323–334.
19. Delsuc F, Brinkmann H, Philippe H. Phylogenomics and the reconstruction of the tree of life. Nat Rev Genet 2005;6:361–375.
20. Edwards SV, Liu L, Pearl DK. High-resolution species trees without concatenation. Proc Natl Acad Sci U S A 2007;104(14):5936–5941.
21. Felsenstein J. Phylip—phylogeny inference package (version 3.2). Cladistics 1989;5:164–166.
22. Giancarlo R, Scaturro D, Utro F. Textual data compression in computational biology: a synopsis. Bioinformatics 2009;25(13):1575–1586.
23. Gusfield D. *Algorithms on Strings, Trees, and Sequences: Computer Science and Computational Biology*. New York, NY: Cambridge University Press; 1997.
24. Huynen MA, Bork P. Measuring genome evolution. Proc Natl Acad Sci U S A 1998;95:5849–5856.
25. Iliopoulos CS, Mchugh J, Peterlongo P, Pisanti N, Rytter W, Sagot M-F. A first approach to finding common motifs with gaps. Int J Found Comput Sci 2005;16(6):1145–1154.
26. Kong S-G, Fan W-L, Chen H-D, Hsu Z-T, Zhou N, Zheng B, Lee H-C. Inverse symmetry in complete genomes and whole-genome inverse duplication. PLoS ONE 2009;4(11):e7553.
27. Kopelowitz T, Lewenstein M. Dynamic weighted ancestors. Proceedings of the 18th Annual ACM-SIAM Symposium on Discrete Algorithms (SODA 2007). New Orleans, LA SIAM; 2007. p 565–574.
28. Kulhavý R. A Kullback-Leibler distance approach to system identification. Annu Rev Control 1996;20:119–130.
29. Kullback S, Leibler RA. On information and sufficiency. Teh Ann Math Stat 1951;22(1):79–86.
30. Lempel A, Ziv J. On the complexity of finite sequences. IEEE Trans Inform Theory 1976;22(1):75–81.
31. Lin J. Divergence measures based on the Shannon entropy. IEEE Trans Inform Theory 1991;37:145–151.

32. Martinsen ES, Perkins SL, Schall JJ. A three-genome phylogeny of malaria parasites (*Plasmodium* and closely related genera): evolution of life-history traits and host switches. Mol Phylogenet Evol 2008;47:261–273.

33. Shiino T, Okabe N, Yasui Y, Sunagawa T, Ujike M, Obuchi M, Kishida N, Xu H, Takashita E, Anraku A, Ito R, Doi T, Ejima M, Sugawara H, Horikawa H, Yamazaki S, Kato Y, Oguchi A, Fujita N, Odagiri T, Tashiro M, Watanabe H. Molecular evolutionary analysis of the influenza A(H1N1)pdm, May–September, 2009: temporal and spatial spreading profile of the viruses in Japan. PLoS ONE 2010;5(6):e11057.

34. Sims GE, Jun SR, Wu GA, Kim SH. Alignment-free genome comparison with feature frequency profiles (FFP) and optimal resolutions. Proc Natl Acad Sci U S A 2009;106(8):2677–2682.

35. Sims GE, Jun S-R, Wu GA, Kim S-H. Whole-genome phylogeny of mammals: evolutionary information in genic and nongenic regions. Proc Natl Acad Sci U S A 2009;106(40):17077–17082.

36. Smith GJD, Vijaykrishna D, Bahl J, Lycett SJ, Worobey M, Pybus OG, Ma SK, Cheung CL, Raghwani J, Bhatt S, Malik Peiris JS, Guan Y, Rambaut A. Origins and evolutionary genomics of the 2009 swine-origin H1N1 Influenza A epidemic. Nature 2009;459(7250):1122–1125.

37. Thompson JD, Higgins DG, Gibson TJ. CLUSTAL W: improving the sensitivity of progressive multiple sequence alignment through sequence weighting, position-specific gap penalties and weight matrix choice. Nucleic Acids Res 1994;22:4673–4680.

38. Ukkonen E. Maximal and minimal representations of gapped and non-gapped motifs of a string. Theor Comput Sci 2009;410(43):4341–4349.

39. Ulitsky I, Burstein D, Tuller T, Chor B. The average common substring approach to phylogenomic reconstruction. J Comput Biol 2006;13(2):336–350.

40. Venter CJ et al. The sequence of the human genome. Science 2001;291:1305–1350.

41. Verzotto D, Comin M. Alignment-free phylogeny of whole genomes using underlying subwords. BMC Algorithms Mol Biol 2012;7:34.

42. Wildman DE, Uddin M, Opazo JC, Liu G, Lefort V, Guindon S, Gascuel O, Grossman LI, Romero R, Goodman M. Genomics, biogeography, and the diversification of placental mammals. Proc Natl Acad Sci U S A 2007;104:14395–14400.

43. Wu T-J, Huang Y-H, Li L-A. Optimal word sizes for dissimilarity measures and estimation of the degree of dissimilarity between DNA sequences. Bioinformatics 2005;21(22):4125–4132.

44. Ziv J, Lempel A. A universal algorithm for sequential data compression. IEEE Trans Inform Theory 1977;23(3):337–343.

CHAPTER 4

A MAXIMUM LIKELIHOOD FRAMEWORK FOR MULTIPLE SEQUENCE LOCAL ALIGNMENT

Chengpeng Bi[1,2]

[1]*Bioinformatics and Intelligent Computing Lab, Division of Clinical Pharmacology, Children's Mercy Hospitals, Kansas City, USA*
[2]*School of Medicine, University of Missouri, Kansas City, USA*

4.1 INTRODUCTION

With the advance of high throughput sequencing technology, it is easy for molecular biologists to generate several terabytes of biological sequences in few days. However, efficient algorithms, to analyze and interpret these data, are still lacking; in particular, efficient motif finding algorithms in a set of biological sequences. Multiple sequence local alignment is often used to solve the motif finding problem. Figure 4.1 represents an example of multiple sequence local alignment used to find motifs in DNA sequences.

This chapter first constructs a maximum likelihood framework for multiple sequence local alignment, and then compares a set of sequence pattern or motif finding algorithms by consolidating them under the framework. Based on the likelihood modeling framework, several motif finding algorithms are then developed

Pattern Recognition in Computational Molecular Biology: Techniques and Approaches,
First Edition. Edited by Mourad Elloumi, Costas S. Iliopoulos, Jason T. L. Wang, and Albert Y. Zomaya.

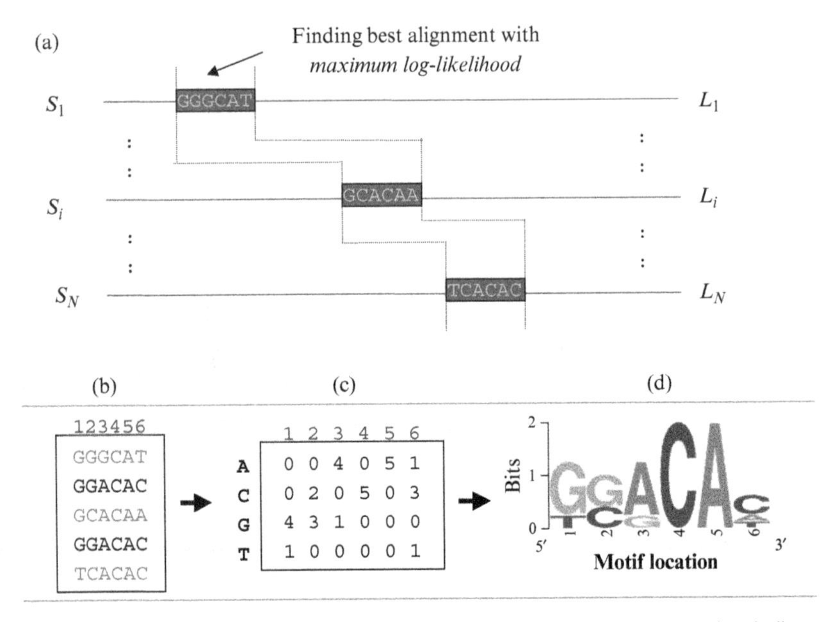

Figure 4.1 Sequence local alignment and motif finding. (a) Multiple sequence local alignment of protein–DNA interaction binding sites. The short motif sequence shown in a highlighted box represents the site recognized and bound with a regulatory protein. (b) Motif sequences (sites) aligned to discover a pattern. (c) Nucleotide counting matrix is derived from the aligned sequences in (b) and counting column-wise. (d) Sequence logo representation of the motif alignment, which visually displays the residue conservation across the alignment positions. Sequence logo is plotted using *BiLogo* server (http://bipad.cmh.edu/bilogo.html). Notice, sequence logo plots the motif conservation across positions. Each column consists of stacks of nucleotide letters (A, C, G, T). The height of a letter represents its occurrence percentage on that position. The total information content on each column is adjusted to 2.0 bits as a maximum.

to maximize the log-likelihood function by imputing the unobserved data. This will facilitate the evaluation and comparison of different algorithms under the same objective function. In the case of DNA *cis*-regulatory sequence motif finding problem, the binding sites or motif locations are unobserved or missing data. In discovery of protein structural domains, the domain locations are missing data. The missing data space is the search space of multiple sequence local alignment. Two different types of optimization algorithms are investigated to explore the missing data space:

The first type is represented by the deterministic algorithms, Deterministic Expectation Maximization (DEM) and Winner-take-all Expectation Maximization (WEM) [1–11], that are to optimize a specified objective function by performing iteratively optimal local search in the alignment space.

The second one is represented by the stochastic sampling algorithms, *Gibbs* [12, 15–18, 22], *pGibbs* [3–5], and *Metropolis* [13, 21], that are to iteratively draw motif location samples according to a conditional probability distribution via Monte Carlo simulation.

The rest of this chapter is organized as follows: Section 4.2 briefly describes the framework for multiple sequence local alignment. Section 4.3 is devoted to develop a suite of motif finding algorithms. Section 4.4 compares the developed algorithms' time complexities. Section 4.5 presents case studies to evaluate and compare the developed algorithms through simulated and real biological examples. Section 4.6 concludes the chapter.

4.2 MULTIPLE SEQUENCE LOCAL ALIGNMENT

Multiple sequence local alignment can be used to locate over-represented sites in a given set of sequences. The aligned over-represented sites are then used to build a frequency matrix that depicts a conserved domain or motif (Figure 4.1. Let $\mathbf{S} = \{S_1, \ldots, S_i, \ldots, S_N\}$ be a set of sequences. Let L_i be the length of a sequence S_i, and S_{ij} be the jth residue (character) of S_i. If S_i codes a DNA or RNA sequence then S_{ij} will be a four-letter sequence, and if S_i codes a protein then S_{ij} will be a 20-letter sequence. If only one motif per sequence (i.e., *oops*) model is assumed, there are N motifs in total for N sequences. A zero or one motif per sequence (i.e., *zoops*) model is also frequently used. Nonetheless, both *oops* and *zoops* models assume that sequence data come from a two-component multinomial mixture model: (i) the background model, assuming that residues at nonmotif positions follow an independent and identical multinomial distribution $\vec{\theta}_0$; and (ii) the w-mer motif model (or w-motif), assuming that residues within the motif are independent but not identical, in other words, residues at different motif positions come from different multinomial distributions $\vec{\theta}_j$.

Let A_i be an indicator variable drawn from the location space $\{0, 1\}^{(L_i - w+1)}$ of sequence S_i, $\mathbf{A} = [A_1, \ldots, A_i, \ldots, A_N]^T$ be the set of indicator variables representing the motif start sites (i.e., a local alignment) in the sequences, and w be the motif width. The total number V of local alignments can be generally expressed as $V = \prod_{i=1}^{N} \binom{L_i - w + 1}{|A_i|}$, where $|A_i| = \sum_l A_{il}$ is the number of motif sites on sequence S_i. Therefore, if $|A_i| = 1$ for all i, it is an *oops* model, otherwise it is a *zoops* or multiple-site model. The total number of motif sites is $|\mathbf{A}| = \sum_i |A_i|$. Alternatively, an indicator variable $a_i = l$ is used to represent the motif starting at position l on sequence S_i, which is equivalent to $A_{il} = 1$. Note that $a_i = 0$ means no motifs found on sequence S_i. If multiple sites occur on a sequence, a vector $\vec{a}_i$ is used to store all the positions. Obviously a motif indicator vector $\vec{a}_i \in \mathcal{P}_i(\{1, 2, \ldots, L_i - w + 1\})$, where $\mathcal{P}_i$ is the power set of the ith sequence motif sites. The alignment of motif sites is initialized by randomly generating a set of motif start sites (i.e., $\mathbf{A}^{(0)}$ or equivalently $[\vec{a}_1^{(0)}, \ldots, \vec{a}_N^{(0)}]^T$) and then it is progressively refined until convergence.

4.2.1 Overall Objective Function

Suppose each alignment is considered of as a hidden state in the alignment space, the motif finding problem can be formulated as finding the optimized alignment state

υ^* among the entire alignment space. Index a state by $\upsilon \equiv [\vec{a}_1, \ldots, \vec{a}_i, \ldots, \vec{a}_N]^T = \mathbf{A}^{(\upsilon)}$, and let the energy of state υ be $E(\upsilon) = E(\mathbf{S}, \mathbf{A}^{(\upsilon)})$ where $\mathbf{A}^{(\upsilon)}$ is the alignment corresponding to the state υ. The energy may be related to an alignment score or the motif sequence specificity/binding energy [2]. Then at equilibrium the probability of state υ, $p(\upsilon)$, is calculated as:

$$p(\upsilon) = \frac{1}{Z(T)} \exp(-E(\mathbf{S}, \mathbf{A}^{(\upsilon)})/k_B T) \tag{4.1}$$

where k_B is the Boltzmann constant, T is the temperature, and $Z(T)$ is defined as:

$$Z(T) = \int_{\vec{a}_1 \in \mathcal{P}_1} \cdots \int_{\vec{a}_N \in \mathcal{P}_N} \exp(-E(\mathbf{S}, [\vec{a}_1, \ldots, \vec{a}_N]^T)/k_B T) d\mathbf{A}$$

Therefore, the optimized alignment state υ^* is the one with the maximum probability p^*. If υ^* is found, then the parameter estimation $\mathbf{\Theta}^*$ is done. However, computing the partition function or normalized constant (the denominator in the above equation) is commonly intractable, because the alignment problem has been proven to be NP-complete [23].

4.2.2 Maximum Likelihood Model

The *de novo* motif finding is a data missing problem, that is, DNA or protein sequences $\mathbf{S}$ are observed, but the motif locations $\mathbf{A}$ are unobserved. Based on maximum likelihood modeling, motif algorithms can be used to estimate the model parameters $\mathbf{\Theta}$ and $\vec{\theta}_0$ given the observed sequences $\mathbf{S}$. The complete data for motif sequence model is $(\mathbf{S}, \mathbf{A}) = \{(S_i, A_i) \text{ s.t. } i \in \{1, \cdots, N\}\}$. The conditional likelihood of sequence S_i, given the hidden variable a_i, is as follows:

$$p(S_i | a_i = l, \mathbf{\Theta}, \vec{\theta}_0) = \prod_{y \in A_{il}^c} \prod_{k=1}^{K} \theta_{0k}^{I(S_{iy}=k)} \prod_{m=1}^{w} \prod_{k=1}^{K} \theta_{mk}^{I(S_{i,l+m-1}=k)} \tag{4.2}$$

where A_{il}^c denotes the background sites and $I(\cdot)$ is the indicator function. Although Equation 4.2 is for an *oops* model, it is easy to extend to other motif models [3]. To simplify the notation, $\mathbf{\Theta}$ is used to contain the background parameters $\vec{\theta}_0$ in the following derivation. Let $\mathbf{\Theta}^{(t)}$ be the parameter estimates after tth iteration, then given the observed data, the conditional expected complete data log-likelihood (often referred to as the Q-function) is defined as:

$$Q(\mathbf{\Theta}; \mathbf{\Theta}^{(t)}) = E[\log p(\mathbf{S}, A | \Theta) | \mathbf{S}, \mathbf{\Theta}^{(t)}] = \int_{\vec{a} \in \mathbf{A}} \log p(\mathbf{S}, \vec{a} | \mathbf{\Theta}) p(\vec{a} | \mathbf{S}, \mathbf{\Theta}^{(t)}) d_{\vec{a}}$$

The above Q-function can be reduced to:

$$Q(\mathbf{\Theta}, \mathbf{\Theta}^{(t)}) = \sum_{i=1}^{N} \sum_{l=1}^{L_i - w + 1} p(a_i = l | S_i, \mathbf{\Theta}^{(t)}) \log p(S_i | a_i = l, \mathbf{\Theta}) \tag{4.3}$$

The conditional probability of a potential binding site a_i is computed as:

$$p(a_i = l|S_i, \mathbf{\Theta}^{(t)}) = \frac{p(S_i|a_i = l, \mathbf{\Theta}^{(t)})g(a_i = l)}{\sum_{j=1}^{L_i-w+1} p(S_i|a_i = j, \mathbf{\Theta}^{(t)})g(a_i = j)} \tag{4.4}$$

where $g(a_i)$ is the probability of the motif positional preference, and it is often treated as a constant, that is, all positions are equally likely or $g(a_i) = 1/(L_i - w + 1)$. By combining Equations 4.2 and 4.4, one can derive an explicit formula for the conditional density function as:

$$p(a_i = l|S_i, \mathbf{\Theta}^{(t)}) = \frac{\prod_{j=1}^{w}\prod_{k=1}^{K}\left(\frac{\theta_{jk}^{(t)}}{\theta_{0k}^{(t)}}\right)^{I(S_{i,l+j-1}=k)}}{\sum_{q=1}^{L_i-w+1}\left\{\prod_{j=1}^{w}\prod_{k=1}^{K}\left(\frac{\theta_{jk}^{(t)}}{\theta_{0k}^{(t)}}\right)^{I(S_{i,q+j-1}=k)}\right\}} \tag{4.5}$$

To facilitate the computation of the Q-function and implementation of the algorithms, one can define two quantities depending on if it is a motif ($j \neq 0$) or background sequence ($j = 0$) as follows:

$$N_{jk} = \begin{cases} \sum_{i=1}^{N}\sum_{l=1}^{L_i-w+1} p(a_i = l|S_i, \mathbf{\Theta}^{(t)}) I(S_{i,(l+j-1)} = k) & \text{if } j \neq 0 \\ \sum_{i=1}^{N}\sum_{l=1}^{L_i-w+1} p(a_i = l|S_i, \mathbf{\Theta}^{(t)}) \sum_{y \in A_{il}^c} I(S_{iy} = k) & \text{if } j = 0 \end{cases}$$

After some modifications, Equation 4.3 can be nicely expressed in its compact form as:

$$Q(\mathbf{\Theta}, \mathbf{\Theta}^{(t)}) = \sum_{k=1}^{K} N_{0k} \log \theta_{0k} + \sum_{j=1}^{w}\sum_{k=1}^{K} N_{jk} \log \theta_{jk} \tag{4.6}$$

Since a local alignment $\mathbf{A}$ is uniquely mapped to an estimate of the Position Weight Matrix (PWM) $\mathbf{\Theta}$, a new objective function $h(\cdot)$ corresponding to the above alignment log-likelihood is defined as:

$$h(\mathbf{A}^{(t)}) = h(\mathbf{\Theta}^{(t)}) = |\mathbf{A}| \sum_{k=1}^{K} \theta_{0k} \log \theta_{0k} + |\mathbf{A}| \sum_{j=1}^{w}\sum_{k=1}^{K} \theta_{jk}^{(t)} \log \theta_{jk}^{(t)} \tag{4.7}$$

where $|\mathbf{A}|$ is the number of aligned motif sites. Note that h relates to a quantity that may have its real biological meaning, for example, it may indicate the protein–DNA binding affinity [2] or energy of a protein three-dimensional configuration [18]. Assuming the Boltzmann distribution, one can compute the probability $p(v)$ of such a configuration. For a large data set, $\vec{\theta}_0$ can be computed from the training sequences and then treated as constants. The maximum likelihood-based alignment is to derive a local alignment together with its motif model that can maximize the function defined as earlier.

4.3 MOTIF FINDING ALGORITHMS

In this section, five motif finding algorithms will be developed that are derived from the framework described as earlier, in particular, on the basis of the Equations 4.5 and 4.7. The deterministic motif finding algorithms, DEM and WEM [4–15], are to maximize the likelihood function h through iteratively greedy local search. The stochastic sampling algorithms, *Gibbs*, *pGibbs*, and *Metropolis*, are to iteratively draw motif location samples through Monte Carlo simulation, and at the same time keep track of alignments with the best local likelihoods.

4.3.1 DEM Motif Algorithm

The DEM algorithm merely iterates E- and M-steps a number of times until it converges: (i) The E-step computes $p(a_i = l|S_i, \mathbf{\Theta}^{(t)})$; and (ii) the M-step updates the parameters as:

$$\theta_{jk} = N_{jk} / \sum_{k \in K} N_{jk} \tag{4.8}$$

where $j = 0, 1, \ldots, w$. The stopping criterion is controlled by a threshold or precision ξ. When the likelihood difference between two consecutive iterations is less than the precision ξ, the DEM algorithm stops.

The E-step is a scanning procedure as in Equation 4.5, and it calculates each potential site probabilities across all sequences. The M-step is an averaging procedure that is to do position-wide (j) weighted summation over all sites by each residue types ($k \in K$) divided by the total weight. In each run, the DEM motif finding algorithm starts with an initial alignment that is randomly produced or received from an input, and then it either goes through I_{max} rounds of refinement or ends up with a covergence. Note that I_{max} is useful in performing truncated search, however, the DEM motif finding algorithm does not enforce the number of iterations but rather watches the convergence condition: $h(\mathbf{A}^{(t+1)}) - h(\mathbf{A}^{(t)}) < \xi$, where ξ is a small quantity.

The pseudo-code of the DEM motif algorithm is given as follows:

4.3.2 WEM Motif Finding Algorithm

The WEM motif finding algorithm replaces the conventional E-step with a series of so-called *partial winning* or W-step, each of which computes a conditional probability $\tilde{p}^{(t)}(\vec{a}_i|S_i, \mathbf{\Theta}^{(t+(i-1)/N)})$. It then determines a winning site $a_i^{(t+i/N)}$ on sequence S_i such that it gets the maximum likelihood (i.e., the ith winner):

$$a_i^{(t+i/N)} = \arg \max_{a \in \mathcal{P}_i} \{\tilde{p}^{(t+i/N)}(a|S_i, \mathbf{\Theta}^{t+(i-1)/N})\} \forall i \tag{4.9}$$

Algorithm 4.1 DEM

Input: $\mathbf{S}, \mathbf{A}^{(0)}, w, I_{max}$
Initialize $\xi \leftarrow 10^{-5}, t \leftarrow 0$
Build $\mathbf{\Theta}^{(0)}$ from $\mathbf{A}^{(0)}$
repeat
  for $i = 1$ to N **do**
    E-step: compute $p(a_i|\cdot)$ by Equation 4.5;
    M-step: update parameters $\{\theta_{jk}\}$ by Equation 4.8;
  end for
  $t \leftarrow t + 1$;
until $h(\mathbf{A}^{(t+1)}) - h(\mathbf{A}^{(t)}) < \xi$ or $t > I_{max}$
return The final motif alignment $\mathbf{A}^*$ and model $\mathbf{\Theta}$.

Notice that although one winner per sequence is presumed here, it can be easily generalized. Based on the winning site WEM performs a partial update from $\mathbf{\Theta}^{t+(i-1)/N}$ to $\mathbf{\Theta}^{t+i/N}$, that is, a partial U-step. These partial W- and U-steps iterate until all winning sites among N sequences are determined which completes one WEM round. Each full round is followed by a complete parameter update from $\mathbf{\Theta}^{(t)}$ to $\mathbf{\Theta}^{(t+1)}$. A number of WEM rounds are often required until a convergence. To simplify the W-step, the following formula is used as the target density function of the unobserved data a:

$$\tilde{p}^{(t+i/N)}(a|S_i, \hat{\mathbf{\Theta}}^{t+i-1/N}) \propto \prod_{j=1}^{w} \prod_{k \in K} \left(\frac{\hat{\theta}_{jk}^{(t)}}{\hat{\theta}_{0k}^{(t)}} \right)^{I(S_{i,a+j-1}=k)} \quad \forall i \tag{4.10}$$

where $a \in \mathcal{P}_i$. Equation 4.10 is nothing but the numerator of Equation 4.5, and it is applied to scan sequence S_i for all potential motif sites. The scanned site with the largest p-value is selected as the new winner, which is in turn used for the next partial U-step.

A partial W-step is immediately followed by a partial U-step. The partial U-step is reduced as follows:

$$\hat{\theta}_{jk}^{(t+i/N)} = \hat{\theta}_{jk}^{(t+(i-1)/N)} + \lambda(a^{(t-1)+i/N}, a^{t+i/N}) \forall i, j, k \tag{4.11}$$

where $a^{(t-1)+i/N}$ and $a^{t+i/N}$ are the winning sites on sequence S_i of two consecutive rounds, respectively, and $\lambda(\cdot, \cdot)$ is a self-tuning function that adjusts the residue type changes at the same motif location between two partial alignments of consecutive rounds. For example, if in the previous alignment, a DNA motif position j on sequence S_i appears as a base "A" (i.e., adenine) and in the current alignment it changes to a "C" (i.e., cytosine), then the tuning function first decreases the "A" count by 1 and then increases the "C" count by 1 at the same motif location. Of course, one can skip such tuning operation if $\lambda = 0$ (i.e., two winning sites in consecutive rounds are exactly

Algorithm 4.2 WEM

Input: $\mathbf{S}, \mathbf{A}^{(0)}, w$
Initialize $\xi \leftarrow 10^{-5}, t \leftarrow 0$
Build $\mathbf{\Theta}^{(0)}$ from $\mathbf{A}^{(0)}$
repeat
  for $i = 1$ to N **do**
    W-step: determine a new winner by Equation 4.9;
    U-step: follow the rule by Equation 4.11;
  end for
  $t \leftarrow t + 1$;
until $h(\mathbf{A}^{(t+1)}) - h(\mathbf{A}^{(t)}) < \xi$
return The final motif alignment $\mathbf{A}^*$ and model $\mathbf{\Theta}$.

the same). Therefore, each partial update costs a linear time $O(L + w)$, that includes locating the winning position in $O(L)$ time, and here w is usually very small, but a sequence can be very long, that is, L is large. Note that the standard DEM algorithm carries out overall averaging update in $O(wL)$ time. The estimation of $\hat{\theta}_{jk}^{(t+i/N)}$ relies on the estimation of $\hat{\theta}_{jk}^{(t+(i-1)/N)}$, and $\hat{\theta}_{jk}^{(0)}$ is randomly initialized. The pseudo-code of the WEM motif finding algorithm is given as follows:

In practice, the function $h(\cdot)$ is evaluated after a complete WEM round, that is, to save time and ensure a sufficient increase of h-value.

4.3.3 Metropolis Motif Finding Algorithm

This algorithm starts with any alignment $\mathbf{A}^{(0)}$ by randomly initializing a seed, note that there exists a one–one map: $\mathbf{A}^{(t)} \rightarrow \mathbf{\Theta}^{(t)}$), and then it iterates the following two steps: (i) propose a random perturbation of the current state t, that is, $\mathbf{A}^{(t)} \rightarrow \mathbf{A}'$, where $\mathbf{A}'$ is generated from a proposal distribution $P(\mathbf{A}^{(t)} \rightarrow \mathbf{A}')$ by a series of independent random sampling; calculate the change $\Delta h = h(\mathbf{\Theta}') - h(\mathbf{\Theta}^{(t)})$; and (ii) define the acceptance probability α_M as:

$$\alpha_M(\mathbf{A}', \mathbf{A}^{(t)}) = \min\{1, \exp\ (\Delta h)\} \tag{4.12}$$

One can toss a Bernoulli coin with a probability α_M coming up heads: if heads show up, accept the new move: $\mathbf{A}^{(t+1)} = \mathbf{A}'$; otherwise stay as it was: $\mathbf{A}^{(t+1)} = \mathbf{A}^{(t)}$. The proposal distribution is a product of N independent site probabilities defined in Equation 4.5, which results in:

$$P(\mathbf{A}^{(t)} \rightarrow \mathbf{A}') = \prod_{i=1}^{N} p(a_i | S_i, \mathbf{\Theta}^{(t)}) \tag{4.13}$$

Algorithm 4.3 *Metropolis*

Input: $\mathbf{S}, \mathbf{A}^{(0)}, w, I_{max}$
Initialize $t \leftarrow 0$
Build $\boldsymbol{\Theta}^{(0)}$ from $\mathbf{A}^{(0)}$, $\boldsymbol{\Theta}^* = \boldsymbol{\Theta}^{(0)}$
repeat
  Draw a new alignment $\mathbf{A}'$;
  Accept or reject $\mathbf{A}'$ according to equation 4.12 or 4.14;
  if $h(\boldsymbol{\Theta}^*) < h(\boldsymbol{\Theta}^{(t)})$ **then** $\boldsymbol{\Theta}^* \leftarrow \boldsymbol{\Theta}^{(t)}$ **end if**
  $t \leftarrow t + 1$;
until $t > I_{max}$
return The final motif alignment $\mathbf{A}^*$ and model $\boldsymbol{\Theta}$.

The Hastings acceptance rate α_H can be defined as follows:

$$\alpha H(\mathbf{A}', \mathbf{A}^{(t)}) = \min \left\{ 1, \exp\ (\Delta h) \prod_{i=1}^{N} \frac{p(a_i' | S_i, \boldsymbol{\Theta}^{(t)})}{p(a_i^{(t)} | S_i, \boldsymbol{\Theta}')} \right\} \tag{4.14}$$

The pseudo-code of the *Metropolis* motif finding sampler is given as follows:

Notice that *Metropolis* updates the alignment after a whole round of N samplings, whereas both *Gibbs* and *pGibbs* iteratively update every single sampling.

4.3.4 Gibbs Motif Finding Algorithm

The *Gibbs* motif finding sampler first initializes $\mathbf{A}^{(0)}$ in the same way as *Metropolis* and then systematically chooses a sequence S_i, $i \in \{1, \ldots, N\}$ and updates it with a new sample drawn from the site conditional density distribution $\pi(\cdot | S_i, \mathbf{A}^{(t)}_{[-i]})$ that can be approximated by Equation 4.5, here $\mathbf{A}^{(t)}_{[-i]}$ is the alignment with exclusion of motif sites on sequence S_i. A slight difference is that $\boldsymbol{\Theta}^{(t)}$ is built from $\mathbf{A}^{(t)}_{[-i]}$ rather than $\mathbf{A}^{(t)}$. Consequently, a little extra work is put on removing the motif sites ($\mathbf{a}_i^{(t)}$) on sequence S_i from the current alignment $\mathbf{A}^{(t)}$ first and then rebuilding a new PWM matrix $\boldsymbol{\Theta}^{(t)}_{[-i]}$ from $\mathbf{A}^{(t)}_{[-i]}$. Such a reconstruction of PWM matrix is called *exclusion* (of sites on sequence S_i) step. Note that both *Metropolis* and *pGibbs* do not have the exclusion step and thus save a little bit time depending on sequence size. The new matrix shall be used in Equation 4.5 to scan the sequence S_i. The pseudo-code of the *Gibbs* motif finding sampler is given as follows:

As a special case of *Metropolis*, *Gibbs* constructs a Markov chain whose equilibrium distribution is π. The motif optimization problem is solved throughout the Monte Carlo simulation by keeping track of all the best log-likelihood solutions along the sampling.

Algorithm 4.4 *Gibbs*

Input: $\mathbf{S}, \mathbf{A}^{(0)}, w, I_{max}$
Initialize $t \leftarrow 0$
Build $\mathbf{\Theta}^{(0)}$ from $\mathbf{A}^{(0)}$, $\mathbf{\Theta}^* = \mathbf{\Theta}^{(0)}$
repeat
  for $i = 1$ to N **do**
    Draw a new sample: $a_i^{(t+i/N)} \sim \pi(\cdot|S_i, \mathbf{\Theta}_{[-i]}^{(t+(i-1)/N)})$;
    Update the motif parameters: $\mathbf{A}^{(t+i/N)} \mapsto \mathbf{\Theta}^{(t+i/N)}$;
    if $h(\mathbf{\Theta}^*) < h(\mathbf{\Theta}^{(t+i/N)})$ **then** $\mathbf{\Theta}^* \leftarrow \mathbf{\Theta}^{(t+i/N)}$ **end if**
  end for
  $t \leftarrow t + 1$;
until $t > I_{max}$
return The final motif alignment $\mathbf{A}^*$ and model $\mathbf{\Theta}$.

Algorithm 4.5 *pGibbs*

Input: $\mathbf{S}, \mathbf{A}^{(0)}, w, I_{max}$
Initialize $t \leftarrow 0$
Build $\mathbf{\Theta}^{(0)}$ from $\mathbf{A}^{(0)}$, $\mathbf{\Theta}^* = \mathbf{\Theta}^{(0)}$
repeat
  for $i = 1$ to N **do**
    Draw a new sample: $a_i^{(t+i/N)} \sim \pi(\cdot|S_i, \mathbf{\Theta}^{(t+(i-1)/N)})$;
    Update the motif parameters: $\mathbf{A}^{(t+i/N)} \mapsto \mathbf{\Theta}^{(t+i/N)}$;
    if $h(\mathbf{\Theta}^*) < h(\mathbf{\Theta}^{(t+i/N)})$ **then** $\mathbf{\Theta}^* \leftarrow \mathbf{\Theta}^{(t+i/N)}$ **end if**
  end for
  $t \leftarrow t + 1$;
until $t > I_{max}$
return The final motif alignment $\mathbf{A}^*$ and model $\mathbf{\Theta}$.

4.3.5 Pseudo-Gibbs Motif Finding Algorithm

The *pseudo-Gibbs* (*pGibbs*) motif finding sampler first initializes $\mathbf{A}^{(0)}$ in the same way as the DEM motif finding algorithm, and it systematically chooses a sequence S_i, $i \in (1, \ldots, N)$, and then updates the motif parameters with a new sample drawn from the site conditional density distribution $\pi(\cdot|S_i, \mathbf{\Theta}^{(t)})$ built from $\mathbf{A}^{(t)}$. This process iterates a number of times (I_{max}). A slight difference between *Gibbs* and *pGibbs* consists in the fact that *Gibbs* draws from $\pi(\cdot|S_i, \mathbf{\Theta}_{[-i]}^{(t)})$ built from $\mathbf{A}_{[-i]}^{(t)}$ rather than $\mathbf{A}^{(t)}$. In other words, *pGibbs* can be implemented directly based on Equation 4.5, whereas *Gibbs* requires the exclusion step to rebuild a new PWM matrix, that is, $\mathbf{\Theta}_{[-i]}^{(t)}$. The pseudo-code of the *pGibbs* motif finding sampler is given as follows:

Since *pGibbs* is a special case of the Monte Carlo EM algorithm [24], it indeed constructs a Markov chain and converges to the equilibrium distribution π [9, 10, 18]. The motif optimization problem is solved throughout the Monte Carlo simulation algorithm by recording all the best likelihood solutions along the stochastic sampling. The *Metropolis*, *Gibbs*, and *pGibbs* samplers are quite time-consuming due to the burn-in time issue [12, 18], whereas the deterministic algorithms such as DEM and WEM usually converge very fast [4–15].

4.4 TIME COMPLEXITY

In one iteration, DEM, WEM, *Metropolis*, *Gibbs*, and *pGibbs* have the same scanning time $O(\overline{L}N)$, where $\overline{L}$ is the average length of a sequence. However, they vary in their update step. DEM takes $O(w\overline{L}N)$ time to update a PWM table, whereas WEM takes only $O(wN + \overline{L}N)$ time. Both *Metropolis* and *Gibbs* need $O(\overline{L}N)$ time to draw new samples in an iteration. *Gibbs* updates a PWM table in $O(|K|wN)$ time because it carries out update sequence by sequence. *Metropolis* takes $O(|K|w)$ time since it updates only once at each iteration.

DEM and WEM automatically determine the number of iterations I_{auto} needed to converge. On the other hand, *Gibbs*, *pGibbs*, and *Metropolis* consume the specified number of iterations ($I_{max} = 100$ by default). The running time difference between the deterministic algorithms DEM and WEM, and the stochastic algorithms *Metropolis*, *pGibbs*, and *Gibbs* is largely attributed to the number of iterations per run. In practice, DEM and WEM converge much faster than their stochastic counterparts, that is to say, $\overline{I}_{auto} < I_{max}$. Such a difference becomes significant when the sequence is getting longer. Although DEM and WEM may be highly time efficient, they are more likely subject to local optima, as exhibited in their performance (see Table 4.1).

4.5 CASE STUDIES

The five motif finding algorithms presented in this chapter were implemented in C++ on the same computer, and thus ensure that their comparisons are fair in both running time and prediction accuracy.

Table 4.1 Values of *pla*, obtained for the different motif finding algorithms, for the CRP data set ($w = 22, N = 18$)

L	DEM	WEM	*Metropolis*	*pGibbs*	*Gibbs*
100	1.00 ± 0.00	0.77 ± 0.07	0.96 ± 0.05	0.98 ± 0.05	0.99 ± 0.03
200	0.96 ± 0.05	0.65 ± 0.06	0.94 ± 0.06	0.97 ± 0.05	0.98 ± 0.05
400	0.41 ± 0.03	0.41 ± 0.06	0.69 ± 0.18	0.62 ± 0.13	0.88 ± 0.08
800	0.22 ± 0.04	0.24 ± 0.03	0.31 ± 0.11	0.27 ± 0.08	0.43 ± 0.15
1600	0.18 ± 0.02	0.16 ± 0.03	0.18 ± 0.03	0.20 ± 0.03	0.23 ± 0.09
$\overline{pla}$	0.55	0.44	0.62	0.61	0.70

4.5.1 Performance Evaluation

Given a known alignment **O** and a predicted one **A**, we define the following functions to measure the prediction accuracy of a motif finding algorithm:

- α_j: Let o_j be the position of the *j*th known motif in a sequence S_i, and a_j be the position of the *j*th predicted motif in the same sequence. To measure the prediction accuracy of a motif finding algorithm on the *j*th predicted motif in S_i, we define the α_j function:

$$\alpha_j(o_j, a_j) = \frac{w_{a_j}^{o_j}}{w_j} \tag{4.15}$$

  where w_j is the width of the *j*th known motif in S_i, and $w_{a_j}^{o_j}$ is the width of the overlapping submotif between the *j*th known motif and the *j*th predicted one in S_i. Hence, when the jth predicted motif and *j*th known one match exactly, we have $\alpha_j = 1$. When they match partially, we have $0 < \alpha_j < 1$. And when they completely mismatch, we have $\alpha_j = 0$.
- pla_i: To measure the prediction accuracy of a motif finding algorithm on a sequence S_i, we define the pla_i function:

$$pla_i(O_i, A_i) = \frac{1}{|A_i|} \sum_{j=1}^{|A_i|} \alpha_j(o_j, a_j) \tag{4.16}$$

  where O_i is the observed motif sites on sequence S_i and A_i is the predicted motif sites on sequence S_i. Clearly, we have $0 \leq pla_i(O_i, A_i) \leq 1$.
- pla: To measure the prediction accuracy of a motif finding algorithm on a set of N sequences, we define the function pla:

$$pla(\mathbf{O}, \mathbf{A}) = \frac{1}{N} \sum_{i=1}^{N} pla_i(O_i, A_i) \tag{4.17}$$

  Clearly, we have $0 \leq pla(\mathbf{O}, \mathbf{A}) \leq 1$.

4.5.2 CRP Binding Sites

The first example used is the gold standard CRP binding data set [1, 3, 15]. CRP is a prokaryotic DNA sequence binding protein, and it is a positive regulatory factor necessary for the expression of catabolite repressible genes. Each of the 22 base pairs (bp) binding sequences is embedded in a 105 bp DNA sequence, that is, $L = 105$. The motif finding algorithms assume the locations of these sites a_i, $i \in \{1, 2, \ldots, N\}$, are unknown. The motif width is set as 22 bp long according to experimentally verified site length. Figure 4.2a displays the 22-mer motif sequence logo. The local alignment

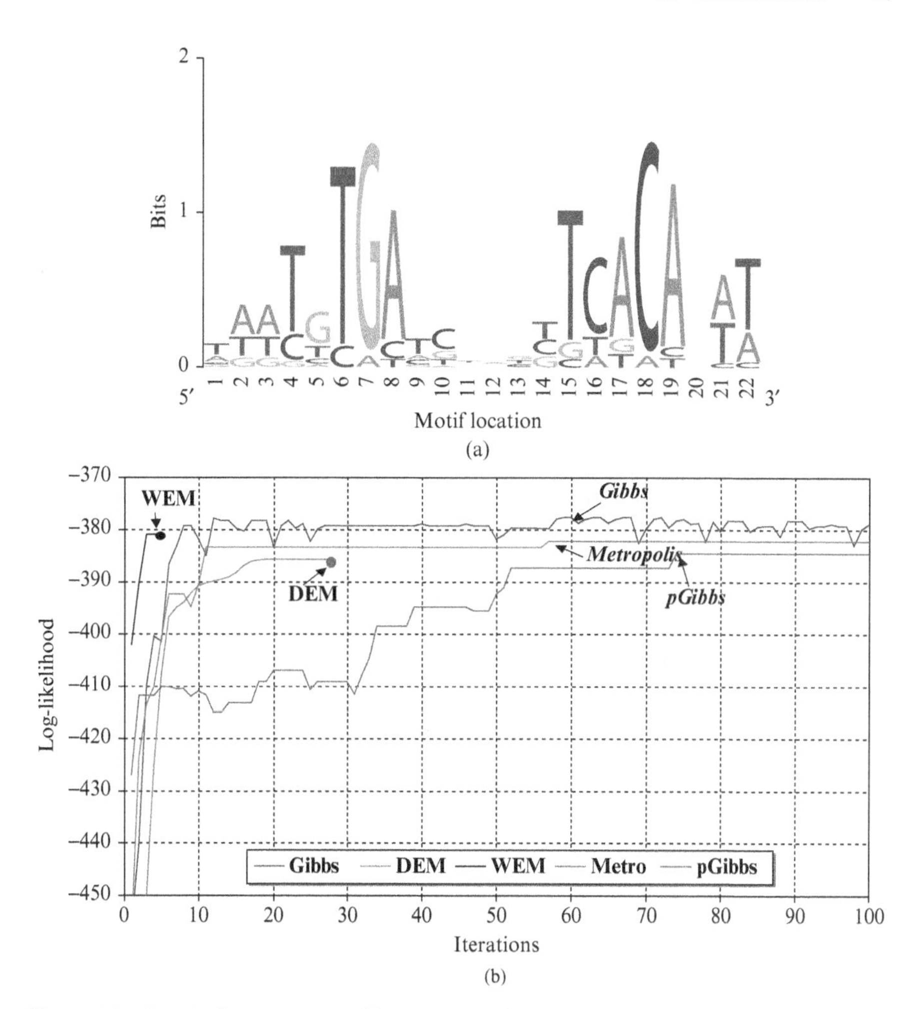

Figure 4.2 Results for sequences of CRP binding sites. (a) Motif sequence logo of CRP binding sites (22 bp). Sequence logos are plotted using *BiLogo* server (http://bipad.cmh.edu/bilogo.html). (b) Comparison of the curves along 100 iterations at maximum.

is performed on both forward and reverse DNA strands. The motif finding problem becomes more difficult as the sequence length gets longer. While testing on the CRP binding data, DEM took 15 iterations on average ($\overline{I}_{auto} = 15.0$) to converge, whereas WEM took only two iterations on average ($\overline{I}_{auto} = 2.0$) to converge. The stochastic algorithms *Metropolis*, *Gibbs*, and *pGibbs* often take much more iterations to converge.

Figure 4.2b shows converging curves of the five motif finding algorithms. The local greedy search algorithms DEM and WEM are monotonically increasing the likelihoods in each iteration until there is no significant improvement. Needless to say, WEM converges much faster than DEM, and requires the shortest time to finish up.

Among the three stochastic motif finding samplers, *Gibbs* exhibited the longest fluctuating curve. This indicates not only that *Gibbs* may converge slower than *Metropolis* and *pGibbs*, but also that it is more likely to efficiently explore local alignment space as compared with *Metropolis* and *pGibbs* (Figure 4.2b). Therefore, *Gibbs* is more able to escape from a local optimum, and this is exemplified by its best performance shown in Table 4.1.

To test the length L effect, the 18 verified CRP binding sites were planted into different data sets, each being different sequence length, that is, $L = 100, 200, 400, 800$, and 1600 bp long. The simulated background sequences were generated based on the zero-order Markov model ($\vec{\theta}_0 = [0.302, 0.183, 0.209, 0.306]^T$), which was estimated according to the original CRP data set. Table 4.1 summarizes the values of *pla* obtained for the different motif finding algorithms. Each algorithm was run 100 times. Overall, *Gibbs* is the best motif finding algorithm ($pla = 0.70$ on average). *Metropolis* and *pGibbs* are second to *Gibbs*. DEM is ranked the third ($pla = 0.55$ on average), however, much better than WEM ($pla = 0.44$ on average). When the sequence length is short (e.g., 100 or 200 bp), the five algorithms were able to find the correct, although with some different accuracy. DEM achieved pretty high accuracy, similar to the three stochastic algorithms *Metropolis*, *Gibbs*, and *pGibbs*, whereas WEM was the one with the lowest performance on average. In the 400 bp group, only the three stochastic algorithms can correctly locate the CRP motif with $pla > 0.62$ (*Gibbs* is the best with $pla = 0.88$).

When the sequence becomes longer, the motif finding task turns to very hard. For example, with $L \geq 800$ bp, only *Gibbs* was able to correctly locate the motif sites, we obtained $pla = 0.43$. Because a longer sequence will contain more decoy signals with large log-likelihoods, and thus multi-mode becomes a serious issue. In such a scenario, an algorithm is more easily getting trapped in a local or false optimum.

4.5.3 Multiple Motifs in Helix–Turn–Helix Protein Structure

The Helix–Turn–Helix (HTH) is composed of two almost perpendicular α helices linked by a several-residue β turn. The HTH motif is a common recognition element used by transcription regulators in prokaryotes and eukaryotes. Many bacterial transcription regulators which bind DNA through an HTH motif can be classified into subfamilies on the basis of sequence similarities. IclR is one of the subfamilies, and it is the repressor of the acetate operon in bacteria. The NCBI Conserved Domain Database (CDD) stores 94 IclR sequences with conserved HTH_ICLR domains (conserved domains ID: pfam09339.1) [20]. However, there are two protein sequences containing some unknown amino acids and thus these sequences are removed from the data set. The length of the 92 available HTH_ICLR sequences ranges from 219 to 572 amino acids and the total size is 25,107 residues.

The algorithms developed in this chapter correctly detected the 29-residue long core motif in the 39 most diverse sequences. The core motif alignment is partially shown in the middle of Figure 4.3a. The core motif sequence logo is also displayed in the middle of Figure 4.3b. Structure data show that three helixes are separated with transition structure (e.g., loops or backbone) in between (see highlighted part in

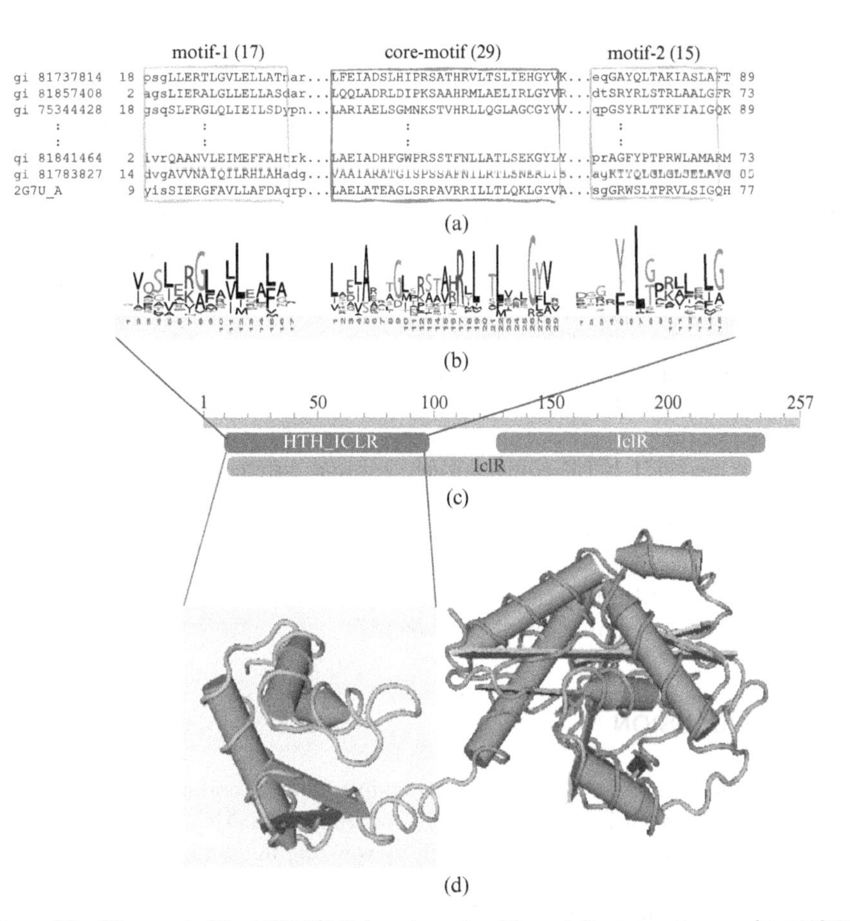

Figure 4.3 Alignment of the HTH_ICLR domains using 39 most diverse sequences from NCBI CDD database [20]. The 2G7U chain A structure data (mmdb: 38,409, Zhang et al. [20], unpublished) is used to plot the 3D structure using Cn3D software downloaded from NCBI [20]. (a) Alignment of the HTH_ICLR domains. The core motif is aligned in the center, the left subunit (motif-1) is aligned with three residues shift to the left compared to the same alignment reported in the CDD [20]. The right subunit (motif-2) is aligned with two residues shift to the left. (b) Sequence logos of three motifs: motif-1, core-motif, and motif-2 displayed from left to right. The sequence logos are plotted using *WebLogo*. (c) The organization of a typical protein in the IclR family (chain A) from NCBI CDD. (d) The 3D structure of the 2G7U chain A. The highlighted structure part is the HTH_ICLR domains.

Figure 4.3d). This flexible alignment divides the whole conserved domain into three sub-domains and thus it may maximize the total log-likelihood by allowing some spacer. Such refinement would presumably optimize the configuration energy while predicting the 3D structure of a unknown protein.

To locate the three subunits in the HTH_ICLR conserved domain, one can use the common erasing technique [1, 14] to perform motif finding recursively. The first or core motif sites are detected as discussed earlier, and then these core motif sites are masked and perform another round of motif finding. The three subunits are recursively found by the motif algorithms through motif erasing. Their alignments are in part presented in Figure 4.3a and the corresponding sequence logos are illustrated in Figure 4.3b. Among the top likelihood solutions, the core motif is aligned in the center which is consistent with those annotated in CDD whereas the left subunit (motif-1) is aligned with three residues left-shift compared to the same alignment reported in the CDD [20]. The right subunit (motif-2) is aligned with two residues shift to the left as well. Figure 4.3d illustrates the 3D structure of the IclR protein 2G7U chain A. The highlighted structure part of 2G7U_A in Figure 4.3d is the reconstructed HTH_ICLR conserved domain where three helixes are clearly observed: motif-1 covers one helix object, core motif involves one helix–turn–helix structure, and motif-2 indicates a conserved protein backbone.

In addition, the spacer distributions can be deduced from the alignment of the three subunits in the HTH_ICLR domain. There are four different distances between the motif-1 and the core motif: 4, 5, 6, and 7 residues with frequencies of 2, 2, 33, and 2, respectively. The 6-residue is the dominant spacer. On the other hand, there are three different distances between the motif-2 and the core-motif: 1, 2, and 3 residues with frequencies of 10, 10, and 19, respectively. Therefore, 3-residue is the dominant spacer between the motif-2 and the core-motif.

4.6 CONCLUSION

In this chapter, we presented two kinds of optimization algorithms for motif finding, which are built under the maximum likelihood framework. This framework unifies a suite of motif finding algorithms through maximizing the same likelihood function by imputing the unobserved data, that is, motif locations. The optimization algorithms presented are either deterministic or stochastic sampling. The deterministic algorithms, that is, DEM and WEM, maximize the likelihood function by performing iteratively optimal local search in the alignment space. The stochastic algorithms, that is, *Metropolis*, *Gibbs*, and *pGibbs*, iteratively draw motif location samples through Markov chain Monte Carlo (MCMC) simulation. Experimental results show that deterministic algorithms DEM and WEM are faster than their stochastic counterparts, *Metropolis*, *Gibbs*, and *pGibbs*, however, they are less performant than the stochastic algorithms in most cases. More runs can significantly improve the performance of DEM and WEM in highly conserved motif finding cases. Further studies also show that subtle motif finding challenges the five algorithms, DEM, WEM, *Metropolis*, *Gibbs*, and *pGibbs*. New strategies are needed to deal with this challenge.

Like the deterministic algorithms DEM and WEM, the stochastic ones *Metropolis*, *Gibbs*, and *pGibbs*, may also suffer from local optima. Therefore, running a population of MCMC chains, that is, multiple runs, is necessary to improve the mixing and proposal mechanisms [18]. Alternative strategies might be applied to the stochastic algorithms in order to alleviate the computing burden owing to burn-in time and slow convergence. For example, one can use either a single long chain or multiple short chains, each starting from different initial values. However, with parallel processing machines, using multiple short chains seems to be computationally more efficient than a single long chain. In motif finding practice, using several short chains, for example, each being 100 iterations per run, is likely to evolve a better solution. However, if long burn-in periods are required, or if the chains have very high autocorrelations, using a number of smaller chains may result in each not being long enough to be of any value.

A key issue in the design of MCMC algorithms is to improve the proposal mechanism and the mixing behaviour [18]. The attempts to improve the issue have been made in two aspects. The first one is to simply run a population of MCMC chains as discussed in this chapter. This indeed increases the chance of getting a better solution. Furthermore, one can take advantage of such a pool of solutions and make information exchange among population possible. That is the idea of hybrid MCMC technique. There are quite a few of this kind of motif finding algorithms developed recently, such as, the DEM algorithm coupled with a simple genetic algorithm [7], a population-based *Metropolis* sampler [6], and the *pGibbs* sampler improved by memetic algorithms [8]. For example, the PMC motif finding algorithm [6] runs a population of Markov chains by *Metropolis* sampler and exchanges information through population-based proposal distributions. It adaptively evolves the whole population of individual chains towards a global maximum. It would be very interesting to design a unified framework that could integrate multiple algorithms to produce synergistic motif alignment in the future.

REFERENCES

1. Bailey TL, Elkan C. Unsupervised learning of multiple motifs in biopolymers using expectation maximization. Mach Learn 1995;21:51–80.
2. Berg OG, von Hippel PH. Selection of DNA binding sites by regulatory proteins: statistical-mechanical theory and application to operators and promoters. J Mol Biol 1987;193:723–750.
3. Bi C. SEAM: a Stochastic EM-type Algorithm for Motif-finding in biopolymer sequences. J Bioinform Comput Biol 2007;5:47–77.
4. Bi C. Data augmentation algorithms for detecting conserved domains in protein sequences: a comparative study. J Proteome Res 2008;7:192–210.
5. Bi C. A Monte Carlo EM algorithm for de novo motif discovery in biomolecular sequences. IEEE/ACM Trans Comput Biol Bioinform 2009;6:370–386.
6. Bi C. DNA motif alignment through evolving a population of Markov chains. BMC Bioinformatics 2009;10:S13.
7. Bi C. Deterministic local alignment methods improved by a simple genetic algorithm. Neurocomputing 2010;73:2394–2406.
8. Bi C. Memetic algorithms for de novo motif-finding in biomedical sequences. Artif Intell Med 2012;56:1–17.
9. Celeux G, Diebolt J. Stochastic versions of the EM algorithm: an experimental study in the mixture case. J Stat Comput Simul 1996;55:287–314.
10. Delyon B, Lavielle M, Moulines E. Convergence of a stochastic approximation version of the EM algorithm. Ann Stat 1999;27:94–128.

11. Dempster AP, Laird AM, Rubin DB. Maximum likelihood from incomplete data via the EM algorithm (with discussion). J R Stat Soc B 1977;39:1–38.
12. Geman S, Geman D. Stochastic relaxation, Gibbs distribution and Bayesian restoration of images. IEEE Trans Pattern Anal Mach Intell 1984;6:721–741.
13. Hastings WK. Monte Carlo sampling methods using Markov chains and their applications. Biometrika 1970;57:97–109.
14. Lawrence CE, Altschul SF, Boguski MS, Liu JS, Neuwald AF, Wootton JC. Detecting subtle sequence signals: a Gibbs sampling strategy for multiple alignment. Science 1993;262:208–214.
15. Lawrence CE, Reilly AA. An expectation maximization algorithm for the identification and characterization of common sites in unaligned biopolymer sequences. Proteins 1990;7:41–51.
16. Liu JS. The collapsed Gibbs sampler with applications to a gene regulation problem. J Am Stat Assoc 1994;89:958–966.
17. Liu JS. Bayesian modeling and computation in bioinformatics research. In: Jiang T, Xu Y, Zhang M, editors. *Current Topics in Computational Biology*. Cambridge (MA): MIT Press; 2002. p 11–44.
18. Liu JS. *Monte Carlo Strategies in Scientific Computing*. New York: Springer-Verlag; 2002.
19. Liu JS, Neuwald AF, Lawrence CE. Bayesian models for multiple local sequence alignment and Gibbs sampling strategies. J Am Stat Assoc 1995;90:1156–1170.
20. Marchler-Bauer A, Anderson JB, Derbyshire MK, DeWeese-Scott C, Gonzales NR, Gwadz M, Hao L, He S, Hurwitz DI, Jackson JD, Ke Z, Krylov D, Lanczycki CJ, Liebert CA, Liu C, Lu F, Lu S, Marchler GH, Mullokandov M, Song JS, Thanki N, Yamashita RA, Yin JJ, Zhang D, Bryant SH. CDD: a conserved domain database for interactive domain family analysis. Nucleic Acids Res 2007;35:D237–D240.
21. Metropolis N, Rosenbluth AW, Rosenbluth MN, Teller A, Teller H. Equations of state calculations by fast computing machines. J Chem Phys 1953;21:1087–1091.
22. Tanner MA, Wong WH. The calculation of posterior distributions by data augmentation. J Am Stat Assoc 1987;82:528–550.
23. Wang L, Jiang T. On the complexity of multiple sequence alignment. J Comput Biol 1994;1:337–348.
24. Wei GCG, Tanner MA. A Monte Carlo implementation of the EM algorithm and the poor man's data augmentation algorithms. J Am Stat Assoc 1990;85:699–704.

CHAPTER 5

GLOBAL SEQUENCE ALIGNMENT WITH A BOUNDED NUMBER OF GAPS

Carl Barton[1], Tomáš Flouri[2], Costas S. Iliopoulos[3], and Solon P. Pissis[3]

[1]*The Blizard Institute, Barts and The London School of Medicine and Dentistry, Queen Mary University of London, London, UK*
[2]*Heidelberg Institute for Theoretical Studies, Heidelberg, Germany*
[3]*Department of Informatics, King's College London, London, UK*

5.1 INTRODUCTION

Alignments are a commonly used technique to compare strings and are based on notions of distance [9] or of similarity between strings; for example, similarities among biological sequences [15]. Alignments are often computed by dynamic programming [17].

A *gap* is a sequence of consecutive insertions or deletions of characters in an alignment. The extensive use of alignments on biological sequences has shown that it can be desirable to penalize the formation of long gaps rather than penalizing individual insertions or deletions of characters [18].

A gap in a biological sequence can be regarded as the absence (respectively presence) of a region, which is (respectively is not) present in another sequence, because

Pattern Recognition in Computational Molecular Biology: Techniques and Approaches,
First Edition. Edited by Mourad Elloumi, Costas S. Iliopoulos, Jason T. L. Wang, and Albert Y. Zomaya.

of the natural diversity among individuals. The gap concept in sequence alignment is therefore important in many biological applications because the insertion or deletion of an entire region (particularly in DNA) often occurs as a single event. Many of these single mutational events can create gaps of varying sizes with almost equal likelihood—within a wide, but bounded, range of sizes.

There are several mechanisms that can give rise to gaps in DNA sequences (cf. Reference [8]): repetitive DNA can be generated by single mutational events that copy and reinsert (at a different position of the gene/genome) long pieces of DNA; unequal cross-over in meiosis can cause an insertion in one sequence and a reciprocal deletion in another; DNA slippage during replication, where a portion of the DNA is repeated on the replicated copy because the replication machinery loses its place on the template, can initiate backward slipping and section repetitions; insertions of transposable elements—jumping genes—into a DNA sequence can also occur; insertion of DNA by retroviruses may be observed; and, finally, translocations of DNA between chromosomes can cause gaps.

Continuous advances in sequencing technology are turning whole-genome sequencing into a routine procedure, resulting in massive amounts of sequence data that need to be processed. The goal of producing large amounts of sequencing data, from closely related organisms, is driving the application known as *resequencing* [4]—the assembly of a genome that is directed by using a reference genome. Tens of gigabytes of data, in the form of short sequences (reads), need to be mapped (aligned) back to reference sequences, a few gigabases long, to infer the genomic location from which the read derived. This is a challenging task because of the high data volume and the large genome sizes. In addition, the performance, in terms of speed, sensitivity, and accuracy, deteriorates in the presence of inherent genomic variability and sequencing errors, particularly so, for relatively short consecutive sequences of deletions or insertions in the reference sequence or in the reads.

Concerning the length of these gaps, a broad range of lengths is possible. In practice, however, the length of the short reads, that is, 25–150 base pairs (bp), is too small to confidently and directly detect a large gap. In Figure 5.1, the distribution of lengths of gaps in *Homo sapiens* exome sequencing is demonstrated.[1] The shape of the gap length distribution is consistent with other studies (cf. References [12, 13, 16]). The presented data reflect a gap occurrence frequency of approximately 5.7×10^{-6} across the *exome*—the part of the genome formed by exons that codes portions of genes in the genome that are expressed. Similar gap occurrence frequencies were observed for other organisms: 1.7×10^{-5} in *Beta vulgaris*; 2.4×10^{-5} in *Arabidopsis thaliana*; and 3.2×10^{-6} in *bacteriophage PhiX174* [10].

During the procedure of resequencing, when performing end-to-end short-read alignment, the Needleman–Wunsch algorithm [11] with scoring matrices and affine gap penalty scores is used to identify these gaps. However, due to the observed gap occurrence frequencies, for short reads the presence of many gaps is unlikely. Hence, applying a standard dynamic programming approach for pairwise global sequence

[1]Data generated by the Exome Sequencing Programme at the NIHR Biomedical Research Centre at Guy's and St Thomas' NHS Foundation Trust in partnership with King's College London.

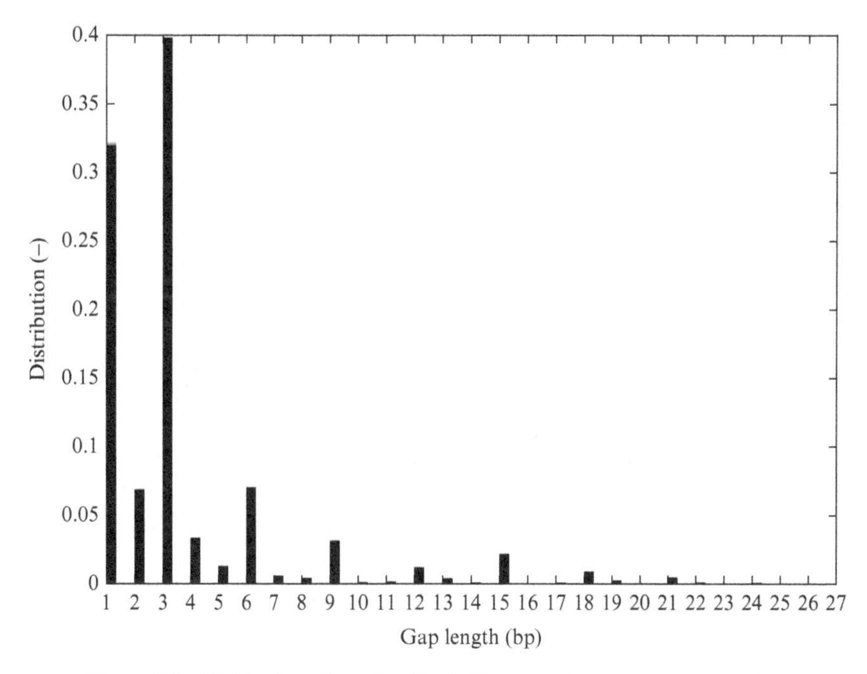

Figure 5.1 Distribution of gap lengths in *Homo sapiens* exome sequencing.

alignment [9, 11], which cannot bound the number of gaps in the alignment, can deteriorate the mapping confidence of those reads [1, 2, 6].

In this chapter, motivated by the aforementioned observations, we present `GapMis` [6, 7], an algorithm for pairwise global sequence alignment with a *single* gap. The algorithm requires $\Theta(mk)$ time and space, where m is the length of the shortest sequence, and k is the maximum allowed edit distance between the two sequences. Moreover, we present `GapsMis` [2, 3], an algorithm for pairwise global sequence alignment with a *variable*, but bounded, number of gaps. The algorithm requires $\Theta(mk\ell)$ time and $\Theta(mk)$ space, where ℓ is the maximum allowed number of gaps inserted in the alignment. These algorithms can be directly applied with an *alignment score* scheme [14] such as scoring matrices and affine gap penalty scores.

5.2 DEFINITIONS AND NOTATION

In this section, we give a few definitions, generally following Ref. [5].

An *alphabet* Σ is a finite nonempty set whose elements are called *letters*. A *string* on an alphabet Σ is a finite, possibly empty, sequence of elements of Σ. The zero-letter sequence is called the *empty string* and is denoted by ϵ. The *length* of a string x is defined as the length of the sequence associated with the string x and is denoted by $|x|$. We denote by $x[i]$, for all $1 \leq i \leq |x|$, the letter at index i of x. Each index i, for all $1 \leq i \leq |x|$, is a position in x when $x \neq \epsilon$. It follows that the ith letter of x is the letter at position i in x and that

$$x = x[1 \ldots |x|]$$

A string x is a *factor* of a string y if there exist two strings u and v, such that $y = uxv$. Let x, y, u, and v be strings, such that $y = uxv$. If $u = \epsilon$, then x is a *prefix* of y. If $v = \epsilon$, then x is a *suffix* of y.

Let x be a nonempty string and y be a string. We say that there exists an *occurrence* of x in y, or, more simply, that x *occurs in* y, when x is a factor of y. Every occurrence of x can be characterized by a position in y. Thus, we say that x occurs at the *starting position* i in y when $y[i \ldots i + |x| - 1] = x$. It is sometimes more suitable to consider the *ending position* $i + |x| - 1$.

An *aligned pair* is a pair (a, b) such that $(a, b) \in \Sigma \cup \{\epsilon\} \times \Sigma \cup \{\epsilon\}/\{\epsilon, \epsilon\}$.

An *alignment* between string x and string y is a string of aligned pairs whose projection on the first component is x and the projection on the second component is y.

A *gap sequence*, or simply *gap*, is a finite nonempty maximal sequence of length p of aligned pairs

$$(a_0, b_0), (a_1, b_1), \ \ldots \ , (a_{p-1}, b_{p-1})$$

such that *either*

- $a_0 = a_1 = \ \ldots \ = a_{p-1} = \epsilon$ or
- $b_0 = b_1 = \ \ldots \ = b_{p-1} = \epsilon$ holds.

A *gap-free sequence* is a string of length p of aligned pairs

$$(a_0, b_0), (a_1, b_1), \ \ldots \ , (a_{p-1}, b_{p-1})$$

such that $a_i, b_i \in \Sigma$, for all $0 \leq i < p$.

The *edit distance*, denoted by $\delta_E(x, y)$, is defined for two strings x and y as the minimum total cost of operations required to transform one string into the other. The allowable operations are as follows:

- *Insertion*: Insert a letter in y, not present in x. $(\epsilon, b), b \neq \epsilon$.
- *Deletion*: Delete a letter in y, present in x. $(a, \epsilon), a \neq \epsilon$.
- *Substitution*: Substitute a letter in y with a letter in x. $(a, b), a \neq b$, and $a, b \neq \epsilon$.

Note that, for simplicity, we only count the number of edit operations, considering the cost of each operation equal to 1: $Ins(b) = Del(a) = Sub(a, b) := 1$.

An *optimal* global alignment between two strings x and y is an alignment whose cost is the edit distance between x and y.

5.3 PROBLEM DEFINITION

Let $\delta_E^\ell(x, y)$, defined for two strings x and y, denote the minimum number of operations required to transform one string into the other such that their alignment consists of at most ℓ gap sequences.

The aforementioned ideas are the basis of the pairwise *global sequence alignment with k-differences and ℓ-gaps* problem, formally defined as follows.

Problem 5.1 *Given a string x of length n, a string y of length $m \leq n$, an integer k, such that $0 \leq k < n$, and an integer ℓ, such that $0 \leq \ell \leq k$, find a prefix of x, say x', such that $\delta_E^\ell(x', y)$ is minimum, $\delta_E^\ell(x', y) \leq k$, and for the corresponding alignment $z = z_0 g_1 z_1 g_2, \ldots, z_{\beta-1} g_\beta z_\beta$: $\beta \leq \ell$; z_0 and z_β are possibly empty gap-free sequences; $g_1, \ldots, g_\beta$ are gap sequences; and $z_1, \ldots, z_{\beta-1}$ are nonempty gap-free sequences.*

Finding a prefix of x that satisfies the above restrictions is equivalent to the notion of semi-global alignment between x and y, which is relevant to the application of resequencing (see Section 5.1 in this regard).

■ **EXAMPLE 5.1** Let $x =$ GCGACGTCCGAA, $y =$ GCAAGTAGA, $k = 4$, and $\ell = 2$. Consider the following alignment between the prefix $x' = x[1 \ldots 11] =$ GCGACGTCCGA of x and y.

1	2	3	4	5	6	7	8	9	10	11
G	C	G	A	C	G	T	C	C	G	A
|	|	.	|		|	|		.	|	|
G	C	A	A	-	G	T	-	A	G	A

With the above problem definition, this alignment is a solution to this problem instance, since $\delta_E^\ell(x', y)$ is minimum, $\delta_E^\ell(x', y) = 4 \leq k$, $\beta = 2 \leq \ell$, positions 1–4 are z_0, position 5 is g_1, positions 6 and 7 are z_1, position 8 is g_2, and positions 9–11 are z_2.

Let $G_s[0 \ldots n, 0 \ldots m]$ be a matrix, for all $1 \leq s \leq \ell$, where $G_s[i, j]$ contains the minimum number of operations required to transform factor $x[1 \ldots i]$ of x into factor $y[1 \ldots j]$ of y allowing for the insertion of at most s gaps. More formally, $G_s[i, j] = \delta_E^s(x[1 \ldots i], y[1 \ldots j])$.

In order to compute the exact location of the inserted gaps, we also need to maintain matrix $H_s[0 \ldots n, 0 \ldots m]$, such that

$$H_s[i,j] = \begin{cases} -b & \text{if a gap of length } b \text{ is inserted after } x[i] \\ a & \text{if a gap of length } a \text{ is inserted after } y[j] \\ 0 & \text{if no gap is inserted} \end{cases}$$

		0	1	2	3	4	5	6	7	8	9
		ϵ	C	A	T	T	C	G	A	C	G
0	ϵ	0	1	2	3	4	5	6	7	8	9
1	A	1	1	1	3	4	5	6	6	8	9
2	C	2	1	2	2	4	4	6	7	6	9
3	A	3	3	1	3	3	5	5	6	8	7
4	T	4	4	4	1	3	4	5	6	7	8
5	C	5	4	5	5	2	3	4	5	6	7
6	G	6	6	5	6	5	3	3	4	5	6
7	A	7	7	6	6	6	5	4	3	4	5
8	C	8	7	8	7	7	6	5	4	3	4
9	G	9	9	8	9	8	7	6	5	4	3

Matrix G_1

(a)

		0	1	2	3	4	5	6	7	8	9
		ϵ	C	A	T	T	C	G	A	C	G
0	ϵ	0	−1	−2	−3	−4	−5	−6	−7	−8	−9
1	A	1	0	0	0	0	0	0	0	0	0
2	C	2	0	0	0	0	0	0	0	0	0
3	A	3	0	0	0	0	0	0	0	0	0
4	T	4	0	0	0	0	0	−2	0	0	−5
5	C	5	0	0	0	0	0	−1	−2	0	−4
6	G	6	0	0	0	2	0	0	−1	−2	0
7	A	7	0	0	0	3	2	0	0	−1	−2
8	C	8	0	0	0	0	0	2	1	0	−1
9	G	9	0	0	0	0	4	0	2	1	0

Matrix H_1

(b)

Figure 5.2 Matrices G_1 and H_1 for x = ACATCGACG and y = CATTCGACG.

		0	1	2	3	4	5	6	7	8	9
		ϵ	C	A	T	T	C	G	A	C	G
0	ϵ	0	1	2	3	4	5	6	7	8	9
1	A	1	1	1	2	3	4	5	6	7	8
2	C	2	1	2	2	3	3	5	6	6	7
3	A	3	2	1	2	3	4	4	5	7	7
4	T	4	3	2	1	2	3	4	5	6	7
5	C	5	4	3	2	2	2	4	5	5	7
6	G	6	5	4	3	3	3	2	4	5	5
7	A	7	6	5	4	4	4	4	2	4	5
8	C	8	7	6	5	5	4	5	4	2	4
9	G	9	8	7	6	6	6	4	5	4	2

Matrix G_2

(a)

		0	1	2	3	4	5	6	7	8	9
		ϵ	C	A	T	T	C	G	A	C	G
0	ϵ	**0**	−1	−2	−3	−4	−5	−6	−7	−8	−9
1	A	**1**	0	0	−1	−2	−3	−4	0	−6	−7
2	C	2	**0**	0	0	0	0	0	0	0	−1
3	A	3	1	**0**	**−1**	0	0	0	0	0	0
4	T	4	2	1	0	**0**	−2	−3	0	0	−6
5	C	5	0	2	1	0	**0**	0	0	0	0
6	G	6	4	3	2	0	0	**0**	−1	−2	0
7	A	7	5	0	3	0	0	0	**0**	−1	−2
8	C	8	0	5	4	0	0	0	1	**0**	−1
9	G	9	7	6	5	0	0	0	2	1	**0**

Matrix H_2

(b)

Figure 5.3 Matrices G_2 and H_2 for x = ACATCGACG and y = CATTCGACG.

The computation of matrix H_s, for all $1 \leq s \leq \ell$, denotes the direction of the gap inserted. The direction of the gap is identified by defining insertions in x as negative integers and insertions in y as positive.

■ **EXAMPLE 5.2** Let x = ACATCGACG and y = CATTCGACG. Figure 5.2a illustrates matrix G_1 and Figure 5.2b illustrates matrix H_1. Figure 5.3a illustrates matrix G_2 and Figure 5.3b illustrates matrix H_2.

5.4 ALGORITHMS

Algorithm GapsMis is a generalization of algorithm GapMis, first introduced in Reference [6], for solving Problem 5.1. Algorithm GapsMis computes matrices $G_{1 \ldots \ell}$ and matrices $H_{1 \ldots \ell}$. It takes as input the string x of length n, the string y

Algorithm 5.1 `GapMis` (x, n, y, m)

```
    {Initializing matrix G_1 and matrix H_1}
    for i ← 0 to n do
      G_1[i, 0] ← i;
      H_1[i, 0] ← i;
    end for
 5: for j ← 0 to m do
      G_1[0, j] ← j;
      H_1[0, j] ← −j;
    end for
    {Computing matrix G_1 and matrix H_1}
    for i ← 1 to n do
10:   for j ← 1 to m do
        if i < j then
          u ← G_1[i − 1, j − 1] + δ_E(x[i], y[j]);
          v ← G_1[i, i] + (j − i);
          G_1[i, j] ← min{u, v};
15:       if v < u then
            H_1[i, j] ← i − j;
          else
            H_1[i, j] ← 0;
          end if
20:     end if
        if i > j then
          u ← G_1[i − 1, j − 1] + δ_E(x[i], y[j]);
          v ← G_1[j, j] + (i − j);
          G_1[i, j] ← min{u, v};
25:       if v < u then
            H_1[i, j] ← i − j;
          else
            H_1[i, j] ← 0;
          end if
30:     end if
        if i = j then
          G_1[i, j] ← G_1[i − 1, j − 1] + δ_E(x[i], y[j]);
          H_1[i, j] ← 0;
        end if
35:   end for
    end for
    return G_1 and H_1;
```

of length m, and ℓ. It assumes that matrices G_1 and H_1 are computed by algorithm `GapMis`.

Proposition 5.1 [6] *Algorithm* `GapMis` *correctly computes matrix* G_1 *and matrix* H_1 *in* $\Theta(mn)$ *time.*

The idea of Algorithm `GapsMis` is to iteratively compute the minimum cost alignment with at most ℓ gaps by allowing the insertion of at most one gap each time we compute the matrix G_s. The first gap insertion is handled by algorithm `GapMis`. In order to allow for the insertion of a new gap, we consider the matrix G_{s-1}, computed for at most $s-1$ gaps, and take the minimum cost alignment from $G_{s-1}[0,j]$ to $G_{s-1}[i,j]$ (lines 18 and 19 in Algorithm `GapsMis`), from $G_{s-1}[i,0]$ to $G_{s-1}[i,j]$ (lines 24 and 25 in Algorithm `GapsMis`), along with the possibility of extending the alignment from $G_s[i-1,j-1]$ to $G_s[i,j]$ (line 29 in Algorithm `GapsMis`). This minimum includes the respective cost of forming the new alignment (e.g., the cost of any inserted gap), forming variables u, v, and w. We, finally, take the minimum value of these three, which is the new minimum cost alignment with at most s gaps (line 30 in Algorithm `GapsMis`). Taking these minima in a naive way would lead to a poor runtime of $\Theta(mn^2)$ per matrix; however, we can improve this to $\Theta(mn)$ by reusing the previous minima we have already computed.

Assume we compute $G_s[i,j]$ after we have computed $G_s[i-1,j]$. Clearly to compute $G_s[i-1,j]$, we must have computed the following:

$$r \in \{0, \ldots, i-2\} : G_{s-1}[r,j] + i - r \leq G_{s-1}[c,j]$$

for all, $0 \leq c \leq i-2$.

If we store the value of r, we can compute $G_s[i-1,j]$, then we can easily update the new value of r in $G_s[i,j]$ as follows:

$$\textbf{if } G_{s-1}[i-1,j] < G_{s-1}[r,j] + i - r \textbf{ then } r \leftarrow i-1$$

The above check can easily be done in constant time. In Algorithm `GapsMis` we store array *minI*, which maintains the current minimum value for each column. Similar to index r, we define index q to maintain the current minimum value for each row. Clearly the same optimization can be made for q (*minJ* in Algorithm `GapsMis`); however, we only need to store one value for q at a time as we compute the matrix row by row.

The effect of the above optimization means all computation at each cell will take constant time. This reduces the time complexity of the algorithm to $\Theta(mn)$ per matrix.

Trivially, the computation of $H_s[i,j]$ depends only on the minimum value selected from $\{u, v, w, G_{s-1}[i,j]\}$ (lines 31–43 in Algorithm `GapsMis`).

$$H_s[i,j] = \begin{cases} H_{s-1}[i,j] & \text{if } G_{s-1}[i,j] = G_s[i,j] \\ i-r & \text{if } u = G_s[i,j] \\ -(j-q) & \text{if } v = G_s[i,j] \\ 0 & \text{if } w = G_s[i,j] \end{cases}$$

Algorithm 5.2 `GapsMis`(x, n, y, m, ℓ)

$G_1, H_1 \leftarrow$ `GapMis`(x, n, y, m);
{Initializing matrices $G_2, \dots, G_\ell$ and matrices $H_2, \dots, H_\ell$}
for $s \leftarrow 2$ to ℓ **do**
  for $i \leftarrow 0$ to n **do**
    $G_s[i, 0] \leftarrow i$;
5:    $H_s[i, 0] \leftarrow i$;
  end for
  for $j \leftarrow 0$ to m **do**
    $G_s[0, j] \leftarrow j$;
    $H_s[0, j] \leftarrow -j$;
10: **end for**
end for
{Computing matrices $G_2, \dots, G_\ell$ and matrices $H_2, \dots, H_\ell$}
for $s \leftarrow 2$ to ℓ **do**
  $minI[0 \dots m] \leftarrow 0$;
  for $i \leftarrow 1$ to n **do**
15:    $minJ \leftarrow 0$;
    for $j \leftarrow 1$ to m **do**
      $newminI \leftarrow 0$;
      if $G_{s-1}[i, j] < G_{s-1}[minI[j], j] + i - minI[j]$ **then**
        $minI[j] \leftarrow i$;
20:        $newminI \leftarrow 1$;
      end if
      $u \leftarrow G_{s-1}[minI[j], j] + i - minI[j]$;
      $newminJ \leftarrow 0$;
      if $G_{s-1}[i, j] < G_{s-1}[i, minJ] + j - minJ$ **then**
25:        $minJ \leftarrow j$;
        $newminJ \leftarrow 1$;
      end if
      $v \leftarrow G_{s-1}[i, minJ] + j - minJ$;
      $w \leftarrow G_s[i-1, j-1] + \delta_E(x[i], y[j])$;
30:      $G_s[i, j] \leftarrow \min\{u, v, w\}$;
      if $u = \min\{u, v, w\}$ **and** $newminI = 1$ **then**
        $H_s[i, j] \leftarrow i - minI[j]$;
      else
        $H_s[i, j] \leftarrow H_{s-1}[i, j]$;
35:      **end if**
      if $v = \min\{u, v, w\}$ **and** $newminJ = 1$ **then**
        $H_s[i, j] \leftarrow -(j - minJ)$;
      else
        $H_s[i, j] \leftarrow H_{s-1}[i, j]$;
40:      **end if**
      if $w = \min\{u, v, w\}$ **then**
        $H_s[i, j] \leftarrow 0$;
      end if
    end for
45: **end for**
end for
return $G_1, \dots, G_\ell$ and $H_1, \dots, H_\ell$;

The first line in the above definition corresponds to taking the same alignment that is in the previous matrix and the others represent either inserting an additional gap or extending an alignment with at most s gaps.

Theorem 5.1 [3] *Algorithm* `GapsMis` *correctly computes matrices* $G_2, \ldots, G_\ell$ *and matrices* $H_2, \ldots, H_\ell$ *in* $\Theta(mn\ell)$ *time.*

For solving Problem 5.1, it is sufficient to locate a value in $G_{1 \ldots \ell}[0 \ldots n, m]$, say $G_s[i, m]$, for some $0 \leq i \leq n$, such that $G_s[i, m]$ is minimum and $G_s[i, m] \leq k$. In order to determine the corresponding alignment with cost $G_s[i, m]$, we find the minimum $j, j \leq s \leq \ell$, such that $G_j[i, m] = G_s[i, m]$. This ensures that a representation of this alignment is stored in matrix H_j.

Algorithm `GapsPos` determines this alignment by finding the positions of the inserted gaps. It takes as input the matrix H_s, an integer $0 \leq i \leq n$, the length m of y, and the maximum allowed number s of inserted gaps. It produces as output the exact number $\beta \leq s$ of inserted gaps and the following three arrays:

- Array *gap_pos* of size s, such that *gap_pos*$[i]$, for all $0 \leq i \leq s-1$, gives the position of the ith inserted gap or 0 if no gap is inserted.
- Array *gap_len* of size s, such that *gap_len*$[i]$, for all $0 \leq i \leq s-1$, gives the length of the ith inserted gap.
- Array *where* of size s, such that if *gap_len*$[i] > 0$, then *where*$[i]$, for all $0 \leq i \leq s-1$, is equal to 1 if the ith gap is inserted in y or equal to 2 if the ith gap is inserted in x.

■ **EXAMPLE 5.3** Let $x =$ ACATCGACG, $y =$ CATTCGACG, $k = 2$, and $\ell = 2$. Starting the trace-back from cell $H_2[9, 9]$ (see bold terms in Figure 5.3b), that is, $i = 9$, gives a solution since $G_2[9, 9] = 2$ is minimum and $G_2[9, 9] = 2 \leq k$ (see Figure 5.3a). Finally, we can determine the corresponding alignment by finding the positions of the inserted gaps, which are at $H_2[3, 3]$ and $H_2[1, 0]$ ($\beta = 2 \leq \ell$), using Algorithm `GapsPos`. A solution to this problem instance is the alignment in Figure 5.4.

However, since the threshold k is given, pruned versions $G^{\mathrm{P}}_{1 \ldots \ell}$ and $H^{\mathrm{P}}_{1 \ldots \ell}$ of matrices $G_{1 \ldots \ell}$ and matrices $H_{1 \ldots \ell}$, respectively, can be computed in $\Theta(mk)$ time per matrix, similarly as shown in Reference [8] for computing the traditional dynamic programming matrix [9, 11] for global sequence alignment.

1	2	3	4	5	6	7	8	9	10
A	C	A	-	T	C	G	A	C	G
	|	|		|	|	|	|	|	|
-	C	A	T	T	C	G	A	C	G

Figure 5.4 Solution for $x =$ ACATCGACG, $y =$ CATTCGACG, $k = 2$, and $\ell = 2$.

Algorithm 5.3 `GapsPos`(H_s, i, m, s)

```
    {Initializing variables}
    j ← m;
    β ← 0;
    gap_pos[0 ... s − 1] ← 0;
    gap_len[0 ... s − 1] ← 0;
 5: where[0 ... s − 1] ← 0;
    {Trace-back}
    while i ≥ 0 and j ≥ 0 do
      if H_s[i,j] = 0 then
        i ← i − 1;
        j ← j − 1;
10:   else
        if H_s[i,j] < 0 then
          gap_pos[s − 1 − β] ← j;
          gap_len[s − 1 − β] ← −H_s[i,j];
          where[s − 1 − β] ← 1;
15:       j ← j + H_s[i,j];
        else
          gap_pos[s − 1 − β] ← i;
          gap_len[s − 1 − β] ← H_s[i,j];
          where[s − 1 − β] ← 2;
20:       i ← i − H_s[i,j];
        end if
        β ← β + 1;
      end if
    end while
25: return β, gap_pos, gap_len, where;
```

Lemma 5.1 [3] *There exist at most* $k + 1$ *cells of matrix* G_s, $1 \leq s \leq \ell$, *that give a solution to Problem 5.1.*

Hence we only need to compute a diagonal band of width $2k + 1$ in matrices $G_{1 \dots \ell}$ and in matrices $H_{1 \dots \ell}$. As a result, Algorithm `GapsMis` can be easily modified to compute $G^{\mathrm{P}}_{1 \dots \ell}$ and $H^{\mathrm{P}}_{1 \dots \ell}$ in $\Theta(mk\ell)$ time, by replacing lines 14 and 16 of Algorithm `GapsMis` by the following:

```
for i ← 1 to min{n, m + k} do
  ...
  for j ← max{1, i − k} to min{m, i + k} do
    ...
  end for
end for
```

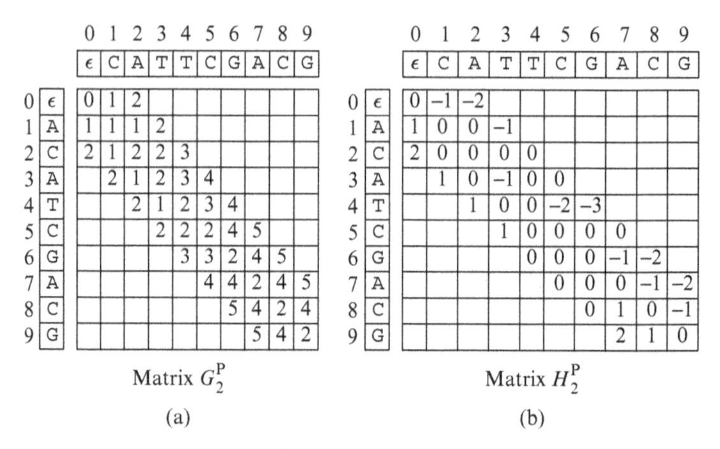

Figure 5.5 Matrices G_2^P and H_2^P for x = ACATCGACG, y = CATTCGACG, and $k = 2$.

■ **EXAMPLE 5.4** Let $x =$ ACATCGACG, $y =$ CATTCGACG, $k = 2$, and $\ell = 2$. Figure 5.5a and b illustrates matrix G_2^P and matrix H_2^P, respectively.

The same can be applied to Algorithm GapMis [6]. For any cell providing a valid solution, Algorithm GapsPos requires an additional time of $\mathcal{O}(m)$.

The computation of matrices G_s and H_s, for all $2 \leq s \leq \ell$, depends only on matrix G_{s-1}. Trivially, the space complexity can be reduced to $\Theta(mk)$. Therefore, we obtain the following result.

Theorem 5.2 [3] *Problem 5.1 can be solved in $\Theta(mk\ell)$ time and $\Theta(mk)$ space.*

Alternatively, we could compute matrices $G^P_{1\,\dots\,\ell}$ and matrices $H^P_{1\,\dots\,\ell}$ based on a simple *alignment scoring* scheme depending on the application of the algorithm, and compute the *maximum score* among all possible alignments of x and y in $\Theta(k)$ time by Lemma 5.3.

5.5 CONCLUSION

In this chapter, we presented GapMis and GapsMis, two algorithms for pairwise global sequence alignment with a variable, but bounded, number of gaps. They compute different versions of the standard dynamic programming matrix [9, 11]. GapMis requires $\Theta(mk)$ time and space, where m is the length of the shortest sequence and k is the maximum allowed edit distance between the two sequences. GapsMis requires $\Theta(mk\ell)$ time and $\Theta(mk)$ space, where ℓ is the maximum allowed number of gaps inserted in the alignment. These algorithms can be directly applied with an alignment score scheme such as scoring matrices and affine gap penalty scores.

REFERENCES

1. Alachiotis N, Berger S, Flouri T, Pissis SP, Stamatakis A. libgapmis: extending short-read alignments. BMC Bioinformatics 2013;14(Suppl 11):S4.
2. Barton C, Flouri T, Iliopoulos CS, Pissis SP. GapsMis: flexible sequence alignment with a bounded number of gaps. In: *Proceedings of the International Conference on Bioinformatics, Computational Biology and Biomedical Informatics, BCB'13*. Washington, DC: ACM; 2013. p 402–411.
3. Barton C, Flouri T, Iliopoulos CS, Pissis SP. Global and local sequence alignment with a bounded number of gaps. Theor Comput Sci 2015. doi: 10.1016/j.tcs.2015.03.016.
4. ten Bosch JR, Grody WW. Keeping up with the next generation: massively parallel sequencing in clinical diagnostics. J Mol Diagn 2008;10(6):484–492.
5. Crochemore M, Hancart C, Lecroq T. *Algorithms on Strings*. New York, NY: Cambridge University Press; 2007.
6. Flouri T, Frousios K, Iliopoulos CS, Park K, Pissis SP, Tischler G. GapMis: a tool for pairwise sequence alignment with a single gap. Recent Pat DNA Gene Seq 2013;7(2):102–114.
7. Flouri T, Park K, Frousios K, Pissis SP, Iliopoulos CS, Tischler G. Approximate string-matching with a single gap for sequence alignment. In: *Proceedings of the 2nd ACM Conference on Bioinformatics, Computational Biology and Biomedicine, BCB '11*. Chicago, IL: ACM; 2011. p 490–492.
8. Gusfield D *Algorithms on Strings, Trees, and Sequences: Computer Science and Computational Biology*. New York, NY: Cambridge University Press; 1997.
9. Levenshtein VI. Binary codes capable of correcting deletions, insertions, and reversals. Technical Report 8. Soviet Physics Doklady; 1966.
10. Minosche AE, Dohm JC, Himmelbauer H. Evaluation of genomic high-throughput sequencing data generated on Illumina HiSeq and Genome Analyzer systems. Genome Biol 2011;12:R112.
11. Needleman SB, Wunsch CD. A general method applicable to the search for similarities in the amino acid sequence of two proteins. J Mol Biol 1970;48(3):443–453.
12. Ng SB, Turner EH, Robertson PD, Flygare SD, Bigham AW, Lee C, Shaffer T, Wong M, Bhattacharjee A, Eichler EE, Bamshad M, Nickerson DA, Shendure J. Targeted capture and massively parallel sequencing of 12 human exomes. Nature 2009;461(7261):272–276.
13. Ostergaard P, Simpson MA, Brice G, Mansour S, Connell FC, Onoufriadis A, Child AH, Hwang J, Kalidas K, Mortimer PS, Trembath R, Jeffery S. Rapid identification of mutations in GJC2 in primary lymphoedema using whole exome sequencing combined with linkage analysis with delineation of the phenotype. J Med Genet 2010;48(4):251–255.
14. Rice P, Longden I, Bleasby A. EMBOSS: the european molecular biology open software suite. Trends Genet 2000;16(6):276–277.
15. Sellers PH. On the theory and computation of evolutionary distances. SIAM J Appl Math 1974;26(4):787–793.
16. Simpson MA, Irving MD, Asilmaz E, Gray MJ, Dafou D, Elmslie FV, Mansour S, Holder SE, Brain CE, Burton BK, Kim KH, Pauli RM, Aftimos S, Stewart H, Ae Kim C, Holder-Espinasse M, Robertson SP, Drake WM, Trembath RC. Mutations in NOTCH2

cause Hajdu-Cheney syndrome, a disorder of severe and progressive bone loss. Nat Genet 2011;43(4):303–305.

17. Wagner RA, Fischer MJ. The string-to-string correction problem. J ACM 1974;21(1):168–173.
18. Zachariah MA, Crooks GE, Holbrook SR, Brenner SE. A generalized affine gap model significantly improves protein sequence alignment accuracy. Proteins 2004;58(2):329–338.

PART II

PATTERN RECOGNITION IN SECONDARY STRUCTURES

CHAPTER 6

A SHORT REVIEW ON PROTEIN SECONDARY STRUCTURE PREDICTION METHODS

Renxiang Yan[1], Jiangning Song[2,3], Weiwen Cai[1], and Ziding Zhang[4]

[1] *Institute of Applied Genomics, College of Biological Science and Engineering, Fuzhou University, Fuzhou, China*
[2] *National Engineering Laboratory for Industrial Enzymes and Key Laboratory of Systems Microbial Biotechnology, Tianjin Institute of Industrial Biotechnology, Chinese Academy of Sciences, Tianjin, China*
[3] *Department of Biochemistry and Molecular Biology, Faculty of Medicine, Monash University, Melbourne, VIC, Australia*
[4] *State Key Laboratory of Agrobiotechnology, College of Biological Sciences, China Agricultural University, Beijing, China*

6.1 INTRODUCTION

Proteins are large biological molecules or macromolecules, consisting of one or more polypeptide chains of amino acid residues. Depending on the complexity of the molecular structures, protein structure can be distinguished into four distinct levels, namely, *primary*, *secondary*, *tertiary*, and *quaternary structures* [3, 4, 28]. The *primary structure* of a protein refers to the linear sequence of amino acids. The protein *secondary structures* are the regular local substructures. Protein *tertiary structure* refers to the *three-dimensional* (3D) structure of a protein, while protein *quaternary structure* is the multi-subunit 3D structure.

Pattern Recognition in Computational Molecular Biology: Techniques and Approaches,
First Edition. Edited by Mourad Elloumi, Costas S. Iliopoulos, Jason T. L. Wang, and Albert Y. Zomaya.

Proteins play a wide range of cellular functions in many biological processes, including catalyzing metabolic reactions [26], replicating DNA [36], responding to stimuli [41], and transporting molecules from one location to another [22]. The function of a protein is directly determined by and mainly dependent on its 3D structure. Although the number of experimentally determined protein structures has been increasing at a fast rate in recent years, the gap between the available protein sequences and structurally known proteins continues to increase. Therefore, there is a great need for accurate protein 3D structure prediction methods [16, 42]. Protein secondary structure prediction is an essential step to model 3D structure of a protein. For example, secondary structure predictions based on Neural Network (NN)-trained models are used by almost all the major contemporary threading/alignment programs for protein 3D structure prediction [19, 38, 39, 44].

In 1951, protein secondary structure was initially defined on the basis of hydrogen bonding patterns by Pauling and Corey [27]. At present, there are several publicly available programs for defining/assigning protein secondary structures (e.g., DSSP [21], STRIDE [13], and SST [23]). These programs accept protein PDB structures as the input [2]. Among them, DSSP and STRIDE are the most widely used in the community. Proteins adopt a limited number of unique secondary structures. For example, there are seven states of secondary structure defined by STRIDE, which utilizes both hydrogen bond energy and main chain dihedral angles to determine secondary structures for structurally known proteins. The seven states are G (3-turn helix), H (4-turn helix), I (5-turn helix), T (hydrogen bonded turn), E (extended strand in parallel/antiparallel β-sheet conformation), B (residue in isolated β-bridge), and C (coil). According to the Critical Assessment of Protein Structure Prediction (CASP) competition standard [35], the seven states can be simplified to three states by the following transformation: H, G, and I are transformed to H (α-helix), E, and B are transformed to E (β-strand), while the rest states are transformed to C (coil).

Broadly, protein secondary structure prediction is to assign one type of secondary structure for each residue in a protein by computational methods. A series of elegant protein secondary structure prediction algorithms have been developed in the past decades. The existing methods can be categorized in different ways. According to the adopted algorithms, the existing protein secondary structure prediction algorithms can be grouped into two main categories: simple statistical- and pattern recognition-based methods. The highlight of the simple statistical-based methods is that the biological meanings of the established statistical models are comprehensible. However, the accuracies of the simple statistical-based methods for protein secondary structure prediction are only around 65%. By contrast, a major advantage of the pattern recognition-based methods is that various information/features can readily be incorporated into a prediction model. State-of-the-art machine learning algorithms, such as NNs and Support Vector Machines (SVMs) have been employed to develop protein secondary structure prediction methods and their accuracies can usually go beyond 65%. Although some critics argue pattern recognition-based methods are "blackbox," the best-performing methods have been consistently developed using the pattern recognition-based algorithms.

Early efforts on protein secondary structure prediction can be traced back to the 1960s when Pauling et al. pioneered an explanation for the formation of certain local conformations (e.g., α-helix and β-strand) [28]. Later, Chou and Fasman found that individual amino acids prefer certain type of secondary structures over others [5]. They subsequently proposed a method termed *Chou–Fasman* to predict protein secondary structure based on such observation. The Chou–Fasman method yielded average accuracy of 50–60%. In 1978, Garnier, Osguthorpe, and Robson (GOR) further extended the Chou–Fasman algorithm and developed a method, called *GOR*, which took into account not only the propensities of individual amino acids to form particular secondary structures but also the conditional probability of an amino acid to form a secondary structure given that its immediate neighbors have formed that structure [14]. The accuracy achieved by the GOR method is approximately 65%. Both Chou–Fasman and GOR methods can be considered as simple statistical-based methods. However, it seems that the Q_3 prediction accuracy (percentage of correctly predicted residues) [30] of Chou–Fasman and GOR stalled at levels slightly above 60%. In the 1990s, the accuracies of protein secondary structure prediction methods were enhanced significantly by combining evolutionary profiles and pattern recognition techniques. Evolutionary profiles are extracted from the Multiple Sequence Alignments (MSAs), which represent the divergence of proteins in the same family, and contain important information to infer the protein secondary structure. For example, the PHD method developed by Rost et al. is a multi-layer NN-based method and can reach an accuracy of 70% [30]. Jones further enhanced the NN-based *protein secondary structure prediction* (PSIPRED) with an accuracy of ~75% [20]. Perhaps the main difference between PHD and PSIPRED is that PSI-BLAST [1] for MSAs was included in PSIPRED [20], whereas PHD used *MaxHom* [10, 33] to generate MSAs. However, PSIPRED achieves a better accuracy than PHD. The accuracy of PSIPRED is probably the highest in the past 10 years and as a result PSIPRED is the most widely used protein secondary structure prediction program in the community. In recent years, evolutionary information obtained from improved and iterative database search techniques (e.g., PSI-BLAST) as well as newly sequenced genome data have further boosted prediction accuracy. After more than three decades of efforts, the prediction accuracies have been improved from ~50% to ~80%. A brief development of protein secondary structure prediction methods is summarized in Figure 6.1. In the

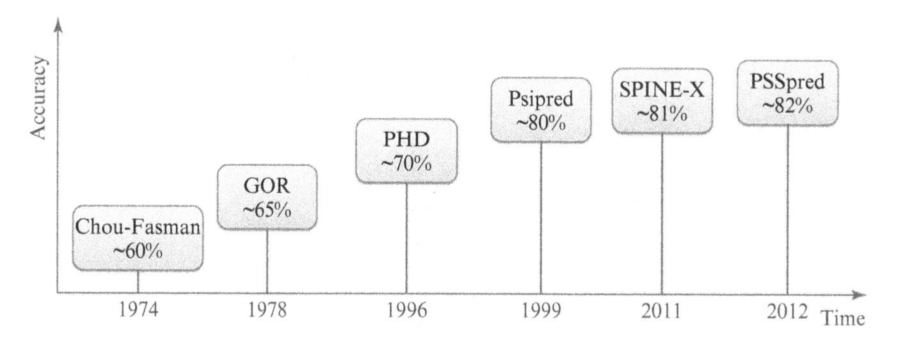

Figure 6.1 The development history of protein secondary structure prediction methods.

following sections of this chapter, we review some classical and popular protein secondary structure prediction methods and also benchmark the top methods to inform readers regarding their updated accuracies.

6.2 REPRESENTATIVE PROTEIN SECONDARY STRUCTURE PREDICTION METHODS

Seven protein secondary structure prediction methods, covering simple statistical- and pattern recognition-based techniques, are discussed in detail in this chapter. Comparatively, the methods using pattern recognition are more accurate. Typical pattern recognition algorithms comprise of the NN [9, 27], SVM [25], and random forest [24] methods. For protein secondary structure prediction, NN is most widely used compared with other pattern recognition methods. An NN generally consists of three components, namely, an *input layer*, a *hidden layer*, and *an output layer*. There are connections called *weights* between the *input* and *hidden layers*, and between the *hidden* and *output layers*. The input features are fed to the *input layer* while the prediction results are obtained in the *output layer*. In the model training process, features of structurally known proteins are used to train the weights of an NN. Prior to model training, the architectures of NNs should be designed; there are usually three output nodes for an NN, where each node represents one protein secondary structure state. Considering there are three states (i.e., *H*, *E*, and *C*), most methods usually employ the following scheme to encode the output nodes before training, that is, (1, 0, 0) for *H*, (0, 1, 0) for *E*, and (0, 0, 1) for *C* (Figure 6.2). Following the design of the NN architecture and labels, features of structurally known protein can be used to tune the weights of NNs. This process is called *training*. For each residue, a sliding window is used to extract the information of the residue and its neighbors, which is fed to the NN. As such, the prediction accuracy of protein secondary structure prediction methods highly relies on the selection of a subset of optimal input features.

In Figure 6.3, we summarize a general flowchart for the development of typical protein secondary structure prediction methods. In the first step, a protein is iteratively threaded through the NCBI NR database with an *e*-value of 0.01 or lower to generate

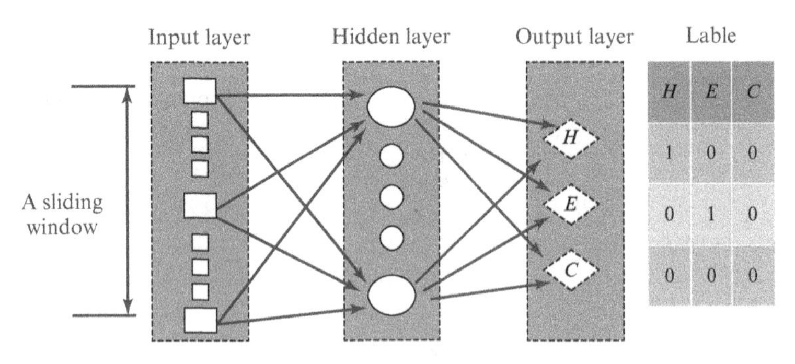

Figure 6.2 The architecture of an NN for protein secondary structure prediction.

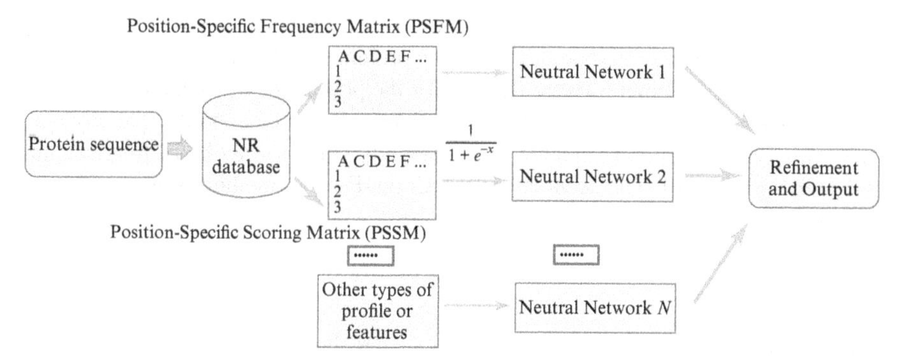

Figure 6.3 A general flowchart for the development of protein secondary structure prediction methods.

different types of sequence profiles, including the Position-Specific Frequency Matrix (PSFM) and Position-Specific Scoring Matrix (PSSM) profiles. In addition, other types of sequence profiles can also be employed as inputs. For example, sequence profiles can be constructed by different weighted schemes or constructed on the basis of the consideration of gaps or not. Given that some values in the sequence profiles, for example, PSSM profile, may be negative, the following standard logistic function is usually used to scale the value to the range of 0–1:

$$f(x) = \frac{1}{1 + e^{-x}} \quad (6.1)$$

where x is the element value in the sequence profiles. It is noteworthy that other physical or biological indices can also be used here.

In the second step, the trained NN models are used to calculate the probabilities of three-state secondary structures. The state-of-the-art methods generally utilize two sets of NNs to improve the prediction accuracy. The outputs of the first set of NNs are fed to the second set. In general, the outputs of the latter can obtain higher prediction accuracy than the former. Moreover, in order to further improve the prediction accuracy, some refinement or smoothing techniques may also be used to correct potential prediction errors. Another critical issue is how to accurately estimate in advance the reliability of a prediction when applied to a new sequence. The confidence score for each residue is generally reported by the protein secondary structure prediction programs. The prediction is relatively reliable if half of the residues are predicted with high confidence scores. Refer to Figure 6.3 for details of the general flowchart of the prediction method development. In the following sections, we discuss in detail several classical methods of protein secondary structure prediction.

6.2.1 Chou–Fasman

The Chou–Fasman method was developed on the basis of the observation that amino acids vary in their abilities to form different secondary structure elements. Chou and

Fasman first analyzed the relative frequencies of each amino acid in α-helix, β-strand, and coil using a data set of structurally known proteins [5]. From these frequencies, a set of probability parameters were derived for the preference of each amino acid for each secondary structure type and these parameters were used to predict the probability at which a given protein sequence of amino acids would form α-helix, β-strand, and coil in the protein [6]. The accuracy of the Chou–Fasman method is about 60–65% in terms of the Q_3 measure.

6.2.2 GOR

The GOR method was developed in the late 1970s shortly after the development of the Chou–Fasman method. It is an information theory-based method for prediction of protein secondary structure [14]. Similar to Chou–Fasman, the GOR method is based on probability parameters derived from empirical studies of structurally known proteins. However, unlike Chou–Fasman, the GOR method takes into account not only the propensities of individual amino acids to form particular secondary structures but also the conditional probabilities of amino acids to form a secondary structure given that its immediate neighbors have already formed that structure [15]. The Q_3 accuracy of the GOR method is around 65%.

6.2.3 PHD

The PHD method uses evolutionary information contained in MSAs and NNs to predict protein secondary structure [30]. In PHD, a feed-forward NN was trained on the basis of structurally known proteins [29]. The back-propagation algorithm is employed by PHD [9]. Compared with previous methods, a key aspect of PHD is the use of evolutionary information used as input in place of single sequences. To generate the sequence profiles, the query sequence is threaded through the sequence database using *MaxHom* [10, 33] for collecting MSAs. It is reported that inclusion of protein family information in this form improved the prediction accuracy by 6–8%. In conclusion, NNs are also used to make predictions. A combination of three levels of NNs resulted in an overall Q_3 of around 70%. This prediction accuracy was once regarded as the best at the time when the PHD method was developed.

6.2.4 PSIPRED

The PSIPRED method is similar to PHD [20]. To obtain the sequence profiles, the query sequence is iteratively BLASTed against the NCBI NR database for three repeats with an *e*-value cutoff of 0.001 for generating the PSSM in PSIPRED. The *Henikoff weight scheme* [17] was used to reduce the redundancy of MSAs during the construction of PSSM. The elements in the PSSM matrix were scaled to the required 0–1 range using the standard logistic function. Two standard feed-forward back-propagation NN architectures with a single hidden layer were used in PSIPRED. A sliding window containing 15 residues was used as the input for the first NN. The output, that is, the possibilities of three secondary structures, of the first NN was then fed into the second NN. It is reported that the Q_3 score of PSIPRED could reach 80%.

6.2.5 SPINE-X

The SPINE-X method is a multistep NN algorithm for protein secondary structure, solvent accessibility, and backbone torsion angles prediction in an iterative way [12]. SPINE-X not only takes sequence profiles into account but also uses several representative physical parameters to improve prediction accuracy. Meanwhile, SPINE-X uses an iterative way to enhance the prediction through the construction of several layers of NNs. SPINE-X has its own way to train NN, particularly focusing on setting the initial weights of NN. It is reported that its own NN training algorithm performs better than that of the standard backpropagation training. The accuracy of protein secondary structure prediction by SPINE-X is slightly higher than 80%.

6.2.6 PSSpred

Protein Secondary Structure Prediction (PSSpred) is a multiple backpropagation NN predictor from PSI-BLAST profiles for protein secondary structure prediction based on PSI-BLAST profiles [43]. The method was developed on the basis of multiple weights NNs training. PSSpred also uses the standard PSSM profile. The frequency file with the Henikoff weighting scheme is also used. In the meantime, it also uses the frequency profile by considering gap information in the profile. Amino acid frequency and log-odds data with the Henikoff weights based on the Rumelhart backpropagation of errors method are then additionally used in training [9]. The final prediction is a combination of seven NN secondary structure predictors. Its accuracy is also slightly higher than 80%.

6.2.7 Meta Methods

Meta methods are developed by combining well-established programs, aiming to further improve the prediction performance by taking into account the outputs of several existing methods. For example, JPred [8] is regarded as one of the best consensus-based methods, which utilizes several popular methods [5, 27], including NNSSP [8], PHD, PSIPRED, and other programs. Its accuracy is reported to be better than any of its component methods. Other meta-based servers include Meta-PP [11], among others The development of meta-based methods requires careful tuning and training of the meta-server algorithms based on all the individual programs. As a consequence, the accuracy of a method depends on the individual component programs and their optimization.

Meanwhile, it is valuable for readers to know the publicly available and state-of-the-art web servers/tools for protein secondary structures prediction. Here, we recommend four tools and their web servers that are listed in Table 6.1.

Table 6.1 Some state-of-the-art of protein secondary structure prediction tools

Method	Web link	References
PSIPRED	http://bioinf.cs.ucl.ac.uk/psipred/	[20]
SPINE-X	http://sparks.informatics.iupui.edu/SPINE-X/	[12]
PSSpred	http://zhanglab.ccmb.med.umich.edu/PSSpred/	[43]
JPred	http://www.compbio.dundee.ac.uk/www-jpred/	[8]

6.3 EVALUATION OF PROTEIN SECONDARY STRUCTURE PREDICTION METHODS

It is valuable to know the real performance of the state-of-the-art of protein secondary structure prediction methods. Therefore, we designed an experiment in this work to systematically benchmark various methods. The details are discussed in the following subsections.

6.3.1 Measures

In general, the measures Q_3, SOV, and RELINFO are used in the community. The Q_3 score is the total number of correctly predicted residue states divided by the total number of residues [20, 30]. SOV is a segment overlap measure defined by Zemla et al. [15]. RELINFO is an information-theory-based measure [29]. Details of SOV and RELINFO can be found in References [15, 29]. In this chapter, we assess the performance of different methods using the Q_3 measure. In addition, other three measures Q_H, Q_E, and Q_C, which describe the fractions of correctly predicted residues out of the total numbers of residues in α-helix, β-strand, and coil, are also commonly used to evaluate the performance [20, 30].

6.3.2 Benchmark

In our previous work, 538 nonredundant proteins were randomly collected from the PDB library to assess the performance by 20 alignment/threading methods [40]. In this work, we also used the same data set to benchmark the prediction performance of different protein secondary structure prediction methods. This data set was named the Y538 data set. The 538 sequences in the Y538 data set were divided into three categories, Easy, Medium, and Hard targets, based on the consensus confidence score of the meta-threading LOMETS program, which consists of nine protein threading programs (dPPAS, MUSTER, HHsearch-I, HHsearch-II, PPAS, PROSPECT, SAM, SPARKS, and SP3) [40]. Details of the scoring functions of these threading programs can be found in Reference [40]. A target is defined as "Easy" if at least one strong template hit is detected for the target by each program with the *Z*-score [40] higher than the confidence cutoff; a target is defined as "Hard" if none of the threading programs has a strong template hit; otherwise, it is considered as a "Medium" target. In total, the 538 proteins were selected to reflect a balanced category distribution of the prediction difficulty with 137 "Easy," 177 "Medium," and 224 "Hard" targets. Benchmarking experiments using data set with balanced distribution can indicate the performances of different methods at different prediction difficulty levels. Moreover, if a protein is classified as "Medium" or "Hard" by LOMETS, it means that it is not easy to find its homologous sequences in the PDB database. In other words, it is probably not easy to accurately predict their secondary structures if they are classified as "Medium" or "Hard."

6.3.3 Performances

Table 6.2 presents a summary of the prediction performances of three community-wide protein secondary structure prediction tools, PSSpred, PSIPRED, and SPINE-X. To make a comparison, we downloaded and installed these tools in our local computer cluster. The 538 proteins were then fed into the three programs and the outputs of possibilities of three types of protein secondary structures were generated. The STRIDE program was used to annotate the structurally determined protein secondary structures from PDB structures. As shown in Table 6.2, PSSpred, SPINE-X, and PSIPRED achieved the Q_3 scores of 0.831, 0.811, and 0.806, respectively. Among them, PSSpred performed better than PSIPRED and SPINE-X. Nevertheless, PSSpred did not perform the best across all the measures. For example, the highest scores of Q_H and Q_C were attained by SPINE-X and PSIPRED, respectively. When the scores of Q_H, Q_E, and Q_C are compared, Q_E is consistently the lowest (Table 6.2). This suggests that the β-strand is more difficult to predict than the α-helix or coil. From the biological perspective, the formation of an α-helix is determined by short-range interactions, whereas the formation of a β-strand is strongly influenced by long-range interactions. The prediction for long-range interactions is theoretically difficult. That is probably the reason the accuracy of Q_E is relatively lower. Although the 538 proteins were divided into three difficulty prediction levels (i.e., Easy, Medium, and Hard), PSSpred consistently generated the highest Q_3 values across all the three levels (Table 6.3). Considering that the sequence profiles were used by all three methods PSSpred, PSIPRED, and SPINE-X, it indicates that sequence profiles can help identify the conserved/variable positions in the sequence, which are important in protein evolution and critical for the maintenance of 3D structures. This might explain why the sequence profile is particularly effective for protein secondary structure prediction.

Table 6.2 Performance comparison of protein secondary structure prediction methods[a]

Methods	Q_3	Q_H	Q_E	Q_C
PSSpred	**0.831**	0.912	**0.748**	0.816
PSIPRED	0.806	0.866	0.695	**0.821**
SPINE-X	0.811	**0.913**	0.685	0.806

[a]Values in bold are the best results.

Table 6.3 Performance comparison of protein secondary structure prediction methods at three prediction difficulty levels[a]

	Easy				Medium				Hard			
Methods	Q_3	Q_H	Q_E	Q_C	Q_3	Q_H	Q_E	Q_C	Q_3	Q_H	Q_E	Q_C
PSSpred	**0.839**	0.922	**0.763**	0.819	**0.837**	**0.910**	**0.748**	**0.832**	**0.817**	**0.902**	**0.735**	0.801
PSIPRED	0.824	0.886	0.731	0.825	0.805	0.857	0.693	0.829	0.788	0.853	0.661	**0.812**
SPINE-X	0.823	**0.932**	0.691	0.812	0.818	0.905	0.696	0.820	0.794	0.897	0.671	0.787

[a]Values in bold are the best results.

Table 6.4 Misclassification rates of residue secondary structure states based on the benchmark data set with 538 proteins[a]

Native	Misclassified	Predicted	PSSpred	PSIPRED	SPINE-X
H	⟶	*E*	0.012	0.012	0.012
H	⟶	*C*	0.076	0.119	0.074
E	⟶	*H*	0.032	0.037	0.035
E	⟶	*C*	0.219	0.267	0.278
C	⟶	*H*	0.106	0.106	0.113
C	⟶	*E*	0.076	0.072	0.081

[a]Misclassification rate is calculated using $E(i)/N(i)$, where $E(i)$ is the number of misclassified protein secondary state i, that is, α-helix, β-strand or coil, and $N(i)$ is the total number of secondary state i in the benchmark data set of the 538 proteins.

To better understand the prediction error generation, it is important to know the misclassification rates between different secondary structure types. As can be seen from Table 6.4, the largest misclassification states were E to C and then C to H, which are consistent for the three protein secondary structure predictors. This result is also in agreement with the viewpoint of Huang and Wang that protein secondary structures can easily wobble between β-strand/coil and α-helix/coil [18].

Here, we also investigated the accuracy of secondary structure prediction for target proteins by the alignment/threading programs. In our previous work, 20 alignment/threading programs that were installed in our local computer cluster were used to build the 3D models of the 538 proteins in the benchmark data set. Likewise, the STRIDE program was used to derive the protein secondary structures of the models built by MODELLER [32] and these alignment/threading programs. The prediction performance of these methods is listed in Table 6.5. It is useful to know how accurately these alignment/threading programs can perform in terms of protein secondary structure prediction, as they represent the best alignment-based protein structure prediction methods available to the community. It should be pointed out that the templates in the library were filtered by an identity of 30% for each target. Accordingly, the accuracies of these alignment/threading methods were lower than that of the sequence-based predictors. Among them, MUSTER obtained the highest Q_3 (0.734) score compared with HHpred, FFAS, and PPA. In fact, the predicted

Table 6.5 Performance comparison of alignment/threading methods in the prediction of protein secondary structure

	Q_3	Q_H	Q_E	Q_C
HHpred [34]	0.703	0.591	0.511	0.888
MUSTER [37]	0.734	0.713	0.562	0.842
FFAS [31]	0.648	0.525	0.464	0.837
PPAS [40]	0.722	0.677	0.525	0.860
PPA [40]	0.663	0.569	0.486	0.827

Table 6.6 Compositions of the predicted and actual secondary structure types in the Y538 data set

Methods	%*H*	%*E*	%*C*
PSSpred	34.7	21.5	43.7
PSIPRED	33.3	20.1	46.5
SPINE-X	35.1	20.2	44.6
Native[a]	32.1	23.7	44.1

[a]The native secondary structures were derived using the program STRIDE.

secondary structure by PSIPRED has already been incorporated into the programs HHpred, MUSTER, and PPAS. The main difference between PPA and PPAS is that the prediction by PSIPRED is also included in PPAS, whereas PPA does not use the secondary structure term in its scoring function. The results show that PPAS attained a Q_3 measure ~8% higher than PPA, suggesting that the quality of 3D models can be significantly improved using the predicted secondary structure information. Furthermore, the reason that MUSTER generated a higher Q_3 score than PPAS could be ascribed to the fact that other informative structural terms (e.g., solvent accessibility and dihedral torsion angles) were also used in the scoring function of MUSTER.

Another critical issue for protein secondary structure prediction is the reasonable secondary structure content prediction. Therefore, it is essential to know whether a method over- or underpredicted a certain secondary structure state, for example, α-helix, when compared with the native structure. It is interesting to note that all three secondary structure predictors have overpredicted the α-helix content but underpredicted the content of β-strand as well as coil (Table 6.6). This provides us with some useful clues as to how to further refine the prediction results when developing improved methods in future studies.

The Y538 data set used in this study is publicly available at http://genomics.fzu.edu.cn/SSlist.txt. For a new method, the test data sets can be directly selected from PDB [2] and are required to share low similarity with proteins used as a training data set. Nevertheless, several other data sets exist as shown in Table 6.7.

Table 6.7 Data sets for protein secondary structure prediction

Data sets	Web	References
RS126[a]	Null[b]	[29]
CB513[c]	Null	[7]
Data set of SPINE-X	Null	[12]
Data set of Psipred	http://bioinfadmin.cs.ucl.ac.uk/downloads/psipred/old/data/tdbdata.tar.gz	[20]

[a]RS126 data set contains 126 proteins with known secondary structures selected by Rost and Sander.
[b]Null means the proteins are not publicly avaiable on the web. The proteins can be found in the corresponding references.
[c]CB513 data set contains 513 proteins selected by Cuff and Barton.

6.4 CONCLUSION

In this chapter, we have reviewed the development history of protein secondary structure prediction methods. The performance of these methods had been improved substantially in the 1990s through the incorporation of evolutionary profiles, which were extracted from the MSAs of proteins from the same structural family, as well as powerful pattern recognition techniques, for example, NNs. In recent years, the evolutionary information generated from iterative PSI-BLAST searches and enlarged NCBI NR databases have substantially enhanced the prediction accuracy. The literature review and our benchmarking tests have revealed that accuracy by the best methods could reach ~80%.

The top-performing methods, for example, PSSpred, PSIPRED, and SPINE-X, are consistently developed using NNs, which suggests that NNs are one of the most suitable pattern recognition algorithms to infer protein secondary structure from sequence profiles. Considering that there exist wobbles between helix/coil and strand/coil in native proteins, the current prediction accuracy (~80%) may have well reached the upper limit, especially in terms of Q_H and Q_C. As the accuracy of Q_E is relatively lower (~70%), which might be attributed to the fact that formation of a β-strand is strongly influenced by long-range interactions, the development of specific β-strand predictors represents a promising direction for further improvement of the prediction accuracy. It is also our hope that the development of such methods will be helpful for the exploration of the protein sequence/structural landscape. In addition, the protein secondary structure by the alignment/threading programs has been also discussed in this chapter. We found that the secondary structure prediction by the alignment/threading methods that combined PSIPRED with other informative structural features, such as solvent accessibility and dihedral torsion angles, was more accurate.

ACKNOWLEDGMENTS

We thank Jun Lin for valuable suggestions in the preparation of the manuscript. This work was supported by National Natural Science Foundation of China (31500673), the Education and Science Foundation for Young teachers of Fujian (JA14049), and Science Development Foundation of Fuzhou University (2013-XY-17)

REFERENCES

1. Altschul SF, Madden TL, Schaffer AA, Zhang J, Zhang Z, Miller W, Lipman DJ. Gapped BLAST and PSI-BLAST: a new generation of protein database search programs. Nucleic Acids Res 1997;25(17):3389–3402.
2. Berman HM. The protein data bank: a historical perspective. Acta Crystallogr 2008;64(1):88–95.
3. Brocchieri L, Karlin S. Protein length in eukaryotic and prokaryotic proteomes. Nucleic Acids Res 2005;33(10):3390–3400.
4. Chiang YS, Gelfand TI, Kister AE, Gelfand IM. New classification of supersecondary structures of sandwich-like proteins uncovers strict patterns of strand assemblage. Proteins 2007;68(4):915–921.
5. Chou PY, Fasman GD. Prediction of protein conformation. Biochemistry 1974;13(2):222–245.
6. Chou PY, Fasman GD. Prediction of the secondary structure of proteins from their amino acid sequence. Adv Enzymol Relat Areas Mol Biol 1978;47:45–148.
7. Cuff JA, Barton GJ. Evaluation and improvement of multiple sequence methods for protein secondary structure prediction. Proteins 1999;34(4):508–519.
8. Cuff JA, Clamp ME, Siddiqui AS, Finlay M, Barton GJ. JPred: a consensus secondary structure prediction server. Bioinformatics (Oxford, England) 1998;14(10):892–893.
9. David ER, Geoffrey EH, Ronald JW. Learning representations by back-propagating errors. In: James AA, Edward R, editors. *Neurocomputing: Foundations of Research*. MIT Press; 1988. p 696–699.
10. Dodge C, Schneider R, Sander C. The HSSP database of protein structure-sequence alignments and family profiles. Nucleic Acids Res 1998;26(1):313–315.
11. Eyrich VA, Rost B. META-PP: single interface to crucial prediction servers. Nucleic Acids Res 2003;31(13):3308–3310.
12. Faraggi E, Zhang T, Yang Y, Kurgan L, Zhou Y. SPINE X: improving protein secondary structure prediction by multistep learning coupled with prediction of solvent accessible surface area and backbone torsion angles. J Computat Chem 2012;33(3):259–267.
13. Frishman D, Argos P. Knowledge-based protein secondary structure assignment. Proteins 1995;23(4):566–579.
14. Garnier J, Osguthorpe DJ, Robson B. Analysis of the accuracy and implications of simple methods for predicting the secondary structure of globular proteins. J Mol Biol 1978;120(1):97–120.
15. Garnier J, Gibrat JF, Robson B. GOR method for predicting protein secondary structure from amino acid sequence. Methods Enzymol 1996;266:540–553.
16. Ginalski K, Grishin NV, Godzik A, Rychlewski L. Practical lessons from protein structure prediction. Nucleic Acids Res 2005;33(6):1874–1891.
17. Henikoff S, Henikoff JG. Position-based sequence weights. J Mol Biol 1994;243(4): 574–578.

18. Huang JT, Wang MT. Secondary structural wobble: the limits of protein prediction accuracy. Biochem Biophys Res Commun 2002;294(3):621–625.
19. Jones DT. GenTHREADER: an efficient and reliable protein fold recognition method for genomic sequences. J Mol Biol 1999;287(4):797–815.
20. Jones DT. Protein secondary structure prediction based on position-specific scoring matrices. J Mol Biol 1999;292(2):195–202.
21. Kabsch W, Sander C. Dictionary of protein secondary structure: pattern recognition of hydrogen-bonded and geometrical features. Biopolymers 1983;22(12):2577–2637.
22. Klingenberg M. Membrane protein oligomeric structure and transport function. Nature 1981;290(5806):449–454.
23. Konagurthu AS, Lesk AM, Allison L. Minimum message length inference of secondary structure from protein coordinate data. Bioinformatics (Oxford, England) 2012;28(12):i97–i105.
24. Leo B. Random forests method. Machine Learning 2001;45(1):15–32.
25. Lin CJ. Formulations of support vector machines: a note from an optimization point of view. Neural Computation 2001; 13(2):307–317.
26. Margolis RN. The nuclear receptor signaling atlas: catalyzing understanding of thyroid hormone signaling and metabolic control. Thyroid 2008;18(2):113–122.
27. Pauling L, Corey RB. Configurations of polypeptide chains with favored orientations around single bonds: two new pleated sheets. Proc Natl Acad Sci USA 1951; 37(11):729–740.
28. Pauling L, Corey RB, Branson HR. The structure of proteins; two hydrogen-bonded helical configurations of the polypeptide chain. Proc Natl Acad Sci USA 1951;37(4):205–211.
29. Rost B, Sander C. Prediction of protein secondary structure at better than 70% accuracy. J Mol Biol 1993;232(2):584–599.
30. Rost B, Sander C, Schneider R. PHD – an automatic mail server for protein secondary structure prediction. Comput Appl Biosci 1994;10(1):53–60.
31. Rychlewski L, Jaroszewski L, Li W, Godzik A. Comparison of sequence profiles. Strategies for structural predictions using sequence information. Protein Sci 2000;9(2):232–241.
32. Sali A, Blundell TL. Comparative protein modelling by satisfaction of spatial restraints. J Mol Biol 1993;234(3):779–815.
33. Schneider R, de Daruvar A, Sander C. The HSSP database of protein structure-sequence alignments. Nucleic Acids Res 1997;25(1):226–230.
34. Soding J. Protein homology detection by HMM-HMM comparison. Bioinformatics (Oxford, England) 2005;21(7):951–960.
35. Sternberg MJ, Bates PA, Kelley LA, MacCallum RM. Progress in protein structure prediction: assessment of CASP3. Curr Opin Struct Biol 1999;9(3):368–373.
36. Stillman BW, Bellett AJ. An adenovirus protein associated with the ends of replicating DNA molecules. Virology 1979;93(1):69–79.
37. Wu S, Zhang Y. MUSTER: Improving protein sequence profile-profile alignments by using multiple sources of structure information. Proteins 2008;72(2):547–556.
38. Yan RX, Si JN, Wang C, Zhang Z. DescFold: a web server for protein fold recognition. BMC Bioinformatics 2009;10:416.
39. Yan RX, Chen Z, Zhang Z. Outer membrane proteins can be simply identified using secondary structure element alignment. BMC Bioinformatics 2011;12(1):76.

40. Yan R, Xu D, Yang J, Walker S, Zhang Y. A comparative assessment and analysis of 20 representative sequence alignment methods for protein structure prediction. Sci Rep 2013;3.
41. Yoshida R, Sanematsu K, Shigemura N, Yasumatsu K, Ninomiya Y. Taste receptor cells responding with action potentials to taste stimuli and their molecular expression of taste related genes. Chem Senses 2005;30(Suppl 1):i19–i20.
42. Zhang Y. Protein structure prediction: when is it useful? Curr Opin Struct Biol 2009;19(2):145–155.
43. Y. Zhang: http://zhanglab.ccmb.med.umich.edu/PSSpred 2012.
44. Zhang Z, Kochhar S, Grigorov MG. Descriptor-based protein remote homology identification. Protein Sci 2005;14(2):431–444.

CHAPTER 7

A GENERIC APPROACH TO BIOLOGICAL SEQUENCE SEGMENTATION PROBLEMS: APPLICATION TO PROTEIN SECONDARY STRUCTURE PREDICTION

Yann Guermeur and Fabien Lauer

LORIA, Université de Lorraine-CNRS, Nancy, France

7.1 INTRODUCTION

We are interested in problems of bioinformatics that can be stated as follows: given a biological sequence and a finite set of categories, split the sequence into consecutive segments each assigned to a category different from those of the previous and next segments. Many problems of central importance in biology fit in this framework, such as protein secondary structure prediction [40], solvent accessibility prediction [14], splice site/alternative splicing prediction [7, 45], or the search for the genes of noncoding RNAs [15]. Our aim is to devise a global solution for the whole class of these problems. On the one hand, it should be generic enough to allow a fast implementation on any instance. On the other hand, it should be flexible enough to make

Pattern Recognition in Computational Molecular Biology: Techniques and Approaches,
First Edition. Edited by Mourad Elloumi, Costas S. Iliopoulos, Jason T. L. Wang, and Albert Y. Zomaya.

it possible to incorporate efficiently the knowledge available for a specific problem, so as to obtain state-of-the-art performance.

The basic principle of our solution is derived from a simple observation: whatever the problem is, all the state-of-the-art prediction methods currently available are made up of combinations of machine learning tools. These tools are either discriminative or generative models. The quality of the classification primarily depends on the choice of the descriptions, the selection of the building blocks of the prediction method, and the design of the global architecture. In that framework, discriminative and generative models appear to exhibit complementary advantages and drawbacks. However, strangely enough, they are seldom combined. Thus, we advocate the use of a hybrid architecture that combines discriminative and generative models in the framework of a modular and hierarchical approach. This architecture is inspired from the seminal works in speech processing that combine Neural Networks (NNs) [2] and Hidden Markov Models (HMMs) [38] (see especially References [29, 34]). To the best of our knowledge, the first example of a transcription of this idea in the field of bioinformatics is provided in Reference [16]. In that article, standard protein secondary structure prediction methods are combined with Multiclass Support Vector Machines (M-SVMs) [18]. Their outputs, after an appropriate postprocessing, provide class posterior probability estimates that are exploited by a Hidden Semi-Markov Model (HSMM) [47]. Other dedicated applications followed [33, 36].

The organization of the chapter is as follows. Section 7.2 reformulates the biological problems of interest as pattern recognition problems. Section 7.3 presents the bottom part of the hierarchy: MSVMpred. Section 7.4 introduces the whole hybrid architecture and focuses on the features of the upper part of the hierarchy, that is, the specification and implementation of the generative model. Its dedication for protein secondary structure prediction is exposed in Section 7.5. At last, we draw conclusions and outline ongoing research in Section 7.6.

7.2 BIOLOGICAL SEQUENCE SEGMENTATION

We make the assumption that the biological problems of interest can be expressed as pattern recognition problems [13] where the data take the form of ordered pairs of sequences of finite length. The first elements of the pairs are written on a common finite alphabet $\mathcal{A}$. The second ones share the same property, for a different alphabet $\mathcal{A}'$. In a pair, both sequences have the same length. Given a sequence written on $\mathcal{A}$, the task to be performed consists in finding a sequence on $\mathcal{A}'$ completing the pair. In the sequel, without loss of generality, $\mathcal{A}'$ is considered as a set of categories devoid of structure, and consequently identified with the set of integers from 1 to Q, denoted $[\![1, Q]\!]$, where $Q = |\mathcal{A}'|$. The problem then appears as a classification (discrimination) one, consisting in assigning a category to each position of the first sequence. We make the assumption, standard in the field of pattern recognition, that the dependency between the two kinds of sequences is of probabilistic nature. Let $L \in \mathbb{N}^*$ be the maximal length of a sequence. Let $\mathcal{X} = \bigcup_{l=1}^{L} \mathcal{A}^l$, $\mathcal{Y} = \bigcup_{l=1}^{L} \mathcal{A}'^l$, and let $\mathcal{B}$ and $\mathcal{B}'$ be the corresponding discrete σ-algebras. The probabilistic dependency takes the form of an unknown probability measure P on the measurable space $(\mathcal{X} \times \mathcal{Y}, \mathcal{B} \otimes \mathcal{B}')$. In short, each biological problem of interest is utterly characterized by an ordered pair of alphabets $(\mathcal{A}, \mathcal{A}')$ and a measure P. Following once more the main path of pattern recognition, we assume by default that training is restricted to pure empirical inference. Let (X, Y) be an ordered pair of random variables with values in $\mathcal{X} \times \mathcal{Y}$, distributed according to P. The only knowledge source available is a data set $d_N = ((x_i, y_i))_{1 \leqslant i \leqslant N}$ obtained as a realization of an N-sample $D_N = ((X_i, Y_i))_{1 \leqslant i \leqslant N}$ made up of independent copies of (X, Y). For a sequence z of either $\mathcal{X}$ or $\mathcal{Y}$, let $\ell(z)$ be its length. For j in $[\![1, \ell(z)]\!]$, let $z^{(j)}$ be the symbol at its jth position. With these notations at hand, ideally, the segmentation problem consists in finding a function from $\mathcal{X} \times [\![1, L]\!]$ into $[\![1, Q]\!]$ minimizing the risk

$$R(f) = \sum_{(x,y) \in \mathcal{X} \times \mathcal{Y}} \left(\frac{1}{\ell(x)} \sum_{j=1}^{\ell(x)} \mathbb{1}_{\{f(x,j) \neq y^{(j)}\}} \right) P((X, Y) = (x, y))$$

Note that the corresponding loss function is Bayes consistent in the sense that the Bayes classifier [13] f_B given by

$$\forall (x, j) \in \mathcal{X} \times [\![1, L]\!],\ j \leqslant \ell(x) \Rightarrow f_B(x, j) \in \underset{1 \leqslant k \leqslant Q}{\operatorname{argmax}} \sum_{y \in \mathcal{Y}:\ y^{(j)} = k} P(Y = y | X = x)$$

is a minimizer of R. However, this functional cannot be used directly as the objective function of a learning problem since the measure P is unknown. This is where the training set comes into action. Basically, P is replaced with the uniform probability measure on d_N.

It is noteworthy that fitting this way the biological problems of interest into a standard statistical framework can prove restrictive from a biological point of view. Indeed, one can think of several problems for which the sequence of categories cannot be inferred from the sole base sequence. A famous example is provided by chaperonin-mediated protein folding. Furthermore, technical reasons can be put forward to justify additional restrictions the consequences of which will have to be dealt with. A major problem springs from the fact that using x as description to predict each of the labels $y^{(j)}$ can be intractable with the standard models from the machine learning toolbox. More precisely, these models can fail to exploit properly distant information. They can treat part of it as noise, which raises well-known problems in pattern recognition such as overfitting. As a consequence, all the classifiers (discriminative models) considered in the sequel except the recurrent NNs apply a standard strategy in sequence segmentation: a local approach based on the use of an analysis window sliding on the sequence to be segmented. Obviously, the loss incurred by this last resort can be significant. The bottom-up approach implemented by our hybrid architecture is precisely designed to overcome this difficulty. For the sake of simplicity, unless otherwise specified, the comments, figures, and formulas provided below do not take into account the specificities of the implementation of the recurrent NNs.

7.3 MSVMpred

MSVMpred is a three-layer cascade of classifiers. The base classifiers use different sets of descriptors all of which include the content of an analysis window. The components of the global output vector are estimates of the class posterior probabilities (the vector belongs to the unit $(Q-1)$-simplex). More precisely, M-SVMs and NNs produce, after a possible postprocessing, initial estimates of the class posterior probabilities. These estimates are exploited by ensemble methods of different capacities that generate new (improved) estimates. At last, these outputs are combined by means of a convex combination to produce the global outputs. The topology of MSVMpred is depicted in Figure 7.1.

7.3.1 Base Classifiers

The base classifiers are of two types: M-SVMs with a dedicated kernel and NNs estimating the class posterior probabilities. The descriptions presented in the input of the M-SVMs are derived from the sole content of an analysis window. Let n_1 be the integer such that $2n_1 + 1$ is the size of this window. Then, neglecting for now the

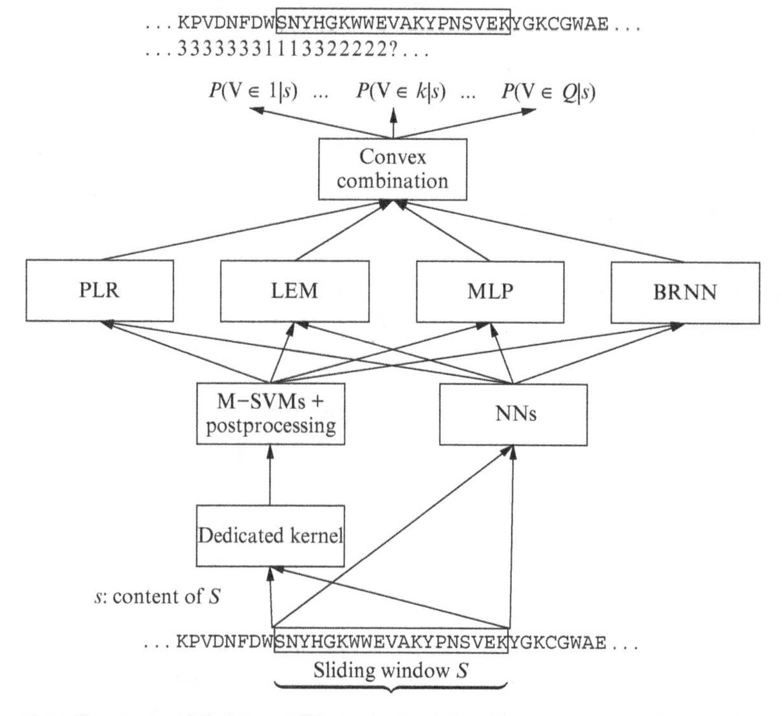

Figure 7.1 Topology of MSVMpred. The cascade of classifiers computes estimates of the class posterior probabilities for the position in the sequence currently at the center of the analysis window S.

problems raised by the extremities of the sequences, the description they use to infer $y^{(j)}$ is simply the subsequence/segment $s^{(j)} = (x^{(j+\delta j)})_{-n_1 \leqslant \delta j \leqslant n_1}$. Among the NNs, the feed-forward ones use the same descriptions as the M-SVMs, whereas the recurrent ones make use of an additional context. A model particularly well suited for the tasks of interest is the Bidirectional Recurrent Neural Network (BRNN) [6]. Indeed, as its name suggests, the additional context it exploits comes from both sides of the sequence processed. It obtains the highest prediction accuracy among all the neural networks assessed so far in protein secondary structure prediction [10, 22, 35].

Both types of classifiers take their values in $\mathbb{R}^Q$ instead of $[\![1, Q]\!]$. Let $h = (h_k)_{1 \leqslant k \leqslant Q}$ be a function computed by one of them and let $\mathcal{Z}$ be its domain (if h is computed by an M-SVM, then $\mathcal{Z}$ is simply $\mathcal{A}^{2n_1+1}$). The function h is associated with the decision rule d_h from $\mathcal{Z}$ into $[\![1, Q]\!] \bigcup \{*\}$ defined as follows:

$$\forall z \in \mathcal{Z}, \quad \begin{cases} \text{if } \exists k \in [\![1, Q]\!] \, : \, h_k(z) > \max_{l \neq k} h_l(z), \text{ then } d_h(z) = k \\ \text{else } d_h(z) = * \end{cases}$$

where $*$ denotes a dummy category introduced to deal with the cases of ex æquo. Contrary to BRNNs, M-SVMs do not output class posterior probability estimates. In order to introduce homogeneity among the outputs of the different base classifiers, the outputs of the M-SVMs are postprocessed by a Polytomous (multinomial) Logistic Regression (PLR) model [24]. In that way, we also ensure that the outputs of all classifiers precisely belong to the probability simplex, a requirement for the use of some of the ensemble methods. In the sequel, N_1 denotes the number of base classifiers. The one of index l is denoted $h^{(l)} = (h_k^{(l)})_{1 \leqslant k \leqslant Q}$.

7.3.2 Ensemble Methods

The second layer of the cascade is made up of ensemble methods. They combine the outputs of the first-layer classifiers $h^{(l)}$ (for l ranging from 1 to N_1) and output new estimates of the class posterior probabilities. Once more, an analysis window is used. The reason for this choice is to take benefit from the fact that the categories of consecutive positions in a sequence are correlated. Basically, the combiners filter the initial predictions. Let $2n_2 + 1$ be the size of the window. Then, the description processed by the combiners to estimate the probabilities associated with the position of index j in the sequence x is

$$z^{(j)} = (h_k^{(l)}(s^{(j+\delta j)}))_{-n_2 \leqslant \delta j \leqslant n_2, 1 \leqslant k \leqslant Q, 1 \leqslant l \leqslant N_1} \in U_{Q-1}^{(2n_2+1)N_1}$$

where U_{Q-1} is the unit $(Q-1)$-simplex.

Several options are available at this level. The four discriminative models currently implemented are a PLR, a Linear Ensemble Method (LEM) [19], a Multi-Layer Perceptron (MLP) [2], and a BRNN. They have been listed in order of increasing capacity [17]. Indeed, the PLR and LEMs are linear separators. An MLP using a softmax activation function for the output units and the cross-entropy loss (a sufficient condition for its outputs to be class posterior probability estimates) is an extension of a PLR obtained by adding a hidden layer. The boundaries it computes are nonlinear in its input space. At last, a BRNN can be seen roughly as an MLP operating on an extended description space.

7.3.3 Convex Combination

The availability of classifiers of different capacities for the second level of the cascade is an important feature of MSVMpred. It makes it possible to cope with one of the main limiting factors to the performance of modular architectures: overfitting. The capacity control is implemented by a convex combination combining the four ensemble methods from the second layer. The behavior of this combination is predictable: it assigns high weights to the combiners of low complexity when the training set size is small (and the combiners of higher complexity tend to overfit the training set). On the contrary, if the problem at hand appears to be a complex one, the latter combiners are favored when this size is large (see Reference [9] for an illustration of the phenomenon).

7.4 POSTPROCESSING WITH A GENERATIVE MODEL

In the field of sequence segmentation, postprocessing the outputs of a discriminative model with a generative one basically consists in using the class posterior probability estimates provided by the first model to derive the emission probabilities of the second one, thanks to Bayes' formula. Then, the segmentation is inferred from the labels associated with the states of the generative model, by means of an algorithm computing the path of highest probability. The rationale for this postprocessing is twofold: widening the context exploited for the prediction (see Section 7.2) and incorporating high-level knowledge on the task of interest (mainly in the topology of the generative models).

The generative model at the top of the hybrid architecture is an HSMM, that is, a model that differs from an HMM through the fact that the underlying (hidden) stochastic process is a semi-Markov chain. This model, introduced for machine recognition of speech, has already found applications in bioinformatics [30]. In the field of biological sequence segmentation, it is indeed useful to relax the Markovian hypothesis, since the distributions of the lengths of the different types of segments seldom exhibit an exponential decay. In practice, the model used by default has Q states, one per category. The global topology of the hierarchy is depicted in Figure 7.2.

For a given biological sequence, the final prediction is thus obtained by means of the dynamic programming algorithm computing the single best sequence of states, that is, the variant of the Viterbi algorithm [38] dedicated to the HSMM implemented. The fact that MSVMpred provides estimates of the class posterior probabilities, rather than emission probabilities, only calls for a minor adaptation of the formulas. Since the category assigned to each position of a training sequence is known, if the model has exactly one state per category, then the best state sequence is known for all the

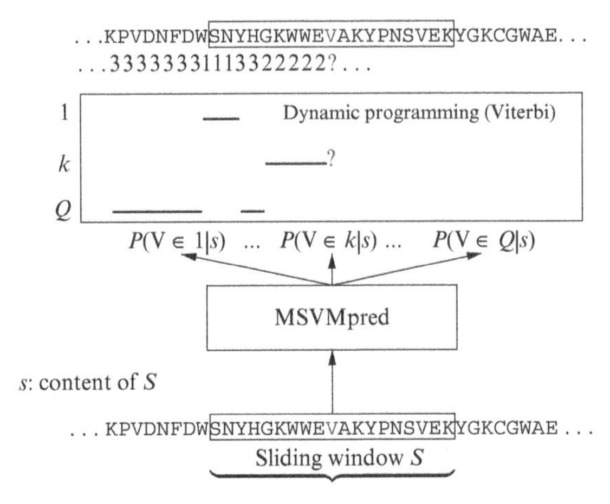

Figure 7.2 Topology of the hybrid prediction method. Thanks to the postprocessing with the generative model, the context effectively available to perform the prediction exceeds that resulting from the combination of the two analysis windows of MSVMpred.

sequences of the training set. As a consequence, applying the maximum likelihood principle to derive the initial state distribution and the transition probabilities boils down to computing the corresponding frequencies on the training set.

Implementing a hybrid approach of the prediction is fully relevant only if the quality of the probability estimates computed by the discriminative models is high enough for the generative model to exploit them efficiently. Our hybrid architecture incorporates an optional treatment specifically introduced to address this issue: a basic postprocessing of the outputs of MSVMpred. This postprocessing aims at constraining the final prediction so as to keep it in a vicinity of that of MSVMpred. For each position in a sequence, the vector of probability estimates is replaced with a vector that is close to the binary coding of the predicted category. Precisely, given a small positive value ϵ ($\epsilon \ll 1/Q$), the highest of the class membership probability estimates is replaced with $1 - (Q - 1)\epsilon$, the $(Q - 1)$ other estimates being replaced with ϵ. In this setting, the influence of the Viterbi algorithm on the path selected vanishes when ϵ reaches zero.

7.5 DEDICATION TO PROTEIN SECONDARY STRUCTURE PREDICTION

Among all the problems of biological sequence segmentation, protein secondary structure prediction emerges as one of those for which the highest number of prediction methods have been developed. Over almost half a century, many different types of approaches have been applied to it, resulting in many marginal improvements and a few significant advances. Even though the methods that constitute the state of the art are quite complex [4, 28, 36], it remains an open problem (the recognition rate currently stagnates around 80%). An interesting point is the fact that among these methods, one can already identify hybrid methods. They will provide us with the baseline performance.

7.5.1 Biological Problem

With the multiplication of genome-sequencing projects, the number of known protein sequences is growing exponentially. Knowing their (three-dimensional/tertiary) structure is a key in understanding their detailed function. Unfortunately, the experimental methods available to determine the structure, X-ray crystallography and Nuclear Magnetic Resonance (NMR), are highly labor-intensive and do not ensure the production of the desired result (e.g., some proteins simply do not crystallize). As a consequence, the gap between the number of known protein sequences and the number of known protein structures is widening rapidly. To bridge this gap, one must resort to empirical inference. The prediction of protein structure from amino acid sequence, that is, *ab initio*, is thus a hot topic in molecular biology. Due to its intrinsic difficulty, it is ordinarily tackled through a divide and conquer approach in which a critical first step is the prediction of the secondary structure, the local, regular structure defined by hydrogen bonds. Considered from the viewpoint of pattern recognition, this prediction is a three-category discrimination task consisting in assigning a conformational state, α-helix (H), β-strand (E), or aperiodic/loop/coil (C), to each residue (amino acid) of a sequence.

7.5.2 MSVMpred2

The instantiation of MSVMpred dedicated to protein secondary structure prediction is named MSVMpred2 [23]. In that field, the idea of an architecture based on a cascade of discriminative models can be traced back to the very first prediction method using pattern-recognition tools [37]. A straightforward transposition of NETtalk [43] provided the first significant improvement in the prediction accuracy, raising it from less than 60% up to more than 64%. In the following years, progress was made at a slow pace, until Rost and Sander [41] obtained the most important advance ever in the field, with a recognition rate passing the 70% mark. It resulted from substituting to the protein sequence the profile of a multiple alignment. The idea was developed by Jones [25], who gave to the profile its current form: that of a Position-Specific Scoring Matrix (PSSM). The prediction accuracy of his prediction method, PSIPRED, is over 76%. MSVMPred2 uses PSSM profiles provided by PSI-BLAST [1] and HHMAKE [44].

The base classifiers of MSVMpred2 are said to belong to the sequence-to-structure level with the combiners belonging to the structure-to-structure level. The sole NNs used are BRNNs. The kernel of the M-SVMs is an elliptic Gaussian kernel function applying a weighting on the predictors as a function of their position in the window. This weighting is learned by application of the principle of multiclass kernel target alignment [8, 20]. At the sequence-to-structure level, the values of the predictors associated with empty positions in the analysis window (at the extremities of the sequence) are set equal to zero. At the structure-to-structure level, these values are set equal to the corresponding *a priori* probabilities (frequencies of the different conformational states computed on the training set).

7.5.3 Hidden Semi-Markov Model

The HSMM selected here is the "*inhomogeneous HMM*" (IHMM) introduced in Reference [39]. It has three states, one for each of the three conformational states. The main advantage of this model compared to an HMM is that already pointed out in Section 7.4: its state transition probabilities are time dependent. This makes it possible to exploit a more suitable model of state durations, a necessary condition to get a high prediction accuracy at the conformational segment level.

7.5.4 Experimental Results

We now present initial experimental results that primarily illustrate the potential of our approach.

7.5.4.1 Protein Data Sets. The instance of the hybrid model dedicated to protein secondary structure prediction was assessed on two data sets. The first one is the well-known CB513 data set, fully described in Reference [12], whose 513 sequences are made up of 84,119 residues. The second one is the newly assembled CM4675 data set. It contains 4675 sequences, for a total of 85,1523 residues. The corresponding maximum pairwise percentage identity is 20%, that is, it is low enough to meet the standard requirements of *ab initio* secondary structure prediction.

To generate the PSI-BLAST PSSMs, the version 2.2.25 of the BLAST package was used. Choosing BLAST in place of the more recent BLAST+ offers the facility to extract more precise PSSMs. Three iterations were performed against the NCBI *nr* database. The E-value inclusion threshold was set to 0.005 and the default scoring matrix (BLOSUM62) was used. The *nr* database, downloaded in May 2012, was filtered by *pfilt* [26] to remove low complexity regions, transmembrane spans, and coiled coil regions. To generate the PSSMs with the HHMAKE algorithm, we used the protocol detailed in Reference [4], with slight differences. The multiple alignments were produced by the version of HHblits provided in the version 2.0.15 of the HH-suite [44] (ftp://toolkit.genzentrum.lmu.de/pub/HH-suite/). This program was applied to the *nr20* database (ftp://toolkit.genzentrum.lmu.de/pub/HH-suite/databases/hhsuite_dbs/), that is, the filtered version of *nr* at 20% identity threshold provided with the HH-suite. The version of HHMAKE used to generate the HMM profiles from these multiple alignments is also the one in the version 2.0.15 of the HH-suite.

As for the labeling of the residues, the initial secondary structure assignment was performed by the DSSP program [27], with the reduction from 8 to 3 conformational states following the Critical Assessment of protein Structure Prediction (CASP) method, that is, $H + G \rightarrow H$ (α-helix), $E + B \rightarrow E$ (β-strand), and all the other states in C (coil).

7.5.4.2 Experimental Protocol. The configuration chosen for MSVMpred2 includes the four main models of M-SVMs: the models of Weston and Watkins [46], Crammer and Singer [11], Lee et al. [32], and the M-SVM^2 [21]. At the sequence-to-structure level, they are used in parallel with four BRNNs. The programs implementing the different M-SVMs are those of MSVMpack [31], while the 1D-BRNN package is used for the BRNNs. The sizes of the first and second analysis windows are, respectively, 13 and 15 ($n_1 = 6$ and $n_2 = 7$).

To assess the accuracy of our prediction method, we implemented a distinct experimental protocol for each of the data sets. The reason for this distinction was to take into account the difference in size of the two sets. For CB513, the protocol was basically the sevenfold cross-validation procedure already implemented in References [9, 23] (with distinct training subsets for the sequence-to-structure level and the structure-to-structure level). At each step of the procedure, the values of the parameters of the IHMM that had to be inferred were derived using the whole training set. As for CM4675, it was simply split into the following independent subsets: a training set for the kernel of the M-SVMs (500 sequences, 98,400 residues), a training set for the sequence-to-structure classifiers (2000 sequences, 369,865 residues), a training set for the postprocessing of the M-SVMs with a PLR (300 sequences, 52,353 residues), a training set for the structure-to-structure classifiers (1000 sequences, 178,244 residues), a training set for the convex combination (200 sequences, 34,252 residues), and a test set (675 sequences, 118,409 residues). Once more, the missing values of the parameters of the IHMM were derived using globally all the training subsets (4000 sequences, 733,114 residues).

It can be inferred from the description of the problem (see Section 7.5.1) that a secondary structure prediction method must fulfill different requirements in order to be useful for the biologist. Thus, several standard measures giving complementary indications must be used to assess the prediction accuracy [5]. We consider the three most popular ones: the per-residue recognition rate Q_3, Pearson–Matthews correlation coefficients $C_{\alpha/\beta/\text{coil}}$, and the Segment overlap (Sov) measure in its most recent version (Sov'99).

Among the most accurate prediction methods are two methods based on a hybrid architecture: YASPIN [33] and the unified multitask architecture introduced in Reference [36]. YASPIN is simply made up of an MLP postprocessed with an HMM. In this cascade, the dedication to the task of interest mainly rests in the topology of the generative model. It assigns three (hidden) states per periodic conformational state. The authors of Reference [33] applied an experimental protocol utterly compatible with ours, and chose the same conversion rule to reduce the number of DSSP conformational states from 8 to 3. As a consequence, the results reported for YASPIN, including a recognition rate slightly superior to 77%, can be directly compared to those of our architecture.

7.5.4.3 Results. The experimental results obtained with MSVMpred2 and the two variants of the hybrid model (with and without postprocessing of the outputs of MSVMpred2) are reported in Table 7.1.

Table 7.1 Prediction accuracy of MSVMpred2 and the hybrid model

Data set		MSVMpred2	Hybrid	Hybrid ($\epsilon = 0.01$)
CB513	Q_3 (%)	78.4	77.5	78.4
	Sov	74.5	73.2	**75.6**
	C_α	0.74	0.74	0.74
	C_β	0.64	0.64	0.64
	C_{coil}	0.61	0.58	0.61
CM4675	Q_3 (%)	81.8	81.0	81.8
	Sov	79.0	77.6	**80.1**
	C_α	0.79	0.78	0.79
	C_β	0.73	0.71	0.73
	C_{coil}	0.65	0.64	0.65

[a] Results in the last column were obtained with the optional treatment of the outputs of MSVMpred2 described in Section 7.4.

In applying the two sample proportion test (the one for large samples), one can notice than even when using CB513, the superiority of MSVMpred2 over YASPIN appears statistically significant with confidence exceeding 0.95. Of course, such a statement is to be tempered since the figures available for YASPIN correspond to a different set of protein sequences. The recognition rate is always significantly above 80% when CM4675 is used. This highlights a fact already noticed in Reference [4]: the complexity of the problem calls for the development of complex modular prediction methods such as ours. The feasibility of their implementation increases with the growth of the protein data sets available. The hybrid method is only superior to MSVMpred2 when it implements the postprocessing described in Section 7.4. In that case, the gain is emphasized, as expected, by means of the Sov. This measure increases by the same amount (1.1 point) on both data sets.

The fact that this postprocessing is necessary in order to improve the performance can be seen as a lukewarm result. It shows that the class posterior probability estimates provided by MSVMpred2 are satisfactory in the sense not only that the hybrid model can yield an improved prediction, but also that they are not yet accurate enough in the sense that they cannot be used directly. Meanwhile, a positive result concerns the correlation between the quality of these estimates and the performance of the overall hybrid model. Note that we cannot measure the quality of the probability estimates (by means of the cross-entropy) without knowledge of the true probabilities. But, as a proof of concept, Figure 7.3 highlights the almost linear dependency between the recognition rate (Q_3) of MSVMpred2 and the one of the hybrid model (without postprocessing). The points on this plot correspond to the performance obtained with various versions of MSVMpred2, including different combinations of input data (PSSMs generated by PSI-BLAST, HHMAKE, or both), base classifiers, and ensemble methods. These results support the thesis that hybrid models can be used to increase the performance of discriminative models while directly benefiting from improvements of the latter.

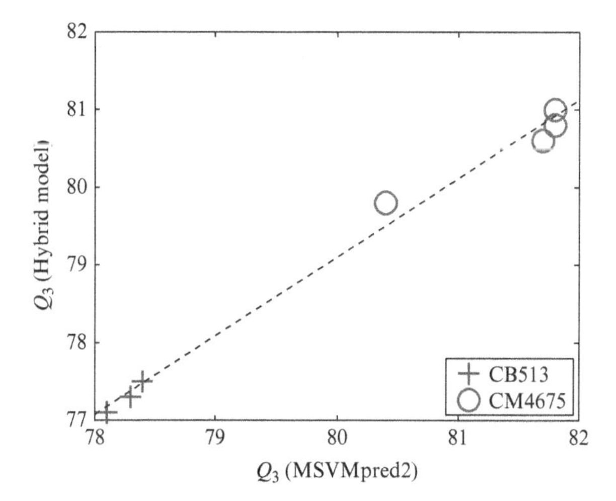

Figure 7.3 Correlation between the recognition rate (Q_3) of MSVMpred2 and the one of the hybrid model. The dashed line is obtained by linear regression.

7.6 CONCLUSIONS AND ONGOING RESEARCH

We have introduced a generic architecture for biological sequence segmentation. It applies a hierarchical approach where low-level treatments are performed by discriminative models whereas higher-level treatments are performed by generative models. Obviously, the aim of a hybrid approach is to make the best of two worlds. From that point of view, the introduction of a postprocessing with a generative model appears as the best solution currently available to overcome the limitations induced by the implementation of the standard local approach, based on the exploitation of the content of an analysis window sliding on the sequence to be segmented. The architecture has been conceived so as to satisfy two contradictory requirements: be easily applicable to all the problems of the class of interest and be flexible enough to adapt to the specificities of any of them. The main characteristic that accounts for this property is modularity. In fact, the modularity of our generic architecture should be as high as to make it possible to incorporate the most efficient features of the methods exhibiting state-of-the-art performance.

The variant of the architecture dedicated to protein secondary structure prediction already comes close to state-of-the-art performance, which is all the more promising as it can be significantly improved. Indeed, it provides us with an ideal example of an architecture that can take benefit in a straightforward way from the ideas at the basis of the most recently developed prediction methods.

We are currently working on the application of the architecture on several biological problems, including those mentioned in Section 7.1. Clearly, there are many options one can think of to improve it. Regarding MSVMpred, a promising approach to be implemented is multitask learning [36]. As for the implementation of the generative model, an appealing option consists in making use of the N-best paradigm [42], in a way similar to that in Reference [3].

ACKNOWLEDGMENTS

The authors thank Christophe Magnan for providing them with the CM4675 data set and the 1D-BRNN package.

REFERENCES

1. Altschul SF, Madden TL, Schäffer AA, Zhang J, Zhang Z, Miller W, Lipman DJ. Gapped BLAST and PSI-BLAST: a new generation of protein database search programs. Nucleic Acids Res 1997;25(17):3389–3402.
2. Anthony M, Bartlett PL. *Neural Network Learning: Theoretical Foundations*. Cambridge: Cambridge University Press; 1999.
3. Aydin Z, Altunbasak Y, Erdogan H. Bayesian protein secondary structure prediction with near-optimal segmentations. IEEE Trans Signal Process 2007;55(7):3512–3525.
4. Aydin Z, Singh A, Bilmes J, Noble WS. Learning sparse models for a dynamic Bayesian network classifier of protein secondary structure. BMC Bioinformatics 2011;12:154.
5. Baldi P, Brunak S, Chauvin Y, Andersen CAF, Nielsen H. Assessing the accuracy of prediction algorithms for classification: an overview. Bioinformatics 2000;16(5):412–424.
6. Baldi P, Brunak S, Frasconi P, Soda G, Pollastri G. Exploiting the past and the future in protein secondary structure prediction. Bioinformatics 1999;15(11):937–946.
7. Barash Y, Calarco JA, Gao W, Pan Q, Wang X, Shai O, Blencowe BJ, Frey BJ. Deciphering the splicing code. Nature 2010;465(6):53–59.
8. Bonidal R. Analyse des systèmes discriminants multi-classes à grande marge [PhD thesis]. Université de Lorraine; 2013. (in French).
9. Bonidal R, Thomarat F, Guermeur Y. Estimating the class posterior probabilities in biological sequence segmentation. SMTDA'12; Crete, Greece; 2012.
10. Chen J, Chaudhari NS. Cascaded bidirectional recurrent neural networks for protein secondary structure prediction. IEEE/ACM Trans Comput Biol Bioinfomatics 2007;4(4):572–582.
11. Crammer K, Singer Y. On the algorithmic implementation of multiclass kernel-based vector machines. J Mach Learn Res 2001;2:265–292.
12. Cuff JA, Barton GJ. Evaluation and improvement of multiple sequence methods for protein secondary structure prediction. Proteins 1999;34(4):508–519.
13. Devroye L, Györfi L, Lugosi G. *A Probabilistic Theory of Pattern Recognition*. New York: Springer-Verlag; 1996.
14. Gianese G, Pascarella S. A consensus procedure improving solvent accessibility prediction. J Comput Chem 2006;27(5):621–626.

15. Gothié E, Guermeur Y, Muller S, Branlant C, Bockmayr A. Recherche des gènes d'ARN non codant. Research Report RR-5057. INRIA; 2003. (in French). Available at https://hal.inria.fr/inria-00071526/

16. Guermeur Y. Combining discriminant models with new multi-class SVMs. Pattern Anal Appl 2002;5(2):168–179.

17. Guermeur Y. VC theory of large margin multi-category classifiers. J Mach Learn Res 2007;8:2551–2594.

18. Guermeur Y. A generic model of multi-class support vector machine. Int J Intell Inform Database Syst 2012;6(6):555–577.

19. Guermeur Y. Combining multi-class SVMs with linear ensemble methods that estimate the class posterior probabilities. Commun Stat Theory Methods 2013;42(16):3011–3030.

20. Guermeur Y, Lifchitz A, Vert R. A kernel for protein secondary structure prediction. In: Schölkopf B, Tsuda K, Vert J-P, editors. *Kernel Methods in Computational Biology*. Chapter 9. Cambridge (MA): The MIT Press; 2004. p 193–206.

21. Guermeur Y, Monfrini E. A quadratic loss multi-class SVM for which a radius-margin bound applies. Informatica 2011;22(1):73–96.

22. Guermeur Y, Pollastri G, Elisseeff A, Zelus D, Paugam-Moisy H, Baldi P. Combining protein secondary structure prediction models with ensemble methods of optimal complexity. Neurocomputing 2004;56:305–327.

23. Guermeur Y, Thomarat F. Estimating the class posterior probabilities in protein secondary structure prediction. PRIB'11; Delft, The Netherlands; 2011. p 260–271.

24. Hosmer DW, Lemeshow S. *Applied Logistic Regression*. London: John Wiley & Sons, Ltd; 1989.

25. Jones DT. Protein secondary structure prediction based on position-specific scoring matrices. J Mol Biol 1999;292(2):195–202.

26. Jones DT, Swindells MB. Getting the most from PSI-BLAST. Trends Biochem Sci 2002;27(3):161–164.

27. Kabsch W, Sander C. Dictionary of protein secondary structure: pattern recognition of hydrogen-bonded and geometrical features. Biopolymers 1983;22(12):2577–2637.

28. Kountouris P, Hirst JD. Prediction of backbone dihedral angles and protein secondary structure using support vector machines. BMC Bioinformatics 2009;10:437.

29. Krogh A, Riis SK. Hidden neural networks. Neural Comput 1999;11(2):541–563.

30. Kulp D, Haussler D, Reese MG, Eeckman FH. A generalized hidden Markov model for the recognition of human genes in DNA. ISMB; St-Louis, MO; 1996. p 134–142.

31. Lauer F, Guermeur Y. MSVMpack: a multi-class support vector machine package. J Mach Learn Res 2011;12:2293–2296.

32. Lee Y, Lin Y, Wahba G. Multicategory support vector machines: theory and application to the classification of microarray data and satellite radiance data. J Am Stat Assoc 2004;99(465):67–81.

33. Lin K, Simossis VA, Taylor WR, Heringa J. A simple and fast secondary structure prediction method using hidden neural networks. Bioinformatics 2005;21(2):152–159.

34. Morgan N, Bourlard H, Renals S, Cohen M, Franco H. Hybrid neural network/hidden Markov model systems for continuous speech recognition. Int J Pattern Recognit Artif Intell 1993;7(4):899–916.

35. Pollastri G, Przybylski D, Rost B, Baldi P. Improving the prediction of protein secondary structure in three and eight classes using recurrent neural networks and profiles. Proteins 2002;47(2):228–235.

36. Qi Y, Oja M, Weston J, Noble WS. A unified multitask architecture for predicting local protein properties. PLoS ONE 2012;7(3):e32235.

37. Qian N, Sejnowski TJ. Predicting the secondary structure of globular proteins using neural network models. J Mol Biol 1988;202(4):865–884.

38. Rabiner LR. A tutorial on hidden Markov models and selected applications in speech recognition. Proc IEEE 1989;77(2):257–286.

39. Ramesh P, Wilpon JG. Modeling state durations in hidden Markov models for automatic speech recognition. ICASSP-92, Volume I; San Francisco, CA; 1992. p 381–384.

40. Rost B. Review: protein secondary structure prediction continues to rise. J Struct Biol 2001;134(2):204–218.

41. Rost B, Sander C. Prediction of protein secondary structure at better than 70% accuracy. J Mol Biol 1993;232(2):584–599.

42. Schwartz R, Chow YL. The N-best algorithm: an efficient and exact procedure for finding the N most likely sentence hypotheses. ICASSP-90, Volume 1; Albuquerque, NM; 1990. p 81–84.

43. Sejnowski TJ, Rosenberg CR. Parallel networks that learn to pronounce english text. Complex Syst 1987;1:145–168.

44. Söding J. Protein homology detection by HMM-HMM comparison. Bioinformatics 2005;21(7):951–960.

45. Sonnenburg S, Schweikert G, Philips P, Behr J, Rätsch G. Accurate splice site prediction using support vector machines. BMC Bioinformatics 2007;8 Suppl 10:S7.

46. Weston J, Watkins C. Multi-class support vector machines. Technical Report CSD-TR-98-04. Royal Holloway, University of London, Department of Computer Science; 1998. Available at http://citeseerx.ist.psu.edu/viewdoc/summary?doi=10.1.1.50.9594;

47. Yu S-Z. Hidden semi-Markov models. Artif Intell 2010;174(2):215–243.

CHAPTER 8

STRUCTURAL MOTIF IDENTIFICATION AND RETRIEVAL: A GEOMETRICAL APPROACH

Virginio Cantoni, Marco Ferretti, Mirto Musci, and Nahumi Nugrahaningsih

Department of Electrical, Computer and Biomedical Engineering, University of Pavia, Via Ferrata 3, 27100 Pavia, Italy

8.1 INTRODUCTION

The 3D structure of a protein is strictly related to its function. For example, antibodies (*Immunoglobulin* protein family) serve the function of capturing foreign objects such as bacteria and viruses. Although belonging to a number of different classes, all antibodies share the Y-like structure: the arms of the Y (*Fab* region) contain the sites that capture the antigens, that is, the foreign objects to be neutralized. The leg of the Y (*Rc* region) serves for the immune system cells to bind to the antibody, in order to destroy the captured pathogen.

In 1972, the Nobel Prize Christian B. Anfinsen postulated (see References [2] and [3]) that the 3D structure of a fold is determined *solely* by its amino acid sequence. This postulate is traditionally referred to as "Anfinsen's Dogma." In other words, given a polypeptide chain, there is a *unique* 3D structure it can fold into. Therefore, the secondary and tertiary structures of a protein are uniquely determined by its primary

Pattern Recognition in Computational Molecular Biology: Techniques and Approaches,
First Edition. Edited by Mourad Elloumi, Costas S. Iliopoulos, Jason T. L. Wang, and Albert Y. Zomaya.

structure. However, it is important to remark that, even if a given amino acid sequence results in a unique tertiary structure, the reverse does not necessarily apply.

Moreover, proteins showing significant similarities in their tertiary structure, and therefore possibly sharing a common ancestor in their evolution, may present little homology in their primary structure [9]. This pushes for the need of methods and tools to find similarities at *each* level of the protein structure, even if the primary structure determines the secondary and the tertiary.

Many algorithms have been described that search for similarities in proteins at each level. Here, we present a novel approach to detect similarities at the secondary level, based on the Generalized Hough Transform (GHT) [4]. However, we do not simply want to assign a score of similarity to a pair of proteins, nor we want to develop yet another (even if more efficient) alignment tool.

As we will show, our main goal is to look for previously unknown common geometrical structures in so-called "unfamiliar" proteins. To do that, we need to abandon the traditional topological description of structural motifs. In other words, our main concern is the precise, flexible, and efficient identification of geometrical similarities among proteins; that is the retrieval of geometrically defined structural motifs.

This chapter is organized as follows. In Section 8.2, we introduce some basic concepts of biochemistry related to the problem at hand. In Section 8.3, we briefly analyze the state of the art regarding the retrieval of structural motifs. The core of this contribution is Section 8.4. Here we fully describe our proposals from both a computer science and a mathematical point of view. In Section 8.5, we briefly discuss our implementation strategy and the available parallelism of our proposals, reviewing the result of some of the benchmarks we performed. Finally, in Section 8.6, we lay down some conclusions and present our future research work.

8.2 A FEW BASIC CONCEPTS

In this section, we provide a brief introduction to some key concepts of biochemistry, which will come in handy for the rest of the chapter. The first concept is that of proteins, of course. Put in a very simple manner, proteins are essential for the functioning of every living organism. Different proteins perform different functions, ranging from oxygen transport (*hemoglobin*) to metals storage (*ferritin*), and from structural support (*keratin*) to immunological functions (*immunoglobulin*). The basic protein unit is the amino acid or residue. Indeed, a protein is a polymer, that is, a set of one or more polypeptides, where a polypeptide is just a long (10–100) chain of amino acids. Noncovalent interactions, such as hydrogen bonds, between the atoms composing the chain maintain the 3D shape of protein. The process through which a polypeptide is folded into a particular 3D shape is called *protein folding*.

8.2.1 Hierarchy of Protein Structures

Proteins are large and complex polymers, and it is convenient for scientists to describe their structure at several levels of abstraction, in order to better understand their spatial arrangement and behavior (see Figure 8.1).

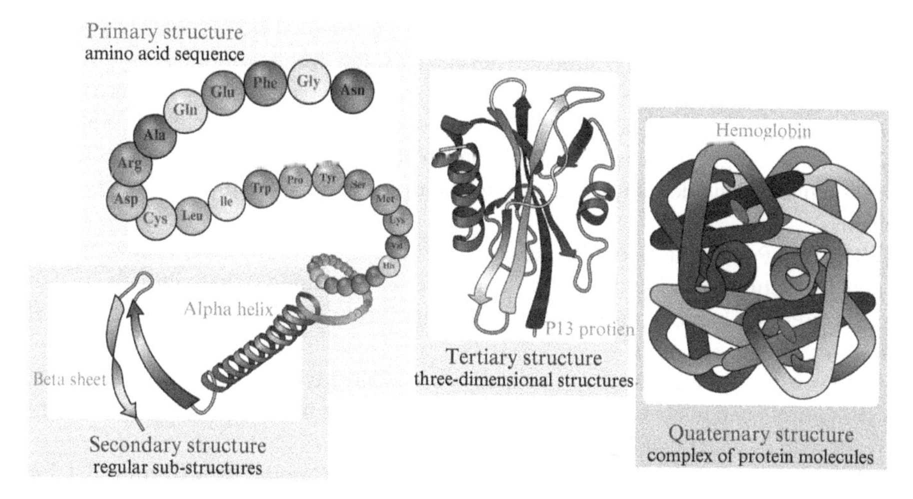

Figure 8.1 Simplified diagram of the multilevel protein hierarchy. *Source*: Modified from a public-domain image created by Mariana Ruiz Villarreal (http://commons.wikimedia.org/wiki/File:Main_protein_structure_levels_en.svg).

1. The *primary structure* simply describes a protein as the exact sequence of all the residues composing its polypeptide(s).
2. The *secondary structure* describes a protein in terms of its Secondary Structure Elements (SSEs) (see Section 8.2.2 for more details).
3. The *tertiary structure* describes the overall 3D shape of a single polypeptide, comprising the atoms it is made up of and their exact position.
4. The *tertiary structure* completely identifies the structure of those proteins that are made up of a single polypeptide. For proteins that are made up of more than one polypeptide, the spatial arrangement of the polypeptides (also called protein subunits) is referred to as *quaternary structure*.

8.2.2 Secondary Structure Elements

An SSE is a small segment of the protein chain that can be identified as a recurring pattern among a multitude of proteins. Each SSE is made up of multiple residues shaped in a peculiar 3D structure.

The first two and the most commonly recurring SSEs are the α-helix and the β-sheet as described by Pauling et al. [23].

The α-helix is a helical structure in which the carbon skeleton of each amino acid is wound around an imaginary line that lies in the middle of the helix and the functional groups of each residue protrude outward. A single turn of the helix extends approximately 5.4 Å (3.6 residues) along the axis.

A β-*sheet* consists of a number of parallel or antiparallel β-strands connected by hydrogen bonds. A β-strand is a polypeptide chain of typically 3–10 amino acids characterized by a zigzag-like structure, where the functional groups of each residue protrude on alternate sides of the strand.

Extracting SSE information from a protein is an essential step for each algorithm that works at a secondary structure level. This job has traditionally been done manually; fortunately, automatic approaches to the recognition method have been developed in the past. The Dictionary of Protein Secondary Structure (DSSP) by Kabsch and Sander [17] is both a dictionary of the secondary structures recurring in proteins and the de facto standard algorithm for identifying them in polypeptide chains.

The DSSP is able to identify, besides α-helices and β-strands, the following secondary structures: 3_{10}-*helices* and π-*helices*, helical structures akin to the α-helix, but much rarer and harder to identify; and less significant SSEs such as *turns*, *isolated residues*, *bends*, and *coils*.

However, for the purpose of this study, we decided to ignore the less significant SSEs entirely and to classify all three helices in the same macro-category, that is, they are all seen as generic helices.

It is important to note that a number of *structural motifs* (see Section 8.4 for a precise definition of this term), including the β-*hairpin* and the *Greek key* (see Section 8.2.3), are composed of few β-strands that can be part of a β-sheet comprising many other β-strands. This poses a problem of consistency: even if the β-sheet is considered a secondary structure and structural motifs are defined as "groups of secondary structures," a single β-sheet can contain, either in whole or in part, one or more structural motifs made up of few β-strands, plus a number of other β-strands that are not part of any structural motif. This would qualify the β-sheet as a sort of super-motif rather than a secondary structure. To further complicate the situation, the β-sheet is traditionally referred to as an SSE and this convention is widely accepted among biochemistry scholars, even if it violates the hierarchy of protein structures that has been described so far. Our solution is to ignore the old convention and consider β-strands instead of β-sheets as the basic secondary structures; this choice is based on the behavior of the DSSP algorithm, which identifies singular strands instead of whole sheets.

Finally, we note that although the DSSP is the oldest and most Common algorithm, it is not the only known method to extract secondary structure information from protein data. Other methods, such as STRIDE [14] and *Predictive Assignment of Linear Secondary Structure Elements* (PALSSE) [21], have been described in the literature.

8.2.3 Structural Motifs

Protein folding is not a random process and is subject to a number of physical and chemical constraints. Some recurring "folding patterns,"that is, small groups of SSEs, have been identified at the secondary level, which are called *super-secondary structures* or *structural motifs*. Structural motifs and the methods to retrieve and identify them are the central topic of this chapter.

Structural motifs can be defined through either a *topological* description or a *geometrical* one. The first description is the most widespread, and it is easier to understand for a human, that is, a biologist. We can say that a topological description is mostly concerned with the kind of connections existing between SSEs in a

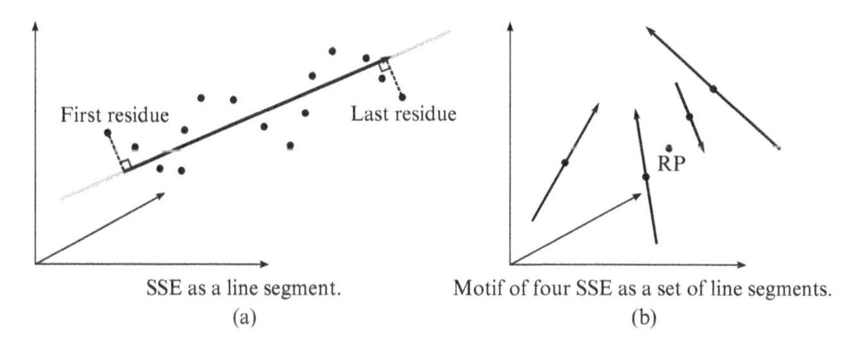

SSE as a line segment.
(a)

Motif of four SSE as a set of line segments.
(b)

Figure 8.2 Approximation of the secondary structure elements and motifs.

motif. In the literature, there are various tools that deal with topological motifs, for example, ProSMoS (Section 8.3.1). Our proposals, instead, are tailored to the retrieval of geometrically defined structural motifs of secondary structures, or simply geometrical motifs. A geometrical motif, instead, is seen as a collection of geometrically defined elements, such as points or line segments.

8.2.3.1 *Representing SSEs and Motifs.* A single SSE can be described at different levels of abstraction, from a simple list of all its amino acids and their absolute positions, to a more compact representation. In the view of our geometrical approach, we simply approximate an SSE by an oriented line segment in the 3D space: the direction of the line segment can be calculated by fitting a line to the positions of the residues composing the SSE, and the endpoints can be chosen as the projections of the first and last residues over the fitted line [12] (Figure 8.2a).

As a consequence, a structural motif can be approximated by a set of oriented line segments, and the *Reference Point* (RP) for the GHT (Section 8.4) can be chosen as the centroid of the midpoints of the line segments (Figure 8.2b). All relevant information about a motif is then stored in a conveniently designed XML format.

8.2.3.2 *A Brief List of Common Structural Motifs.* This section will briefly enumerate some of the most commonly recurring structural motifs. Note that these descriptions are intrinsically topological: obtaining a valid geometrical model for a structural motif is not an easy task. Indeed, as it will become clearer later, we do not try to model any of these motifs, even if we are able to find almost all of them.

- The β-*hairpin* is the simplest structural motif that can be observed. It is made up of two adjacent antiparallel β-strands connected by a short chain of two to five residues. This motif is often part of an antiparallel β-sheet.
- The β-*meander* is a simple super-secondary structure made up of three or more adjacent antiparallel β-strands, each connected to the next by a hairpin.
- First described by Richardson [27], the *Greek key* (Figure 8.3a) is one of the most studied structural motifs. It is composed of four adjacent antiparallel β-strands, where the first three are connected by hairpins, and the fourth one

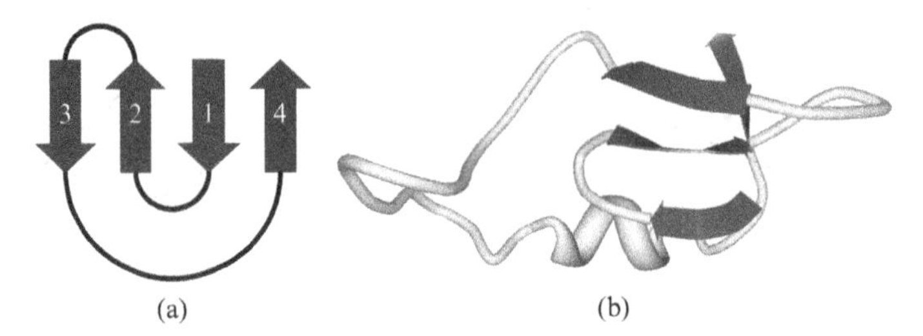

Figure 8.3 Greek key in theory and in practice: (a) idealized structure and (b) noncoplanar Greek key in staphylococcal nuclease [5].

is close to the first and connected to the third with a long loop. Some authors (see, e.g., Branden and Tooze [5]) identify Greek keys among quadruples of strands that are not all hydrogen-bonded and sometimes not even coplanar (see Figure 8.3b for an example). This allows to classify many quadruples of β-strands as Greek keys, but makes it hard to design an algorithm that is able to determine whether four β-strands actually form a Greek key.

- A β-*barrel* consists of several antiparallelβ-strands arranged in a cylindrical structure. The simplest kind of β-barrel is the up and down barrel, where each strand is connected to its two nearest neighbors in the sequence. Richardson [26] also identified a particular kind of β-barrel, the Greek key barrel, which contains one or more Greek keys.
- The β–α–β motif is composed of two parallel β-strands connected by a noncoplanar α-helix. This motif can be found in α/β domains such as the *Triosephosphate Isomerase* (TIM) barrel domain, made up of eight parallel β-strands and eight α-helices arranged in two concentric cylindrical structures.

8.2.4 Available Sources for Protein Data

Large structured online repositories are available that provide information about a great number of proteins, specifying their structures (from primary to quaternary) and relationships with other proteins.

The *Protein Data Bank* (PDB)[1] is a database of protein 3D structures, mainly obtained by X-ray crystallography or *Nuclear Magnetic Resonance* (NMR) spectroscopy. As of March 28, 2014, this database contains 98,900 entries, each one identified by a four-character alphanumeric ID (e.g., `7fab`), which is automatically assigned and has no particular meaning.[2] Contributed by researchers from all around the world, the information provided by the PDB is not restricted to protein 3D structures: each entry also comprises the protein's general description and taxonomy, residue sequence (primary structure), known similarities with other proteins, and references to the relevant literature. An interactive Java applet allows to display and inspect 3D structures.

[1] http://www.pdb.org/.
[2] http://www.rcsb.org/pdb/staticHelp.do?p=help/advancedsearch/pdbIDs.html.

Many entries in the PDB have structural similarities, and some of them are close variants (e.g., `2imm` and `2imn`). According to Orengo et al. [22], "nearly three-quarters of the entries in the current PDB are practically identical" (as of 1997). The PDB is similar to an encyclopedia of protein structures: a specific entry may be found by specifying its ID or name, but the database itself does not provide any hierarchical classification. The *Structural Classification of Proteins* (SCOP) [15] database was designed precisely to facilitate access to the information contained in the PDB. It provides a "detailed and comprehensive description of the relationships of all known proteins structures."

The SCOP classification has been constructed manually by visual comparison of protein sequences. Its hierarchical levels are the following, from higher to lower: *structural classes*, such as all-α (entirely composed of α-helices), all-β (entirely composed of β-strands), α/β (composed of alternating α-helices and β-strands), and α + β (composed of α-helices and β-strands that occur separately); *folds*: *super-families*; *families*; *protein domains*; *species*; and *domains*.

Again the details are not strictly relevant to the discussion; we can say, in a qualitative manner, that the lower one goes down the SCOP hierarchy, the more similar proteins become. Again, qualitatively, we can state that as proteins in the same family are extremely similar from a structural point of view, precise discerning of their areas of similitude become pointless, and alignment tools are much more suited for the task of comparing two "familiar" proteins.

Finally, we note that the SCOP is not the only available classification. Among the others, we mention the *CATH Protein Structure Classification* [22].

8.3 STATE OF THE ART

This section will provide a brief analysis of previous approaches (both topological and geometrical). For more details, adequate references are provided. Note that this section is far from being exhaustive: only some of the most relevant approaches are introduced.

8.3.1 Protein Structure Motif Search

Protein Structure Motif Search (ProSMoS) [28, 29] is a web server application that emulates the behavior of a human expert when looking for a motif. It takes as input a motif meta-matrix and returns the list of proteins in PDB that contains the given motif. It consists of two main steps: the data preprocessing and the query process of the meta-matrix.

The data preprocessing is done off-line and is made of two phases. First, the SSE assignment of protein (taken from PDB) is carried out using PALSSE [21] (see Section 8.2.2). This step generates a vector representation of the SSE, similar to the one described in Section 8.2.3.1. Each vector is characterized by its starting and ending coordinates, and also some biological features such as the type of SSE (e.g., α-helix or β-strand). Second, for every protein, ProSMoS generates a

```
1      3br8.ssd                      7
2      EA  2  -  11   10  10.351 -5.501 31.243  2.850   8.486 53.603
3      HA 16  -  29   14  -4.939 10.501 43.067  4.421   9.638 26.106
4      EA 30  -  38    9   1.644 -0.587 28.320 -6.521   2.466 49.636
5      EA 40  -  48    9  -2.034  3.077 51.419  7.046  -1.407 27.495
6      HA 49  -  63   15  14.153  4.628 26.624  6.094  13.691 44.115
7      EA 68  -  78   11   7.829 10.944 53.440 14.309  -9.307 32.671
8      EA 85  -  89    5  -1.626  2.711 27.563 -7.195   1.581 37.639
9      sheet     1     5   1       3     4      6       7
10     *v-tut-
11      *vuvuv
12       *tu-c
13        *v--
14         *v-
15          *-
16           *
```

Figure 8.4 ProSMoS: the meta-matrix of the `3br8` protein. Lines 2–8 is the list of SSEs inside the protein with their geometrical and biological properties, and lines 10–16 describe the relation matrix among the SSEs of `3br8`.

compact meta-matrix [25] that contains the relation among SSEs inside the protein (Figure 8.4). There are six types of relations, namely: *c*, *t*, -, *u*, *v*, and *N* [29]. With this representation, ProSMoS does not need any geometrical information to perform its topological motif search. At the end of preprocessing, ProSMoS generates an off-line database.

After completing the preprocessing phase, the user who wants to conduct a search makes an online query to the meta-matrices database. The user must describe the expected motif in the ProSMoS meta-matrix format. That is, the user must describe a model for all the relations between the SSEs in the expected motif. Then ProSMoS tries to match the sub-matrix in the meta-matrices database using the Weiss's graph path search algorithm.

8.3.2 Promotif

PROMOTIF [16] is an algorithm that retrieves structural motif information found inside a protein. The user supplies data in the PDB format, and the program produces a series of postscript files containing the identified structural elements. The structural blocks that can be recognized (as of PROMOTIF 2.0) are turns (both β and γ), disulphide bridges, helical geometry and interactions, β-strands and β-sheet topologies, β-bulges, β-hairpins, β–α–β motif, ψ-loops, and main-chain hydrogen-bonding patterns. It also produces a Ramachandran et al. plot [24] in color and a schematic diagram for recognized protein structures. Note that PROMOTIF does not compare multiple proteins, or even a pair of proteins together: it defines a topological model for all the recognizable motifs and looks directly for them in the target protein.

In PROMOTIF, SSE assignment is done using a slightly modified DSSP algorithm. The definition of a structural motif is based on a combination of criteria defined in the literature. Therefore, the algorithm and the features used during the searching process can vary depending on the consensus on a given motif definition. For example, the search for β-turns is based on the following criteria: first PROMOTIF looks for a peculiar topology, that is four consecutive residues (i, $i+1$, $i+2$, $i+3$) where $i+1$

and $i + 2$ are not helical, and the distance between the C_α atoms of residues i and $i + 3$ does not exceed 7 Å, as defined by Lewis et al. [20]; then, to identify the type of a given β-turn, PROMOTIF operates according to the works by Venkatachalaman [30] and Richardson [27], that is, it looks for specific combinations of the ϕ and ψ angles of the $(i + 1)$ and $(i + 2)$ residues.

8.3.3 Secondary-Structure Matching

Secondary-Structure Matching (SSM) [19] is an algorithm that assesses the similarity of a protein with respect to the collection of proteins in a given database. This tool aims to prune the computation time in protein alignment compared to the existing protein alignment web servers (e.g., DALI), with high-quality alignment results. To achieve this goal, it applies protein comparison at the SSE geometrical level, and then does the traditional pair residues alignment. The similarity results can be sorted by their various scores, for example, by the quality similarity value, the statistical significance, and the number of matched SSEs. SSM represents a protein as a graph with SSEs as vertices, and uses graph-matching algorithms to assess the similarity. The SSEs' assignment is calculated by the PROMOTIF algorithm (Section 8.3.2). Each vertex has two labels: (i) the type of the SSE, which can be helix with its subtype, or a β-strand; and (ii) the number of residues. Each edge is labeled with various information: the distance between the two vertices (ρ_{ij}), the angles between the edge and the two vertices (α_1, α_2), the torsion angles (α_3, α_4), the serial number of each vector in the chain, and the chain identifier as shown in Figure 8.5a. As we will see in Section 8.4.1.1, the geometric model of SSM has many resemblances to our own.

SSM offers five flexibility degrees in comparing vertices and edges, namely: highest, high, normal, low, and lowest. Each degree has certain defined tolerances for each label value. It also offers flexibility in deciding whether the connectivity between SSEs is taken into account in the matching process, by providing three options: (i) the connection is ignored; (ii) soft connectivity; and (iii) strict connectivity. Figure 8.5b shows an example of the flexibility of SSM in handling different sets of SSEs having the same graph representation.

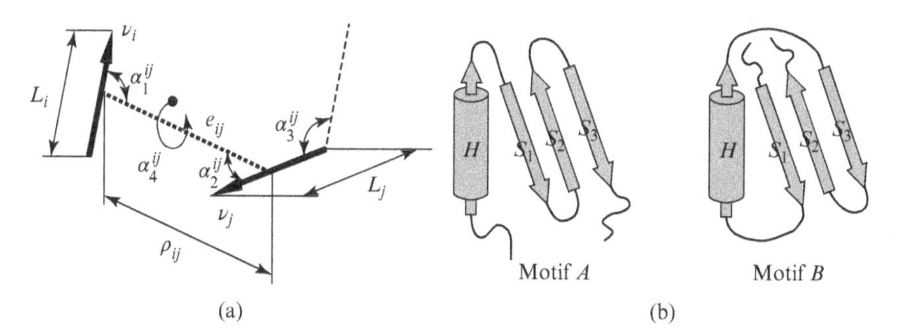

Figure 8.5 (a) The geometric model of an SSM. Vertices and edges are labeled. (b) Two SSEs with the same SSM graph representation but with different connections.

8.3.4 Multiple Structural Alignment by Secondary Structures

Multiple Structural Alignment by Secondary Structures (MASS) [12] is an efficient method for multiple alignments of protein structures and detection of structural motifs. MASS disregards the sequential order of the SSE. Thus, it can find non-sequential and even nontopological structural motifs. MASS is subset alignment detection: it does not require that all the input molecules be aligned. Rather, MASS is capable of detecting structural motifs shared only by a subset of the molecules.[3]

MASS is based on a two-level alignment process, in both secondary and primary structures. Its main feature is the capability to perform alignment of a whole set of proteins at the same time, as opposed to comparing a pair of proteins one at the time.

Going in more detail, the algorithm builds a reference frame, that is, a new basis, from each pair of SSEs in the search space, then tries to align the frames using geometrical hashing [1]. A reference frame for a pair is constructed as follows: (i) the first axis goes along the first SSE; (ii) the second axis goes along the distance vector, that is, a vector that connects the closest point between two SSEs; and (iii) the third axis is obtained through a cross product between the first and the second axes. After this step, the positions of all remaining SSEs are defined on the basis of this reference frame. The list of the reference frames is saved in a *Reference Table* (RT) that is the main input for the alignment phase of the algorithm.

MASS can detect possible multiple alignments. Groups of "similar" alignment are clustered together to obtain a smaller set of good global alignments. Then MASS uses a scoring system to detect the most promising one(s).

8.4 A NOVEL GEOMETRICAL APPROACH TO MOTIF RETRIEVAL

In this section, we will describe our novel approach to the retrieval and the identification of geometrically described structural motifs. All of these methods are based on the GHT [4].

8.4.1 Secondary Structures Cooccurrences

The GHT describes a geometric pattern by selecting an RP in a coordinate system local to the pattern and a set of feature elements (points, segments, etc.) that make up the pattern and that are geometrically referenced to this point. This structural description is encoded in an RT. The transform (voting rule) consists of producing, for each feature element in the *source space* (the object under scrutiny), a number of candidates for the RP of the geometric pattern; these candidates are collected in a*voting space*where evidence of cooccurrences of candidates is accumulated. An analysis of this space (search space) looking for peaks of accumulated cooccurrences allows identifying places where there exist instances of the looked-for geometric pattern. This method is often used in pattern recognition. Its computational complexity can be quite relevant since it depends on the number of feature elements used to describe

[3] http://bioinfo3d.cs.tau.ac.il/MASS/.

the geometric pattern (i.e., the cardinality of the RT), the number of feature elements present in the search space, and the resolution at which the voting space is quantized.

The *Secondary Structure Cooccurrences* (SSC) algorithm [6, 7] is a new approach for protein structural motifs extraction based on a modified GHT. Put in other words, the SSC is able to search for motifs (geometric patterns) of SSEs (feature elements) inside a given protein (search space). However, the GHT algorithm normally operates on a bitmap preprocessed by an edge detection operator that provides the gradient direction for each edge point of the image. The differences between a bitmap and a protein structure are apparent. First, the former is two-dimensional, while the latter is three-dimensional. Second, the basic feature element of a bitmap is clearly the pixel, while the primitive element of a protein structure can be considered as any of the following: a single atom, a residue, an SSE, or a tuple of SSEs. The modified GHT of the SSC adopts pairs of SSEs as the primitive element, represented as line segments (see Section 8.2.3.1). In this way, it is possible to be both orientation- and translation-agnostic during the search. The construction of the RT and the voting procedure of the GHT are still present, but are modified accordingly. Indeed, they are based on the *geometric* features of the SSEs' vector representation. A more detailed explanation is provided in the following sections, which is partially drawn from Reference [10].

8.4.1.1 Algorithm Description. The RT construction and the voting procedure are based on the fact that a pair of oriented line segments (not necessarily representing SSEs) in the 3D space can be identified regardless of roto-translations, by considering the following geometric parameters:

1. the lengths L_1 and L_2 of the segments;
2. the distance Md between their midpoints;
3. the angle ϕ between the two segments (considered as vectors applied to the origin);
4. the shortest distance Ma between the lines the two segments lie on [6, 7].

An alternative set of geometric parameters for identifying a pair of oriented line segments is proposed by Krissinel and Henrick, and is used in the SSM approach (Section 8.3.3). Their alternative method requires two additional parameters, but does not imply the calculation of the shortest distance between two lines, which is non-trivial.

Let $\vec{v}_1$ and $\vec{v}_2$ be the two segments considered as vectors, let $\vec{m}_1$ and $\vec{m}_2$ be the midpoints of the two segments, and let $\vec{d} = \vec{m}_2 - \vec{m}_1$. Regardless of how the two segments are identified, they can be used to construct a local coordinate system:

- the origin is the midpoint $\vec{m}_1$ of the first segment;
- the x axis is $\hat{v}_1$;
- the y axis is taken on the plane defined by $\hat{v}_1$ and $\hat{d}$, and orthonormal to $\hat{v}_1$;
- the z axis is taken orthonormal to the x and y axes.

Table 8.1 The reference table for the SSC algorithm

SSE Pair	SSE Types	Geometric Parameters	Relative RP Position
$(1,2)$	(t_1,t_2)	$(L_1,L_2,Md,\phi,Ma)_{1,2}$	$\vec{RP}'_{1,2}=T^{-1}_{1,2}\vec{RP}$
$(1,3)$	(t_1,t_3)	$(L_1,L_2,Md,\phi,Ma)_{1,3}$	$\vec{RP}'_{1,3}=T^{-1}_{1,3}\vec{RP}$
...	...	...	...
$(1,n)$	(t_1,t_n)	$(L_1,L_2,Md,\phi,Ma)_{1,n}$	$\vec{RP}'_{1,n}=T^{-1}_{1,n}\vec{RP}$
$(2,3)$	(t_2,t_3)	$(L_1,L_2,Md,\phi,Ma)_{2,3}$	$\vec{RP}'_{2,3}=T^{-1}_{2,3}\vec{RP}$
$(2,4)$	(t_2,t_4)	$(L_1,L_2,Md,\phi,Ma)_{2,4}$	$\vec{RP}'_{2,4}=T^{-1}_{2,4}\vec{RP}$
...	...	...	...
$(2,n)$	(t_2,t_n)	$(L_1,L_2,Md,\phi,Ma)_{2,n}$	$\vec{RP}'_{2,n}=T^{-1}_{2,n}\vec{RP}$
...	...	...	...
$(n-1,n)$	(t_{n-1},t_n)	$(L_1,L_2,Md,\phi,Ma)_{n-1,n}$	$\vec{RP}'_{n-1,n}=T^{-1}_{n-1,n}\vec{RP}$

Let us consider a structural motif of c secondary structures, that is, with a cardinality c, with the RP chosen as the centroid of the midpoints of its SSEs. The modified RT is constructed as in Table 8.1: for each SSE pair, the stored RP position is in the local coordinate system of the pair. T represents the transformation matrix from the local coordinate system to the standard basis. The two main differences between this RT formulation and a standard GHT reference table are the usage of pairs of feature elements instead of single elements, and the addition of the geometric parameters into the table.

Let us consider a protein secondary structure with N elements approximated by line segments. The voting procedure is detailed in Procedure 8.1.

Procedure 8.1 SSC voting procedure

Input:
The RT of the motif to be searched
A protein structure with N SSEs
Output: Vote points
for $i = 1$ **to** $N-1$ **do**
  for $j := i+1$ **to** N **do**
    Let (t_i, t_j) be the types of SSEs (i,j)
    Let (L_1, L_2, Md, ϕ, Ma) be the geometric parameters of (i,j)
    Let $\vec{m}_i$ be the midpoint of i
    Let $\{\hat{u}, \hat{v}, \hat{w}\}$ be the local coordinate system of (i,j)
    Let T be the transformation matrix from $\{\hat{u}, \hat{v}, \hat{w}\}$ to the std basis
    for each entry in the **RT do**
      if (t_i, t_j) and (L_1, L_2, Md, ϕ, Ma) match with the entry **then**
        Let $\vec{RP}'$ be the relative RP position stored in the RT entry
        Cast a vote in $\vec{m}_i + T\vec{RP}'$
      end if

end for
end for
end for

Assuming that operations with matrices and vectors can be done in constant time, the time complexity of the RT construction procedure is $O\left(\binom{c}{2}\right) = O(c^2)$, and the time complexity of the voting procedure is

$$O\left(\binom{c}{2} \cdot \binom{N}{2}\right) = O\left(c^2 N^2\right) \tag{8.1}$$

If we neglect the procedure to find vote clusters (Section 8.4.1.2), Equation 8.1 represents the overall time complexity of the SSC algorithm.

Obviously, perfect matches between the (L_1, L_2, Md, ϕ, Ma) parameters of the motif's SSE pairs and the protein's SSE pairs are practically impossible in real-world scenarios, and appropriate tolerances have to be used.

A cluster of $\binom{c}{2}$ "near" (according to a given tolerance) votes around a point $\vec{p}$ reveals an instance of the motif in the protein, with $\vec{p}$ being the (approximate) position of its RP.

To assess the performance and robustness of the SSC algorithm, the following questions should be answered:

1. What are the *minimum* tolerances for a motif to be found even if not identical with its prototype?
2. What are the *maximum* tolerances that do not give rise to false positives?

Answering these questions is not trivial. By modeling the "deformation" of a motif with the application of random errors to the endpoints of its SSEs, we could perform a formal error propagation analysis, and eventually find a closed-form expression relating the tolerances being used with the probability of recognizing the deformed motif [10].

8.4.1.2 *Finding Clusters of Vote Points.* In Section 8.4.1.1, the SSC algorithm has been described but the procedure to find a cluster of $\binom{c}{2}$ vote points, where c is the cardinality (number of SSEs) of the motif being searched, is still to be detailed.

The problem can be solved with two different approaches:

1. Restricting the voting space to a set of quantized values, as in typical implementations of the GHT for image bitmaps. The search for a cluster of vote points consists of finding a region in the quantized space where at least $\binom{c}{2}$ votes have been cast.
2. Allowing the coordinates of the vote points to take any value in R,[4] and employ a trivial algorithm that, for each vote point p, calculates the distance

[4]In practice, the values that the coordinates can take are limited to the resolution of the floating point representation.

between p and each other vote point, in order to establish if at least $\binom{c}{2}$ vote points (including p) lie in a sphere of given radius centered in p.

In the first approach, the *cluster tolerance*, that is, the maximum distance between any pair of vote points that constitute a cluster, is proportional to the quantization step. In the second approach, the cluster tolerance is given by twice the radius, that is, the diameter of the sphere.

The number of cast votes is $m = O(c^2N^2)$, where c is the motif cardinality and N is the number of SSEs in the protein. This upper bound for m is given by the worst case in which all the SSE pairs of the protein match against those of the motif. As a consequence, the second approach has a time complexity of $O(m^2) = O(c^4N^4)$. Cantoni and Mattia [8] suggest the use of a range tree to reduce the time complexity to $O(m\log^3 m) = O(c^2N^2\log^3 nN)$.

Both above-mentioned approaches work under the assumption that any cluster of $\binom{c}{2}$ vote points that are near to each other is "valid." In other words, they assume that any pair of points p_1 and p_2 is allowed to be included in the same cluster, provided that the Euclidean distance between p_1 and p_2 is below a given value.

Actually, this does not produce exact results. We are looking for clusters of $\binom{c}{2}$ vote points because $\binom{c}{2}$ is the number of SSE pairs in a motif of cardinality n and, consequently, the number of rows in the RT. Let us consider the situation in which *two* different SSE pairs of the protein match against the *same* row of the RT, and the corresponding vote points happen to be close enough to each other to be included in the same cluster. In this case, since the cluster contains exactly $\binom{c}{2}$ vote points and two of them are relative to the same RT row, there is at least one RT row that lacks a corresponding vote point in that cluster. This kind of situation may result in false positives.

Moreover, let us consider the situation in which an SSE pair of the protein matches against *two* different rows of the RT. If the corresponding vote points are grouped in the same cluster, then the cardinality of the motif being found may be smaller than the cardinality of the motif being searched, which makes no sense.

A possible solution for this issue is to model the problem as a search for a *clique* of size $\binom{n}{2}$ in an undirected graph, where the vertexes represent the vote points, and an edge between two vertexes is present if and only if the corresponding vote points are (i) close enough according to a given tolerance; (ii) not originated from the same RT row; and (iii) not originated from the same protein SSE pair.

The k-clique problem is known to be NP-complete [18], having a time complexity that is exponential with respect to k. This implies that, in order to perform accurate cluster finding in the voting space by this method, the overall time complexity of the SSC algorithm becomes exponential with respect to the cardinality of the searched motif.

In practice, this does not represent a serious problem because (i) the number of cast votes is usually in the order of tens or few thousands, and (ii) the votes are usually sparse enough to allow for a simple yet effective optimization—the graph is partitioned into a number of small connected components and the clique finder algorithm is applied separately to each of them.

Note, however, that if we increase the accepted tolerances for the geometric parameters, the number of votes increases, so the cost of the precise clique finder is unbearable. Fortunately, we were able to devise an approximated clique finder algorithm that uses domain information to find the cluster of vote points in polynomial time. A more detailed description of this polynomial algorithm is yet to be published and will not be further discussed here.

8.4.2 Cross Motif Search

Cross Motif Search (CMS) [11] is an extension of the SSC algorithm, which is able to enumerate all super-SSEs (groups of SSEs) that two given proteins have in common. The idea behind CMS is not to develop another structural alignment algorithm but to perform an exhaustive enumeration of the common structures between two proteins.

8.4.2.1 Brute-Force Approach. With CMS, every possible super-SSE of a given protein P_1, called *source protein*, is searched in another protein P_2, called *target protein*.

The simplest, straightforward approach is to actually enumerate each super-SSE of the source protein and use SSC to search it inside the target protein. The cardinality of the super-SSEs ranges from 3 to N_1, where N_1 is the number of SSEs in the source protein P_1. Note that super-SSEs of cardinality lower than 3 are nonsignificant; super-SSEs of cardinality 2 are SSE pairs, and super-SSEs of cardinality 1 are single SSEs.

The algorithm is detailed below. For the sake of simplicity, the pseudo-code neglects the fact that the positions in P_2 of the found super-SSEs instances have to be returned by the procedure.

Procedure 8.2 Brute-force CMS

```
Input: Source protein P1, target protein P2
Output: List L of common super-SSEs, initially empty
  N1 := number of SSEs in P1
  // Search super-SSEs of cardinalities 3 to N1
  for c := 3 to N1 do
      for each super-SSE M of cardinality c in P1 do
          if SSC finds M in P2 then
              Append M to L
          end if
      end for
  end for
```

This approach is effective but dreadfully inefficient. The number of super-SSEs of cardinality c is $\binom{N_1}{c}$. As a consequence, the time complexity of a single iteration of the outermost loop in Procedure 8.2 is

$$O\left(\binom{N_1}{c}\cdot c^2N_2^2\right)=O\left(\frac{N_1^c}{c!}\cdot c^2N_2^2\right)=O\left(\frac{cN_1^cN_2^2}{(c-1)!}\right)$$

where c is the current cardinality, N_1 the number of SSEs in the source protein, and N_2 the number of SSEs in the target protein.

The number of iterations of the outermost loop is exactly $N_1 - 2$, therefore the overall complexity of Procedure 8.2 is

$$O\left(\sum_{c=3}^{N_1}\binom{N_1}{c}\cdot c^2N_2^2\right)=O\left(N_2^2\sum_{c=3}^{N_1}c^2\binom{N_1}{c}\right) \tag{8.2}$$

Considering that $N_1 < c$, Equation 8.2 can be rewritten as follows:

$$O\left(N_2^2\sum_{c=3}^{N_1}c^2\binom{N_1}{c}\right)=O\left(N_1^2N_2^2\sum_{c=3}^{N_1}\binom{N_1}{c}\right)=O\left(N_1^2N_2^22^{N_1}\right)$$

Note that this upper bound does not take into account the procedure for finding clusters of vote points.

The search can be limited to super-SSEs of cardinality k, with $3 < k < N_1$. In this case, the overall time complexity is

$$O\left(N_1^2N_2^2\sum_{c=3}^{k}\binom{N_1}{c}\right).$$

8.4.2.2 *Greedy Approach.* The logic of the brute-force CMS is simple: each possible super-SSEs of the source protein P_1 should be searched inside the target protein P_2, with the cardinality of the super-SSEs ranging from 3 to $k < N_1$, where N_1 is the number of SSEs in P_1.

Each iteration of the outermost loop in Procedure 8.2 is completely independent of the previous ones. In other words, iteration c does not take any advantage of the results obtained by iteration $c-1$. By contrast, the greedy CMS approach, rather than constructing super-SSEs of cardinality c from scratch, constructs them by combining super-SSEs of cardinality $c-1$.

The key idea is that a super-SSE of cardinality c is present in a protein only if its subsets of cardinality $c-1$ are present as well.

For example, one cannot find the super-SSEs $\{1;2;3;4\}$ during iteration $c=4$ if one has not found $\{1;2;3\}$, $\{1;2;4\}$, $\{1;3;4\}$, and $\{2;3;4\}$ during iteration $c=3$. This is illustrated in Figure 8.6.

However, it is important to remark that this condition is necessary, but not sufficient. In fact, it is perfectly possible that all the subsets of a super-SSE M are present in the protein but do not give rise to M when considered as a whole, because they are independent of each other and possibly lie in completely different parts of the protein.

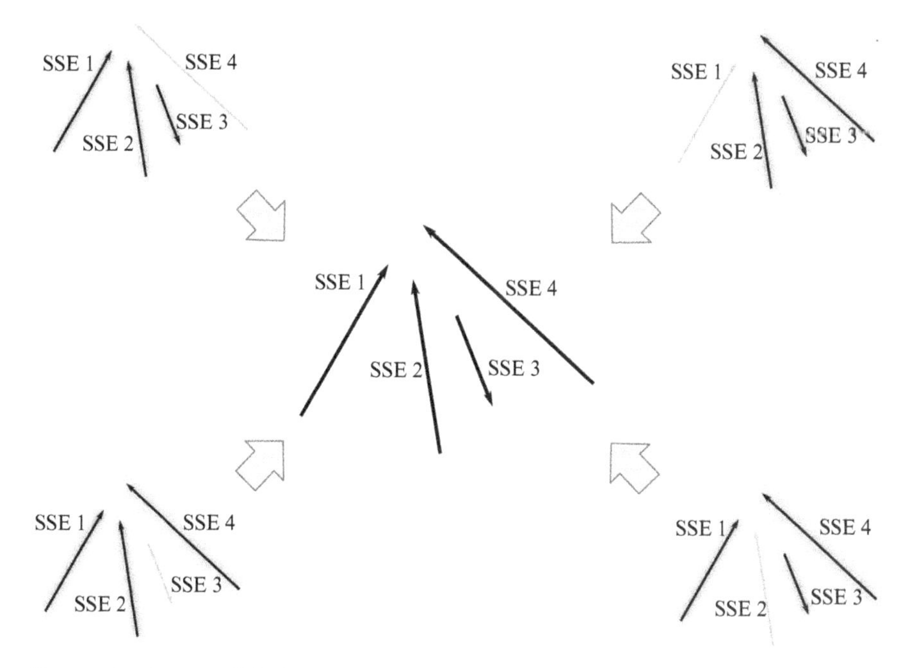

Figure 8.6 A super-SSE of cardinality c is present in a protein only if its subsets of cardinality $c - 1$ are present as well.

The greedy CMS algorithm [11] is detailed in Procedure 8.3. As in Procedure 8.2, the pseudo-code neglects the fact that the positions of the found super-SSEs instances have also to be returned.

Procedure 8.3 Greedy CMS

Input: Source protein P_1, target protein P_2
Output: List L of common super-SSEs, initially empty

```
N_1 := number of SSEs in P_1
Let S be a vector of sets, where the set S[c] contains the common super-SSEs of
cardinality c

// Search super-SSEs of cardinalities 3 to N_1
for c := 3 to N_1 do
    if c = 3 then
        for each super-SSE M of cardinality 3 in P_1 do
            if SSC finds M in P_2 then
                Add M to S[3]
            end if
        end for
    else
```

```
        // Break the loop if no super-SSEs of cardinality c − 1 have been found
        if S[c − 1] = ∅ then
            break
        end if
        for each M̄ in S[c − 1] do
            for i := 1 to N₁ do
                M ← M̄ ∪ {i}
                if ∀M′ : M′ ⊂ M ∧ |M′| = c − 1, M′ ∈ S[c − 1] then
                    if SSC finds M in P₂ then
                        Add M to S[c]
                    end if
                end if
            end for
        end for
    end if
    Append all the elements of S[c] to L
end for
```

Even if the CMS greedy variant has been designed precisely for improving the performance of the brute-force variant, its time complexity is slightly worse. The iteration $c = 3$ of the outermost loop is identical to the iteration $c = 3$ of the brute-force variant. Each subsequent iteration ($c > 3$) processes the results of the previous iteration that are $O\left(\binom{N_1}{c-1}\right)$; more precisely, each super-SSE found during the previous iteration is extended by adding another SSE of P_1: this introduces an N_1 factor.

Therefore, the overall time complexity of the CMS greedy variant is

$$O\left(\underbrace{\binom{N_1}{3} \cdot N_2^2}_{\substack{\text{First iteration}\\ c=3}} + \underbrace{\sum_{c=4}^{N_1} N_1 \binom{N_1}{c-1} \cdot c^2 N_2^2}_{\substack{\text{Subsequent iterations}\\ c>3}}\right) = O\left(N_1^3 N_2^2 2^{N_1}\right) \tag{8.3}$$

However, benchmarks (see Section 8.4.3.1) clearly show that, in the average case, the greedy variant brings a dramatic performance improvement.

8.4.3 Complete Cross Motif Search

Given a set K of proteins, for example, a subset of a protein database such as PDB or SCOP, *Complete Cross Motif Search* (CCMS) [11] applies CMS to all the $\binom{K}{2}$ possible pairs of proteins in K. The outcomes of these pairwise comparisons can be analyzed in order to identify known motifs inside the proteins in K and possibly discover entirely new motifs.

Table 8.2 CCMS Benchmarking Results

Algorithm Variant	Tolerances	Overall Execution Time (s)	Average Execution Time For a Comparison (s)
Brute-force	Strict	647.78	3.409
Brute-force	Loose	653.49	3.439
Greedy	Strict	15.00	0.079
Breedy	Loose	19.67	0.104

As of the end of 2013, a lot of data have been collected, but no in-depth analysis of the CCMS results on a large protein set has been performed: this kind of analysis is the main aim of forthcoming research. Preliminary benchmarks and results are presented in the next section.

8.4.3.1 Benchmark and Results. To assess the performance of the CMS algorithm, several CCMS runs have been performed on a small protein set[5] by using different configurations of CMS: brute-force versus greedy algorithm variant, and strict versus loose tolerances.

The testing machine is an Intel Core 2 Quad Q6600 2.40 GHz (4 cores, 4 threads) with 4 GB of RAM. As strict tolerances we used 0.5 Å for midpoints distance and line distance tolerances, 5° for angle tolerance, and 1.0 Å for the cluster radius tolerance. On the other hand, as loose tolerances, we used 1.0 Å for midpoints distance and line distance tolerances, 7° for angle tolerance, and 2.0 Å for the cluster radius tolerance.

The results are reported in Table 8.2. The maximum cardinality of the common super-SSEs being searched has been set to 5. Because the set contains 20 elements, each CCMS run comprises $\binom{20}{2} = 190$ pairwise comparisons. The fourth column of Table 8.2 reports the average execution time for a single pairwise comparison, calculated as the overall execution time (third column) divided by 190.

The benchmarks clearly show that the performance impact of the greedy variant is indeed remarkable, with an average speedup of 38 with respect to the brute-force variant (43 using strict tolerances, 33 using loose tolerances). The speedup of the greedy variant decreases as looser tolerances are employed. This happens because by using looser tolerances, CMS finds more results, and the more results are found during iteration c, the more super-SSEs have to be processed during iteration $c + 1$.

The protein set contains pairs of proteins that are very similar to one another. For example, PDB entries `2qmy`, `2qx8`, and `2qx9` are basically the same protein. As already said, CMS is not a structural alignment algorithm; it performs an exhaustive enumeration of the common super-SSEs between two proteins. When two proteins are very similar to one another, there is no point in using CMS to have a list of all the super-SSEs they have in common because the outcome would be an enormous number of not very useful results. For example, if we apply CMS to a pair of

[5]PDB entries: `1fnb`, `2prl`, `2pzn`, `2qmy`, `2qx8`, `2qx9`, `2z7g`, `2z7k`, `2z98`, `2z9b`, `2z9c`, `3c3u`, `3c94`, `3c95`, `3dc7`, `3dhp`, `3e9l`, `3e9o`, `4gcr`, `7fab`.

identical proteins having 100 SSEs each, the number of common super-SSEs being found would be almost 2^{100}, a 31-digit number! In fact, processing the pair (`2qmy`, `2qx9`) takes 2.050 s, whereas processing the pair (`7fab`, `2qx9`) takes only 0.079 s (greedy variant, loose tolerances, cardinality from 3 to 5), even if `7fab` contains 10 more SSEs than `2qmy`.

Table 8.3 reports a histogram of the common super-SSEs that have been found, grouped by cardinality. The greedy variant has been employed, and the maximum cardinality of the common super-SSEs being searched has been set to 10. The outstandingly high number of common structures being found is obviously biased by the choice of the protein set: by removing `2qx8` and `2qx9` from the set, we get very different results (see Table 8.4).

The conclusion we can draw is that CMS should be used with pairs of proteins that are not similar to one another (this can be assessed by using a structural

Table 8.3 Number of common super-SSEs found by CCMS in the testing data set, listed by their cardinality

Strict Tolerance		Loose Tolerance	
Cardinality	Number of Super-SSEs Found	Cardinality	Number of Super-SSEs Found
3	1026	3	2481
4	793	4	3489
5	383	5	3546
6	105	6	2642
7	12	7	1506
8	0	8	656
9	0	9	208
10	0	10	42

Table 8.4 Number of common super-SSEs found by CCMS removing `2qx8` and `2qx9` from the testing data set, listed by their cardinality

Strict Tolerance		Loose Tolerance	
Cardinality	Number of Super-SSEs Found	Cardinality	Number of Super-SSEs Found
3	170	3	491
4	79	4	431
5	17	5	237
6	1	6	68
7	0	7	9
8	0	8	0
9	0	9	0
10	0	10	0

alignment tool like PDBeFold or looking at their SCOP classification), in order to detect common structures that are not easy to identify with unaided eye, and are not considered significant by structural alignment tools.

8.5 IMPLEMENTATION NOTES

In this section, we briefly discuss the issues that arise in the implementation of the SSC and the CMS algorithms, with particular regard to the efficient greedy variant (Section 8.4.2.2). The analysis is motivated by the ever-increasing size of the PDB, currently in the order of 10^5 proteins (Section 8.2.4), and by the high asymptotical complexity of our algorithms.

Indeed, without careful implementation strategies and without a good distribution of the computation on the available hardware platforms, the algorithms can turn out to be practically unfeasible. The spectrum of hardware resources available for the computation is wide: one can map the algorithms on a modern desktop system, or even conceive some parallel implementation in a medium-size multiprocessor server or on a massively parallel system.

As always, any implementation is first checked for optimizations that can reduce the complexity and the runtime, almost independently of the hardware.

This section is mainly based on Reference [13].

8.5.1 Optimizations

Many serial optimizations have been devised to speed up both SSC and CCMS. Here we present the two that have the highest impact on performance. The first consists in avoiding waste in computation due to a symmetric analysis of pairs of SSEs thanks to a presorting scheme that introduces a linear, one-way only, processing phase on each pair; the second consists in reducing the computational cost for obtaining the geometric parameters of each couple of SSEs in the search space, by caching partial computations. The two optimizations are briefly described in the following sections.

8.5.1.1 Presorting. The reference table of the SSC algorithm (Section 8.4.1) creates an entry for each possible pair of SSEs in the source protein, including their associated permutations. Since the cardinality of the reference table is one of the main sources of complexity of the kernel of the algorithms, its reduction is of the utmost importance. To avoid missing any valid combination of SSEs, it is necessary to ensure that (i) either the voting rule is independent of the order of SSE in the pair or (ii) both the source space (source protein) and the target space (protein to be analyzed) list a given pair of SSEs in the same order. The first approach places a considerable load in the computation of the voting rule, while the latter is only feasible if one sorts the set of SSEs in both spaces. The real difference between the two approaches is that the voting rule is applied to each pair, while sorting can be done once and off-line. The added cost of sorting (whose asymptotic complexity is $O(N \log N)$) is only paid once and can be distributed over all SSC runs made by the external CCMS algorithm. With a rough approximation, we can say that presorting cuts the computational time almost by half. So sorting is by far the correct choice, provided it does not disrupt the process of motif identification. The RT shown in Table 8.1 employs presorting so that its cardinality is reduced.

A detailed discussion of the presorting process is beyond the scope of this contribution. It may suffice to state that one can sort the SSEs on the basis of a few different criteria (geometrical, biological, mixed-mode); unless the SSEs in a pair are very similar in their geometrical 3D shape, a simple length-based criterion turns out to be the easiest to implement. The approach can no longer suffice if the tolerances required in the method make the pair of SSEs no longer distinguishable. If this is the case, it is mandatory to use more advanced geometrical sorting keys. The interested reader is referred to Reference [13].

8.5.1.2 Caching. The second optimization tries to minimize the number of computations by avoiding duplicate operations. While analyzing a pair of SSEs, the SSC algorithm (Section 8.4.1) computes the geometrical parameters for the Hough transform, from scratch, each time it considers a pair; since the combinations of SSEs are quadratic in the size of the protein, the waste in computation is clear. A consistent speedup (20–40% depending on protein size) can be obtained by precomputing the geometric parameters of all pairs of SSEs in proteins in the search space and by storing them in a cache, to be used later by the actual SSC run. It can also be shown that a caching mechanism can be applied to the voting phase. So, a unified, shared cache can be designed in the contest of all optimizations. This cache is accessed very frequently, and it can introduce itself inefficiencies, especially in a parallel contest, unless it is properly designed. A very efficient memory scheme has been devised for the allocation of data items in the cache [13], in order to be able to use a linear addressing mode that leverages data spatial locality.

8.5.2 Parallel Approaches

An efficient serial implementation of the algorithms (SSC and especially CCMS) may result unfeasible, even with the greedy approach. Preliminary attempt to run a brute-force SSC (Section 8.4.2.1) on a serial machine with the target of finding all motifs (that is motifs with a cardinality c between 3 and N) without using the greedy approach aborted for memory space problems at $c = 4$, even with a small database of 20 proteins. Just to assess the problem, the first version of SSC took some 581 s to detect all structural similarities of a protein with $N = 46$ SSEs [13]. The CCMS problem poses even worse problems, as soon as the size of search domain increases: indeed, remember that the CCMS case compares each protein with its SSEs to all other protein in the selected set. A testing run on a very small set (20 proteins randomly chosen from the PDB) with the further limitation of maximum cardinality $c = 5$ shows that a brute-force approach takes approximately 2800 s [11]. The greedy approach dramatically reduces the runtime, as it was shown in Section 8.4.3.1.

Yet, by taking into account the current cardinality of the PDB, one has to carefully consider whether the optimized serial implementation is efficient enough to allow for repetitive runs, as it might be necessary when new proteins are added to the database.

However, it is possible to prune significantly the search space, remembering that, beyond the family level, proteins share a great deal of SSEs and tertiary structures as well (see Section 8.2.4). So it is quite evident that a CCMS problem within a family or even a super-family produces a wealth of matches that one can likely consider "useless," in that they bear little if none new insight into the structures of "familiar" protein.

As the SCOP classifies protein in approximately 1800 super-families, we can think to choose only one protein per super-families. However, a CCMS run on this reduced search domain still generates a tremendous amount of work. Increasing motif cardinality and especially tolerances adds further complexity so that the computations become quickly unfeasible on any desktop system, even a modern one.

Parallel processing is clearly a possible solution to manage this high computational load. To assess the way in which one can profitably use a parallel processing mode for the SSC and CCMS problems, one should consider the available parallelism (data versus task) at various granularities: the data set to be analyzed (number of proteins), the elements to be searched (motifs, themselves a set of SSEs), and the intrinsic processing mode of the algorithm.

As to the first dimension, data parallelism is clearly the dominant theme: either at the coarse level (proteins) or at the fine level (SSEs in a single protein), there is a huge space for data distribution among processing units. The workload, however, can be unevenly distributed, unless some care is taken in assigning tasks to the processors. Indeed in comparing two dissimilar proteins, the computations in the greedy algorithm can complete much earlier than in a pair of fairly similar proteins. At a finer level, that is at the level of a single protein and of its constituent SSEs, the Hough scheme underlining the SSC kernel does offer a further level of data decomposition. One can distribute the N SSEs of the proteins to multiple tasks that work independently on different motifs; however, the possible gain at this level is surely smaller than that at the coarse one since the computation is definitely shorter.

A set of experiments in parallel processing have been designed and partially carried out [13]. The approach is based on OpenMP since that mode easily matches the data parallel processing characteristics of SSC and CCMS. Preliminary results have been collected on a multicore desktop system, and others are being collected on a small-sized server (a *Xeon* machine with 32 cores). At this stage, no firm conclusion can be drawn on the scalability of the problem. However, preliminary testing indicates that an instance of CCMS on the 1800 proteins can consume in excess of 14 days on such a server, so a true parallel implementation requires a system with at least 1024 processing cores and a hybrid approach based on OpenMP and MPI.

8.6 CONCLUSIONS AND FUTURE WORK

In this chapter, we have described SSC, CMS, and CCMS, a set of novel algorithms for identifying structural motifs inside proteins. These algorithms adopt a geometric approach both in motif description and in motif search since they are based on a well-established structural method of pattern recognition, the GHT.

Comparing our approach to other existing algorithms, interesting similarities and differences can be highlighted. Most approaches described in the literature are either alignment tools (MASS) or are based on a topological motif description (ProSMoS, PROMOTIF). Our approach is based on a geometrical model for the secondary structure level; the same holds true for very few other approaches, such as SSM. A key point, however, is that instead of generating similarity values, as SSM, our approach retrieves the motifs inside proteins and identifies them in a 3D space, thus offering useful information on the inner structure of proteins. In terms of flexibility, even if still limited to a single pair of proteins at a time, our approach provides the user with the capability of applying a wide degree of tolerance to the geometrical comparison.

Being a very precise and flexible set of tools, our algorithms have relatively higher execution times when compared to other approaches (especially alignment tools). Note, however, that a direct comparison is unfair, as our tools have been designed with a very different goal in mind: to look previously unknown geometrical similarities in "unfamiliar" proteins, that is, proteins that are known to be quite dissimilar from the beginning.

Much effort has been directed to consistently reduce both computational complexity and execution times of our algorithms, as handling tolerances is especially demanding in terms of computations. The goal to carry out extensive cross searches, coupled with a wide range of geometrical tolerances, brought us to sophisticated parallel approaches. Preliminary small-scale experiments show that a hybrid OpenMP–MPI parallel processing paradigm is a good solution. Currently, the port of a CCMS problem on a very large subset of the SCOP database is being prepared for execution on an HPC facility available in the Italian Supercomputing Center at CINECA.[6]

Further work is undergoing to devise new algorithms that, while preserving the overall geometrical approach, are able to exploit biological considerations and a hierarchical search approach to prune the searching space and reach new levels of efficiency.

ACKNOWLEDGMENT

The authors thank Giacomo Drago for his significant contribution to the research work and to the writing of this chapter.

REFERENCES

1. Alesker V, Nussinov R, Wolfson HJ. Detection of non-topological motifs in protein structures. Protein Eng 1996;9(12):1103–1119. PMID: 9010924.
2. Anfinsen CB. The formation and stabilization of protein structure. Biochem J 1972;128: 737–749.

[6]http://www.hpc.cineca.it.

3. Anfinsen CB. Principles that govern the folding of protein chains. Science 1973;181 (4096):223–230.
4. Ballard DH. Generalizing the Hough transform to detect arbitrary shapes. Pattern Recognit 1981;13(2):111–122.
5. Branden C, Tooze J. *Introduction to Protein Structure*. 2nd ed. New York: Garland Science; 1999.
6. Cantoni V, Ferone A. 21st International Conference on Pattern Recognition, ICPR'12; Nov 11–15; Tsukuba, Japan: IEEE Computer Society Press; 2012.
7. Cantoni V, Ferone A. 9th International Meeting on Computational Intelligence Methods for Bioinformatics and Biostatistics, CIBB'12; July 12–14; Texas; 2012.
8. Cantoni V, Mattia E. Protein structure analysis through Hough Transform and Range Tree. Il Nuovo Cimento 2012;25(5):39–45.
9. Chothia C, Lesk AM. The relation between the divergence of sequence and structure in proteins. EMBO J 1986;5(4):823–826.
10. Drago G. Pattern recognition techniques for proteins structural motifs extraction: an optimization study [Master's thesis]. Pavia: Department of Electrical, Computer and Biomedical Engineering, University of Pavia; 2013.
11. Drago G, Ferretti M, Musci M. CCMS: a greedy approach to motif extraction. In: *New Trends in Image Analysis and Processing—ICIAP 2013*. Volume 8158, Lecture Notes in Computer Science. Berlin: Springer-Verlag; 2013. p 363–371.
12. Dror O, Benyamini H, Nussinov R, Wolfson H. MASS: multiple structural alignment by secondary structures. Bioinformatics 2003;19(1):i95–i104. PMID: 12855444.
13. Ferretti M, Musci M. Entire motifs search of secondary structures in proteins: a parallelization study. Proceedings of the International Workshop on Parallelism in Bioinformatics (Pbio 2013); Sep 17; Madrid. ACM Digital Library; 2013. p 199–204.
14. Frishman D, Argos P. Knowledge-based protein secondary structure assignment. Proteins 1995;23:566–579.
15. Hubbard TJP, Ailey B, Brenner SE, Murzin AG, Chothia C. SCOP: a structural classification of proteins database. Nucleic Acids Res 1999;27(1):254–256.
16. Hutchinson G, Thornton JM. PROMOTIF—a program to identify and analyze structural motifs in proteins. Protein Sci 1996;5:212–220.
17. Kabsch W, Sander C. Dictionary of protein secondary structure: pattern recognition of hydrogen-bonded and geometrical features. Biopolymers 1983;22(12):2577–2637.
18. Knuth DE. *The Art of Computer Programming*. Boston, MA: Addison-Wesley; 2005.
19. Krissinel E, Henrick K. Secondary-structure matching (SSM), a new tool for fast protein structure alignment in three dimensions. Acta Crystallogr D 2004;60:2256–2268.
20. Lewis PN, Momany FA, Scheraga HA. Chain reversals in proteins. Biochim Biophys Acta, Protein Struct 1973;303(2):211–229.
21. Majumdar I, Sri Krishna S, Grishin NV. PALSSE: a program to delineate linear secondary structural elements from protein structures. BMC Bioinformatics 2005;6(1):202. PMID: 16095538.
22. Orengo CA, Michie AD, Jones S, Jones D, Swindells MB, Thornton JM. CATH—a hierarchic classification of protein domain structures. Structure 1997;5(8):1093–1108.
23. Pauling L, Corey RB, Branson HR. The anatomy and taxonomy of protein structure. Proc Natl Acad Sci U S A 1951;37(4):205–211.

24. Ramachandran GN, Ramakrishnan CT, Sasisekharan V. Stereochemistry of polypeptide chain configurations. J Mol Biol 1963;7(1):95–99.
25. Richards FM, Kundrot CE. Identification of structural motifs from protein coordinate data: secondary structure and first-level supersecondary structure. Proteins 1988;3(2):71–84.
26. Richardson JS. β-sheet topology and the relatedness of proteins. Nature 1977;268 (5620):495–500.
27. Richardson JS. The anatomy and taxonomy of protein structure. Adv Protein Chem 1981; 34:167–339.
28. Shi S, Chitturi B, Grishin NV. ProSMoS server: a pattern-based search using interaction matrix representation of protein structures. Nucleic Acids Res 2009;37(Web Server issue):W526–531. PMID: 19420061.
29. Shi S, Zhong Y, Majumdar I, Sri Krishna S, Grishin NV. Searching for three-dimensional secondary structural patterns in proteins with ProSMoS. Bioinformatics 2007; 23(11):1331–1338.
30. Venkatachalam CM. Stereochemical criteria for polypeptides and proteins. V. Conformation of a system of three linked peptide units. Biopolymers 1968;6(10): 1425–1436.

CHAPTER 9

GENOME-WIDE SEARCH FOR PSEUDOKNOTTED NONCODING RNAs: A COMPARATIVE STUDY

Meghana Vasavada, Kevin Byron, Yang Song, and Jason T.L. Wang

Department of Computer Science, New Jersey Institute of Technology, Newark, USA

9.1 INTRODUCTION

Noncoding RNAs (ncRNAs) are functional RNA molecules, which are involved in many biological processes including gene regulation, chromosome replication, and RNA modification. Researchers have developed computational methods for searching genomes for ncRNAs [3, 6, 7]. Many of these methods are based on matching RNA secondary structures with genomic sequences using sequence–structure alignment. When the secondary structure is pseudoknot-free, the methods work well. However, challenges arise when the structure contains pseudoknots. To our knowledge, there are two tools, *RNA via Tree decOmPoSition* (RNATOPS) [5, 11] and *INFERence of RNA ALignment* (Infernal) [4, 9], which are capable of performing genome-wide pseudoknotted ncRNA search.

Pattern Recognition in Computational Molecular Biology: Techniques and Approaches,
First Edition. Edited by Mourad Elloumi, Costas S. Iliopoulos, Jason T. L. Wang, and Albert Y. Zomaya.

In this chapter, we conduct a comparative study on RNATOPS and Infernal. We perform genome-wide search for pseudoknotted ncRNAs in 13 genomes using the two tools. Through our experiments, we analyze and compare the performances of the two tools with respect to their ncRNA detection accuracies.

The rest of the chapter is organized as follows. In Section 9.2, we present some background materials related to this study. In Section 9.3, we describe the methodology used in the analysis of pseudoknotted ncRNA search. In Section 9.4, we present some experimental results and an interpretation of these results. Section 9.5 concludes the chapter.

9.2 BACKGROUND

In this section, we present basic concepts and terms concerning pseudoknotted ncRNAs, and give an introductory description of the two pseudoknotted ncRNA search tools, RNATOPS and Infernal.

9.2.1 Noncoding RNAs and Their Secondary Structures

An ncRNA is a functional RNA molecule that is not translated into proteins. The DNA sequence from which an ncRNA is transcribed is called ncRNA gene. There are many types of ncRNA genes, including *transfer* RNA (tRNA), *ribosomal* RNA (rRNA), *transfer-messenger* RNA (tm-RNA), *small nuclear* RNA (snRNA), *Ribonuclease* P (RNase P), *telomerase* RNA, and microRNA, among others.

Pseudoknot is an important structural motif in secondary structures of ncRNAs, which perform their functions through both their sequences and secondary structures. These secondary structures are defined by the interacting base pairs, namely guanine (G)-cytosine (C) and adenine (A)-uracil (U), which form three covalently hydrogen bonded and two hydrogen bonded base pairs, respectively. Base pairs in an RNA structure are approximately coplanar and are stacked into other base pairs. Such contiguous stacked base pairs are called *stems* or *helices*. Single-stranded subsequences bounded by base pairs are called *loops*. Single-stranded bases occurring at the end of a stem form a *hairpin loop*. Single-stranded bases occurring within a stem form a bulge or *bulge loop* if the single-stranded bases are on one side of the stem, or an *interior loop* if there are single-stranded bases interrupting both sides of the stem. Finally, there are *multibranched loops* or *junctions* from which three or more stems radiate.

A secondary structure can be represented in several ways, among which arc-based and dot–bracket representations are commonly used. In the arc-based representation, nucleotides and hydrogen bonds are represented by vertices and arcs, respectively. For pseudoknot-free secondary structures, all arcs are either nested or in parallel whereas for pseudoknotted structures crossover arcs indicate pseudoknots. In the dot–bracket notation, "." represents unpaired bases and matching brackets "<" and ">" indicate base-pairing nucleotides. The base-pairing nucleotides forming pseudoknots are represented by upper–lower case character pairs, such as A..a or B..b, or matching parentheses "(" and ")" as shown in Figure 9.1.

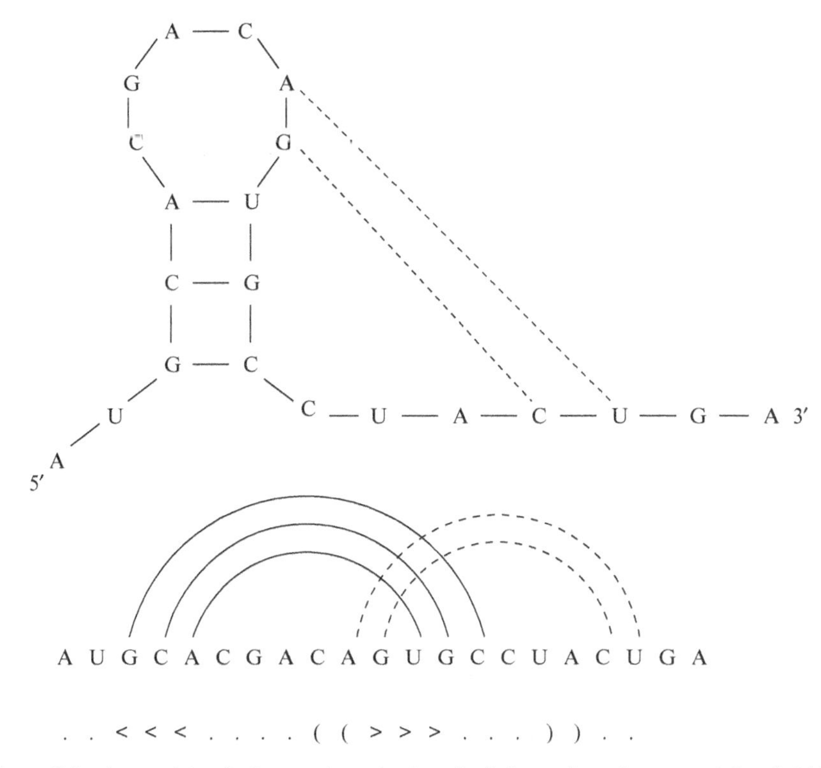

Figure 9.1 A pseudoknotted secondary structure (top), its arc-based representation (middle), and dot–bracket notation (bottom).

9.2.2 Pseudoknotted ncRNA Search Tools

There are two widely used tools, RNATOPS [5, 11] and Infernal [4, 9], which are capable of performing genome-wide search for pseudoknotted ncRNAs. RNATOPS is a profile-based RNA structure search program, which is able to detect RNA pseudoknots in genomes [11]. Infernal is a software package that applies stochastic context-free grammars to genome-wide ncRNA search. Infernal accepts a Stockholm alignment, which is a multiple alignment of RNA sequences together with the consensus secondary structure of the sequences, and creates a covariance model, which is a statistical representation of the Stockholm alignment [2, 8–10]. Infernal then uses the covariance model to identify ncRNAs in a genome by detecting RNA secondary structures on the genome that are similar to the covariance model. Figure 9.2 illustrates a hypothetical Stockholm alignment with a pseudoknot.

9.3 METHODOLOGY

Figure 9.3 illustrates the methodology used in our analysis of genome-wide pseudoknotted ncRNA search. The steps of the methodology are described in detail below.

```
O. princeps.1        AUGCACGUCGACGGACGUGCCUCGAGG
M. musculus.1        AUGCAAGUCGACGGACGUGCGUCGGGG
H. sapiens.1         AUGCACGUCGACGAACGUGCGCGGGGG
#=GC SS_cons         ...<..<<..>>.AA.>.<<.aa.>>.
```

Figure 9.2 A hypothetical Stockholm alignment with a pseudoknot.

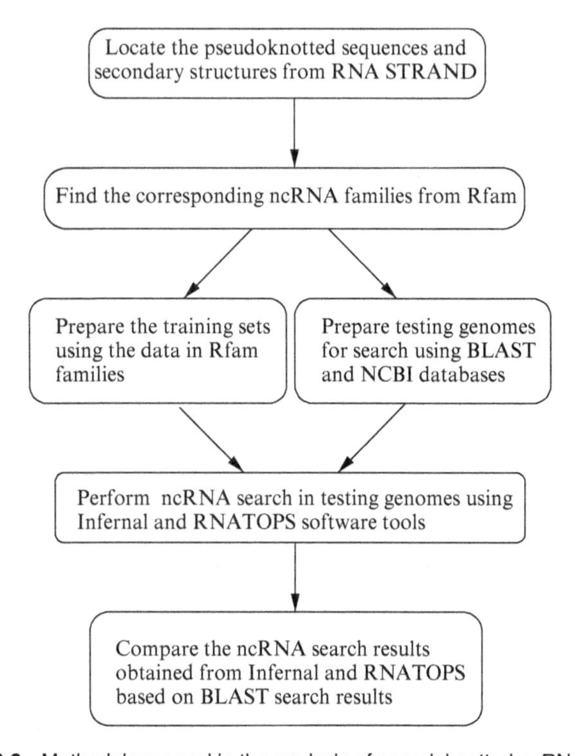

Figure 9.3 Methodology used in the analysis of pseudoknotted ncRNA search.

These steps include (i) preparation of training sets, (ii) preparation of testing sets (genomes), (iii) ncRNA search with RNATOPS, (iv) ncRNA search with Infernal, and (v) comparative analysis.

1. Preparation of training sets
   Thirteen pseudoknotted ncRNA sequences, out of 2333 sequences, from different organisms were located and retrieved from the RNA STRAND 2.0 database [1]. This database contains known RNA secondary structures of prokaryotic and eukaryotic organisms. We then found the ncRNA families (ncRNA types) corresponding to the retrieved pseudoknotted ncRNA sequences in the Rfam database [6]. The Rfam database contains a wide variety of ncRNA families (ncRNA types). Table 9.1 lists the 6 Rfam

Table 9.1 The six ncRNA families used in our study

Rfam ID	Rfam Accession	Description
RNaseP_bact_a	RF00010	Bacterial RNase P class A
RNaseP_bact_b	RF00011	Bacterial RNase P class B
RNaseP_arch	RF00373	Archaeal RNase P
IRES_Cripavirus	RF00458	Cripavirus *Internal Ribosome Entry Site* (IRES)
Intron_gpI	RF00028	Group I catalytic intron
HDV_ribozyme	RF00094	Hepatitis delta virus ribozyme

families corresponding to the 13 pseudoknotted ncRNA sequences. For each ncRNA family in Table 9.1, we randomly selected 10 seed sequences of roughly the same length from the family and obtained the multiple alignment of the sequences in Stockholm format together with the consensus structure of the sequences. Thus, we obtained six Stockholm alignments, which were used as training data.

2. Preparation of testing sets (genomes)
   The 13 pseudoknotted ncRNA sequences described in Step 1 above were then used in BLAST searches for homologous genome sequences containing the ncRNA sequences. We identified 13 genome sequences, each of which contained a pseudoknotted ncRNA sequence described in Step 1 above. Table 9.2 shows the 13 genomes, organisms they belong to, and their RNA STRAND IDs. Table 9.3 shows the GenBank accession numbers of the 13 genomes. These genome sequences were used as testing data. Note that training sets

Table 9.2 The 13 genomes used in our study

Organism	RNA Strand ID	Genome
Cricket paralysis virus	PDB_01157	*Cricket paralysis virus nonstructural polyprotein*
Plautia stali intestine virus	PDB_01129	*Plautia stali intestine virus RNA*
Pyrococcus horikoshii	ASE_00253	*Pyrococcus horikoshii OT3 DNA*
Pyrococcus furiosus	ASE_00249	*Pyrococcus furiosus COM1*
Thermotoga maritima	ASE_00424	*Thermotoga maritima MSB8*
Deinococcus radiodurans	ASE_00101	*Deinococcus radiodurans strain R1*
Thermus thermophilus	ASE_00428	*Thermus thermophilus HB27*
Methanococcus jannaschii	ASE_00196	*Methanococcus jannaschii DSM 2661*
Archaeoglobus fulgidus	ASE_00009	*Archaeoglobus fulgidus strain DSM 4304*
Bacillus subtilis	ASE_00032	*Bacillus subtilis BEST7003*
Tetrahymena thermophila	PDB_00082	*Tetrahymena thermophila strain ATCC 30382 18S ribosomal RNA*
Azoarcus sp. BH72	PDB_00908	*Azoarcus sp. BH72*
Hepatitis Delta Virus	PDB_00335	*Hepatitis delta virus isolate 59045-CAR delta antigen gene*

Table 9.3 The GenBank accession numbers of the 13 genomes

Genome	GenBank Acession
Cricket paralysis virus nonstructural polyprotein	AF218039.1
Plautia stali intestine virus RNA	AB006531.1
Pyrococcus horikoshii OT3 DNA	BA000001.2
Pyrococcus furiosus COM1	CP003685.1
Thermotoga maritima MSB8	AE000512.1
Deinococcus radiodurans, strain R1	AE000513.1
Thermus thermophilus HB27	AE017221.1
Methanocaldococcus jannaschii DSM 2661	L77117.1
Archaeoglobus fulgidus strain DSM 4304	AE000782.1
Bacillus subtilis BEST7003	AP012496.1
Tetrahymena thermophila strain ATCC 30382 18S ribosomal RNA	JN547815.1
Azoarcus sp. BH72	AM406670.1
Hepatitis delta virus isolate 59045-CAR delta antigen gene	JX888110.1

and testing sets were separated. That is, the genomes from which the seed sequences (training data) were taken differed from the 13 genomes in the testing sets.

3. ncRNA search with RNATOPS
   Each Stockholm alignment was converted into the *Pasta* formatted data [11] using the *RNApasta* program in RNATOPS. The *Pasta* formatted training data and the genome sequences (in FASTA format) were uploaded to the RNATOPS-W web server. The option of "automatic filter," and output options of "filtered" and "whole structure search" were chosen. Once a query is submitted, the web server provides a reference number that can be used to access the result obtained from the server.
4. ncRNA search with Infernal
   The *Covariance Model* (CM) of each training Stockholm alignment was built using the "cmbuild" function in Infernal v1.0.2. The model of the structure profile obtained was saved in a "cm" file. Thus, the model for each of the six ncRNA families was created. The "cmsearch" function in the Infernal tool was then used to search a given genome based on each covariance model constructed to detect ncRNA sequences in the genome that are similar to the covariance model.
5. Comparative analysis
   The ncRNA search results obtained from Infernal and RNATOPS were compared and verified with those from BLAST. Note that BLAST is used as a validation tool here. Since the 13 pseudoknotted ncRNA sequences of interest are known in advance (cf. Step 1 above), we can use BLAST to find the positions of the sequences on the genomes that contain the sequences through BLAST's local alignment functionality. The positions identified by BLAST are referred to as *true positions* of the ncRNA sequences in genomes. On

the other hand, Infernal and RNATOPS are predictive tools, which employ covariance models and *Pasta* formatted training data, respectively, to predict the positions of pseudoknotted ncRNA sequences in a genome. In general, when the positions of pseudoknotted ncRNA sequences are not known in advance, we can use these two tools to predict the positions of the pseudoknotted ncRNA sequences in a genome. The purpose of this study is to see which predictive tool is more accurate compared with the search results from BLAST.

9.4 RESULTS AND INTERPRETATION

It was observed that RNATOPS detected ncRNAs with less computation time than Infernal due to the filter functions incorporated in the RNATOPS tool. On the other hand, Infernal yields more hits than RNATOPS.

Table 9.4 summarizes the search results from RNATOPS. It can be seen from Table 9.4 that RNATOPS had three hits where the positions of the ncRNAs detected by RNATOPS are close to the true positions found by BLAST. For example, RNATOPS detected an ncRNA of IRES_Cripavirus (Rfam Accession: RF00458) in the genome of *Cricket Paralysis virus* (GenBank Accession: AF218039.1) at the positions 6027–6215, compared to the true positions 6030–6219 found by BLAST.

On the other hand, with the filter functions, RNATOPS detected six ncRNAs whose positions are quite different from the true positions found by BLAST. These included RNaseP_arch (Rfam accession: RF00373) ncRNAs in the genomes of *Pyrococcus horikoshii* (GenBank accession: BA000001.2), *Pyrococcus furiosus* (GenBank accession: CP003685.1), *Methanococcus jannaschii* (GenBank accession: L77117.1), and *Archaeoglobus fulgidus* (GenBank accession: AE000782.1), and

Table 9.4 Results of ncRNA search with RNATOPS

GenBank Accession	Rfam Accession	True Positions	RNATOPS Positions
AF218039.1	RF00458	6030–6219	6027–6215
AB006531.1	RF00458	6003–6146	6002–6191
BA000001.2	RF00373	168172–168500	Filtered: 168471–168496
CP003685.1	RF00373	1382446–1382117	Filtered: 527680–527705
L77117.1	RF00373	643507–643746	Filtered: 643428–643762
AE000782.1	RF00373	86275–86047	Filtered: 2092331–2092356
AE000512.1	RF00010	752885–753222	NIL
AE000513.1	RF00010	904571–905017	NIL
AE017221.1	RF00010	695567–695957	NIL
AP012496.1	RF00011	2163512–2163112	NIL
JN547815.1	RF00028	2960–3116	Filtered: 2831–3596
AM406670.1	RF00028	3559264–3559459	Filtered: 2141126–2142199
JX888110.1	RF00094	687–731	684–768

Table 9.5 Results of ncRNA search with Infernal

GenBank Accession	Rfam Accession	True Positions	Infernal Positions
AF218039.1	RF00458	6030–6219	+ 6028–6228
AB006531.1	RF00458	6003–6146	+ 6003–6204
BA000001.2	RF00373	168172–168500	+ 168175–168497
CP003685.1	RF00373	1382446–1382117	– 1382443–1382120
L77117.1	RF00373	643507–643746	+ 643504–643761
AE000782.1	RF00373	86275–86047	– 86278–86044
AE000512.1	RF00010	752885–753222	+ 752885–753222
AE000513.1	RF00010	904571–905017	+ 904571–905017
AE017221.1	RF00010	695567–695957	+ 695570–695939
AP012496.1	RF00011	2163512–2163112	– 2163505–2163125
JN547815.1	RF00028	2960–3116	+ 2848–3189
AM406670.1	RF00028	3559264–3559459	+ 3559255–3559454
JX888110.1	RF00094	687–731	+ 685–773

Intron_gpI (Rfam accession: RF00028) ncRNAs in the genomes of *Tetrahymena thermophila* (GenBank accession: JN547815.1) and *Azoarcus* (GenBank accession: AM406670.1).

It was also observed that RNATOPS failed to detect four ncRNAs, which included RNaseP_bact_a (Rfam accession: RF00010) ncRNAs in the genomes of *Thermotoga maritima* (GenBank accession: AE000512.1), *Deinococcus radiodurans* (GenBank accession: AE000513.1), *Thermus thermophilus* (GenBank accession: AE017221.1), and RNaseP_bact_b (Rfam accession: RF00011) ncRNA in the genome of *Bacillus subtilis* (GenBank accession: AP012496.1).

Table 9.5 summarizes the search results from Infernal. The "+" sign in the column of "Infernal positions" means the plus strand of a genome, while the "–" sign means the minus strand of a genome. It can be seen from Table 9.5 that Infernal successfully detected pseudoknotted ncRNAs in all of the 13 genomes analyzed here. However, when the alignments produced by Infernal were carefully examined, it was observed that the Infernal tool generated structural misalignments between the consensus structure of a covariance model and the detected RNA structure in a genome. These structural misalignments occurred when a single base represented by a dot "." was aligned with (part of) a stem represented by "<" or ">".

9.5 CONCLUSION

Analysis of ncRNA search in 13 genomes using six different pseudoknotted ncRNA families in Rfam was carried out using RNATOPS and Infernal software tools. Based on our experimental results, we conclude that Infernal successfully detects pseudoknotted ncRNAs in all the genomes, although the tool generates structural misalignments. On the other hand, RNATOPS is faster, but is unable to detect some of the pseudoknotted ncRNAs in the genomes. It would be useful to develop new tools capable of both quickly detecting all pseudoknotted ncRNAs in genomes and producing correct structural alignments. This would require the design of new filters and RNA sequence–structure alignment algorithms.

REFERENCES

1. Andronescu M, Bereg V, Hoos HH, Condon A. RNA STRAND: the RNA secondary structure and statistical analysis database. BMC Bioinformatics 2008;9:340.
2. Durbin R, Eddy SR, Krogh A, Mitchison G. *Biological Sequence Analysis: Probabilistic Models of Proteins and Nucleic Acids*. Cambridge: Cambridge University Press; 1998. p 233–295.
3. Eddy SR. A memory-efficient dynamic programming algorithm for optimal alignment of a sequence to an RNA secondary structure. BMC Bioinformatics 2002;3:18.
4. Eddy SR, Durbin R. RNA sequence analysis using covariance models. Nucleic Acids Res 1994;22:2079–2088.
5. Huang Z, Wu Y, Robertson J, Feng L, Malmberg RL, Cai L. Fast and accurate search for non-coding RNA pseudoknot structures in genomes. Bioinformatics 2008;24(20):2281–2287.
6. Jones SG, Moxon S, Marshall M, Khanna A, Eddy SR, Bateman A. Rfam: annotating non-coding RNAs in complete genomes. Nucleic Acids Res 2005;33:D121–D124.
7. Khaladkar M, Liu J, Wen D, Wang JTL, Tian B. Mining small RNA structure elements in untranslated regions of human and mouse mRNAs using structure-based alignment. BMC Genomics 2008;9:189.
8. Nawrocki EP, Eddy SR. Query-dependent banding (QDB) for faster RNA similarity searches. PLoS Comput Biol 2007;3(e56):3.
9. Nawrocki EP, Kolbe D, Eddy SR. Infernal User's Guide: Sequence analysis using profiles of RNA secondary structure consensus. Available at: http://infernal.janelia.org. Accessed 2015 Aug 4.
10. Rivas E, Eddy SR. A dynamic programming algorithm for RNA structure prediction including pseudoknots. J Mol Biol 1999;285:2053–2068.
11. Wang Y, Huang Z, Wu Y, Malmberg RL, Cai L. RNATOPS-W: a web server for RNA structure searches of genomes. Bioinformatics 2009;25(8):1080–1081.

PART III

PATTERN RECOGNITION IN TERTIARY STRUCTURES

CHAPTER 10

MOTIF DISCOVERY IN PROTEIN 3D-STRUCTURES USING GRAPH MINING TECHNIQUES

Wajdi Dhifli[1,2,3] and Engelbert Mephu Nguifo[1,2]

[1]*LIMOS – Blaise Pascal University – Clermont University, Clermont-Ferrand, France*
[2]*LIMOS – CNRS UMR, Aubière, France*
[3]*Department of Computer Science, University of Quebec At Montreal, Downtown station, Montreal (Quebec), Canada*

10.1 INTRODUCTION

Proteins are biological macromolecules that play crucial roles in almost every biological process. They are responsible in one form or another for a variety of physiological functions. Proteins are made of complex structures composed of a number of amino acids that are interconnected in space. The amino acids themselves are composed of a set of interconnected atoms. The advances in both computational and biological techniques of protein studies have yielded enormous online databases. However, for comparative studies, the complexity of protein structures requires adequate bioinformatics methods to mine these databases.

For many years, proteins have been mainly studied on the basis of their primary structures. This is because a primary structure (represented by string of characters)

Pattern Recognition in Computational Molecular Biology: Techniques and Approaches,
First Edition. Edited by Mourad Elloumi, Costas S. Iliopoulos, Jason T. L. Wang, and Albert Y. Zomaya.

can be analyzed by extensive and efficient string and sequence computational concepts. In addition, there has been a huge gap between the number of unique protein primary structures in databases and that of protein tertiary structures. However, the tertiary structure that represents the native form that controls basic protein function could be more relevant for most studies. In addition to embedding the primary structure, it also contains the connections between distant amino acids. Tertiary structures capture homology between proteins that are distantly related in evolution. With the availability of more protein 3D-structures owing to techniques such as X-ray crystallography, increasing efforts have been devoted to directly deal with them.

The most known online database allowing the acquisition of protein structures in computer readable format is the *Protein Data Bank* (PDB) [4]. It represents an online repository of large biological molecules, including proteins and nucleic acids. PDB was created in 1971 at Brookhaven National Laboratory and is continuously expanding. Toward the end of October 2014, it contained already 104,125 structures, with the repository getting updated every week. Protein 3D-structures are available in the PDB website in a special data format also called PDB, which is simply a textual format describing the coordinates of atoms of a molecule in 3D-space. With the high complexity of protein 3D-structures and their availability in large amounts in online databases and in computer-analyzable formats rises an urgent need to exploit automatic techniques from data mining and pattern recognition to study them.

Pattern mining is one of the most important tasks in pattern recognition. The main purpose behind pattern mining is to find hidden relations and behaviors in data in order to better analyze them and to help in understanding the observed phenomena. Pattern mining has been extensively addressed during the last two decades for different types of patterns including association rules and itemsets. In the recent years, many efforts have been devoted to mine patterns from graph data. In this context, the combinatorial nature of graphs makes the search space exponential and usually leads to generating a massive number of patterns. In the literature, this problem is referred to as *information overload* [1, 66]. Moreover, in case the generated pattern are used as descriptors in further analysis and mining tasks, their huge number can cause a deterioration of the task quality. This problem is known as the *curse of dimensionality* [1, 66]. One of the widest domains of application of pattern mining is motif discovery from protein 3D-structures. The discovered motifs are very important in characterizing groups of protein tertiary structures and can be used as descriptors for further analysis such as classification and clustering of proteins [27, 54], functional sites discovery [14, 51], and fingerprint identification [58, 68].

The main goal of this chapter is to bridge the gap between pattern mining in pattern recognition and motif discovery from protein 3D-structures. It precisely shows how to use graph mining techniques to mine protein 3D-structure motifs. The rest of the chapter is organized as follows: Section 10.2 reviews existing techniques to transform protein tertiary structures into graphs that enable exploiting the wide range of graph mining techniques to study them. Section 10.3 presents graph mining. Sections 10.4 and 10.5 review existing techniques for frequent subgraph mining. Section 10.6 presents and discusses feature selection and existing subgraph selection approaches as a way to tackle the problem of information overload encountered in frequent subgraph mining.

10.2 FROM PROTEIN 3D-STRUCTURES TO PROTEIN GRAPHS

A crucial step in the computational study of protein 3D-structures is to look for a convenient representation of their spatial conformation. Since a protein can be seen as a set of connected elements (amino acids and atoms), it can then be easily transformed into a graph where the elements are the nodes and the connections are the edges.

10.2.1 Parsing Protein 3D-Structures into Graphs

In most existing works, a protein is transformed into a graph, as previously explained. There exist different transformation techniques.

10.2.1.1 Transformation Techniques. Some approaches have been proposed in the literature for transforming protein 3D-structures into graphs of amino acids [52]. These approaches use different techniques. In the following, we present the most known approaches. In all these approaches, nodes of the graphs represent the amino acids. However, they differ in the way the edges are considered in an attempt to reflect the truly existing interactions. Some of them express the edges by the strength of interaction between side chains of the amino acids, while others express the edges based on the distance between pairs of amino acids.

- **Triangulation**: Triangulation is a form of tessellation. It is used to transform an object, represented by a set of points in a plane or in a 3D-space, into a set of triangles. It is possible to have multiple triangulations for the same object. The Delaunay triangulation [12] is a special way of triangulation. It was used to build protein graphs in several works [6, 21, 57]. The main idea is to consider the amino acids as a set of points in space, then to iteratively try to create tetrahedrons such that no point is inside the *circumsphere*[1] of any *tetrahedron*,[2] that is, these are empty spheres (see the example in Figure 10.1).
- **Main Atom**: This is the technique mainly used in the literature. The main idea is to abstract each amino acid in only one of its main atoms, M_A. This main atom, M_A, can be real, like the α carbon (C_α) or the β carbon (C_β), or fictive like the amino acid centroid or the side chain centroid [21, 42]. Two amino acids u and v are linked by an edge $e(u, v) = 1$, if the Euclidean distance between their two main atoms $\Delta(M_A(u), M_A(v))$ is below a threshold distance δ. Formally,

$$e(u,v) = \begin{cases} 1, & \text{if } \Delta(M_A(u), M_A(v)) \leq \delta \\ 0, & \text{otherwise} \end{cases} \tag{10.1}$$

[1]The *circumsphere* of a polyhedron is a sphere that contains the polyhedron and touches each of its vertices.
[2]A *tetrahedron* is a polyhedron composed of four triangular faces that meet at each corner.

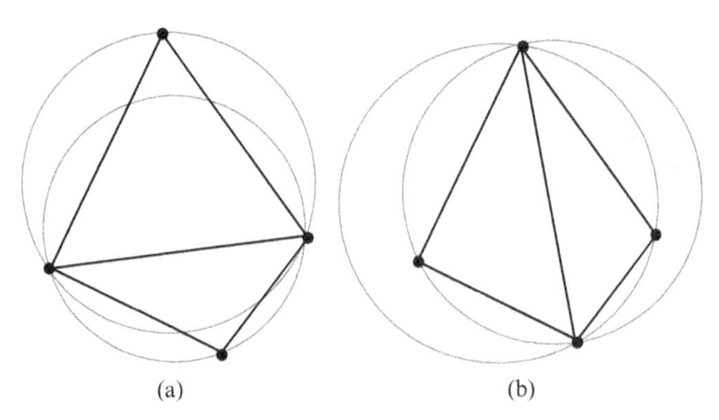

Figure 10.1 Triangulation example in a 2D-space. (a) Triangulation meets the Delaunay condition. (b) Triangulation does not meet the Delaunay condition.

Most studies use the C_α atom with $\delta \geq 7$ Å usually [21] because of its presence in every amino acid. In addition, it is an atom near the middle of the amino acid. It defines the overall shape of the protein conformation.

- **All Atoms**: Instead of using one atom per amino acid, the All Atoms technique [52] considers the distances between all the atoms (A_A) of the amino acids. Two nodes representing the amino acids u and v are linked by an edge $e(u, v)$:

$$e(u, v) = \begin{cases} 1, & \text{if } \Delta(A_A(u), A_A(v)) \leq \delta \\ 0, & \text{otherwise} \end{cases} \tag{10.2}$$

10.2.1.2 Discussion. The correctness of each technique is measured by its ability to reflect in the edges of the graph it generates, the actually existing connections between the amino acids of the protein. The Delaunay triangulation technique has two main drawbacks because of the empty circumspheres condition. First, it can produce many false links between very distant nodes in the protein, especially at the surface of the protein where the circumspheres easily get out of the cloud of atoms. Second, the empty sphere condition does not allow a node to make connection with any other node outside of its tetrahedron sphere. This makes it omit many edges even in the presence of real interactions.

The Main Atom technique suffers a drawback. Since it abstracts the amino acids into one main atom, it may omit possible edges between other atoms in the amino acids that are closer than their main atoms. Moreover, in case the centroids of the amino acids are considered as the main atoms, there may be two problems. In the case where the amino acids are big, if their centroids are farther than the given distance threshold, they will be considered with no links, while a real connection could be established between other close atoms (other than the centroids). In the case where the amino acids are small, if the distance between their centroids is smaller than the given distance threshold, then they will be considered as connected while in reality they may be disconnected. The All Atoms technique overcomes both limitations by theoretically considering the distance between all the atoms in the amino acids; this highly increases the runtime and complexity of the technique. However, the authors propose some heuristics to alleviate the complexity. For instance, they consider only the distance between the centroids of the side chains to decide whether their amino acids are connected or not, without regard to their chemical properties. This reduces the runtime but it may engender many false edges.

A real example of three protein tertiary structures, an all-α protein, an all-β protein, and an α–β protein,[3] transformed into graphs are respectively illustrated in the Figures 10.2–10.4. Although using such a representation may omit some edges and may contain some false ones, it opens new challenges and perspectives by allowing the use of graph mining techniques to study proteins. Besides, it enables extending previous works from the literature on primary sequences such as pattern discovery [16, 31].

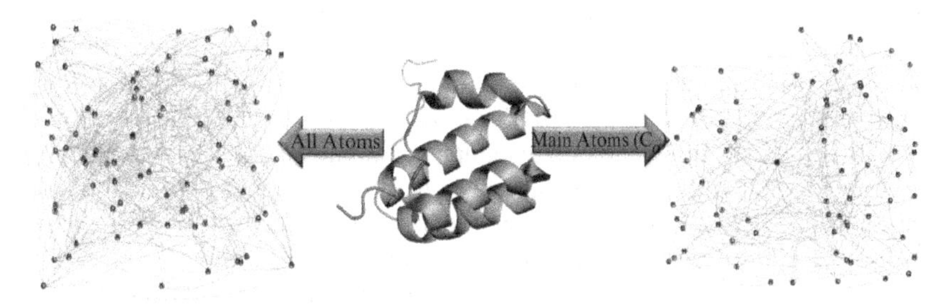

Figure 10.2 Example of an all-α protein 3D-structure (lipid-binding protein, PDB identifier: 2JW2) transformed into graphs of amino acids using the All Atoms technique (81 nodes and 562 edges), then using the Main Atom technique with C_α as the main atom (81 nodes and 276 edges). Both techniques are used with a distance threshold δ of 7 Å.

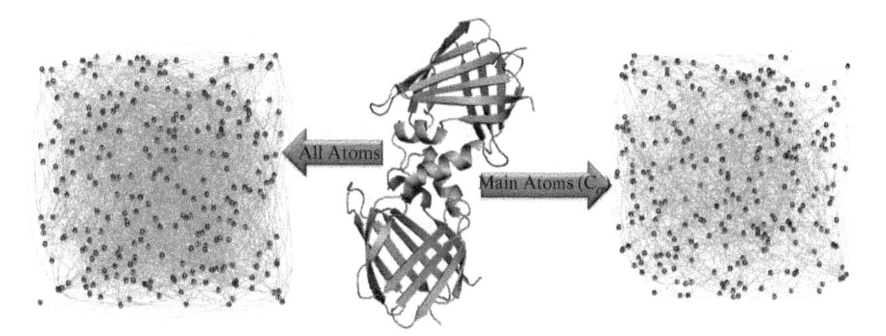

Figure 10.3 Example of an all-β protein 3D-structure (for retinoic acid transport, PDB identifier: 1CBI) transformed into graphs of amino acids using the All Atoms technique (272 nodes and 2168 edges), then using the Main Atom technique with C_α as the main atom (272 nodes and 1106 edges). Both techniques are used with a distance threshold δ of 7 Å.

[3]According to the structural classification of SCOP [2].

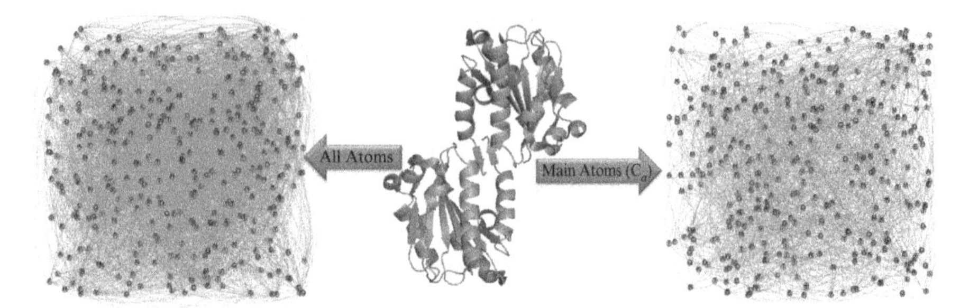

Figure 10.4 Example of an α–β protein 3D-structure (for cell adhesion, PDB identifier: 3BQN) transformed into graphs of amino acids using the All Atoms technique (364 nodes and 3306 edges), then using the Main Atom technique with C_α as the main atom (364 nodes and 1423 edges). Both techniques are used with a distance threshold δ of 7 Å.

10.2.1.3 Protein Graph Repository. The previously mentioned transformation methods from protein 3D-structures to graphs were implemented in a JAR (Java ARchive) file. This file is used as the core program of a web repository for protein graphs and is termed the *Protein Graph Repository* (PGR) [13]. PGR accepts protein PDB files as input and outputs graphs of amino acids under several formats. See Reference [13] for further details. It is worth mentioning that there also exist other tools and online web servers to help in performing analysis on protein tertiary structures. For instance, *ProFunc* [36] is a server for protein function prediction. It performs a set of analyses over protein tertiary structures including scans against the PDB and motif databases, determination of surface morphology and conserved residues, and potential ligand-binding sites.

The preceding part of the chapter showed that proteins are complex data that need to be analyzed and mined using automatic methods because of the increasingly large amounts of available data. It reviewed existing methods to transform protein 3D-structures into graphs. Such representation allows to fully exploit the potential of graph mining and pattern-recognition algorithms to perform complex studies. The next part of the chapter focuses on the discovery of important substructures in protein families that could be performed through frequent subgraph mining, pattern recognition, and motif discovery. Indeed, it reviews the problem of subgraph discovery as a way to mine motifs from protein 3D-structures in a graph perspective. Feature selection for graph patterns is also introduced as a way to tackle the information overload because of the high number of retrieved frequent subgraphs, which is the case for protein tertiary structures.

10.3 GRAPH MINING

Graphs are one of the most powerful structures to model complex data [11]. In fact, any data composed of entities having relationships can be represented by a graph. The entities will be seen as the graph nodes and the relationships as the graph edges.

Graph-processing algorithms are now becoming ubiquitous because of the increasingly lower costs of storage devices and the availability of high processing power. Graph-based algorithms are increasingly used in the modeling and the analysis of many real-world applications such as the World Wide Web, blogs, cell phone communications, XML documents, and even electronic circuits. In chemoinformatics [5, 8, 49, 62] and bioinformatics [6, 14, 21], graphs are used to model various types of molecules and biological data such as chemical compounds, gene and metabolic networks, protein structures, and protein–protein interaction networks.

One of the most powerful techniques to analyze and study graph data is to look for interesting subgraphs. Subgraphs are considered interesting if they obey to one or multiple constraints. These constraints can be structural and topological, based on frequency, coverage, discrimination, or even semantics if the graphs are labeled.

10.4 SUBGRAPH MINING

One of the main and most challenging tasks in graph mining is to look for recurrent substructures, that is, to extract frequent subgraphs [9]. In fact, there exist two types of frequent subgraph discovery problems, namely,

- From a single large graph: In this case, we wish to determine all subgraphs that occur at least a certain number of times in one large graph (e.g., the World Wide Web graph or a protein–protein interaction network).
- From a database of many graphs: In this case, we have a database of graphs (e.g., a family of proteins represented by graphs) and we wish to determine all subgraphs that occur at least in a certain number of graphs of the database.

In different applications, we may be interested in different kinds of subgraphs such as subtrees, cliques (or complete graphs), bipartite cliques, and dense subgraphs. These subgraphs are used as patterns to describe the data under consideration. They may represent communities in social networks, hubs, and authority pages on the web, clusters of proteins involved in similar biochemical functions in protein–protein interaction networks, and so on. In the most common case, subgraphs are mined from data based on their frequency.

10.5 FREQUENT SUBGRAPH DISCOVERY

The problem of frequent pattern mining has been widely addressed in data mining. In the case of graph data, mining frequent patterns is more challenging mainly because of the combinatorial nature of graphs [66]. Indeed, in the case of graphs, the process of computing frequency is different. As in this chapter, we are more interested in the mining of frequent subgraphs from a graph database representing a family (or group in general) of protein tertiary structures, the following subsection defines and gives the formal statement of the problem of frequent subgraph discovery in graph databases.

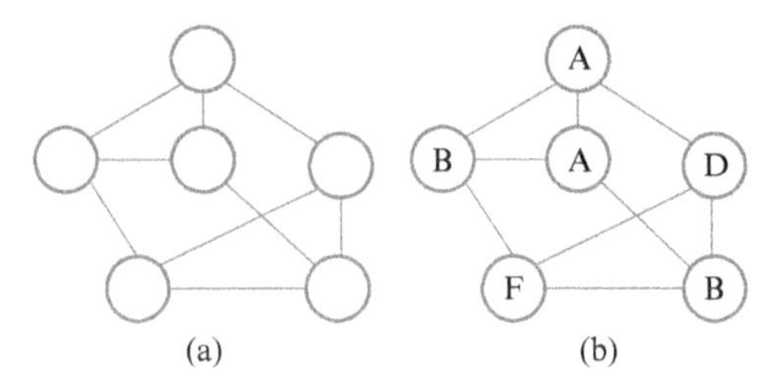

Figure 10.5 Example of an unlabeled graph (a) and a labeled graph (b).

10.5.1 Problem Definition

Let $\mathcal{G}$ be a graph database. Each graph $G = (V, E)$ of $\mathcal{G}$ is given as a collection of nodes V and edges E. We denote by $|V|$ the number of nodes of G (also called *graph order*) and by $|E|$ the number of edges of G (also called *graph size*). If two nodes u and $v \in V$ and $\{u, v\} \in E$, then u and v are said to be *adjacent* nodes. The nodes and edges of G can be labeled within an alphabet Σ, such that G becomes $G = (V, E, \Sigma, L)$, where $\Sigma = \Sigma_V \cup \Sigma_E$ and L is the label function that maps a node or an edge to a label (Figure 10.5 shows an example of an unlabeled graph (a) and a labeled graph (b)). G is called *labeled graph* and the labels of nodes and edges are denoted respectively by $L(u)$ and $L\{u, v\}$.

Definition 10.1 *(Graph isomorphism) A graph isomorphism exists between two graphs $G = (V, E, \Sigma, L)$ and $G' = (V', E', \Sigma', L')$ if there is a bijective function f: $V \to V'$ such that*

- $\forall u, v \in V : \forall\{u, v\} \in E \iff \{f(u), f(v)\} \in E'$
- $\forall v \in V : L(v) = L'(f(v))$
- $\forall\{u, v\} \in E : L\{u, v\} = L'\{f(u), f(v)\}$

where L and L′ are, respectively, the label functions of G and G′.

Definition 10.2 *(Subgraph isomorphism) A labeled graph G is a subgraph of a labeled graph G′, denoted by $G \subseteq G'$, if and only if there exists an injective function $f : V \to V'$, such that*

- $\forall u, v \in V : \forall\{u, v\} \in E \to \{f(u), f(v)\} \in E'$
- $\forall v \in V, L(v) = L'(f(v))$
- $\forall\{u, v\} \in E : L\{u, v\} = L'\{f(u), f(v)\}$

Under these conditions, the function f represents an embedding of G in G′, G is called a subgraph of G′, and G′ is called the supergraph of G.

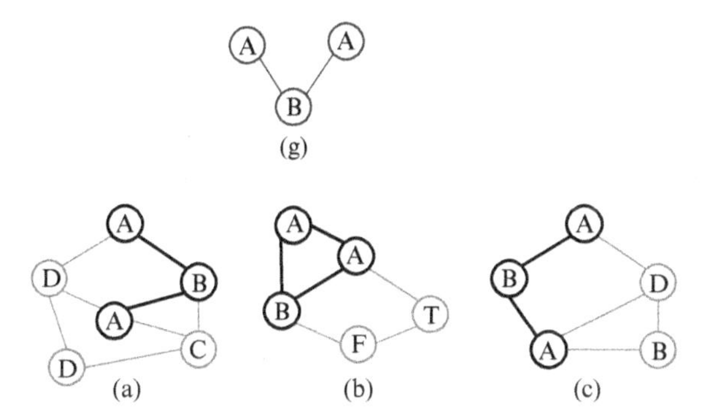

Figure 10.6 Example of a subgraph (g) that is frequent in the graphs (a), (b), and (c) with a support of 100%.

Definition 10.3 *(Frequent subgraph) Given a subgraph g, a graph database $\mathcal{G}$, and a minimum frequency threshold τ (minimum support), let $\mathcal{G}_g$ be the subset of $\mathcal{G}$ where g appears (i.e., g has a subgraph isomorphism in each graph in $\mathcal{G}_g$). The number of graphs where g occurs is denoted by $\mid \mathcal{G}_g \mid$. The subgraph g is considered as frequent if*

$$support(g) = \frac{\mid \mathcal{G}_g \mid}{\mid \mathcal{G} \mid} \geq \tau \ \ (or \frac{\mid \mathcal{G}_g \mid}{\mid \mathcal{G} \mid} * \ 100 \ \textit{if in percentage}) \qquad (10.3)$$

Figure 10.6 shows an example of a subgraph (g) that is frequent in the graphs (a), (b), and (c) with a support of 100%, that is, (g) is frequent for all minimum frequency threshold $\tau \in [0..100]$.

Frequent subgraph mining is mainly performed over two steps. The first step consists of generating the candidate subgraphs. In the second step, each one of the candidate subgraphs is checked for frequency. The challenge in frequent subgraph mining comes from the expensive graph isomorphism and subgraph isomorphism that are known to be NP-hard [66]. In the first step, graph isomorphism is used to avoid the generation of duplicate candidate subgraphs. In the second step, subgraph isomorphism checks are needed to determine the support of each candidate subgraph in all graphs in the database.

Since graphs are combinatorial by nature, the number of possible extensions of candidate subgraphs is exponential. Thus, there is an urgent need to limit the search space.

Property 10.1 *(Antimonotonicity) All subgraphs of a frequent subgraph are themselves frequent. Consequently, all supergraphs of an infrequent subgraph are infrequent.*

This property is used in pruning the candidate subgraphs such that if a candidate subgraph is infrequent, then all its supergraphs can directly be pruned with no need to compute their frequency.

10.5.2 Candidates Generation

The generation of candidates is an essential step in the discovery of frequent subgraphs. Generally, a candidate's search space is iteratively constructed, in a tree-like form, by extending the candidate subgraphs where each node in the tree represents a candidate subgraph and each child node of the latter represents one of its possible extensions, that is, one of its supergraphs. Two approaches are mainly used for generating candidate subgraphs, namely, *Breadth First Search* (BFS) and *Depth First Search* (DFS) (see the illustrative example below). The BFS approach follows a level-wise strategy where candidates of size k are generated by merging and fusing candidates of size $k - 1$; that is to say, no candidate of size k is generated before all candidates of size $k - 1$ are generated and explored. In a DFS strategy, a candidate subgraph is iteratively checked for frequency and then arbitrarily extended until it is no longer frequent or no further extension is possible. Most recent approaches (see Section 1.5.3) prefer DFS over BFS because it uses less memory and is faster as it avoids the expensive merger and fusion of subgraphs used in BFS.

10.5.2.1 Illustrative example. Figure 10.7 shows an illustrative example of a candidates search tree in a BFS manner (a) and a DFS manner (b). Each node in the search tree represents a candidate subgraph. Since BFS (Figure 10.7a) follows a level-wise strategy, the tree is constructed using the following order:

1. Level 1: n_0
2. Level 2: n_1, n_2, n_3
3. Level 3: n_4 (constructed from n_1 and n_3), n_5 (constructed from n_2 and n_3).

The DFS (Figure 10.7b) follows the same branch until the bottom of the tree before visiting the other branches. Suppose that nodes on the right are visited before those on the left, candidates in the tree are constructed using the following order:

– $n_0, n_1, n_4, n_2, n_5, n_3$ (n_4: but pruned for duplication as already generated from n_1) (n_5: but pruned for duplication as already generated from n_2).

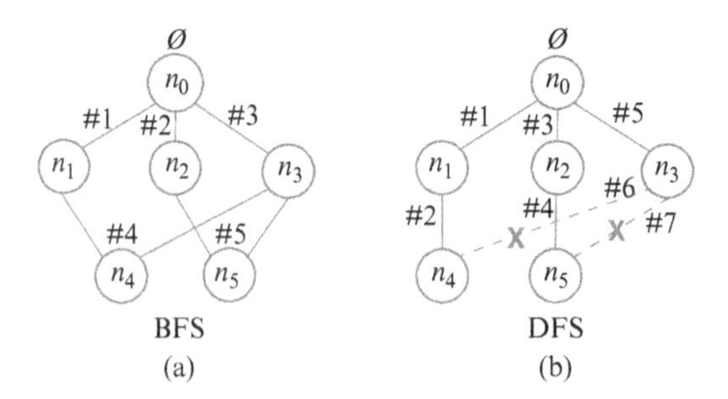

Figure 10.7 Example of a candidates search tree in a BFS manner (a) and a DFS manner (b). Each node in the search tree represents a candidate subgraph.

10.5.3 Frequent Subgraph Discovery Approaches

Many approaches for frequent subgraph mining have been proposed [25, 32, 35]. A pioneer work is Reference [10], where the authors proposed an approximate and greedy search algorithm named *SUBDUE* for discovering frequent graph patterns based on a minimum description length and background knowledge. Other works based on the principles of artificial intelligence have been proposed, such as *WARMR* [30] and *FARMER* [43]. They successfully mined frequent subgraphs from the data of chemical compounds. Although these approaches allow to completely discover all frequent subgraphs, they suffer from the high consumption in terms of time and computational resources. In addition, the discovered subgraphs are semantically very complex because the graphs were initially transformed into datalog facts.

Besides these studies, there exist two main categories of approaches of frequent subgraph discovery, namely, the Apriori-based approaches and the pattern-growth approaches.

10.5.3.1 Apriori-Based Approaches. Generally, Apriori-based approaches start from subgraphs of small sizes. Then, in a bottom-up manner, they generate candidate subgraphs by adding a node or an edge to an existing frequent subgraph. The main idea behind Apriori-based approaches is that candidate subgraphs of size $k+1$ are generated by means of a join operation on two frequent subgraphs of size k having a common subgraph of size $k-1$. Thus, in order to be able to generate candidate subgraphs of level $k+1$, all subgraphs of size k have to be already generated. Consequently, Apriori-based approaches use the BFS strategy as they follow a level-wise candidates generation.

The main algorithms that have been proposed in this category are AGM [24], FSG [33], and DPMine [60]. AGM [24] and FSG [33] are very similar, but the main difference between them is that AGM generates a candidate by extending a frequent subgraph with a node, while FSG generates a candidate by extending a frequent subgraph with an edge. DPMine [60] uses edge-disjoint paths as the expansion units for candidates generation. It starts by identifying all frequent paths, then all subgraphs with two paths. After that, it starts generating subgraphs with k paths by merging pairs of frequent subgraphs of $k-1$ paths having $k-2$ paths in common.

10.5.3.2 Pattern-Growth Approaches. Pattern-growth approaches extend an already discovered frequent subgraph by adding an edge in every possible position. Adding an edge may result in adding a new node. Extensions are recursively performed until no more frequent subgraphs are generated. In contrast to Apriori-based approaches, pattern-growth approaches are more flexible on the search method, where both BFS and DFS can be applied. Pattern-growth approaches do not need the expensive join operations used in the Apriori-based approaches; nevertheless, they suffer from the problem of duplicates generation. Indeed, the edge extension strategy can result in the generation of the same subgraph multiple times from different extensions. Hence, existing pattern-growth approaches have tried to propose ways to avoid or at least minimize the generation of duplicate candidate subgraphs.

The main algorithms that have been proposed in this category are MoFa [5], gSpan [64], FFSM [22], and GASTON [44]. MoFa [5] is mainly proposed to mine frequent subgraphs in a set of molecules. In order to accelerate the mining process, MoFa stores the embedding list of previously found subgraphs such that the extensions will be restricted only to these embeddings. Even though MoFa also uses structural and background knowledge for pruning, it still generates many duplicates.

The gSpan algorithm [64] addresses the problem of duplication differently. It first starts generating candidates using the right-most extension technique. In this technique, according to a DFS on the graph, the straight path from the starting node to the target node is called the *right-most path* [64]. Only extensions on the right-most path are allowed. It was proved that candidates generation using the right-most extension technique is complete. To alleviate the cost of isomorphism between subgraphs, gSpan uses a canonical representation where each subgraph is simply represented by a unique code called the *minimum DFS code*, allowing an easy detection of isomorphic subgraphs.

FFSM [22] also uses a canonical representation, in a matrix form, called the *Canonical Adjacency Matrix* (CAM) to represent graphs and to detect isomorphism. It generates new candidate subgraphs either by an extension of a CAM or by joining two CAMs using a set of adapted operators.

In many contexts, GASTON [44] is considered the fastest subgraph mining algorithm. In contrast to all existing approaches, it exploits the fact that a wide range of the discovered frequent patterns are paths and trees, and only a portion (that is sometimes very small) represents subgraphs with cycles. Hence, GASTON considers them differently by splitting the frequent subgraph mining into path mining, then subtree mining, and finally subgraph mining. Consequently, the subgraph isomorphism is only performed in the final step. GASTON also records the embedding list to save unnecessary isomorphism detection by extending only patterns that appear in the list.

10.5.4 Variants of Frequent Subgraph Mining: Closed and Maximal Subgraphs

According to the antimonotonicity (Property 10.1), all subgraphs of a frequent subgraph are also frequent. This results in the problem of information overload. Indeed, this problem becomes even more serious with large subgraphs as they contain an exponential number of smaller frequent subgraphs. To overcome this problem, variants of frequent subgraph mining have been proposed, namely, closed subgraph mining and maximal subgraph mining.

Definition 10.4 (Closed subgraph) *A frequent subgraph g is said to be closed, if it has no supergraph g' ($g \subset g'$) that is also frequent and has the same support.*

Definition 10.5 (Maximal subgraph) *A frequent subgraph g is said to be maximal, if it has no supergraph g' ($g \subset g'$) that is also frequent.*

According to Definitions 10.4 and 10.5, both closed and maximal frequent subgraphs present a compact representation of the frequent subgraphs. Closed subgraph compactness is lossless as it contains all the information about the frequent subgraphs and their supports. However, maximal subgraph compactness does not consider the whole information as although all frequent subgraphs can be restored, the exact support of each subgraph is lost. The main approach that has been proposed in the literature for closed subgraph mining is CloseGraph [65] and those for maximal subgraph mining are SPIN [23] and MARGIN [59]. Although the set of closed or maximal subgraphs is much smaller than the set of frequent ones, the number of subgraphs is still very high in real-world cases that alleviate the problem of information overload but do not resolve it. The most used technique to tackle this problem is feature selection.

10.6 FEATURE SELECTION

Feature selection is also known in the literature as pattern selection, attribute selection, variable selection, or variable subset selection [18, 34, 39, 40, 50]. The task of feature selection has been widely addressed in pattern recognition and data mining not only for subgraphs but also for other types of patterns such as association rules, itemsets, and sequential motifs. Although many approaches have been proposed in the literature, allowing an efficient computation of frequent patterns, the number of discovered patterns is often very high. This is an obvious impact of the high dimensional nature of many types of data. Besides, the most frequent pattern discovery approaches were not originally designed to consider the relevance of features.

The main goal of feature selection is to reduce the number of features by removing the *redundant* and *irrelevant* ones such that only a subset of interesting features is retained.

10.6.1 Relevance of a Feature

Given an interestingness criterion, a feature is redundant if it does not bring any additional information over the currently selected features and thus it can be replaced by at least one of the already selected features. A feature is considered *irrelevant* if it does not provide any useful information in any context so that it does not have any influence on the output of the prediction.

Many approaches have been proposed for feature selection to resolve the information overload problem when the number of patterns is high. It is possible to categorize the existing approaches in different ways depending on the criterion used for classification. The most conventional classification of feature selection approaches comprises three categories, namely, wrapper, embedded, or filter approaches [40]. Wrapper approaches [7] divide the feature set into subsets using a sampling technique, generally cross-validation, and each one of the feature subsets is evaluated by training and testing the learning model. The selected subset of features is the one having the best predictive score. Similarly to wrappers, embedded approaches [34] are also related to a predictive model. In contrast, the selection process is performed as part of the model construction by searching in the combined space of feature subsets. Filter approaches [37] assess the relevance of features according to a significance criterion (statistical test), independently of the chosen predictor.

10.7 FEATURE SELECTION FOR SUBGRAPHS

Because of the combinatorial nature of graphs, subgraph mining is one of the fields most concerned with the information overload problem. To be able to handle the exponential number of frequent subgraphs, many feature selection approaches have been proposed for selecting significant subgraphs. Although the main goal of these approaches is to obtain a smaller yet more informative subset of subgraphs, each approach has a different way of evaluation.

10.7.1 Problem Statement

In general, the process of feature selection for subgraphs can be formulated as follows. Given a graph database $\mathcal{G} = \{G_1, \cdots, G_n\}$ and an evaluation function F, find all significant subgraphs $g^* \in \mathcal{G}$ such that

- g^* are the set of subgraphs that maximize the evaluation function F, that is, $g^* = argmax_g(F(g))$; or,
- g^* are the subgraphs having an evaluation score that is greater or equal to a given threshold, that is, $g^* = \forall g, F(g) \geq \tau$, if F is a threshold-based function; or,
- g^* are the k subgraphs having successively the best score with F, that is, $g^* = Top_k F(g)$.

In all three cases, the best scenario is when the evaluation function F is embedded within the subgraph mining. In such a case, significant subgraphs g^* are mined from $\mathcal{G}$ without exhaustively generating the entire set of subgraphs. However, a direct mining of g^* from $\mathcal{G}$ can be very difficult especially if the evaluation function F is not antimonotonic. An example of an antimonotonic evaluation function is frequency (see Definition 10.3 and Property 10.1). Considering the second case, g^* is the set of all frequent subgraphs, that is, any subgraph in g^* has a frequency score greater than or equal to a user-defined threshold τ ($\forall g \in g^*$, frequency$(g) \geq \tau$). If considering the third case, a user can select the k most frequent subgraphs, that is, the Top_k frequent subgraphs.

Many approaches have been (and are being) proposed for significant subgraph selection. The following subsections present and discuss some of the recent and most interesting subgraph selection methods in the literature. These methods are grouped into categories based on their selection strategy: generating top-k subgraphs as in Reference [61], clustering-based subgraph selection as in Reference [67], sampling-based approaches [1], approximate subgraph mining as in Reference [14], discriminative subgraphs as in Reference [68], and other significant subgraph selection approaches [45, 51].

10.7.2 Mining Top-k Subgraphs

Top-k selection approaches accept a parameter k and a criterion F, then return the best k subgraphs that are ranked according to F. The main motivation behind top-k

selection approaches is that, in many application domains, a user may be interested in finding a specific number of patterns that best qualify for a given evaluation criterion. Redundancy-aware top-k patterns [61] and TGP [38] are two of the most interesting approaches for top-k subgraph selection. Although top-k selection approaches are usually very fast and simple, the user still has to specify a proper number of patterns to select, that is, the parameter k, which may be difficult to define in some applications.

10.7.3 Clustering-Based Subgraph Selection

Clustering-based subgraph selection aims generally to obtain a set of representative patterns, where each representative resembles a cluster centroid. In fact, clustering is the process of bringing together a set of objects into groups of similar ones. The definition of similarity between the input objects varies from one clustering model to another. In most of these models, the concept of similarity is based on distances, such as Euclidean distance or cosine distance. Clustering-based subgraph selection generally uses a clustering technique to group similar subgraphs, then each group of similar subgraphs is summarized by one representative, that is, its centroid. RP–FP, RP–GD [41], and RING [67] are some of the most interesting clustering-based subgraph selection approaches. Even though the latter generally give good representatives, most known approaches suffer from high computational complexity because of the combinatorial nature of clustering algorithms.

10.7.4 Sampling-Based Approaches

In statistics, sampling can simply be defined by the task of selecting a subset of individuals among a given statistical population to estimate the characteristics of the whole set. In our context (i.e., pattern selection), sampling is a method for selecting a subset of n patterns out of N such that the sampled n patterns allows (or approximately) estimating the characteristics of all the N patterns. Sampling-based approaches are mainly proposed because of the assumption that, in many real application contexts, it is very expensive or even impossible to generate the entire set of frequent patterns. Thus, sampling-based selection approaches try to approximately generate the set of significant patterns by only considering a sample of the entire set. ORIGAMI [19], output space sampling [1], and MCSs [55] are among the most interesting sampling-based approaches for subgraphs. This selection technique is fast and more efficient in large-scale applications than most of the other known techniques. However, the sampling is usually driven by a probability distribution. Thus, the latter should be defined very carefully as it highly affects the quality of the generated subgraphs.

10.7.5 Approximate Subgraph Mining

Approximation is usually used when the exact result is unknown or difficult to obtain such that the obtained inexact results are within required limits of accuracy. In many applications, mining the whole set of frequent subgraphs is very difficult. Besides, in domains of application such as protein–protein interaction networks, slight differences between subgraphs may not be important. In such cases, approximation is a way to handle such issues by only enumerating an approximate set of subgraphs such that similar but slightly different subgraphs will collapse into one representative. Approximation in subgraph mining is performed by structural, topological, or label approximation. Examples of very interesting works on approximation are smoothing–clustering [8] on structural approximation, TRS [15] on topological approximation, and UnSubPatt [14] on label approximation. Existing approximation-based approaches for subgraph mining effectively alleviate redundancy caused by structural, topological, and/or label similarity in the input subgraph set. However, the general tendency of existing approaches is to define similarity and approximation algorithms that are computationally expensive.

10.7.6 Discriminative Subgraph Selection

Supervised classification is one of the important applications that use frequent subgraphs and, in general, frequent patterns. In the case where graphs in the database are labeled (i.e., each graph is affiliated to a class), it is possible to extract discriminative frequent subgraphs. These subgraphs have recently been used as attributes for machine learning classifiers to help building models that best discriminate between classes. Several selection methods and algorithms have been proposed for mining discriminative frequent graph patterns. Most of existing approaches are filter methods or boosting methods. GAIA [27], GraphSig [48], CORK [58], LEAP [63], and D&D [68] are some of the most known filter methods. They differ in the way of considering the discrimination power of patterns, but they all tend to select patterns that best discriminate between classes; then they consider them as descriptors of the original graphs for classification. In filter methods, the training and building of classification models are performed separately, that is, after mining the set of discriminative features. In boosting methods, the search for discriminative subgraphs is performed at the same time as the construction of the classification models. Some of the most known boosting methods are LPGBCMP [17], COM [26], gPLS [53], and gBoost [54].

A common drawback of all these methods is that in the case where graphs in the database have no class labels, they become useless. Besides, they consider the discrimination power of patterns individually. This may fail if no individual pattern has high discrimination power; yet, jointly some patterns may have higher discrimination. Moreover, the selected patterns may be individually discriminative but redundant if there exists a significant overlap in their supporting graphs. This makes them more vulnerable to overfitting.

10.7.7 Other Significant Subgraph Selection Approaches

It is worth mentioning that other selection approaches have been proposed for mining significant subgraphs (SkyGraph [45], MIPs [47], Ant-motifs [51], etc.), not necessarily for a specific application context, but only for extracting relevant or hidden information or to simply characterize a given data set. Many of these approaches have their own original and unique selection technique, and thus it is difficult to classify all of them under one of the above subgraph-selection categories.

10.8 DISCUSSION

All the previously mentioned approaches are mainly designed for graph data. In order to adopt them to mine motifs from protein 3D-structures, it is only necessary to parse the proteins into structure graphs, which was the subject of Section 10.2.1. Since there are many subgraph selection approaches at present, it is difficult to compare them in general because many of the approaches were originally designed for a particular application context. Hence, the choice of an appropriate selection method highly depends on the users' needs and the application constraints. In order to help assisting such choice, Table 10.1 lists and characterizes all the subgraph selection approaches that have been investigated in this chapter.

Many important functional and evolutionary relationships among proteins cannot be efficiently discovered through sequence analysis alone, but only through tertiary structure comparison. Such a case represents one of the very interesting domains of application of motif discovery where, for instance, recurrent motifs can represent conserved features that describe the functional and evolutionary relationships that are sought [28, 29, 56]. Top-k subgraphs can be used to discover a specific number of patterns that best score for one or more given quality functions. For instance, in drug prediction, the number of possible target molecules is usually exponential [49, 62]. To minimize the search space, it is possible to look for molecules that best qualify for a given set of chemical properties. The best k molecules are then selected and used as targets to predict the new drug.

Clustering-based, sampling-based, and approximate subgraph mining are very handy in scenarios where the search space or the number of discovered patterns is very high, which may hinder or even make impossible further explorations. This is the case of mining motifs from protein tertiary structures [14, 19, 51] as the latter are known to be dense macromolecules and with an increase in their availability in online databases, using such selection methods is becoming a necessity rather than a choice.

Protein classification and fingerprints identification are among the most important tasks in the study of protein 3D-structures. Both tasks can efficiently be performed through mining-discriminative subgraph motifs. The aim is to identify motifs that best discriminate a given family of proteins (for instance, functional or structural family) from other proteins. Motifs that frequently appear in a considered group of proteins and not in others are usually referred to as *fingerprints*. Such motifs can be used to identify functional sites [20], substructures that characterize a given protein family, and also to identify the function or, in general, the affiliation of an unknown protein [54].

Table 10.1 Characteristics of existing subgraph selection approaches

Subgraph selection approach	Descriptor		
	Post Processing	Learning-Task Dependent	Selected Subgraphs
TGP [38]	No	No	Top-*k* frequent closed
Redundancy aware top-*k* [61]	Yes	No	Top-*k* significant and non redundant
RP-FP [41]	Yes	No	Frequent closed representatives
RP-GD [41]	No	No	Frequent closed representatives
RING [67]	No	No	Frequent representatives
ORIGAMI [19]	Yes	No	α-Orthogonal β-representative
Output space sampling [1]	No	No	Sample of frequent subgraphs
MCSs [55]	No	No	Maximum common subgraphs
Smoothing–clustering [8]	Yes	No	Approximate structural representatives
TRS [15]	Yes	No	Topological representatives
UnSubPatt [14]	Yes	No	Unsubstituted (label) representatives
D&D [68]	No	Yes	Diverse discriminative
LPGBCMP [17], COM [26], GAIA [27], CORK [58], GraphSig [48], LEAP [63], gBoost [54], gPLS [53]	No	Yes	Discriminative
MIPs [47]	No	No	Most informative closed
SkyGraph [45]	No	No	Undominated
Ant-motifs [51]	No	No	Ant-like shape

10.9 CONCLUSION

Proteins are complex data that need to be analyzed and mined using automatic methods because of their large numbers that keep increasing by the day. Motif discovery is one of the core tasks in protein tertiary structure analysis. Pattern mining is also one of the most important tasks in pattern recognition. The main aim of this chapter was to review how to bridge the gap between both tasks and both domains of research. The first part of the chapter reviewed existing methods to transform protein 3D-structures into graphs, enabling the usage of graph mining and pattern-recognition techniques to perform protein 3D-structure analysis. The second part of the chapter focused on subgraph discovery from graph databases that can be used to extract motifs from protein 3D-structures in a graph perspective. It presented the context and formalization of frequent subgraph discovery: problem definition, candidates generation, the main approaches proposed in the literature (Apriori and pattern growth-based approaches), and variants of frequent subgraph mining (closed and maximal subgraphs). It also presented the general framework of feature selection and the complexity of adopting the old techniques for frequent subgraphs. The chapter further investigated a panoply of subgraph selection approaches proposed in the literature that can be used easily with protein 3D-structure motifs.

Even though great advances have been noticed in recent years in all the reviewed three tasks (i.e., protein graph transformation, subgraph extraction, and subgraph selection techniques), many enhancements are still needed, especially in terms of scalability in order to match the increasing needs of analyzing larger protein 3D-structure databases. It is also worth mentioning some of the other existing techniques for mining protein 3D-structure motifs, such as *geometric 3D-motifs* as in Reference [46] and *3D-templates* as in Reference [3]. The main difference between geometric 3D-motifs and the motifs discussed in this chapter (i.e., in the form of subgraphs) is that besides connections between amino acids in 3D-space, geometric 3D-motifs also consider the geometry of residues around each other, that is, the exact position and orientation of the residue in space. 3D-templates are very similar to geometric 3D-motifs except that they obey constraints such as the number of atoms, a specific geometric conformation, and the atom type. Mining and matching motifs from graph representations of protein structures allow more flexibility than geometric 3D-motifs and templates.

ACKNOWLEDGMENTS

This work was supported by the BREEDWHEAT project and a MENRT grant from the French Ministry of Higher Education and Research.

REFERENCES

1. Al Hasan M, Zaki MJ. Output space sampling for graph patterns. Proceedings of the VLDB Endowment (35th International Conference on Very Large Data Bases), Volume 2(1); Lyon, France 2009. p. 730–741.
2. Andreeva A, Howorth D, Chandonia J-M, Brenner SE, Hubbard TJP, Chothia C, Murzin AG. Data growth and its impact on the SCOP database: new developments. Nucleic Acids Res 2008;36(1):D419–D425.
3. Barker JA, Thornton JM. An algorithm for constraint-based structural template matching: application to 3D templates with statistical analysis. Bioinformatics 2003;19(13):1644–1649.
4. Berman HM, Westbrook JD, Feng Z, Gilliland G, Bhat TN, Weissig H, Shindyalov IN, Bourne PE. The protein data bank. Nucleic Acids Res 2000;28(1):235–242.
5. Borgelt C, Berthold MR. Mining molecular fragments: finding relevant substructures of molecules. IEEE International Conference on Data Mining (ICDM); 2002. p. 51–58.
6. Bostick DL, Shen M, Vaisman II. A simple topological representation of protein structure: implications for new, fast, and robust structural classification. Proteins 2004;56(3):487–501.
7. Cadenas JM, Garrido MC, Martínez R. Feature subset selection filter-wrapper based on low quality data. Expert Syst Appl 2013;40(16):6241–6252.
8. Chen C, Lin CX, Yan X, Han J. On effective presentation of graph patterns: a structural representative approach. Proceedings of the 17th ACM Conference on Information and Knowledge Management. ACM; Napa Valley, CA; 2008. p. 299–308.
9. Cheng H, Yan X, Han J. Mining graph patterns. In: *Managing and Mining Graph Data.* Springer-Verlag; 2010. p. 365–392.
10. Cook DJ, Holder LB. Substructure discovery using minimum description length and background knowledge. J Artif Intell Res 1994;1:231–255.
11. Cook DJ, Holder LB. *Mining Graph Data.* John Wiley & Sons; 2006.
12. Delaunay B. Sur la sphère vide. À la mémoire de Georges Voronoï. Bull Acad Sci URSS 1934;6:793–800.
13. Dhifli W, Saidi R. Protein graph repository. Extraction et gestion des connaissances (EGC); 2010. p. 641–642.
14. Dhifli W, Saidi R, Nguifo EM. Smoothing 3D protein structure motifs through graph mining and amino-acids similarities. J Comput Biol 2014;21(2):162–172.
15. Dhifli W, Moussaoui M, Saidi R, Nguifo EM. Mining topological representative substructures from molecular networks. Proceedings of the 13th International Workshop on Data Mining in Bioinformatics (BIOKDD); New York City; 2014.
16. Doppelt O, Moriaud F, Bornot A, de Brevern AG. Functional annotation strategy for protein structures. Bioinformation 2007;1(9):357–359.
17. Fei H, Huan J. Boosting with structure information in the functional space: an application to graph classification. ACM Knowledge Discovery and Data Mining Conference (KDD); 2010. p. 643–652.
18. Guyon I, Elisseeff A. An introduction to variable and feature selection. J Mach Learn Res 2003;3:1157–1182.

19. Hasan MA, Chaoji V, Salem S, Besson J, Zaki MJ. ORIGAMI: mining representative orthogonal graph patterns. Proceedings of the IEEE International Conference on Data Mining. IEEE Computer Society; 2007. p. 153–162.

20. Helma C, Cramer T, Kramer S, De Raedt L. Data mining and machine learning techniques for the identification of mutagenicity inducing substructures and structure activity relationships of noncongeneric compounds. J Chem Inf Model 2004;44(4):1402–1411.

21. Huan J, Bandyopadhyay D, Wang W, Snoeyink J, Prins J, Tropsha A. Comparing graph representations of protein structure for mining family-specific residue-based packing motifs. J Comput Biol 2005;12(6):657–671.

22. Huan J, Wang W, Prins J. Efficient mining of frequent subgraphs in the presence of isomorphism. IEEE International Conference on Data Mining (ICDM); Melbourne, FL; 2003. p. 549–552.

23. Huan J, Wang W, Prins J, Yang J. SPIN: Mining maximal frequent subgraphs from graph databases. Proceedings of the 10th ACM SIGKDD International Conference on Knowledge Discovery and Data Mining. ACM; Seattle, WA; 2004. p. 581–586.

24. Inokuchi A, Washio T, Motoda H. An Apriori-based algorithm for mining frequent substructures from graph data. The European Conference on Machine Learning and Principles and Practice of Knowledge Discovery in Databases; Lyon, France; 2000. p. 13–23.

25. Jiang C, Coenen F, Zito M. A survey of frequent subgraph mining algorithms. Knowl Eng Rev 2013;28:75–105.

26. Jin N, Young C, Wang W. Graph classification based on pattern co-occurrence. ACM International Conference on Information and Knowledge Management; Hong Kong, China; 2009. p. 573–582.

27. Jin N, Young C, Wang W. GAIA: graph classification using evolutionary computation. Proceedings of the 2010 ACM SIGMOD International Conference on Management of Data, SIGMOD. ACM; Indianapolis, IN; 2010. p. 879–890.

28. Jonassen I. Efficient discovery of conserved patterns using a pattern graph. Comput Appl Biosci 1997;13(5):509–522.

29. Jonassen I, Eidhammer I, Conklin D, Taylor WR. Structure motif discovery and mining the PDB. Bioinformatics 2002;18(2):362–367.

30. King RD, Srinivasan A, Dehaspe L. WARMR: a data mining tool for chemical data. J Comput Aided Mol Des 2001;15(2):173–181.

31. Kleywegt GJ. Recognition of spatial motifs in protein structures. J Mol Biol 1999;285(4):1887–1897.

32. Krishna V, Ranga Suri NNR, Athithan G. A comparative survey of algorithms for frequent subgraph discovery. Curr Sci 2011;100(2):190–198.

33. Kuramochi M, Karypis G. Frequent subgraph discovery. IEEE International Conference on Data Mining (ICDM); San Jose, CA 2001. p. 313–320.

34. Ladha L, Deepa T. Feature selection methods and algorithms. Int J Comput Sci Eng (IJCSE) 2011;5(5):1787–1797.

35. Lakshmi K, Meyyappan T. Frequent subgraph mining algorithms - a survey and framework for classification. Int J Inf Technol Conver Serv (IJITCS) 2012;2:189–202.

36. Laskowski RA, Watson JD, Thornton JM. ProFunc: a server for predicting protein function from 3D structure. Nucleic Acids Res 2005;33(Web-Server-Issue):89–93.

37. Lazar C, Taminau J, Meganck S, Steenhoff D, Coletta A, Molter C, de Schaetzen V, Duque R, Bersini H, Nowe A. A survey on filter techniques for feature selection in gene expression microarray analysis. IEEE/ACM Trans Comput Biol Bioinform (TCBB) 2012;9(4):1106–1119.
38. Li Y, Lin Q, Li R, Duan D. TGP: Mining top-k frequent closed graph pattern without minimum support. Proceedings of the 6th International Conference on Advanced Data Mining and Applications (ADMA). Chongqing, China; Springer-Verlag; 2010. p. 537–548.
39. Liu H, Motoda H. Feature Extraction, Construction and Selection: A Data Mining Perspective. Norwell (MA): Kluwer Academic Publishers; 1998.
40. Liu H, Motoda H. *Computational Methods of Feature Selection*, Data Mining and Knowledge Discovery Series. Chapman and Hall/CRC; 2007.
41. Liu Y, Li J, Zhu J, Gao H. Clustering frequent graph patterns. J Digit Inf Manage 2007;5(6):335–346.
42. Lovell SC, Davis IW, Arendall WB, de Bakker PIW, Word JM, Prisant MG, Richardson JS, Richardson DC. Structure validation by Cα geometry: ϕ, ψ and Cβ deviation. Proteins 2003;50(3):437–450.
43. Nijssen S, Kok J. Faster association rules for multiple relations. Proceedings of the 17th International Joint Conference on Artificial Intelligence, IJCAI. Seattle, WA; Morgan Kaufmann Publishers; 2001. p. 891–896.
44. Nijssen S, Kok JN. A quickstart in frequent structure mining can make a difference. ACM Knowledge Discovery and Data Mining Conference (KDD); Seattle, WA; 2004. p. 647–652.
45. Papadopoulos AN, Lyritsis A, Manolopoulos Y. SkyGraph: an algorithm for important subgraph discovery in relational graphs. Data Min Knowl Discov 2008;17(1):57–76.
46. Pennec X, Ayache N. A geometric algorithm to find small but highly similar 3D substructures in proteins. Bioinformatics 1998;14(6):516–522.
47. Pennerath F, Napoli A. The model of most informative patterns and its application to knowledge extraction from graph databases. Proceedings of the European Conference on Machine Learning and Knowledge Discovery in Databases: Part II, ECML PKDD '09. Bled, Slovenia; Springer-Verlag; 2009. p. 205–220.
48. Ranu S, Singh AK. GraphSig: A scalable approach to mining significant subgraphs in large graph databases. IEEE 25th International Conference on Data Engineering; Shanghai, China; 2009. p. 844–855.
49. Ranu S, Singh AK. Indexing and mining topological patterns for drug discovery. Proceedings of the 15th International Conference on Extending Database Technology, EDBT. ACM; Berlin, Germany; 2012. p. 562–565.
50. Saeys Y, Inza I, Larra naga P. A review of feature selection techniques in bioinformatics. Bioinformatics 2007;23(19):2507–2517.
51. Saidi R, Dhifli W, Maddouri M, Nguifo EM. A novel approach of spatial motif extraction to classify protein structures. Proceedings of the 13th Journées Ouvertes en Biologie, Informatique et Mathématiques (JOBIM); Rennes, France 2012. p. 209–216.
52. Saidi R, Maddouri M, Nguifo EM. Comparing graph-based representations of protein for mining purposes. Proceedings of the ACM KDD Workshop on Statistical and Relational Learning in Bioinformatics; Paris, France 2009. p. 35–38.
53. Saigo H, Krämer N, Tsuda K. Partial least squares regression for graph mining. ACM Knowledge Discovery and Data Mining Conference (KDD); Las Vegas, NV 2008. p. 578–586.

54. Saigo H, Nowozin S, Kadowaki T, Kudo T, Tsuda K. gBoost: a mathematical programming approach to graph classification and regression. Mach Learn 2009;75(1):69–89.
55. Schietgat L, Costa F, Ramon J, Raedt L. Effective feature construction by maximum common subgraph sampling. Mach Learn 2011;83(2):137–161.
56. Spriggs RV, Artymiuk PJ, Willett P. Searching for patterns of amino acids in 3D protein structures. J Chem Inf Comput Sci 2003;43(2):412–421.
57. Stout M, Bacardit J, Hirst JD, Smith RE, Krasnogor N. Prediction of topological contacts in proteins using learning classifier systems. Soft Comput 2008;13:245–258.
58. Thoma M, Cheng H, Gretton A, Han J, Kriegel H-P, Smola A, Song L, Yu PS, Yan X, Borgwardt KM. Discriminative frequent subgraph mining with optimality guarantees. Stat Anal Data Min 2010;3(5):302–318.
59. Thomas LT, Valluri SR, Karlapalem K. MARGIN: Maximal frequent subgraph mining. 6th International Conference on Data Mining (ICDM); 2006. p. 1097–1101.
60. Vanetik N, Gudes E, Shimony SE. Computing frequent graph patterns from semistructured data. Proceedings of the IEEE International Conference on Data Mining. Melbourne, FL IEEE Computer Society; 2002. p. 458–465.
61. Xin D, Cheng H, Yan X, Han J. Extracting redundancy-aware top-k patterns. Proceedings of the 12th ACM SIGKDD International Conference on Knowledge Discovery and Data Mining. ACM; Philadelphia; 2006. p. 444–453.
62. Yamanishi Y, Kotera M, Kanehisa M, Goto S. Drug-target interaction prediction from chemical, genomic and pharmacological data in an integrated framework. Bioinformatics 2010;26(12):246–254.
63. Yan X, Cheng H, Han J, Yu PS. Mining significant graph patterns by leap search. Proceedings of the ACM SIGMOD International Conference on Management of Data, SIGMOD. Vancouver, Canada; ACM; 2008. p. 433–444.
64. Yan X, Han J. gSpan: Graph-based substructure pattern mining. Proceedings of the 2002 IEEE International Conference on Data Mining (ICDM), Volume 02. Melbourne, FL IEEE Computer Society; 2002. p. 721–724.
65. Yan X, Han J. CloseGraph: Mining closed frequent graph patterns. ACM Knowledge Discovery and Data Mining Conference (KDD); Washington, DC; 2003. p. 286–295.
66. Zaki MJ, Meira W Jr. *Data Mining and Analysis: Fundamental Concepts and Algorithms*. Cambridge University Press; 2014.
67. Zhang S, Yang J, Li S. RING: an integrated method for frequent representative subgraph mining. Proceedings of the IEEE International Conference on Data Mining. Miami, FL; IEEE Computer Society; 2009. p. 1082–1087.
68. Zhu Y, Yu JX, Cheng H, Qin L. Graph classification: a diversified discriminative feature selection approach. Proceedings of the ACM International Conference on Information and Knowledge Management, CIKM. Sheraton, Maui Hawaii; ACM; 2012. p. 205–214.

CHAPTER 11

FUZZY AND UNCERTAIN LEARNING TECHNIQUES FOR THE ANALYSIS AND PREDICTION OF PROTEIN TERTIARY STRUCTURES

Chinua Umoja, Xiaxia Yu, and Robert Harrison

Department of Computer Science, Georgia State University, Atlanta, GA, USA

11.1 INTRODUCTION

Proteins, the building blocks of life, are considered among the most important molecules because of the roles they play in organisms. A protein is an organic compound, made entirely from different combinations of amino acids; proteins provide a gamut of functions to the organisms that use them. At a very basic level, a protein is a linear chain of amino acids, ranging from a few tens up to thousands of amino acids. Proteins derive their structure from folding of this chain; under the influence of several chemical and physical factors, proteins fold into three-dimensional structures, which determine their biological functions.

The application for the determination of protein structure has given rise to many advances in the field of the biotechnology, for example, the design of new proteins and folds [5], disease study [36], drug design projects [17], refinement of theoretical

Pattern Recognition in Computational Molecular Biology: Techniques and Approaches,
First Edition. Edited by Mourad Elloumi, Costas S. Iliopoulos, Jason T. L. Wang, and Albert Y. Zomaya.

models obtained by comparative modeling [14], and obtaining experimental structures from incomplete nuclear magnetic resonance data [39]. Furthermore, the function of a protein is a consequence of its structure, and predicting the native structures of proteins would help to take advantage of the large amount of biological information that is being generated by genome-sequencing projects.

Determining the protein structure, the functional conformation of a protein, is viewed by many as one of the most interesting challenges of modern computational biology [3]. Not only has determining protein structures from their respective amino acid sequence remained a central problem in computational biology but this determination has also found its way into computer science and machine learning because of the scope and complexity of the problem. Even some of the most simplistic reductions of the structure prediction problem have been deemed to be NP-hard [60, 61].

The common methodologies that are used are either comparative modeling or *ab initio* methods; these methodologies present different complexities and, in many cases, different solutions. In comparative modeling, databases of known structures are referred to for the development of the protein structure. Comparative modeling methods include homology modeling, threading, and fold recognition. *Ab initio* methods are those that use the amino acid sequence and the interactions between them to determine structure. In *ab initio* methods, correct structure is generally calculated by determining the overall energy of the structure; the lower the energy, the more stable the structure, which increases the probability of its being in the correct conformation and structure.

The major goal when developing protein structure development strategies is to develop more reliable ways to determine correct and incorrect structure and the creation or improvements of methods that find functional conformations from those correct structures. To determine whether a structure is correct or incorrect in the definition of energy conformations, one must account for an array of physical properties: Gibbs free energy, hydrophobicity, electrostatic potential, van der Waal's force, and so on. Physically, it is understood that conformations with the lowest Gibbs free energy are the most stable structures and are, by definition, the natural conformations. Protein folding may be altered by outside factors such as chaperones and electrostatic conditions of the environment, and therefore isolated and refolded proteins may be nonnative. Different machine learning algorithms have been proposed for searching through possible conformations as well as efficiently evaluating the quality of protein models; these algorithms include, but are not limited to, evolutionary algorithms such as *Genetic Algorithms* (GAs) and Monte Carlo Methods, and supervised learning methods such as *Artificial Neural Networks* (ANNs) and *Support Vector Machines* (SVMs). In this chapter, we cover the machine learning methods that are currently being used to solve the challenge of protein structure prediction as well as their benefits and drawbacks (Figures 11.1–11.9).

The rest of this chapter is organized as follows. In Section 11.2, GAs and the factors that should be taken into account when applying them are described. Then, Section 11.3 describes the application of supervised machine learning. Section 11.4 describes the use of fuzzy logic for protein structure prediction. Section 11.5 summarizes and concludes the chapter.

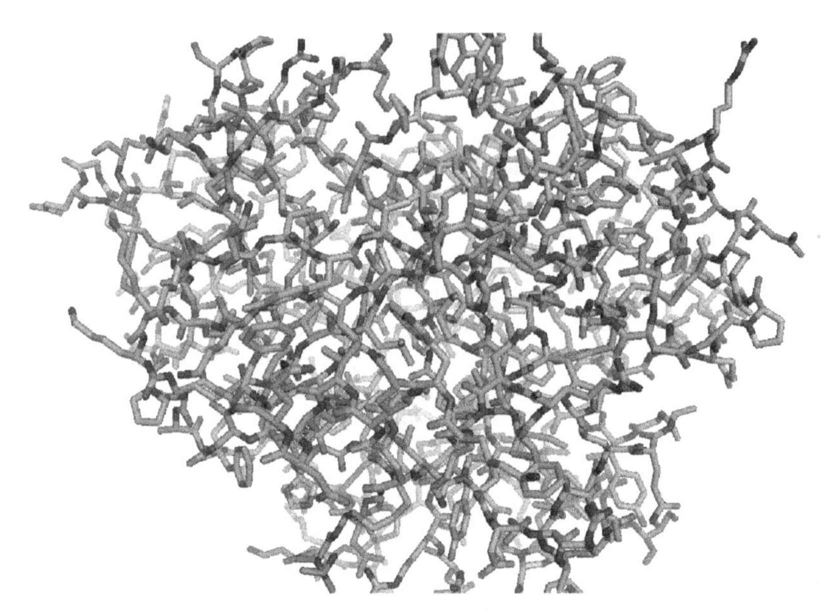

Figure 11.1 The figure of the chemical structure of a protein, *Protein Data Bank* (PDB) code 1AIX [6].

```
IVEGSDAEIGMSPWQVMLFRKSPQELLCGASLISDRWVLT
AAHCLLYPPWDKNFTENDLLVRIGKHSRTRYERNIEKISM
LEKIYIHPRYNWRENLDRDIALMKLKKPVAFSDYIHPVCL
PDRETAASLLQAGYKGRVTGWGNLKETWTANVGKGQPSVL
QVVNLPIVERPVCKDSTRIRITDNMFCAGYKPDEGKRGDA
CEGDSGGPFVMKSPFNNRWYQMGIVSWGEGCDRDGKYGFY
THVFRLKKWIQKVIDQFGE
```

Figure 11.2 Primary sequence of protein 1AIX [6].

11.2 GENETIC ALGORITHMS

One of the most common applications of machine learning for *ab initio* protein structure prediction is the use of the evolutionary algorithms, specifically GAs. GAs are general optimization algorithms that mimic the process of evolution through Darwinian natural selection, gene mutation, and reproduction. This mimicry comes into play with the major steps of this type of algorithm: fitness test, crossover, and mutation. Recent studies from Reference [16] show that GAs are said to be superior to the other evolutionary algorithms.

The benefits of GAs lie in the distinct characteristic of the maintenance of a large population of potential solutions characterized as chromosomes, genes, or individuals depending on the author. These chromosomes are all in competition with each other

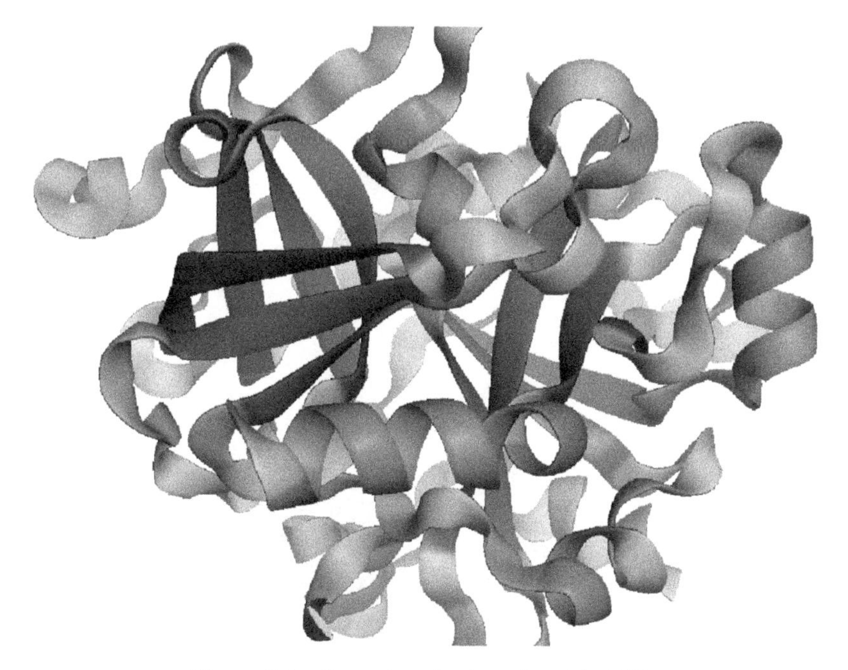

Figure 11.3 Secondary structure of protein 1AIX.

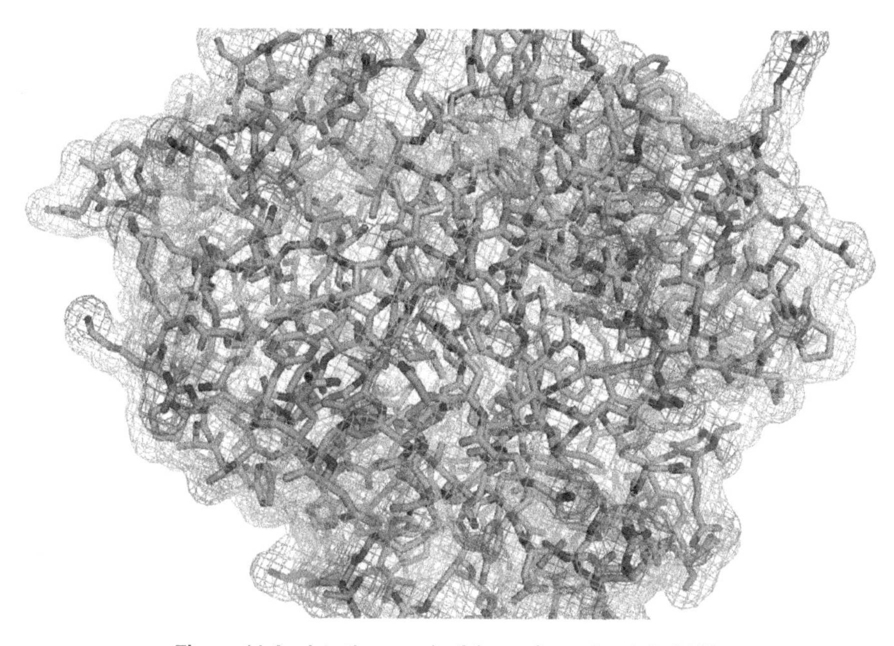

Figure 11.4 A tertiary mesh of the surface of protein 1AIX.

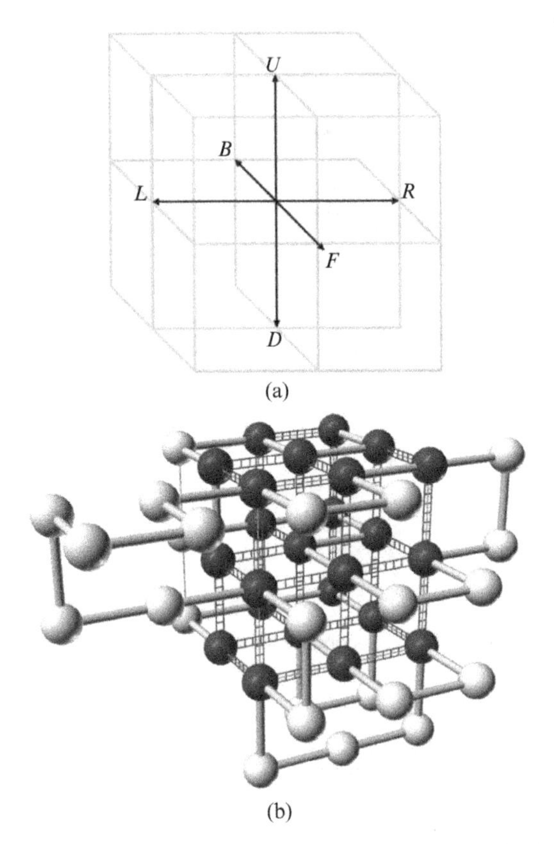

Figure 11.5 An example of the possible moves in the HH model (a), as well as a sequence with 48 monomers [HHHHPHHPHHHHHPPHPPHHPPHPPPPPPHPPHPPPHPPHH-PPHHHPH]. *Source*: Custódio et al. [16]. Reproduced with permission of Elsevier (b).

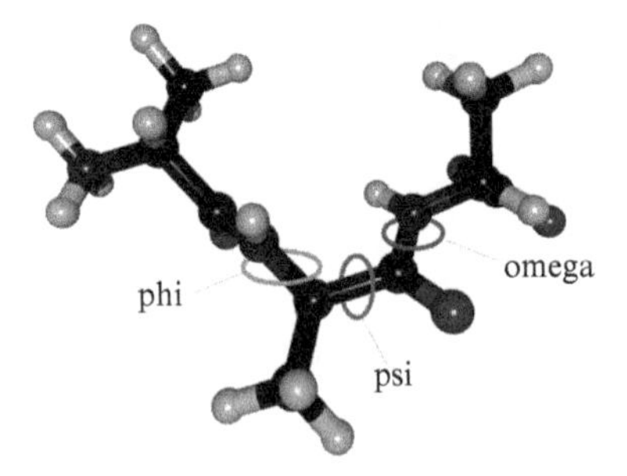

Figure 11.6 A molecule used to depict the ϕ, labeled phi, ψ, labeled psi, and ω, labeled omega, dihedral angles.

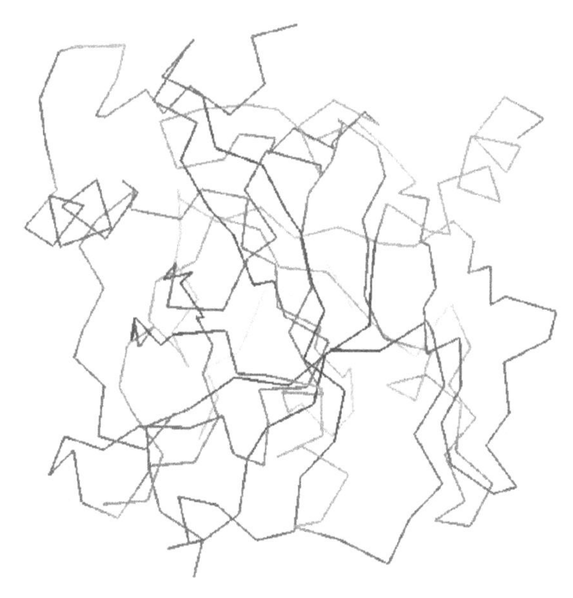

Figure 11.7 Illustration of the final product of an all-atom model. This is the actual backbone of protein 1AIX [6].

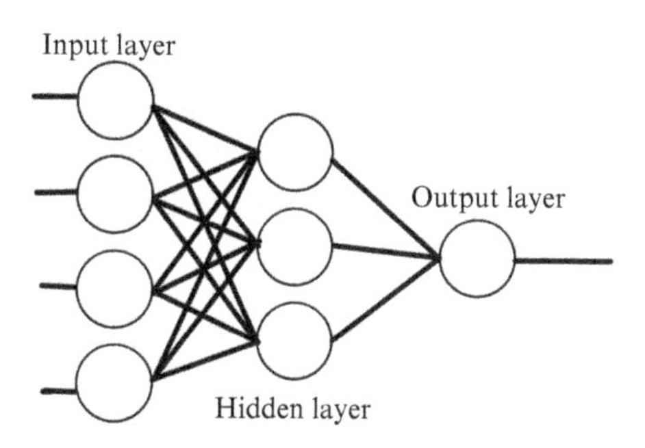

Figure 11.8 Basic structure of an ANN.

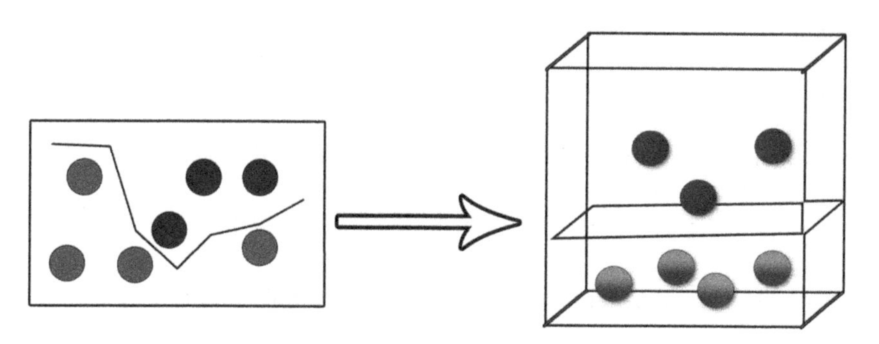

Figure 11.9 Basic structure of an SVM.

in the form of a fitness function that is used to determine the strength of a possible solution. Chromosomes are written as strings of information (in basic cases, binary string where each bit represents a specific value). The initial population is usually a random set of possible solutions. In each generation, including the initial population, every individual is evaluated and ranked by their respective fitness. In most common GAs, the individuals with the highest fitness are very likely to be chosen to be members of the next generation, whereas individuals with minimal fitness values are very unlikely to be chosen to go on to the next generation.

Mutation and crossover are additional GA operations that are applied to the sets of chromosomes to generate new individuals from generation to generation. The mutation operation represents the random and slight genetic changes that happen in nature, while the crossover operations represent the transfer of information in reproduction. There are multiple types of both operations, and depending on one's problem set, some versions of the operation work better than others.

In mutation, the algorithm is given percentage likelihood to select random bits of information to be changed. This step usually happens in one of the following three ways: single-point mutation (bit inversion), segment mutation, or order-changing mutation. In single-point mutation, only a random bit of information is changed to another possible value. In segment mutation, a block of information is mutated; sometimes, this presents itself as a single bit being mutated and the rest of the gene adjusting to make it a valid solution. In the order-changing method, two or more bits of information are exchanged.

In crossover, the algorithm is given a percentage likelihood that crossover will occur in the current generation of individuals; the determination of which individuals will be paired to perform a crossover is usually influenced by the ranking provided by the fitness function. Crossovers usually occur in one of the four following ways: single-point crossover, multipoint crossover, uniform crossover, and arithmetic crossover. In the single-point crossover, a point is selected in both individuals; this point is called the *crossover point*; the new genes are created by separating both individuals at the crossover point and combining those halved genes with the opposite half of the other individual. The multipoint crossover is similar to the single-point crossover but one selects multiple crossover points that split each individual into segments and these are combined with another individual's respective segments. In uniform crossover, the child gene is created by choosing random genes from both parents, and in arithmetic crossover, the child gene is the product of some function of both of the parents, for example, intersection-AND or exclusive-OR.

In each GA, the functions covered are applied in some way, hoping that after a certain number of generations and with a large enough initial population, the individuals will converge on an optimal solution or solution set. The effectiveness of a GA is in the determination of how the fitness test, crossover, and mutation are applied.

11.2.1 GA Model Selection in Protein Structure Prediction

When applying GAs to *ab initio* protein structure prediction, the major goal is to find the structural conformation with the lowest energy conformation. This is not an easy task and one of the simplest and common applications used to find possible low energy conformations (the hydrophobic–hydrophilic model) was determined by Unger and Moult to be an NP-hard problem [60, 61]. But before looking at possible ways to determine energy conformation, when designing a GA for this problem, one must first decide what type of model to use to formulate the problem. Currently, the types of models one has to choose are cubic lattice models and all-atom models.

The cubic lattice model is a discretization of the conformational space of the protein, because of which predicting the exact protein structure is impossible [16]. However, this system is still very widely used because it is significantly less complex and still captures many of the aspects of protein structures [30]. Using the cubic lattice, one is able to predict folding patterns, but it still presents the issues that arise from predicting the energy landscape of protein, namely, are massive multimodality, multiple conformations with equal energy, and large regions of unfeasible conformations [16]. With the cubic lattice model, each amino acid of the protein is given a position on a three-dimensional grid and linked together in a chain with backbone dihedrals.

Most authors who use the cubic lattice model [16, 30, 36, 61] use a specific type of cubic lattice model called the *Hydrophobic–Hydrophilic* (HH) or *Hydrophobic–Polar* (HP) model. This is a simplified model in which each amino acid is recognized as hydrophobic or hydrophilic. This is because hydrophobic amino acids tend to migrate toward each other to avoid polar solvents; this migration causes the formation of a hydrophobic core. The only requirement for this type of system is that there can be no collisions of amino acids in the conformation. In this system, the energy conformation is calculated by counting the number of HH contacts in adjacent nonbonded sites in the lattice, so the equation for a valid conformation would be the following:

$$E(c) = -n$$

In this function for energy conformation, c represents the conformation and n represents the total number of HH contacts in the conformation. There are other equations that may calculate energy conformations for the cubic lattice model, some of which will be further discussed in the next section. The all-atom model is designed to be much more complex in behavior than the cubic lattice model. In this model, elements are no longer the amino acids in the protein but all the atoms that comprise the protein. The chain is no longer bound by a lattice or backbone dihedrals in the all-atom model. One reason the all-atom model is considered more complex is that an experimentally observed native structure may have a higher potential energy conformation than an unnatural but optimized structure. This is because in all-atom models the energy conformation is not the only factor that matters; it is also necessary to account for the effects on the entropy of the system as a whole (Figure 11.5) [16].

In all-atom models, energy conformations are calculated by using the physics of the atoms themselves. The potential energy is evaluated by calculating force fields that are empirical and classical physics approximations for the properties of atoms and chemical bonds. The Lennard-Jones potential and Coulomb's law are used to describe the electrostatic interaction between atoms, which take into account the polarity of every atom in the system. The uses of both the Lennard-Jones potential and Coulomb's law are discussed in more detail in the next section, as well as the equations used to represent them. While the force field can be directly optimized, a better approach is to use molecular dynamics simulations to model the motions of the atoms and search conformational space for minimums.

11.2.2 Common Methodology

In a GA, the methodology comes down to a few key points: individual representation, fitness test, mutation, and crossover rates and attributes. Although population size and number of generations are critical for the performance of a GA, there is little general guidance from theory. Constants are empirically adjusted to maximize the effectiveness of a GA on a specific problem. It is noteworthy to mention that even though usually each individual in the initial population is randomly generated, it is better to have a diverse set of individuals to cover a wider range of potential solutions.

11.2.2.1 Individual Representation. In the models previously discussed, current work only has the representation of the individuals in a very particular way. For the cubic lattice model, whether it is an HP model or something slightly more complicated, the representation of individuals is usually very similar. Unger and Moult [61] stated the concept of giving amino acids in the individual an absolute direction on the square lattice, that is, an individual is coded by a sequence $(U, D, L, R, F, B)^{n-1}$, where U, D, L, R, F, and B code, respectively, up, down, left, right, forward, and backward movement on the lattice, and n is the length of the protein. In Reference [40], the backward move was removed and the remaining moves were given in terms of relative movement from current position; this was done to further avoid invalid conformations. This representation is still in use for these types of models [16].

In the representation for all-atom models, each atom is given three dihedral angles usually characterized by (ϕ, ψ, ω) (Figure 11.6). In some cases, ω is excluded because even though the torsion angle ω are explicit, they do not change during execution and are generally kept in the *trans*(180°) configuration models (Figure 11.7) [16].

11.2.2.2 Fitness Test. The fitness functions of most GAs for protein structure prediction are directly related to the energy functions or force fields of proteins. Energy functions can be measured by molecular mechanics and force field components such as bond length potential, bond angle potential, torsion angle potential, improper torsion angle potential, dihedral angles, van der Waals pair interactions, electrostatic potential, and hydrogen bonds [36, 50]. As mentioned in the previous section, a very simplistic method for the cubic lattice model is counting the number of HH contacts, where a model with more contacts is in lower energy conformation than one with fewer contacts.

Because of the natural complexity of the system in an all-atom model, many more functions can be used but this increases the overall complexity of the GA. Some fitness functions measure the summation of the different potentials [50] but newer ones use equations that account for instances. An example of a fitness function based on Reference [16], is

$$E_{\text{total}}(r_i) = E_{\text{torc}} + E_{\text{LJ}} + E_{\text{coul}}$$

$$E_{\text{torc}} = \sum_{n}^{(N_\phi)} K_{\phi_n}[1 + \cos(n_{\text{n}}\phi_{\text{n}} - \delta_{\text{n}})]$$

$$E_{\text{LJ}} = \sum_{i \leq j}^{N_{\text{atoms}}} -\left(\frac{A_{ij}}{r_{ij}}\right)^6 + \left(\frac{B_{ij}}{r_{ij}}\right)^{12}$$

$$E_{\text{coul}} = \sum_{i<j}^{N_{\text{atoms}}} \frac{q_i q_j}{4\pi\varepsilon_0\varepsilon_r(r_{ij})r_{ij}}$$

Here, E_{torc} is the torsion energy, E_{LJ} is the Lennard-Jones potential, and E_{coul} is the Coulomb or electrostatic potential. In the previous equations, K_{ϕ_n} is an energy constant based on the torsion of a bond; n_n is the period; ϕ_n is the torsion angle; δ_n is the phase angle; values A_{ij} and B_{ij} are Lennard-Jones parameters that depend on the atomic type and the distance, r_{ij}, between atoms i and j at which the interparticle potential is 0; q_i and q_j are the charges of atoms i and j, respectively; and $\varepsilon_r(x)$ is a dielectric function that depends on that distance represented by x [16].

11.2.2.3 Crossover and Mutation. Applying the proper type and amount of both crossover and mutation is very important to the speed at which the optimal solutions arise. Although fitness functions are alsovery important, mutation and crossover determine how the individuals change and new solutions arise. In an early paper, Krasnogor et al. determined in the HP model that the single-point crossover was not able to transfer blocks of information well, but segment mutations acted as a powerful local search, which led them to state the best combination of parameters that had a small crossover probability and a high mutation and segment mutation probability [30].

Custòdio et al. [16] used combinations of six different genetic operators to get solutions in both their HP model and their all-atom model. They are a two-point crossover, a multipoint crossover, a segment mutation, an exhaustive search mutation, a local move, and a loop move. In exhaustive search mutation, Custòdio et al. test all possible directions for a randomly selected base and keep only the best solution; in their paper, this mutation demands four fitness functions and it was developed to hybridize the GA and an exhaustive local search. Local moves and loop moves are order-changing mutations, and where local changes swap the movements of two random locations, loop movements swap locations from two units that are five bases apart. Custòdio et al. also mention a repair function when collisions happen that fixes invalid conformations created by crossovers and mutations.

Different opinions have been expressed in the field on the selection of which parents perform set crossover. In most methods, one would have a random selection with a bias toward the individuals with higher fitness; this is to simulate keeping the stronger individuals throughout generation. Some believe in the tournament method where you take a constant number of random individuals and use crossover between them to create new individuals. Neither of these methods preserves the diversity of the individuals in a given population from generation to generation. This can be an issue because if one does not have a diverse population of highly fit individuals, one reduces the chance of converging to the optimum solution. A solution, called *crowding*, to this problem is given in Reference [16] based on the works of References [18] and [35]. In this method, the new individual created runs a search over the parent generation for the most similar parent. Either the parent or child is kept for the next generation, depending on which individual has the lower energy. If the energies are the same, then either individual is given a 50% chance of being represented.

11.2.2.4 Tracking Results. The quality of the model found with a GA depends both on the speed of the calculations and the quality of the energy function for which an optimum is being sought. The speed of the calculation can be estimated from the number of energy evaluations the GA has to perform as the system evolves from generation to generation. In addition, from Reference [17], we know when testing against a known structure, a prediction is considered similar if (i) secondary structure elements agree with the known structure; (ii) the orientation of those secondary structures is within 20° of the ones in the known structure; and (iii) the length of the secondary structure elements differ by no more than two residues of the known structure.

We know from foundational papers such as References [60, 61] that applying GAs to the HP model outperforms the Monte Carlo method in terms of finding a global minimum of two-dimensional lattices [61]. This concept was improved upon later in Reference [40], achieving a high number of HH contacts with less energy evaluations. We must mention that a number of early prediction methods are said to show experimental bias [41]. Pederson and Moult described in their paper that potentials in Reference [17] were parameterized to favor experimental structures, and earlier papers including Reference [53] relied on experimental secondary structures for success. They even mention that their own earlier works were only effective on small peptides.

More recent papers have shown less bias and continued success on furthering the belief that GAs are a great way to tackle the protein structure prediction issue. In the most recent studies, Cusòdio et al. [16] prove that their method is not only an efficient way to investigate the complex energy landscape of proteins but it also shows an increase in the probability of finding native structures. For the cubic lattice method, they developed tests that were performed against three HP benchmark sequences, and showed comparable solutions to models specifically designed for the HP model and superior to other evolutionary algorithms such as the Hydrophobic Zipper strategy [19], the Contact Interactions method [56], and Constraint-Based Hydrophobic Core Construction [66]. Even the all-atom implementation they developed found native structures with a success rate of 100% when they tested it against the polyalanine peptides.

11.3 SUPERVISED MACHINE LEARNING ALGORITHM

Referring back to Section 11.1, in terms of the protein structure prediction methods such as homology modeling and threading, the most common machine learning algorithms applied are those under the umbrella of supervised learning. Supervised learning algorithms are those in which inferences are made from labeled training sets. By labeled training set, we mean a provided set of examples where we know exactly what group or solution each example provides; the set itself is in general called a *training set* and the labels that correspond to the training set are called *training labels*. In supervised learning algorithms, examples are usually provided in the form of attribute vectors, and after one's algorithm is trained to predict a label hypothesis for each example provided, one runs it on the sample data to generate the answer. Traditional statistics methods as well as methods created to mimic the way neurological networks work are generally the functions one uses to create these label hypotheses. In this section, we will discuss the two most popular applications of supervised learning algorithms to the protein structure prediction problem, specifically, the homology modeling and threading techniques. The applications of supervised machine learning algorithms include ANNs and SVMs.

11.3.1 Artificial Neural Networks

ANNs are computational models that were first created by Rosenblatt in 1959 [46], inspired by the networks of biological neurons that contain multiple layers of simple computing nodes. In an ANN, each node operates as a nonlinear summing device and is represented as the neuron in the brain that could accept input values from feeding information and sending the processed outputs through the whole network. ANNs are very robust and have been successfully applied to solving many complex problems including prediction, classification, and pattern recognition. The structure of an ANN is given by a number of layers. These layers are divided into three groups: the input layer, the hidden layer, and the output layer (Figure 11.8). The input layer is the level at which the input data is fed into the ANN. The hidden layer is the level at which most of the processing and calculations are performed; the hidden layer is also the level at which the input values are subjected to the weighted values that are used to represent learned behavior. The output layer is the level where results are provided. The most commonly used ANN in the protein structure prediction problem, and one of the most commonly used in general, is the feed-forward back-propagated ANN. Feed-forward ANNs are those where the nodes do not form a directed cycle; information moves only in one direction. Back-propagation ANNs adjust the weights in the network to minimize the error in the output. In a feed-forward back-propagated ANN, the activation state of each node X_i is a real value between 0 and 1. The strength of the connection, or weight, between a node i and another node j is represented by a real number W_{ij}. The activation of a given node in the ANN can be calculated by summing the products of every input node's output Y_j and weight W_{ij}, and then adding a bias term, b_j

$$X_i = \sum_j W_{ij} Y_j + b_j$$

Once the activation X_i for a node has been calculated, the output of that node Y_i can be computed using the logistic sigmoid function and then propagated to the next layer of the ANN.

$$Y_i = \frac{1}{1 + e^{-X_i}}$$

During each cycle, the input is given to the ANN. The weight of the nodes is adjusted at the end of the cycle, and this procedure is repeated. This form of supervised training, in which the desired output is presented to the ANN along with the inputs, has been used to train the ANN [32]. In protein structure prediction, ANNs are used to determine the conformation of the protein backbone using ϕ and ψ dihedral angles.

11.3.2 ANNs in Protein Structure Prediction

In 1988, Qian and Sejnowski [44] were the first to use ANNs to predict protein secondary structure. The model they used is a fully connected *Multi-Layer Perceptron* (MLP) with one hidden layer [4] (Haykin). The input is a small window of 13 amino acids, and the three output nodes are the probabilities of an amino acid that belongs to a helix, sheet, or loop region. They randomly weighted each input and used back-propagation [4] to train the system on a PDB file chosen from the PDB. The whole system achieved up to 64% accuracy, and the authors suggested that a significant improvement could not be made. However, in spite of this, a number of papers using ANNs to predict protein secondary structure were published later using similar techniques [24, 29, 52]. Scientists and researchers are working hard to improve the performance of ANNs. One method is to include more information in combination of the inputs, for example, the evolutionary information and multiple sequence alignments [47, 48]. Moreover, owing to the rapid growth of the PDB database, the training sets for ANNs have also increased. This has therefore enhanced ANNs' performances [8, 42, 43]. Furthermore, other algorithms were combined with ANNs in order to improve their performances [32, 34]. Servers were built on the basis of these algorithms to provide a service to the structural biology community. The PSIPRED protein structure prediction server based on three developed methods, PSIPRED, GenTHREADER, and MEMSAT 2, was built by McGuffin and his colleagues in 2000 [37]. In 2008, the Jpred 3 secondary structure prediction server [11] was finalized on the basis of the Jnet algorithm, which was proposed by Cuff and Barton [15]. Since these artificial intelligence methods are based on an empirical risk-minimization principle, they have some disadvantages, for example, finding the local minimal instead of global minimal; having low convergence rate; more prone to overfitting; and when the size of fault samples is limited, it might cause poor generalization. Even though scientists are trying to overcome these disadvantages [45], most of them still remain.

11.3.3 Support Vector Machines

The theory of SVMs was developed by Cortes and Vapnik [12]. Although they can expound to multiple classes, at a very basic level, SVMs are designed for two-class classification problems (Figure 11.9). What is provided to the machine is a set of nD-dimensional vectors x_i and their associated classes y_i:

$$\{x_1, y_1\}, \ldots, \{x_n, y_n\} \in \mathrm{IR}^D \times \{\pm 1\}$$

The main idea is to construct a hyperplane to separate the two classes (labeled $y \in \{-1, 1\}$) so that margin (the distance between the hyperplane and the nearest point) is maximal. It is equivalent to solving the following optimization problem:

$$\min \frac{1}{2}(w \cdot w) + C \sum_{i=1}^{n} \xi_i$$

with constraints

$$y_i((w \cdot x_i) + b) \geq 1 - \xi_i$$

$$\xi_i \geq i = 1, \ldots, n$$

where ξ_i is a slack variable giving a soft classification boundary and C is a constant corresponding to the value $\|w\|^2$ [13]. The boundary is also called the *optimal separating hyperplane*, which is chosen by maximizing the distance from the closest training data [55].

SVMs are considered the most efficient supervised machine learning methods not only because of their well-developed learning theory [62, 63] but also because of their superior performances in practical applications in pattern-recognition problems [12, 22, 49, 57, 58]. SVMs have many good features, including avoiding of overfitting with the use of structural risk minimization, the capability of handling large feature spaces, the assurance that the training converges to a global optimal solution, and the condensation of information in a given data set that is made without loss of useful information [26].

SVMs have been applied to solve protein structure prediction problems since 2001 [26]. In Reference [26], the authors used frequency profiles with evolutionary information as input. The goal was to improve the performances of the SVM. Their research issues were to define the type of kernels to be used and what data to represent or encode for the training.

One method is to study different SVM kernels, for example, using a quadratic kernel function to increase the accuracy of helix prediction [64], dual-layered SVMs with a profile [23], two layers of SVMs with weighted cost functions for balanced training [7], and mixed-modal SVMs [65].

Another way to increase the accuracy of SVMs is to include different encoding methods as input, for example, SVMpsi incorporated PSI-BLAST *Position-Specific Scoring Matrix* (PSSM) profiles as an input vector [28], mixed encoding schemes of an orthogonal matrix, hydrophobicity matrix, and BLOSUM62 substitution matrix together with SVMs [25], while Chen et al. [10] used PseAAC [51] as the protein sequence representation and the optimal length of local protein structure was used as input [21]. Furthermore, in an attempt to improve the performances of SVMs [9], combinations of the above two methods have been tried out. Normally, SVMs are binary classifiers. In some studies, scientists also extend this classification method to predict multiclass structures [38, 64].

11.4 FUZZY APPLICATION

A not entirely novel concept, but less explored, is the application of fuzzy logic to protein structure prediction. Fuzzy logic systems incorporate a system of many-valued logic in which an element in a set is given a value to describe its membership to different groupings; the basic example of this can be understood by visualizing an item that is partially true and partially false. Biological systems are not always rigid; this is to say that, in some protein structures, torsion angles change and amino acids are not always in constant position relative to the main structure. It has already been shown that computationally efficient implementations of *Fuzzy Decision Trees* (FDTs) [2] are competitive with other efficient learning algorithms to analyze and predict protein structure, and have been effective in recognizing and extracting sections in RNA structures. Another example is in the homology modeling and threading methods that are so commonly used in the supervised learning techniques; the application of fuzzy logic would assign a confidence value to the structures that the algorithms would be comparing as done in Reference [68].

11.4.1 Fuzzy Logic

Fuzzy logic and fuzzy sets operate under the notion of uncertainty: the idea that an element may not completely belong to one set and the operations that impact that belief. As a reminder, in common logic systems (referred to as *crisp*), elements of a set are either inside the set or outsideit, true or false, 1 or 0. A fuzzy set A is defined as a set $A = \{x_1, x_2, ..., x_n\}$ for $n = \| A \|$ and there is a membership function $\mu_A : X \rightarrow [0, 1]$, where $0 \leq \mu_A(x_i) \leq 1$ for $x_i \in A$, $i = 1, ..., n$. This logic extends to the same sets of operations given to crisp sets (e.g., intersection and union) [67].

11.4.2 Fuzzy SVMs

The fuzzification of SVMs replaces data as a function of parameters with data as a function of belief in class memberships. This happens by applying a membership value to each input point of the SVM as well as reformulating the machine so that two inputs can make different contributions to the learning. The basic definition of an SVM is changed in the following manner: let S be a set of training points of labeled data, containing nD-dimensional vectors x_i and their associated classes y_i, as well as their associated fuzzy membership value s_i:

$$x_1, y_1, s_1, ..., x_n, y_n, s_n \in \text{IR}^D \times \{\pm 1\}$$

where $x_i \in \text{IR}^D, y_i \in \{\pm 1\}$, and $0 \leq \sigma \leq s_i \leq 1$ with $i = 1, ..., n$. σ is treated as a minimum confidence value used to remove small membership function values. Let $z = \varphi(x)$ denote the corresponding feature space vector with a mapping φ from IR^D to a feature space Z. With the incorporation of the fuzzy membership value s_i defined over x_i, the measure of error of that point ξ_i is adjusted by the membership. This is to say that the fuzzy membership must be included in the SVM to maximize the margin, and it is done with this function [33]:

$$\min \frac{1}{2}(w \cdot w) + C \sum_{i=1}^{n} s_i \xi_i$$

with constraints

$$y_i((w \cdot x_i) + b) \geq 1 - \xi_i$$
$$\xi_i \geq i = 1, ..., n$$

11.4.3 Adaptive-Network-Based Fuzzy Inference Systems

An *Adaptive-Network-based Fuzzy Inference System* or *Adaptive Neural Fuzzy Inference System* (ANFIS) is an ANN described in Reference [27] based on the Takagi–Sugeno fuzzy inference system [54]. This method consists of five layers (Figure 11.10):

- Layer 1: Each node is this layer is a membership function

  $$O_i^1 = \mu_{A_i}(x)$$

  where x is the input to node i and A_i is the group or label of membership associated with this node.
- Layer 2: This layer consists of nodes, which multiply the incoming signals and send the product out to the next layer. Each node output represents the firing strength of a specific fuzzy rule.
- Layer 3: Each node normalizes the firing strength based on node i over the summation of all of the nodes.

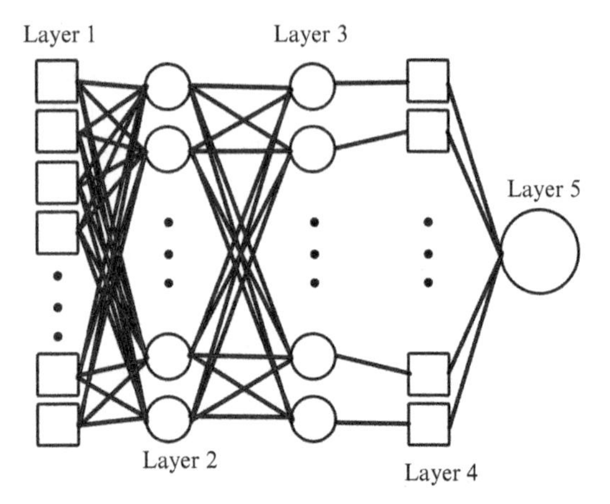

Figure 11.10 Structure of an ANFIS.

- Layer 4: Each node i in this layer is given the following node function

$$O_i^4 = \overline{w}_i f_i = \overline{w}_i(p_i x + q_i y + r_i)$$

where $\overline{w}_i$ is the output of Layer 3 and p_i, q_i, r_i is the consequent parameter set.
- Layer 5: This layer consists of a single node that computes the overall output as the summation of all of the incoming signals from the previous layer

$$O_i^5 = \text{overall output} = \sum_i \overline{w}_i f_i$$

11.4.4 Fuzzy Decision Trees

The FDT is another method of machine learning that is yet to be discussed in this chapter. Being considered one of the most popular methods, FDTs represent classification of knowledge with a fuzzy rule set to make its classifications more robust and immune to data uncertainties [1]. In Reference [1], there is an extension of the methodology described in Reference [59] to develop an FDT that gives better accuracy than the original. The algorithm developed in Reference [1] is as follows:

1. Generate the root node with a fuzzy set of all training data with the membership value set to 1.
2. A node is a leaf node, if the fuzzy set of data at the node satisfies any of the following conditions:
   a. The number of objects in the node's data set is less than a given threshold.
   b. All the objects in the node's fuzzy data set belong to one class.
   c. No more attributes are available for classification.

3. The class assigned to a leaf node is the name of the class with the greatest membership value in the node's fuzzy data set.
4. If the node does not satisfy any of the above conditions, then do the following:
   a. If the current node is the root node of the tree, calculate the classification ambiguity of each attribute and select the attribute with the smallest classification ambiguity as the test attribute.
   b. Else if the current node is not a root node, then for each attribute not selected so far as a test attribute up in the path from the current node to the root node of the tree, calculate the information gain and select the attribute with the maximum information gain as the test attribute.
   c. Divide the fuzzy data set at the node into fuzzy subsets using the selected test attribute, with the membership value of an object in a subset set to the product of the membership value in the parent node data set and the value of the selected attribute fuzzy term in the data at the node.
   d. For each of the subsets, generate a new node with the branch labeled with the fuzzy term of the selected attribute.
5. For each newly generated node, repeat recursively from step 2.

11.5 CONCLUSION

The prediction and analysis of correctness of protein tertiary structure have proven to be arduous tasks in the fields of biology and computer science. There is a wealth of research on the attempts from different groups to define the best methods of finding correct conformations of proteins, as well as to determine whether or not a conformation is infeasible or incorrect. These attempts include a wide range of machine learning techniques, the most successful being GAs, ANNs, and SVMs. Through the application of fuzzy methodologies to these machine learning techniques, favorable results have been obtained.

Zhong et al. [68] discovered that cluster membership functions in SVMs may not reveal the nonlinear sequence-to-structure relationship. They improved this method by proposing the use of fuzzy SVMs. These machines assume that the confidence value from the SVM's decision function plays more important roles in assigning the final class for sequence segments than the clustering membership score. Ding et al. [20] showed that the basic fuzzy SVMs are an efficient approach for predicting protein structure and can also be used to predict the quality of other protein attributes. By using ANFIS, Riis et al. [45] showed in 1996 that the application of a hybrid neuro-fuzzy GA generated decision models that were precise, were generalizable, and provided better overall results than many crisp methods. The method also showed a rapid convergence in fuzzy rule generalization. A similar work in 2013 [31] showed similar advantages with a comparable methodology. Abu-Halaweh and Harrison [2] showed that FDTs are reliable methods to determine essential features in a similar problem, the prediction and classification of real and pseudo-microRNAs, as well as to determine important features for the classification problem that could be further used in other methodologies such as SVMs.

REFERENCES

1. Abu-Halaweh NM, Harrison RW. Practical fuzzy decision trees. Computational Intelligence and Data Mining, 2009. CIDM'09. IEEE Symposium on. Nashville, TN; IEEE; 2009. p 211–216.
2. Abu-Halaweh NM, Harrison RW. Identifying essential features for the classification of real and pseudo micrornas precursors using fuzzy decision trees. Computational Intelligence in Bioinformatics and Computational Biology (CIBCB), 2010 IEEE Symposium on. IEEE; Montreal, Canada; 2010. p 1–7.
3. Ben-David M, Noivirt-Brik O, Paz A, Prilusky J, Sussman JL, Levy Y. Assessment of CASP8 structure predictions for template free targets. Proteins 2009;77(S9):50–65.
4. Bishop CM. *Neural Networks for Pattern Recognition*. New York: Oxford University Press; 1995.
5. Blundell TL, Sibanda BL, Sternberg MJE, Thornton JM. Knowledge-based prediction of protein structures. Nature 1987;326:347–352.
6. Bode W, Turk D, Karshikov A. The refined 1.9-a x-ray crystal structure of d-phe-pro-arg chloromethylketone-inhibited human alpha-thrombin: structure analysis, overall structure, electrostatic properties, detailed active-site geometry, and structure-function relationships. Protein Sci. 1992;1(4):426–471.
7. Casbon J. *Protein secondary structure prediction with support vector machines* [Master's thesis]. University of Sussex; 2002.
8. Chandonia J-M, Karplus M. New methods for accurate prediction of protein secondary structure. Proteins 1999;35(3):293–306.
9. Chatterjee P, Basu S, Kundu M, Nasipuri M, Plewczynski D. PSP_MCSVM: brainstorming consensus prediction of protein secondary structures using two-stage multiclass support vector machines. J Mol Model 2011;17(9):2191–2201.
10. Chen C, Chen L, Zou X, Cai P. Prediction of protein secondary structure content by using the concept of Chou's pseudo amino acid composition and support vector machine. Protein Pept Lett 2009;16(1):27–31.
11. Cole C, Barber JD, Barton GJ. The jpred 3 secondary structure prediction server. Nucleic Acids Res 2008;36 Suppl 2:W197–W201.
12. Cortes C, Vapnik V. Support-vector networks. Mach Learn 1995;20(3):273–297.
13. Cristianini N, Shawe-Taylor J. An introduction to support vector machines; 2000.
14. Crivelli S, Eskow E, Bader B, Lamberti V, Byrd R, Schnabel R, Head-Gordon T. A physical approach to protein structure prediction. Biophys J 2002;82(1):36–49.
15. Cuff JA, Barton GJ. Application of multiple sequence alignment profiles to improve protein secondary structure prediction. Proteins 2000;40(3):502–511.
16. Custódio FL, Barbosa HJC, Dardenne LE. A multiple minima genetic algorithm for protein structure prediction. Appl Soft Comput 2014;15:88–99.
17. Dandekar T, Argos P. Ab initio tertiary-fold prediction of helical and non-helical protein chains using a genetic algorithm. Int J Biol Macromol 1996;18(1):1–4.

18. De Jong KA. Analysis of the behavior of a class of genetic adaptive systems; 1975.
19. Dill KA, Fiebig KM, Chan HS. Cooperativity in protein-folding kinetics. Proc Natl Acad Sci U S A 1993;90(5):1942–1946.
20. Ding Y-S, Zhang T-L, Chou K-C. Prediction of protein structure classes with pseudo amino acid composition and fuzzy support vector machine network. Protein Pept Lett 2007;14(8):811–815.
21. Fai CY, Hassan R, Mohamad MS. Protein secondary structure prediction using optimal local protein structure and support vector machine. Int J Bio-Sci Bio-Technol 2012;4(2):35.
22. Furey TS, Cristianini N, Duffy N, Bednarski DW, Schummer M, Haussler D. Support vector machine classification and validation of cancer tissue samples using microarray expression data. Bioinformatics 2000;16(10):906–914.
23. Guo J, Chen H, Sun Z, Lin Y. A novel method for protein secondary structure prediction using dual-layer svm and profiles. Proteins 2004;54(4):738–743.
24. Holley LH, Karplus M. Protein secondary structure prediction with a neural network. Proc Natl Acad Sci U S A 1989;86(1):152–156.
25. Hu H-J, Pan Y, Harrison R, Tai PC. Improved protein secondary structure prediction using support vector machine with a new encoding scheme and an advanced tertiary classifier. IEEE Trans NanoBiosci 2004;3(4):265–271.
26. Hua S, Sun Z. A novel method of protein secondary structure prediction with high segment overlap measure: support vector machine approach. J Mol Biol 2001;308(2):397–407.
27. Jang J-SR. ANFIS: adaptive-network-based fuzzy inference system. IEEE Trans Syst Man Cybern 1993;23(3):665–685.
28. Kim H, Park H. Protein secondary structure prediction based on an improved support vector machines approach. Protein Eng 2003;16(8):553–560.
29. Kneller DG, Cohen FE, Langridge R. Improvements in protein secondary structure prediction by an enhanced neural network. J Mol Biol 1990;214(1):171–182.
30. Krasnogor N, Hart W, Smith J, Pelta D. Protein structure prediction with evolutionary algorithms; 1999.
31. Krishnaji A, Rao AA. An improved hybrid neuro fuzzy genetic system (I-HNFGS) for protein secondary structure prediction from amino acid sequence. Advances in Computing, Communications and Informatics (ICACCI), 2013 International Conference on. IEEE; Mysore, India; 2013. p 1218–1223.
32. Kushwaha SK, Shakya M. Neural network: a machine learning technique for tertiary structure prediction of proteins from peptide sequences. Advances in Computing, Control, & Telecommunication Technologies, 2009. ACT'09. International Conference on. IEEE; Kerala, India; 2009. p 98–101.
33. Lin C-F, Wang S-D. Fuzzy support vector machines. IEEE Trans Neural Netw 2002;13(2):464–471.
34. Lin K, Simossis VA, Taylor WR, Heringa J. A simple and fast secondary structure prediction method using hidden neural networks. Bioinformatics 2005;21(2):152–159.
35. Mahfoud SW. Crowding and preselection revisited. Urbana 1992;51:61801.
36. Mansour N, Kanj F, Khachfe H. Enhanced genetic algorithm for protein structure prediction based on the HP model. Search Algorithms Appl 2011;3:69.
37. McGuffin LJ, Bryson K, Jones DT. The psipred protein structure prediction server. Bioinformatics 2000;16(4):404–405.

38. Nguyen MN, Rajapakse JC. Multi-class support vector machines for protein secondary structure prediction. Genome Inform Ser 2003;14:218–227.

39. Olszewski KA, Piela L, Scheraga HA. Mean field theory as a tool for intramolecular conformational optimization. 3. Test on melittin. J Phys Chem 1993;97(1):267–270.

40. Patton AL, Punch WF III, Goodman ED. A standard GA approach to native protein conformation prediction. ICGA; Pittsburgh, PA; 1995. p 574–581.

41. Pedersen JT, Moult J. Genetic algorithms for protein structure prediction. Curr Opin Struct Biol 1996;6(2):227–231.

42. Pollastri G, Mclysaght A. Porter: a new, accurate server for protein secondary structure prediction. Bioinformatics 2005;21(8):1719–1720.

43. Pollastri G, Przybylski D, Rost B, Baldi P. Improving the prediction of protein secondary structure in three and eight classes using recurrent neural networks and profiles. Proteins 2002;47(2):228–235.

44. Qian N, Sejnowski TJ. Predicting the secondary structure of globular proteins using neural network models. J Mol Biol 1988;202(4):865–884.

45. Riis SK, Krogh A. Improving prediction of protein secondary structure using structured neural networks and multiple sequence alignments. J Comput Biol 1996;3(1):163–183.

46. Rosenblatt F. Two theorems of statistical separability in the perceptron, mechanization of thought processes. Proceedings of a Symposium on the Mechanisation of Thought Processes; London; 1959. p 421–456.

47. Rost B, Sander C. Prediction of protein secondary structure at better than 70% accuracy. J Mol Biol 1993;232(2):584–599.

48. Rost B, Sander C. Combining evolutionary information and neural networks to predict protein secondary structure. Proteins 1994;19(1):55–72.

49. SchiilkopP B, Burgest C, Vapnik V. Extracting support data for a given task; 1995.

50. Schulze-Kremer S. Genetic algorithms for protein tertiary structure prediction. Machine Learning: ECML-93. Springer; Vienna, Austria; 1993. p 262–279.

51. Shen H-B, Chou K-C. Pseaac: a flexible web server for generating various kinds of protein pseudo amino acid composition. Anal Biochem 2008;373(2):386–388.

52. Stolorz P, Lapedes A, Xia Y. Predicting protein secondary structure using neural net and statistical methods. J Mol Biol 1992;225(2):363–377.

53. Sun S. Reduced representation model of protein structure prediction: statistical potential and genetic algorithms. Protein Sci 1993;2(5):762–785.

54. Takagi T, Sugeno M. Fuzzy identification of systems and its applications to modeling and control. IEEE Trans Syst Man Cybern 1985;15(1):116–132.

55. Tang X, Zhuang L, Cai J, Li C. Multi-fault classification based on support vector machine trained by chaos particle swarm optimization. Knowl Based Syst 2010;23(5):486–490.

56. Toma L, Toma S. Contact interactions method: a new algorithm for protein folding simulations. Protein Sci 1996;5(1):147–153.

57. Tong S, Chang E. Support vector machine active learning for image retrieval. Proceedings of the 9th ACM International Conference on Multimedia. Ontario, Canada; ACM; 2001. p 107–118.

58. Tong S, Koller D. Support vector machine active learning with applications to text classification. J Mach Learn Res 2002;2:45–66.

59. Umano M, Okamoto H, Hatono I, Tamura H, Kawachi F, Umedzu S, Kinoshita J. Fuzzy decision trees by fuzzy ID3 algorithm and its application to diagnosis systems. Fuzzy Systems, 1994. IEEE World Congress on Computational Intelligence., Proceedings of the 3rd IEEE Conference on. IEEE; Orlando, FL; 1994. p 2113–2118.
60. Unger R, Moult J. Finding the lowest free energy conformation of a protein is an NP-hard problem: proof and implications. Bull Math Biol 1993;55(6):1183–1198.
61. Unger R, Moult J. Genetic algorithms for protein folding simulations. J Mol Biol 1993;231(1):75–81.
62. Vapnik V. Statistical learning theory; 1998.
63. Vapnik V. *The Nature of Statistical Learning Theory*. New York: Springer-Verlag; 2000.
64. Ward JJ, McGuffin LJ, Buxton BF, Jones DT. Secondary structure prediction with support vector machines. Bioinformatics 2003;19(13):1650–1655.
65. Yang B, Wu Q, Ying Z, Sui H. Predicting protein secondary structure using a mixed-modal SVM method in a compound pyramid model. Knowl Based Syst 2011;24(2):304–313.
66. Yue K, Dill KA. Forces of tertiary structural organization in globular proteins. Proc Natl Acad Sci U S A 1995;92(1):146–150.
67. Zadeh LA, logic F. Neural networks, and soft computing. Commun ACM 1994;37(3):77–84.
68. Zhong W, He J, Pan Y. Multiclass fuzzy clustering support vector machines for protein local structure prediction. Bioinformatics and Bioengineering, 2007. BIBE 2007. Proceedings of the 7th IEEE International Conference on. IEEE; Boston, MA; 2007. p 21–26.

CHAPTER 12

PROTEIN INTER-DOMAIN LINKER PREDICTION

Maad Shatnawi[1], Paul D. Yoo[3,5], and Sami Muhaidat[2,4]

[1]*Higher Colleges of Technology, Abu Dhabi, UAE*
[2]*ECE Department, Khalifa University, Abu Dhabi, UAE*
[3]*Centre for Distributed and High Performance Computing, University of Sydney, Sydney, Australia*
[4]*EE Department, University of Surrey, Guildford, UK*
[5]*Department of Computing and Informatics, Bournemouth University, UK*

12.1 INTRODUCTION

The accurate and stable prediction of protein domains is an important stage for the prediction of protein structure, function, evolution, and design. Predicting inter-domain linkers is very important for accurate identification of structural domains within a protein sequence. Downsizing of proteins without loss of their function is one of the major targets of protein engineering [47, 50]. This has an important effect in reducing computational cost. Many domain prediction methods first detect domain linkers, and in turn predict the location of domain regions. The knowledge of domains is used to classify proteins, predict *Protein–Protein Interaction* (PPI), and understand their structures, functions, and evolution. Therefore, efficient computational methods for splitting proteins into structural domains are gaining practical importance in proteomics research [20].

This chapter provides a comprehensive comparative study of the existing approaches in domain linker prediction and discusses the technical challenges and

Pattern Recognition in Computational Molecular Biology: Techniques and Approaches,
First Edition. Edited by Mourad Elloumi, Costas S. Iliopoulos, Jason T. L. Wang, and Albert Y. Zomaya.

unresolved issues in this field. It also presents two different domain linker prediction approaches.

The rest of this chapter is organized as follows. Section 12.2 provides a brief overview of protein structure. Section 12.3 addresses the key technical challenges that face inter-domain linker prediction and the open issues in this field. Section 12.4 discusses the evaluation measures that are typically used in domain linker prediction. Section 12.5 provides a comprehensive description and comparison for the most current research approaches in the field of inter-domain linker prediction. Section 12.6 presents a *Domain Boundary Prediction Using Enhanced General Regression Network* (DomNet) approach. Section 12.7 presents a domain linker prediction approach using *Amino Acid* (AA) compositional index and simulated annealing. Concluding remarks are presented in Section 12.8.

12.2 PROTEIN STRUCTURE OVERVIEW

A *protein* is a polypeptide that is a linear polymer of several AAs connected by peptide bonds. The *primary structure* of a protein is the linear sequence of its AA units starting from the amino-terminal residue (N-terminal) to the carboxyl-terminal residue (C-terminal). Although protein chains can become cross-linked, most polypeptides are unbranched polymers, and therefore, their primary structure can be presented by the AA sequence along their main chain or backbone [3]. AAs consist of carbon, hydrogen, oxygen, and nitrogen atoms that are clustered into functional groups. Two of these functional groups define the AAs: the amino group (NH_2) and the carboxyl group (COOH).

There are over 300 naturally occurring AAs on earth, but the number of different AAs in proteins is only 20. All AAs have the same general structure, but each has a different R-group. The carbon atom to which the R group is connected is called *alpha carbon*. These AAs are alanine, arginine, asparagine, aspartic acid, cysteine, glutamic acid, glutamine, glycine, histidine, isoleucine, leucine, lysine, methionine, phenylalanine, proline, serine, threonine, tryptophan, tyrosine, and valine. These 20 AAs are represented by one-letter abbreviation as A, R, N, D, C, Q, E, G, H, I, L, K, M, F, P, S, T, W, Y, and V. AAs are connected to make proteins by a chemical reaction in which a molecule of water is removed, leaving two AA residues connected by a peptide bond.

Connecting multiple AAs in this way produces a polypeptide. This reaction leaves the C of the carboxyl group directly linked to the N of the amino group. This linked group of atoms (CONH) is called *peptide bond*. Polypeptides can be thought of as a string of alpha carbons alternating with peptide bonds. Since each alpha carbon is attached to an R-group, a given polypeptide is distinguished by the sequence of its R-groups.

The *secondary structure* of a protein is the general *three-dimensional* (3D) form of its local parts without displaying specific atomic positions, which are considered to be a *tertiary structure*. The most common secondary structures are alpha (α) helices and beta (β) sheets. The α-helix is a right-handed spiral array while the β-sheet is made up of beta strands connected crosswise by two or more hydrogen bonds, forming a twisted pleated sheet. These secondary structures are linked together by tight turns and loose flexible loops [3].

Domains are the basic functional units of protein tertiary structures. A protein *domain* is a conserved part of a protein sequence that can evolve, function, and exist independently. Each domain forms a compact 3D structure and mostly can be independently stable and folded. Several domains are joined together in several combinations forming multi-domain protein sequences [8, 45]. Each domain in a multi-domain protein has its own functions and mutually works with its neighboring domains. One domain may exist in a variety of different proteins. Domains vary in length from 25 to 500 AA residues. Accurate detection of domain boundaries is significant for many protein research activities as it will allow the breakdown of a protein into its functional units [42].

12.3 TECHNICAL CHALLENGES AND OPEN ISSUES

There are several technical challenges that face computational analysis of protein sequences in general and domain boundary prediction in particular. First, there have been a huge amount of newly discovered protein sequences in the postgenomic era. Second, protein chains are typically large and contain multiple domains that are difficult to characterize by experimental methods. Third, the availability of large, comprehensive, and accurate benchmark data sets is required for the training and evaluation of prediction methods.

One of the challenges of protein prediction methods is the protein sequences representation. Protein prediction methods vary in protein sequences representation and feature extraction in order to build their classification models. Two kinds of models were generally used to represent protein samples: the sequential model and the discrete model. The most and simplest sequential model for a protein sample is its entire AA sequence. However, this approach does not work well when the query protein did not have high sequence similarity to any attribute-known proteins [9]. Several nonsequential models, or discrete models, were proposed. The simplest discrete model is the protein AA composition that contains the normalized occurrence frequencies of the 20 native AAs in a protein. However, all the sequence-order knowledge will be lost using this representation that, in turn, will negatively affect the prediction quality [9]. Some approaches such as in Reference [30] use physicochemical properties of AAs. Other approaches use AA flexibility such as CHOPnet [24], gene ontology, solvent accessibility information, and/or evolutionary information such as DOMpro [7]. Protein secondary structure information has also been broadly used for domain boundary prediction such as SSEP-Domain [18]. Several approaches have used the 3D coordinates of protein structure while other methods identify domains in globular proteins from one-dimensional atomic coordinates [16, 45].

There are various challenges that face machine-learning protein prediction methods. Selecting the best machine-learning approach is a great challenge. There are a variety of techniques that diverse in accuracy, robustness, complexity, computational cost, data diversity, overfitting, and dealing with missing attributes and different features. Most machine-learning approaches of protein sequences are computationally expensive and often suffer from low prediction accuracy. They are further susceptible to overfitting. In other words, after a certain point, adding new features or new training examples can reduce the prediction quality [28]. Furthermore, protein chain data are imbalanced as domain segments are much longer than linkers and, therefore, classifiers will usually be biased toward the majority of the classes. This raises the challenge of choosing the appropriate evaluation metrics. For example, a technique that fails to predict any linker in a protein sequence, which has, respectively, 95% and 5% of its AAs as domains and linkers, achieves a high prediction accuracy of as much as 95%.

Another significant issue in domain boundary prediction is that the prediction accuracy is considerably low in multi-domain proteins compared to single-domain proteins. For example, the experiments of Liu and Rost [24] on subsets of 1187 proteins of CATH [31] and SCOP [29] data sets showed a domain prediction accuracy of 69% while the accuracy for multi-domain proteins alone was only 38% [45].

12.4 PREDICTION ASSESSMENT

There are several assessment metrics that are used in the evaluation of domain linker prediction methods. The most frequently used evaluation measures in this field are *accuracy*, *sensitivity*, *precision*, *specificity*, *F-measure*, *Matthews Correlation Coefficient* (MCC), *N-score*, and *C-score.*

Accuracy is defined as the proportion of correctly predicted linkers and domains to all of the structure-derived linkers and domains listed in the data set. *Sensitivity*, or *Recall*, is defined as the proportion of correctly predicted linkers to all of the structure-derived linkers listed in the data set. *Precision* is defined as the proportion of correctly predicted linkers to all of the predicted linkers. *Specificity* is defined as the proportion of correctly predicted domains to all the structure-derived domains listed in the data set. These four metrics can be represented mathematically as follows:

$$Accuracy = \frac{TP + TN}{TP + TN + FN + FP} \tag{12.1}$$

$$Recall = \frac{TP}{TP + FN} \tag{12.2}$$

$$Precision = \frac{TP}{TP + FP} \tag{12.3}$$

$$Specificity = \frac{TN}{TN + FP} \tag{12.4}$$

where TP (true positive) is the number of AAs within the known linker segment predicted as "Linkers," TN (true negative) is the number of AAs within the known domain segment predicted as "Domains," FN (false negative) is the number of AAs within the known linker segments predicted as "Domains," and FP (false positive) is the number of AAs within the known domain segment predicted as "Linkers."

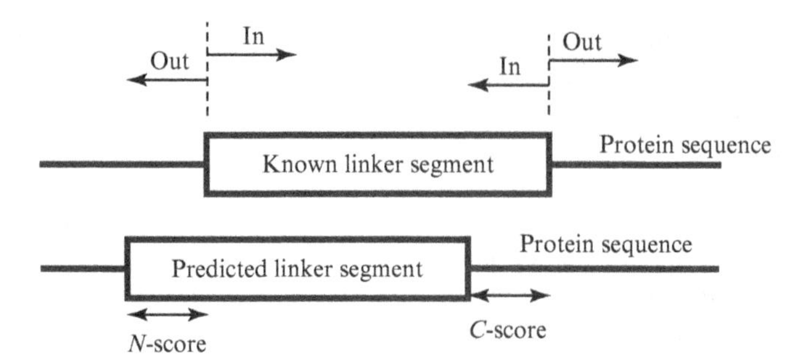

Figure 12.1 *N-score* and *C-score* are the number of AA residues that do not match when comparing the predicted linker segment with the known linker segment. A lower score means a more accurate prediction. For exact match, the *N-score* and *C-score* should be equal to zero.

The *F-measure* is an evaluation metric that combines *Precision* and *Recall* in a single value. It is defined as the harmonic mean of *Precision* (P) and *Recall* (R) [11, 37]:

$$F1 = \frac{2PR}{P + R} \tag{12.5}$$

MCC [27] is defined as

$$MCC = \frac{(TP \times TN) - (FP \times FN)}{\sqrt{(TP + FP)(TP + FN)(FP + TN)(TN + FN)}} \tag{12.6}$$

The MCC varies between −1 and 1. As it becomes closer to 1, the model is regarded as a better prediction system. *N-score* and *C-score* are the numbers of AA residues that do not match when comparing the predicted linker segment with the known linker segment as shown in Figure 12.1. A lower score means a more accurate prediction. If the prediction of this segment is an exact match, then the *N-score* and *C-scores* should be equal to zero [38, 48].

12.5 CURRENT APPROACHES

This section provides an overview and discussion of some of the current prediction approaches of inter-domain linker regions.

12.5.1 DomCut

Suyama and Ohara [42] developed DomCut as a simple method to predict linker regions among functional domains based on the difference in AA composition between domain and linker regions. The prediction is made by calculating the linker index from the SWISS-PROT database [2] of domain and linker segments. A sequence segment is considered a linker if it satisfies three conditions: connecting two adjacent domains, in the range from 10 to 100 residues, and not containing membrane spanning regions. As a result, they got a nonredundant sequence set that is composed of 273 sequences (486 linker and 794 domain segments). The average lengths of linker and domain segments were 35.8 and 122.1 AAs, respectively. To represent the preference for AA residues in linker regions, the authors introduced the linker index, which is defined as

$$S_i = -\ln\left(\frac{f_i^{linker}}{f_i^{domain}}\right) \tag{12.7}$$

where f_i^{linker} and f_i^{domain} are the frequencies of AA residue i in the linker and domain regions, respectively.

A linker preference profile was generated by plotting the averaged linker index values along an AA sequence using a siding window of size 15. A linker was predicted if there was a trough in the linker region and the averaged linker index value at the minimum of the trough was lower than the threshold value. The regions without domain or linker assignment were taken to be uncertain regions. At the threshold value of −0.09, the sensitivity and selectivity of DomCut were 53.5% and 50.1%, respectively. The web server with additional information is available at http://www.kazusa.or.jp/tech/suyama/domcut.

The information contained in the linker index will not be alone sufficient to precisely predict linker regions. It was reported by Dong et al. [12] that DomCut has low sensitivity and specificity in comparison to other methods. However, the integration of linker index with other protein sequence features will enhance the production. Therefore, DomCut has been used by many researchers such as Pang et al. [32] and Zaki et al. [49], and therefore, it can be considered as the basis for the development of several domain boundary predictors.

12.5.2 Scooby-Domain

SequenCe hydrOphOBicitY predicts DOMAINs (Scooby-Domain) web application was developed by George et al. [17] and extended by Pang et al. [32] to visually identify foldable regions in a protein sequence. Scooby-Domain uses the distribution of observed lengths and hydrophobicities in domains with known 3D structure to predict novel domains and their boundaries in a protein sequence. It utilizes a multilevel smoothing window to determine the percentage of hydrophobic AAs within a putative domain-sized region in a sequence. Each smoothing window calculates the fraction of hydrophobic residues it encapsulates along a sequence and places the value at its central position. This creates a triangular-shaped 2D matrix where the value at cell (i,j) is the average hydrophobicity encapsulated by a window of size j that is centered at residue position i. Matrix values are converted to probability scores by referring to the observed distribution of domain sizes and hydrophobicities. Using the observed distribution of domain lengths and percentage hydrophobicities, the probability that the region can fold into a domain or be unfolded is then calculated.

Scooby-Domain employs an A* search algorithm to search through a large number of alternative domain annotations. The A* search algorithm considers combinations of different domain sizes, using a heuristic function to conduct the search. The corresponding sequence stretch for the first predicted domain is removed from the sequence. The search process is repeated until there are less than 34 residues remaining, which is the size of the smallest domain, or until there are no probabilities greater than 0.33, which is an arbitrary cutoff, to prevent nondomain regions from being predicted as a domain.

Two linker prediction scoring systems, DomCut [42] and PDLI [12], were used separately to complement Scooby-Domain's prediction. The performance of Scooby-Domain was evaluated with the inclusion of homology information. Homologues of the query sequence were detected using PSI-BLAST [1] searches of the SWISS-PROT database [2] and *Multiple Sequence Alignments* (MSAs) were generated using PRALINE [41]. On a test set of 173 proteins with consensus CATH [31] and SCOP [29] domain definitions, Scooby-Domain has a sensitivity of 50% and an accuracy of 29%.

The advantages of Scooby-Domain include its ability to predict discontinuous domains and successful predictions are not limited by the length of the query sequence. A* search is a very flexible method, and it may be easily adapted and improved to include more sophistication in its predictions. . However, A* search algorithm has an exponential computational time complexity in its worst case [33, 36]. Furthermore, domains that are connected by small linkers may not be identifiable by Scooby-Domain because window averaging may lose any signal at the linker.

12.5.3 FIEFDom

Bondugula et al. [4] presented *Fuzzy Integration of Extracted Fragments for Domains* (FIEFDom) as the method to predict domain boundaries of a multi-domain protein from its AA sequence using a *Fuzzy Mean Operator* (FMO). Using the *nonredundant* (nr) sequence database together with a *Reference Protein Set* (RPS) containing known domain boundaries, the operator is used to assign a likelihood value for each residue of the query sequence as belonging to a domain boundary. FMO represents a special case of the fuzzy nearest neighbor algorithm [21] with the number of classes set to one. The approach is a three-step procedure. First, the *Position-Specific Scoring Matrix* (PSSM) of the query sequence is generated using a large database of known sequences. Second, the generated profile is used to search for similar fragments in the RPS. Third, the matches with the proteins in RPS are parsed, and the domain *Boundary Propensity* (PB) of the query protein is predicted using an FMO. For SCOP 1.65 data set with a maximal sequence identity of 30%, the average domain prediction accuracy of FIEFDom is 97% for one-domain proteins and 58% for multi-domain proteins.

The advantages of FMO include its simplicity, ease of updating, and its asymptotic error bounds. The choice of the program to designate a region as a domain boundary can be traced back to all proteins in the local database that contributed to the decision. The model need not be trained or tuned whenever new examples of domain boundaries become available. In addition, the users can choose the domain definitions such as CATH [31] and SCOP [29] to suit their needs by replacing the RPS. FIEFDom works well for protein sequences with many close homologs and that with only remote homologs. On the other hand, this approach did not address the issue of predicting domains with noncontiguous sequences and therefore it discarded such proteins.

12.5.4 Chatterjee et al. (2009)

Chatterjee et al. [6] proposed the physicochemical properties as additional features to train a *Support Vector Machine* (SVM) classifier for improved prediction of multi-domains in protein chains. Three kernel functions were examined: linear, cubic polynomial, and *Radial Basis Function* (RBF). The feature set consists of six different features: predicted secondary structure, predicted solvent accessibility, predicted conformational flexibility profile, AA composition, PSSM, and AA physicochemical properties. A 13 AA residue window is slided over the protein chain every time by one residue position. The efficiency of this approach was evaluated on CATH data sets [31]. The SVM classifier with a cubic polynomial kernel had shown the best performances in terms of accuracy and precision. These two measures were 76.46% and 86.82%, respectively.

12.5.5 DROP

Ebina et al. [14] developed *Domain linker pRediction using OPtimal features* (DROP) as an SVM, with an RBF kernel, domain linker predictor trained by 25 optimal features. The optimal combination of features was selected from a set of 3000 features using a random forest algorithm, which calculates the *Mean Decrease Gini Index* (MDGI), complemented with a stepwise feature selection. The selected features were primarily related to secondary structures, PSSM elements of hydrophilic residues and prolines.

For each residue, a 3000-dimensional real-valued feature vector was extracted. These features are as follows; 544 AA indices describing physicochemical properties, 20 PSSM elements, 3 *Probabilities of Secondary Structure* (PSS), 2 α-helix/β-sheet core propensities, 1 sequential hydrophobic cluster index, sequence complexity as defined by Shannon's entropy, 1 expected contact order, 20 elements of AA compositions, 3 domain/coil/linker propensity indices, 2 linker likelihood scores and three newly defined scores quantifying the AA composition similarity between domain and linker regions. Vector elements were averaged with windows of ±5, ±10, ±15, or ±20 residues around the considered residue to include local and semi-local information into the vectors. The total number of vectors for linkers and domains were 2230 and 52,335, respectively.

The efficiency of DROP was evaluated by two domain linker data sets: DS-All [13, 43] and CASP8 FM. [1] DS-All contains 169 protein sequences, with a maximum sequence identity of 28.6% and 201 linkers. DROP achieved a prediction sensitivity and precision of 41.3% and 49.4%, respectively, with more than 19.9% improvement by the optimal features.

[1] http://predictioncenter.org/casp8/.

DROP does not use sequence similarity to domain databases. One of the advantages of this approach is the use of random forest approach for feature selection. Instead of exhaustively searching all feature combination, random forest is based on random sampling that provides a quick and inexpensive screening for the optimal features. However, random forest can be trapped in local minima and severs from overprediction. As a result, DROP overpredicts domain linkers in single-domain targets of *Benchmarking Data Set* (BDS) [13] and CAFASP4.[2] This can be decreased by increasing the default threshold level or by including nonlocal features such as C? density and foldability index.

The use of SVM as a predictor is another advantage. SVM is the state-of-the-art machine-learning technique and has many benefits, and it overcomes many limitations of other techniques. SVM is having strong foundations in statistical learning theory [10] and has been successfully applied in various classification problems [51]. SVM offers several related computational advantages such as the lack of local minima in the optimization [44].

Research in inter-domain boundary prediction is a wide field in terms of protein sequence extracted features, prediction technique, evaluation measures, and data sets used for training and testing. Some prediction approaches use secondary structure features, others use physiochemical properties, PSSM, AA composition, and/or solvent accessibility information, while others use linker index and/or domain lengths and hydrophobicities. Some approaches uses SVM as a prediction technique, others use random forest, whereas others use FMO, *Enhanced General Regression Network* (EGRN), or A* search algorithm. Some approaches measure their performance based on prediction accuracy while others use sensitivity, selectivity, precision, and/or correlation coefficient.

12.6 DOMAIN BOUNDARY PREDICTION USING ENHANCED GENERAL REGRESSION NETWORK

The domain boundary prediction in Domain boundary prediction using enhanced general regression Network (DomNet) consists of four consecutive steps. First, a comprehensive multi-domain data set was built for the purpose of benchmarking structure-based domain identification methods. Second, a compact protein sequence profile was constructed from conserved domain database. Third, an *Enhanced General Regression Network* (EGRN), specially designed for the high dimensional problem of protein sequence data is employed. In the final step, the performance of the proposed model is compared to other existing machine-learning models as well as the widely known domain boundary predictors on two benchmark data sets.

12.6.1 Multi-Domain Benchmark Data Set

Two data sets namely, Benchmark_2 and CASP7, were used in this study. Benchmark_2 is a comprehensive data set for the purpose of benchmarking structure-based

[2]http://www.cs.bgu.ac.il/ dfischer/CAFASP4/.

domain identification methods. This data set comprises 315 polypeptide chains, 106 one-domain chains, 140 two-domain chains, 54 three-domain chains, 8 four-domain chains, 5 five-domain chains, and 2 six-domain chains. The data set is nonredundant in a structural sense: each combination of topologies occurs only once per data set. Sequences of protein chains are taken from the *Protein Data Bank* (PDB).[3]

Critical Assessment of Techniques for Protein Structure Prediction (CASP)[4] is one of the most widely known benchmark data sets in domain boundary prediction. Annually, most of the well-known domain predictors participate in the CASP competition. The data set contains 66 one-domain chains, 26 two-domain chains, and 2 three-domain chains.

The secondary structure information and the solvent accessibility were predicted for each chain in both data sets using SSpro [35] and ACCpro [34]. Evolutionary information for each chain is obtained by PSSM, which was constructed using PSI-BLAST [1] and RPS-BLAST for the conserved domain database [25]. The inter-domain linker index was taken from DomCut [42].

12.6.2 Compact Domain Profile

The EGRN model uses a nonlinear auto-associative network for filtering noise or less discriminative features. In the process of filtering, the model finds the optimal data dimension for prediction. The reduced data dimension of 15 features (dm_15) provides better prediction accuracy than its original data dimensions that consists of 23 features (dm_23). If we assume that the additional structural information in the data set such as the secondary structure, solvent accessibility, and domain linker index is proven to be useful for the domain boundary prediction, the best data dimension (dm_15) observed in the experiments may divulge that not all the information in the sequence profile (PSSM) may be useful for accurate prediction. Thus, if we can additionally exploit useful information (i.e., mid- or long-range interaction information) that does not exist in the sequence profile from other profiles and it is combined with the compact data set (dm_15 data set), a new profile that contains more structural information can be constructed.

We found the additional information from the conserved domain database [25]. In molecular evolution, domains may have been utilized as building blocks and may have been recombined in different arrangements to modulate protein function. In addition, protein sequences are severely modified in molecular evolution as mutation, insertion, and deletion occur in their residues. Although many rearrangements and modifications occur during the evolution process, some fundamental structures are conserved and these regions can be identified as a conserved domain. The conserved domain profile contains PSSMs for proteins with similar domain architectures and these can be searched by *Reserved Position-Specific BLAST* (RPS-BLAST). Conserved domains contain conserved sequence patterns or motifs, which allow for their detection in polypeptide sequences so that PSSMs in conserved domain database can

[3]http://www.ncbi.nlm.nih.gov/Structure/VAST/nrpdb.html.
[4]http://predictioncenter.org/casp7.

be useful to find remote homology. We believe that these PSSMs should be useful in boosting up the weak signal of mid- or long-range interactions that reside in AA sequences. DomNet adopts the idea above and uses the novel *Compact Domain* profile (CD profile) that consists of PSSMs extracted from a conserved domain database and existing sequence profiles generated by PSI-BLAST. We extracted 157 additional PSSMs of proteins with similar domain architectures from the conserved domain database and these were combined with the existing Benchmark_2 training set. Finally, the training set was dimensionally normalized using the auto-associative network used in EGRN.

12.6.3 The Enhanced Semi-Parametric Model

Protein sequence data can be mathematically viewed as points in a high-dimensional space. Learning in the high-dimensional space causes several important problems. First, the large network training requires a large data set of known examples, which leads to an exponential rise in computational complexity and susceptibility to the overfitting problem. Second, with nonparametric models such as ANNs, better generalization and faster training can be achieved when they have fewer weights to be adjusted by fewer inputs. In other words, beyond a certain point, adding new features can actually lead to a reduction in the performance of the classification system [28].

One solution to the problems above can be semi-parametric modeling. Semi-parametric models take assumptions that are stronger than those of non-parametric models, but are less restrictive than those of parametric model. In particular, they avoid most serious practical disadvantages of nonparametric methods but at the price of an increased risk of specification errors. The proposed model, EGRN, takes a form of the semi-parametric model and finds the optimal tradeoff between parametric and nonparametric models. So, it can have advantages of both models while effectively avoiding the curse of dimensionality. The EGRN contains the evolutionary information represented within the local model. Its global model also works as a collaborative filter that transfers the knowledge among local models in formats of hyper-parameters. The local model contains an efficient vector quantization method. Consider a vector training sequence consisting of M vectors, $T = \{x_1, x_2, \ldots, x_m\}$, and N code vectors $P = \{p_1, p_2, \ldots, p_n\}$. The whole region is partitioned by vector quantization function into a set of *Vorinoi Regions* $V = \{S_1, S_2, \ldots, S_n\}$, then each local region is represented by the code vector p_i of the noise-free vectors generated by an efficient dimensionality reduction method. The centroid vector within a cluster can be expressed as

$$Q_i(X_m) = p_i = \frac{\sum_{X_m \in S_i} X_m}{N}, i = 1, 2, \ldots, N \tag{12.8}$$

where N is a number of observations in the cluster S_i.

As a collaborative filter, we use *General Regression Neural Network* (GRNN). In the literature, GRNN was shown to be effective in noisy environments as it deals with sparse data effectively. Generally, it provides more accurate predictive performance than other existing neural network models. However, one drawback is that as it incorporates every training vector pair $\{x_i \to y_i\}$ into its architecture (x_i is the single training vector in the input space and y_i is the associated desired scalar output), it requires extensive computer resources by performing very large computations. Moreover, like general kernel methods, it severely suffers from the curse of dimensionality. It is unable to ignore irrelevant inputs without major modifications to the basic algorithm. A standard version of the GRNN equation is as follows:

$$\hat{y}(x) = \frac{\sum_{i=1}^{NV} y_i \exp \dfrac{-(x-x_i)^T(x-x_i)}{2\sigma^2}}{\sum_{i=1}^{NV} \exp \dfrac{-(x-x_i)^T(x-x_i)}{2\sigma^2}} \tag{12.9}$$

where x_i is the training vector for class i in the input space; σ, the single learning or smoothing parameter chosen during network training, y_i, the scalar output related to x_i; and NV, the total number of training vectors.

To construct a semi-parametric model, we substitute $Q_i(X)$ of the local model for each training sample x_i used in the GRNN decision function. Hence, the approximation of EGRN can be written as

$$Z_i \exp \frac{-(x-Q_i(x))^T(x-Q_i(x))}{2\sigma^2} = \sum_{j=1}^{Z_i} \exp \frac{-(x-x_j')^T(x-x_j')}{2\sigma^2} \tag{12.10}$$

The solution for noise reduction and finding center vectors by quantizing input vectors is proposed by Yoo et al. [46]. It involves the embedding of an efficient dimensionality reduction method into its architecture, quantizing the noiseless desired input vectors x_j', having positive and negative class groups, and associating them with the center of the input vectors mapping to each group p_j. This approach not only achieves the network reduction by efficiently reducing the dimensions of input vectors but also creates a more compact network. This method was theoretically proven to be effective for high-dimensionality problems dealing with protein data. The general algorithm for EGRN is

$$\hat{y}(x) = \frac{\sum_{i=0}^{NI} y_i \exp \dfrac{-(x-p_i)^T(x-p_i)}{2\sigma^2}}{\sum_{i=0}^{NI} \exp \dfrac{-(x-p_i)^T(x-p_i)}{2\sigma^2}} \tag{12.11}$$

where p_i is the center of each quantized region for class i in the input space; σ, the single learning or smoothing parameter chosen during network training; y_i, the scalar output related to p_i; and NI, the number of unique centers p_i.

GRNN is considered as an overfitting and purely nonparametric model, whereas EGRN can be considered as a semi-parametric model as it contains the evolutionary information found in the local model. As a key collaborator, the global model transfers the knowledge among the local models. Hence, the performance of the proposed model has some advantages in comparison to the pure parametric models and pure nonparametric models in terms of learning bias and generalization variance, especially on high-dimensional protein data sets.

This method that includes the selection of the output y_i and finding the optimal size of the associated input regions in each class i is shown to be effective in finding an optimal tradeoff between model bias and variance. However, if we assume that a centroid vector is calculated by its associated cluster and the cluster contains noise vector, the centroid vectors may incorrectly represent its associated cluster or regions. Hence, in this method, an auto-associative network [23] was used to perform nonlinear mapping from x to $x\prime$ to eliminate noise or less-discriminatory features. The basic architecture of EGRN is depicted in Figure 12.2.

The auto-associative network contains three units, namely, encoding, bottleneck, and decoding units. In the input units, the first transfer function, f_1 (generally nonlinear, i.e., hyperbolic tangent, sigmoidal function) maps x, the input column vector of length l, to the encoding layer so that it produces h^x, a column vector of length n by using the following equation:

$$h_k^x = f_1((W^x x + b^x)_k) \tag{12.12}$$

where W^x is an $n \times l$ weight matrix; b^x, a column vector of length n containing the bias parameters; and $k = 1, 2, \ldots, n$.

Afterwards, the second transfer function f_2, maps the output of the encoding layer to the bottleneck layer. Let m represent a nonlinear principal component.

$$m = f_2(w^x \cdot h^x + \overline{b}^x) \tag{12.13}$$

Then, the third transfer function, f_3, maps from m to the decoding layer h^m,

$$h_k^x = f_3((w^x m + b^m)_k) \tag{12.14}$$

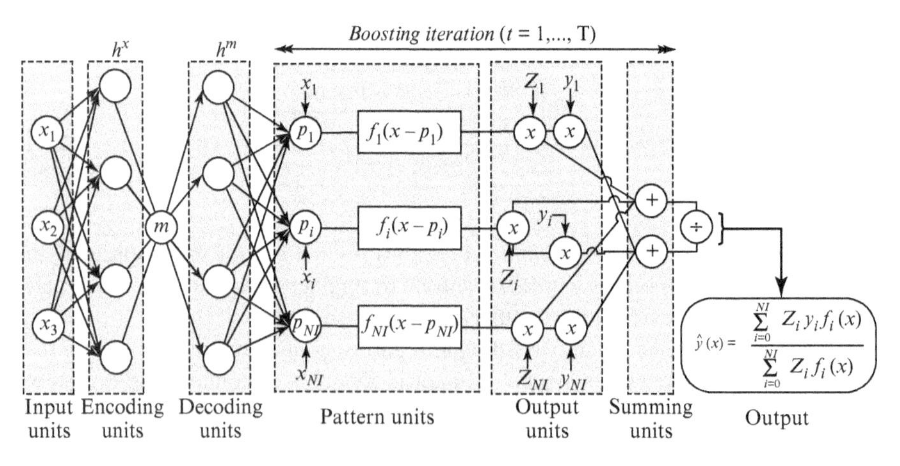

Figure 12.2 Basic EGRN architecture.

Finally, the last transfer function, f_4, produces $x_i\prime$ vectors by mapping from h^m, the output column vector of length l, to the pattern units.

$$x_i\prime = f_4((w^m \cdot h^m + \overline{b}^m)_i) \tag{12.15}$$

In order to maximize the performance of EGRN, a network boosting method called *Adaptive Boosting* (AdaBoost) [15] was utilized and modified for the EGRN for the network boosting. As observed in our experiments, the modified AdaBoost was tested with the EGRN and showed that it can fit into its architecture for the more accurate delineation of domain boundaries. The boosting algorithm takes the training set $(x_1, y_1), \ldots, (x_{NV}, y_{NV})$, where each x_i belongs to a domain X, and each label y_i is in a label set $Y = \{-1, +1\}$ as domain boundaries are indicated as positive (+1) or negative (−1) values only. AdaBoost calls the base learning algorithm (EGRN) repeatedly in a series of rounds $t = 1, \ldots, T$. And it maintains a distribution or set of weights over the training set. The weight of this distribution in training example i in round t is denoted $D_t(i)$.

All weights are set equally in the initial state, but, in each round, the weights of incorrectly classified examples are increased so that the learner is forced to focus on the hard examples in the training set. The weak learner's job is to find a weak hypothesis $h_t : X \rightarrow Y$ appropriate for the distribution D_t. The goodness of a weak hypothesis is measured by its error:

$$err_t = Pr_{i\sim D_t}[h_t(x_i) \neq y_i] = \sum_{i:h_t(x_i)\neq y_i} D_t(i) \tag{12.16}$$

In the above equation, the error is measured with respect to distribution D_t on which the learner was trained. In some cases where a learner cannot use the weights D_t in the training examples, a subset of the training examples can be sampled according to D_t, and these unweighted examples can be used to train the learner.

12.6.4 Training, Testing, and Validation

In order to thoroughly test the performance of the EGRN model, two extensive experiments on two benchmark data sets were conducted. For the experiment on the Benchmark_2 data set, a sevenfold cross-validation scheme for the model evaluation was adopted. Random data set selection and testing were conducted seven times for each different window size data set. When multiple random training and testing experiments were performed, a model was formed from each training sample. The estimated prediction accuracy is the average of the prediction accuracy for the models and each window size, derived from the independently and randomly generated test divisions. We tested four different window sizes (7, 11, 19, and 27) for each model.

During the postprocessing of the network output, as the network generates the raw outputs that have many local peaks, we filtered these raw outputs by using Liu and Rost's [24] method. We determined the threshold for each network outputs according to the length (L) of the protein and to the distribution of raw output values for all residues in that protein. We compiled the 92nd percentile of the raw output T_1 and set the threshold T to

$$T = \begin{cases} \max(T_1, 60) & \text{for } L \leq 100 \\ \max(T_1, 30) & \text{for } L \leq 200 \\ T_1 & \text{for } L > 200 \end{cases}$$

T was set to the threshold that divides domain boundaries and others. If the value of a residue is above the threshold, the residue is regarded as domain boundary. Second, we assigned the central residue as a domain boundary if three or more residues were predicted as a domain boundary. And all parameters for these filters were developed using the validation set only.

As for the experiment on CASP7 data set, we trained each network using the window size of 7 training sets obtained from the experiment on the Benchmark_2 data set. Each protein chain in CASP7 data set was fed into each predictor and a prediction was made. The identical postprocessing procedure was conducted as performed in the experiment on the Benchmark_2 data set. With the CASP7 data set, if the predicted domain boundary is in the range of [−15, 15] residues of the true domain boundary, then we consider it as a correct prediction as it used the window size 7 training set. The performance of this approach was measured by accuracy, sensitivity, specificity, and MCC.

12.6.5 Experimental Results

For the optimal learning of EGRN, the most suitable data dimension for the network using its embedded auto-associative network should first be sought. The data dimensions (dm_23) of Benchmark_2 data set were reduced to the range of 11–19. Table 12.1 shows the prediction accuracy obtained by EGRN for the different data dimensions as well as for the different window sizes. As observed in Table 12.1, a data dimension of 15 shows the best accuracy for all window sizes by showing the average accuracy of 65.55%. In addition, the window sizes of 7 and 11 provide better accuracy than other window sizes and correspond with a previous study. For further comparison of EGRN to other existing models, EGRN will use a data dimension of 15 and window sizes of 7 and 11.

Table 12.1 Accuracy of domain boundary placement on the CASP7 benchmark data set

win_size	dm_11	dm_13	dm_15	dm_17	dm_19	Average accuracy
win_7	60.7	63	68.3	62.4	61.4	63.16
win_11	63.3	62.2	64.9	63.5	63.2	63.42
win_19	60.1	60.4	65.9	58.1	58	60.50
win_27	62.5	62.4	63.1	62.7	63.5	62.84
Average accuracy	61.65	62	65.55	61.33	60.1	62.50
Variance	2.25	1.25	4.70	5.17	14.98	1.80

With the result obtained from the above experiment, the predictive performance of DomNet is compared with the best contemporary domain predictors, namely, DOM-pro [7], DomPred [26], DomSSEA [26], DomCut [42], and DomainDiscovery [40]. DOMpro [7] uses machine-learning algorithms in the form of recursive neural networks and predicts protein domains using a combination of evolutionary information in the form of profiles, predicted secondary structure, and predicted relative solvent accessibility. DomPred [26] uses a combined homology and fold-recognition-based approach. The sequence homology approach simply attempts to distinguish boundaries from overlapping edges in PSI-BLAST MSAs. The fold-recognition approach relies on secondary structure element alignments, using the DomSSEA method [26], in order to find domain boundaries in more distant homologs. The method has an accuracy of 49% at predicting the domain boundary location within 20 residues using a representative set of two domain chains. DomCut [42] predicts linker regions based on sequence alone and relies solely on AA propensity. This method simply defines a linker region to be one that has lower linker index values than a specified threshold value. DomainDiscovery [40] uses SVM and additional sequence-based information and domain linker index during its training.

Table 12.2 shows the comparison of this model with the contemporary domain predictors on the CASP7 benchmark data set. EGRN model has shown the best overall accuracy among the state-of-the-art predictors by reaching a prediction accuracy of 69.6%. DomNet correctly predicted accuracy at 82.3%, 34.7%, and 29.2% for one-domain, two-domain, and three-domain chains, respectively.

12.7 INTER-DOMAIN LINKERS PREDICTION USING COMPOSITIONAL INDEX AND SIMULATED ANNEALING

The compositional index of protein chains is used along with the simulated annealing optimization technique to predict inter-domain linker regions within these chains. The method consists of two main steps: calculating the compositional index and then refining the prediction by detecting the optimal set of threshold values that distinguish inter-domain linkers from nonlinkers. In the first step, linker and nonlinker segments are extracted from the training data set and the differences in AA appearances in linker segments and nonlinker segments are computed. Then, the AA composition of the test protein sequence is computed, and finally the AA compositional index is calculated.

Table 12.2 Comparison of prediction accuracy of EGRN with other predictors

Predictor	One-domain	Two-domain	Three-domain	Average accuracy
DomNet	82.3	34.7	29.2	69.60
DOMpro	84.4	0.0	0.0	62.64
DomPred	85.9	9.5	33	66.28
DomSSEA	80.5	19.1	33.0	62.06
DomCut	85.3	9.5	7.5	65.21
DomainDiscovery	80.5	31.0	29.2	67.34

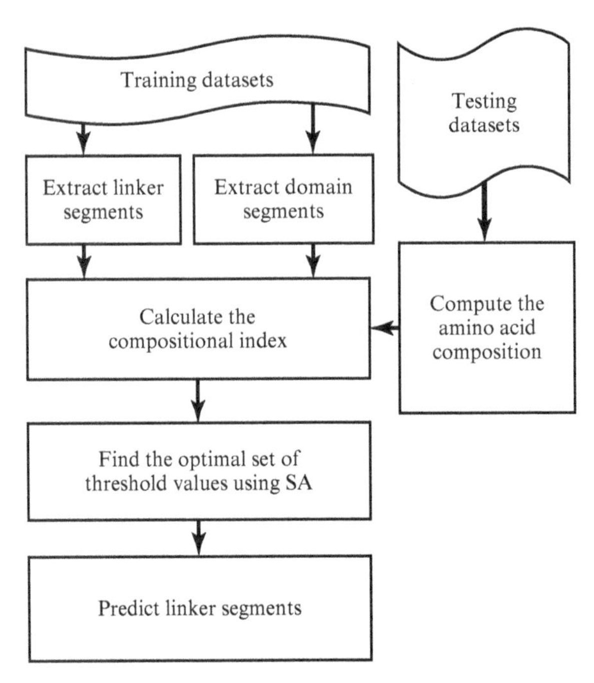

Figure 12.3 Method overview. The method consists of two main steps: calculating the compositional index and then refining the prediction by detecting the optimal set of threshold values that distinguish inter-domain linkers from nonlinkers using simulated annealing.

In the second step, a simulated annealing algorithm [22] is applied to find the optimal set of threshold values that will separate linker segments from nonlinker segments. In the following sections, we describe both steps. An overview of the proposed method is illustrated in Figure 12.3.

12.7.1 Compositional Index

Following Zaki et al. [49], we denote by S^* the enumerated set of protein sequences in the database. From each protein sequence s_i in S^*, we extract known linker segments and nonlinker segments and store them in data sets S_1 and S_2, respectively. To represent the preference for AA residues in linker segments, the compositional index r is calculated. The compositional index r_i for the AA residue i will be calculated as follows:

$$r_i = -\ln\left(\frac{f_i^{linker}}{f_i^{domain}}\right) \cdot a_i \quad (12.17)$$

where f_i^{linker} and f_i^{domain} are the frequencies of AA residue i in the linker and domain regions, respectively. This is somewhat analogous to DomCut method [42]. However, the information contained in the index values proposed by Suyama and Ohara [42] has no sufficient information to accurately predict the linker segments, thus we follow the improved index proposed by Zaki et al. [49] in which AA compositional knowledge was incorporated. The typical AA composition contains 20 components, each of which reflects the normalized occurrence frequency for one of the 20 natural AAs in a sequence. The AA composition in this case is denoted by a_i. Each residue in every testing protein sequence is represented by its corresponding compositional index r_i. Subsequently, the index values are averaged over a window that slides along the length of each protein sequence. To calculate the average compositional index values m_J^w along a protein sequence s, using a sliding window of size w, we apply the following formula:

$$m_j^w = \begin{cases} \dfrac{\sum_{i=1}^{j+((w-1)/2)} r_{si}}{j+((w-1)/2)} & \text{if } 1 \le j \le (w-1)/2 \\ \dfrac{\sum_{i=j-((w-1)/2)}^{j+((w-1)/2)} r_{si}}{j+((w-1)/2)} & \text{if } (w-1)/2 < j \le L-((w-1)/2) \\ \dfrac{\sum_{i=j}^{L} r_{si}}{L-j+1+((w-1)/2)} & \text{if } L-((w-1)/2) < j \le L \end{cases}$$

where L is the length of the protein and s_j is the AA at position j in protein sequence s.

A linker preference profile will be generated by plotting thc averaged linker index values along the AA sequence using a siding window of size 15. As shown in Figure 12.4, an optimized threshold value (0 in this case) separates linker regions. AAs with a linker score less than the set threshold value are considered as linker region.

12.7.2 Detecting the Optimal Set of Threshold Values Using Simulated Annealing

Simulated annealing [22] is a probabilistic searching method for the global optimization of a given function in a large search space. It is inspired by the annealing technique, that is, the heating and controlled cooling of a metal to increase the size of its crystals and reduce their defects. The major advantage of simulated annealing over other optimization techniques is its ability to avoid being trapped in local optima. This is because the algorithm applies a random search that does not only accept changes that increase the objective function f (assuming a maximization problem), but also some changes that reduce it [5, 19].

In this case, the values m_j are used in conjunction with the simulated annealing algorithm to improve the prediction by detecting linkers and structural domains. This is done by first dividing each protein sequence into segments. The segment size was set to the standard linker size among the data sets. Then, starting from a random threshold value for each segment, simulated annealing is applied to predict the optimal threshold for each segment that maximizes both the recall and precision of the linker segment prediction.

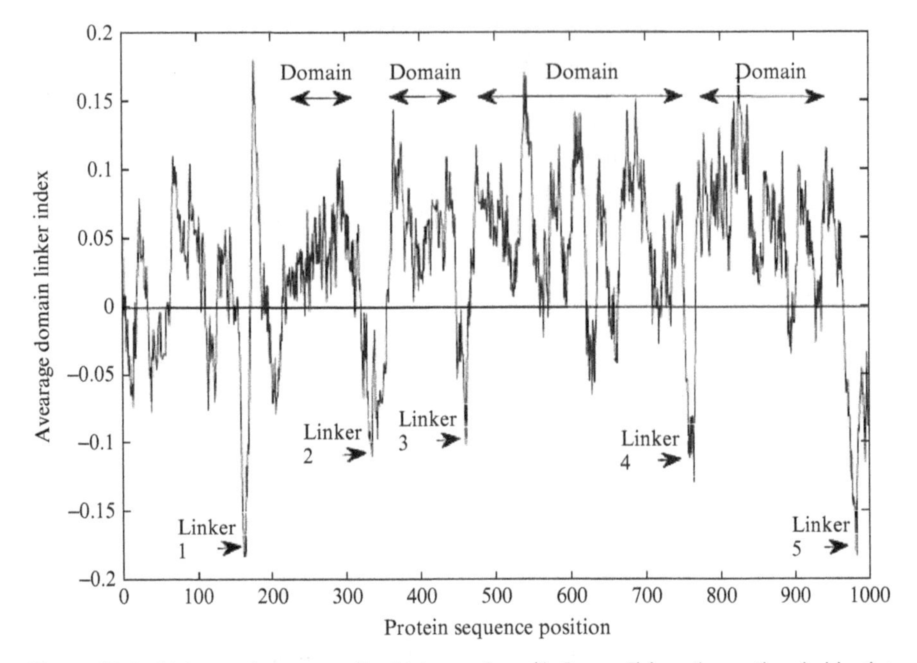

Figure 12.4 Linker preference profile. Linker regions (1, 2, …, 5) less than a threshold value of 0 are identified from the protein sequence.

Recall and *Precision* were selected to be the performance measures to evaluate the performance of the method due to several reasons. First, both evaluation measures were used to evaluate the performance of most of the current approaches that will allow comparing the accuracy of this method to the state-of-the-art methods. Secondly, *Recall* and *Precision* can handle unbalanced data situation where data points are not equally distributed among classes [39].

12.7.3 Experimental Results

The performance of the method was evaluated on two protein sequence data sets. The first data set was used by Suyama and Ohara [42] to evaluate the performance of DomCut, which was extracted from the SWISS-PROT database [2]. The data set contains nonredundant set of 273 protein sequences (486 linker and 794 domain segments). The average numbers of AA residues in linker and domain segments were 35.8 and 122.1, respectively. The second data set is DS-All [13, 43], which was used to evaluate the performance of DROP [14] in predicting inter-domain linker segments.

When the performance was evaluated on DomCut/SWISS-PROT protein sequence data set, an average *Recall* of 0.452 and *Precision* of 0.761 were achieved with the segment size of 36 AA (average linker size in the data set). With 18 AA segment size (half of the average linker size), we achieved a *Recall* of 0.558 and *Precision* of 0.836. It is worth to mention that *Recall* and *Precision* of DomCut were, respectively, 0.535 and 0.501.

When the performance was evaluated on 151 protein sequences of DS-All data set (182 linker and 332 domain segments), setting the segment size to 13 AA (average linker size in DS-All data set), we achieved an average *Recall* of 0.592 and *Precision* of 0.595. The comparison of the performance of this approach against the currently available domain linker prediction approaches [14] is reported in Table 12.3. It is clear to see that the proposed method outperformed the state-of-the-art domain linker prediction approaches in both *Recall* and *Precision.*

To demonstrate the performance of this method, in Figure 12.5, we show the optimal threshold values for an example (protein XYNA_THENE) in DomCut data set as predicted by this method, while, in Figure 12.6, we show the optimal threshold

Table 12.3 Prediction performance of publicly domain linker prediction approaches

Approach	Recall	Precision
Proposed method	0.592	0.595
DROP	0.413	0.494
DROP-SD5.0	0.428	0.515
DROP-SD8.0	0.418	0.503
SVM-PeP	0.403	0.491
SVM-SD3.0	0.373	0.446
SVM-SD2.0	0.353	0.420
SVM-Af	0.214	0.257
Random	0.050	0.060

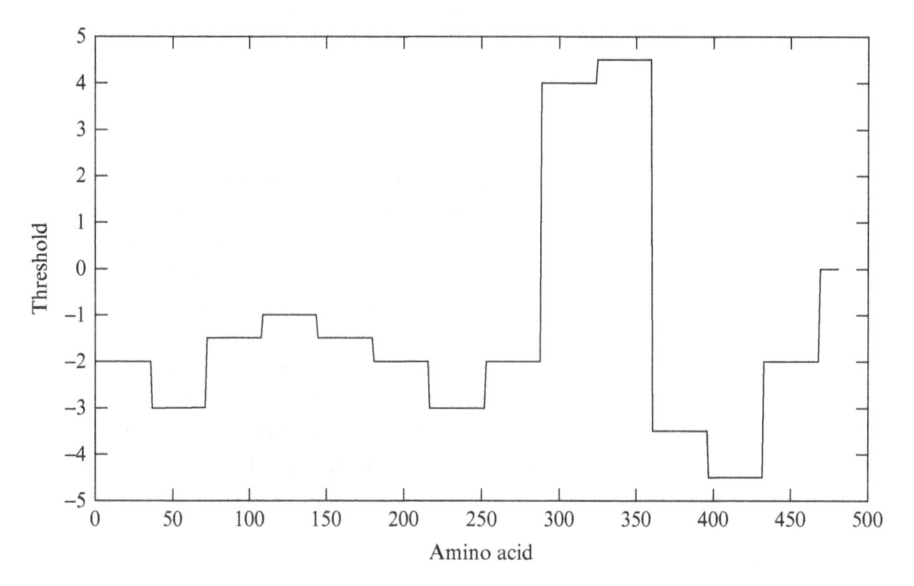

Figure 12.5 Optimal threshold values for XYNA_THENE protein sequence in DomCut data set.

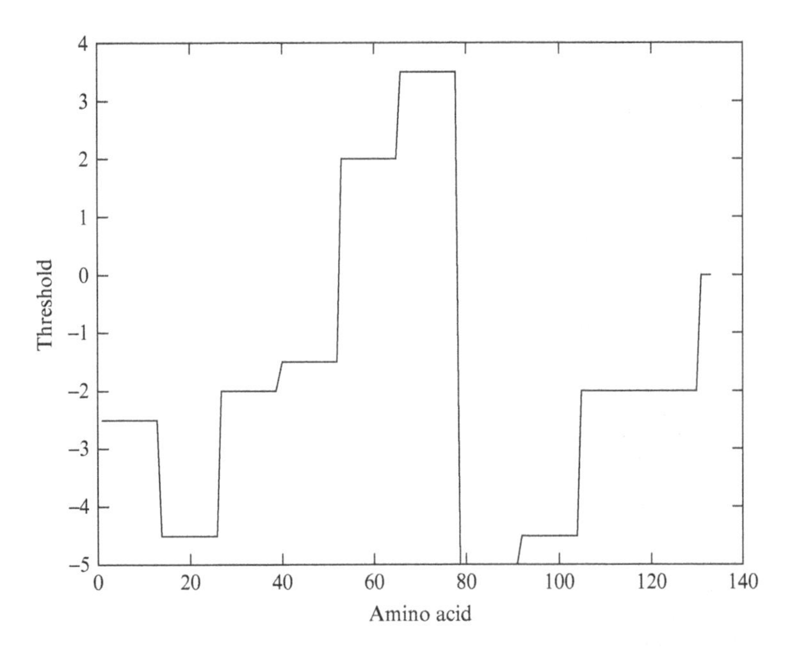

Figure 12.6 Optimal threshold values for 6pax_A protein sequence in DS-All data set.

values for another example (protein 6pax_A) in DS-All data set. It is clearly shown that the compositional index threshold values at linker segments are increased by the algorithm, whereas the threshold values of domains are reduced.

12.8 CONCLUSION

This chapter is focused on the prediction of protein inter-domain linker regions. The main challenges faced by domain linker prediction were presented. We investigated most of the relevant existing approaches carried out toward this perspective and provided a comparison of these approaches. It is clearly noted that the prediction of inter-domain linkers still needs much research effort in order to achieve reasonable prediction accuracy.

This chapter also presents two domain boundary prediction approaches. The first approach uses EGRN that is trained by a novel CD profile, secondary structure, solvent accessibility information, and inter-domain linker index to detect possible domain boundaries for a target sequence. The performance of this model was compared to several machine-learning models on the Benchmark_2 data set and achieved the best performance.

The second approach uses the AA compositional index to predict protein inter-domain linker segments from AA sequence information. Then, a simulated annealing algorithm is employed to improve the prediction by finding the optimal set of threshold values that separate domains from linker segments. The performance of the proposed method was compared to the state-of-the-art approaches on two well-known protein sequence data sets. Experimental results show that the proposed method outperformed the currently available approaches for inter-domain linker prediction in terms of *Recall* and *Precision.*

REFERENCES

1. Altschul SF, Madden TL, Schäffer AA, Zhang J, Zhang Z, Miller W, Lipman DJ. Gapped BLAST and PSI-BLAST: a new generation of protein database search programs. Nucleic Acids Res 1997;25(17):3389–3402.
2. Bairoch A, Apweiler R. The SWISS-PROT protein sequence database and its supplement TrEMBL in 2000. Nucleic Acids Res 2000;28(1):45–48.
3. Berg JM, Tymoczko JL, Stryer L. *Biochemistry*. 5th ed. New York: W.H. Freeman; 2002.
4. Bondugula R, Lee MS, Wallqvist A. FIEFDom: a transparent domain boundary recognition system using a fuzzy mean operator. Nucleic Acids Res 2009;37(2):452–462.
5. Busetti F. Simulated annealing overview; 2003.
6. Chatterjee P, Basu S, Kundu M, Nasipuri M, Basu DK. Improved prediction of multi-domains in protein chains using a support vector machine; 2009.
7. Cheng J, Sweredoski MJ, Baldi P. DOMpro: protein domain prediction using profiles, secondary structure, relative solvent accessibility, and recursive neural networks. Data Min Knowl Discov 2006;13(1):1–10.
8. Chothia C. Proteins. One thousand families for the molecular biologist. Nature 1992;357(6379):543.
9. Chou K-C. Some remarks on protein attribute prediction and pseudo amino acid composition. J Theor Biol 2011;273(1):236–247.
10. Cristianini N, Shawe-Taylor J. *An Introduction to Support Vector Machines and other Kernel-Based Learning Methods*. Cambridge University Press; 2000.
11. DMW Powers. Evaluation: from precision, recall and F-measure to ROC, informedness, markedness & correlation. J Mach Learn Technol 2011;2(1):37–63.
12. Dong Q, Wang X, Lin L, Xu Z. Domain boundary prediction based on profile domain linker propensity index. Comput Biol Chem 2006;30(2):127–133.
13. Ebina T, Toh H, Kuroda Y. Loop-length-dependent SVM prediction of domain linkers for high-throughput structural proteomics. Pept Sci 2009;92(1):1–8.
14. Ebina T, Toh H, Kuroda Y. DROP: an SVM domain linker predictor trained with optimal features selected by random forest. Bioinformatics 2011;27(4):487–494.
15. Freund Y, Schapire RE. Experiments with a new boosting algorithm. Bari, Italy; ICML, Volume 96; 1996. p 148–156.
16. Galzitskaya OV, Melnik BS. Prediction of protein domain boundaries from sequence alone. Protein Sci 2003;12(4):696–701.
17. George RA, Lin K, Heringa J. Scooby-domain: prediction of globular domains in protein sequence. Nucleic Acids Res 2005;33 Suppl 2:W160–W163.
18. Gewehr JE, Zimmer R. SSEP-Domain: protein domain prediction by alignment of secondary structure elements and profiles. Bioinformatics 2006;22(2):181–187.

19. Henderson D, Jacobson SH, Johnson AW. The theory and practice of simulated annealing. In: *Handbook of Metaheuristics*. Springer-Verlag; 2003. p 287–319.
20. Hondoh T, Kato A, Yokoyama S, Kuroda Y. Computer-aided NMR assay for detecting natively folded structural domains. Protein Sci 2006;15(4):871–883.
21. Keller JM, Gray MR, Givens JA. A fuzzy K-nearest neighbor algorithm. IEEE Trans Syst Man Cybern 1985;15(4):580–585.
22. Kirkpatrick S, Gelatt D Jr., Vecchi MP. Optimization by simulated annealing. Science 1983;220(4598):671–680.
23. Kramer MA. Nonlinear principal component analysis using autoassociative neural networks. AIChE J 1991;37(2):233–243.
24. Liu J, Rost B. Sequence-based prediction of protein domains. Nucleic Acids Res 2004;32(12):3522–3530.
25. Marchler-Bauer A, Anderson JB, Derbyshire MK, DeWeese-Scott C, Gonzales NR, Gwadz M, Hao L, He S, Hurwitz DI, Jackson JD, Ke Z, Krylov D, Lanczycki CJ, Liebert CA, Liu C, Lu F, Lu S, Marchler GH, Mullokandov M, Song JS, Thanki N, Yamashita RA, Yin JJ, Zhang D, Bryant SH. CDD: a conserved domain database for interactive domain family analysis. Nucleic acids research 2007;35 Suppl 1:D237–D240.
26. Marsden RL, McGuffin LJ, Jones DT. Rapid protein domain assignment from amino acid sequence using predicted secondary structure. Protein Sci 2002;11(12):2814–2824.
27. Matthews BW. Comparison of the predicted and observed secondary structure of T4 phage lysozyme. Biochim Biophys Acta, Protein Struct 1975;405(2):442–451.
28. Melo JCB, Cavalcanti GDC, Guimaraes KS. PCA feature extraction for protein structure prediction. Neural Networks, 2003. Proceedings of the International Joint Conference on, Volume 4. IEEE; Portland, OR; 2003. p 2952–2957.
29. Murzin AG, Brenner SE, Hubbard T, Chothia C. SCOP: a structural classification of proteins database for the investigation of sequences and structures. J Mol Biol 1995;247(4):536–540.
30. Nagarajan N, Yona G. Automatic prediction of protein domains from sequence information using a hybrid learning system. Bioinformatics 2004;20(9):1335–1360.
31. Orengo CA, Michie AD, Jones S, Jones DT, Swindells MB, Thornton JM. CATH–a hierarchic classification of protein domain structures. Structure 1997;5(8):1093–1109.
32. Pang CNI, Lin K, Wouters MA, Heringa J, George RA. Identifying foldable regions in protein sequence from the hydrophobic signal. Nucleic Acids Res 2008;36(2):578–588.
33. Pearl J. Heuristics: intelligent search strategies for computer problem solving; 1984.
34. Pollastri G, Baldi P, Fariselli P, Casadio R. Prediction of coordination number and relative solvent accessibility in proteins. Proteins 2002;47(2):142–153.
35. Pollastri G, Przybylski D, Rost B, Baldi P. Improving the prediction of protein secondary structure in three and eight classes using recurrent neural networks and profiles. Proteins 2002;47(2):228–235.
36. Russell S, Norvig P. *Artificial Intelligence: A Modern Approach*. 3rd ed. Upper Saddle River (NJ): Prentice Hall Press; 2009.
37. Sasaki Y. The truth of the F-measure. Teaching and Tutorial materials, 2007;1–5.
38. Shatnawi M, Zaki N. Prediction of inter-domain linker regions in protein sequences: a survey. The 9th International Conference on Innovations in Information Technology (IIT), Al Ain: UAE; 2013. p 237–242.

39. Shatnawi M, Zaki N. Prediction of protein inter-domain linkers using compositional index and simulated annealing. Proceeding of the 15th Annual Conference Companion on Genetic and Evolutionary Computation Conference Companion, GECCO '13 Companion. New York: ACM; 2013. p 1603–1608.
40. Sikder AR, Zomaya AY. Improving the performance of domaindiscovery of protein domain boundary assignment using inter-domain linker index. BMC Bioinformatics 2006;7 Suppl 5:S6.
41. Simossis VA, Heringa J. Praline: a multiple sequence alignment toolbox that integrates homology-extended and secondary structure information. Nucleic Acids Res 2005;33 Suppl 2:W289–W294.
42. Suyama M, Ohara O. DomCut: prediction of inter-domain linker regions in amino acid sequences. Bioinformatics 2003;19(5):673–674.
43. Tanaka T, Yokoyama S, Kuroda Y. Improvement of domain linker prediction by incorporating loop-length-dependent characteristics. Pept Sci 2006;84(2):161–168.
44. Vapnik VN. Statistical learning theory. Wiley; 1998.
45. Yoo PD, Sikder AR, Taheri J, Zhou BB, Zomaya AY. Domnet: protein domain boundary prediction using enhanced general regression network and new profiles. IEEE Trans NanoBiosci 2008;7(2):172–181.
46. Yoo PD, Sikder AR, Zhou BB, Zomaya AY. Improved general regression network for protein domain boundary prediction. BMC Bioinformatics 2008;9 Suppl 1:S12.
47. Zaki N. Prediction of protein-protein interactions using pairwise alignment and inter-domain linker region. Eng Lett 2008;16(4):505.
48. Zaki N, Bouktif S, Lazarova-Molnar S. A combination of compositional index and genetic algorithm for predicting transmembrane helical segments. PloS ONE 2011;6(7):e21821.
49. Zaki N, Bouktif S, Lazarova-Molnar S. A genetic algorithm to enhance transmembrane helices prediction. Proceedings of the 13th Annual Conference on Genetic and Evolutionary Computation. ACM, Dublin, Ireland; 2011. p 347–354.
50. Zaki N, Campbell P. Domain linker region knowledge contributes to protein-protein interaction prediction. Proceedings of International Conference on Machine Learning and Computing (ICMLC 2009), Perth, Australia; 2009.
51. Zaki N, Wolfsheimer S, Nuel G, Khuri S. Conotoxin protein classification using free scores of words and support vector machines. BMC Bioinformatics 2011;12(1):217.

CHAPTER 13

PREDICTION OF PROLINE *CIS–TRANS* ISOMERIZATION

Paul D. Yoo[4,5], Maad Shatnawi[2], Sami Muhaidat[1,3], Kamal Taha[1], and Albert Y. Zomaya[4]

[1]*ECE Department, Khalifa University, Abu Dhabi, UAE*
[2]*Higher Colleges of Technology, Abu Dhabi, UAE*
[3]*EE Department, University of Surrey, Guildford, UK*
[4]*Centre for Distributed and High Performance Computing, University of Sydney, Sydney, Australia*
[5]*Department of Computing and Informatics, Bournemouth University, UK*

13.1 INTRODUCTION

The Cis–Trans *Isomerization* (CTI) is a very important step for rate-determining in protein-folding reactions [5, 15, 22]. Prolyl CTI can be catalyzed by prolyl isomerizes that are enzymes found in both prokaryotes and eukaryotes that inter-converts the *cis–trans* isomers of peptide bonds with the *Amino Acid* (AA) proline [17]. These enzymes are involved not only in the catalysis of folding [17] but also in regulatory process [7, 25]. The CTI of prolyl peptide bonds has been suggested to dominate the folding of the α subunit of tryptophan synthase from *Escherichia coli* (aTS) [24]. A CTI, which is necessary to achieve the final conformational state of the prolyl bonds for such proteins, has often been found as the rate-limiting step in in vitro protein folding [16]. Therefore, it is very important to accurately predict proline *cis–trans* isomers of proteins based on their AA sequences.

Pattern Recognition in Computational Molecular Biology: Techniques and Approaches,
First Edition. Edited by Mourad Elloumi, Costas S. Iliopoulos, Jason T. L. Wang, and Albert Y. Zomaya.

The Human Genome Project has produced a massive biological data, and therefore, accurate and efficient computational modeling methods that can find useful patterns from this data have gained high attention. The first computational model for proline CTI prediction from AA sequences was made in 1990 by Frömmel and Preissner [8]. The authors had taken the physicochemical properties of adjacent local residues (±6) of prolyl AAs and found six different patterns that allow one to assign correctly approximately 72.7% (176 *cis*-prolyl residues in their relatively small data set of 242 Xaa-Pro bonds of known *cis*-prolyl residues), whereby no false-positive one is predicted.

Since the seminal work of Frömmel and Preissner [8], *Support Vector Machines* (SVMs) proved to be the most suitable proline CTI prediction method. The first SVM-based computational predictor was developed by Wang et al. [21]. They constructed an SVM model with a polynomial kernel function and used AA sequences as input, and achieved the Q_2 accuracy of 76.6%. Song et al. [20] also built an SVM with *Radial Basis Function* (RBF) and used the evolutionary information through the *Position-Specific-Scoring Matrix* (PSSM) scores generated by PSI-BLAST [1] and predicted secondary structure information obtained from PSI-PRED [14] as input. They reached the Q_2 accuracy of 71.5%, and *Mathews Correlation Coefficient* (MCC) of 0.40. Pahlke et al. [13] showed the importance of protein secondary structure information in the proline CTI prediction. Their computational algorithm called COPS- the first attempt to predict for all 20 naturally occurring AAs whether the peptide bond is a protein is in *cis* or *trans* conformation- used secondary structure information of AA triplets only. Most recently, Exarchos et al. [6] used an SVM with a wrapper feature selection algorithm trained by the PSSM scores, and predicted secondary structure information, real-valued solvent, accessibility level for each AA, and the physicochemical properties of the neighboring residues. They achieved a 70% accuracy in the prediction of the peptide bond conformation between any two AAs only.

As seen above, the recent development of computational modeling for proline CTI prediction has mostly been based on SVM and its variants, and evolutionary information (PSSM scores), and secondary structure information. These models showed approximately 70–73% Q_2 accuracies and 0.40 MCC. This observation is compatible with the results of other computational biology studies. SVM models showed a great prediction performance in computational biology and bioinformatics tasks [26–30].

In this chapter, we introduce a prediction approach that utilizes biophysically motivated intelligent voting model with a powerful randomized meta-learning technique through the use of AA sequence information only for the accurate and efficient proline CTI prediction. The proposed model has been developed based on the Random Forest data modeling [4] and evolutionary information. The predictive performance is compared with the most widely used SVM model and its variants on the same data set used in Reference [20]. The rest of this chapter is organized as follows. Section 13.2 provides the methods overview. Section 13.3 discusses the model evaluation and analysis. Concluding remarks are presented in Section 13.4.

13.2 METHODS

Our experiments consist of data construction, model development, and model validation and testing. In the model construction phase, SVM-based voting models with two different methods are constructed through a series of experiments in order to select a proper kernel function and tune the parameters. The predictive performance of the proposed methods is compared with those of SVM_{LIB} (Lib-SVM) and SVM_{ADA} (Adaboosted-Lib-SVM) for Q_2 accuracy, Sensitivity (Sn), Specificity (Sp), MCC, type I and II error rates, and STandard DEViation (STDEV) intervals over 7–11 sub-folds of each classifier for model stability and generalization ability.

13.2.1 Evolutionary Data Set Construction

To make a fair comparison with existing proline CTI prediction models, we have used Song et al. [20] data set. The data set has 2424 nonhomologous protein chains, obtained from the culled PDB list provided by PSICES server.[1] All the tertiary structures in the data set were determined by X-ray crystallography method with resolution better than 2.0 Å and R-factor less than 0.25. In addition, the sequence identity of each pair of sequences is less than 25%, and the protein chains with sequence length less than 60 residues were excluded from the data set. In total, there are 609,182 residues, and every sequence contains at least one proline residue.

The evolutionary information in the form of PSSMs was included in the windows as direct input. It is the most widely used input form for protein structure prediction in 1D, 2D, and 3D, as well as other computational structural proteomic prediction tasks [6, 13, 14, 26–30]. The idea of using evolutionary information in the form of PSSMs was first proposed by Jones et al. [10] and their work has significantly improved prediction accuracy.

To generate PSSM scores, we used the nr (*nonredundant*) database and blastpgp program obtained from NCBI website.[2] We ran blastpgp program to query each protein in our data set against the nr database to generate the PSSMs. The blastpgp program was set up to three iterations with a cutoff e-value of 0.001. Finally, the PSSM scores were scaled to the range of 0–1 using the standard logistic function:

$$f(x) = \frac{1}{1 + \exp(-x)} \tag{13.1}$$

where x is the raw profile matrix value. The scaled PSSM scores were used as the direct input to the learning models. A $M \times 20$ PSSM matrix is generated for each protein sequence, where M is the target sequence length and 20 is the number of AA types. Each element of the matrix represents the log-odds score of each AA at one position in the multiple alignments. The window size $(2l + 1)$ indicates the scope of the vicinity of the target prolyl peptide bonds, determining how much neighboring

[1] http://dunbrack.fccc.edu/PISCES.php.

[2] ftp://ftp.ncbi.nlm.nih.gov/blast/db/.

sequence information is included in the prediction. We selected the windows size l of 9 and built our models as it produced the best predictive results as reported in Reference [20].

Since our data set is composed of 27,196 *trans* residues (negative samples) and only 1265 *cis* residues (positive samples), it is important to reduce this data imbalance. There are two general approaches to reduce such imbalance problem: increasing the number of undersamples by random resampling and decreasing the number of oversamples by random removal. In this study, we adopted the first approach and made 1–1 ratios between the sizes of positive and negative training samples.

13.2.2 Protein Secondary Structure Information

The recent computational proteomic studies report that protein secondary structure information is useful in various protein sequence-based predictions [6, 13, 14, 26–30]. Although the mutations at sequence level can obscure the similarity between homologs, the secondary structure patterns of the sequence remain conserved. That is because changes at the structural level are less tolerated. Most of recent studies use the probability matrix of secondary structure states predicted from PSI-PRED [14]. PSI-PRED is a well-known computational predictor and predicts protein secondary structures in three different states (α-helix, β-sheet, and loops). However, there is one significant limitation in using predicted secondary structure information. The best secondary structure prediction model still cannot reach the upper boundary of its prediction accuracy. In other words, it is not good enough yet to be used as a confirmation tool. It shows approximately 75–80% Q_3 accuracies only. Incorrectly predicted secondary structure information within the training data leads to poor learning and inaccurate prediction. Although predicted secondary information may be useful to some extent, it should not be used if one attempts to reach better than 80% Q_2 accuracy. We therefore used evolutionary information in the form of PSSMs obtained from protein AA sequences only. To achieve more than 80% Q_2 accuracy, accurate, and correct information encoding presented in input data set is critical.

13.2.3 Method I: Intelligent Voting

This new intelligent voting approach of Random Forest data modeling [4] combines a number of methods and procedures to exploit homology information effectively. If we take a large collection of weak learners, each performing only better than chance, then by putting them together, it is possible to make an ensemble learner that can perform arbitrarily well. Randomness is introduced by bootstrap resampling [3] to grow each tree in the ensemble learner, and also by finding the best splitter at each node within a randomly selected subset of inputs. Method I grows many *Decision Trees* (DTs) [2] as in Figure 13.1. To classify a new input vector x, put the input vector down each of the DTs in the ensemble learner. Each DT is trained on a bootstrap sample of the training data.

To estimate the performance of the ensemble learner, Method I performs a kind of cross-validation by using *Out-Of-Bag* (OOB) data. Since each DT in the ensemble

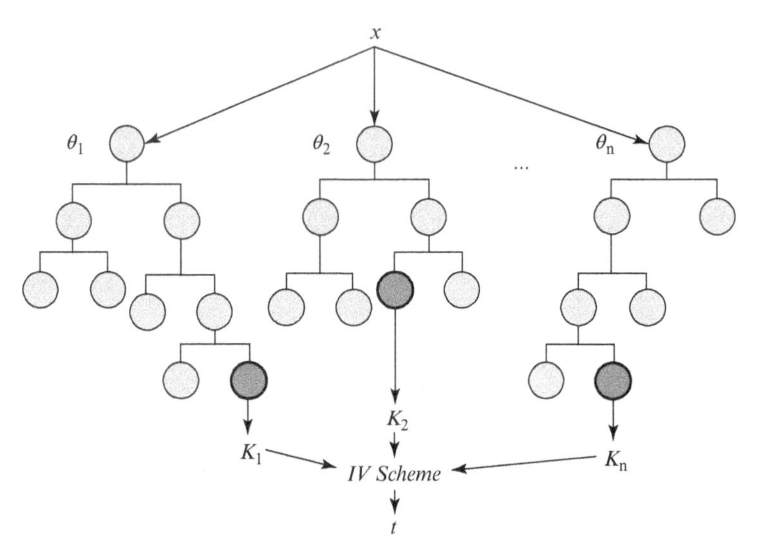

Figure 13.1 A general architecture of Method I ensemble. The collection of DTs $\{h(x, \theta_k), k = 1, 2, \cdots\}$, where the θ_k are independently, identically distributed random DTs, and each DT cast "a unit vote" for the final classification of input x.

grows on a bootstrap sample of the data, the sequences left out of the bootstrap sample, the OOB data, can be used as legitimate test set for that tree. On average one-third of the training data will be OOB for a given tree. Consequently, each PSSM in the training data set will be left out of one-third of the trees in the ensemble and uses these OOB predictions to estimate the error rate of the full ensemble.

Like CART [23], Method I uses the Gini index for determining the final class in each DT. The Gini index of node impurity is the most commonly used measure for classification problems. If a data set T contains examples from n classes, Gini index $G(T)$ is calculated as

$$G(T) = 1 - \sum_{j=1}^{n} (P_j)^2 \tag{13.2}$$

where P_j is the relative frequency of class j in T. If a data set T is split into two subsets T_1 and T_2 with sizes N_1 and N_2, respectively, the Gini index $G(T)$ of the split data containing examples from n classes will be

$$G_{split}(T) = \frac{N_1}{N} G(T_1) + \frac{N_2}{N} G(T_2) \tag{13.3}$$

The attribute value that provides the smallest $G_{split}(T)$ is chosen to split the node.

Figure 13.2 shows the three key steps of the Method I ensemble. First, a random seed is chosen that randomly pulls out a collection of samples from the training data set while maintaining the class distribution. Second, with this selected data set, a set of attributes from the original data set is randomly chosen based on user-defined values. All the input variables are not considered because of enormous computation and high chances of overfitting. In a data set where M is the total number of input attributes

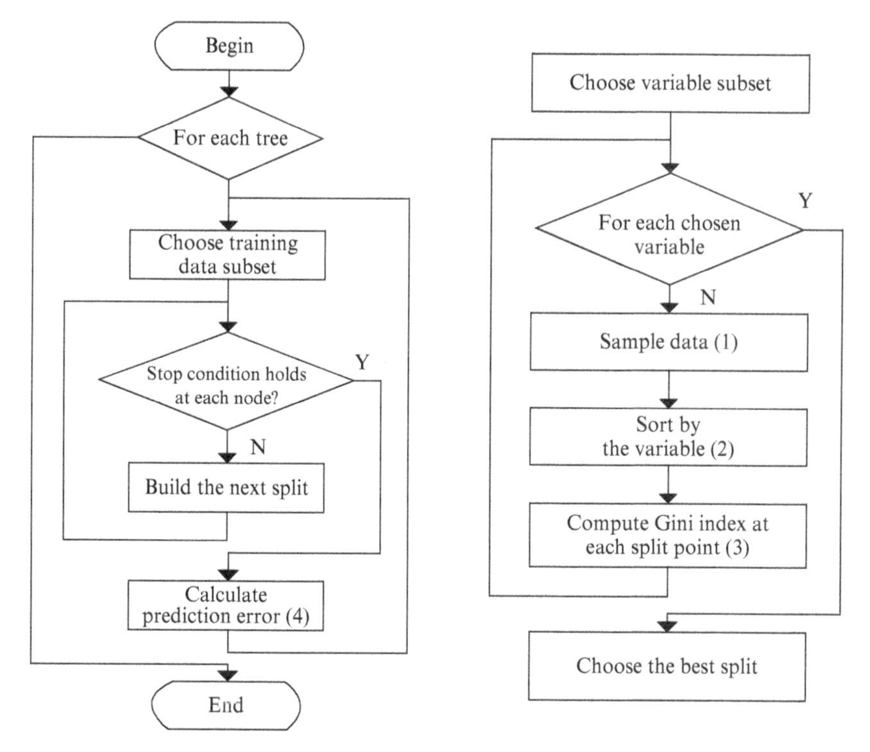

Figure 13.2 A flowchart of Method I ensemble. The left-hand side shows the main flow of the Method I ensemble while the right-hand side flowchart is the expansion of the process, to build the next split, of the main flowchart of left-hand side.

in the data set, only R attributes are chosen at random for each tree where $R < M$. Third, the attributes from this set create the best possible split using the Gini index to develop a DT model. The process repeats for each of the branches until the termination condition stating that leaves are the nodes that are too small to split. In this study, Method I ensemble was constructed and implemented in the Weka RF package [9].

13.2.4 Method II: Randomized Meta-Learning

Method II builds an ensemble of randomized base classifiers (i.e., Method I) and averages their classification. Each one is based on the same input data, but uses a different random number seed. Some learning algorithms already have a built-in random component. For example, when learning *Multi-Layer Perceptrons* (MLPs) using the back-propagation algorithm, the initial network weights are set to small random values. The learned classifier depends on the random numbers because the algorithm may find a different local minimum of the error function. One way to make the outcome of classification more stable is to run the learner several times with different random number seeds (initial weights) and combine the classifiers' predictions by voting or averaging. Learning in Method I builds a randomized DT in each iteration of the bagging algorithm and often produces excellent predictors. Although

Table 13.1 Model performance comparisons

Model	Q_2	*Sp*	*Sn*	MCC	Type I	Type II	STDEV
SVM_{LIB}	0.7672	0.5722	0.9622	0.5813	0.4278	0.0378	0.0253
SVM_{AB}	0.7653	0.5656	0.9650	0.5796	0.4344	0.0350	0.0226
Method I	0.8150	0.6698	0.9599	0.6592	0.3302	0.0401	0.0182
Method II	0.8658	0.7816	0.9500	0.7443	0.2183	0.0500	0.0227

The parameters of each model were given the following values: Method I (debug: false, maxDepth: 0, numExecutionSlots: 1, numFeatures: 0, numTrees: 13, printTrees: false, seed: 1), Method II (the same setting for Method I, and for Method II, seed: 3 and iteration: 10), SVM_{LIB} (SVMType: C-SVM, cacheSize: 40.0, coef0: 0.0, cost: 13, debug: false, degree: 3, eps: 0.0010, gamma: 0.0, kernelType: rbf, loss: 0.1, normalize: false, nu: 0.5, seed: 1, shrinking: true), and SVM_{AB} (the same as SVM_{LIB}, and for Adaboost, numIterations: 14, seed: 1, weightThreshold: 100).

bagging [3] and randomization yield similar results, it sometimes pays to combine them because they introduce randomness in different, perhaps complementary, ways. Randomization demands more work than bagging because the learning algorithm must be modified, but it can profitably be applied to a greater variety of learners.

The Method II input vector is formed by 3 seeds and 10 number of iterations. The steps involved in decision making are depicted in Figure 13.2. Table 13.1 presents experimental results related to Methods I and II predictions. Using Method II to combine all the DTs, the system reduced type I error rate significantly as depicted in Table 13.1.

13.2.5 Model Validation and Testing

To accurately assess the predictive performance of each model, we adopted a cross-validation scheme for our model evaluation. First, we apply the holdout method in our proline CTI data set. The training set is given to the inducer and the induced classifier is tested on the test set. The function approximator fits a function using the training set only. The function approximator is then asked to predict the output values for the data in the testing set. The errors it makes are accumulated as before to give the mean absolute test set error, which is used to evaluate the model. However, the holdout method has a key drawback in that the single random division of a sample into training and testing sets may introduce bias in model selection and evaluation. Since the estimated classification rate can be very different depending on the characteristic of the data, the holdout estimate can be misleading if we get an unfortunate split. Hence, in our experiment, we adopted multiple train-and-test experiments to overcome the limitation of the holdout method. We created 7- to 11-fold cross-validation.

13.2.6 Parameter Tuning

In our experiments, we used a semi-autonomous approach through Weka meta-learner, *CVParameterSelection* searches, and then checked the neighboring values of the best parameters found by the search. The list of the full parameters that we have used in our experiments is provided in the footnote of Table 13.1.

13.3 MODEL EVALUATION AND ANALYSIS

The performance of the models used in this study are measured by the accuracy (Q_2: the proportion of true-positive and true-negative residues with respect to the total positive and negative residues), the sensitivity (*Sn*: also called *Recall*, the proportion of correctly predicted isomerization residues with respect to the total positively identified residues), the specificity (*Sp*: the proportion of incorrectly predicted isomerization residues with respect to the total number of proline isomerization residues), and MCC, which is a correlation coefficient between the observed and predicted binary classifications that has a value in the range of ?1–+1. An MCC coefficient of +1 represents a perfect prediction, 0 no better than random prediction, and ?1 indicates total disagreement between prediction and observation. Hence, a high value of MCC means that the model is considered more robust prediction model. The above measures can be obtained using the following equations:

$$Q_2 = \frac{TP + TN}{TP + TN + FN + FP} \tag{13.4}$$

$$Sn = \frac{TP}{TP + FN} \tag{13.5}$$

$$Sp = \frac{TN}{TN + FP} \tag{13.6}$$

$$MCC = \frac{(TP \times TN) - (FP \times FN)}{\sqrt{(TP + FP)(TP + FN)(FP + TN)(TN + FN)}} \tag{13.7}$$

where *TP* is the number of true postives, *FN* is the number of false negatives or underpredictions, *TN* is the number of true negatives, and *FP* is the number of false positives or overpredictions [18]. *Sn* and *Sp* can handle imbalanced data situation where data points are not equally distributed among classes [19].

We adopted the polynomial kernel and RBF kernel to construct the SVM classifiers, which is aligned with the existing proline CTI prediction studies [20].

$$K(\vec{x}_i \cdot \vec{x}_j + 1)^d \tag{13.8}$$

$$K(\vec{x}_i \cdot \vec{x}_j) = \exp\left(-r\|\vec{x}_i - \vec{x}_j\|^2\right) \tag{13.9}$$

where the degree d needs to be tuned as for polynomial function, and the gamma and the regulator parameters for RBF need to be regulated. See the footnote of Table 13.1 for the parameter settings used for this study. For the optimal learning of the prediction models, the most suitable data fold for each model should be sought.

Table 13.1 shows the comparison of our proposed methods results with those of SVM_{LIB} and its variant, SVM_{AB}. The proposed methods (Methods I and II) outperformed other SVM models. Methods I and II achieved 81.5% and 86.58% Q_2 accuracies, respectively, while SVM models achieved approximately 76% Q_2 accuracy only. Table 13.1 also shows that our proposed methods are superior to SVM models in terms of specificity and MCC, which indicate the model robustness and stableness. The Q_2 accuracy of 86.58% that we achieved on proline CTI prediction is far better than those of any existing computational proline CTI predictors reported in the literature. The best Q_2 accuracy that we have found in the literature was approximately 73% on the same data set as used in this research.

The performance of each model is measured by types I and II error rates as well since incorrectly predicted residues can be as valuable as the correctly predicted residues for further modification of the model. Type I errors mean experimentally verified *trans* residues that are predicted (incorrectly) to be *cis* residues; type II errors indicate experimentally verified *cis* residues that are predicted (incorrectly) to be *trans* residues. Method II shows the lowest type I error rate (0.21833) while SVM_{AB} reaches the lowest type II error rate (0.035). Although our proposed methods seem to be not very useful in improving type II error rate, it reduces type I error rate effectively. Interestingly, type I error rates are worse with SVMs while type II error rates are worse with our proposed methods.

STDEV provides a good idea on model generalization ability. Although nonparametric machine-learning models have been proved to be useful in many different applications, their generalization capacity has often been shown to be unreliable because of the potential of overfitting. The symptom of overfitting is that the model fits the training sample too well, and thus the model output becomes unstable for prediction. On the other hand, a more stable model, such as a linear model, may not learn enough about the underlying relationship, resulting in underfitting the data. It is clear that both underfitting and overfitting will affect the generalization capacity of a model. The underfitting and overfitting problems in many data-modeling procedures can be analyzed through the well-known bias-plus-variance decomposition of the prediction error. The best generalization ability came from Method I, where a multiple DTs have been built. The idea used in Method II using a different random-number seed seems to be useful in better learning, however, not enough to improve its generalization ability. Method I shows the best STDEV value of 0.0182, while other models reach approximately 0.022–0.025. Method II outperforms other models in Q_2, Sp, and MCC, and no significant differences observed in Sn, while Methods I and II improve Sp and MCC significantly. Moreover, type II error rates are much lower than type I error rates in general, and Methods I and II effectively reduce type I error rate. In addition to that, Method II improves in Q_2, Sp, and MCC and type I error rate. However, no significant improvement observed in Sn and type II error rate and STDEV. This result indicates that the proposed methods could be improved by reducing the errors of experimentally verified *cis* residues that are incorrectly predicted to be *trans* residues. Method I clearly outperforms other models in terms of generalization ability and stability. Again, using a different random-number seed does not really make the outcome of classification more stable, which contradicts the findings in Reference [11].

Although the two proposed methods have shown to be useful for proline CTI prediction tasks, we suggest the following to be taken into account for the sake of improvement. First, since our methods use PSSMs only as input, homology information presented in PSSMs may not have enough information to reach the upper-boundary accuracy. Recent studies suggest that global sequence homology is seen as a strong indicator for the occurrence of prolyl *cis* residues [12], meaning that accurate descriptors of CTI residues and their corresponding encoding schemes must be identified. Second, solvent accessibility as a new possible input feature of proline CTI must be well examined as proline *cis* residues are more frequently found in surface-accessible areas compared to *trans* residues [12].

13.4 CONCLUSION

In this chapter, we have presented two methods for proline CTI residue prediction in proteins. The models are trained using a Random Forest ensemble method, which grows multiple trees and classifies according to the most votes over all the trees in the forest. The two methods are able to predict proline CTI with the Q_2 accuracy of 86.58% outperforming any existing proline CTI predictors reported in the literature. Experimental results demonstrate that the proposed methods can achieve a test error better than the most widely used SVM models. It has also demonstrated that pure evolutionary information in the format of PSSM scores as input works greatly in reducing the error rate during the model learning process, meaning that noise presented (i.e., predicted secondary information) in input data set may lead to significant degrading in the performance of the models.

REFERENCES

1. Altschul SF, Gish W, Miller W, Myers EW, Lipman DJ. Basic local alignment search tool. J Mol Biol 1990;215(3):403–410.
2. Argentiero P, Chin R, Beaudet P. An automated approach to the design of decision tree classifiers. IEEE Trans Pattern Anal Mach Intell 1982;4(1):51–57.
3. Breiman L. Bagging predictors. Mach Learn 1996;24(2):123–140.
4. Breiman L. Random forests. Mach Learn 2001;45(1):5–32.
5. Eckert B, Martin A, Balbach J, Schmid FX. Prolyl isomerization as a molecular timer in phage infection. Nat Struct Mol Biol 2005;12(7):619–623.
6. Exarchos KP, Papaloukas C, Exarchos TP, Troganis AN, Fotiadis DI. Prediction of *cis/trans* isomerization using feature selection and support vector machines. J Biomed Inform 2009;42(1):140–149.
7. Fischer G, Aumüller T. Regulation of peptide bond *cis/trans* isomerization by enzyme catalysis and its implication in physiological processes. In: *Reviews of Physiology, Biochemistry and Pharmacology*. Volume 148, Reviews of Physiology, Biochemistry and Pharmacology. Berlin Heidelberg: Springer-Verlag; 2003. p 105–150.
8. Frömmel C, Preissner R. Prediction of prolyl residues in *cis*-conformation in protein structures on the basis of the amino acid sequence. FEBS Lett 1990;277(1):159–163.
9. Holmes G, Donkin A, Witten IH. WEKA: A machine learning workbench. Intelligent Information Systems, 1994. Proceedings of the 1994 2nd Australian and New Zealand Conference on. Brisbane, Australia; IEEE; 1994. p 357–361.

10. Jones DT. Protein secondary structure prediction based on position-specific scoring matrices. J Mol Biol 1999;292(2):195–202.
11. Lira MMS, de Aquino RRB, Ferreira AA, Carvalho MA, Neto ON, Santos GSM. Combining multiple artificial neural networks using random committee to decide upon electrical disturbance classification. Neural Networks, 2007. IJCNN 2007. International Joint Conference on. IEEE; Orlando, FL; 2007. p 2863–2868.
12. Lorenzen S, Peters B, Goede A, Preissner R, Frömmel C. Conservation of *cis* prolyl bonds in proteins during evolution. Proteins 2005;58(3):589–595.
13. Pahlke D, Leitner D, Wiedemann U, Labudde D. COPS-*cis/trans* peptide bond conformation prediction of amino acids on the basis of secondary structure information. Bioinformatics 2005;21(5):685–686.
14. Petersen TN, Lundegaard C, Nielsen M, Bohr H, Bohr J, Brunak S, Gippert GP, Lund O. Prediction of protein secondary structure at 80% accuracy. Proteins 2000;41(1):17–20.
15. Reimer U, Scherer G, Drewello M, Kruber S, Schutkowski M, Fischer G. Side-chain effects on peptidyl-prolyl *cis/trans* isomerisation. J Mol Biol 1998;279(2):449–460.
16. Schmid FX. Prolyl isomerase: enzymatic catalysis of slow protein-folding reactions. Annu Rev Biophys Biomol Struct 1993;22(1):123–143.
17. Schmid FX. Prolyl isomerases. Adv Protein Chem 2001;59:243–282.
18. Shatnawi M, Zaki N. Prediction of inter-domain linker regions in protein sequences: A survey. The 9th International Conference on Innovations in Information Technology (IIT); Al Ain, UAE; 2013. p 237–242.
19. Shatnawi M, Zaki N. Prediction of protein inter-domain linkers using compositional index and simulated annealing. Proceeding of the 15th Annual Conference Companion on Genetic and Evolutionary Computation Conference Companion, GECCO '13 Companion. New York: ACM; 2013. p 1603–1608.
20. Song J, Burrage K, Yuan Z, Huber T. Prediction of *cis/trans* isomerization in proteins using PSI-BLAST profiles and secondary structure information. BMC Bioinformatics 2006;7(1):124.
21. Wang M-L, Li W-J, Xu W-B. Support vector machines for prediction of peptidyl prolyl *cis/trans* isomerization. J Pept Res 2004;63(1):23–28.
22. Wedemeyer W, Welker E, Scheraga HA. Proline *cis–trans* isomerization and protein folding. Biochemistry 2002;41(50):14637–14644.
23. Wu X, Kumar V. *The Top Ten Algorithms in Data Mining*. Boca Raton (FL): CRC Press; 2010.
24. Wu Y, Matthews CR. A *cis*-prolyl peptide bond isomerization dominates the folding of the α subunit of Trp synthase, a TIM barrel protein. J Mol Biol 2002;322(1):7–13.
25. Yaffe MB, Schutkowski M, Shen M, Zhou XZ, Stukenberg PT, Rahfeld J-U, Xu J, Kuang J, Kirschner MW, Fischer G. Sequence-specific and phosphorylation-dependent proline isomerization: a potential mitotic regulatory mechanism. Science 1997;278(5345): 1957–1960.
26. Yoo P, Ho YS, Ng J, Charleston M, Saksena N, Yang P, Zomaya A. Hierarchical kernel mixture models for the prediction of AIDS disease progression using HIV structural gp120 profiles. BMC Genomics 2010;11 Suppl 4:S22.
27. Yoo PD, Ho YS, Zhou BB, Zomaya AY. SiteSeek: post-translational modification analysis using adaptive locality-effective kernel methods and new profiles. BMC Bioinformatics 2008;9(1):272.

28. Yoo PD, Sikder AR, Taheri J, Zhou BB, Zomaya AY. DomNet: protein domain boundary prediction using enhanced general regression network and new profiles. IEEE Trans NanoBiosci 2008;7(2):172–181.
29. Yoo PD, Zhou BB, Zomaya AY. Machine learning techniques for protein secondary structure prediction: an overview and evaluation. Curr Bioinform 2008;3(2):74–86.
30. Yoo P, Zhou B, Zomaya A. A modular kernel approach for integrative analysis of protein domain boundaries. BMC Genomics 2009;10 Suppl 3:S21.

PART IV

PATTERN RECOGNITION IN QUATERNARY STRUCTURES

CHAPTER 14

PREDICTION OF PROTEIN QUATERNARY STRUCTURES

Akbar Vaseghi[1], Maryam Faridounnia[2], Soheila Shokrollahzade[3], Samad Jahandideh[4], and Kuo-Chen Chou[5,6]

[1]*Department of Genetics, Faculty of Biological Sciences, Tarbiat Modares University, Tehran, Iran*
[2]*Bijvoet Center for Biomolecular Research, Utrecht University, Utrecht, The Netherlands*
[3]*Department of Medicinal Biotechnology, Iran University of Medical Sciences, Tehran, Iran*
[4]*Bioinformatics and Systems Biology Program, Sanford-Burnham Medical Research Institute, La Jolla, CA, USA*
[5]*Department of Computational Biology, Gordon Life Science Institute, Belmont, MA, USA*
[6]*Center of Excellence in Genomic Medicine Research (CEGMR), King Abdulaziz University, Jeddah, Saudi Arabia*

14.1 INTRODUCTION

In all living organisms, a protein sequence built up by 20 main amino acid residues and linked through peptide bonds, is encoded by genes and is called the *primary structure*. Depending on the sequence of amino acids, environmental conditions, and a number of other factors, the polypeptide chain folds into very well-defined conformational segments called the *secondary structure* (β-strands, α-helices, etc.). These segments fold into larger conformations with different arrangements creating a *tertiary structure*. In order to be functional and stable, many of these polypeptide chains

Pattern Recognition in Computational Molecular Biology: Techniques and Approaches,
First Edition. Edited by Mourad Elloumi, Costas S. Iliopoulos, Jason T. L. Wang, and Albert Y. Zomaya.

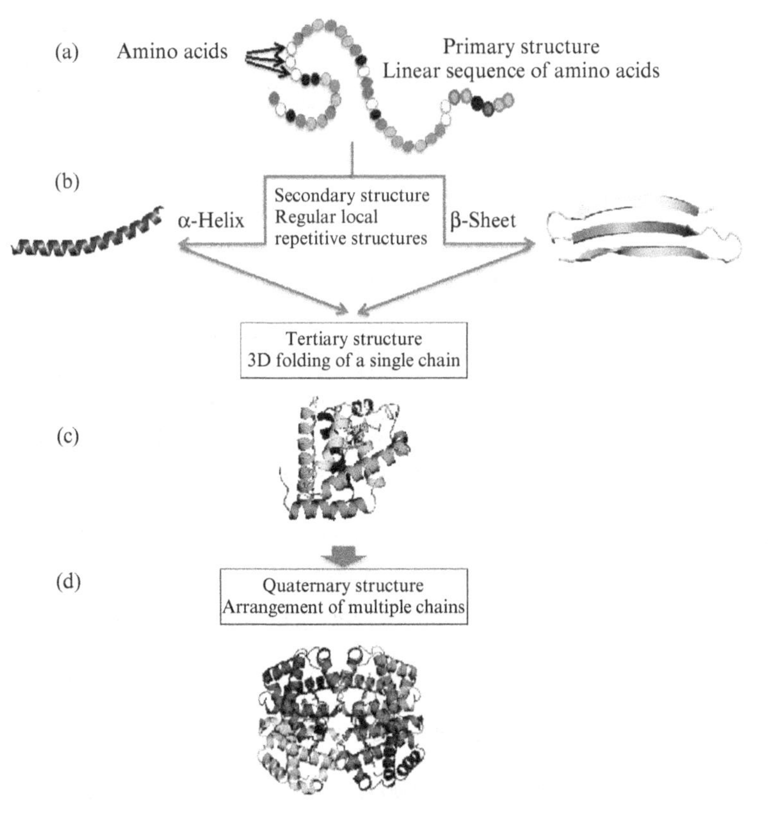

Figure 14.1 The different protein structures: (a) primary structure (polypeptide chain), (b) secondary structure (α-helix), (c) tertiary structure (example: myoglobin), and (d) quaternary structure (example: hemoglobin).

with tertiary structures assemble together as *quaternary structures* (Figure 14.1). These protein assemblies are involved in a variety of biological processes such as cellular metabolism, signal transduction, chromosome transcription, replication, and DNA damage repair.

Quaternary structures are the spatial arrangements including multiple copies of one or multiple types of polypeptide chains, assembling through noncovalent interactions to enable a biological function, which depending on the conditions can be very specific or multipurpose. This creates the protein universe with various classes of subunit constructions, such as monomer, dimer, trimer, tetramer, pentamer, and hexamer [22] (Figure 14.2).

Knowledge of the protein quaternary structure is important because it enables to discover the biological function of the protein and, hence, it enables to target this function during drug development [18]. For instance, the hemoglobin [54], potassium channel [19, 30], and the M2 proton channel [69] are tetramers; while the phospholamban [53], Gamma-AminoButyric Acid type A (GABAA) receptor [20, 73]), and

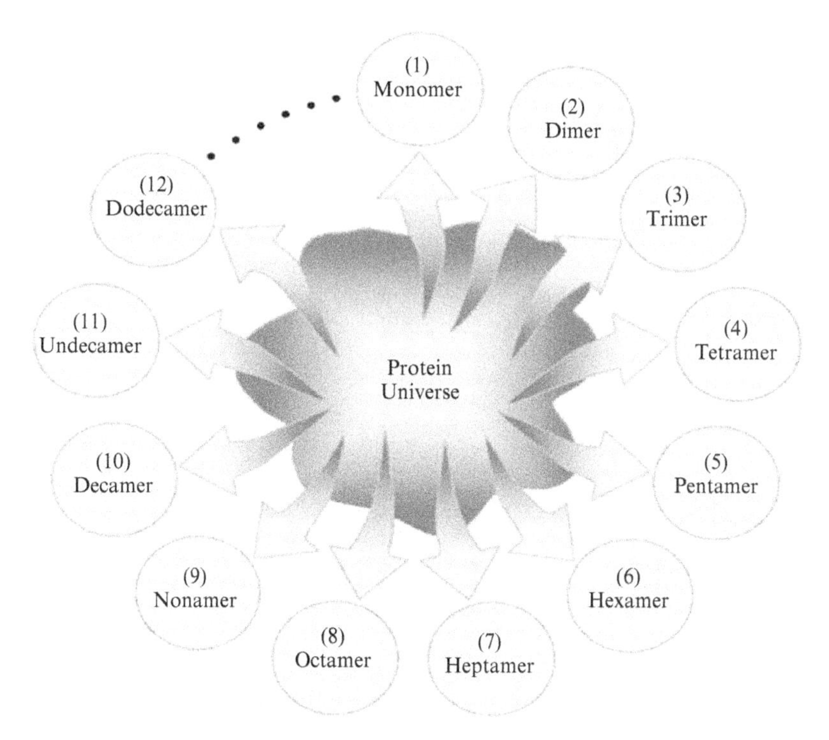

Figure 14.2 The schematic representation of different polypeptide chains that form various oligomers. *Source*: Chou and Cai (2003) [22]. Reproduced with permission of John Wiley and Sons, Inc.

alpha7 nicotinic acetylcholine receptor [20] are pentamers. The sodium channel is a monomer [11], whereas the p7 hepatitis C virus is a hexamer [52].

Among the aforementioned proteins, hemoglobin is a classical example that is made up of two α-chains and two β-chains, and these four chains aggregate into one structure to perform the protein's cooperative function during the oxygen-transportation process, as elucidated from the viewpoint of low-frequency collective motion [13–15, 54]. In addition, recent findings have revealed that the novel molecular wedge allosteric drug-inhibited mechanism [56] for the M2 proton channel can be understood [39] through a unique left-handed twisting packing [23] arrangement of four transmembrane helices from four identical protein chains [69].

There are three categories of rotational symmetry for oligomeric proteins, including (i) cyclic symmetry C_n, which has an n-fold symmetry by $360°/n$ rotations ($n = 2, 3, 4,...$); for example, Figure 14.3a shows an object which has C_5 symmetry with a fivefold rotational axis. (ii) Dihedral symmetry, D_n, is generated when an n-fold symmetry intersects with a twofold axis of symmetry. For example, Figure 14.3b shows D_4 symmetry. (iii) Polyhedral symmetries such as cubic, tetrahedrons, and icosahedrons have 12, 24, or 60 identical subunits (Figure 14.3c, e, and d). The most common symmetry for soluble proteins is C_2 in homodimers. Among other common symmetries, D_2 tetramers are more common than C_4 (potassium

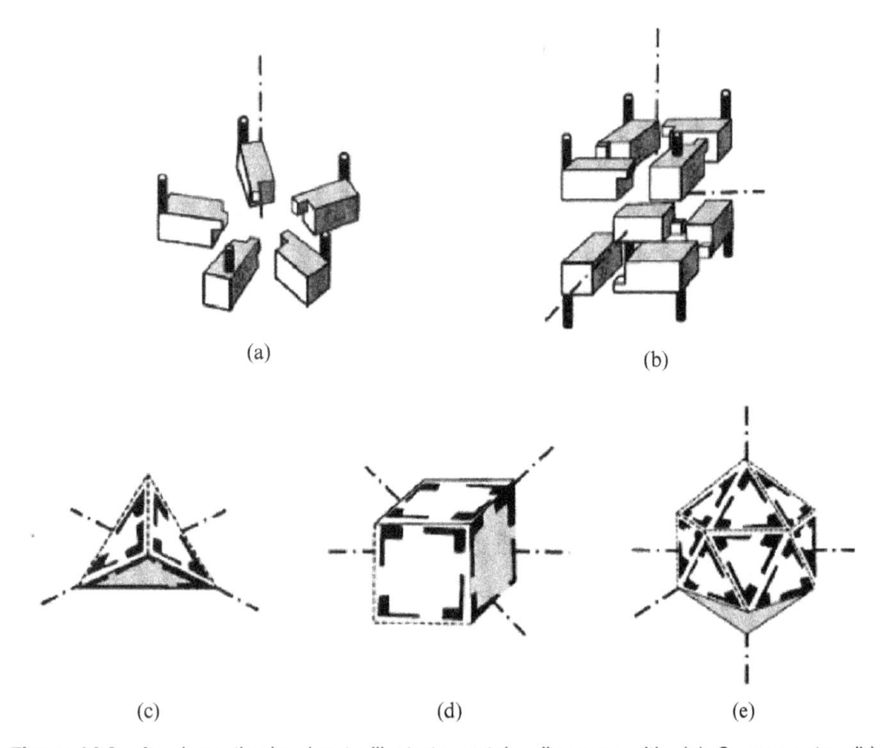

Figure 14.3 A schematic drawing to illustrate protein oligomers with: (a) C_5 symmetry, (b) D_4 symmetry, (c) tetrahedron symmetry, (d) cubic symmetry, and (e) icosahedrons symmetry. *Source*: Chou and Cai (2003) [22]. *Reproduced with permission of John Wiley and Sons, Inc.*

channel) and C_5 (pentameric acetylcholine receptor), and D_3 (of bacteriorhodopsin) hexamers are more common than C_6 ([40]).

Determination of the protein structure and understanding its function is essential for any relevant medical, engineering, or pharmaceutical applications. Therefore, the study of quaternary structure of proteins, despite all the obstacles in acquiring data from large macromolecular assemblies, is one of the major goals in biomolecular sciences. Although the structure of a large number of proteins is experimentally solved, there exist many proteins with unknown fold and without any obvious homology. The protein structure can be investigated by experimental [4, 43, 45, 71, 75, 79] and computational methods, namely, Comparative Modeling (CM) [1, 2, 57, 60, 67, 83]. Considering the fact that determination of protein structure using the experimental methods is expensive, labor intensive, and time consuming, the Structural Genomics (SG) project has been introduced with the goal of developing and integrating new technologies that cover the gene-to-structure process. This SG approach facilitates high-throughput structural biology. However, regarding the considerably higher speed and lower cost of computing, the development of computational methods for protein structure prediction is very promising for prediction of unsolved protein structures [65].

The rest of this chapter is organized as follows. In Sections 14.2–14.5, we discuss respectively protein structure prediction, template-based predictions, critical assessment of protein structure prediction, and quaternary structure prediction. Finally, in Section 14.6, we present the conclusion of this chapter.

14.2 PROTEIN STRUCTURE PREDICTION

Development of large-scale sequencing techniques speeded up sequencing and resulted in a huge gap between the number of known protein sequences and the number of solved structures (Figure 14.4). Accurate prediction of protein structure is a difficult task due to (i) limited knowledge about stability of protein structure, (ii) the role of chaperons in the folding process of proteins from different families, (iii) multiple folding pathways to attain the native state being evidenced for a single protein, and (iv) an enormous quantity of possible conformations that can be suggested for each protein. Nevertheless, the goal of protein structure prediction projects is to reduce this gap. During its early stage, the problem was simplified by mimicking one possible mechanism of protein folding, which includes the two folding steps: (i) formation of local secondary structures and (ii) arrangement of secondary structures to achieve the folded conformation. Later, before trying to tackle tertiary structure prediction, for decades, many secondary structure prediction methods were developed. Here, we briefly introduce the three generations of secondary structure prediction methods and continue with popular methods for prediction of tertiary structure.

14.2.1 Secondary Structure Prediction

Up to now, different methods have been developed for prediction of secondary structure. Here, we introduce three different generations of these methods in separate sections.

14.2.1.1 First Generation. In the first generation, a few popular methods were developed by simple analysis of amino acids distribution in α-helices and β-strands. The first method that was widely used for prediction of secondary structures was the Chou–Fasman method [24]. This method was developed in 1974 on the basis of statistical analysis of preference and avoidance of all 20 amino acids in α-helices and β-strands. They simply calculated the propensity of all 20 amino acids for α-helices and β-strands in a small database of solved protein structures. The Chou–Fasman method consists of a number of simple rules. For example, if the property in a window of six amino acids (for a helix) or five amino acids (for a strand) is above a predefined threshold value, this segment is the nucleation point of a possible secondary structure.

The next standard method was the Garnier, Osuguthorpe, and Robson (GOR) method [33]. The GOR method, developed four years after the Chou–Fasman method, was slightly more sophisticated and more accurate. Secondary structure prediction was primarily based on a sliding window of 17 residues and assigning a value to each residue that expresses the likelihood of it being in a particular secondary structure.

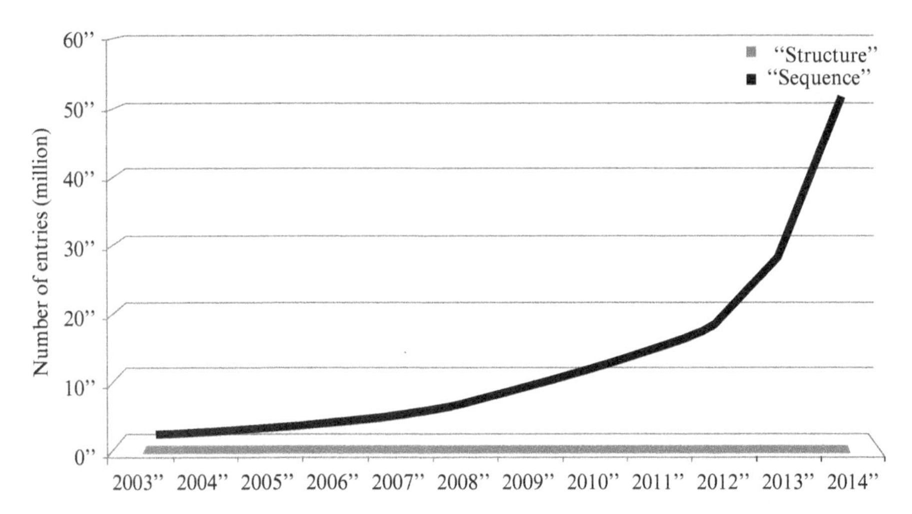

Figure 14.4 Comparison of determined number of protein sequences and protein structures based on statistical data from PDB and UniProtKB.

14.2.1.2 Second Generation. The second generation of secondary structure prediction methods covers a combination of methodology development and feature generation. Different predictor methods including statistical learning and machine learning are applied for prediction of secondary structures [62, 63, 72]. Furthermore, a class of features including physicochemical properties and sequence patterns was defined to fit into the predictor models [34]. In this generation, accuracy of secondary structure prediction had 10% improvement compared to the methods in the first generation with an accuracy of 50–60%.

14.2.1.3 Third Generation. These methods outperformed developed methods in the earlier generations in terms of accuracy [10, 64]. Their accuracy reached close to 80% [55]. The main cause for improvement in accuracy was adding evolutionary information to the list of features generated from multiple sequence alignments. For example, Rost and Sander in 1993 developed the PHD method that is composed of several cascading neural networks [61]. In the neural network, homologous sequences that are determined by BLAST [55] are aligned using the MaxHom alignment [68] of known structures; then, generated conservation scores are used to train the network, which then can be used to predict the secondary structure of the aligned sequences of the unknown protein.

In addition to prediction of secondary structure, many other methods are developed for prediction of local attributes of proteins, such as disordered regions, membrane-spanning beta-barrels, and transmembrane helices.

14.2.2 Modeling of Tertiary Structure

Despite the advances in protein structure determination in recent decades, some classes of proteins such as membrane proteins, that is, proteins integrated into a complex lipid bilayer environment—that are of eminent importance in biological sciences and medicinal biology—and other proteins with large quaternary structure arrangements, are not easy to crystallize or are too large for determination with Nuclear Magnetic Resonance (NMR) [12, 44]. Therefore, developing accurate protein structure prediction tools is very important, as many computational methods that have been developed until now comprise machine-learning-based methods [2], SCWRL and MolIDE [74], Rosetta [1, 60], MODELLER [67], homology modeling, fold recognition, threading, and *ab initio* prediction. In this section, we briefly introduce more popular methodologies for tertiary structure prediction.

14.3 TEMPLATE-BASED PREDICTIONS

Homology modeling and threading methods are two types of template-based approaches. The homology modeling method needs to have the homologous protein structure as template and threading methods are a new approach in fold recognition, in which the tool attempts to fit the sequence in the known structures.

14.3.1 Homology Modeling

Considering that high sequence similarity results in similar three-dimensional (3D) structures, a large number of methods (homology modeling, comparative modeling, or template-based modeling) use a properly known structure as a seeding point. This type of method has five stages, including (i) finding a proper template, (ii) multiple sequence alignment of the query protein to the templates, (iii) modeling the target structure including main chain modeling, loop modeling, and side chain modeling, (iv) energy refinement, and (v) model evaluation. *Swiss-model* [36] and *3D Jigsaw* [3] are some of the most commonly used servers for homology modeling.

14.3.2 Threading Methods

The threading method is a protein structure prediction technique applied when there is not enough sequence similarity between the target and the template. In this method, the target sequence is pasted onto a known fold and the compatibility of the sequence and the fold is evaluated by calculating a score. The local secondary structure, the environment, and the pairwise interaction of side chains of close amino acids are the structural properties that are used to evaluate the fit. FFAS03 [41], 3DPSSM [46], and I-TASSER [66] are the most popular threading programs.

14.3.3 *Ab initio* Modeling

Anfinsen's hypothesis implies that the native state of the protein represents the global free energy minimum. Considering this hypothesis, *ab initio* methods try to find these global minima of the protein [29] and predict tertiary structure without reference to a specific template with homologous structure [51]. Finding the correct conformation using *ab initio* modeling requires an efficient search method for exploring the conformational space to find the energy minima and an accurate potential function that calculates the free energy of a given structure. *Rosetta* is the most popular program in this category [42].

14.4 CRITICAL ASSESSMENT OF PROTEIN STRUCTURE PREDICTION

The best method for evaluation of the performance of structure prediction methods is *in silico* test on unknown structures, which is the focus of the Critical Assessment of protein Structure Prediction (CASP) meetings [50]. CASP invites the scientists that developed methods for protein structure prediction to model structure of proteins with solved structure but not released yet. The models are submitted and compared to the experimentally determined structures. The most accurate methods are introduced and the progression of the field is discussed during these biannual meetings. Evaluation of the results is carried out in different categories including tertiary structure prediction, residue–residue contact prediction, disordered regions prediction, function prediction, model quality assessment, model refinement, and high-accuracy template-based prediction.

14.5 QUATERNARY STRUCTURE PREDICTION

The Protein Data Bank (PDB) contains more than 100,000 protein structures mostly determined by X-ray crystallography and NMR [5]. By convention, a crystallographic PDB entry reports atomic coordinates for the crystal ASymmetric Unit (ASU). ASU includes the subunit contacts and intermolecular contacts that define the quaternary structure. In addition to PDB, many databases exist such as Protein Biological unit Database (ProtBuD) [78], Protein InTerfaces and Assemblies (PITA) [58], Probable Quaternary Structure (PQS) [37], Protein Quaternary Structure investigation) PiQSi) [48], and Protein Interfaces, Surfaces, and Assemblies (PISA) [47] designed for quaternary structure studies. PQS, PISA, and PITA are developed at the European Bioinformatics Institute (EBI-EMBL, Hinxton, UK). These databases also include the ones that are not limited to biological units. PITA and PQS apply crystal symmetries to the molecules in the ASU, select neighbors, and score each pairwise interface on the basis of the buried area, plus a statistical potential in PITA or a solvation energy term in PQS [6]. Nevertheless, PiQSi facilitated the visual inspection of the quaternary structure of protein complexes in PDB [49]. PiQSi characterizes a query protein as belonging to monomer, homo-oligomer, or hetero-oligomer according to its sequence information alone. The first version of PiQSi annotated over 10,000 structures from the PDB biological unit and corrected the quaternary state of approximately 15% of them. All data are also available at http://www.PiQSi.org [49].

Using the above-mentioned databases, a few sequence-based computational methods have been developed for the prediction of protein quaternary structure using statistical models or machine learning methods. The earliest method for prediction of quaternary structure was a decision tree model that uses simple sequence-based features to discriminate between the primary sequences of homodimers and non-homodimers [31]. Subsequently, new algorithms were developed for the prediction of different types of homomeric structures ([16], Chou, 2004, [21]) and heteromeric structures [9].

Features, databases, and prediction models are three important elements in developing a reliable and accurate prediction methodology. Different combinations of predictor models and features have been examined for quaternary structure prediction. For example, pseudo amino acid composition, originally introduced by Kou-Chen Chou at The Gordon Life Science Institute, was a feature for improving the prediction of protein subcellular location [17] (Figure 14.5). Pseudo amino acid composition was introduced to get the desired results when trying sequence-similarity-search-based prediction. This kind of approach fails when a query protein does not have significant homology to the known protein(s). Thus, various discrete models were proposed that do not rely on sequence-order (see Reference [17] for more details). Later, pseudo amino acid composition was used for the prediction of quaternary structure [17, 22]. Moreover, the functional domain composition information, which has been very useful in identifying enzyme functional classes [25] and protease types [26], was fed into the Nearest Neighbor (NN) algorithm [28] that is proved to be very successful in prediction of protein subcellular localization [26] as prediction engine for prediction of quaternary structure [76].

Training of models using small database selected from the population increases the error rate in testing. In such circumstances, predictor performance can be improved by

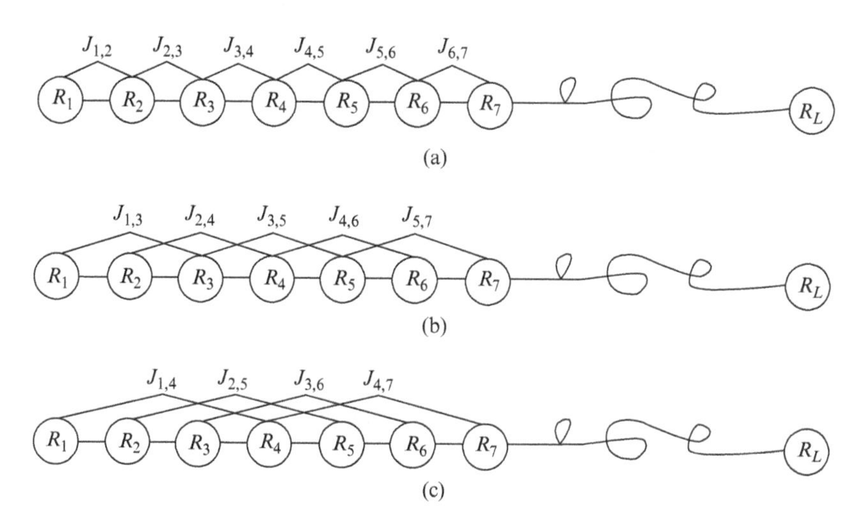

Figure 14.5 Definition procedure for the huge number of possible sequence order patterns as a set of discrete numbers. Panels (a–c) represent the correlation mode between all the most contiguous residues, all the second-most contiguous residues, and all the third-most contiguous residues, respectively. *Source*: Chou and Cai (2003) [22]. *Reproduced with permission of John Wiley and Sons, Inc.*

using larger a data set in the training step. Moreover, redundancy of protein sequences can result in overfitting. In this case, overfitting refers to receiving higher weights by nonrelevant features that are redundantly repeated in the data set. However, there is overfitting in the use of features, models, or evaluation procedures that violate parsimony, that is, they use more complicated approaches than are necessary. A general rule is that the best predictor model is the simplest. For example, if a simple model, for example, the linear regression model, perfectly fits to a database, then applying an overfitting-prone complicated model like an artificial neural network results in adding a level of complexity without any corresponding benefit in performance or, even worse, with poorer performance than the simpler model.

After the first publication on classification of protein quaternary structure, the Support Vector Machine*s* (SVMs) were introduced to classify quaternary structure properties from the protein primary sequences [81]. SVM is a new type of data mining method applied with success to different problems in bioinformatics and computational biology, such as translation initiation sites [84], membrane protein types [8], protein–protein interactions [8], and protein subcellular localization [38]. In this study, a binary SVM was applied to discriminate between the primary sequences of homodimers and non-homodimers, and the obtained results were similar to the previous investigation by Garian in terms of performance [32]. Subsequently, different methodologies were applied for classification and prediction of protein quaternary structure—for example, the NN algorithm [80], threading [35], and the Function for the Degree Of Disagreement (FDOD) are applied to discriminate between homodimers and other homo-oligomeric proteins from the primary structure [70]. By extending the problem to different types of homo-oligomer and hetero-oligomer quaternary structures, recently Quat-2L was developed for predicting protein quaternary structure and identifying the query protein as monomer, homo-oligomer, or hetero-oligomer for the first step (with a success rate of 71.14%), and for different types of homo-oligomers and hetero-oligomers for the second step (with success rates of 76.91% and 82.52%, respectively). Quat-2L is available as a web-server at http://icpr.jci.jx.cn/bio-info/Quat-2L [77]. Hybridization of pseudo amino acids composition and Quat-2L mode can be used to predict protein quaternary structural attribute of a protein chain according to its sequence information alone with higher success rate [77]. The sequence-segmented pseudo amino acids composition approach can capture essential information about the compositions and hydrophobicity of residues in the surface patches buried in the interfaces of associated subunits [82].

Three conventional procedures for performance evaluation of predictor methods are defined as self-consistency, cross-validation, and jackknife [30]. The self-consistency procedure is as simple as training and testing of the predictor model on the same data set that only evaluates correlation between input features and outputs. On the other hand, during cross-validation and jackknife tests, none of the test samples is presented in training procedures. The jackknife test is a type of cross-validation in which only one sample is fed into the model as testing sample after each training procedure; therefore, the number of iterations is equal to the total number of samples in data set. However, in the cross-validation procedure, the data set is divided into some subsets and for each iteration one subset is eliminated during training and the total number of iteration is equal to the number of subsets [76].

14.6 CONCLUSION

The function of a protein is related to its quaternary structure and thus prediction of quaternary structure from the protein sequence is very useful and can be obtained by studying the relation of the quaternary structural attribute of a protein chain and its sequence feature. Predicting quaternary structure of a protein is a difficult task. However, the development and growth of novel statistical learning and machine learning methods, the increase in the number of solved structures, and the improvement of computational resources, have together increased the chances of development of more accurate prediction methods in the near future. Among protein structure prediction problems, prediction of quaternary structure is fairly new and the methods that have been developed are still immature. More efforts are needed to increase the accuracy of these methods to a higher level of reliability to become applicable for biologists. We are very optimistic that accuracy of quaternary structure prediction will be improved by using recently developed state-of-the-art machine learning methods, such as the random forest method [7] and multi-class SVM [59].

ACKNOWLEDGMENTS

We express our thanks to all members of "The First Protein Structural Bioinformatics Online Workshop." This chapter is extracted from our literature review and discussion on protein structure prediction in this workshop.

REFERENCES

1. Baker D, Sali A. Protein structure prediction and structural genomics. Science 2001;294:93–96.
2. Baldi P. *Bioinformatics: The Machine Learning Approach*. The MIT Press; 2001.
3. Bates PA, Kelley LA, MacCallum RM, Sternberg MJ. Enhancement of protein modeling by human intervention in applying the automatic programs 3D-JIGSAW and 3D-PSSM. Proteins 2001;5:39–46.
4. Benesch JL, Robinson CV. Mass spectrometry of macromolecular assemblies: preservation and dissociation. Current Opin Struct Biol 2006;16:245–251.
5. Berman HM, Westbrook J, Feng Z, Gilliland G, Bhat TN, Weissig H, Shindyalov IN, Bourne PE. The protein data bank. Nucleic Acids Res 2000;28:235–242.
6. Berman HM, Battistuz T, Bhat T, Bluhm WF, Bourne PE, Burkhardt K, Feng Z, Gilliland GL, Iype L, Jain S. The protein data bank. Acta Crystallogr Sect D: Biol Crystallogr 2002;58:899–907.
7. Breiman L. Random forests. Mach Learn 2001;45:5–32.

8. Cai YD, Liu XJ, Xu XB, Chou KC. Support vector machines for predicting membrane protein types by incorporating quasi-sequence-order effect. Internet Electron J Mol Des 2002;1:219–226.
9. Carugo O. A structural proteomics filter: prediction of the quaternary structural type of hetero-oligomeric proteins on the basis of their sequences. J Appl Crystallogr 2007;40:986–989.
10. Chandonia JM, Karplus M. New methods for accurate prediction of protein secondary structure. Proteins 1999;35:293–306.
11. Chen Z, Alcayaga C, Suarez-Isla BA, O'Rourke B, Tomaselli G, Marban E. A "minimal" sodium channel construct 2002.
12. Cheng J, Tegge AN, Baldi P. Machine learning methods for protein structure prediction. Biomed Eng IEEE Rev 2008;1:41–49.
13. Chou K-C. The biological functions of low-frequency phonons. 4. Resonance effects and allosteric transition. Biophys Chem 1984;20:61–71.
14. Chou K-C. Review: Low-frequency collective motion in biomacromolecules and its biological functions. Biophys Chem 1988;30:3–48.
15. Chou K-C. Low-frequency resonance and cooperativity of hemoglobin. Trends Biochem Sci 1989;14:212–213.
16. Chou K-C. Prediction of protein subcellular locations by incorporating quasi-sequence-order effect. Biochem Biophys Res Commun 2000;278:477–483.
17. Chou KC. Prediction of protein cellular attributes using pseudo-amino acid composition. Proteins 2001;43:246–255.
18. Chou K-C. Structural bioinformatics and its impact to biomedical science. Curr Med Chem 2004;11:2105–2134.
19. Chou K-C. Modelling extracellular domains of GABA-A receptors: subtypes 1, 2, 3, and 5. Biochem Biophys Res Commun 2004;316:636–642.
20. Chou K-C. Insights from modelling the 3D structure of the extracellular domain of alpha7 nicotinic acetylcholine receptor. Biochem Biophy Res Commun 2004;319:433–438.
21. Chou K-C. Using amphiphilic pseudo amino acid composition to predict enzyme subfamily classes. Bioinformatics 2005;21:10–19.
22. Chou KC, Cai YD. Predicting protein quaternary structure by pseudo amino acid composition. Proteins 2003;53:282–289.
23. Chou KC, Maggiora GM, Nemethy G, Scheraga HA. Energetics of the structure of the four-alpha-helix bundle in proteins. Proceedings of the National Academy of Sciences of the United States of America (PNAS USA)1988;85:4295–4299.
24. Chou PY, Fasman GD. Prediction of protein conformation. Biochemistry 1974;13:222–245.
25. Chou K-C, Shen H-B. Euk-mPLoc: a fusion classifier for large-scale eukaryotic protein subcellular location prediction by incorporating multiple sites. J Proteome Res 2007;6:1728–1734.
26. Chou K-C, Shen H-B. ProtIdent: A web server for identifying proteases and their types by fusing functional domain and sequential evolution information. Biochem Biophys Res Commun 2008;376:321–325.
27. Chou KC, Zhang CT. Review: Prediction of protein structural classes. Critical Reviews in Biochemistry and Molecular Biology 1995;30:275–349.

28. Cover T, Hart P. The nearest neighbor decision rule. IEEE Trans Inform Theory 1967;13:21–27.
29. Defay T, Cohen FE. Evaluation of current techniques for ab initio protein structure prediction. Proteins 1995;23:431–445.
30. Doyle DA, Morais CJ, Pfuetzner RA, Kuo A, Gulbis JM, Cohen SL, Chait BT, MacKinnon R. The structure of the potassium channel: molecular basis of K+ conduction and selectivity. Science 1998;280:69–77.
31. Garian R. Prediction of quaternary structure from primary structure. Bioinformatics 2001;17:551–556.
32. Garian R. Prediction of quaternary structure from primary structure. Bioinformatics 2001;17:551–556.
33. Garnier J, Gibrat JF, Robson B. GOR method for predicting protein secondary structure from amino acid sequence. Methods Enzymol 1996;266:540–553.
34. Geourjon C, Deleage G. SOPMA: significant improvements in protein secondary structure prediction by consensus prediction from multiple alignments. Comput Appl Biosci 1995;11:681–684.
35. Grimm V, Zhang Y, Skolnick J. Benchmarking of dimeric threading and structure refinement. Proteins 2006;63:457–465.
36. Guex N, Peitsch MC. SWISS-MODEL and the Swiss-Pdb viewer: an environment for comparative protein modeling. Electrophoresis 1997;18(15):2714–2723.
37. Henrick K, Thornton JM. PQS: a protein quaternary structure file server. Trends Biochem Sci 1998;23:358–361.
38. Hua S, Sun Z. Support vector machine approach for protein subcellular localization prediction. Bioinformatics 2001;17:721–728.
39. Huang RB, Du DS, Wang CH, Chou KC. An in-depth analysis of the biological functional studies based on the NMR M2 channel structure of influenza A virus. Biochem Biophys Res Commun 2008;377:1243–1247.
40. Janin J, Bahadur RP, Chakrabarti P. Protein–protein interaction and quaternary structure. Q Rev Biophys 2008;41:133–180.
41. Jaroszewski L, Rychlewski L, Li Z, Li WZ, Godzik A. FFAS03: a server for profile–profile sequence alignments. Nucleic Acids Res 2005;33:W284–W288.
42. Jauch R, Yeo HC, Kolatkar PR, Clarke ND. Assessment of CASP7 structure predictions for template free targets. Proteins 2007;69:57–67.
43. Blundell TL, Johnson LN. *Protein Crystallography*. New York: Academic; 1976.
44. Jones DT. GenTHREADER: an efficient and reliable protein fold recognition method for genomic sequences. J Mol Biol 1999;287:797–815.
45. Karlsson R. SPR for molecular interaction analysis: a review of emerging application areas. J Mol Recogn 2004;17:151–161.
46. Kelleya LA, MacCalluma RM, Sternberga MJE. Enhanced genome annotation using structural profiles in the program 3D-PSSM1. J Mol Biol 2000;299(2):501–522.
47. Krissinel E, Henrick K. Inference of macromolecular assemblies from crystalline state. J Mol Biol 2007;372:774–797.
48. Levy ED. PiQSi: protein quaternary structure investigation. Structure 2007;15: 1364–1367.
49. Levy ED. PiQSi: protein quaternary structure investigation. Structure 2007b;15: 1364–1367.

50. Moult J, Fidelis K, Kryshtafovych A, Schwede T, Tramontano A. Critical assessment of methods of protein structure prediction (CASP) – round X. Proteins 2014;82 (Suppl 2):1–6.
51. Ortiz AR, Kolinski A, Rotkiewicz P, Ilkowski B, Skolnick J. *Ab initio* folding of proteins using restraints derived from evolutionary information. Proteins 1999;37:177–185.
52. OuYang B, Xie S, Berardi MJ, Zhao XM, Dev J, Yu W, Sun B, Chou JJ. Unusual architecture of the p7 channel from hepatitis C virus. Nature 2013;498:521–525.
53. Oxenoid K, Chou JJ. The structure of phospholamban pentamer reveals a channel-like architecture in membranes. Proc Natl Acad Sci USA 2005;102:10870–10875.
54. Perutz MF. The hemoglobin molecule. Sci Am 1964;211:65–76.
55. Petersen TN, Lundegaard C, Nielsen M, Bohr H, Bohr J, Brunak S, Gippert GP, Lund O. Prediction of protein secondary structure at 80% accuracy. Proteins 2000;41:17–20.
56. Pielak RM, Jason R, Schnell JR, Chou JJ. Mechanism of drug inhibition and drug resistance of influenza A M2 channel. Proc Natl Acad Sci USA 2009;106:7379–7384.
57. Pieper U, Eswar N, Davis FP, Braberg H, Madhusudhan MS, Rossi A, Marti-Renom M, Karchin R, Webb BM, Eramian D. MODBASE: a database of annotated comparative protein structure models and associated resources. Nucleic Acids Res 2006;34:D291–D295.
58. Ponstingl H, Kabir T, Thornton JM. Automatic inference of protein quaternary structure from crystals. J Appl Cryst 2003;36:1116–1122.
59. Rashid M, Saha S, Raghava GP. Support vector machine-based method for predicting subcellular localization of mycobacterial proteins using evolutionary information and motifs. BMC Bioinf 2007;8:337.
60. Rohl CA, Strauss CE, Chivian D, Baker D. Modeling structurally variable regions in homologous proteins with rosetta. Proteins 2004;55:656–677.
61. Rost B, Sander C. Improved prediction of protein secondary structure by use of sequence profiles and neural networks. Proc Natl Acad Sci USA 1993a;90:7558–7562.
62. Rost B, Sander C. Prediction of protein secondary structure at better than 70% accuracy. J Mol Biol 1993b;232:584–599.
63. Rost B, Sander C. Secondary structure prediction of all-helical proteins in two states. Protein Eng 1993c;6:831–836.
64. Rost B, Sander C. Third generation prediction of secondary structures. Methods Mol Biol 2000;143:71–95.
65. Rost B, Liu J, Przybylski D, Nair R, Wrzeszczynski KO, Bigelow H, Ofran Y. Prediction of protein structure through evolution. In: *Handbook of Chemoinformatics: From Data to Knowledge*4 vols. Wiley; 2003. p 1789–1811.
66. Roy A, Kucukural A, Zhang Y. I-TASSER: a unified platform for automated protein structure and function prediction. Nat Protoc 2010;5:725–738.
67. Sali A, Blundell T. Comparative protein modelling by satisfaction of spatial restraints. Protein Struct Distance Anal 1994;64:C86.
68. Sander C, Schneider R. Database of homology-derived structures and the structural meaning of sequence alignment. Proteins 1991;9:56–68.
69. Schnell JR, Chou JJ. Structure and mechanism of the M2 proton channel of influenza A virus. Nature 2008;451:591–595.
70. Song J, Tang H. Accurate classification of homodimeric vs other homooligomeric proteins using a new measure of information discrepancy. J Chem Inform Comput Sci 2004;44:1324–1327.

71. Sund H, Weber K. The quaternary structure of proteins. Angew Chem Int Edit Engl 1966;5:231–245.
72. Taylor WR, Thornton JM. Prediction of super-secondary structure in proteins. Nature 1983;301:540–542.
73. Tretter V, Ehya N, Fuchs K, Sieghart W. Stoichiometry and assembly of a recombinant GABAA receptor subtype. J Neurosci 1997;17:2728–2737.
74. Wang Q, Canutescu AA, Dunbrack RL. SCWRL and MolIDE: computer programs for side-chain conformation prediction and homology modeling. Nat Protoc 2008;3:1832–1847.
75. Wuthrich K. *NMR of Proteins and Nucleic Acids*. New York: Wiley; 1986.
76. Xiao X, Wang P, Chou K-C. Predicting the quaternary structure attribute of a protein by hybridizing functional domain composition and pseudo amino acid composition. J Appl Crystallogr 2009;42:169–173.
77. Xiao X, Wang P, Chou K-C. Quat-2L: a web-server for predicting protein quaternary structural attributes. Mol Diversity 2011;15:149–155.
78. Xu Q, Canutescu A, Obradovic Z, Dunbrack RL Jr. ProtBuD: a database of biological unit structures of protein families and superfamilies. Bioinformatics 2006;22:2876–2882.
79. Yan Y, Marriott G. Analysis of protein interactions using fluorescence technologies. Curr Opin Chem Biol 2003;7:635–640.
80. Yu X, Wang C, Li Y. Classification of protein quaternary structure by functional domain composition. BMC Bioinform 2006;7:187.
81. Zhang S-W, Pan Q, Zhang H-C, Zhang Y-L, Wang H-Y. Classification of protein quaternary structure with support vector machine. Bioinformatics 2003;19:2390–2396.
82. Zhang S-W, Chen W, Yang F, Pan Q. Using Chou's pseudo amino acid composition to predict protein quaternary structure: a sequence-segmented PseAAC approach. Amino Acids 2008;35:591–598.
83. Zhou HX, Shan Y. Prediction of protein interaction sites from sequence profile and residue neighbor list. Proteins 2001;44:336–343.
84. Zien A, Rätsch G, Mika S, Schölkopf B, Lengauer T, Müller K-R. Engineering support vector machine kernels that recognize translation initiation sites. Bioinformatics 2000;16:799–807.

CHAPTER 15

COMPARISON OF PROTEIN QUATERNARY STRUCTURES BY GRAPH APPROACHES

Sheng-Lung Peng[1] and Yu-Wei Tsay[2]

[1]Department of Computer Science and Information Engineering, National Dong Hwa University, Hualien, Taiwan
[2]Institute of Information Science, Academia Sinica, Taipei, Taiwan

15.1 INTRODUCTION

Measuring protein structural similarity attempts to establish an equivalence relation between polymer structures based on their shapes and three-dimensional conformations. This procedure is a valuable tool for the evaluation of proteins with low sequence similarity, especially when relations between proteins cannot be easily detected by alignment techniques. In this section, we give a brief overview on protein structure comparison and discuss the advantages and disadvantages of various approaches.

As a result of the tremendous progress in biotechnology, comparison or alignment of protein structures has become a fundamental and significant task in computational biology. Structural comparison refers to the analysis of the similarities and differences between two or more structures. Structural alignment refers to establishing which amino acid residues are equivalent among them [8]. In general, these approaches

Pattern Recognition in Computational Molecular Biology: Techniques and Approaches,
First Edition. Edited by Mourad Elloumi, Costas S. Iliopoulos, Jason T. L. Wang, and Albert Y. Zomaya.

provide a measure of structural similarity between proteins. They are used to identify similar folds and evolutionarily related proteins.

When aligned proteins are assumed to share a common ancestor, a structural alignment supports identification of evolutionarily equivalent residues [34]. The structural alignment of functionally related proteins provides insights into the functional mechanisms and has been successfully applied in the functional annotation of proteins whose structures have been determined [58].

As a rule, proteins with high sequence identity and high structure similarity tend to possess functionally conserved regions [57]. In most cases, the relationships between the sequence and structures of homologous proteins are well recognized from these functionally conserved regions. However, the bias of the expected sequence and structure similarity relationship still remains unexplored [18]. Numerous amino acid sequences may yield various structures, while similar sequences may sometimes yield dissimilar structures. Reference [24] illustrates an asymmetric relationship between protein sequence and structure. Thus, it is believed that comparison of protein structures can reveal distant evolutionary relationships that would not be detected by sequence information alone.

The Root Mean Square Deviation (RMSD) value [30] gives a direction to measure the average distance between the given structural proteins. Let A and B be two proteins. Assume that $(a_1, a_2, \dots, a_n)$ (respectively, $(b_1, b_2, \dots, b_n)$) be the corresponding atoms of A (respectively, B). Then the RMSD value of A and B is defined as follows:

$$\mathrm{RMSD} = \sqrt{\frac{1}{n}\sum_{i=1}^{n}\delta_i^2} \tag{15.1}$$

where δ_i is the distance between a_i and b_i for $1 \leq i \leq n$. The minimum value of RMSD obtained by translation and rotation of two superimposed proteins indicated the divergence between the two proteins. RMSD is also an excellent measure for nearly identical structures. The smaller the RMSD value of the two structures, the more similar is their conformation. In addition, there is fairly general agreement that the RMSD requires performing an exhaustive search against whole proteins. Although the RMSD protocol is most popularly implemented, it suffers from a few drawbacks. This point has not received much attention.

1. The RMSD approach ignores many significant global alignments that may result in misjudgment. Because all atoms are equally weighted during the RMSD calculation, it becomes biased to a local structure rather than to a global one rendering it incapable of being performed over all numbers of atoms. Only selected backbone atoms between two residues are chosen for the least-squares fitting.
2. Another drawback is that the protein environment of physicochemical properties, such as ligand complexity, compound reaction, and molecule bonding, is completely ignored [59].

3. Finally, although an RMSD value seeks to reach a global minimum between two proteins, there is no standard criteria for an RMSD value. Unrelated proteins may have a large RMSD, but this may also happen for two relevant proteins consisting of identical substructures [51]. Once the shapes of two proteins turn divergent, the RMSD loses its effectiveness. This may lead to a confusing analysis.

DALI, another popular structural alignment method, breaks the input structures into hexapeptide fragments and calculates a distance matrix by evaluating the contact patterns between successive fragments [23]. Secondary structure features that involve residues that are contiguous in sequence appear on the main diagonal of the matrix; other diagonals in the matrix reflect spatial contacts between residues that are not in proximity to each other in the sequence. When these diagonals are parallel to the main diagonal, the features they represent are parallel. On the other hand, when they are perpendicular, their features are antiparallel. This representation is memory-intensive because the features in the square matrix are symmetrical to the main diagonal.

When the distance matrices of two proteins share the same or similar features in approximately the same positions, they can be said to have similar folds with loops of similar length connecting their secondary structure elements. The actual alignment process of DALI requires a similarity search after the distance matrices of two proteins are built; this is normally conducted via a series of overlapping 6×6 submatrices. Submatrix matches are then reassembled into a final alignment via a standard score-maximization algorithm—the original version of DALI used a Monte Carlo simulation to maximize the structural similarity score, which is a function of the distances between putative corresponding atoms. In particular, more distant atoms within corresponding features are exponentially downweighted to reduce the effects of noise introduced by loop mobility, helix torsions, and other minor structural variations. Because DALI relies on an all-to-all distance matrix, it can account for the possibility that structurally aligned features might appear in different orders within the two sequences being compared.

In this chapter, two methods based on graph theory are surveyed to measure the similarity of given proteins. Instead of finding the minimum RMSD between pairwise superimposed proteins, the method explores a remodeling of one protein to one graph, via a graph-theoretic approach, thus attempting to overcome the inherent drawback of RMSD. The rest of the chapter is organized as follows. In Section 15.2, we introduce various types of models on graph matching and protein graph remodeling. Next, in Section 15.3, a method to measure the similarity of two given proteins by a maximal common edge subgraph is proposed. Section 15.4 gives an alternative inspection to graph matching by using graph spectra.

15.2 SIMILARITY IN THE GRAPH MODEL

Graph theory defines the all-against-all relations (called *edges*) between nodes (or *vertices*), that is, objects. The powerful combinatorial methods of graph properties have been mainly applied in information theory, combinatorial optimization, structural biology, chemical molecules, and many other fields. Measurement of similarity using graph theory is a practical approach in various fields. When graphs are used to represent structured objects, the problem of measuring object similarity turns into the problem of computing the similarity of graphs [11]. This approach mainly consists of two phases, namely, *graph matching* and *graph scoring*. Zager and Verghes [60] divide the problem of graph similarity matching into four types, namely, *graph isomorphism*, *graph edit distance*, *maximum common subgraph*, and *statistical comparison* depending on their levels.

- *Graph isomorphism*
  Many similarity measures are based on isomorphic relations [54]. However, it is known that the graph isomorphism problem is difficult. Although certain cases of graph isomorphism may be solved in polynomial time, the complexity of the graph similarity approach is considered impractical for most applications.
- *Graph edit distance*
  To find a desirable graph matching, it should be more effective to examine another subject—the *graph edit distance*—to determine the minimum cost transformation from one graph to another [10]. The approach of the graph edit distance consists of two phases, graph matching (edit operation) and graph scoring (cost function). By the operation of addition or deletion of nodes and edges, one graph may bijectively map to another.
- *Maximal common subgraph*
  In addition, the Maximal Common Subgraph (MCS), identifying the largest isomorphic subgraphs of two graphs, is a generalization of the maximum clique problem and subgraph isomorphism problem [16]. Although Bunke [10] shows in detail that graph edit distance computation is equivalent to the MCS problem, most of the approximation efforts are surveyed on MCS algorithms. The MCS approach has been applied frequently in the area of cheminformatics [48]; this involves finding similarities between chemical compounds from their structural equation.
- *Statistical comparison*
  There are also other methods, known as *statistical comparison* methods, which are capable of graph matching by assessing aggregate measures of graph structure, such as degree distribution, diameter, and betweenness measures [56]. Here, *graph spectra* are another such illustration. In mathematics, graph spectral theory is the study of the properties of a graph in relationship to the characteristic polynomial, eigenvalues, and eigenvectors of its adjacency matrix or Laplacian matrix. Since graph isomorphism is regarded as complicated and time-consuming, graph spectra are a good choice for graph matching [14].

Let $G = (V, E)$ denote a graph, where V and E are the vertex and edge sets of G, respectively. Given two graphs $G_A = (V_A, E_A)$ and $G_B = (V_B, E_B)$, where $|V_A| \leq |V_B|$, the graph-matching problem is to find a 1–1 mapping function $f : V_A \rightarrow V_B$ to minimize (or maximize) a desired objective. For example, assume that $|V_A| = |V_B|$. If we want to know whether there exists a function f such that for any two vertices $u, v \in V_A$ are adjacent in G_A if and only if $f(u)$ and $f(v)$ are adjacent in G_B; then this becomes the graph isomorphism problem. In general, the graph-matching problem tries to determine the similarity between two graphs. In recent years, therefore, numerous attempts have been made on graph matching to show its efficiency. As already mentioned, Zager and Verghese [60] have shown that the problem of graph matching may be divided into different types depending on their levels. Graph scoring qualitatively measures the mutual dependence of two graphs [47]. Generally, the value of similarity ranges between 0 and 1, from dissimilar to identical. The Jaccard similarity coefficient [26], a statistic commonly used for comparing similarity, can be rewritten as follows:

$$J_{sim}(G_A, G_B) = \frac{(|V(G_{AB})| + |E(G_{AB})|)^2}{(|V(G_A)| + |E(G_A)|) \cdot (|V(G_B)| + |E(G_B)|)} \tag{15.2}$$

In Equation 15.2, the denominator is the product of the sizes of graphs G_A and G_B. The numerator is the square of the size of the graph G_{AB}, that is, the common subgraph of G_A and G_B [28]. Similarly, Dice's coefficient is also a measure guide for the diversity of samples. It can be rewritten as Equation 15.3

$$D_{sim}(G_A, G_B) = \frac{2 \cdot (|V(G_{AB})| + |E(G_{AB})|)}{(|V(G_A)| + |E(G_A)|) + (|V(G_B)| + |E(G_B)|)} \tag{15.3}$$

15.2.1 Graph Model for Proteins

The structure of a protein can be regarded as a conformation with various local elements (helix and sheet) and forces (Van der Waal's forces and hydrogen bonds), folding into its specific characteristic and functional structure. With the help of *graph transformation*, folded polypeptide chains are represented as a graph by several mapping rules. Proteins contain complex relationships between their polymers, reaction of residues, interaction of covalent bonds, bonding of peptides, and hydrophobic packing, which are essential in structure determination. The intention here is to transform a protein structure into a graph. Formally, a graph-transforming system consists of a set of rewritten graph rules: $L \rightarrow R$, where L is called the *pattern graph* and R the *replacement graph* [15]. This is the key operation in the graph transformation approach.

Vishveshwara et al. [55] describe the protein graph in detail in their description of protein remodeling. For instance, the transformation of nodes and edges is motivated by various purposes as illustrated in Figure 15.1. Graph components are constructed in such a way that they switch protein structures. In Table 15.1, we outline some categories of protein graph approaches to a set of graphs, representing each specific graph-rewriting and graph-measuring skill. It would be useful to begin with a summary of the common research in this area:

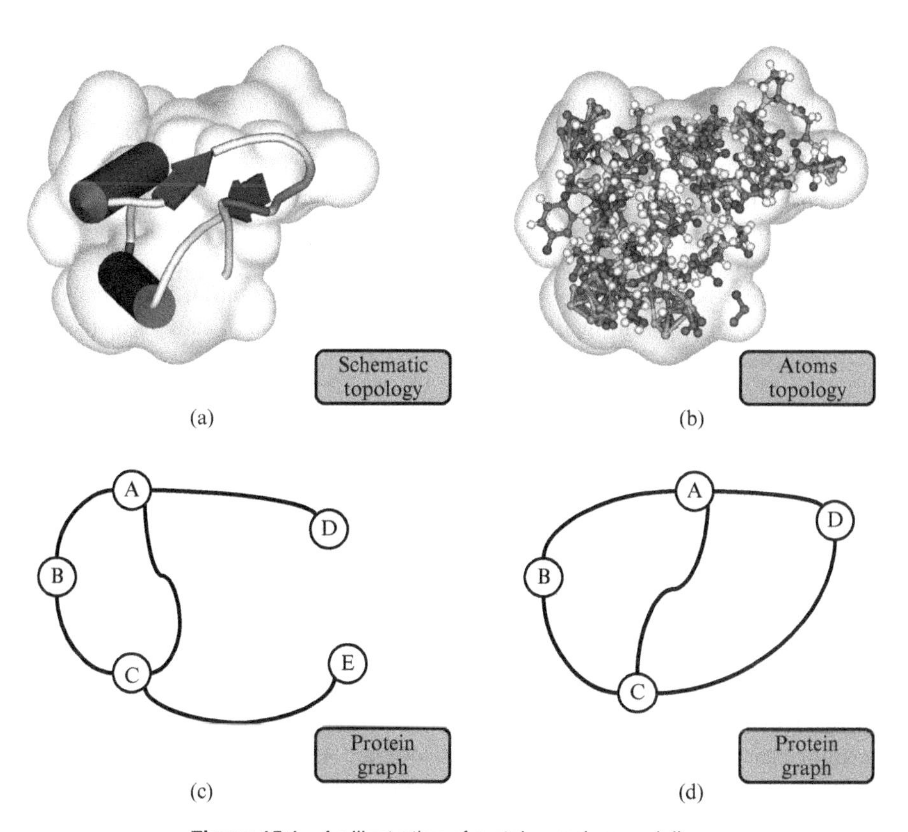

Figure 15.1 An illustration of protein graph remodeling.

Table 15.1 Summary of related studies in protein graph building by graph-theoretic approach

	Graph transformation		Graph similarity	
References	Nodes	Edges	Matching	Scoring
[25]	C_α atoms	Labeled	Subgraph mining	Mutual information
[24]	DSSP*	Attributed	MCES[b]	Jaccard coefficient
[12]	Side chains	Interacted energy	Biconnected graph	Energy function
[52]	Residues	Geometrical constraints	Clique finding	Mutual information
[7]	SSE[c]	Labeled	SVM[d]	Statistics

[a] Dictionary of secondary structures of proteins.
[b] Maximum common edge subgraph.
[c] Secondary structure elements.
[d] Support vector machine.

- *Geometric relation*
  Basically, proteins have been represented in various ways using different levels of detail. It has been showed that conformation of a protein structure is determined geometrically by using various constraints [32]. The most common method for protein modeling is to reserve its topological relation in graphs. From the aspect of graph theory, a simplified representation of a protein structure directs attention toward connectivity patterns. This helps in going into details regarding interaction relations within a polypeptide folding. In brief, geometric-based protein modeling involves the refinement of its vertices and edges, adapting information obtained from interatomic distances of pairs of atoms.
- *Chemical relation*
  Compared with geometric relationship, chemical relations result in a more complicated description of the protein graph model. This is due to the various chemical properties of amino acids, including electrostatic charge, hydrophobicity, bonding type, size, and specific functional groups [29]. By giving values to edges and nodes in graph, the variation of each labeled components can be easily discovered.

Once a problem has been formalized in graph-theoretical language, the concept of graph theory can be used to analyze specific problems [50]. Proteins are made up of elements such as carbon, hydrogen, nitrogen, and oxygen. To be able to perform their biological function, proteins fold into specific spatial conformations, driven by a number of non-covalent interactions. Therefore, methods adopted in graph transformation vary from divergent chemical structures, including molecular geometry, electronic structure, and so on. With different viewpoints, relational features in protein structures can be regarded as vertices and edges of a corresponding graph.

Although the exact graph-matching problem attempts to find graph or subgraph isomorphisms, it is known that finding a bijective relation between two graphs is difficult [19]. Another point that needs to be noted in real applications is that the chances of locating an isomorphic relation between two graphs are very small. A more relaxed inexact graph matching is required. However, the inexact graph-matching problem has also proved to be NP-complete [1]. When the graph grows larger, the efficiency of graph-theoretic techniques may drop.

15.3 MEASURING STRUCTURAL SIMILARITY VIA MCES

In this section, a method is proposed to measure the similarity of given proteins [24]. Instead of finding the minimum RMSD of superimposed pairwise proteins, the purpose here is to explore the remodeling of one protein to one graph: a graph-theoretic approach that attempts to overcome the inherent drawback of RMSD. This section begins with a simple observation on a small protein—crambin—a hydrophobic and

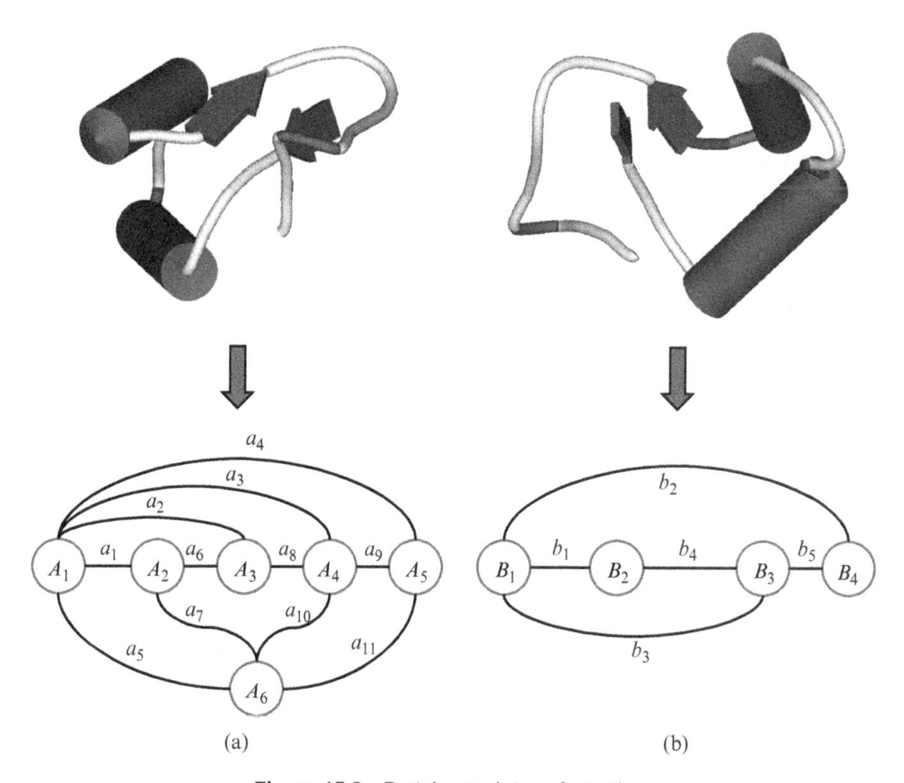

Figure 15.2 Protein graph transformation.

water-insoluble molecule. Crambin was originally isolated from the seeds of the plant *Crambe abyssinica*, an oilseed crop native to the Mediterranean area and one of the richest known sources of erucic acid. Crambin has also been used widely in developing the methodology for the determination of protein structure from NMR data [2].

15.3.1 Problem Formulation

Given two proteins A and B as shown in Figure 15.2, let $G_A = (V_A, E_A)$ and $G_B = (V_B, E_B)$ be their graphs (called *P-graphs*) , respectively. Without loss of generality, we assume that $|V_A| \geq |V_B|$. The goal is to find the largest edge subgraph of G_B isomorphic to an edge subgraph of G_A, that is, the Maximum Common Edge Subgraph (MCES) of these two graphs. We then use the MCES of G_A and G_B to measure the similarity of proteins A and B. It is known that finding the MCES of two graphs is NP-hard [19]. However, the constructed P-graphs are labeled. Thus some useful properties can be used to avoid an exhaustive search. Since the graphs are labeled, the proposed algorithm is efficient in practice.

15.3.2 Constructing P-Graphs

As already mentioned, for remodeling a protein structure to a graph, a consistent model is required to represent each protein substructure. First of all, each vertex in the graph is created according to the Dictionary of Secondary Structures of Proteins (DSSP) [40]. Under this metric, a protein secondary structure is represented by a single letter code, for example, H for helix (containing G, H, and I), T for hydrogen turn (containing T, E, and B), and C for coil (containing only C), which can be written as $\hat{H} = \{G, H, I\}$, $\hat{T} = \{T, E, B\}$, and $\hat{C} = \{C\}$. For a protein, we construct a labeled P-graph. Each vertex corresponds to a secondary structure. With every vertex, we associate two attributes stored in a table, called the *P-table*. The first attribute denotes the type of the secondary structure as in Reference [7]. Depending on the polarity of a side chain, a polypeptide chain behaves according to the hydrophilic or hydrophobic characters of its amino acids [13]. If most amino acids of a substructure are polar, then it is marked as polar, for example, amino acids in $\hat{P} = \{R, N, D, E, Q, H, K, S, T, Y\}$ are polar; else, the chain is marked as nonpolar, for example, amino acids in $\hat{N} = \{A, G, I, L, M, F, W, V\}$ are nonpolar.

In addition, some amino acids exhibit special properties on structural stabilization, such as *cysteine* (CYS), which forms covalent disulfide bonds with other cysteine residues and *proline*, (PRO), which has a cyclic formation with the polypeptide backbone. Hence, we include another attribute based on the chemical bonds, that is, the set $\hat{B} = \{C, P\}$, with the numbers under each labeling indicating their differences incrementally.

These characteristics are recorded as the second attribute in the P-table. The definition of the vertex set for a P-graph can be set outas follows:

- Vertex set $V = \{v_1, v_2, \dots, v_n\}$:
  - v_i is the vertex corresponding to the ith *Secondary Structure* (SS) within the given protein.
  - DSSP label:

$$D[v_i] = \begin{cases} 1 & \text{the } i\text{th SS is a helix } (\hat{H}) \\ 2 & \text{the } i\text{th SS is a sheet } (\hat{T}) \\ 4 & \text{the } i\text{th SS is a coil } (\hat{C}) \end{cases} \tag{15.4}$$

  - Polarity label:

$$P[v_i] = \begin{cases} 1 & \text{the majority of the } i\text{th SS is polar } (\hat{P}) \\ 2 & \text{the majority of the } i\text{th SS is nonpolar } (\hat{N}) \\ 4 & \text{the } i\text{th SS contains CYS or PRO residue } (\hat{B}) \end{cases} \tag{15.5}$$

  Note that the priority for an SS containing CYS or PRO is higher than the priority that an SS is polar or nonpolar.

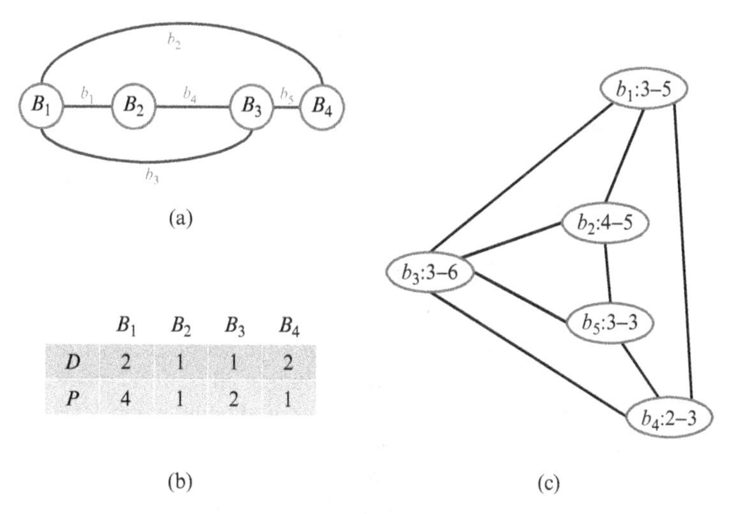

	B_1	B_2	B_3	B_4
D	2	1	1	2
P	4	1	2	1

Figure 15.3 (a) a P-graph G_B with $V_B = \{B_1, B_2, B_3, B_4\}$ and $E_B = \{b_1, b_2, b_3, b_4, b_5\}$. (b) The P-table of G_B. (c) The line graph $L(G_B)$ of G_B.

Since a local segment of a biopolymer is treated as one vertex in its P-graph, the relations among the segments are considered as edges. In proteins, a secondary structure is described by patterns of hydrogen bonds between the backbone amide and carboxyl groups. Thus, an edge can be defined by a correlated hydrogen bonding. In such a case, a connection is established between two neighboring secondary structures, namely, a *Physical Edge*. As shown in Figure 15.3a, four substructures, that is, B_1, B_2, B_3, and B_4, are in P-graph G_B. Edges $b_1 = (B_1, B_2)$, $b_4 = (B_2, B_3)$, and $b_5 = (B_3, B_4)$ are defined as the physical edges.

Obviously, there are other issues involved, for example, proteins are conformed by various chemical bonds. For example, in Figure 15.3a, B_1 is adjacent to B_4 because they are in close proximity. These edges are called *Geometric Edges*, for example, $b_2 = (B_1, B_4)$ and $b_3 = (B_1, B_3)$. In addition, in order to avoid connecting similar attributes in a P-graph label, the *Constrained Edge* attempts to delete those edges with matching properties due to the repulsive interaction among different regions of the polypeptide chains [5]. These approaches operate when the same forces are opposition or the atoms are not closely bonded. The definition of the edge set for a P-graph G is set out in the following.

Edge set $E = \{e_1, e_2, \ldots, e_k\}$:

- Physical edge: $(v_i, v_j) \in E$ if $j = i + 1$.
- Geometric edge: $(v_i, v_j) \in E$ if the distance $dist(v_i, v_j) \leq d$ in the three-dimensional space for a given threshold d. We usually use the coordinate of the center of v_i to denote the coordinate of v_i. Thus the distance of v_i and v_j is well defined. Moreover, d is usually determined as the average of the distances between all pairs of atoms in the protein.

- Constrained Edge: An edge is removed if both $D[v_i] = D[v_j]$ and $P[v_i] = P[v_j]$.

By using the technique proposed in Reference [3], the MCES of two P-graphs can be found by finding the largest clique from the modular graph of the line graphs of the two P-graphs.

15.3.3 Constructing Line Graphs

For an undirected graph G, the line graph of G, denoted as $L(G)$, is the graph obtained from G by letting the vertex set of $L(G)$ be the edge set of G and with every two vertices of $L(G)$ being adjacent if the corresponding edges in G share a common vertex. As illustrated in Figure 15.3c, each vertex in the line graph $L(G_B)$ is labeled with the corresponding edge in the original graph G_B. For example, the edge b_1 in G_B is adjacent to edges b_2 and b_3 via vertex B_1, and is adjacent to edge b_4 through vertex B_2. Therefore, there are edges connecting vertex b_1 to vertices b_2, b_3, and b_4 in $L(G_B)$. For constructing the modular graph, the chemical relation of G_B is also annotated in each vertex in $L(G_B)$ from its P-table in Figure 15.3b. For example, vertex b_1 in $L(G_B)$ can be viewed as a group of chemical substances obtained by B_1 and B_2 in G_B. Thus b_1 will be labeled as (3,5) which is computed according to (2 + 1, 4 + 1). In this manner, b_1 is composed of the substructures of a helix and a sheet, and one of which is the polar component.

15.3.4 Constructing Modular Graphs

A *modular graph*, also termed a *compatibility graph*, is widely used in graph isomorphism. The modular graph can be regarded as a technique of finding the maximum common subgraph of two given labeled graphs [31]. For line graphs $L(G_A)$ and $L(G_B)$, the vertex set of the modular product is the Cartesian product $V(L(G_A)) \times V(L(G_B))$. Any two vertices (i,j) and (m,n) are adjacent in the modular product of $L(G_A)$ and $L(G_B)$ if and only if the relations of i to m and j to n are exclusive disjunctions (XOR). In other words, i is adjacent to m and j is adjacent to n, or i is not adjacent to m and j is not adjacent to n if and only if vertices (i,j) and (m,n) are adjacent in the modular product of $L(G_A)$ and $L(G_B)$. Let G_A and G_B be the graphs depicted in Figure 15.2. Owing to page constraints, we consider only a_1, a_2, and a_9 from $L(G_A)$ and b_1, b_2, and b_3 from $L(G_B)$ in Figure 15.4a to show the construction of the corresponding modular graph. If the chemical properties of vertices are identical, then they will be merged as in Figure 15.4b and will be denoted as a vertex in the modular graph. For example, a_1 is composed of 3–5 the same as b_1, so a_1 and b_1 join together as a single vertex in Figure 15.4c. The remaining vertices are determined similarly. By definition, there is an edge x from a_1b_1 to a_2b_2 as $(a_1, a_2) \in E(L(G_A))$ and $(b_1, b_2) \in E(L(G_B))$. The validating of this edge is shown in Figure 15.4d. However, edges y and z will not exist in the modular graph as they do not satisfy the definition. Figure 15.4e and f shows the reason. Finally, the construction of the modular graph M_{AB} of $L(G_A)$ and $L(G_B)$ can be determined.

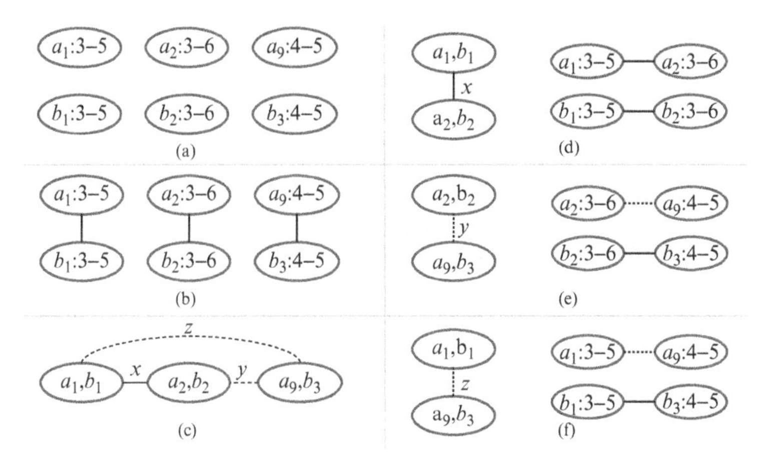

Figure 15.4 The process of constructing modular graph from $L(G_A)$ and $L(G_B)$.

15.3.5 Maximum Clique Detection

The problem of finding a clique with the maximum size for a given graph is NP-hard [41]. It can be solved by a well-known branch-and-bound algorithm [39]. The key issue is to find a proper upper bound size of maximum clique. As the modular graph M_{AB} has been constructed, to find an MCES between G_A and G_B simply involves locating the maximum clique in M_{AB}, then backtracking to the original graphs G_A and G_B. Since clique detection is an NP-hard problem, we use the degree of the vertex to prune those sparsely connected vertices, enumerating each candidate vertex in order of decreasing degree.

Once the MCES G_{AB} has been found from G_A and G_B, a formula is needed to measure the similarity of proteins A and B. In view of the MCES, it is required to consider both vertices and edges in subgraph G_{AB}. Thus, the Jaccard similarity coefficient is adopted for our graph scoring function.

15.3.6 Experimental Results

To validate the protein structure comparison of the graph-based approach, [24] tested some small proteins of plant crambin. Eight protein structures were selected as materials, namely, 2EYA, 1CRN, 1OKH, 1ED0, 1ORL, 1WUW, 1NBL, and 1H5W. As illustrated in Table 15.2 and Figure 15.5 from *Protein Data Bank* (PDB) [4], all the molecular descriptions are listed. For further explanation of the proposed results, it may be helpful to refer the protein database of *Class, Architecture, Topology, and Homology* (CATH) [37] in Table 15.3. It is a hierarchical domain classification of protein structures in the PDB, using a combination of automated and manual procedures, where **C** stands for secondary structure content, **A** for general spatial arrangement of secondary structures, **T** for spatial arrangement and connectivity of secondary structures, **H** for manual curation of evidence of evolutionary relationship, **S** indicates $\geq$ 35% sequence similarity, **O** indicates $\geq$ 60% sequence similarity, **L** indicates

Table 15.2 Annotations of eight selected macromolecules in the PDB

PID	Fold	Domain	Species	Description
2EYA	Crambin-like	Crambin	*Abyssinian cabbage*	Plant thionin
1CRN	Crambin-like	Crambin	*Abyssinian cabbage*	Plant thionin
1OKH	Crambin-like	Viscotoxin A3	*European mistletoe*	Plant thionin
1ED0	Crambin-like	Viscotoxin A3	*European mistletoe*	Plant thionin
1ORL	Crambin-like	Viscotoxin C1	*European mistletoe*	Plant thionin
1WUW	Crambin-like	Beta-Hordothionin	*Barley*	Plant thionin
1NBL	Crambin-like	Hellethionin D	*Helleborus purpurascens*	Plant thionin
1H5W	Upper collar protein gp10	Upper collar protein gp10	Bacteriophage phi-29	Phage connector

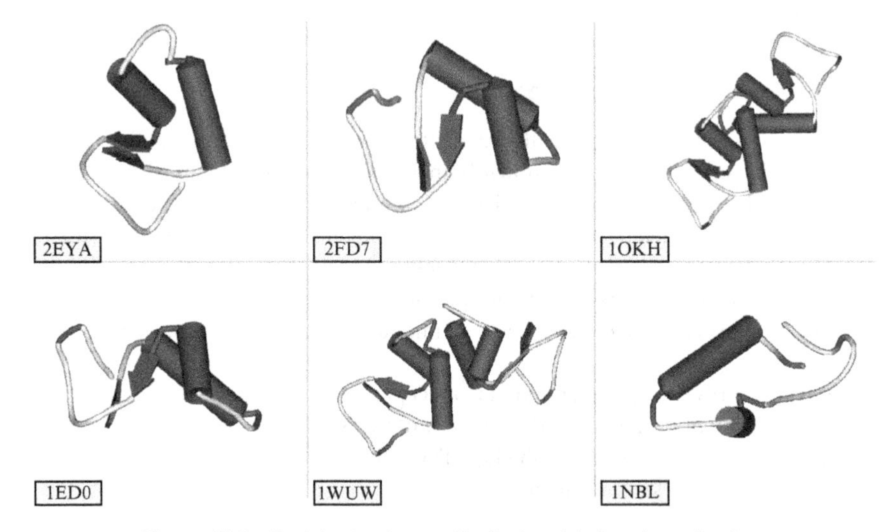

Figure 15.5 Protein structures with display style in schematic view.

Table 15.3 Entries of selected macromolecules and their CATH codes

PID	C	A	T	H	S	O	L	I	D
2EYA	3	30	1350	10	1	1	1	1	15
1CRN	3	30	1350	10	1	1	1	3	1
1OKH	3	30	1350	10	2	1	3	1	1
1ED0	3	30	1350	10	2	1	3	1	3
1ORL	3	30	1350	10	2	1	5	1	1
1WUW	3	30	1350	10	2	2	2	1	1
1NBL	3	30	1350	10	2	3	1	1	1
1H5W	3	30	1350	20	1	1	1	1	1

$\geq$ 95% sequence similarity, **I** indicates 100% sequence similarity, and **D** means unique domains. Table 15.4 summarized the structural comparison from the viewpoint of RMSD to P-graph. In the following, we discuss it in detail from two aspects.

The proposed method presents a convincing viewpoint, showing inferred evolutionary relationships among various strains. There are several examples, according to P-graph matching—one may notice that 1OKH, 1ED0, and 1ORL are mutually shape-related in Table 15.2. The method does not say that RMSD cannot be used in protein structural comparison; instead, the value of RMSD should be described together with other parameters, for example, sequence identity, number of alpha carbons, backbone atoms, and so on. It follows from what has been said that the result is more competitive with the existing RMSD approach, although it relies on more information than RMSD. Besides, there was no structure info on 1CRN and 1ORL, and 2FD7 for reference.

In addition, results suggest one further viewpoint, *graph model construction*. Attributes of edges are graded into three types, physical, geometric, and constrained edges. Adopting from several properties on vertices and edges, protein comparison could be improved by incorporating more specific parameters. It may be worth pointing out that the construction of P-graphs is limited by the number of secondary structures.

15.4 PROTEIN COMPARISON VIA GRAPH SPECTRA

The graph-matching problem attempts to find graph or subgraph isomorphisms. It is already known that finding an optimal bijective relation for graph isomorphism of two graphs is NP-hard [19]. However, in real applications, there is a small chance to locate an isomorphic relation between graphs. Further, the more relaxed inexact graph matching is required. It has also been found that there is no efficient way for determining graph isomorphism. Hence, as already mentioned, *graph spectra* give an alternative solution to graph matching. The properties of topological relationships can be deduced from the graph spectra of the proteins being compared. In this section, we survey the recent studies in this area [42, 43].

15.4.1 Graph Spectra

In general, the topology of a graph is complicated. One practical way to tackle this is to symbolize the graph as a matrix, turning it into numbers and vectors. As already mentioned, there is no efficient way of determining the graph isomorphism or finding the MCES. The *graph spectrum* gives an alternative solution to graph matching. By definition, the spectrum of a finite graph G can be defined according to its adjacency matrix A_G and diagonal degree matrix D_G, whose entries $a_{i,j}$ and $d_{i,j}$ can be written as follows.

$$a_{i,j} = \begin{cases} 1, & \text{if } (i,j) \in E \\ 0, & \text{otherwise} \end{cases} \quad (15.6)$$

Table 15.4 Comparison of the proposed method with DALI RMSD

<table>
<tr><th>PID</th><th>1CRN</th><th>1OKH</th><th>1ED0</th><th>1ORL</th><th>1WUW</th><th>1NBL</th><th>1H5W</th></tr>
<tr><td>2EYA</td><td>1.1Å | 0.33</td><td>1.4Å | 0.30</td><td>1.3Å | 0.30</td><td>1.6Å | 0.30</td><td>1.2Å | 0.25</td><td>2.1Å | 0.00</td><td>3.6Å | 0.07</td></tr>
<tr><td>1CRN</td><td>—</td><td>0.6Å | 0.00</td><td>1.0Å | 0.00</td><td>1.0Å | 0.00</td><td>0.6Å | 0.00</td><td>1.9Å | 0.00</td><td>5.7Å | 0.01</td></tr>
<tr><td>1OKH</td><td>—</td><td>—</td><td>0.9Å | 1.00</td><td>0.9Å | 1.00</td><td>0.6Å | 0.83</td><td>1.7Å | 0.00</td><td>5.7Å | 0.01</td></tr>
<tr><td>1ED0</td><td>—</td><td>—</td><td>—</td><td>1.2Å | 1.00</td><td>0.9Å | 0.83</td><td>1.5Å | 0.00</td><td>5.3Å | 0.01</td></tr>
<tr><td>1ORL</td><td>—</td><td>—</td><td>—</td><td>—</td><td>0.9Å | 0.83</td><td>1.7Å | 0.00</td><td>5.2Å | 0.01</td></tr>
<tr><td>1WUW</td><td>—</td><td>—</td><td>—</td><td>—</td><td>—</td><td>1.7Å | 0.00</td><td>1.7Å | 0.08</td></tr>
<tr><td>1NBL</td><td>—</td><td>—</td><td>—</td><td>—</td><td>—</td><td>—</td><td>3.8Å | 0.00</td></tr>
</table>

$$d_{i,j} = \begin{cases} \deg(v_i), & \text{if } i = j \\ 0, & \text{if } i \neq j \end{cases} \qquad (15.7)$$

Given a matrix M, the spectrum of M is a multiset of *eigenvalues* [14]. An ordinary spectrum of a finite graph G is the spectrum of its *adjacency matrix* A_G. The *Laplacian spectrum* of G is defined in the matrix $L_G = D_G - A_G$. Let the Laplacian matrices of graphs X and Y be defined as follows:

$$L_X = \begin{pmatrix} 3 & -1 & -1 & -1 \\ -1 & 2 & -1 & 0 \\ -1 & -1 & 3 & -1 \\ -1 & 0 & -1 & 2 \end{pmatrix} \qquad L_Y = \begin{pmatrix} 3 & -1 & -1 & -1 & 0 \\ -1 & 2 & -1 & 0 & 0 \\ -1 & -1 & 3 & 0 & -1 \\ -1 & 0 & 0 & 1 & 0 \\ 0 & 0 & -1 & 0 & 1 \end{pmatrix}$$

By definition, the Laplacian spectra of X and Y are $\lambda_X = [4, 4, 2, 0]$ and $\lambda_Y = [4, 4, 1, 1, 0]$, respectively. In other words, $\{4, 4, 2, 0\}$ (respectively, $\{4, 4, 1, 1, 0\}$) is the eigenvalue multiset of λ_X (respectively, λ_Y). To measure the similarity of X and Y, these two vectors will be normalized into the same dimension, that is, λ_X will become $[4, 4, 2, 0, 0]$. The distance between X and Y is defined as the *Euclidean distance* between λ_X and λ_Y, that is, $\sqrt{2} = 1.414$. In Reference [43], the authors applied the minimum weight bipartite matching on the Laplacian spectra to determine the distance. For example, the bipartite matching distance between $\lambda_X = [4, 4, 2, 0]$ and $\lambda_Y = [4, 4, 1, 1, 0]$ is 1 as $[4, 4, 2, 0]$ will match $[4, 4, 1, 0]$ of $[4, 4, 1, 1, 0]$. Peng and Tsay's [43] shows that the determination of similarity among some proteins of the MHC by applying bipartite matching distance is better than the one using Euclidean distance. In the following, we discuss the issue of which matrix is better for use in determining the similarity of two protein graphs.

15.4.2 Matrix Selection

Recall that the *spectrum* of a matrix M is the multiset that contains all the eigenvalues of M [14]. If M is a matrix obtained from a graph G, then the spectrum of M is also called the *spectrum* of G. If two graphs have the same spectrum, then they are *cospectral* (or *isospectral*) graphs. In this section, we compare four spectra of matrices to examine their accuracy in protein structural comparison. These four spectra are derived from the *adjacency matrix, Laplacian matrix, signless Laplacian matrix,* and *Seidel adjacency matrix.*

An ordinary spectrum of a finite graph G is the spectrum of its *adjacency matrix* A. It is known as the *characteristic polynomial* of graph G [6]. The *Laplacian matrix* L_G of G is $D_G - A_G$ [9]. Apparently, comparing with the binary relation of G, the spectrum of G tends to enrich information on the adjacent relation.

Let C_n (respectively, K_n and S_n) denote the cycle (respectively, complete and star) graph of n vertices. Collatz and Sinogowitz found the smallest pair of cospectral graphs as shown in Figure 15.6 [6]. Assume that the graph depicted in Figure 15.6a

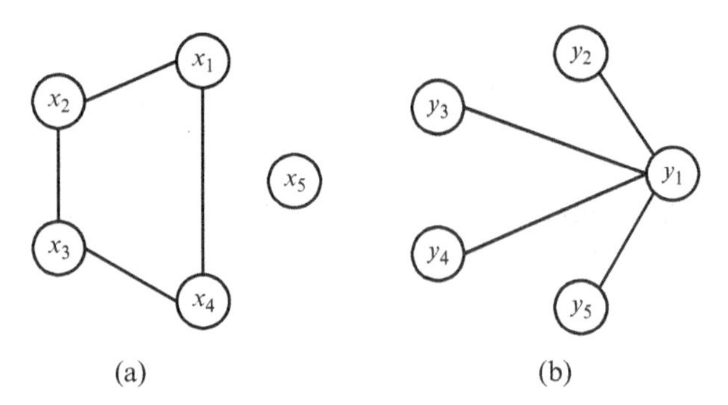

Figure 15.6 An illustration of cospectral graphs.

is X which is the union of C_4 and K_1 and the graph in Figure 15.6b is Y which is S_5. The spectra of X and Y (by their adjacency matrices) are identical, that is, $\lambda_X = \lambda_Y = [2, 0, 0, 0, -2]$.

From the point of view of a topological relation, the spectra of adjacency matrices from X and Y should not be equal owing to their incompatible topologies. Therefore, the Laplacian matrix is proposed to improve the factors of the graph relation. It is the difference of the degree diagonal matrix and the adjacency matrix of the given graph. In this case, the spectra of X and Y are $\lambda_X = [4, 2, 2, 0, 0]$ and $\lambda_Y = [5, 1, 1, 1, 0]$, respectively. Apparently, the Laplacian matrix gives a distinct outcome of its eigenvalues. On the other hand, the matrix of $D_X + A_X$ is called the *signless Laplacian matrix* Q_X of X. In this case, the spectra of X and Y are still $\lambda_X = [4, 2, 2, 0, 0]$ and $\lambda_Y = [5, 1, 1, 1, 0]$, respectively, that is, the spectra of the Laplacian and signless Laplacian matrices are equal for these two graphs.

Although studies have found that the signless Laplacian matrix spectrum performs better than other commonly used graph spectra [14], most of the surveys were confined to combinatorial, graph-theoretical analysis, and linear algebra problems. As of now, exploration in biological application problems is seldom carried out. In addition, there is another matrix, called the *Seidel adjacency matrix S*. It is modified from the adjacency matrix such that each entry $s_{i,i}$ is 0, $s_{i,j}$ is -1 if the vertices i and j are adjacent, and $s_{i,j}$ is 1 if they are not adjacent [53] (see Eq. 15.8). Besides, the spectra of Seidel adjacency matrices of Figure 15.6 are $\lambda_X = [3, 2, -1, -2, -3]$ and $\lambda_Y = [4, -1, -1, -1, -1]$.

$$S_{i,j} = \begin{cases} 1, & \text{if } a(i,j) = 0 \\ -1, & \text{if } a(i,j) = 1 \\ 0, & \text{if } i = j \end{cases} \tag{15.8}$$

It is easy to observe that the similarity of X and Y should be diverse owing to their topologies. Therefore, the difference of their spectra should also be large. Note that in Figure 15.6, according to the Euclidean distance, the spectra of Seidel adjacency matrices do not give a better result than those of Laplacian matrices or signless Laplacian matrices.

15.4.3 Graph Cospectrality and Similarity

Graphs with the same spectra are called *cospectral* or *isospectral* [9]. The relation between graph cospectra and graph isomorphism is examined in the following propositions.

1. Isomorphism → Cospectrality (Yes)
2. Cospectrality → Isomorphism (No)
3. Non-isomorphism → Non-cospectrality (No)
4. Non-cospectrality → Non-isomorphism (Yes)

In Proposition (1), do isomorphic graphs imply cospectral graphs? Of course, this is trivial to explain from the isomorphic graphs as sketched in Figure 15.7. Thus, the spectra of their adjacency matrices are identical. It is also helpful to explain Proposition (4). As a result, isomorphic graphs are also cospectral. However, graph cospectrality does not imply graph isomorphism. In Figure 15.6, X and Y are cospectral graphs but they are not isomorphic. It supports the arguments of Propositions (2) and (3). In summary, determining cospectral graphs is an easy problem in graph matching, while determining graph isomorphism is an NP-hard problem.

15.4.4 Cospectral Comparison

Basically, the scoring of two spectra is measured by the distance equation of *Euclidean distance*. However, there are several drawbacks in this graph spectra scoring:

- Size issue: Vectors of graph spectra are required to have the same dimension.
- Zero padding issue: Euclidean distance requires 1–1 mapping in the axis; therefore, the step of zero padding is essential but it is easy to confuse the padding zero value with the real one.
- Scoring issue: If two nonrelevant graphs are matched, then they could be mutually incompatible by using Euclidean distance.

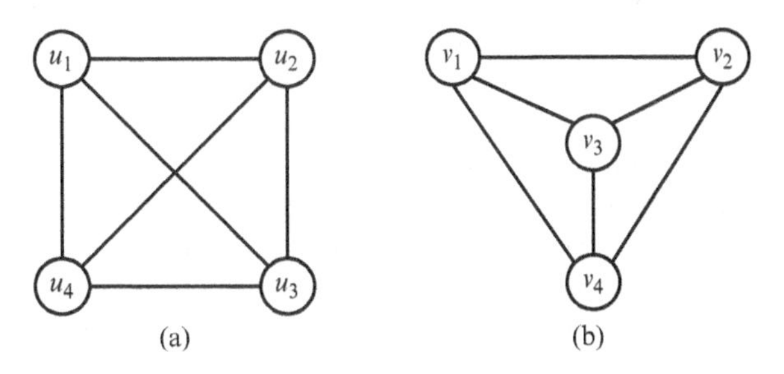

Figure 15.7 An example of isomorphic graphs.

Peng and Tsay [43] give a dependable procedure to evaluate spectra similarity based on minimum weight maximal bipartite matching (using the *Hungarian* algorithm). In comparing spectra efficiency with the mentioned matrices, this method is adapted to inspect their similarity.

In general, proteins are composed of multiple polypeptide chains, with units of multimeric proteins joining together. Each subunit is a single polypeptide chain and forms a stable folded structure by itself [49]. The assembly of subunits provides an efficient way of constructing large structures. In Reference [42], the *Major Histocompatibility Complex* (MHC) is chosen as the material for experiments. MHC, as an immune system in most vertebrates, encodes for a small complex cell surface protein. The MHC protein is known for the *Human Leukocyte Antigen* (HLA), one of the most intensively studied genes in humans [20]. Generally, there are three classes of MHC molecules—Class I, Class II, and Class III. In humans, the HLA complex consists of more than 200 genes located close together in Chromosome 6 [27]. If the HLA protein recognizes a peptide as foreign, it responds by triggering the infected cell to self-destruct. Owing to a great diversity of microbes in the environment, the MHC gene widely varies its peptide through several mechanisms [38]. Table 15.5 shows the annotation and description of the MHC gene family and Table 15.6 shows their CATH codes.

15.4.5 Experimental Results

The graph spectrum provides a useful tool for graph matching, especially when an isomorphic relation is difficult to determine. Although many studies have been made on graph spectra analysis, it appears that little attention has been directed to protein graph matching. In the experiment [42], for each P-graph G, the adjacency matrix A_G, Laplacian matrix L_G, signless Laplacian matrix Q_G, and Seidel adjacency matrix S_G are considered. The goal is to find a practical matrix representation for graph spectra, evaluating similarity according to their spectra in an appropriate approach. At first, in a protein structure of PDB form, we simply adopt each secondary structure as a vertex, spatially closed as an edge; transforming a protein structure into an undirected simple graph. Next, the four matrices are generated to evaluate their performance by the graph scoring equation. Finally, the results of proposed method are compared with RMSD and CATH code.

Various types of MHC are chosen as the material, for the verification of the proposed method, including 1MY7, 1GYU, 3F04, 2E5Y, 2ENG, 1BXO, 1OG1, 1OJQ, and 1A6Z. Obviously, most MHC antigens contain multi-chains and multi-domains, with intracellular assembly in the immune system of processed proteins on the cell surface. The purpose here is to explore the criteria of various matrix representations for the graph spectra, measuring the spectrum decomposition of the Laplacian matrix from the protein graph. The tested antigen molecules are summarized in Table 15.5. CATH codes of the tested proteins [37] are listed in Table 15.6. Besides, the pairwise comparison of selected materials are computed by the network tool DALI server [21]. DALI is widely used for carrying out automatic comparisons

Table 15.5 The annotation and description of the MHC gene family

Protein ID	1MY7	1GYU
Class	All beta	All beta
Fold	Immunoglobulin-like	Immunoglobulin-like
Architecture	Sandwich	Sandwich
Domain	p65 subunit of NF-kappa B	Gamma1-adaptin domain
Protein ID	**3F04**	**2E5Y**
Class	Mainly beta	Mainly beta
Fold	Immunoglobulin-like	—
Architecture	Sandwich	Sandwich
Domain	C2-domain Calcium	ATP synthase
Protein ID	**2ENG**	**1BXO**
Class	Mainly beta	Mainly beta
Fold	Double psi beta-barrel	Acid proteases
Architecture	Beta barrel	Beta barrel
Domain	Endoglucanase V	Acid protease
Protein ID	**1OG1**	**1OJQ**
Class	Alpha–beta	Alpha–beta
Fold	ADP-ribosylation	ADP-ribosylation
Architecture	Alpha–beta complex	Alpha–beta complex
Domain	Eukaryotic mono-ADP	Exoenzyme c3
Protein ID	**1K5N**	**1A6Z**
Class	Alpha–beta	Alpha–beta
Fold	Immunoglobulin-like	Immunoglobulin-like
Architecture	2-Layer sandwich	2-Layer sandwich
Domain	Class I MHC, alpha-3 domain	Hemochromatosis protein Hfe

Table 15.6 Entries of selected macromolecules and their CATH codes

PID	Domain	C	A	T	H	S	O	L	I	D
1MY7	A1	2	60	40	10	16	1	1	1	1
1GYU	A1	2	60	40	1230	1	1	1	1	2
3F04	A1	2	60	40	150	1	1	1	1	1
2E5Y	A1	2	60	15	10	1	1	1	1	1
2ENG	A1	2	40	40	10	1	1	1	1	1
1BXO	A2	2	40	70	10	1	1	1	1	1
1OG1	A1	3	90	176	10	6	1	1	1	3
1OJQ	A1	3	90	176	10	4	1	1	1	2
1K5N	A1	3	30	500	10	1	1	1	1	1
1A6Z	A1	3	30	500	20	1	1	1	1	1

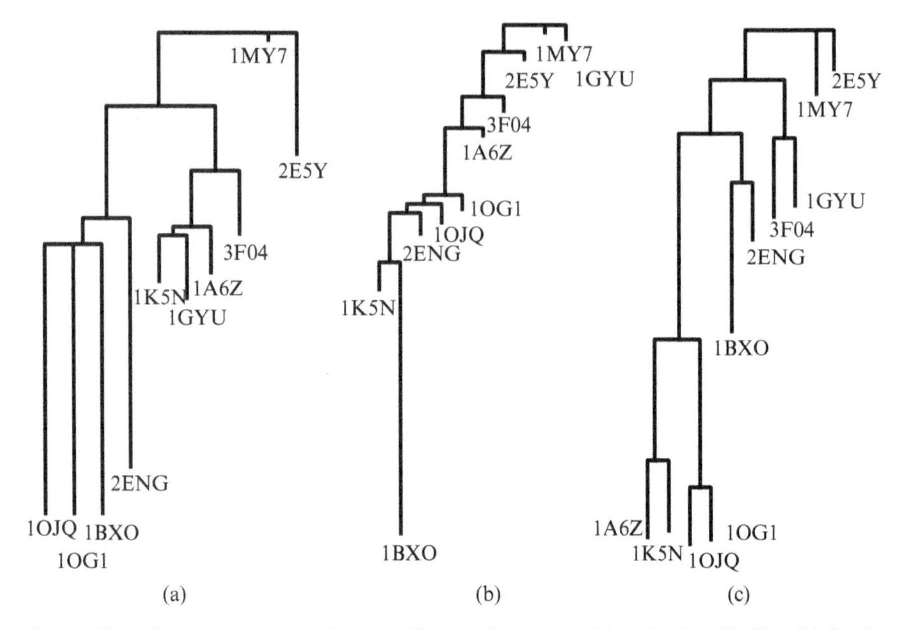

Figure 15.8 Clustering constructed by different distance metrics. (a) For RMSD, (b) for the Seidel spectrum, and (c) for the CATH code.

of protein structures determined by X-ray crystallography or NMR. It is able to measure structural relation according to intramolecular distance similarity.

Entities of MHC-related proteins are inferred as evolutionary relationships for symbolic expression of topologies. Their clusterings are constructed in Figure 15.8 by *PHYLIP* [17] using *Neighbor-Joining* [36]. In this experiment, the authors find that the spectra of the adjacency matrix, Laplacian matrix, signless Laplacian matrix, and Seidel adjacency matrix are seldom different. Results show that when the Seidel adjacency matrix is applied, it gives more accurate relatedness of leaf nodes in cluster shape. Owing to page constraints, we only concentrate on the capability of the spectrum approach and pay scant attention to matrix comparison.

Clusters are provided to expect their accuracy by different similarity scoring issues. Comparatively, CATH may result in a better classification on protein domain and its subunits. Thus, the comparison is according to the authority of the CATH code. Basically, the selected MHC proteins can be roughly classified into the following groups:

- A: A_1 {1MY7, 1GYU, 3F04}, A_2 {2E5Y}, A_3 {2ENG}, A_4 {1BXO}
- B: B_1 {1OG1, 1OJQ}, B_2 {1K5N, 1A6Z}

Let us now look at Figure 15.8a in detail. With the RMSD approach, the topological distance is quite unsatisfactory. It should be noticed that 1BXO, 1OG1, and 1OJQ are unable to estimate their superimposing structures within selected proteins by pairwise DALI. The clusters in Figure 15.8a are apt to mislead one into making wrong assumptions. In addition, protein 1GYU is confused with 1K5N and 1A6Z in B_2, and 2ENG also cannot be calculated in RMSD. In brief, the performance of RMSD value is unacceptable because of the principle underlying the method. As mentioned in the proposed spectrum method, 1OG1, 1OJQ, 3F04, and 1K5N are in the correct group, A_3 and A_4 are placed unsuitably in B but still in the same group. This is because the spectra length of 1BXO is too long for comparison with the other spectra. Apart from this, these MHC-related structure taxa are justly placed.

15.5 CONCLUSION

In this chapter, we survey some methods to determine the structural similarity of two proteins based on graph theory. In particular, we focus on the protein graphs that are constructed from the secondary structures. Earlier, a protein graph was constructed according to the protein atoms, that is, an atom is a vertex. However, the resulting graph becomes huge and it takes a lot of computing time for running graph algorithms. By using a secondary structure as a vertex, the resulting graph becomes small. Therefore, many graph algorithms can be applied. With the help of a computer simulation, the relative similarity of protein structures can be efficiently located. On comparison with the general RMSD approach and its ability, these methods provide an alternative conception and promotive advantage in terms of efficiency. However, the similarity accuracy of two proteins depends on their corresponding graphs. Recently, Peng and Tsay used the idea of graph entropy to adjust protein graphs [44]. In the future, these methods may be useful for constructing protein graphs by considering more chemical properties of the proteins.

Recently, Pržulj proposed the concept of graphlets as another direction for applying graph theory to measure the similarity of proteins [46]. Graphlets are small, induced subgraphs. By extending the degree distribution of a graph to the graphlet degree distribution [45], the similarity of two protein graphs can be determined. Results of this direction are also positive [22, 33, 35] although the graphs used are different from the protein graphs of this chapter. In conclusion, the graph-based approach offers a practical direction for protein structural comparison.

REFERENCES

1. Abdulkader AM. *Parallel algorithms for labelled graph matching* [PhD thesis]. Colorado School of Mines; 1998.
2. Ahn HC, Jurani N, Macura S, Markley JL. Three-dimensional structure of the water-insoluble protein crambin in dodecylphosphocholine micelles and its minimal solvent-exposed surface. J Am Chem Soc 2006;128(13):4398–4404.
3. Barrow HG, Burstall RM. Subgraph isomorphism, matching relational structures and maximal cliques. Inform Process Lett 1976;4(4):83–84.

4. Berman HM, Westbrook J, Feng Z, Gilliland G, Bhat TN, Weissig H, Shindyalov IN, Bourne PE. The protein data bank. Nucleic Acids Res 2000;28(1):235–242.
5. Bhagavan NV. *Medical Biochemistry*. Waltham, MA; Academic Press; 2002.
6. Biggs NL. *Algebraic Graph Theory*. 2nd ed. Cambridge: Cambridge University Press; 1993.
7. Borgwardt KM, Ong CS, Schönauer S, Vishwanathan SVN, Smola AJ, Kriegel H-P. Protein function prediction via graph kernels. Bioinformatics 2005;21 Suppl 1:i47–i56.
8. Bourne PE, Shindyalov IN. Structure comparison and alignment. In: *Structural Bioinformatics*. Hoboken, NJ: John Wiley & Sons, Inc.; Hoboken, NJ; 2005.
9. Brouwer AE, Haemers WH. The Gewirtz graph: an exercise in the theory of graph spectra. Eur J Combin 1993;14:397–407.
10. Bunke H. On a relation between graph edit distance and maximum common subgraph. Pattern Recognit Lett 1997;18(8):689–694.
11. Bunke H. Graph matching: theoretical foundations, algorithms, and applications. In Proceedings of Vision Interface, Volume 21; Montreal 2000.
12. Canutescu AA, Shelenkov AA, Dunbrack RL. A graph-theory algorithm for rapid protein side-chain prediction. Protein Sci 2003;12(9):2001–2014.
13. Creighton TE. *Proteins: Structures and Molecular Properties*. W. H. Freeman & Co.; New York, NY; 1993.
14. Cvetkovic DM, Doob M, Sachs H, Cvetkovi&cacute M. *Spectra of Graphs: Theory and Applications*. 3rd Revised and Enlarged Edition. Vch Verlagsgesellschaft Mbh; New York, NY; 1998.
15. Ehrig H, Engels G, Kreowski HJ. *Handbook of Graph Grammars and Computing by Graph Transformation: Applications, Languages and Tools*. Hackensack (NJ): World Scientific Publishing Company; 1997.
16. Eppstein D. Subgraph isomorphism in planar graphs and related problems. J Graph Algorithms Appl 1999;3:1–27.
17. Felsenstein J. PHYLIP - phylogeny inference package (Version 3.2). Cladistics 1989;5:164–166.
18. Gan HH, Perlow RA, Roy S, Ko J, Wu M, Huang J, Yan S, Nicoletta A, Vafai J, Sun D, Wang L, Noah JE, Pasquali S, Schlick T. Analysis of protein sequence/structure similarity relationships. Biophysical 2002;83:2781–2791.
19. Garey MR, Johnson DS. *Computers and Intractability: A Guide to the Theory of NP-Completeness*, Series of Books in the Mathematical Sciences; New York, NY: W. H. Freeman & Co; 1979.
20. Goodsell DS. *The Machinery of Life*. 2nd ed. Göttingen, German: Springer-Verlag; Göttingen; 2009.
21. Hasegawa H, Holm L. Advances and pitfalls of protein structural alignment. Curr Opin Struct Biol 2009;19(3):341–348.
22. Hayes W, Sun K, Przulj N. Graphlet-based measures are suitable for biological network comparison. Bioinformatics 2013;29:483–491.
23. Holm L, Kääriäinen S, Rosenström P, Schenkel A. Searching protein structure databases with DaliLite v.3. Bioinformatics 2008;24(23):2780–2781.
24. Hsu C-H, Peng S-L, Tsay Y-W. An improved algorithm for protein structural comparison based on graph theoretical approach. Chiang Mai J Sci 2011;38 (Special):71–81.

25. Huan J, Bandyopadhyay D, Wang W, Snoeyink J, Prins J, Tropsha A. Comparing graph representations of protein structure for mining family-specific residue-based packing motifs. J Comput Biol 2005;12(6):657–671.
26. Jaccard P. Étude comparative de la distribution florale dans une portion des alpes et des jura. Bull Soc Vaud Sci Nat 1901;37:547–579.
27. Janeway CA. *Immunobiology*. 5th ed Garland Science; New York, NY; 2004.
28. Johnson M. Relating metrics, lines and variables defined on graphs to problems in medicinal chemistry. In: Alavi Y, Chartrand G, Lesniak L, Lick DR, Wall CE, editors. *Graph Theory with Applications to Algorithms and Computer Science*. New York: John Wiley & Sons, Inc.; 1985. p 457–470.
29. Lehninger A, Nelson DL, Cox MM. *Lehninger, Principles of Biochemistry*. 5th ed. New York, NY: W. H. Freeman; New York, NY; 2008.
30. Lesk AM. Detection of three-dimensional patterns of atoms in chemical structures. Commun ACM 1979;22(4):219–224.
31. Levi G. A note on the derivation of maximal common subgraphs of two directed or undirected graphs. Calcolo 1973;9(4):341–352.
32. Lund O, Hansen J, Brunak S, Bohr J. Relationship between protein structure and geometrical constraints. Protein Sci 1996;5(11):2217–2225.
33. Malod-Dognin N, Przulj N. GR-Align: fast and flexible alignment of protein 3D structures using graphlet degree similarity. Bioinformatics 2014;30:1259–1265.
34. Mayr G, Domingues F, Lackner P. Comparative analysis of protein structure alignments. BMC Struct Biol 2007;7(1):50.
35. Milenkovic T, Wong WL, Hayes W, Przulj N. Optimal network alignment with graphlet degree vectors. Cancer Inform 2010;9:121–137.
36. Nei M, Li WH. Mathematical model for studying genetic variation in terms of restriction endonucleases. Proc Natl Acad Sci U S A 1979;76(10):5269–5273.
37. Orengo CA, Michie AD, Jones S, Jones DT, Swindells MB, Thornton JM. CATH–a hierarchic classification of protein domain structures. Structure 1997;5(8):1093–1109.
38. Pamer E, Cresswell P. Mechanisms of MHC class I–restricted antigen processing. Annu Rev Immunol 1998;16(10):323–358.
39. Pardalos P. A branch and bound algorithm for the maximum clique problem. Comput Oper Res 1992;19(5):363–375.
40. Pauling L, Corey R, Branson H. The structure of proteins; two hydrogen-bonded helical configurations of the polypeptide chain. Proc Natl Acad Sci U S A 1951;37(4):205–211.
41. Pemmaraju S, Skiena S. *Computational Discrete Mathematics: Combinatorics and Graph Theory with Mathematica*. Cambridge, England; Cambridge University Press; 1990.
42. Peng S-L, Tsay Y-W. On the usage of graph spectra in protein structural similarity. J Comput 2012;23:95–102.
43. Peng S-L, Tsay Y-W. Using bipartite matching in graph spectra for protein structural similarity. Proceedings of BioMedical Engineering and Informatics; Hangzhou, China; 2013. p 497–502.
44. Peng S-L, Tsay Y-W. Adjusting protein graphs based on graph entropy. BMC Bioinformatics 2014;15 Suppl 15:S6.
45. Przulj N. Biological network comparison using graphlet degree distribution. Bioinformatics 2007;23:e177–e183. (Erratum in Bioinformatics 2010;26:853–854.)

46. Przulj N, Corneil DG, Jurisica I. Efficient estimation of graphlet frequency distributions in protein-protein interaction networks. Bioinformatics 2006;22:974–980.
47. Rand WM. Objective criteria for the evaluation of clustering methods. J Am Stat Assoc 1971;66(336):846–850.
48. Raymond JW, Willett P. Maximum common subgraph isomorphism algorithms for the matching of chemical structures. J Comput Aided Mol Des 2002;16:521–533.
49. Richardson JS. The anatomy and taxonomy of protein structure. In: Anfinsen CB, Edsall JT, Richards FM, editors. *Advances in Protein Chemistry*. Volume 34. New York: Academic Press; 1981.
50. Roberts FS. *Graph Theory and Its Applications to Problems of Society*. Philadelphia (PA): Society for Industrial Mathematics; 1987.
51. Rogen P, Fain B. Automatic classification of protein structure by using Gauss integrals. Proc Natl Acad Sci U S A 2003;100(1):119–124.
52. Samudrala R, Moult J. A graph-theoretic algorithm for comparative modeling of protein structure. J Mol Biol 1998;279:279–287.
53. Seidel JJ. A survey of two-graphs. Coll Int Teorie Combin 1973;17:481–511.
54. Ullmann JR. An algorithm for subgraph isomorphism. J ACM (JACM) 1976;23(1):42.
55. Vishveshwara S, Brinda KV, Kannan N. Protein structure: insights from graph theory. J Comput Chem 2002;1:187–211.
56. Watts DJ. *Small Worlds: The Dynamics of Networks between Order and Randomness*, Princeton Studies in Complexity. Princeton (NJ): Princeton University Press; 2003.
57. Wilson CA, Kreychman J, Gerstein M. Assessing annotation transfer for genomics: quantifying the relations between protein sequence, structure and function through traditional and probabilistic scores. J Mol Biol 2000;297(1):233–249.
58. Yakunin AF, Yee AA, Savchenko A, Edwards AM, Arrowsmith CH. Structural proteomics: a tool for genome annotation. Curr Opin Chem Biol 2004;8(1):42–48.
59. Yusuf D, Davis AM, Kleywegt GJ, Schmitt S. An alternative method for the evaluation of docking performance: RSR vs RMSD. J Chem Inform Model 2008;48(7):1411–1422.
60. Zager LA, Verghese GC. Graph similarity scoring and matching. Appl Math Lett 2008;21(1):86–94.

CHAPTER 16

STRUCTURAL DOMAINS IN PREDICTION OF BIOLOGICAL PROTEIN–PROTEIN INTERACTIONS

Mina Maleki[1], Michael Hall[2], and Luis Rueda[1]

[1]*School of Computer Science, University of Windsor, Windsor, Canada*
[2]*Department of Computing Science, University of Alberta, Edmonton, Canada*

16.1 INTRODUCTION

Proteins are large molecules that constitute the bulk of the cellular machinery of any living organism or biological system. They play important roles in fundamental and essential biological processes such as DNA synthesis, transcription, translation, and splicing. Proteins perform their functions by interacting with molecules such as DNA, RNA, and other proteins. Interactomics aims to study the main aspects of protein interactions in living systems. To understand the complex cellular mechanisms involved in a biological system, it is necessary to study the nature of these interactions at the molecular level, in which prediction of *Protein–Protein Interactions* (PPIs) plays a significant role [24].

Although prediction of PPIs has been studied from many different perspectives, the main aspects that are studied include sites of interfaces (where), arrangement of

Pattern Recognition in Computational Molecular Biology: Techniques and Approaches,
First Edition. Edited by Mourad Elloumi, Costas S. Iliopoulos, Jason T. L. Wang, and Albert Y. Zomaya.

proteins in a complex (how), type of protein complex (what), molecular interaction events (if), and temporal and spatial trends (dynamics) [46]. This chapter focuses mostly on the prediction of types of PPIs, more specifically, on obligate (permanent) and nonobligate (transient) complexes. These problems have been investigated in various ways, involving both experimental (in vivo or in vitro) and computational (in silico) approaches. Experimental approaches such as yeast two-hybrid and affinity purification followed by mass spectrometry tend to be costly, labor-intensive, and suffer from noise. Nonetheless, these techniques have been successfully used to produce high-throughput protein interaction data for many organisms [49, 84]. Typically, structural information in the main databases such as the *Protein Data Bank* (PDB) [6] is derived through costly techniques such as X-ray crystallography or *Nuclear Magnetic Resonance* (NMR) for smaller proteins. Therefore, using computational approaches for the prediction of PPIs is a good choice for many reasons [76]. So far, a variety of machine-learning approaches have been used to predict PPIs, mostly based on combinations of classification, clustering, and feature selection techniques. These systems, in general, represent objects (complexes, sites, patches, protein chains, domains, or motifs) as features or properties.

Domains are the minimal, structural, and fundamental units of proteins. These functional units often have a biological role and serve some specific purpose, such as signal binding or manipulation of a substrate within cells [25, 39]. Using the functional domain information to formulate protein samples for statistical predictions was originally proposed in References [10, 21], and quite encouraged results were achieved. The structural domain information, particularly the functional domain information [22], has not only been used to enhance the success rates in predicting the structural classes for a very low homology data set in which none of proteins included has more than 20% pairwise sequence identity to any other in a same class, but also been used (either independently or combined with other information) to predict protein subcellular location [21, 23, 28, 29], proteases and their types [27], membrane protein type [10], network of substrate–enzyme–product triads [15], drug–target interaction networks [43], regulatory pathways [44], enzyme family class [71], protein fold pattern [72], protein quaternary structural attribute [73, 80, 81], and others. Recent studies focus on employing domain knowledge to predict PPIs [1, 14, 42, 54, 75, 82, 83]. This is based on claims that only a few highly conserved residues are crucial for PPIs [32, 40] and that most domains and *Domain–Domain Interactions* (DDIs) are evolutionarily conserved [74]. We consider that two proteins interact if a domain in one protein interacts with a domain in the other protein [11, 62].

Our recent work on the prediction of PPIs focuses on studying the temporal and stability aspects of the interactions using structural and sequence domains [42, 52, 54, 55]. The problem of predicting obligate versus nonobligate protein complexes, as well as transient versus permanent [45] has been addressed. For this, we use a wide range of physicochemical and sequence-based properties, including desolvation [2, 53, 67] and electrostatic energies [56, 78], amino acid properties, solvent accessibility, statistical measures, and short-linear motifs [33, 61].

In this chapter, computational domain-based models to predict obligate and nonobligate PPIs are described. In these models, desolvation energies of structural domain pairs present in the interface are considered as features for prediction. The rest of the chapter is organized as follows. First, a brief description of domains and different types of domains is presented. Then, a general domain-based framework for the prediction of PPI types is presented. Within this framework, we discuss different subproblems including feature extraction, feature selection, classification, and evaluation. Next, we explain the data sets used and discuss some experimental results. We conclude the chapter with a brief conclusion.

16.2 STRUCTURAL DOMAINS

As mentioned earlier, proteins are composed of one or more minimal functional units called *domains*. The importance of functional domain information is typically reflected by the fact that it is the core of a protein that plays the major role for its function. For this reason, determining the three-dimensional structure of a protein has been done by using experimental approaches [5, 26, 70, 79] or by computational models [16, 17, 19, 20, 30], and the focus has mostly been on considering functional domains. There are several repositories of domain information available, including databases such as *Protein families* (Pfam) [41] and *Class, Architecture, Topology, and Homologous superfamily* (CATH) [31]. The Pfam database contains domains that are derived from sequence homology with other known structures, whereas CATH domains are based on structural homology.

The sequence domains in Pfam are identified using multiple sequence alignments and *Hidden Markov Model*s (HMMs). Other classifications of Pfam entries are (i) families of related protein regions, (ii) short unstable units that can make a stable structure when repeated multiple times, and (iii) short motifs present in disordered parts of a protein.

In contrast, the structural domains in the CATH database are organized in a hierarchical manner, which can be visualized as a tree with levels numbered from 1 to 8, hereafter referred to as L1–L8. Domains at upper levels of the tree represent more general classes of structure than those at lower levels. For example, domains at level 1 represent mainly alpha ($c1$), mainly beta ($c2$), mixed alpha–beta ($c3$), and few secondary structures ($c4$), whereas those at level 2 represent more specific structures. As shown in Figure 16.1, roll, beta barrels, and two-layer sandwich are three different sample architectures of domains in class $c3$. Domains at level 3 are even more specific, and so on.

In this review, structural CATH domains are considered as the basis for our predictions.

16.3 THE PREDICTION FRAMEWORK

A general domain-based model to predict PPI types is shown in Figure 16.2. The data set is a list of complexes (interacting chains) with their predefined PPI types (classes). To predict PPI types, first, the prediction properties (features) of each complex in the

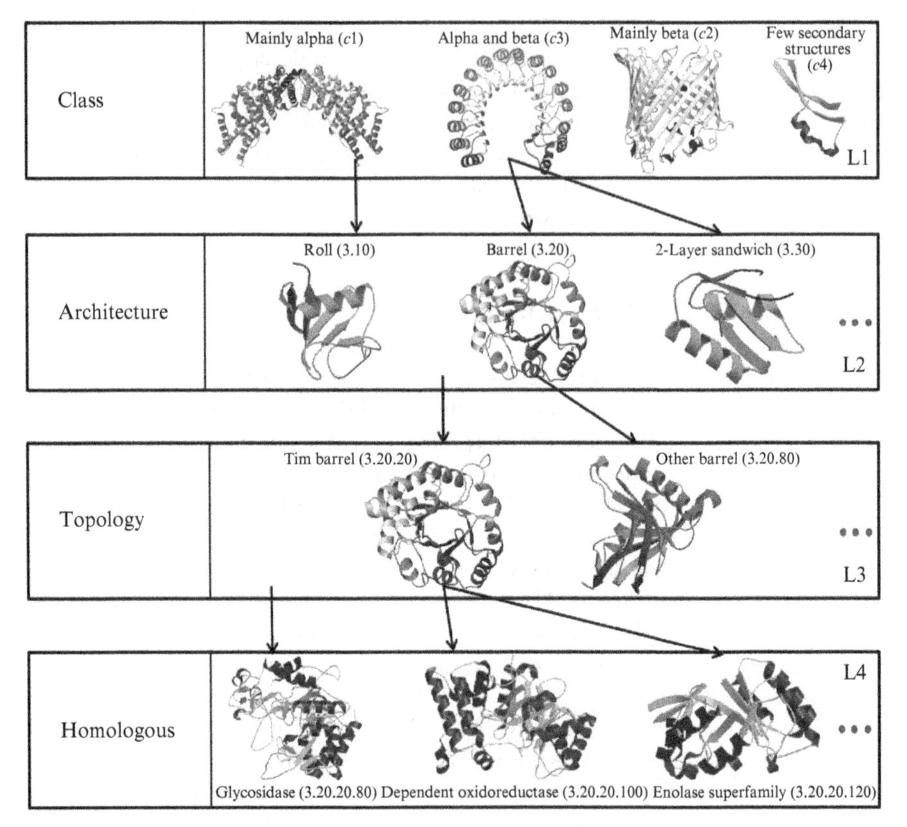

Figure 16.1 Four levels of the *Class, Architecture, Topology, and Homologous superfamily* (CATH) hierarchy.

data set are extracted employing different PPI databases. Then, the extracted features are passed through a feature selection module to remove noisy, irrelevant, and redundant features and select the most powerful and discriminative ones for prediction. After that, the selected features are used for classification and the outputs of the classification module are the predicted PPI types. Finally, the performance of the prediction model can be evaluated using different numerical performance metrics and visual analysis tools.

More details about the four main subproblems of feature extraction, feature selection, classification, and evaluation and analysis of the presented prediction model are discussed below.

16.4 FEATURE EXTRACTION AND PREDICTION PROPERTIES

Features are the observed properties of each sample that are used for prediction. Using the most relevant features is very important for successful prediction. Recently, the

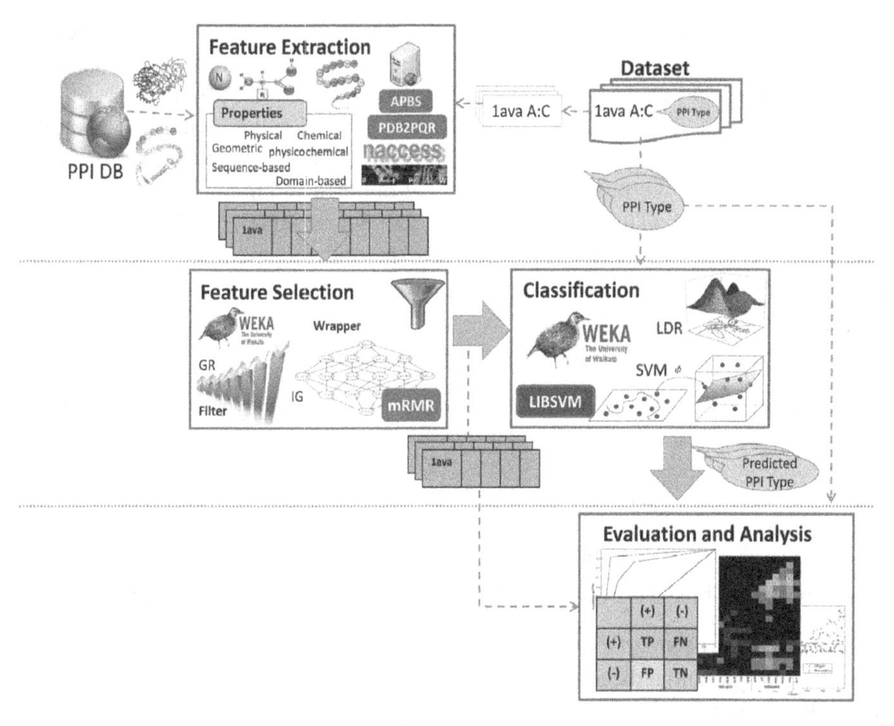

Figure 16.2 Domain-based framework used to predict PPI types.

pseudo amino acid composition [18] or PseAAC [13, 37, 86] has been increasing and widely used to extract various protein key features for analyzing various attributes of proteins, for example, the structural domain information has been incorporated into pseudo amino acid composition for the prediction of protein subcellular location [7, 23], enzyme subclass [8], membrane protein type [9], and protein quaternary structural attribute [80].

To extract properties for the prediction of PPI types, first, the 3D structures of each complex in the data set should be collected from the PDB [6]. Then, unnecessary information is filtered out from the downloaded PDB files and each PDB file is separated into two files according to its side-chain information. After collecting the domain information for each complex from domain databases, that information is added to each atom present in the chain. Complexes that do not have domain information in at least one of their subunits are discarded.

Before computing the prediction properties, we must identify which pairs of atoms are part of the interface. For this, we use a cutoff distance such that a pair of atoms within this distance is considered to be interacting. While in most studies this cutoff is considered to be less than 7 Å to calculate some types of features such as electrostatic energy, a larger cutoff may be used. In this way, the distances between all atom pairs of interacting chains are computed and those that are lower than the defined cutoff distance are considered to be interface atoms.

Some studies in PPI consider the analysis of a wide range of properties for predicting types of complexes or types of protein–protein interfaces (binding sites), including physical [32, 87], chemical [32, 47, 57, 87], physicochemical [2, 51, 56], geometric [32, 47, 48, 87], sequence-based [50, 58, 87], and domain-based features [32, 40, 52, 55, 63]. A summary of the employed feature types for prediction in different studies along with the characteristics of obligate and nonobligate interactions (or interfaces) based on using those features is shown in Table 16.1. In the following, some of our proposed features for predicting the obligate and nonobligate PPIs are described.

16.4.1 Physicochemical Properties

One of the types commonly used for prediction are physicochemical features, which consider both chemical and physical characteristics of the interacting residues as the prediction properties. *Atomic Contact Vectors* (ACV) [57], solvent accessibility (*Solvent-Accessible Surface Area*, SASA) and its variants [51, 87], and binding free energies [2, 47, 53, 56, 67, 78] are some examples of physicochemical features that are employed to predict obligate and nonobligate PPIs. In this review, we explain two subtypes of binding free energy: electrostatic and desolvation energies.

16.4.1.1 Desolvation Energy. Desolvation energy is defined as the knowledge-based contact potential (accounting for hydrophobic interactions), self-energy change upon desolvation of charged and polar atom groups, and side-chain entropy loss. As in Reference [12], the binding free energy ΔG_{bind} is defined as follows:

$$\Delta G_{bind} = \Delta E_{elec} + \Delta G_{des} \tag{16.1}$$

where ΔE_{elec} is the total electrostatic energy and ΔG_{des} is the total desolvation energy, which for a protein is defined as follows [12]:

$$\Delta G_{des} = g(r)\Sigma\Sigma e_{ij} \tag{16.2}$$

Considering the interaction between the ith atom of a ligand and the jth atom of a receptor, e_{ij} is the *Atomic Contact Potential* (ACP) between them, which is the estimation of the free energy after replacing atom–water contacts by atom–atom contacts for various types of interface protein atoms [85]; $g(r)$ is a smooth function based on their distance. The value of $g(r)$ is 1 for atoms that are less than 5 Å apart [12]. For simplicity, the smooth function can be considered to be linear within the range of 5–7 Å, and the value of $g(r)$ is $(7 - r)/2$. More details about the calculation of desolvation energy are given in Reference [2, 53, 67].

16.4.1.2 Electrostatic Energy. Electrostatic energy involves a long-range interaction that can occur between charged atoms of two interacting proteins or two different molecules [3]. Moreover, these interactions can occur between charged atoms on the protein surface and charges in the environment. In order to compute electrostatic energy values, the PDB2PQR [35] and APBS [4] software packages can be used.

Table 16.1 Properties employed in different studies for the prediction of obligate and nonobligate complexes

Type	Property	Nonobligate	Obligate	References
	Hydrophobicity	Less hydrophobic residues	More hydrophobic residues	
Chemical	Polarity	High	Low	[32, 47, 57, 87]
Physical	*Interface Area* (IA) ratio	Small interfaces < 1500 Å^2	Large and twisted interfaces from 1500 to 10,000 Å^2	[32, 87]
Physicochemical	Atomic contact	Less number of contacts	More number of contacts	[57]
	Association bonds	Salt bridges	Hydrogen bonds and covalent disulphide bridges	[32]
	Binding affinity	Weak PPIs (Kd > 10^{-6}M)	Strong PPIs (Kd < 10^{-6}M)	[47, 60]
	Low-ASA pairs	Small (45 ± 20.6)	Large (83 ± 53.2)	[51]
Geometric	Gap volume	More gap volume	More complementary interfaces	[87]
	Secondary structure	Turns, more helix	B-sheet, less helix	[32, 47]
Sequence-based	B-factor	More flexible (large B-factor)	Rigid (small B-factor)	[50]
Evolutionary	Conservation score sequence profile	Less conserved	Evolve slowly and at similar rates	[36, 50, 57, 87]
Domain-based	Interacting folds	Domain–polypeptide interactions	Domain–domain interactions	[60]

For each complex in the data sets, after extracting the structural data from the PDB, PDB2PQR is used to prepare the structures for electrostatic calculations. The main tasks performed by PDB2PQR include the addition of missing heavy atoms, placement of missing hydrogen atoms, and assignment of charges. After that, APBS is used to compute electrostatic energy values of interactions between solutes in salty and aqueous media. In APBS, the Poisson–Boltzmann equation is solved numerically and electrostatic calculations are performed in the range of 10 atoms to 1 million atoms. More details about the calculation of electrostatic energy properties and the parameters that should be set are given in References [56, 78].

16.4.2 Domain-Based Properties

After identifying all the unique domains present in the interface of at least one complex in the data sets, desolvation (or electrostatic) energies for all domain–domain interactions (pairs of domains) are calculated using Equation 16.2. For each ligand–receptor (protein–protein) pair, the cumulative desolvation (or electrostatic) energy values across all occurrences of that DDI are computed if any duplicate DDIs are found during the calculation step. A domain is considered to be in the interface, if it has at least one residue interacting with a domain in the other chain. In the following, we describe the method used for desolvation energies of two different types of structural domain-based properties from CATH.

16.4.2.1 Domain-Based Properties at the Individual Level. Since the CATH database is organized in a hierarchical scheme, a separate data set of feature vectors can be created for each level of the CATH hierarchy. To speed up computations, after calculating the desolvation energies for all DDIs in level 8, for each DDI in higher levels the desolvation energy can be calculated by taking the sum of the desolvation energies of the corresponding DDIs at the next lower level. For each node in the CATH tree, the set of DDIs associated with it is completely disjoint (with the exception of reflexive pairs, which have been accounted for in postprocessing). Thus, when the desolvation energies of DDIs from one level are combined to find the desolvation energies for the respective parent nodes at the next higher level, no redundancy is introduced into the corresponding features.

Each of these data sets can be used for classification separately in order to measure the predictive power of each individual level in the CATH hierarchy. More details about the extraction of domain-based feature vectors for each level are given in Reference [54].

16.4.2.2 Domain-Based Properties at the Combined Level. The described approach can be extended to consider multiple levels of the CATH hierarchy simultaneously. In this way, different domains at different levels of granularity can be viewed. By allowing arbitrary combinations of nodes, the total number of feature vectors would be exponential, with each feature vector corresponding to a sequence of nodes chosen to represent the domains found in the data set.

In order to maintain computational tractability and eliminate any redundancy in the feature vectors, the following constraints can be imposed: (i) There can be no overlap between nodes. That is, there cannot exist a pair of nodes in a sequence such that one node is an ancestor of the other. (ii) Only combinations of nodes taken from levels 2 and 3 of the hierarchy have been considered. Based on the results presented in Reference [54], it is acceptable to conclude that the optimal combination of nodes can be found somewhere between these two levels. (iii) Nodes at level 3, which are the sole child of their parent node at level 2, are discarded. However, the number of node sequences to be evaluated is still exponential with respect to the number of nodes at level 2.

Even though an exhaustive enumeration of the entire search space is still computationally tractable given the size of data sets, this would be a poor choice in general. Accordingly, a method based on *Sequential Floating Forward Search* (SFFS) has been implemented [66] to find a reasonable approximation to the best combination of nodes between two different levels. For this, SFFS is initialized at the sequence of nodes consisting of the set of all nodes at level 2 as this sequence showed the greatest promise in the previous study by Maleki et al. [54]. Then, the search proceeds downward through the CATH tree toward the sequence of nodes corresponding to the lower level (level 3).

The SFFS algorithm starts with an empty feature set and, sequentially, the best feature that satisfies some criterion function is combined with the features that have already been selected. Also, this algorithm has the capability of eliminating the worst feature from the selected feature set. Therefore, the SFFS proceeds dynamically increasing and decreasing the number of features until the desired subset of features is reached.

16.5 FEATURE SELECTION

Feature selection is the process of choosing the best subset of relevant features that represents the whole data set efficiently after removing redundant and/or irrelevant ones. Applying feature selection before running a classifier is useful in reducing the dimensionality of the data and, thus, reducing the prediction time while improving the prediction performance. There are two different ways of performing feature selection: wrapper methods and filter methods [59].

16.5.1 Filter Methods

In filter-based feature selection methods, the quality of the selected features is scored and ranked independently of the classification algorithm by using some statistical criteria based on their relevance. Although feature selection based on filter methods is fast, it does not consider the dependency of features from each other; a feature that is not useful by itself can be very useful when combined with others.

Some of the most well-known filter methods are the following:

minimum Redundancy Maximum Relevance (mRMR): One of the most widely used feature selection algorithms based on mutual information is mRMR [34, 64]. It uses maximum relevance score as the class separability criterion. In mRMR, the features that have minimal redundancy and are highly relevant to the classes are recursively selected and scored to generate a ranked list of all features.

Information Gain (IG): IG is based on the concept of entropy [59]. The IG value of a feature X with respect to class attribute Y is calculated as follows:

$$IG(Y,X) = H(Y) - H(Y|X) \tag{16.3}$$

Here, $H(Y)$ is the entropy of class Y and $H(Y|X)$ is the conditional entropy of Y given X, which are calculated by means of the following formulas:

$$H(Y) = -\sum_{y \in Y} p(y)\log_2(p(y)) \tag{16.4}$$

and

$$H(Y|X) = -\sum_{x \in X} p(x) \sum_{y \in Y} p(y|x)\log_2(p(y|x)) \tag{16.5}$$

where $p(y)$ is the marginal probability density function for random variable Y and $p(y|x)$ is the conditional probability of Y given X.

Gain Ratio (GR): GR attribute evaluation is a well-known feature selection method that is based on the concept of IG and entropy [59]. The GR value of a feature X with respect to class attribute Y is calculated as follows:

$$GR(Y,X) = \frac{(H(Y) - H(Y|X))}{H(X)} = \frac{IG(Y,X)}{H(X)} \tag{16.6}$$

where $H(Y)$, the entropy of class Y, and $H(Y|X)$, the conditional entropy of Y given X, are calculated using Equations 16.4 and 16.5 , respectively. A value of GR = 1 indicates that feature X is highly relevant and one of the best features to predict class Y, while GR = 0 means that feature X is not relevant at all.

Chi-Square (χ^2): This method measures the degree of independence of each feature to the classes by computing the value of the chi-square statistic [59]. The χ^2 value of a feature X with respect to class attribute Y is calculated as follows:

$$\chi^2(Y,X) = \frac{N \times (AD - CB)^2}{(A+C) \times (B+D) \times (A+B) \times (C+D)} \tag{16.7}$$

where A is the number of times feature X and class Y co-occur, B is the number of times X occurs without Y, C is the number of times Y occurs without X, $D$is the number of times neither X nor Y occurs, and N is the total number of samples.

16.5.2 Wrapper Methods

The aim of this kind of methods is to find the best subset of features using a particular predictive model (classifier) to score feature subsets. As performing an exhaustive search to find the best subset of features is computationally intensive, some heuristic search methods can be employed such as the following.

Forward selection: Starting with an empty set, keep adding features one at a time until no further improvement can be achieved.

Backward elimination: Starting with the full set of features, keep removing features one at a time until no further improvement can be achieved.

Forward–backward wrapping: Interleave the two phases of adding and removing in either forward selection or backward elimination.

Floating forward (backward) selection (elimination): It works in the same way as forward selection, while keeping alternative solutions in memory for future exploration.

Thus, the aim of each of these heuristic search methods is to find an optimal feature subset for a specific data set. Note that the score of each subset of features is calculated based on the employed classification method.

16.6 CLASSIFICATION

After extracting the features and selecting the most discriminating ones, a classifier can be applied in order to assign the class labels. The samples are first divided into train and test sets using different splitting methods such as m-fold cross-validation or leave-one-out methods. All classification methods employ two phases of processing for training and testing. In the training phase, the training samples are used to build a model that is a description of each training class. Then, in testing phase, that model is used to predict the classes of the test samples. There are a variety of classification methods, of which some of the commonly used methods are explained below.

16.6.1 Linear Dimensionality Reduction

Linear Dimensionality Reduction(LDR) methods have been successfully used in pattern recognition for a long time due to their easy implementation and high classification speed [69]. In LDR, the objects (protein complexes in our context) are represented by two normally distributed random vectors $\mathbf{x}_1 \sim N(\mathbf{m}_1, \mathbf{S}_1)$ and $\mathbf{x}_2 \sim N(\mathbf{m}_2, \mathbf{S}_2)$, respectively, with p_1 and p_2 being the a priori probabilities. The aim of LDR is to apply a linear transformation to project large dimensional data onto a lower dimensional space, $\mathbf{y}_i = \mathbf{A}\mathbf{x}_i$, in such a way that the classification in the new space is as efficient as possible, if not better. To obtain the underlying transformation matrix $\mathbf{A}$, $\mathbf{S}_W = p_1\mathbf{S}_1 + p_2\mathbf{S}_2$ and $\mathbf{S}_E = (\mathbf{m}_1 - \mathbf{m}_2)(\mathbf{m}_1 - \mathbf{m}_2)^t$, the within-class and between-class scatter matrices, respectively, are first computed.

Three LDR criteria are the most widely used in the prediction of PPIs:

(a) *Fisher's Discriminant Analysis* (FDA) [38], which aims to maximize

$$J_{FDA}(\mathbf{A}) = tr\{(\mathbf{A}\mathbf{S}_W\mathbf{A}^t)^{-1}(\mathbf{A}\mathbf{S}_E\mathbf{A}^t)\} \tag{16.8}$$

The optimal $\mathbf{A}$ is found by considering the eigenvector corresponding to the largest eigenvalue of $\mathbf{S}_{FDA} = \mathbf{S}_W^{-1}\mathbf{S}_E$.

(b) The *Heteroscedastic Discriminant Analysis* (HDA) approach [69], which aims to obtain the matrix $\mathbf{A}$ that maximizes the following function:

$$\begin{aligned} J_{HDA}(\mathbf{A}) = tr\{&(\mathbf{A}\mathbf{S}_W\mathbf{A}^t)^{-1}[\mathbf{A}\mathbf{S}_E\mathbf{A}^t \\ &- \mathbf{A}\mathbf{S}_W^{1/2}\frac{p_1\log(\mathbf{S}_W^{-(1/2)}\mathbf{S}_1\mathbf{S}_W^{-(1/2)}) + p_2\log(\mathbf{S}_W^{-(1/2)}\mathbf{S}_2\mathbf{S}_W^{-(1/2)})}{p_1p_2}\mathbf{S}_W^{1/2}\mathbf{A}^t]\} \end{aligned} \tag{16.9}$$

This criterion is maximized by obtaining the eigenvectors, corresponding to the largest eigenvalues, of the matrix:

$$\mathbf{S}_{HDA} = \mathbf{S}_W^{-1}\left[\mathbf{S}_E - \mathbf{S}_W^{1/2}\frac{p_1\log(\mathbf{S}_W^{-(1/2)}\mathbf{S}_1\mathbf{S}_W^{-(12)})+p_2\log(\mathbf{S}_W^{-(1/2)}\mathbf{S}_2\mathbf{S}_W^{-(1/2)})}{p_1p_2}\mathbf{S}_W^{1/2}\right] \tag{16.10}$$

(c) The *Chernoff Discriminant Analysis* (CDA) approach [69], which aims to maximize

$$\begin{aligned} J_{CDA}(\mathbf{A}) = tr\{&p_1p_2\mathbf{A}\mathbf{S}_E\mathbf{A}^t(\mathbf{A}\mathbf{S}_W\mathbf{A}^t)^{-1} + \log(\mathbf{A}\mathbf{S}_W\mathbf{A}^t) \\ &- p_1\log(\mathbf{A}\mathbf{S}_1\mathbf{A}^t) - p_2\log(\mathbf{A}\mathbf{S}_2\mathbf{A}^t)\} \end{aligned} \tag{16.11}$$

To solve this problem, a gradient-based algorithm is used [69]. This iterative algorithm needs a learning rate, α_k, which is maximized by using the secant method to ensure that the gradient algorithm converges. The initialization of the matrix $\mathbf{A}$ is also an important issue in the gradient-based algorithm. Using different initializations, the solution for $\mathbf{A}$ that yields the maximum Chernoff distance in the transformed space can be selected. More details about this method can be found in References [68, 69].

Once the dimension reduction takes place, the vectors in the new space of dimension d can be classified using any classification technique. To achieve the reduction, the linear transformation matrix $\mathbf{A}$, which corresponds to the one obtained by one of the LDR criteria, is found independently for every fold in the cross-validation process. Different classifiers can be employed to classify the vectors in the lower dimensional space such as *Quadratic Bayesian* (QB) classifier [69], which is the optimal classifier for normal distributions, and a *Linear Bayesian* (LB) classifier obtained by deriving a Bayesian classifier with a common covariance matrix, $\mathbf{S} = \mathbf{S}_1 + \mathbf{S}_2$.

16.6.2 Support Vector Machines

Support Vector Machines (SMVs) are well-known machine-learning techniques used for classification, regression, and other tasks. The aim of the SVM is to find the support vectors (most difficult vectors to be classified) to derive a decision boundary that separates the feature space into two regions.

Let $\{\mathbf{x}_i\}$ where $i = 1, 2, \ldots, n$, be the feature vectors of a training data set $\mathbf{X}$. These vectors belong to one of two classes, ω_1 and ω_2, which are assumed to be linearly separable. The goal of the SVM is to find a hyperplane that classifies all the training vectors as follows:

$$g(\mathbf{x}) = \mathbf{w}^t\mathbf{x} + w_0 \tag{16.12}$$

This kind of hyperplane is not unique. The SVM chooses the hyperplane that leaves the maximum margin from that hyperplane to the *support vectors*.

The distance from a point to a hyperplane is given by

$$z = \frac{|g(\mathbf{x})|}{||\mathbf{w}||} \tag{16.13}$$

If for each $\mathbf{x}_i$ we denote the corresponding class label by y_i ($+1$ for ω_1, -1 for ω_2), the SVM finds the best hyperplane by computing the parameters $\mathbf{w}$ and w_0 of the hyperplane so that the following formula is minimized:

$$J(\mathbf{w}) = \frac{1}{2}||\mathbf{w}||^2 \tag{16.14}$$

subject to

$$y_i(\mathbf{w}^t x_i + w_0) \geq 1, \quad i = 1, 2, \cdots, n \tag{16.15}$$

The classification by using the SVM is usually inefficient when using a linear classifier because, in general the data are not linearly separable, and hence the use of kernels is crucial in mapping the data onto a higher dimensional space in which the classification is more efficient. The effectiveness of the SVM depends on the selection of the kernel, the selection parameters, and the soft margin [38]. There are a number of different kernels that can be used in SVMs such as polynomial, *Radial Basis Function* (RBF), and sigmoid.

16.6.3 *k*-Nearest Neighbor

k-*Nearest Neighbor* (*k*-NN) is one of the simplest classification methods, in which the class of each test sample can be easily found by a majority vote of the class labels of its neighbors. To achieve this, after computing and sorting the distances between the test sample and each training sample, the most frequent class label in the first k training samples (nearest neighbors) is assigned as the class of the test sample. Determining the appropriate number of neighbors is one of the challenges of this method.

16.6.4 Naive Bayes

One of the simplest probabilistic classifiers is *Naive Bayes* (NB). Assuming independence of features, the class of each test sample can be found by applying Bayes' theorem. The basic mechanism of NB is rather simple. The reader is referred to Reference [77] for more details.

16.7 EVALUATION AND ANALYSIS

To evaluate the performance of a prediction model or to compare the performance of different prediction models, there are a variety of numerical metrics and visual analysis tools. One of the well-known numerical performance metrics is accuracy, which can be computed as follows: $acc = (TP + TN)/(N + P)$, where TP and TN are the total numbers of true-positive (true obligate) and true-negative (true nonobligate) predictions, respectively. P and N are the total numbers of complexes in the positive and negative classes, respectively. For unbalanced class problems, the performance can be analyzed in terms of specificity ($SP = TN/N$), sensitivity ($SN = TP/P$), or geometric mean ($G_m = \sqrt{SE \times SP}$).

Moreover, the *Receiver Operating Characteristic* (ROC) curve is a visual tool that can be plotted based on the *True-Positive Rate* (TPR), also known as "sensitivity," versus the *False-Positive Rate* (FPR), or "1 − specificity," at various threshold settings. To generate the ROC curves, the sensitivity and specificity of each subset of features are determined for different parameter values of the employed classifier. Then, by applying a simple algorithm, the FPR and TPR points are filtered as follows: (i) for the same FPR values, the greatest TPR value (top point) is chosen, and (ii) for the same TPR values, the smallest FPR value (left point) is chosen. A polynomial function with degree 2 is then fitted to the selected points. The ROC analysis is suitable for unbalanced class problems and yields better insight than simple performance metrics.

16.8 RESULTS AND DISCUSSION

In this section, some prediction results using domain-based prediction properties are shown.

For this, two preclassified data sets of protein complexes were obtained from the studies of Zhu et al., [87] (ZH^{Org}) and Mintseris and Weng [58] (MW^{Org}). Then, complexes that did not have domain information in at least one of their subunits were discarded. A summary of the number of obligate and nonobligate complexes before and after filtering these complexes by considering CATH domains is given in Table 16.2, where MW and ZH refer to as the original MW^{Org} and ZH^{Org} data sets after removing the complexes that do not have CATH domains in their interactions.

16.8.1 Analysis of the Prediction Properties

The results of the SVM classifier (maximum accuracies) using domain-based features of the individual levels of CATH for the MW and ZH data sets are depicted

Table 16.2 Data sets included in this review and their number of complexes

Data Set Name	# Complexes	# Obligate	# Nonobligate
MW^{Org}	327	115	212
MW	287	106	181
ZH^{Org}	137	75	62
ZH	127	72	55

Table 16.3 Prediction results of SVM for the MW and ZH data sets

Subset	#D	#DDIs	SVM	Subset	#D	#DDIs	SVM
MW-L1	4	9	72.82	ZH-L1	4	9	74.80
MW-L2	26	106	77.35	ZH-L2	24	67	79.30
MW-L3	237	342	71.25	ZH-L3	136	154	69.76
MW-L4	386	403	70.73	ZH-L4	186	180	69.69
MW-L5	740	563	70.38	ZH-L5	272	228	67.93
MW-L6	803	573	69.69	ZH-L6	278	230	67.72
MW-L7	864	576	69.69	ZH-L7	287	230	67.72
MW-L8	899	576	69.69	ZH-L8	301	236	67.72

in Table 16.3. The names of the subsets show the data set name (MW or ZH) and the level of the domains in the CATH hierarchy (L1–L8). For each subset of features, the number of domains (#*D*) and the number of nonzero DDIs (features) are also shown.

From the table, it is observed that, for both MW and ZH data sets, only domain-based features taken from the first few levels of CATH are more powerful and discriminative to predict obligate and nonobligate complexes, and features taken from other levels, L4–L8, could be ignored because of their low prediction performance. More prediction results and discussions can be found in Reference [54].

As the prediction results using features taken from levels 2 and 3 of the CATH hierarchy are the best in comparison to the features of levels 4–8, it can be concluded that the optimal combination of nodes can be found somewhere between these two levels. In Reference [55], by comparing the prediction results of the LDR, NB, SVM, and *k*-NN classifiers with individual and combined domain-based features for the MW and ZH data sets, it has been concluded that, for both the MW and ZH data sets, (i) domain-based properties at the combined level yield higher accuracies than domain-based properties at the individual levels; (ii) domain-based features related to level 2 of CATH are more powerful than the features from level 3; (iii) SVM is the most powerful classifier for all subsets of features; and (iv) SVM, LDR, NB, and *k*-NN classifiers, however, show a similar trend. For all classifiers, DDIs from L2 are better than those of L3, while DDIs from a combination of L2 and L3 are much better than those of both L2 and L3 individually.

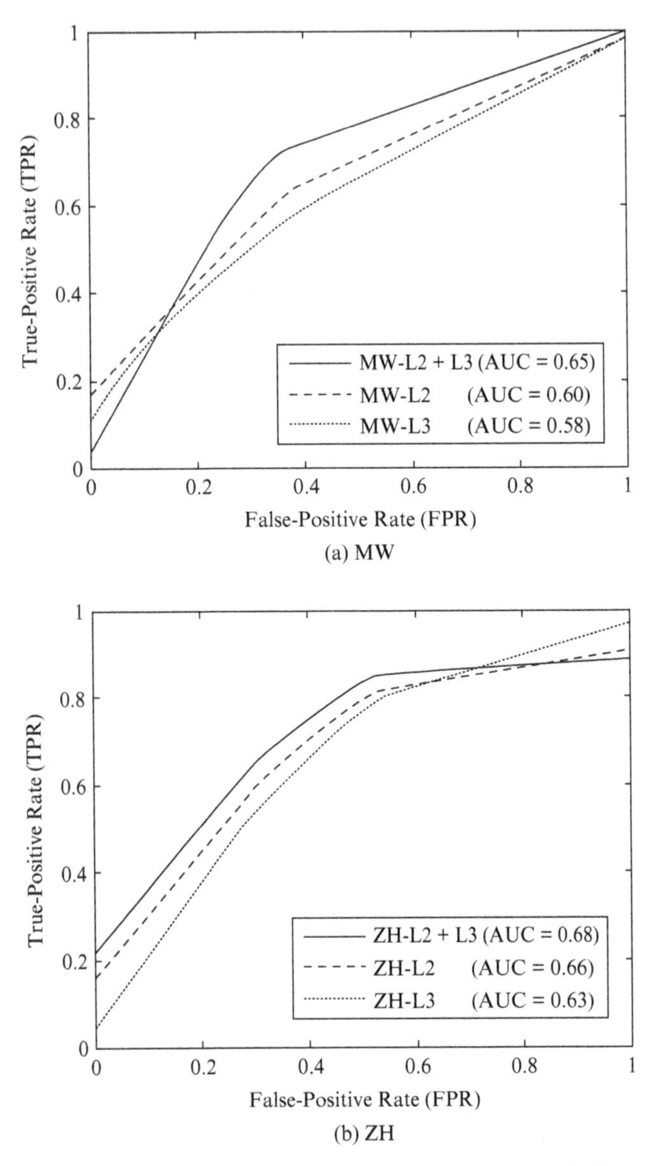

Figure 16.3 ROC curves and AUC values for all subsets of features of (a) MW and (b) ZH data sets.

Similarly, the ROC curves of the MW and ZH data sets using individual and combined DDI properties of the CATH hierarchy, plotted in Figure 16.3a and b, respectively, confirm the classification results. To generate the ROC curves, the sensitivity and specificity of each subset of features were determined for different values of d and β values in the CDA classifier. From the plots, it is clear that, for both data sets, the prediction performances of LDR using DDI properties on the combined level

(ZH-L2 + L3 and MW-L2 + L3) are clearly better than using DDI properties on level 2 (ZH-L2 and MW-L2) and much better than those on level 3 (ZH-L3 and MW-L3).

In addition, the *Area Under the Curve* (AUC) is computed for each of the above ROC curves using the trapezoid rule. The AUC values are also shown in Figure 16.3. The AUC for ZH-L2 + L3 is 0.68, which is greater than the AUCs of both ZH-L2 (0.66) and ZH-L3 (0.63). Similarly, the AUC for the MW data set using DDI properties on the combined level (MW-L2 + L3) is 0.65 while that for MW-L2 is 0.60. Also, the AUC of MW-L2 is greater than that of MW-L3. Generally, by comparing the AUC values, it can be concluded that DDI properties from the combined levels show much better predictive power than those from the individual levels. More prediction results and discussions can be found in Reference [55].

16.8.2 Analysis of Structural DDIs

A summary of the number of DDIs present in both the ZH and MW data sets, categorized by complex type, obligate and nonobligate, is shown in Table 16.4. Domain names show the "class" of each complex, which are determined based on the secondary structure composition of the complexes. In CATH, there are generally four classes of mainly alpha ($c1$), mainly beta ($c2$), mixed alpha–beta ($c3$), and secondary structure content ($c4$) [31].

From the table, it is clear that most of the DDIs are between domains of $c3$ and other classes in which $c3$:$c3$ (DDIs of $c3$ domains with themselves) has the highest rank among all levels. However, domains of $c4$ have the least number of interactions than those of other levels. In addition, the number of obligate DDIs is larger than that of nonobligate DDIs, when considering the interactions of homo-DDIs such as $c1$:$c1$ and $c3$:$c3$.

To provide a visual insight of the distribution of DDIs present in the complexes of the MW and ZH data sets, a schematic view of the DDIs of level 3 is shown in Figure 16.4. In each figure, DDIs that are obligate, nonobligate, and common are

Table 16.4 A summary of the number of CATH DDIs of level 3 present in the ZH and MW data sets

		ZH-CATH-L3		MW-CATH-L3	
Domain 1	Domain 2	# Obligate	# Nonobligate	# Obligate	# Nonobligate
$c1$	$c1$	17	8	29	18
$c1$	$c2$	6	4	18	41
$c1$	$c3$	12	18	63	59
$c1$	$c4$	1	1	1	7
$c2$	$c2$	15	18	28	92
$c2$	$c3$	7	22	27	77
$c2$	$c4$	2	2	10	6
$c3$	$c3$	101	25	88	70
$c3$	$c4$	0	0	4	8
$c4$	$c4$	1	0	0	0

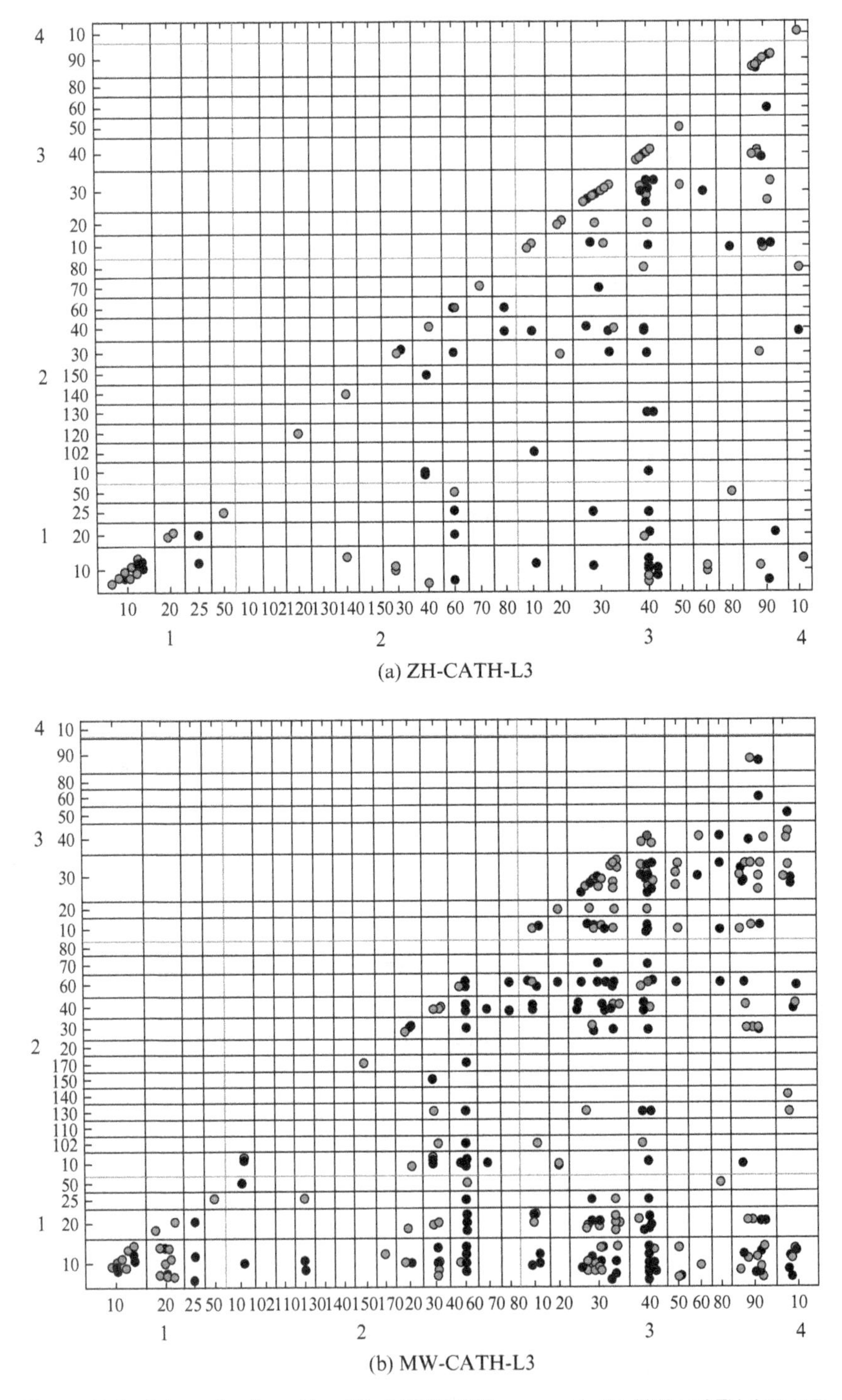

(a) ZH-CATH-L3

(b) MW-CATH-L3

Figure 16.4 Schematic view of level 3 of CATH DDIs present in the MW and ZH data sets.

shown in black, light, and dark gray lines, respectively. Also, the horizontal and vertical light gray and black lines indicate the boundaries of the classes in levels 1 and 2 of CATH, respectively.

Some concentrations of black or light gray dots can be seen in some parts of the plots in both data sets. This suggests that some domains are more likely to occur in obligate, while others in nonobligate complexes. This is more noticeable along the diagonal, in which most of the obligate DDIs (black dots) are concentrated, which indicates that the largest number of homo-domain pairs is in obligate complexes. Moreover, by considering the distribution of DDIs, it can be seen that all DDIs can be divided into three main groups: more obligate (columns 1–20, 2–30, and 3–90), more nonobligate (columns 2–60 and 3–40), or noninteraction (columns 2–102 and 2–150).

16.9 CONCLUSION

The idea of employing a structural domain-based approach for predicting obligate and nonobligate protein complexes is explained in this chapter. The prediction results demonstrate a significant improvement by combining nodes from different levels in the CATH hierarchy, rather than considering DDI features of each level separately. Also, it has been shown that DDIs at upper levels of the CATH hierarchy are more powerful than those at lower levels for prediction based on the obtained prediction results using different classification methods. The plotted ROC curves and AUC values corroborate the prediction results.

Furthermore, by grouping the DDI feature vectors of the CATH hierarchy based on their secondary structures, a numerical analysis shows that while most of the interactions are between domains that have alpha–beta structures, there are fewer interactions between domains of $c4$ and domains of other classes.

In addition, by considering the distribution of DDIs in level 3 of the CATH hierarchy, it has been shown visually that some DDIs in level 3 can be considered less important (noninteraction group), while the others are more important (more obligate and more nonobligate groups) to predict obligate and nonobligate complexes. Also, both visual and numerical analyses indicate that most homo-DDIs are obligate.

The approach described in this review can also be used for the prediction of other types of complexes, including intra- and interdomains and homo- and hetero-oligomers. Other properties can also be used including geometric (e.g., shape, planarity, and roughness) and other statistical and physicochemical properties such as residue and atom vicinity, secondary structure elements, hydrophobicity, and salt bridges, among others. Moreover, performing a biological analysis to find the interacting domains of different types of complexes and also investigating the types of DDIs are some of the worthy research topics in this area.

In addition, user-friendly and publicly accessible web-servers represent the future direction for developing practically more useful models, simulated methods, or predictors. The *Protein–Protein Interaction Prediction At Windsor* (PPIPAW) [65], a Web server developed by authors, focuses on predicting two types of PPIs, namely obligate and nonobligate, providing an additional insight into the characteristics of the SASA of the interface and the desolvation energies for atom-type and amino acids participating in the interface. This server is being updated periodically to account for other types of properties, including domains, electrostatic energies, and others.

REFERENCES

1. Akutsu T, Hayashida M. Domain-based prediction and analysis of protein-protein interactions. *Biological Data Mining in Protein Interaction Networks*. Chapter 3. Medical Information Science Reference; 2009. p 29–44.
2. Aziz M, Maleki M, Rueda L, Raza M, Banerjee S. Prediction of biological protein-protein interactions using atom-type and amino acid properties. Proteomics 2011;11(19): 3802–3810.
3. Baker N. *Continuum Models for Biomolecular Solvation*. Richland (WA): Pacific Northwest National Laboratory; 2008.
4. Baker NA, Sept D, Joseph S, Holst MJ, Mccammon JA. Electrostatics of nanosystems: application to microtubules and the ribosome. Proc Natl Acad Sci U S A 2001;98(18): 10037–10041.
5. Berardi MJ, Shih WM, Harrison SC, Chou JJ. Mitochondrial uncoupling protein 2 structure determined by NMR molecular fragment searching. Nature 2011;479:109–113.
6. Berman H, Westbrook J, Feng Z, Gilliland G, Bhat T, Weissig H, Shindyalov I, Bourne P. The protein data bank. Nucleic Acids Res 2000;28:235–242.
7. Cai YD. Nearest neighbour algorithm for predicting protein subcellular location by combining functional domain composition and pseudo-amino acid composition. Biochem Biophys Res Commun 2003;305:407–411.
8. Cai YD. Predicting enzyme subclass by functional domain composition and pseudo amino acid composition. J Proteome Res 2005;4:967–971.
9. Cai YD. Predicting membrane protein type by functional domain composition and pseudo-amino acid composition. J Theor Biol 2006;238:395–400.
10. Cai YD, Zhou GP. Support vector machines for predicting membrane protein types by using functional domain composition. Biophys J 2003;84:3257–3363.
11. Caffrey D, Somaroo S, Hughes J, Mintseris J, Huang E. Are protein-protein interfaces more conserved in sequence than the rest of the protein surface? Protein Sci 2004;13(1):190–202.
12. Camacho C, Zhang C. FastContact: rapid estimate of contact and binding free energies. Bioinformatics 2005;21(10):2534–2536.
13. Cao D, Xu QS, Liang YZ. propy: a tool to generate various modes of Chou's PseAAC. Bioinformatics 2013;29:960–962.
14. Chandrasekaran P, Doss C, Nisha J, Sethumadhavan R, Shanthi V, Ramanathan K, Rajasekaran R. In silico analysis of detrimental mutations in ADD domain of chromatin remodeling protein ATRX that cause ATR-X syndrome: X-linked disorder. Netw Model Anal Health Inform Bioinform 2013;2(3):123–135.
15. Chen L, Feng KY, Cai Y. Predicting the network of substrate-enzyme- product triads by combining compound similarity and functional domain composition. BMC Bioinformatics 2010;11:293.

16. Chou KC. The convergence-divergence duality in lectin domains of the selectin family and its implications. FEBS Lett 1995;363:123–126.
17. Chou KC. Knowledge-based model building of tetiary structures for lectin domains of the selectin family. J Protein Chem 1996;15:161–168.
18. Chou KC. Prediction of protein cellular attributes using pseudo amino acid composition. Proteins 2001;43:246–255.
19. Chou KC. Insights from modelling the 3D structure of the extracellular domain of alpha7 nicotinic acetylcholine receptor. Biochem Biophys Res Commun 2004;319:433–438.
20. Chou KC. Review: structural bioinformatics and its impact to biomedical science. Curr Med Chem 2004;11:2105–2134.
21. Chou KC, Cai YD. Using functional domain composition and support vector machines for prediction of protein subcellular location. J Biol Chem 2002;277:45767–45769.
22. Chou KC, Cai YD. Predicting protein structural class by functional domain composition. Biochem Biophys Res Commun 2004;321:1007–1009.
23. Chou KC, Cai YD. Predicting subcellular localization of proteins by hybridizing functional domain composition and pseudo-amino acid composition. J Cell Biochem 2004;91:1197–1203.
24. Chou KC, Cai Y. Predicting protein-protein interactions from sequences in a hybridization space. J Proteome Res 2006;5:316–322.
25. Chou KC, Cai Y. *Biomolecular Networks: Methods and Applications in Systems Biology*. John Wiley and Sons, Inc.; 2009.
26. Chou JJ, Li S, Klee CB, Bax A. Solution structure of Ca2+-calmodulin reveals flexible hand-like properties of its domains. Nat Struct Biol 2001;8:990–997.
27. Chou KC, Shen HB. Protident: a web server for identifying proteases and their types by fusing functional domain and sequential evolution information. Biochem Biophys Res Commun 2008;376:321–325.
28. Chou KC, Shen HB. A new method for predicting the subcellular localization of eukaryotic proteins with both single and multiple sites: Euk- mploc 2.0. PLoS ONE 2010;5:e9931.
29. Chou KC, Shen HB. Plant-mPLoc: a top-down strategy to augmen the power for predicting plant protein subcellular localization. PLoS ONE 2010;5:e11335.
30. Chou KC, Watenpaugh kD, Heinrikson RL. A model of the complex between cyclin-dependent kinase 5 (Cdk5) and the activation domain of neuronal Cdk5 activator. Biochem Biophys Res Commun 1999;259:420–428.
31. Cuff A, Sillitoe I, Lewis T, Redfern O, Garratt R, Thornton J, Orengo C. The CATH classification revisited-architectures reviewed and new ways to characterize structural divergence in superfamilies. Nucleic Acids Res 2009;37:310–314.
32. De S, Krishnadev O, Srinivasan N, Rekha N. Interaction preferences across protein-protein interfaces of obligatory and non-obligatory components are different. BMC Struct Biol 2005;5(15). doi: 10.1186/1472-6807-5-15.
33. Deng M, Jiang R, Sun F, Zhang X, editors. Research in computational molecular biology. 17th Annual International Conference on Research in Computational Molecular Biology (RECOMB 2013); Beijing, China; 2013.
34. Ding C, Peng H. Minimum redundancy feature selection from microarray gene expression data. J Bioinform Comput Biol 2005;3(2):185–205.

35. Dolinsky TJ, Czodrowski P, Li H, Nielsen JE, Jensen JH, Klebe G, Baker NA. PDB2PQR: expanding and upgrading automated preparation of biomolecular structures for molecular simulations. Nucleic Acids Res 2007;35:522–525.
36. Dong Q, Wang X, Lin L, Guan Y. Exploiting residue-level and profile-level interface propensities for usage in binding sites prediction of proteins. BMC Bioinformatics 2007;8:147.
37. Du P, Gu S, Jiao Y. PseAAC-General: fast building various modes of general form of Chou's pseudo-amino acid composition for large-scale protein datasets. Int J Mol Sci 2014;15:3495–3506.
38. Duda R, Hart P, Stork D. *Pattern Classification*. 2nd ed. New York: John Wiley and Sons, Inc.; 2000.
39. Dwivedi VD, Arora S, Pandey A. Computational analysis of physico-chemical properties and homology modeling of carbonic anhydrase from cordyceps militaris. Netw Model Anal Health Inform Bioinform 2013:1–4.
40. Eichborn JV, Günther S, Preissner R. Structural features and evolution of protein-protein interactions. Int Conf Genome Inform 2010;22:1–10.
41. Finn R, Mistry J, Tate J, Coggill P, Heger A, Pollington J, Gavin O, Gunasekaran P, Ceric G, Forslund K, Holm L, Sonnhammer E, Eddy S, ateman A. The Pfam protein families database. Nucleic Acids Res 2010;38:211–222.
42. Hall M, Maleki M, Rueda L. Multi-level structural domain-domain interactions for prediction of obligate and non-obligate protein-protein interactions. ACM Conference on Bioinformatics, Computational Biology and Biomedicine (ACM-BCB); Florida; Oct 2012. p 518–520.
43. He Z, Zheng J, Shi XH, Hu LL. Predicting drug-target interaction networks based on functional groups and biological features. PLoS ONE 2010;5:e9603.
44. Huang T, Cheng L, Cai YD. Classification and analysis of regulatory pathways using graph property, biochemical and physicochemical property, and functional property. PLoS ONE 2011;6:e25297.
45. Jones S, Thornton JM. Principles of protein-protein interactions. Proc Natl Acad Sci U S A 1996;93(1):13–20.
46. Kurareva I, Abagyan R. Predicting molecular interactions in structural proteomics. In: Nussinov R, Shreiber G, editors. *Computational Protein-Protein Interactions*. Chapter 10. CRC Press; 2009. p 185–209.
47. La D, Kong M, Hoffman W, Choi Y, Kihara D. Predicting permanent and transient protein-protein interfaces. Proteins 2013;81(5):805–818.
48. Lawrence MC, Colman PM. Shape complementarity at protein/protein interfaces. J Mol Biol 1993;234(4):946–950.
49. Liu Z, Chen L. Proteome-wide prediction of protein-protein interactions from high-throughput data. Protein Cell 2012;3(7):508–520.
50. Liu R, Jiang W, Zhou Y. Identifying protein-protein interaction sites in transient complexes with temperature factor, sequence profile and accessible surface area. Amino Acids 2010;38:263–270.
51. Liu Q, Li J. Propensity vectors of low-ASA residue pairs in the distinction of protein interactions. Proteins 2010;78(3):589–602.

52. Maleki M, Aziz M, Rueda L. Analysis of obligate and non-obligate complexes using desolvation energies in domain-domain interactions. 10th International Workshop on Data Mining in Bioinformatics (BIOKDD 2011) in Conjunction with ACM SIGKDD 2011; San Diego (CA); Aug 2011. p 21–26.
53. Maleki M, Aziz M, Rueda L. Analysis of relevant physicochemical properties in obligate and non-obligate protein-protein interactions. Workshop on Computational Structural Bioinformatics in Conjunction with BIBM 2011; 2011. p 543–547.
54. Maleki M, Hall M, Rueda L. Using structural domain to predict obligate and non-obligate protein-protein interactions. IEEE Symposium on Computational Intelligence in Bioinformatics and Computational Biology (CIBCB2012); San Diego (CA); May 2012. p 9–15.
55. Maleki M, Hall M, Rueda L. Using desolvation energies of structural domains to predict stability of protein complexes. J Netw Model Anal Health Inform Bioinform (NetMahib) 2013;2:267–275.
56. Maleki M, Vasudev G, Rueda L. The role of electrostatic energy in prediction of obligate protein-protein interactions. BMC Proteome Sci 2013;11 Suppl 1:S11.
57. Mintseris J, Weng Z. Atomic contact vectors in protein-protein recognition. Proteins 2003;53:629–639.
58. Mintseris J, Weng Z. Structure, function, and evolution of transient and obligate protein-protein interactions. Proc Natl Acad Sci U S A 2005;102(31):10930–10935.
59. Novakovic J, Strbac P, Bulatovic D. Toward optimal feature selection using ranking methods and classification algorithms. Yugosl J Oper Res 2011;21(1):119–135.
60. Ozbabacan SA, Engin H, Gursoy A, Keskin O. Transient protein-protein interactions. Protein Eng Des Sel 2011;24(9):635–648.
61. Pandit M, Rueda L. Prediction of biological protein-protein interaction types using short, linear motifs. ACM Conference on Bioinformatics, Computational Biology and Biomedicine (ACMBCB 2013); Washington (DC); 2013. p 699–700.
62. Park J, Bolser D. Conservation of protein interaction network in evolution. Genome Inform 2001;12:135–140.
63. Park SH, Reyes J, Gilbert D, Kim JW, Kim S. Prediction of protein-protein interaction types using association rule based classification. BMC Bioinformatics 2009;10(36). doi: 10.1186/1471-2105-10-36.
64. Peng H, Long F, Ding C. Feature selection based on mutual information: criteria of max-dependency, max-relevance, and min-redundancy. IEEE Trans Pattern Anal Mach Intell 2005;27(8):1226–1238.
65. Protein-Protein Interaction Prediction at Windsor (PPIPAW) [Online]. Available at http://sanjuan.cs.uwindsor.ca/ppipaw/. Accessed 2015 Aug 6.
66. Pudil P, Ferri F, Novovicova J, Kittler J. Floating search methods for feature selection with nonmonotonic criterion functions. Proceedings of the 12th international Conference on Pattern Recognition, Volume 2; 1994. p 279–283.
67. Rueda L, Banerjee S, Aziz M, Raza M. Protein-protein interaction prediction using desolvation energies and interface properties. Proceedings of the 2nd IEEE International Conference on Bioinformatics & Biomedicine (BIBM 2010); Hong Kong, China 2010. p 17–22.
68. Rueda L, Garate C, Banerjee S, Aziz M. Biological protein-protein interaction prediction using binding free energies and linear dimensionality reduction. Proceedings of the 5th IAPR International Conference on Pattern Recognition in Bioinformatics (PRIB 2010); Netherlands 2010. p 383–394.

69. Rueda L, Herrera M. Linear dimensionality reduction by maximizing the chernoff distance in the transformed space. Pattern Recognit 2008;41(10):3138–3152.

70. Schnell JR, Chou JJ. Structure and mechanism of the M2 proton channel of influenza a virus. Nature 2008;5(451):591–595.

71. Shen HB. EzyPred: a top-down approach for predicting enzyme functional classes and subclasses. Biochem Biophys Res Commun 2007;364:53–59.

72. Shen HB. Predicting protein fold pattern with functional domain and sequential evolution information. J Theor Biol 2009;256:441–446.

73. Shen HB. Quatident: a web server for identifying protein quaternary structural attribute by fusing functional domain and sequential evolution information. J Proteome Res 2009;8:1577–1584.

74. Singh DB, Gupta MK, Kesharwani RK, Misra K. Comparative docking and ADMET study of some curcumin derivatives and herbal congeners targeting ß-amyloid. Netw Model Anal Health Inform Bioinform 2013;2(1):13–27.

75. Singhal M, Resat H. A domain-based approach to predict protein-protein interactions. BMC Bioinformatics 2007;8(199). doi: 10.1186/1471–2105–8–199.

76. Skrabanek L, Saini HK, Bader GD, Enright AJ. Computational prediction of protein-protein interactions. Mol Biotechnol 2008;38(1):1–17.

77. Theodoridis S, Koutroumbas K. *Pattern Recognition.* 4th ed. Amsterdam: Elsevier Academic Press; 2008.

78. Vasudev G, Rueda L. A model to predict and analyze protein-protein interaction types using electrostatic energies. 5th IEEE International Conference on Bioinformatics and Biomedicine (BIBM 2012); 2012. p 543–547.

79. Wang J, Pielak RM, McClintock MA, Chou JJ. Solution structure and functional analysis of the influenza B proton channel. Nat Struct Mol Biol 2009;16:1267–1271.

80. Xiao X, Wang P. Predicting protein quaternary structural attribute by hybridizing functional domain composition and pseudo amino acid composition. J Appl Crystallogr 2009;42:169–173.

81. Xiao X, Wang P. Quat-2l: a web-server for predicting protein quaternary structural attributes. Mol Divers 2011;15:149–155.

82. Zaki N. Protein-protein interaction prediction using homology and inter-domain linker region information. In: *Advances in Electrical Engineering and Computational Science.* Volume 39. Springer-Verlag; 2009. p 635–645.

83. Zaki N, Lazarova-Molnar PCS, El-Hajj W. Protein-protein interaction based on pairwise similarity. BMC Bioinformatics 2009;10(150). doi: 10.1186/1471–2105–10–150.

84. Zhang Q, Petrey D, Deng L, Qiang L. Structure-based prediction of protein-protein interactions on a genome-wide scale. Nature 2012;490(7421):556–560.

85. Zhang C, Vasmatzis G, Cornette JL, DeLisi C. Determination of atomic desolvation energies from the structures of crystallized proteins. J Mol Biol 1997;267:707–726.

86. Zhong WZ, Zhou SF. Molecular science for drug development and biomedicine. Int J Mol Sci 2014;15:20072–20078.

87. Zhu H, Domingues F, Sommer I, Lengauer T. NOXclass: prediction of protein-protein interaction types. BMC Bioinformatics 2006;7(27). doi: 10.1186/1471-2105-7-27.

PART V

PATTERN RECOGNITION IN MICROARRAYS

CHAPTER 17

CONTENT-BASED RETRIEVAL OF MICROARRAY EXPERIMENTS

Hasan Oğul

Department of Computer Engineering, Başkent University, Ankara, Turkey

17.1 INTRODUCTION

Owing to recent developments in biotechnology and computational biology, there has been a rapid growth in the accumulation of biological and biochemical data in large public databases. These data may appear in distinct formats such as sequences, structures, networks, or experimental measurements. One of the largest repositories for experimental measurements is gene expression databases which usually contain the results of high-throughput microarray or similar experiments. They mainly store very large numerical matrices of experimental results in association with their textual annotations and categorical attributes, which provide brief information to database users for easy access to the experiment queried.

Accessing gene expression data sets is important for several reasons. First, these data sets provide an environment for suggesting and testing biological hypotheses. Researchers usually try to infer valuable knowledge from data set content and produce testable hypotheses to explain relevant phenomena. Therefore, the availability of previous experimental results facilitates the design and analysis of new wet-lab experiments. Second, the knowledge hidden in completed experiments might be a treasure for application of medical practices in diagnostic and prognostic

Pattern Recognition in Computational Molecular Biology: Techniques and Approaches,
First Edition. Edited by Mourad Elloumi, Costas S. Iliopoulos, Jason T. L. Wang, and Albert Y. Zomaya.

evaluation of patients with certain diseases. The experiences gained from these practices are further utilized for drug design and discovery. Third, these data sets can serve as benchmark suites for computational studies in the field. A new release of an algorithm, mathematical model, or a software tool requires a validation through ground-truth rules or real-life simulations. They usually need to be compared against existing approaches to discern their ability to perform better. Existing data sets provide an appropriate environment for these *in silico* evaluations. Not only bioinformaticians but also other technical researchers in computer science, operations research, statistics, and relevant fields benefit from these data sets in their benchmarking works. The computational studies related to gene expression data have enriched exploratory and predictive data analysis, such as clustering, classification, and regression, as well as the research relevant to big data, such as distributed database design, cloud, and parallel computing.

Obtaining a data set of a gene expression experiment can be done through the query engines provided by public databases on the Internet. Users can access the data sets by building structured queries that characterize the experiment to be conducted. A query may contain the combination of more than one attribute. Owing to several limitations of this structured approach, the latest trend in gene expression databases is to include semantic queries that can meet the user information requirements at higher levels. Since it is not a trivial task for a user to build semantic queries, the most user-friendly way is to submit an experiment instead of a query to retrieve similar entries in the collection. This scheme is called *content-based retrieval of experiments* and is the main focus of this chapter.

The rest of the chapter is organized as follows. In Section 17.2, the basic terminology and techniques in conventional information retrieval systems are introduced. Section 17.3 gives an elaborate view of the concept of content-based information retrieval. Sections 17.4 and 17.5 relate these concepts and methods of information retrieval to microarray gene expression analysis. Section 17.6 provides an introduction to the similarity metrics that would be particularly useful for comparing microarray experiments. Section 17.7 outlines briefly the methods for evaluating information retrieval performance. The existing software tools are listed with their brief descriptions in Section 17.8. The chapter concludes with Section 17.9.

17.2 INFORMATION RETRIEVAL: TERMINOLOGY AND BACKGROUND

In broad terms, information retrieval can be defined as the study of searching unstructured queries in an unstructured (or at least nonprecisely structured) collection of data. In this respect, information retrieval is concerned with the strategies for representing, storing, organizing, and accessing information objects stored in local or remote collections. Owing to its unstructured nature, a *query* in a traditional database transaction is replaced by the term *information need* in the information retrieval community. It is usually more difficult to characterize an information need than a simple structured query. It requires interpreting and conceptualizing the input need, and on the other hand, finding a proper representation for stored objects that suit the input format and conception as well. The subsequent task is to match the information need and the relevant object in the collection.

Although it has been one of the very old questions of library and archiving studies, the information retrieval concept received less attention than structured querying and database systems for many years. The explosion of the research in the field started with the rapid dissemination of information by the Internet and spread in the use of web content worldwide. Since its invention in 1991, the Internet has become a global repository of knowledge, which allows unprecedented sharing of information in a size never seen before. While improving the quality of life to some degree, the freedom in the style and content made it difficult to traverse and access the desired knowledge over the years. Beyond satisfying basic information requirements, the users have begun to expect faster and more intelligent access tools that can simplify their lives and communication with the others. With these developments, the science of information retrieval has become more popular and more sophisticated than the old-fashioned library management studies.

An effective information retrieval tool should deal with two common major problems. First, it may retrieve too many objects, of which only a small portion are relevant to the user query. Second, most relevant objects may not necessarily appear at the top of query output order. In fact, an effective information retrieval system aims at retrieving all the objects that are relevant to the user query while retrieving as few nonrelevant objects as possible. Current research on information retrieval goes beyond this primary goal and deals with several other tasks such as modeling, object categorization or classification, information visualization, user interfaces, high-performance computing, and natural language processing. In this manner, it intersects with several other research fields such as data mining, machine learning, and pattern recognition.

A key distinction should be made between *data retrieval* and *information retrieval*. Data, information, and knowledge are three encompassing terms to explain the flow of concepts in representing and understanding things. Data is the lowest unit of concepts. It involves raw measurements of entities. Information is the data that is processed to be useful. Knowledge, on the other hand, is a more structured and high-level representation of information that is used for reasoning and inference. In that sense, data retrieval and information retrieval refer to different tasks. Data retrieval is the process of finding and reporting all objects that satisfy clearly the defined conditions. These conditions are usually given to describe the data in terms of regular or relational algebra expression. The user does not deal with what is meant by retrieved data but its correctness. On the other hand, the information retrieval task is concerned with the relevance, not correctness, of the objects retrieved from the collection. Its primary goal is to explain the meaning of the retrieved object to the user. It is not well structured but contains a level of semantic value.

Information retrieval systems can be distinguished into *Boolean retrieval* and *ranked retrieval* systems. In Boolean systems, the user is allowed to specify his information need in more complex queries in order to get a single best-fit solution. Although Boolean retrieval has an old tradition, it has several shortcomings. First, it has no inherent notion of the relevance, that is, the single outcome is reported as a unique confident result. Second, it is not always possible to form a good structured request by users. The information need is often described by ambiguous semantic requirements. In ranked retrieval, on the other hand, the system returns a list of relevant results ordered by their confidence level on the relevance between query and the outcome. How to score this confidence is another challenging question in information retrieval.

Another issue in distinguishing between information retrieval systems is the *scale* at which they operate. The largest scale is the web. A *web retrieval system* has to search over a tremendous collection of objects in millions of computers. In addition to the crucial need for very efficient indexing techniques, web retrieval systems should deal with the hierarchy and linked structure of the documents. The second class, *personal information retrieval systems* target the documents contained in personal computers. These may include personal files, emails, logs, malicious contents, and so on. The third class can be referred as *domain-specific retrieval systems* in between the former two. Here, the retrieval is performed over a specified collection, either in a local computer or remote locations such as the web or a distributed system. The collection usually comprises a specific type of documents or objects. Therefore, such systems require building of domain-specific index structures and matching strategies. The confidence of relevance should also be defined accordingly.

A key concern in an information retrieval system is *indexing*. It is used to traverse a collection to locate the relevant object to the queried one in a faster way than a brute-force approach. In its naive form, an *index* is a data structure containing the terms or concepts with which are associated pointers to the related object. In traditional information retrieval for text documents, each term corresponds to a word appearing in at least one of the documents in the collection. Since it is not an efficient way to include all possible words in the collection, an appropriate index should include a representative set of keywords that can be deduced by various operations such as stemming, compression, and feature selection. In a trade-off between efficiency and representativeness, creating an optimal index is one of the challenging questions in the information retrieval field. A good indexing is expected to improve the quality of the answer set of an information need. Since the ultimate goal is to satisfy the user's requirement at a higher semantic level, a computationally feasible index is not sufficient to represent the objects in a collection in the long run. Current studies on information retrieval focus also on human interaction and try to develop strategies to enhance the tools for better understanding user behavior.

17.3 CONTENT-BASED RETRIEVAL

Until recently, information retrieval was considered to be a science of texts, in which a document was stored, queried, and accessed by its text content. This trend had been tandem with the happenings and developments in the Internet world. Along with the immense accumulation of other types of data over the Internet, the main focus of information retrieval researchers has shifted from text data to other formats, such as image, audio, and video [21]. Traditional approaches for solely textual information retrieval have been adopted into multimedia and other formats in a variety of ways. It has been, of course, inevasible to develop additional domain-specific models that fit into the nature of new data types.

Multimedia information retrieval is the most attractive field of study as opposed to traditional information retrieval studies. Here, a document is replaced by a multimedia object and the methods used in document information retrieval are transplanted into the new object by redefining the concept of term appearing in these sorts of files. This is not a trivial task, however. There are two alternatives to extract a set of terms from multimedia files. The first approach is to use the textual annotations about the object. These annotations, also called *meta-data*, may contain brief information about the object content or its environment, which are usually provided by the people producing or uploading the file. If the user having a retrieval request is interested in this brief information, the retrieval task is analogous to typical document retrieval as the meta-data annotation can simply be regarded as a text document. This is not usually the case. The user may ask to find details inside the multimedia object just the way he/she perceives when he/she looks or hears. In this stage, *content-based retrieval* comes into prominence. This term is used for the tasks that retrieve the objects using its actual content rather than meta-data annotations. There is no straightforward way of vectorizing a multimedia file to an index using a set of terms. Even if a term concept is abstracted to represent some part of the multimedia content using some domain-specific data exploration techniques, the information need of a user cannot be adopted to a query by these abstract definitions. The latest trend to overcome these problems is *query-by-example* retrieval. In this approach, the user query is maximally abstracted by requesting an example that would resemble the object to be retrieved. The information retrieval framework is then responsible for vectorizing both the query and the objects in the collection. The remaining task is to match the abstract views of the query and the objects in the collection (Figure 17.1). The user is, on the other hand, not aware of what kind of search is being performed behind the system.

Visual objects, such as image and video, contain more versatile components in comparison with text data. Content-based image retrieval methods attempt to extract low-level visual characteristics that may be defined by gradient, color, shape, texture, or spatial relationships and combine them with higher level characteristics inferred from semantic analysis of more general conceptual features of data. In video, sequential attributes of contiguous images are also incorporated. In spite of a great effort, the field is still considered to be in its infancy as satisfactory performance that is close to human perception is yet to be achieved. For audio data, the situation is worse. The characterization of visual data requires low-level signal processing techniques and the real human perception can only be abstracted by high-level features extracted from low-level signal characteristics.

While text documents and multimedia objects dominate the current web content, some other special data types are also subject to information retrieval. Scientific articles, for example, have a special format to be reconsidered for information retrieval purposes. Recommendation systems operate with a strategy of retrieving similar users from the collection of previous user-item rating matrices. Here, the document contains integer matrices instead of text data. A challenging information retrieval

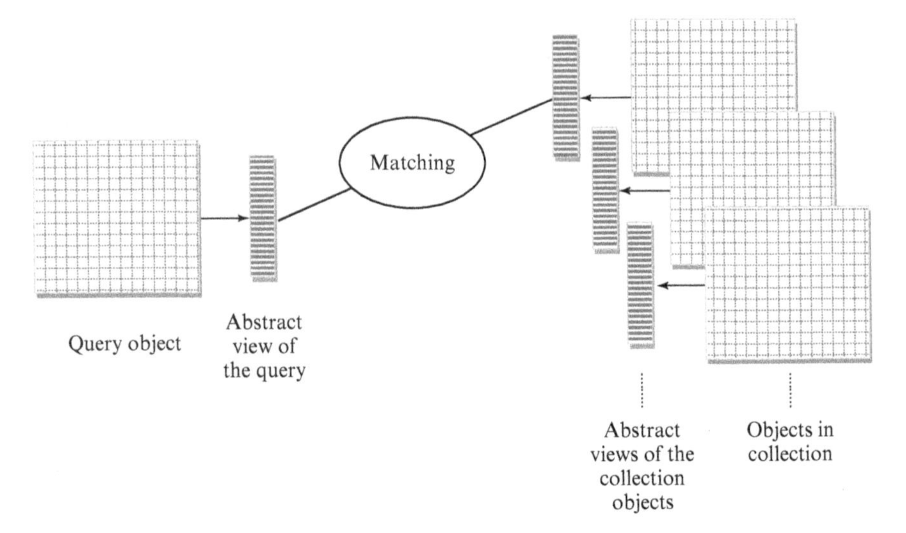

Figure 17.1 Query-by-example framework for content-based information retrieval.

request has recently arisen in the biosciences. Research in biological and biochemical sciences has produced an enormous amount of data accumulated in public databases. Two major data types in this respect are sequences, such as DNA, RNA, and proteins, and gene expression measurements, usually obtained by microarray experiments. While they both need domain-specific treatment for information retrieval studies, we focus on the latter in the following sections.

17.4 MICROARRAY DATA AND DATABASES

Microarray technology enables measuring the expression levels of thousands of genes simultaneously. It has become a major platform for wet-lab analysis of gene expression and regulation since its invention. It is still most widely used tool in spite of several recent developments in related technology such as RNA-Seq [23].

A microarray experiment results from gene expression data in two dimensions. The data typically form a numerical matrix with rows corresponding to the genes under consideration, and columns representing the conditions under which the experiment is done. Each cell quantifies the expression level of a gene in a given condition (see Figure 17.2). The analysis of a gene expression matrix is a challenging task owing to numerous reasons. First, data may contain noise or fluctuations because of platform differences, environmental conditions, and biological instability. Therefore, more than one replicate is used in the experiment to get more reliable results. The second problem is that the data are large in size. In many situations, only a subset of genes, for example, those that are differentially expressed, contribute to the results inferred from a computational analysis. Third, the biological knowledge inherited in one matrix probably repeats in some others. These cross-correlations

Gene expression matrix	Condition 1	Condition 2	......	Condition j	......	Condition m
Gene 1						
Gene 2						
⋮	⋮	⋮	⋮	⋮	⋮	⋮
Gene i				e_{ij}		
⋮	⋮	⋮	⋮	⋮	⋮	⋮
Gene n						

Figure 17.2 Format of a gene expression matrix: the expression of ith gene in jth condition is shown by e_{ij}.

might be related to the conservation between distinct species, experimental factors, and disease or treatment similarities among others. Therefore, the gene expression matrix has attracted computational scientists to design and develop more reliable and efficient data analysis methods.

A major direction enriched in gene expression analysis is *clustering*. From an information and computer science point of view, it is defined as partitioning related objects into previously unlabeled groups. For a gene expression matrix, this can emerge in two ways: clustering genes or clustering conditions [26]. A condition may refer to a distinct subject for which an experiment is performed. Therefore, a clustering may result, for example, with grouping related people on the basis of behavior of their genes. A more attractive application is clustering of genes. In this case, the genes having similar profiles over a set of subjects or conditions are grouped into coherent sets. A gene group may indicate a similarity in their functions, biological processes and pathways included, or cellular localizations. A recent extension of clustering is called *biclustering*, where the genes and conditions are partitioned simultaneously so that the group of genes that behave similarly under a subset (not all) of conditions can be identified [22].

Another computational focus in gene expression analysis is *classification*. In classification, an object is assigned to one of the previously labeled categories. The most common way of doing this is to employ a supervised learning strategy, where a model that maps predefined numerical features extracted from a sample to one of known labels is learned through a training set. In microarray data, this task usually refers to assigning a subject of study to one of biological classes often labeled by disease types and subtypes in human experiments. This is considered to be a promising predictive tool for diagnostic purposes in biomedicine [30].

Since the microarray expression data has a large number of gene entries, *gene selection* is another crucial issue in gene expression analysis studies. It is defined as selecting a representative subset of all genes such that the gene expression matrix comprises a significantly lowered number of rows thereafter. Although the primary goal of gene selection is to reduce the size of data studied for better efficiency, the task also comes in handy for biological knowledge discovery. In a cancer-related microarray experiment, for example, the user can investigate the genes associated with the cancer type in question. The problem can be approached by variety of ways, such as wrapping and filtering techniques [13].

The number of public repositories for gene expression studies has been increasing. Two of them are most popular: GEO (Gene Expression Omnibus) [1] and ArrayExpress [24]. GEO is a National Center of Biotechnology Information (NCBI) utility, which can be accessed from http://www.ncbi.nlm.nih.gov/geo/. It provides diverse facilities to help the users for storing, accessing, downloading, and visualizing curated gene expression data produced in different platforms and technologies. It is a large public repository of standardized biological data [2]. ArrayExpress is a counterpart of GEO provided and maintained by the European Bioinformatics Institute (EBI). It can be accessed from http://www.ebi.ac.uk/arrayexpress/. Several other gene expression collections are available on the Internet with generic or specified content. The Arabidopsis Information Resource (TAIR), for example, maintains a repository of biological data specific to model higher plant *Arabidopsis thaliana*, including the results of microarray experiments done by different research laboratories [25]. TAIR can be accessed from http://www.arabidopsis.org/.

17.5 METHODS FOR RETRIEVING MICROARRAY EXPERIMENTS

Many of the microarray databases are public to serve an *in silico* laboratory for researchers working in related fields [1, 18]. Everyone can freely benefit from the results of others in designing experiments or putting forth new hypotheses. Querying a microarray experiment in any of the above databases broadly involves the following steps:

1. A set of attributes among those provided by the database tool is selected to determine the search scope. These attributes might be the name of the organism, platform type, authors, citation information, description, and so on.
2. The values (or the ranges) of the selected attributes are fixed to describe the main characteristics of the experiment to be searched. The attribute values can be combined to build a hybrid query using logical operators such as AND and OR.
3. Matching experiments are returned in rank by the order of match quality based on the query attributes and the attributes of database entries. While a perfect match is obligated for many of the attribute types, such as organism and author, some more approximate matches can be allowed in performing comparison between attribute values, such as asking for the occurrence of queried keywords in the description part.

This kind of database query is referred to as a *metadata-based search* and poses several limitations. The user information needs may not be met by the possible query combinations that the database engine provides. A structural query limits the flexibility of the user to build semantic needs. Keyword-based search over free text descriptions may provide flexibility to some degree. However, text annotations are not always fully descriptive and sometimes they are not available at all. Furthermore, the text descriptions tell something about the design and implementation of the experiments, but not anything about the results and biological implications of the experiment. These limitations suggest the use of content-based information retrieval techniques applied for multimedia objects in a similar manner. Since it is usually not possible to build semantic content-based information needs in a structural query, the best way to retrieve a microarray experiment is Query by Example, where an experiment on hand is submitted to get similar entries in the collection in terms of their knowledge inherited in the numerical results. The idea of content-based search of microarray experiments was first introduced in Reference [17]. A few groups have used similar ideas in their studies to produce new biological hypotheses by associating their experiments with similar previous experiments. Their results confirm the usability of the approach in practice [6, 14, 15, 19, 29].

Content-based retrieval of microarray experiments requires constructing fingerprints that represent the experiment content in a computational framework. The *fingerprint* here refers analogously to the term of *index* in traditional information retrieval with some domain-specific differences. The comparison of a query experiment is done through its fingerprint over the other fingerprints present in the database. Therefore, a successful implementation of a content-based search strategy should first deal with how to derive a representative fingerprint from given experiment and second, how to compare two fingerprints in an effective and efficient way.

Given a gene expression matrix E, where e_{ij} represents the expression of the ith gene in jth condition, the problem of relevant experiment retrieval is defined as finding the matrix M_k among the collection of matrices $\{M_1, M_2, \ldots, M_t\}$ in the microarray repository, where k achieves the lowest distance $d(E, M_k)$. In other words, the retrieval process involves a trivial task of comparing distances between the query matrix and others in the databases, and reporting the one that attains the lowest score. The problem can also be defined as a ranking problem where some of top experiments are returned in an ascending order by their distance scores with the query. It is often a more plausible output as we usually need similar experiments but not only the most similar one. Regardless of the output format, the nontrivial task here is to infer distance $d(X,Y)$ for any given matrices X and Y. Making a one-to-one comparison between the matched cells is not sensible as the cells are not usually paired. Each experiment in a gene expression database is performed on different condition sets. Therefore, there is no match between column labels of any two matrices in the database. The rows, representing the probes in a microarray chip (or the gene-involved experiments in essence), also differ in each experiment owing to the organism in question or the experimental design. However, it is expected to have a large consensus between the row labels when both experiments are performed on the same organism. The evolutionary homology between the organisms makes the similarity between the experiments performed on them more feasible. Therefore, it is reasonable to assume that the all rows (or a selected subset of them) between compared matrices are paired in a computational model. Then, the information retrieval model can be built on matching the logical views of overall gene expression profiles of two matrices. The logical view can be designed as a single-dimension vector representing the gene behaviors over the conditions pertaining to each individual experiment.

With the basic assumptions above, the problem of comparing two gene expression matrices can be turned into a two-stage process: (i) reducing each matrix into a one-dimensional vector, that is, a fingerprint, preferably of length lower than the number of genes and (ii) comparing their representative vectors in one-dimensional space to evaluate their relevance.

In the literature, we encounter two main approaches for reducing a gene expression matrix into a representative vector. The first idea is to use differential expression profiles [7, 16]. A gene is said to be *differentially expressed* if its expression is significantly altered between two conditions. The fold change between two conditions is called a *differential expression* and is qualified by a single numeric value. The differential expression of a gene is the most straightforward and effective way of describing its behavior, especially for the experiments performed on control and treatment groups. Its generalization for the whole gene expression matrix is referred as the *differential expression profile* of the experiment. The differential expression profile can be simply obtained by concatenating the differential expression of all genes in the experiment into a single vector. Having a differential expression profile for both query and each of the experiments in the repository, the database search can be done over these profiles instead of the expression matrices themselves (see Figure 17.3). This strategy corresponds to the concept of indexing in traditional information retrieval where the frequency of each word in the corpus is evaluated instead of the differential expressions of the genes. To increase computational efficiency, [7] suggested using a dimension reduction technique based on Independent Component Analysis such that the resulting vector has a dimension that is lower than gene count. Another suggestion is to use binary fingerprints that represent only up- or downregulation of genes [3].

A second approach for modeling a gene expression matrix in an information retrieval context is to utilize gene sets instead of individual genes. A gene set is a group of genes labeled for a specific biological function, process, or evolutionary family. A matrix can be represented by the enrichment of predefined gene sets and the representative vector is indexed by the labels of these gene sets. An obvious advantage of this model over differential expression profiles is the profitability of memory space, and eventually the computational efficiency in comparison and database traversal stages. The knowledge contained in gene set profiles further provides a more semantic interpretation to the system. A classical yet powerful method that can be used to form this representation is the so-called Gene Set Enrichment Analysis (GSEA). GSEA is a data exploration and semantic annotation technique for gene expression data [27]. There exist only a couple of examples using GSEA or similar techniques for content-based microarray experiment retrieval in the

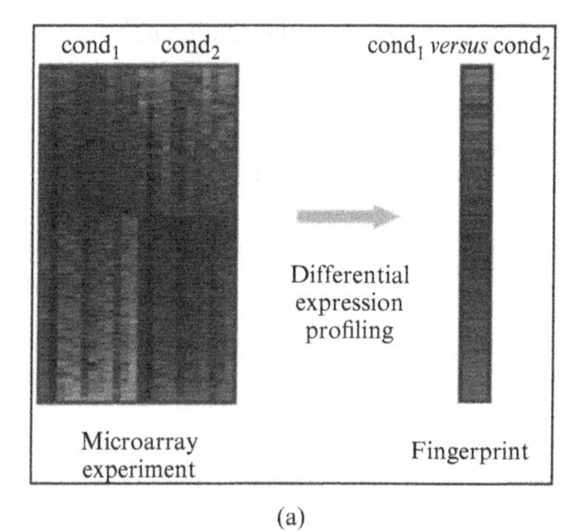

(a)

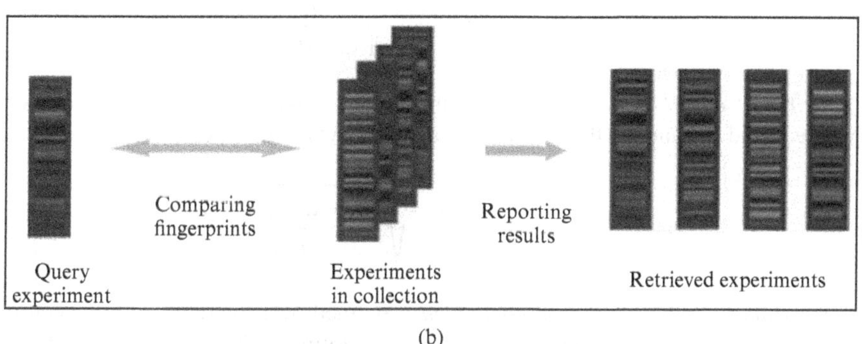

(b)

Figure 17.3 Microarray experiment retrieval framework based on differential expression fingerprints: (a) fingerprint extraction, (b) database search and ranked retrieval.

literature [4, 11, 28]. A major limitation of this approach is the difficulty of finding a reliable gene set collection on which the enrichment analysis can be performed.

17.6 SIMILARITY METRICS

Comparing two entities is an old question in science. It answers the question in what degree they are similar or different. Comparison of two entities represented by two distinct vectors of numerical values is made over a similarity or distance metric. Although such metrics are diverse, we describe here the metrics that are more tailored for comparing the fingerprints of gene expression experiments. In a formal notation, given two distinct experiments represented by their fingerprints A and B of length n, the distance between them, formulated by $d(A,B)$, is calculated by individual entries A_i and B_i, where i lies between 1 and n.

For the vectors having real numbers, geometric and correlation-based distance metrics are frequently used, which are also applicable for microarray fingerprints based on differential expression profiles. The *Euclidean distance* is the common distance metric:

$$d_{\mathrm{Euclidean}}(A,B)=\sqrt{\sum_{i=1}^{n}(A_i-B_i)^2}$$

The *Manhattan distance* is another geometric measure given by

$$d_{\mathrm{Manhattan}}(A,B)=\frac{1}{n}\sum_{i=1}^{n}|A_i-B_i|$$

The *Cosine distance* defines the angle between two vectors, which in turn qualifies a distance between them:

$$d_{\mathrm{Cosine}}(A,B)=1-\frac{A\cdot B}{\|A\|\cdot\|B\|}$$

Pearson's correlation coefficient is a measure of correlation between two vectors (where μ and σ stand for the mean and standard deviation, respectively):

$$d_{\mathrm{Pearson}}(A,B)=1-\frac{1}{n-1}\sqrt{\frac{(A_i-\mu_A)\,(B_i-\mu_B)}{\sigma_A\sigma_B}}$$

Spearman's rank correlation coefficient is defined over the ranks of two vectors. The rank vectors of two fingerprint vectors are compared by Pearson's correlation coefficient:

$$d_{\mathrm{Spearman}}(A,B)=d_{\mathrm{Pearson}}(\mathrm{Rank}(A),\mathrm{Rank}(B))$$

When the vectors have binary entries, as in binary fingerprints for microarray experiments, diverse metrics are used. The *Hamming distance* is a metric borrowed from information theory and adopted for binary fingerprints:

$$d_{\mathrm{Hamming}}(A,B)=1-\frac{\sum_{i=1}^{n}(A_i\wedge B_i)+\sum_{i=1}^{n}(A_i'\wedge B_i')}{n}$$

The *Tanimoto distance* is another applicable metric that has proven to be more favorable in comparing binary gene expression fingerprints [3]:

$$d_{\mathrm{Tanimoto}}(A,B)=1-\frac{\sum_{i=1}^{n}(A_i\wedge B_i)}{\sum_{i=1}^{n}(A_i\vee B_i)}$$

A recent study also suggests the use of adaptive metrics instead of fixed distance functions [20]. An example of an adaptive metric is the Weighted Rank Metric (WRM) given by

$$d_{\mathrm{WRM}}(A,B) = \sqrt{\sum_{i=1}^{n} (w_i(\mathrm{Rank}(A)_i) - w_i(\mathrm{Rank}(B)_i))^2}$$

The vector w comprises the weight of each rank, that is, the significance of each rank in measuring the relevance of two experiments. Adaptive methods require a parameter tuning step but are proven to be more appropriate for expression data [20] as the contribution of all genes or gene sets are not equal in the development of local biological processes.

17.7 EVALUATING RETRIEVAL PERFORMANCE

The performance of an information retrieval system can be evaluated in a variety of techniques borrowed from statistics and machine learning sciences. All are applicable for microarray experiment retrieval. In a binary retrieval system, the performance can be evaluated by a set of scalar measures computed using the number of True Positive (TP), True Negative (TN), False Positive (FP), and False Negative (FN) predictions. A true positive indicates that the microarray experiment retrieved by the system is one of the experiments requested by the user. Each experiment in the collection that is not returned by the system and not requested by the user either is called a *true negative*. A false positive refers to a retrieved experiment that is not required by the user, while a false negative is a relevant object missed by the retrieval system. *Accuracy* is the most straightforward measure in data mining and related sciences; it is defined as the ratio of correct predictions (sum of TP and TN) to all samples in the collection. However, it does not make any sense for assessing a microarray retrieval system as the collection contains a large number of experiments that are not relevant to the user need, that is, an imbalance between positive and negative samples. A high accuracy is eventually achieved even by returning nothing to the user. Two common evaluation metrics for information retrieval studies are *recall* and *precision*. *Recall* is the ratio of the number of relevant experiments retrieved to the total number of all relevant experiments in the collection. It is given by the formula:

$$\mathrm{Recall} = \frac{\mathrm{TP}}{\mathrm{TP} + \mathrm{FN}}$$

Precision is the ratio of the number of relevant experiments retrieved to the total number of experiments retrieved. It is defined by:

$$\mathrm{Precision} = \frac{\mathrm{TP}}{\mathrm{TP} + \mathrm{FP}}$$

Ideally, both recall and precision are desired to be high, that is, an ideal system should retrieve as many relevant experiments as possible (i.e., high recall) and very few nonrelevant experiments (i.e., high precision). In practice, allowing more concession in candidate experiments yields a higher recall at the cost of precision. Therefore, a compromise should be struck in reality. Indeed, a better precision is usually more favorable.

In a ranked retrieval system, simple variants of recall and precision can be used. For example, a modified precision can be adopted for some top percentage of retrieved experiments as the evaluation measure. In this case, the denominator in the formula above is replaced by the number of experiments in this top list. Alternatively, a precision can be measured at a fixed recall level (e.g., 70%). A generalization of this approach is average precision, which is another scalar measure computed by averaging over the precision values at different recall values.

Some measures are developed to evaluate and qualify the trade-off between precision and recall in information retrieval, which are also convenient for assessing the quality of ranked retrieval results. Receiving Operating Characteristic (ROC) curve is tailored for this objective although it is designed to illustrate the performance of a binary classification system as its discrimination threshold is varied. It is created by plotting the fraction of TPs out of all positives versus the fraction of FPs out of all negatives at various threshold settings. In a test of ranked information retrieval system, the ROC curve evaluation model is transplanted by using various rank levels instead of different threshold settings in drawing TP versus FP plot. Among several instances of ROC curves, the best one can be deduced by checking the area under them. This yields a new evaluation metric simply called Area Under Curve (AUC). If we attach importance to retrieve the relevant experiments in earlier hits, the area under the curve can be computed for a top list, say, the first N experiments, and the resulting metric is then referred to as AUC_N.

Another tool for analyzing the performance of a retrieval system is the Cumulative Match Characteristic (CMC) curve, which was originally proposed and commonly used for evaluating biometric identification systems. A CMC curve depicts the increase in the retrieval success of the system with increase in the rank before which a correct match is obtained. It is more convenient when the user needs a specific experiment that should be returned in the top of the resulting list, while the ROC curve is more useful in evaluating the plenty of relevant experiments in the same list. The AUC analysis is still applicable for CMC curves in the same way.

17.8 SOFTWARE TOOLS

Since the content-based microarray experiment retrieval is still its infancy, there are not too many available tools that can work in practice. Here, we briefly introduce a few software applications available on the web in the chronological order of their publications.

CellMontage [10] is the first software application that can search a query experiment in a large repository (69,000 microarray experiments from GEO). It uses Spearman's rank correlation coefficient to compare the differential expression profile of the query with the other such profiles in the collection. The test results have shown that CellMontage can retrieve experiments from same cell lines to a certain extent. A disadvantage of the tool is that the long time taken to retrieve an experiment when the query contains many genes (>10,000).

GeneChaser [5] is a web server that enables a user without any bioinformatics skill to identify the conditions in which a list of genes is differentially expressed. While being favorable owing to its high usability, it can only allow to search up to hundreds of genes at a time.

GEM-TREND [9] is another web service for mining gene expression data sets. The input to GEM-TREND can be a query gene-expression fingerprint having up- or downregulated labels or in the form of gene expression ratio data, which is then translated into the same fingerprint format. It can accept only up to 500 genes in query. GEM-TREND serves also as a tool for network generation and cluster analysis for retrieved gene expression profiles. Its major limitations are the maximum limit on the number of genes in the query and the requirement for preprocessing the input query to obtain a fingerprint in the accepted format.

ProfileChaser [8] is the web implementation of highly accessed and cited methodology introduced by Engreitz et al. [7]. The fingerprint is solely based on differential expression. However, a remarkable speed-up is achieved through dimension reduction by selecting a subset of genes in fingerprint construction in database search. ProfileChaser can perform an online search through the current version of GEO and directly retrieve the relevant experiment from GEO's repository. It is currently the unique example of an online query-by-example microarray retrieval system over GEO.

openSESAME [12] is similar to GEM-TREND in that it requires a list of genes as the input query to search the experiments having similar experimental conditions, diseases, or biological states in its collection obtained from GEO. Although it has a similar objective, it does not account for an exact query-by-example retrieval system.

SPIEDw [31] is another differential-expression-based microarray search engine that performs on its own database. The database contains approximately 200,000 entries for microarray experiments collected from several resources. The software can query a gene list with their expression values possibly relative to a control group in its collection to retrieve the relevant experiment having similar gene expression values.

17.9 CONCLUSION AND FUTURE DIRECTIONS

Large and exponentially growing compendium of public repositories for gene expression experiments offer extensive opportunities for bioscience researchers in discovering new knowledge and going beyond what they envisaged in their current experiments. The Internet makes it easy to access the data for everything in our daily and academic life. The technology does not make any sense without providing automated and user-friendly interfaces to those without any computational skills. In this respect, a content-based search platform serves as an intelligent environment for creating semantic queries with minimal human intervention and manual processing.

It makes the interpretation easier and enhances the human perception in retrieved information.

In parallel to the exponential growth in the databases, we anticipate an increase in the number of attempts to develop more accurate and efficient systems for microarray experiment retrieval in the following decade. A major problem is that using only the differential expression profiles in modeling experiments leads to serious information loss in terms of the biological phenomena hidden behind the large data matrices. Novel encoding techniques are needed to build more effective fingerprints that can better explain biological implications of the experiment. The second challenge is the big data. The enormous size of database contents makes it difficult to traverse whole collection in a reasonable time period. Therefore, the incorporation of high-performance computing techniques should also be considered in practical applications in addition to algorithmic and modeling issues. An improvement in computational efficiency can be achieved through data-parallel implementations or faster data structures for indexing.

ACKNOWLEDGMENT

This study was supported by the Scientific and Technological Research Council of Turkey (TUBITAK) under the Project 113E527.

REFERENCES

1. Barrett T, Suzek TO, Troup DB, Wilhite SE, Ngau WC, Ledoux P, Rudnev D, Lash AE, Fujibuchi W, Edgar R. NCBI GEO: mining millions of expression profiles—database and tools. Nucleic Acids Res 2005;33:D562–D566.
2. Barrett T, Wilhite SE, Ledoux P, Evangelista C, Kim IF, Tomashevsky M, Marshall KA, Phillippy KH, Sherman PM, Holko M, Yefanov A, Lee H, Zhang N, Robertson CL, Serova N, Davis S, Soboleva A. NCBI GEO: archive for high-throughput functional genomic data. Nucleic Acids Res 2009;37:D885–D890.
3. Bell F, Sacan A. Content based searching of gene expression databases using binary fingerprints of differential expression profiles. 7th International Symposium on Health Informatics and Bioinformatics (HIBIT), 107–113, 2012.
4. Caldas J, Gehlenborg N, Faisal A, Brazma A, Kaski S. Probabilistic retrieval and visualization of biologically relevant microarray experiments. Bioinformatics 2009;25:i145–i153.
5. Chen R, Mallelwar R, Thosar A, Venkatasubrahmanyam S, Butte AJ. GeneChaser: identifying all biological and clinical conditions in which genes of interest are differentially expressed. BMC Bioinf 2008;9:548.
6. Dudley JT, Tibshirani R, Deshpande T, Butte AJ. Disease signatures are robust across tissues and experiments. Mol Syst Biol 2009;5:307.
7. Engreitz JM, Morgan AA, Dudley JT, Chen R, Thathoo R, Altman RB, Butte AJ. Content-based microarray search using differential expression profiles. BMC Bioinf 2010;11:603.

8. Engreitz JM, Chen R, Morgan AA, Dudley JT, Mallelwar R, Butte AJ. ProfileChaser: searching microarray repositories based on genome-wide patterns of differential expression. Bioinformatics 2011;27:3317–3318.
9. Feng C, Araki M, Kunimoto R, Tamon A, Makiguchi H, Niijima S, Tsujimoto G, Okuno Y. GEM-TREND: a web tool for gene expression data mining toward relevant network discovery. BMC Genom 2009;10:411.
10. Fujibuchi W, Kiseleva L, Taniguchi T, Harada H, Horton P. CellMontage: similar expression profile search server. Bioinformatics 2007;23:3103–3104.
11. Georgii E, Salojärvi J, Brosché M, Kangasjärvi J, Kaski S. Targeted retrieval of gene expression measurements using regulatory models. Bioinformatics 2012;28:2349–2356.
12. Gower AC, Spira A, Lenburg ME. Discovering biological connections between experimental conditions based on common patterns of differential gene expression. BMC Bioinf 2011;12:381.
13. Guyon I, Weston J, Barnhill S, Vapnik V. Gene selection for cancer classification using support vector machines. Mach Learn 2002;46:389–422.
14. Hassane DC, Guzman ML, Corbett C, Li X, Abboud R, Young F, Liesveld JL, Carroll M, Jordan CT. Discovery of agents that eradicate leukemia stem cells using an in silico screen of public gene expression data. Blood 2008;111:5654–5662.
15. Hibbs MA, Hess DC, Myers CL, Huttenhower C, Li K, Troyanskaya OG. Exploring the functional landscape of gene expression: directed search of large microarray compendia. Bioinformatics 2007;23:2692–2699.
16. Horton PB, Kiseleva L, Fujibuchi W. RaPiDS: an algorithm for rapid expression profile database search. Genome Informatics 2006;17:67–76.
17. Hunter L, Taylor RC, Leach SM, Simon R. GEST: a gene expression search tool based on a novel Bayesian similarity metric. Bioinformatics, 17:S115-S122, 2001.
18. Ivliev AE, 't Hoen PA, Villerius MP, den Dunnen JT, Brandt BW. Microarray retriever: a web-based tool for searching and large scale retrieval of public microarray data. Nucleic Acids Res 2008;36:W327–W331.
19. Lamb J, Crawford ED, Peck D, Modell JW, Blat IC, Wrobel MJ, Lerner J, Brunet JP, Subramanian A, Ross KN, Reich M, Hieronymus H, Wei G, Armstrong SA, Haggarty SJ, Clemons PA, Wei R, Carr SA, Lander ES, Golub TR. The Connectivity Map: using gene-expression signatures to connect small molecules, genes, and disease. Science 2006;313:1929–1935.
20. Le HS, Oltvai ZN, Bar-Joseph Z. Cross-species queries of large gene expression databases. Bioinformatics 2010;26:2416–2423.
21. Lew MS, Sebe N, Djebera C, Jain R. Content-based multimedia information retrieval: State of the art and challenges. ACM T Multimed Comput Commun Appl 2006;2:1–19.
22. Madeira S, Oliveria AL. Biclustering algorithms for biological data analysis: a survey. IEEE Trans Comput Biol Bioinf 2004;1:24–45.
23. Metzker M. Sequencing technologies – the next generation. Nat Rev Genet 2010;11:31–46.
24. Rocca-Serra P, Brazma A, Parkinson H, Sarkans U, Shojatalab M, Contrino S, Vilo J, Abeygunawardena N, Mukherjee G, Holloway E, Kapushesky M, Kemmeren P, Lara GG, Oezcimen A, Sansone SA. ArrayExpress—a public database of microarray experiments and gene expression profiles. Nucleic Acids Res 2007;35:D747–D750.

25. Rhee SY, Beavis W, Berardini TZ, Chen G, Dixon D, Doyle A, Garcia-Hernandez M, Huala E, Lander G, Montoya M, Miller N, Mueller LA, Mundodi S, Reiser L, Tacklind J, Weems DC, Wu Y, Xu I, Yoo D, Yoon J, Zhang P. The Arabidopsis Information Resource (TAIR): a model organism database providing a centralized, curated gateway to Arabidopsis biology, research materials and community. Nucleic Acids Res 2003;31:224–228.
26. Shannon W, Culverhouse R, Duncan J. Analyzing microarray data using cluster analysis. Pharmacogenomics 2003;4:41–52.
27. Subramanian A, Tamayo P, Mootha VK, Mukherjee S, Ebert BL, Gillette MA, Paulovich A, Pomeroy SL, Golub TR, Lander ES, Mesirov JP. Gene set enrichment analysis: a knowledge-based approach for interpreting genome-wide expression profiles. Proc Natl Acad Sci USA 2005;102:15545–15550.
28. Suthram S, Dudley JT, Chiang AP, Chen R, Hastie TJ, Butte AJ. Network-based elucidation of human disease similarities reveals common functional modules enriched for pluripotent drug targets. PLoS Comput Biol, 6:e1000662, 2010.
29. Tanay A, Steinfeld I, Kupiec M, Shamir R. Integrative analysis of genome-wide experiments in the context of a large high-throughput data compendium. Mol Syst Biol 2005;1:2005.0002.
30. Furey TS, Cristianini N, Duffy N, Bednarski DW, Schummer M, Haussler D. Support vector machine classification and validation of cancer tissue samples using microarray expression data. Bioinformatics 2000;16:906–914.
31. Williams G. SPIEDw: a searchable platform-independent expression database web tool. BMC Genom 2013;14:765.

CHAPTER 18

EXTRACTION OF DIFFERENTIALLY EXPRESSED GENES IN MICROARRAY DATA

Tiratha Raj Singh[1], Brigitte Vannier[2], and Ahmed Moussa[3]

[1]*Biotechnolgy and Bioinformatics Department, Jaypee University of Information and Technology (JUIT), Solan, Himachal Pradesh, India*
[2]*Receptors, Regulation and Tumor Cells (2RTC) Laboratory, University of Poitiers, Poitiers, France*
[3]*LabTIC Laboratory, ENSA, Abdelmalek Essaadi University, Tangier, Morocco*

18.1 INTRODUCTION

Gene expression profiling has become an invaluable tool in functional genomics. DNA microarrays have revolutionized the study of gene expression and are now a staple of biological inquiry. By using the microarray, it is possible to observe the expression-level changes in tens of thousands of genes over multiple conditions, all in a single experiment. Microarrays find applications in gene discovery, disease diagnosis (e.g., *Differentially Expressed* (DE) genes may be implicated in cancer), drug discovery, and toxicological research. In a typical eukaryotic gene expression profiling experiment, a mixture of labeled cRNA molecules isolated from a tissue of interest is hybridized to the array. The array is then scanned to determine fluorescent intensity at each probe set, which must be converted to a measure of relative transcript abundance.

Pattern Recognition in Computational Molecular Biology: Techniques and Approaches,
First Edition. Edited by Mourad Elloumi, Costas S. Iliopoulos, Jason T. L. Wang, and Albert Y. Zomaya.

A common task in analyzing microarray data is to determine which genes are DE across two kinds of tissue samples or samples collected from two different experimental conditions. Selection of DE genes across multiple conditions is one of the major goals in many microarray experiments [7]. A wide variety of statistical methods have been employed to analyze the data generated in experiments by using *Affymetrix* GeneChips and/or cDNA Chips for several species including rat and human [20, 22]. Even if it is important to understand the relative performance of these methods in terms of accuracy in detecting and quantifying relative gene expression levels and changes in gene expression, all these techniques suffer from errors in accuracy [5]. The probability that a false identification (type I error) is committed can increase sharply when the number of tested genes gets large. Correlation between the statistical tests attributed to gene coregulation and dependency in the measurement errors of the gene expression levels further complicates the problem. The variation present in microarray data can be caused by biological differences and/or by technical variations. The best way to address this question is to use replicates for each condition.

The rest of the chapter is organized as follows. First, we discuss briefly how to extract signal from microarray data, then, we consider the problem of DE genes detection in a replicated experiment, next we provide a review of statistical methods dealing with DE gene identification and expose our new method for DE gene selection on the basis of comparisons between statistics provided from both experiments and replicates. *Gene Ontology* (GO) enrichment analysis is used to complete the analysis of selected targets. We conclude this chapter by giving an example using publicly available data related to cancer pathology for target research identification.

18.2 FROM MICROARRAY IMAGE TO SIGNAL

Microarray technology reveals an unprecedented view into the biology of RNA. In addition, an increase in its popularity allows us to approach a wide range of complex biological issues. Examples of well-known microarray technologies platforms are *Affymetrix* [22] and *Agilent cDNA* [18]. The manufacturing process of these two technologies and relevant signal extraction from microarrays are presented in the following two subsections.

18.2.1 Signal from Oligo DNA Array Image

The *Affymetrix* gene chip represents a very reliable and standardized technology for genome-wide gene expression screening [15]. In this technology, probe sets of 11–20 pairs with 25-mer oligonucleotides are used to detect a single transcript. Each oligonucleotide pair consists of a probe with *Perfect Match* (PM) to the target and another probe with a single base *MisMatch* (MM) in the 13th position [8]. Each MM is intended to act as a calibration signal for the true value given by the PM. PM and MM pairs are part of probe sets. Each probe set is designed to represent a different gene transcript. Typically, probe sets consist of 11–16 PM and MM pairs. Scanner software supplied by the *Affymetrix* platform store a numeric image data file with a suffix ".DAT" corresponding to PM and MM signals across the chips. Owing to

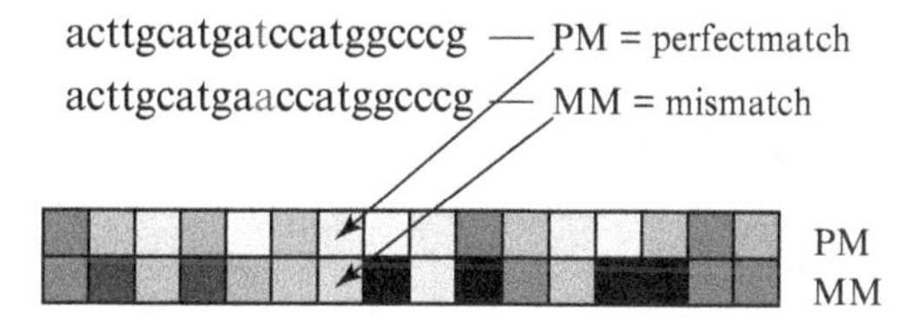

Figure 18.1 Probe set in *Affymetrix* technology.

the big size of ".DAT" file, a reliable algorithm has been developed to transform them into a ".CEL" file that keeps the essential report and provides a summary of the ".DAT" file [22]. The ".CEL" files contain probe sets that represent a mapping of probe cells to aggregates of probe spots (Figure 18.1). Each probe set is related to a gene, the quality control purpose that can be found in the spot file, or a description file with a suffix ".CDF." Intensity expression of the each gene may be computed from PM and MM to its related probe set.

18.2.2 Signal from Two-Color cDNA Array

Two-color microarrays are also widely used to perform transcriptome analysis. They are popular for their variety and wide use as in *Comparative Genome Hybridization* (CGH) [20]. In this technology, targets are obtained for each condition to be studied by reverse transcription of extracted mRNA [8]. Differential analysis of the transcriptome is conducted by taking equal amounts of cDNA targets from each condition, labeling them with different fluorophores (usually Cy5 and Cy3 dyes, which fluoresce at red and green wavelengths, respectively), and then hybridizing them competitively on the same slide. The expression level of all the genes is determined in parallel by measuring and comparing, for each spot, the fluorescence intensity level at the appropriate wavelengths (red and green in this case). Therefore, microarray experiments strongly rely on the quality of the data extraction process, that is, the image acquisition and analysis steps [23]. The signal of cDNA is assessed by scanning the hybridized arrays using a confocal laser scanner capable of interrogating both the Cy3- and Cy5-labeled probes and producing separate TIFF images for each (Figure 18.2). Image segmentation is used to determine, for each spot, the area or set of pixels that are related to the foreground signal, whereas the remaining neighboring ones are usually considered as background [3]. Ratio analysis, associated with quality scores and weights, is commonly employed to determine expression differences between two samples. This ratio reflects differential gene (cDNA) or protein expression or change in DNA copy number (CGH) between two samples.

18.3 MICROARRAY SIGNAL ANALYSIS

The main goal of both intraplatform and interplatform microarray technologies can be achieved in detecting DE genes when the microarray experiment is performed appropriately. The choice of the method, the algorithm, and the tools is always

Figure 18.2 Scanned image of cDNA microarray.

difficult, and each step in microarray analysis, to achieve optimal DE gene detection, is crucial. In this section, we attempt to present some popular methods for DE gene selection. We also present a new method for DE gene selection based on comparison between replicated experiences.

18.3.1 Absolute Analysis and Replicates in Microarrays

The possibility of an absolute analysis (is the gene present or absent in the condition?) using oligonucleotide microarray is an important advantage compared to cDNA technology. In fact, in many situations, we have to just know whether the transcript of a particular gene is present or absent. In this context, we can easily evaluate the gene expression level provided from probe sets and analyze the results by comparing the *p-values* of expression levels of all genes with thresholds $\alpha 1$ and $\alpha 2$. *Affymetrix* technology provides by default two levels of significance, $\alpha 1$ and $\alpha 2$ ($\alpha 1 = 0.04$ and $\alpha 2 = 0.06$). Genes with expression *p-values* under $\alpha 1$ are called *present*, genes with expression *p-values* higher than $\alpha 2$ are called *absent*, and genes with *p-values* between $\alpha 1$ and $\alpha 2$ are called *marginal*.

As the cost of cDNA or oligonucleotides chips production has become lower, researchers have also been encouraged to use more chips in their experiments. Most of the time, experiments dealing with gene expression use several replicates for each condition. There are two types of primary replicates: technical and biological. Technical replicates are a repetition of the same biological samples on multiple microarrays. Biological replicates are multiple samples obtained in the same biological condition.

The use of replicates has three major advantages:

- Replicates can be used to measure experiment variations so that statistical tests can be applied to evaluate differences. This point will be discussed in this chapter.

- Averaging across replicates increases the precision of gene expression measurements and allows detection of the small changes.
- Replicates can be compared to detect outliers due to aberrations within arrays, samples, or experimental procedures. The presence of outlier samples can have a severe impact on data interpretation. Most array platforms have internal controls to detect various problems in an experiment. However, internal controls cannot identify all issues.

In DE gene selection, the hypothesis underlying microarray analysis is that the intensities for each arrayed gene represent its relative expression level. Biologically relevant patterns of expression are typically identified by comparing the expression levels between different states on a gene-by-gene basis. But before an appropriate comparison of the levels, data normalization must be carried out across all chips.

18.3.2 Microarray Normalization

Data normalization minimizes the effects caused by technical variations and, as a result, allows the data to be comparable in order to find actual biological changes [26]. There are a number of reasons for data to be normalized, including unequal quantities of starting RNA, differences in labeling or detection efficiencies between the fluorescent dyes used, and systematic biases in the expression levels. Several normalization approaches have been proposed, most of which derive from studies using two-color microarrays [18]. Some authors proposed normalization of the hybridization intensity ratios; others use global, linear, or nonlinear methods [21]. The use of nonlinear normalization methods [27] is believed to be superior to the above-mentioned approaches. However, these approaches are not appropriate for GeneChip arrays because only one sample is hybridized to each array instead of two (red and green). Bolstad et al. [6] present a review of these methods and propose to deal with the quantile method that performs best in oligonucleotide chips. The goal of quantile normalization is to make the probe intensities distribution the same for all the arrays. Figure 18.3a shows the *MA* plot (*M* for log ratios and *A* for mean average) for four *Affymetrix* chips before normalization using quantiles, while Figure 18.3b shows the *MA* plot for the same chips after normalization. This figure illustrates the impact of the normalization step by correcting the signals of genes across all chips (Figure 18.3).

18.4 ALGORITHMS FOR DE GENE SELECTION

In *two-sample* DE gene selection [6, 9], there is a general assumption that gene expression measurements are normally distributed. The simplest statistical method to detect DE genes is the *two-sample t-test* [9]. The *two-sample t-test* allows us to formulate statements concerning the difference between the means of two normally distributed variables with the assumption that the variances are unknown. When the experiments concern comparison of two conditions (treated vs baseline), the objective of the comparative analysis is to answer the question: does the expression of a

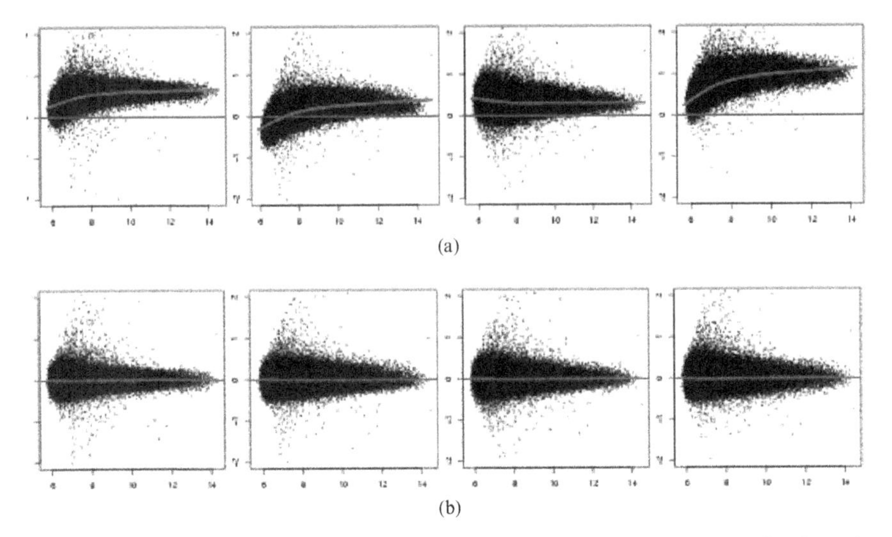

Figure 18.3 The *MA* plot for four *Affymetrix* chips (a) before and (b) after normalization using quantile normalization.

transcript on a chip (treated) change significantly with respect to the other chip (baseline)? In this context, and by using simple *Fold-Change* (FC) or *two-sample t-test* methods, five distinct answers are possible: increase, decrease, marginal decrease, marginal increase, and no change. These detection calls are obtained by comparing *p-values* of the *t-tests* with the given thresholds [14]. Multiple studies, especially in the case of large-scale data, have shown that FC on its own is an unreliable indicator [17], the detection call is a limited tool, and other multiple testing procedures can be used. Some of these procedures, such as the *Bonferroni* procedure [11], control the *Family-Wise Error Rate* (FWER) [13]. The other multiple testing procedures, such as the *Benjamini and Hochberg* (BH) procedure [5], control the *False Discovery Rate* (FDR) [28]. These statistics were originally designed to perform a single test but are not appropriate for large-scale data analysis. This motivates the development of many new statistics that borrow information across multiple genes to identify the DE gene, including a *Significance Analysis of Microarrays* (SAM) offering a random testing approach [25]. All these methods have been widely criticized for their high rate of false positives, especially with a large data variance and/or weak signals [29]. For this, the algorithm proposed in the next section uses more statistics provided from *replicates*. In fact, *replicates* can be used to measure experiment variations so that statistical tests can be applied to evaluate differences.

18.4.1 Within–Between DE Gene (WB-DEG) Selection Algorithm

For simplification, let us consider one experience with two conditions (treated vs control) with three replicates for each condition. We can process two types of comparisons using statistics: Between Condition Comparison (BCC) that performs statistical test between two microarrays not from the same condition and *Within Condition Comparisons* (WCC) where the statistical test is applied on microarrays from the same condition (replicated microarrays).

For the common problem of comparing gene expression between two groups, a standard statistical test is the classical t-statistic, $t_{12i} = mean(x_{i1}) - mean(x_{i2})$, where x_{ij} is the signal of gene i in condition j.

The modulation *p-value* corresponds to the probability that the signal of gene i remains constant (null hypothesis) between treated and control samples. The *p-value* significance expressed as ($p_{12i} = -\log_{10} p\text{-}value_{12i}$) may be computed for each gene i. A volcano plot is a graphical display to study *FC* and *p-value* significance for each gene with two conditions. We will use this plot to quantify the background noise in a replicated array. In fact, the method displays two volcano plots: the first one displays the *p-value* significance and FC related to BCC conditions (Figure 18.4a), while the second displays the *p-value* significance and FC related to WCC conditions (Figure 18.4b). From the latter volcano plot, the user can set a suitable threshold for selecting modulated genes from the first volcano plot.

The *Within–Between DE Gene* (WB-DEG) [16] Selection Algorithm is run as follows:

- Compute the FDR for each type of comparison, namely, FDR^{WCC} and FDR^{BCC}.
- Compute $\text{FDR}^{\text{WB}} = \frac{\text{FDR}^{\text{WCC}}}{\text{FDR}^{\text{BCC}}}$ that controls the selection procedure [16], interpreted as the expected proportion of false positives if genes with the observed statistic are declared DE
- Select genes with $\text{FDR}^{\text{WB}} < \text{cutoff}$.

Figure 18.4 illustrates an example of selection (selected genes located above the cutoff) using two FDR^{WB} cutoff (1% and 10%).

This algorithm was implemented in the R/Bioconductor project [2] as a graphical interface displaying all the options and the menu of the microarray analysis. WB-DEGS application deals with the DE gene selection, especially for experiments with a high background noise. WB-DEGS aimed to improve the gene identification specificity and accuracy.

18.4.2 Comparison of the WB-DEGs with Two Classical DE Gene Selection Methods on Latin Square Data

The Latin Square Data generated by the *Affymetrix* set consist of 14 labeled transcripts spiked at varying concentrations into a labeled mixture of RNA from a tissue in which the spiked transcripts are known not present. Each transcript concentration series or experiment consists of 13 transcripts spiked at successively doubling concentrations between 0.25 and 1024 PM and one transcript not spiked (a concentration of 0 PM). Each RNA sample was hybridized in triplicate [1]. This data set enables the sensitivity and accuracy evaluation of gene selection methods by comparing the trends in calculated transcript abundance with the known trends in transcript

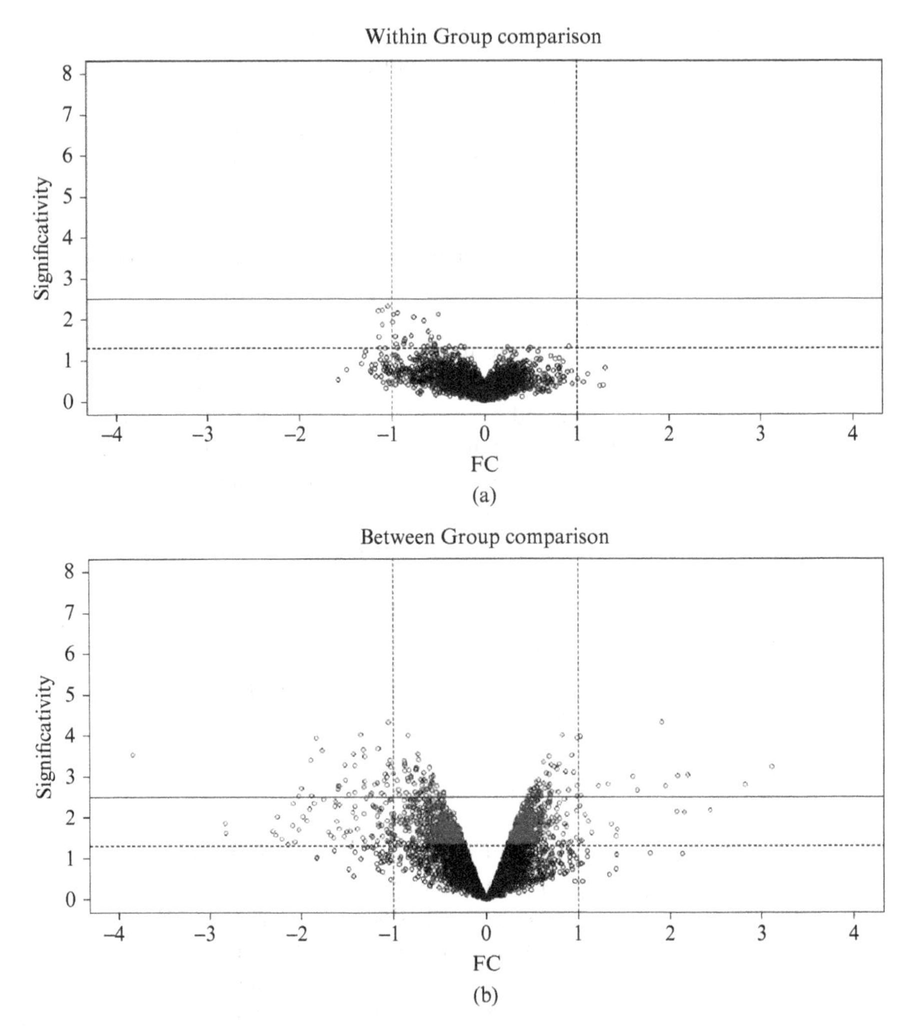

Figure 18.4 Selection of DE gene using FDR^{WB} control. Volcano plots for (a) within-group comparison and (b) between-group comparison.

concentration. Latin square data also allow the comparison between two or more DE gene selection methods. Two parameters are used to evaluate the robustness of gene selection methods. On the one hand, the sensitivity is the proportion of actual positive genes correctly identified as such. On the other hand, the *accuracy* is the degree of closeness of the quantity of selected genes to spiked genes. In this work, we have compared results of the WB-DEGS algorithm with those of the SAM and multiple *t*-test algorithms [29]. In fact, a recent study classifies numerous DE gene selection methods into three classes: algorithms based on FC, algorithms based on *t*-test, and algorithms based on random permutation such as SAM [12]. Table 18.1 summarizes statistical results of the three algorithms using three *p-value* cutoffs (1%, 5%,

Table 18.1 Results of DE gene selection methods using sensitivity and accuracy

	Sensitivity			Accuracy		
FDR^{WB}	1%	5%	8%	1%	5%	8%
t-Test	50.39	45.36	29.87	58.26	66.45	67.35
SAM	65.44	66.21	69.52	67.6	68.84	75.38
t-Test + BH correction	58.59	60.49	68.11	74.32	78.26	80.36
WB-DEGS	60.58	62.47	69.21	75.65	80.25	85.46

and 8%). Regarding the *accuracy*, the WB-DEGS and SAM algorithms provide good results compared with multiple *t*-test algorithms with and without *BH* correction, where the sensitivity states that when using WB-DEGS, true genes are selected.

18.5 GENE ONTOLOGY ENRICHMENT AND GENE SET ENRICHMENT ANALYSIS

Once the statistical analysis is complete and a group of DE genes is selected, GO enrichment annotations can be used to complete the microarray analysis [24]. GO terms are organized hierarchically such that higher level terms are more general and thus are assigned to more genes, and more specific descend terms are related to parents by either "is a" or "part of" relationship. The relationships form a directed acyclic graph, where the terms can have one or more parents and zero or more children. The terms are also separated into three categories/ontologies [4]:

- Cellular component: This describes where in the cell a gene acts, what organelle a gene product functions in, or what functional complex an enzyme is part of.
- Molecular function: This defines the function carried out by a gene product; one product may carry out many functions; set of functions together make up a biological process.
- Biological process: This relates to some biological phenomena affecting the state of an organism. Examples include the cell cycle, DNA replication, and limb formation.

The method derives its power by focusing on gene sets, that is, groups of genes that share common biological function, chromosomal location, or regulation. Using GOrilla [10], the functional enrichment analysis was applied to the 487 selected genes from the above data. Figure 18.5 shows the significantly enriched bar chart of biological process (Figure 18.5a) and cellular component (Figure 18.5b), while Figure 18.6 illustrates the direct acyclic graph for this set of selected genes related to the cellular component analysis. There are several evidences on the studies performed to analyze the microarray data sets for various diseases. One such example is a study being performed on *Alzheimer's Disease* (AD) where the authors presented a comprehensive analysis on GO and other biological annotations [19].

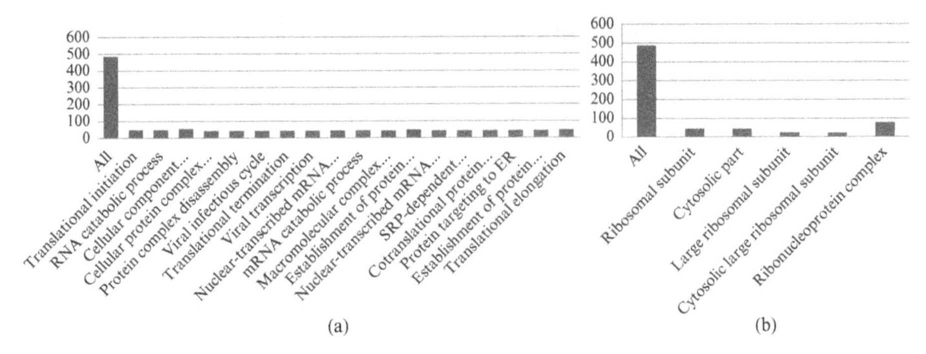

Figure 18.5 Enriched bar charts of selected genes using GO on biological process (a) and cellular components (b) categories.

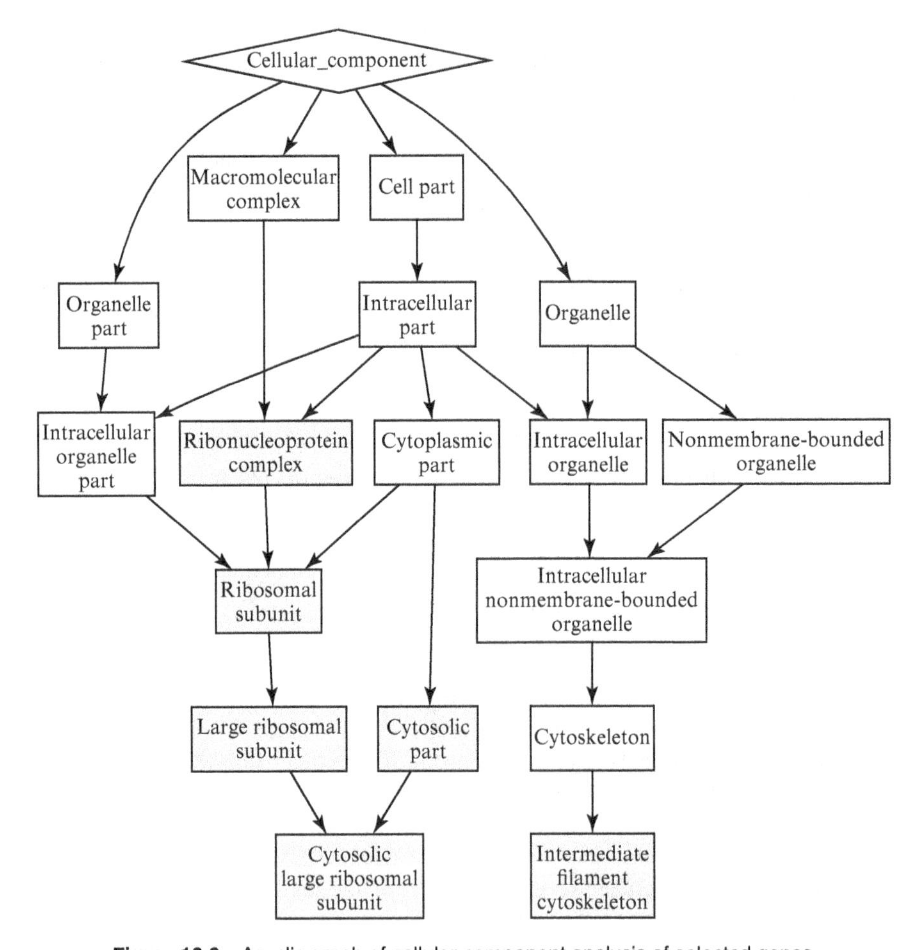

Figure 18.6 Acyclic graph of cellular component analysis of selected genes.

18.6 CONCLUSION

In this chapter, we have presented a suitable workflow of microarray data analysis and presented a limited overview of methods and descriptions available in gene selection methods. The objective was to elucidate whether the DE gene selection process is influenced by gene selection methods and platforms on which the analyses are performed. These several considerations often lead to select wrong DE genes in the majority of cases. To reduce these errors, we present WB-DEGS, a new algorithm that improves the selection process specificity and minimizes the background noise by controlling the FDR. WB-DEGS, implemented in the R programming language, allows a limited number of false positives and provides a comprehensive interface allowing the analysis and visualization of different steps of microarray data in a user-friendly exploration. Our approach has also some impact on the Gene Set Enrichment Analysis. Integration of functional genomic data and analyzing it through several types of computational approaches will help in the generation of biologically meaningful information. Discussed case studies have the potential to greatly facilitate biological discoveries.

REFERENCES

1. http://www.affymetrix.com/support/technical/sample_data/datasets.affx.
2. http://bioinfoindia.org/WB-DEGS/.
3. Ahmed AA, Vias M, Iyer NG, Caldas X, Brenton JD. Microarray segmentation methods significantly influence data precision. Nucleic Acids Res 2004;32:e50.
4. Ashburner M, Ball CA, Blake JA, Botstein D, Butler H, Cherry JM, Davis AP, Dolinski K, Dwight SS, Eppig JT, Harris MA, Hill DP, Issel-Tarver L, Kasarskis A, Lewis S, Matese JC, Richardson JE, Ringwald M, Rubin GM, Sherlock G. Gene ontology: tool for the unification of biology. The Gene Ontology Consortium. Nat Genet 2000;25(1):25–29.
5. Benjamini Y, Hochberg Y. Controlling the false discovery rate: a practical and powerful approach to multiple testing. J R Stat Soc [Ser B] 1995;57:289–300.
6. Bolstad BM, Irizarry RA, Astrand M, Speed TP. A comparison of normalization methods for high density oligonucleotide array data based on variance and bias. Bioinformatics 2003;19(2):185–193.
7. Brunet JP, Tamayo P, Golub TR, Mesirov JP. Metagenes and molecular pattern discovery using matrix factorization. Proc Natl Acad Sci U S A 2004;101(12):4164–4169.
8. Choaglin D, Mosteller F, Tukey F. *Understanding Robust and Exploratory Data Analysis*. Volume 79. New York: John Wiley and Sons; 2000. p 7–32.
9. Cui X, Churchill GA. Statistical tests for differential expression in cDNA microarray experiments. Genome Biol 2003;4:210.
10. Eden E, Navon R, Steinfeld I, Lipson D, Yakhini Z. GOrilla: a tool for discovery and visualization of enriched GO terms in ranked gene lists. BMC Bioinformatics 2009;10:48.
11. Holm S. A simple sequentially rejective bonferroni test procedure. Scand J Stat 1979;6:65–70.

12. Jeffery IB, Higgins DG, Culhane AC. Comparison and evaluation of methods for generating differentially expressed gene lists from microarray data. BMC Bioinformatics 2006;7:359.
13. Jung SH, Bang H, Young S. Sample size calculation for multiple testing in microarray data analysis. Biostatistics 2005;6(1):157–169.
14. Liu WM, Mei R, Di X, Ryder TB, Hubbell E, Dee S, Webster TA, Harrington CA, Ho MH, Baid J, Smeekens SP. Analysis of high density expression microarray with signed-rank calls algorithmes. Bioinformatics 2002;12:1593–1599.
15. Lockhart D, Fodor SP, Gingeras TR, Lockhart DJ. High density synthetic oligonucleotide arrays. Nat Biotechnol 1996;14:1675–1680.
16. Moussa A, Vannier B, Maouene M. New algorithm for gene selection in microarray data analysis workflow. Comput Technol Appl 2012;3(2):169–174.
17. Newton MA, Kendziorski CM, Richmond CS, Blattner FR, Tsui KW. On differential variability of expression ratios: improving statistical inference about gene expression changes from microarraydata. J Comput Biol 2001;8:37–52.
18. Novikov E, Barillot E. An algorithm for automatic evaluation of the spot quality in two-color DNA microarray experiment. BMC Bioinformatics 2005;6:293.
19. Panigrahi PP, Singh TR. Computational studies on alzheimers disease associated pathways and regulatory patterns using microarray gene expression and network data: revealed association with aging and other diseases. J Theor Biol 2013;334C:109–112.
20. Pinkel D, Segraves R, Sudar D, Clark S, Poole I, Kowbel D, Collins C, Kuo WL, Chen C, Zhai Y, Dairkee SH, Ljung BM, Gray JW, Albertson DG. High resolution analysis of DNA copy number variation using comparative genomic hybridization to microarrays. Nat Genet 1999;20:207–211.
21. Quackenbush J. Microarray data normalization and transformation. Nat Genet 2002;32:496–501.
22. Schadt E, Li C, Ellis B, Wong WH. Feature extraction and normalization algorithms for high-density oligonucleotide gene expression data. Cell Biochem Suppl 2001;37:120–125.
23. Schena M, Shalon D, Davis RW, Brown PO. Quantitative monitoring of gene expression patterns with a complementary DNA microarray. Science 1995;270:467–470.
24. Subramanian A, Tamayo P, Mootha VK, Mukherjee S, Ebert BL, Gillette MA, Paulovich A, Pomeroy SL, Golub TR, Lander ES, Mesirov JP. Gene set enrichment analysis: a knowledge-based approach for interpreting genome-wide expression profiles. Proc Natl Acad Sci U S A 2005;102(43):15545–15550.
25. Tusher V, Tibshirani R, Chu G. Significance analysis of microarrays applied to the ionizing radiation response. Proc Natl Acad Sci U S A 2001;98(9):5116–5121.
26. Vannier B, Moussa A. Application of microarray technology in gene expression analysis. *Nanotechnology in Health Care*. Jaipur, India: S.P. Publications; 2011. p 207–213.
27. Yang YH, Dudoit S, Luu P, Lin DM, Peng V, Ngai J, Speed TP. Normalization for cDNA microarray data: a robust composite method addressing single and multiple slide systematic variation. Nucleic Acids Res 2002;30(4):e15.
28. Yekutieli D, Benjamini Y. Identifying differentially expressed genes using false discovery rate controlling procedures. Bioinformatics 2003;19(3):368–375.
29. Zhijin Wu. A review of statistical methods for preprocessing oligonucleotide microarrays stat methods. Med Res 2009;18(6):533–541.

CHAPTER 19

CLUSTERING AND CLASSIFICATION TECHNIQUES FOR GENE EXPRESSION PROFILE PATTERN ANALYSIS

Emanuel Weitschek[1,2], Giulia Fiscon[2,3], Valentina Fustaino[4], Giovanni Felici[2], and Paola Bertolazzi[2]

[1]*Department of Engineering, Uninettuno International University, Rome, Italy*
[2]*Institute of Systems Analysis and Computer Science "A. Ruberti", National Research Council, Rome, Italy*
[3]*Department of Computer, Control and Management Engineering, Sapienza University, Rome, Italy*
[4]*Cellular Biology and Neurobiology Institute, National Research Council, Rome, Italy*

19.1 INTRODUCTION

Because of the advances in microarray technology, especially in the *Next-Generation Sequencing* (NGS) techniques, several biological databases and repositories contain raw and normalized gene expression profiles, which are accessible with up-to-date online services. The analysis of gene expression profiles from microarray/*RNA sequencing* (RNA-Seq) experimental samples demands new efficient methods from statistics and computer science.

Pattern Recognition in Computational Molecular Biology: Techniques and Approaches,
First Edition. Edited by Mourad Elloumi, Costas S. Iliopoulos, Jason T. L. Wang, and Albert Y. Zomaya.

In this chapter, two main types of gene expression data analysis are taken into consideration: *gene clustering* and *experiment classification*. Gene clustering is the detection of gene groups that present similar patterns. Indeed, several clustering methods can be applied to group similar genes in the gene expression experiments. On the other hand, experiment classification aims to distinguish among two or more classes to which the different samples belong (e.g., different cell types or diseased versus healthy samples).

This chapter provides first a general introduction to microarray and RNA-Seq technologies. Then, gene expression profiles are investigated by means of pattern-recognition methods with data-mining techniques such as classification and clustering. In addition, the integrated software packages *GenePattern* [40], *Gene Expression Logic Analyzer* (GELA) [55], TM4 software suite [42], and other common analysis tools are illustrated. For gene expression profile pattern discovery and experiment classification, the software packages are tested on four real case studies:

1. *Alzheimer's disease* versus healthy mice;
2. *Multiple sclerosis* samples;
3. *Psoriasis* tissues;
4. *Breast cancer* patients.

The performed experiments and the described techniques provide an effective overview to the field of gene expression profile classification and clustering through pattern analysis.

The rest of the chapter is organized as follows. In Section 19.2, we introduce the *Transcriptome Analysis*, highlighting the widespread approaches to handle it. In Sections 19.3 and 19.4, we provide an overview of the *microarray* and *RNA-Seq* technologies, respectively, including the description of the technique with its applications and the corresponding data analysis workflow. Then, in Section 19.5, we provide a comparative analysis between microarray and RNA-Seq technologies, by highlighting their advantages and disadvantages. In Section 19.6, we deepen the analysis of gene expression profile data obtained from microarray and RNA-Seq technologies, explaining how to deal with them, describing the widespread normalization techniques that have to be applied to these raw data, and providing an overview of gene clustering, experiment classification methods, and software suites that can be used to analyze them. Furthermore, in Section 19.7, we present the application of the whole described analysis process to four real case studies (i.e., Alzheimer's disease mice, multiple sclerosis samples, psoriasis tissues, and breast cancer patients). Finally, in Section 19.8, we draw the summary of this chapter.

19.2 TRANSCRIPTOME ANALYSIS

An *RNA molecule* is a complementary single-stranded copy of a double-stranded DNA molecule, obtained after the process of transcription. It provides a critical contribution in the coding, noncoding, expression, and regulation of genes. The *transcriptome* is the set and quantity of RNA transcripts that characterize a cell in a certain development stage. Its deep analysis is critical either for detecting genes activity and quantifying their expression levels, or for identifying functional elements, aiming to understand cellular development stages and mostly pathological conditions.

Several methods have been developed to address the transcriptome analysis focusing on its quantification. They can be mainly divided into two groups:

- *Hybridization-based* methods
- *Sequence-based* methods

The first type of methods includes the widespread *microarray*-based methods that exploit the hybridization techniques (i.e., based on the nucleotides property to pair with their complementary fixed on a specific support) to catch information about gene expression levels with a high-throughput and low costs (unless when investigating large genomes), showing, on the other hand, a high background noise, a limited dynamic detection range, and the inability to identify yet unknown transcripts. In contrary, the second type of methods provides directly the desired sequence. The Sanger et al. sequencing [43] is the initial widespread sequencing method. Recently, among the sequence-based approaches, the RNA-Seq [35, 53] stands out with lower time and cost. RNA-Seq performs a mapping and a transcriptome quantification relying on the novel NGS techniques (i.e., massively parallel sequencing). The latter guarantees an unequal high-throughput and quantitative deep sequencing that outperforms the other transcriptome analysis techniques according to several points of view, exhaustively explained in Section 19.4.

19.3 MICROARRAYS

The *microarray technology* (also known as biochips, DNA chips, gene chips, or high-density arrays) is a high-resolution method to detect the expression level of a large set of genes with a unique parallel experiment [6, 45]. Specifically, a microarray is a semiconductor device composed of a grid of multiple rows and columns. A cell of the array is associated to a DNA probe sequence, hybridized by Watson–Crick complementarity [9] to the DNA of a target gene (e.g., the mRNA sequences). The mRNA sequences contain the necessary information for the amino acids to form a given protein. The microarray experimental process is composed of the following steps. First, the mRNA sequences are amplified. Next, the mRNA sequences are fluorescently tagged. Then, the mRNA sequences are poured on the array. Next, the array is hybridized. Finally, the array is scanned with a laser that measures the quantity of fluorescent light in each cell. This measure is the expression level of the current gene.

19.3.1 Applications

Microarrays have several applications ranging from the genomic and transcriptomic areas, including pharmacogenomics, drug discovery, diagnostics, and forensic purposes [23]. The widespread transcriptomic application of this technology is the measure of gene expression in different sample conditions, also called *Gene Expression Profiling* (GEP) [45]. Gene expression microarrays include case–control studies,[1] body maps, tumors profiling, and outcomes prediction. Genomics microarrays are used to screen the sample for mutations, *Single Nucleotide Polymorphisms* (SNPs), *Short Tandem Repeats* (STRs), comparative genome hybridizations, genotyping, resequencing, and pathogen characterization [48].

19.3.2 Microarray Technology

A microarray consists of a rectangular support generally constructed on glass, silicon, or plastic substrates in a 1–2 cm^2 area [22]. It is organized in grids where thousands of different probes are immobilized in fixed locations. Each location is called *spot* and contains a pool of probes. Probes, usually synthetic oligonucleotides or larger DNA/complementary *DNA* (cDNA) fragments, are nucleic acid sequences that hybridize the labeled targets via Watson–Crick duplex formation [9]. The targets are then held into the spots. Each spot can recognize a specific target that may be either a chromosomal region for the genomic microarray, or part of an mRNA in the case of gene expression microarrays. A scanner detects the single-spot fluorescence signals and converts them into digital quantifiable signals.

Microarray devices and systems are commercialized by several companies specialized on manufacturing technology [48]. There are two main platforms that differ in the probe manufacturing: robotic deposition and in situ synthesis on a solid substrate. Section 19.3.3 are focused on in situ synthesized oligonucleotide microarray, which is the microarray technology of *GeneChip* by *Affymetrix* (affymetrix .com), the most widespread company specialized on gene expression microarray technology.

19.3.3 Microarray Workflow

A gene expression microarray experiment is usually composed of four laboratory steps [47]:

1. *Sample preparation and labeling*: The first performed steps on the biological sample are the RNA extraction, purification, reverse transcription, and labeling.
2. *Hybridization*: Probes and targets form a double-hybrid strand according to the Watson–Crick base complementarity [9].
3. *Washing*: The excess material is removed from the array to ensure the accuracy of the experiment by reducing nonspecific hybridizations. The targets remain bounded on the array.

[1] *Case–control studies* are defined as specific studies that aim to identify subjects by outcome status at the outset of the investigation, for example, whether the subject is diagnosed with a disease. Subjects with such an outcome are categorized as *cases*. Once the outcome status is identified, controls (i.e., subjects without the outcome but from the same source population) are selected [46].

4. *Imaging acquisition*: A scanner excites the fluorophores and measures the fluorescence intensity providing a color image representation as the output.
5. *Data extraction*: Specific software converts the image in numeric values by quantifying the fluorescence intensity.

For further details on the microarray experiment protocol, refer [6, 45].

19.4 RNA-Seq

RNA-Seq [35, 53] is a family of methods that takes advantage of NGS technologies for detecting the transcriptome expression levels with flexibility, low cost, and high reproducibility. RNA-Seq allows performing more measurements and consequently a quantitative analysis of the expression levels. Different from the traditional sequence-based approaches, RNA-Seq yields quantitative measures instead of a qualitative and normalized description of the considered sequence. Unlike tag-based techniques, it provides different sequences for each transcript with a single-base resolution. Indeed, RNA-Seq uses directly short fragments of the DNA molecule without any other insertion as input data and provides discrete measures as output, suitable to be simply managed. Furthermore, it is suited for two different tasks: (i) transcriptome reconstruction and analysis; (ii) quantification of the expression levels.

The RNA-Seq technique is characterized by the three following main phases that constitute the RNA-Seq workflow:

- Sample preparation through the building of the fragments library to be sequenced;
- Sequencing;
- Sequenced reads analysis and expression levels quantification.

Even if the RNA-Seq workflow should be characterized by the RNA extraction and its direct sequencing, actually the RNA-Seq techniques require a preprocessing phase in order to make the sample of RNA suitable to the following part of the experimental protocol. The whole detailed procedure is sketched in Figure 19.1, and the three leading phases are thoroughly explained as follows.

- *Sample preparation and library of fragments building*: First, the input RNA sequence is usually purified by the poly-A binding (steps 1 and 2 of Figure 19.1) in order to minimize the copious ribosomial RNA transcripts. A sequence random-size fragmentation (step 3 of Figure 19.1) reduces the RNA sequence into shorter segments ($\sim$ 200–300 bases) because the next steps require smaller strings of RNA. The obtained fragments can be turned into cDNA by the reverse transcription, which performs first a complementary copy of the RNA strand and, secondly, using as primer the remaining (partially degraded) RNA fragments, yields as output the corresponding double-stranded cDNA (step

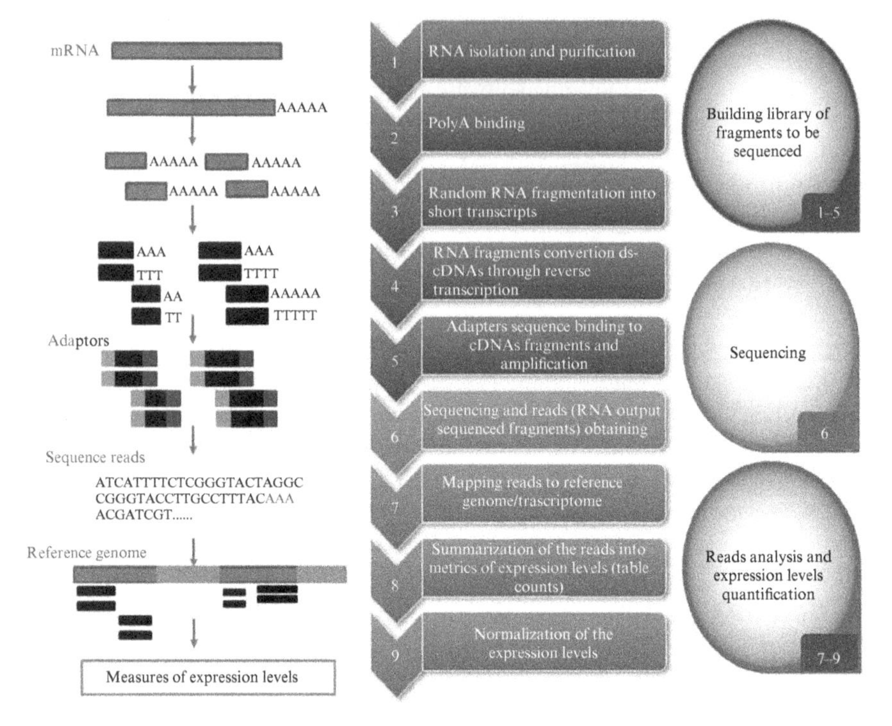

Figure 19.1 Overview of the RNA-Seq workflow.

4 of Figure 19.1). Step 4 is necessary because the sequencers are unable to perform directly a sequencing of an RNA molecule. The cDNAs obtained by step 4 are amplified and the adapters are attached to one or both ends (step 5 of Figure 19.1) in order to build up a suitable library of fragments to be sequenced.

- *Sequencing*: The nucleotide sequence of short cDNA segments has to be turned into a quantified signal. A sequencer processes the amplified fragments of input cDNA (called *reads*) and identifies the successive order of the nucleotides that make up the analyzed sequence (step 6 of Figure 19.1). Finally, the sequencer provides the sequenced reads as output. Today, widespread sequencers are those from Illumina and Roche.
- *Reads analysis and expression levels quantification*: In order to quantify the read referred to each gene, the sequenced reads are aligned and hence mapped to the reference genome (step 7 of Figure 19.1). The mapping aims to look for a unique region where the reads and the reference genome are identical. Starting from the genome, the transcriptome can be reconstructed by removing the intronic regions. In order to estimate the expression levels of each gene, a quantification of all reads, which correspond either to the same exon or transcript or gene, is performed. The reads are summarized in a *table of counts*, where a *count* is the

real output of RNA-Seq, representing the measure of the expression level for a specific gene (step 8 of Figure 19.1). Afterward, there are two issues to be addressed: how to deal with the overlapping reads and how to take into account the RNA isoforms.[2] In fact, RNA-Seq is able to quantify even the expression levels of isoform transcripts. The first issue can be solved assigning to each gene only the reads that are exactly mapped in such a gene, while those that overlap more than one gene are not counted. The second issue can be handled with a scheme that counts the mapped reads shared with all isoforms of a gene. Lastly, step 9 of Figure 19.1 performs the measure normalization required to have comparable measurements.

19.5 BENEFITS AND DRAWBACKS OF RNA-Seq AND MICROARRAY TECHNOLOGIES

In this subsection, the main *advantages* as well as the identified *drawbacks* of the RNA-Seq and microarray technologies are highlighted. As mentioned in the previous section, RNA-Seq shows several advantages, which make it the mainstream tool for deeply profiling the transcriptome and for quantifying the expression levels with an high accuracy.

The supremacy of RNA-Seq, summarized in Figure 19.2, is due to several features explained as follows:

- it is able to detect isoforms, allelic expression, complex transcriptomes, and currently unknown transcripts (unlike the hybridization-based approaches);
- it is characterized by a single-base resolution when determining the transcript boundaries, thanks to which several aspects of the today's genome annotation can be addressed;
- high-throughput quantitative measures are feasible and can be easily reproduced;
- it guarantees low background signal, lack of an upper limit for the quantification, and a large dynamic range of expression levels over which transcripts can be profiled;
- it presents high sensitivity and accuracy to quantify the expression levels;
- it requires a low amount of RNA (unlike microarray or other sequencing methods);
- the costs for mapping transcriptomes are relatively low.

On the other hand, there are some drawbacks to be underlined:

- RNA-Seq data are complex to be analyzed and sensitive to bias;

[2]The different gene transcripts generated from a single gene.

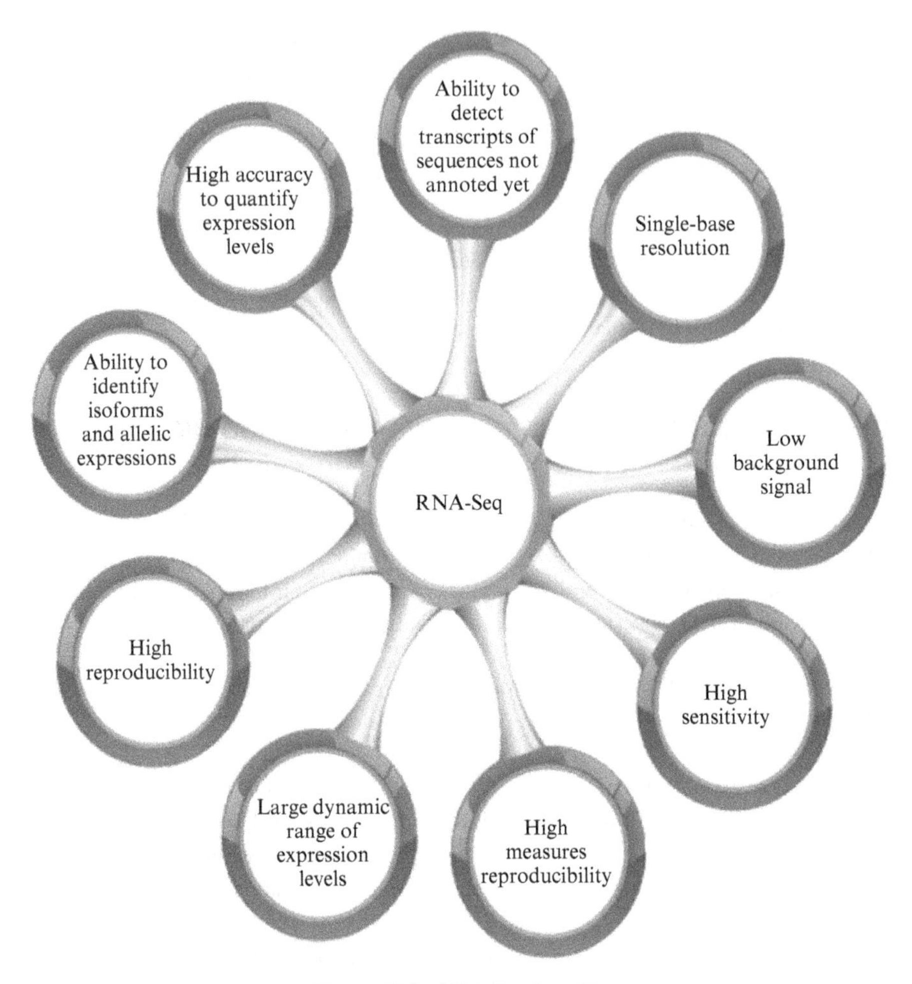

Figure 19.2 RNA-Seq benefits.

- the cost of the sequencing platform can be too high for certain studies and laboratories;
- the requirement of an up-to-date high-level *Information Technology* (IT) infrastructure to store and process the large amount of data (unlike the microarrays).

Following benefits can be identified in microarray technology (summarized in Figure 19.3:

- the costs of a microarray experiment are low, data size is relatively small, and short time is required to perform it;
- the measures are quantitative and obtained with a high-throughput technology;

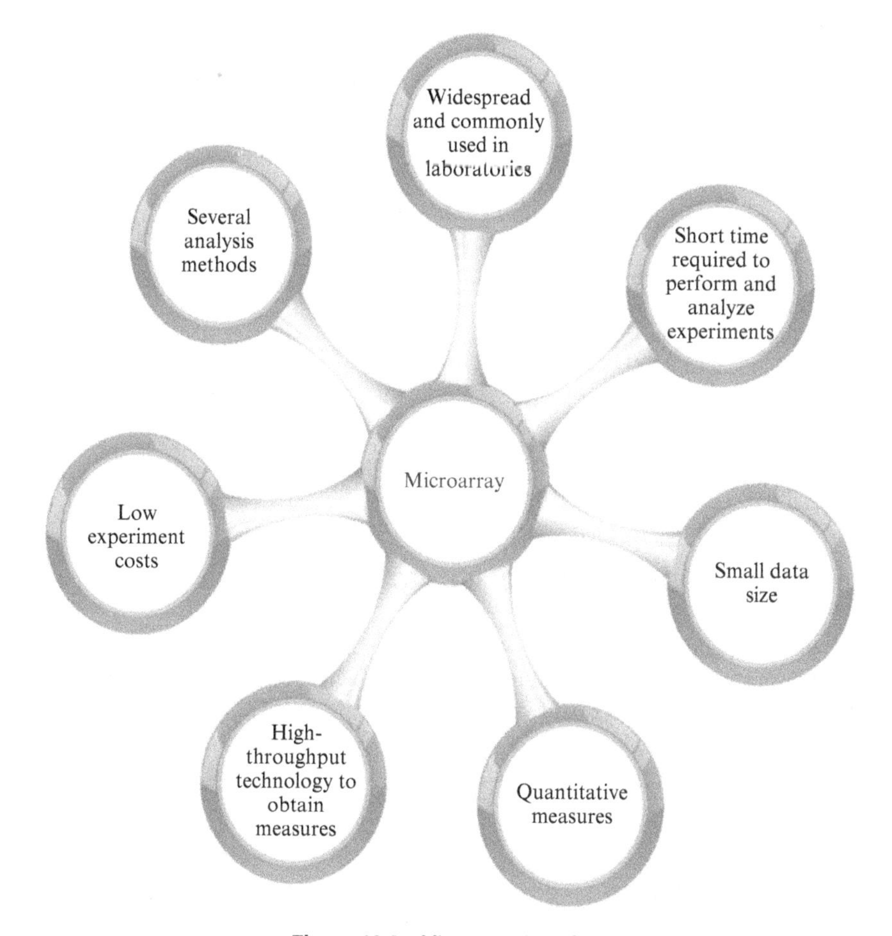

Figure 19.3 Microarray benefits.

- the microarray technologies are widespread, and hence several methods are available for analyzing microarray gene expression profile data;

The drawbacks of the microarray technology are summarized as follows:

- a background signal correction of the gene expression profiles is necessary;
- the quantification of the gene expression intensity signals has upper limits;
- the conservation of the gene chips demands high caution and care;
- the gene expression intensity signals may present high background noise;
- microarrays have a limited dynamic detection range of gene expression profiles and are unable to identify currently unknown transcripts.

Table 19.1 provides a summary of the comparison between benefits and drawbacks of microarrays and RNA-Seq.

Table 19.1 Microarrays versus RNA-Seq

	Microarrays	RNA-Seq
High-throughput	√	√
Quantitative measures	√	√
High measure reproducibility		√
Unknown transcript identification		√
Isoform and allelic expression detection		√
Single-base resolution		√
Low background signal		√
Unlimited quantification		√
Low amount of RNA		√
Widespread	√	
Complexity		√
Handling of data	√	
Low cost	√	

19.6 GENE EXPRESSION PROFILE ANALYSIS

After an adequate experimental setup and execution, either with microarrays technology or with RNA-Seq techniques, the obtained raw data has to be processed and analyzed with effective and efficient statistical, mathematical, and computer science methods. In this section, the whole gene expression profile analysis process is described, focusing on two particular types of analysis: *genes clustering* and *experiment classification*.

19.6.1 Data Definition

In general, a gene expression profile data set is a collection of experiments organized in records. Each experiment has a fixed number of gene expression profiles, which can be discrete or continuous. An experiment is described by a set of gene expression profiles represented as a multidimensional vector. The data set can be stored in a matrix, where each row i represents a gene ($i = 1, \dots, k$), each column j an experimental sample (Exp_j, $j = 1, \dots, n$), and each cell (i,j) the gene expression value ($\exp r_{(i,j)}$), as shown in Table 19.2.

Table 19.2 Gene expression profile data set

Experiment class	Exp_1 *control*	… …	Exp_m *control*	Exp_{m+1} *case*	… …	Exp_n *case*
$gene_1$	$\exp r_{(1,1)}$	…	$\exp r_{(1,m)}$	$\exp r_{(1,m+1)}$	…	$\exp r_{(1,n)}$
$gene_2$	$\exp r_{(2,1)}$	…	$\exp r_{(2,m)}$	$\exp r_{(2,m+1)}$	…	$\exp r_{(2,n)}$
…	…	…	…	…	…	…
$gene_k$	$\exp r_{(k,1)}$	…	$\exp r_{(k,m)}$	$\exp r_{(k,m+1)}$	…	$\exp r_{(k,n)}$

The matrix may present two heading lines containing the experimental sample identifiers and the class labels. In a case–control study, the class labels can be either *case* or *control*. In particular, we define *case* samples as subjects who are diagnosed with a disease, while *control* samples as healthy subjects without the disease, but extracted from the same source population of the cases. Each cell of the matrix (expr) contains the gene expression value.

19.6.2 Data Analysis

A typical gene expression profile analysis, whose aim is to cluster similar genes and to classify the experimental samples, is composed of the following steps:

1. *Normalization and background correction*
2. *Genes clustering*
3. *Experiment classification*

Normalization is a preprocessing step that is required to transform the raw data into reliable and comparable gene expression measurements.

Genes clustering is the detection of gene groups that present similar patterns [41]. In References [26] and [56], different clustering methods that can be applied to group similar genes in microarray experiments are described. In Reference [3], the authors introduce the biclustering technique: here the genes and the experimental samples are clustered simultaneously in order to find groups of genes that are related to a particular class (e.g., in a case–control study); ideally two biclusters should appear, one containing the case samples and the associated genes and the other the control samples and the associated genes.

In *experiment classification*, the aim is to separate either the samples in classes (e.g., case *versus* control) or different cell types [27]. The experimental sample separation should be performed with classification models composed of human interpretable patterns, for example, logic formulas ("if–then rules").

19.6.3 Normalization and Background Correction

In this subsection, we explain the normalization preprocessing step, highlighting the widespread techniques to normalize the gene expression profile data provided by microarray, as well as RNA-Seq technologies.

19.6.3.1 Microarray Normalization Methods. Microarrays are subject to different sources of variation and noise that make data not directly comparable and that originate from array-manufacturing process, sample preparation and labeling, hybridization, and quantification of spot intensities. Normalization methods aim to handle these systematic errors and bias introduced by the experimental platforms. Therefore, normalization is a basic prerequisite for microarray data analysis and for the quantitative comparisons of two or more arrays. Every normalization method includes a specific background correction step that is used to remove the contribution of unspecific background signals from the spot signal intensities. Furthermore, each method computes a measure of the expression level from the probe intensities that represent the amount of the corresponding mRNA [15].

The most commonly used normalization methods are *Microarray Analysis Suite* 5.0 (MAS 5.0) [1], *dChip* [33], and *Robust Multiarray Average* (RMA) [25].

In MAS 5.0, the normalization step operates on the gene-level intensities and applies a simple linear scaling. The intensities between two or more arrays are mathematically represented on a straight line with a zero y-intercept. The slope of the line is multiplied with a scaling factor to let the mean of the experiment chip be equal to the one of the baseline chip (reference).

Also in *dChip*, the normalization is applied on the gene-level intensities, but the intensities are represented as nonlinear smooth curves. A rank-invariant set is used to force a given number of nondifferentially expressed genes to have equal values among the data sets [44]. Otherwise, RMA employs a probe-level quantile normalization in multiple arrays by forcing the quantile values of the probe intensity distribution for each array to be equal [4]. Microarray normalization methods are available at the open-source software for bioinformatics R/Bioconductor (www.bioconductor.org) [20], in *GenePattern* [40] and in *Microarray Data Analysis System* (MIDAS) from the TM4 software suite [42].

19.6.3.2 RNA-Seq Normalization Methods. As mentioned in Section 19.4, RNA-Seq returns the total number of reads aligned to each gene (called *counts*), which hence represent the mapping into the reference transcriptome. The counts are stored in a matrix (Table 19.3), whose rows and columns contain the genes and the sequenced samples, respectively.

Starting from the raw output, a normalization step is required in order to remove the inherent bias of the sequencing (e.g., the length of RNA transcripts and the depth of sequencing for a sample). Indeed, the metrics of normalization make the measurements directly comparable and the expression levels estimation worthwhile. First, the number of reads mapped to a gene can be conditioned by the length of the reference transcripts: trivially, shorter transcripts are more likely to have less mapped reads than longer ones. Moreover, because of the sequencing depth, differences can occur that may make the comparisons between two samples inconsistent.

Table 19.3 Table of RNA-Seq counts

Experiment class	Exp_1 *control*	… …	Exp_m *control*	Exp_{m+1} *case*	… …	Exp_n *case*
$gene_1$	$count_{(1,1)}$	…	$count_{(1,m)}$	$count_{(1,m+1)}$	…	$count_{(1,n)}$
$gene_2$	$count_{(2,1)}$	…	$count_{(2,m)}$	$count_{(2,m+1)}$	…	$count_{(2,n)}$
…	…	…	…	…	…	…
$gene_k$	$count_{(k,1)}$	…	$count_{(k,m)}$	$count_{(k,m+1)}$	…	$count_{(k,n)}$

Among the normalization methods, we focus on the two widespread techniques based on the *Reads Per Kilobase per Million mapped* (RPKM) [36] and the *Fragments Per Kilobase per Million mapped* (FPKM) [51] values. The RPKM normalizes the read counts according to their length and the counts sum for each sample. Specifically, let R be the count of mapped reads into the genes exon, N_r the total number of mapped reads, and L the length of all exons of such a gene (in terms of number of base pairs), then the value is calculated with the following formula:

$$RPKM = \frac{R}{N_r \cdot L} \cdot 10^9 \tag{19.1}$$

The FPKM computes the expected fragments (pairs of reads) per kilobase of the transcript per million sequenced fragments. Specifically, let N_f be the count of mapped fragments, F the total number of the mapped fragments, and L the length of all exons of such a gene (in terms of number of base pairs), then the value is calculated with the following formula:

$$FPKM = \frac{F}{N_f \cdot L} \cdot 10^9 \tag{19.2}$$

The RPKM value is really close to the FPKM value; indeed, if all mapped reads are paired, the two values will coincide. However, the latter is able to handle a higher number of reads from single molecules.

Recently, a further normalization technique based on *RNA-Seq by Expectation Maximization* (RSEM) has been developed [32]. The RSEM technique outputs abundance estimates, 95% credibility intervals, and visualization files. Specifically, it reports a guess of how many ambiguously mapping reads belong to a transcript/gene (i.e., called *raw count* values of the gene expression data) and estimates the frequency of the gene/transcript among the total number of transcripts that were sequenced (i.e., called also *scaled estimate* values). Newer versions of RSEM call the *scaled estimate* value *Transcripts Per Million* (TPM). It is closely related to FPKM and hence it is independent from the transcript length. Finally, it is noteworthy how RSEM appears to provide an accurate transcript quantification without requiring a reference genome.

Finally, it should be underlined that a normalization of an RNA-Seq experiment does not imply an alteration of the raw data, in contrary to the normalization techniques of microarray data. The major RNA-Seq software analysis tools, such as MEV [24] from TM4 software suite, include automatic conversion from raw data counts in RPKM, FPKM, and RSEM normalized values.

19.6.4 Genes Clustering

After the normalization procedure, the real analysis process begins with *gene clustering*. A large number of genes (10,000–40,000) are present in every sample of gene expression profile data, and the extraction of a subset of representative genes, which are able to characterize the data, is a crucial analysis step. Gene cluster analysis is a technique whose aim is to partition the genes into groups, based on a distance function, such that similar or related genes are in same clusters, and different or unrelated genes are in distinct ones [16]. The cluster analysis groups the genes by analyzing their expression profiles, without considering any class label, if available. For this reason, it is also called *unsupervised learning*: it creates an implicit labeling of genes that is given by the clusters and is derived only from their expression profiles.

For defining the genes partition, a *distance measure* (or distance metric) between them has to be defined. The measure has to maximize the similarity of the gene expression in the same clusters and the dissimilarity of the gene expression in different ones. Various metrics and distance functions can be adopted, the widespread ones are *Minkowski distance* [8], that of order 1 is called *Manhattan distance* [49], and that of order 2 is called *Euclidean distance* [11, 12], *Mahalanobis distance* [50], and *Correlation distance* [31]. The adoption of a measure of the similarity of the genes over the samples is already an important tool of analysis; in fact, we can just examine one gene and verify which genes exhibit a similar behavior. In this context, the correlation distances are of particular relevance (the most used is Pearson's correlation) that are designed to measure the relative variations of the genes over the sample and do not take into account the scale of the expression values.

Clustering is indeed a more sophisticated analysis that can be performed, but it is still based on the concept of similarity between genes as measured by a distance function. Clustering algorithms can be divided into two main groups: *partition algorithms* and *hierarchical algorithms*. In the partition algorithms, the objects are divided into a given number of clusters. In the hierarchical algorithms, the objects are grouped in clusters starting from initial clusters (that contains all the objects) or vice versa. The most important partition clustering algorithm is *K-means* [34]. Widely used hierarchical clustering algorithms are *AGglomerative NESting* (AGNES) [29] and *DIvisive ANAlysis* (DIANA) [29]. A new type of clustering algorithm, particularly designed for the gene expression profiles, is presented in the software GELA [54, 55], where the clusters are selected by a discretization procedure.[3]

In Reference [3], a biclustering method is described, where the genes and the experimental samples are clustered simultaneously. This is a very important approach, as the clusters may contain the genes associated to experiments of a particular class (e.g., a cluster may group together diseased experimental samples and the corresponding involved genes).

After the clusters have been computed, a cluster validation step can be performed: in this step, the clusters are validated with statistical measures, such as entropy, purity, precision, recall, and correlation. According to these measures, the clustering algorithm may be adjusted or fine tuned.

Finally, the results have to be presented to the user and knowledge has to be transferred with a graphical interface and cluster lists of similar genes.

Another important step that can be taken into account is the reduction of the dimensions of the space where the data lies. There are several methods that may perform this step based on the projection of the data point onto a subspace of smaller dimension,

[3]Discretization is the conversion of numeric (continuous) values into discrete ones.

with respect to either the number of genes or the number of samples; the most popular among such methods is the *Principal Component Analysis* (PCA) [28]. PCA identifies a projection onto a small number of *components*, or *factors*, that are obtained as linear combination of the original ones. For example, if PCA is applied on the genes, the sample points are represented in a new space where each coordinate is obtained as a linear combination of the original genes (symmetrically, it can be applied to the samples). Such a projection is characterized by the fact that the covariance matrix is preserved at best by few coordinates, and hence it is possible to visualize the points in a small space (2–3 dimensions) loosing very little part of the information contained in the data.

19.6.5 Experiment Classification

Classification is the action of assigning an unknown object into a predefined class after examining its characteristics [16]. The experiment classification aims to differentiate the classes present in the gene expression profile data set. An experimental sample is usually associated with a particular state or class (e.g., control versus case). For classifying the samples, *supervised machine learning* techniques can be used. In supervised learning, the class label of the samples is assigned or known, a classification function or model is learned from a part of the data set (the training set) and applied for verification to the rest of the data set (the test set) for verifying the classification accuracy. A classification method is a systematic approach for building a classification model from the training data set [49], characterized by a learning algorithm that computes a model for representing the relationship between the class and the attributes set of the records. This model should fit to the input data and to new data objects belonging to the same classes of the input data. Moreover, the classification model should contain common patterns for each class and be interpretable by humans.

A general approach for solving a classification problem consists of the following steps [49]: split the data in training set and test set; perform training; compute the classification model; verify the performances of the model on the training and the test sets; and evaluate.

The most popular classification methods that can be applied to gene expression profiles are *C4.5 Classification Tree* (C4.5) [37, 38], *Support Vector Machines* (SVMs) [10, 52], *Random Forest* (RF) [14], *Nearest Neighbor* [13], and *Logic Data Mining* (logic classification formulas or rule-based classifiers) [17].

In a *classification tree*, each node is associated to a predicate that represents the attributes of the objects in the data set; the values of this predicates are then used to iteratively generate new nodes ("growing the tree"). The special class attributes are represented in the tree by the leaves. The most used tree decision classifiers, such as C4.5 [38], rely on rules that create new nodes with the local objective of minimizing the class entropy. The classification tree model can be easily transformed into "if–then rules," which are easily interpretable by humans.

Similar results can be obtained by a class of methods commonly referred to as *Logic Data Mining*, or rule-based classification, where the classifier uses logic propositional formulas in disjunctive or conjunctive normal form ("if–then rules") for classifying the given records. Examples of methods for computing logic classification formulas are RIPPER [7], LSQUARE [17], GELA [54, 55], LAD [5], RIDOR [19], and PART [18]. The major strength of classification formulas is the expressiveness of the models that are very easy to interpret, for example, "if gene Nudt19 < 0.76 then the sample is diseased."

As a success of experiment classification, the work in Reference [2] is cited where the authors are able to distinguish the sample with Alzheimer's disease *versus* the control microarray experimental sample using only few genes and individuating logic formulas in the form of "if–then" classification rules.

SVM maps the data in n-dimensional vectors and tries to construct a separating hyperplane that maximizes the margin (defined as the minimum distance between the hyperplane and the closest point in each class) between the data. Its strength relies on the ability to impose very complex nonlinear transformation on the data, lifting the data in a new space where linear separation is easier to achieve. It normally obtain very good classification performances on numeric data, but the main drawback is that the classification model is not easily interpretable.

RF is a classification method that operates by constructing several classification trees, selecting randomly the features (i.e., genes), and using a special model evaluation named *bagging*.

The *K-Nearest Neighbor* (KNN) algorithm is a classifier based on the closest training data in the feature space [13]. Given a training set of objects whose class is known, a new input object is classified by looking at its k closest neighbors of the training set (e.g., using the Euclidean distance). The main drawback of this approach is that no model is computed to classify the data. Furthermore, when evaluating a classifier, the *cross-validation* data-sampling technique for supervised learning is recommended. Cross-validation is a standard sampling technique that splits the data set in a random way in k disjoint sets, the classification procedure is run k times with different sets. At a generic run k, the k subset is used as the test set and the remaining $k - 1$ sets are merged and used as the training set for building the model. Each of the k sets contains a random distribution of the data. The cross-validation sampling procedure builds k models and each of these models is validated with a different set of data. Classification statistics are computed for every model and the average of those represents an accurate estimation of the data-mining procedure performance.

19.6.6 Software Tools for Gene Expression Profile Analysis

In this subsection, we present four state-of-the-art software packages that can be used to perform the previously described gene expression profile analysis steps.

19.6.6.1 TM4 Software Suite. *TM4 software suite* [42] is a collection of different open-source applications that can be used in the major steps of the gene expression profile analysis pipeline. TM4 comprises four analysis tools, *MicroArray DAta Manager* (MADAM), *Spotfinder*, MIDAS, and *Multi-Experiment Viewer* (MEV).

MADAM is the software tool dedicated to the data management of gene expression profiles. It is able to guide the user in creating a relational database to store the expression values and to keep track of the experimental and analysis workflow.

Spotfinder is the tool designed for the analysis of microarray images and the quantification of gene expressions. It processes the image files that are produced by the microarray scanners and computes the gene expression profile intensities.

After that, the intensity values can be processed and analyzed, but a normalization step is necessary, as explained in Section 19.6.2. MIDAS is the tool that is dedicated to this task, integrating several normalization methods.

MEV is a specific freely available software package able to perform the real gene expression analysis: it provides a collection of clustering, classification, statistical, and visualization methods. MEV can process even normalized data of RNA-Seq [24].

The main advantages of TM4 software suite are its ad hoc analysis tools with user-friendly interfaces and with several integrated analysis methods.

19.6.6.2 GenePattern. *GenePattern* is a user-friendly platform for computational analysis and visualization of genomic data freely available at the Broad Institute website (*broadinstitute.org/cancer/software/genepattern/*) [40]. Its comprehensive environment offers the possibility to perform the entire pipeline of gene expression data analysis from raw data preprocessing, gene selection, and gene clustering to pathway detection [30]. By means of *GenePattern*, the whole pipeline of data analysis can be recorded and rerun also with different data or parameters. Gene expression data can be given as input in *Affymetrix GeneChip* or tab-delimited text formats. Gene expression data from any source can be easily transformed into a tab-separated file. *GenePattern* has an extensible structure that allows to share analysis modules and quickly integrate new methods with existing tools, supporting the steady growing repository of new computational methods. It is also possible to access to *GenePattern* from the R (www.r-project.org) [39], MATLAB (The MathWorks Inc., Natick, MA), or Java programming languages (java.com).

19.6.6.3 WEKA. *Waikato Environment for Knowledge Analysis* (WEKA) [21] is a Java software tool that comprises a collection of open-source data analysis methods, including classification and clustering algorithms. Two main software containers—"weka.classifier" and "weka.cluster"—provide implementations of established classification and clustering algorithms, as the ones described in Section 19.6.5. WEKA strengths are the user-friendly interface, the amount of available algorithms, and the possibility of performing several experiments and comparisons in an integrated software suite. As a general tool for machine learning, WEKA uses an own file format (arff), so a format conversion of the gene expression profile data must be performed.

19.6.6.4 GELA. GELA [54, 55] is a clustering and rule-based classification software, particularly engineered for gene expression analysis. The aims of GELA are to cluster gene expression profiles in order to discover similar genes and to classify the experimental samples. GELA converts the numeric gene expression profiles to discrete by defining intervals, implements a method named *Discrete Cluster Analysis* (DCA) based on the discretization step to cluster the genes, uses an integer-programming method for selecting the characteristic genes for each class, adopts the Lsquare method for computing the logic classification formulas ("if–then rules"), and uses a special weighted sample classification procedure. GELA also integrates standard statistical methods for gene expression profile data analysis: PCA to group similar genes and experiments, and Pearson correlation analysis to find a list of correlated genes to a selected gene. It is available at dmb.iasi.cnr.it/gela.php. GELA also supports the classification and clustering of RNA-Seq data.

19.7 REAL CASE STUDIES

To provide the reader with an example of the analysis process described in the previous sections, we describe in the following some results obtained in our research. In Reference [2], GELA was applied in a microarray case–control study: early (1–3 months) and late stage (6–15 months) experimental samples of *Alzheimer's disease* versus healthy mice had to be distinguished by analyzing their gene expression profiles. In that study, 119 experimental samples and 16,515 gene expression profiles, provided by the *European Brain Research Institute* (EBRI), have been analyzed in order to distinguish the experimental samples with a clear human-interpretable classification model and to detect the genes whose expression or coexpression strongly characterizes the Alzheimer's disease. First, a MAS 5.0 normalization step was performed using the standard Affymetrix Expression Console software (ver. 1.2). Then, clustering of the genes with the DCA individuated each other related gene groups and shrank the whole data set to 3656 genes for 1–3 months and to 3615 for 6–15 months. Finally, the classification model was computed: a small number of classification formulas that are able to distinguish diseased from control mice were extracted from GELA and a small number of genes capable of effectively separating control and diseased mice samples were identified. The logic separating formulas for 1–3 and 6–15 months are reported in Tables 19.4 and 19.5, respectively. Each clause of such tables is able alone to separate the two different classes.

A 30-fold cross-validation and a percentage split sampling technique (10% test and 90% training) were used to validate the logic formulas, obtaining 99% of accuracy (percentage of the correctly classified samples) in both cases. We used 30-fold cross-validation in order to have at each run two samples, one case and one control, as test set.

The same classification analysis was performed with the WEKA [21] software for comparing the results obtained with GELA. A subset of the WEKA classification algorithms, C4.5 Decision Tree algorithm, RF, SVM, and KNN, was used on the Alzheimer's data set. Different parameter settings of the algorithms were used, the best resulting accuracies are reported in Table 19.6. All the algorithms were run by using a 30-fold cross-validation sampling procedure. From the results, it can be seen that all methods (except KNN) perform at a comparable level; GELA and SVM have excellent accuracies, but SVM produces classification models whose interpretation

Table 19.4 Logic formulas in early stage

Alzheimer's disease	(Nudt19 < 0.76) OR
	(Arl16 ≥ 1.31) OR
	(Aph1b ≥ 0.47) OR
	(Slc15a2 ≥ 0.55) OR
	(Agpat5 ≥ 0.73) OR
	(Sox2ot < 0.58 OR Sox2ot ≥ 1.53) OR
	(2210015D19Rik ≥ 0.86) OR
	(Wdfy1 ≥ 1.37)
Control	(Nudt19 ≥ 0.76) OR
	(Arl16 < 1.31) OR
	(Aph1b < 0.47) OR
	(Slc15a2 < 0.55) OR
	(Agpat5 < 0.73) OR
	(0.58 ≥ Sox2ot AND Sox2ot < 1.53) OR
	(2210015D19Rik < 0.86) OR
	(Wdfy1 < 1.37)

is very difficult for humans. On the other hand, GELA and C4.5 are able to extract meaningful and compact models (i.e., classification formulas and trees, respectively).

Other tests have been performed on data sets downloaded from public repositories *ArrayExpress* and *Gene Expression Omnibus* (GEO): *Psoriasis* and *Multiple Sclerosis Diagnostic*. The Psoriasis data set was composed of 54,613 gene expression profiles of 176 experimental samples (85 control and 91 diseased) and was provided from the National Psoriasis Foundation. The Multiple Sclerosis Diagnostic data set contained 22,215 gene expression profiles of 178 experimental samples (44 control and 134 diseased) and was released from the *National Institute of Neurological Disorders and Stroke* (NINDS). All gene expression profile values were normalized using the standard Affymetrix Expression Console software (ver 1.2) by the MAS5 algorithm. The results are reported in Table 19.7. In this case, all methods perform at a comparable level. SVM, KNN, and RF produce classification models that are difficult to understand for humans. On the other hand, GELA and C4.5 obtained clear classification models.

Finally, we describe the application of GELA for a case–control analysis on a public RNA-Seq data set of breast cancer extracted from *The Cancer Genome Atlas* (TCGA) data portal (http://cancergenome.nih.gov/). GELA has been applied to the *breast cancer* data set where 20,531 gene expression profiles of 783 subjects have been analyzed for classifying experimental samples either in control (i.e., healthy subjects) or in breast cancer patients. It was able to detect several clusters of similar genes and to provide a clear, compact, and accurate classification model, composed of small "if–then rules," which also allow detecting a small subset of those genes that characterize the breast cancer. The classification model was tested with a 10-fold cross-validation sampling and performed with a 98% accuracy. For comparing the

Table 19.5 Logic formulas in late stage

Alzheimer's disease	(Slc15a2 ≥ 0.62) OR (Agpat5 < 0.26 OR Agpat5 ≥ 0.55) OR (Sox2ot ≥ 1.78) OR (2210015D19Rik ≥ 0.82) OR (Wdfy1 < 0.75 OR Wdfy1 ≥ 1.29) OR (D14Ertd449e < 0.33 OR D14Ertd449e ≥ 0.52) OR (Tia1 < 0.17 OR Tia1 ≥ 0.49) OR (Txnl4 < 0.74) OR (1810014B01Rik < 0.71 OR 1810014B01Rik ≥ 1.17) OR (Snhg3 < 0.16 OR Snhg3 ≥ 0.35) OR [(1.12 ≥ Actl6a AND Actl6a < 1.42) OR Actl6a ≥ 1.48] OR (Rnf25 < 0.67 OR Rnf25 ≥ 1.26)
Control	(Slc15a2 < 0.62) OR (0.26 ≥ Agpat5 AND Agpat5 < 0.55) OR (Sox2ot < 1.78) OR (2210015D19Rik < 0.82) OR (0.75 ≥ Wdfy1 AND Wdfy1 < 1.29) OR (0.33 ≥ D14Ertd449e AND D14Ertd449e < 0.52) OR (0.17 ≥ Tia1 AND Tia1 < 0.49) OR (Txnl4 ≥ 0.74) OR (0.71 ≥ 1810014B01Rik AND 1810014B01Rik < 1.17) OR (0.16 ≥ Snhg3 AND Snhg3 < 0.35) OR [(0.81 < Actl6a AND Actl6a < 1.12) OR (1.42 < Actl6a AND Actl6a < 1.48)] OR (0.67 ≥ Rnf25 AND Rnf25 < 1.26)

Table 19.6 Classification accuracies [%] on Alzheimer data sets

Method	Setting	Early stage	Late stage	Model
GELA	No settings	100.0	100.0	Yes
SVM	*Polykernel*=2	96.66	100.0	No
RF	*Trees*=100	96.66	94.91	No
C4.5	*Unpruned, minobj*=2	98.33	98.30	Yes
KNN	*k*=2	70.00	86.44	No

results with respect to other supervised machine-learning algorithms, a similar analysis was performed with WEKA [21], using C4.5, RF, SVM, and KNN methods. The accuracy values of a 10-fold cross-validation scheme are depicted in Table 19.8.

Table 19.7 Classification accuracies [%] on multiple sclerosis and psoriasis data sets

Method	Setting	MsDiagnostic	Psoriasis	Model
GELA	No settings	94.94	100.0	Yes
SVM	*Polykernel*=2	90.45	98.86	No
RF	*Trees*=100	91.57	98.86	No
C4.5	*Unpruned, minobj*=2	87.08	97.16	Yes
KNN	*k*=2	87.64	99.43	No

Table 19.8 Classification accuracy [%] on breast cancer data set

Method	Settings	Accuracy	Model
GELA	No settings	98	Yes
C4.5	*Unpruned, minobj*=2	98	Yes
RF	*Trees*=100	99	No
SVM	*Polykernel*=2	99	No
KNN	*k*=2	99	No

ABCA5|23461<5.99 OR ADRB2|154<2.54 ⇒ Breast cancer

Figure 19.4 An example of GELA model for breast cancer.

All methods performed with excellent classification rates showing an accuracy that is even greater than 98%. RF, SVM, and KNN are the best performing methods, but GELA and C4.5 have the advantage to provide a human-readable classification model, such as in Figure 19.4.

19.8 CONCLUSIONS

Thanks to the new advances in microarray and RNA-Seq technologies, a large quantity of gene expression profile data are available both at online open data repositories and at local laboratories. In this chapter, efficient methods, algorithms, and software for performing an effective gene expression profile analysis have been described. In particular, the emphasis was placed on two types of analysis: *gene clustering* and *experiment classifications*, for which several methods have been presented and described. Afterward, gene expression profile analysis tools, which integrate these methods, have been illustrated. For providing the reader with a practical example, the software GELA and WEKA were applied to four real case studies. The performed experiments show a complete knowledge discovery process whose aim is to identify the hidden patterns in the different classes of the experimental samples by analyzing their gene expression profiles.

REFERENCES

1. Affymetrix. *Affymetrix Microarray Suite User Guide*. version 5th ed. Santa Clara (CA): Affymetrix; 2001.
2. Arisi I, D'Onofrio M, Brandi R, Felsani A, Capsoni S, Drovandi G, Felici G, Weitschek E, Bertolazzi P, Cattaneo A. Gene expression biomarkers in the brain of a mouse model for Alzheimer's disease: mining of microarray data by logic classification and feature selection. J Alzheimer's Dis 2011;24(4):721–738.
3. Ayadi W, Elloumi M, Hao J-K. A biclustering algorithm based on a bicluster enumeration tree: application to DNA microarray data. BioData Min 2009;2(1):9.
4. Bolstad BM, Irizarry RA, Åstrand M, Speed TP. A comparison of normalization methods for high density oligonucleotide array data based on variance and bias. Bioinformatics 2003;19(2):185–193.
5. Boros E, Hammer PL, Ibaraki T, Kogan A, Mayoraz E, Muchnik I. An implementation of logical analysis of data. IEEE Trans Knowl Data Eng 2000;12(2):292–306.
6. Chee M, Yang R, Hubbell E, Berno A, Huang XC, Stern D, Winkler J, Lockhart DJ, Morris MS, Fodor SP. Accessing genetic information with high-density DNA arrays. Science 1996;274(5287):610–614.
7. Cohen WW. Fast effective rule induction. Proceedings of the 12th International Conference on Machine Learning. Tahoe City, CA: Morgan Kaufmann Publishers; 1995. p 115–123.
8. Cox TF, Ferry G. Discriminant analysis using non-metric multidimensional scaling. Pattern Recognit 1993;26(1):145–153.
9. Crick FH, Watson JD. The complementary structure of deoxyribonucleic acid. Proc R Soc London, Ser A 1954;223(1152):80–96.
10. Cristianini N, Shawe-Taylor J. *An Introduction to Support Vector Machines and other Kernel-Based Learning Methods*. Cambridge, UK: Cambridge University Press; 2000.
11. Cuisenaire O, Macq B. Fast Euclidean distance transformation by propagation using multiple neighborhoods. Comput Vis Image Underst 1999;76(2):163–172.
12. Danielsson P-E. Euclidean distance mapping. Comput Graph Image Process 1980;14(3):227–248.
13. Dasarathy BV. *Nearest Neighbor (NN) Norms: NN Pattern Classification Techniques*. Los Alamitos (CA): IEEE Computer Society Press; 1990.
14. Díaz-Uriarte R, De Andres SA. Gene selection and classification of microarray data using random forest. BMC Bioinformatics 2006;7(1):3.
15. Do JH, Choi D. Normalization of microarray data: single-labeled and dual-labeled arrays. Mol Cells 2006;22(3):254–261.
16. Dulli S, Furini S, Peron E. *Data Mining*. Milan, Italy: Springer-Verlag; 2009.
17. Felici G, Truemper K. A minsat approach for learning in logic domains. INFORMS J Comput 2002;13(3):1–17.
18. Frank E, Witten IH. Generating accurate rule sets without global optimization; 1998.
19. Gaines BR, Compton P. Induction of ripple-down rules applied to modeling large databases. J Intell Inf Syst 1995;5(3):211–228.

20. Gentleman RC, Carey VJ, Bates DM, Bolstad B, Dettling M, Dudoit S, Ellis B, Gautier L, Ge Y, Gentry J, Hornik K, Hothorn T, Huber W, Iacus S, Irizarry R, Leisch F, Li C, Maechler M, Rossini AJ, Sawitzki G, Smith C, Smyth G, Tierney L, Yang JY, Zhang J. Bioconductor: open software development for computational biology and bioinformatics. Genome Biol 2004;5(10):R80.
21. Hall M, Frank E, Holmes G, Pfahringer B, Reutemann P, Witten IH. The WEKA data mining software: an update. SIGKDD Explor Newsl 2009;11:10–18.
22. Heller MJ. DNA microarray technology: devices, systems, and applications. Annu Rev Biomed Eng 2002;4(1):129–153.
23. Holloway AJ, Van Laar RK, Tothill RW, Bowtell DD. Options available–from start to finish–for obtaining data from DNA microarrays II. Nat Genet 2002;32:481–489.
24. Howe EA, Sinha R, Schlauch D, Quackenbush J. RNA-Seq analysis in MeV. Bioinformatics 2011;27(22):3209–3210.
25. Irizarry RA, Hobbs B, Collin F, Beazer-Barclay YD, Antonellis KJ, Scherf U, Speed TP. Exploration, normalization, and summaries of high density oligonucleotide array probe level data. Biostatistics 2003;4(2):249–264.
26. Jiang D, Tang C, Zhang A. Cluster analysis for gene expression data: a survey. IEEE Trans Knowl Data Eng 2004;16(11):1370–1386.
27. Jirapech-Umpai T, Aitken S. Feature selection and classification for microarray data analysis: evolutionary methods for identifying predictive genes. BMC Bioinformatics 2005;6:148.
28. Jolliffe I. *Principal Component Analysis*. New York, NY: Wiley Online Library; 2005.
29. Kaufman L, Rousseeuw P. *Finding Groups in Data An Introduction to Cluster Analysis*. New York: Wiley Interscience; 1990.
30. Kuehn H, Liberzon A, Reich M, Mesirov JP. Using genepattern for gene expression analysis. Curr Protoc Bioinform 2008;Chapter 7:Unit 7.12.
31. Leow WK, Li R. The analysis and applications of adaptive-binning color histograms. Comput Vis Image Underst 2004;94(1):67–91.
32. Li B, Dewey CN. RSEM: accurate transcript quantification from RNA-Seq data with or without a reference genome. BMC Bioinformatics 2011;12(1):323.
33. Li C, Wong WH. DNA-chip analyzer (dChip). In: Parmigiani G, Garrett ES, Irizarry R, Zeger SL, editors. *The Analysis of Gene Expression Data: Methods and Software*. New York: Springer-Verlag; 2003. p 120–141.
34. MacQueen J. Some methods for classification and analysis of multivariate observations. Proceedings of the 5th Berkeley Symposium on Mathematical Statistics and Probability, Volume 1; California 1967. p 281–297.
35. Marioni JC, Mason CE, Mane SM, Stephens M, Gilad Y. RNA-Seq: an assessment of technical reproducibility and comparison with gene expression arrays. Genome Res 2008;18(9):1509–1517.
36. Mortazavi A, Williams BA, McCue K, Schaeffer L, Wold B. Mapping and quantifying mammalian transcriptomes by RNA-Seq. Nat Methods 2008;5(7):621–628.
37. Quinlan JR. *C4. 5: Programs for Machine Learning*. Volume 1. San Francisco: Morgan Kaufmann Publishers; 1993.
38. Quinlan JR. Improved use of continuous attributes in C4.5. J Artif Intell Res 1996;4:77–90.
39. R Core Team. *R: A Language and Environment for Statistical Computing*. Vienna, Austria: R Foundation for Statistical Computing; 2013.

40. Reich M, Liefeld T, Gould J, Lerner J, Tamayo P, Mesirov JP. Genepattern 2.0. Nat Genet 2006;38(5):500–501.
41. Romdhane LB, Shili H, Ayeb B. Mining microarray gene expression data with unsupervised possibilistic clustering and proximity graphs. Appl Intell 2010;33(2):220–231.
42. Saeed AI, Bhagabati NK, Braisted JC, Liang W, Sharov V, Howe EA, Li J, Thiagarajan M, White JA, Quackenbush J. TM4 microarray software suite. Methods Enzymol 2006;411:134–193.
43. Sanger F, Nicklen S, Coulson AR. DNA sequencing with chain-terminating inhibitors. Proc Natl Acad Sci U S A 1977;74(12):5463–5467.
44. Schadt EE, Li C, Ellis B, Wong WH. Feature extraction and normalization algorithms for high-density oligonucleotide gene expression array data. J Cell Biochem 2001;84(S37):120–125.
45. Schena M, Shalon D, Davis RW, Brown PO. Quantitative monitoring of gene expression patterns with a complementary DNA microarray. Science 1995;270(5235):467–470.
46. Song JW, Chung KC. Observational studies: cohort and case-control studies. Plast Reconstr Surg 2010;126(6):2234.
47. Stekel D. *Microarray Bioinformatics*. Cambridge, UK: Cambridge University Press; 2003.
48. Stoughton RB. Applications of DNA microarrays in biology. Annu Rev Biochem 2005;74:53–82.
49. Tan P, Steinbach M, Kumar V. *Introduction to Data Mining*. Boston: Addison-Wesley; 2005.
50. Torra V, Narukawa Y. On a comparison between Mahalanobis distance and Choquet integral: the Choquet–Mahalanobis operator. Inf Sci 2012;190:56–63.
51. Trapnell C, Williams BA, Pertea G, Mortazavi A, Kwan G, van Baren MJ, Salzberg SL, Wold BJ, Pachter L. Transcript assembly and quantification by RNA-Seq reveals unannotated transcripts and isoform switching during cell differentiation. Nat Biotechnol 2010;28(5):511–515.
52. Vapnik VN. *Statistical Learning Theory*. New York: John Wiley and Sons, Inc.; 1998.
53. Wang Z, Gerstein M, Snyder M. RNA-Seq: a revolutionary tool for transcriptomics. Nat Rev Genet 2009;10(1):57–63.
54. Weitschek E, Felici G, Bertolazzi P. MALA: a microarray clustering and classification software. Database and Expert Systems Applications (DEXA), 23rd International Workshop on Biological Knowledge Discovery Vienna, Austria 2012. pp. 201–205.
55. Weitschek E, Fiscon G, Felici G, Bertolazzi P. *GELA: a software tool for the analysis of gene expression data. Database and Expert Systems Applications (DEXA)*, 26th International Workshop on Biological Knowledge Discovery; Valencia, Spain; 2015. p 31–35.
56. Xu R, Wunsch D. Survey of clustering algorithms. IEEE Trans Neural Netw 2005;16(3):645–678.

CHAPTER 20

MINING INFORMATIVE PATTERNS IN MICROARRAY DATA

Li Teng

Department of Internal Medicine, University of Iowa, Iowa City, IA, USA

20.1 INTRODUCTION

DNA microarray technology has had a central role in biological and biomedical research. It enables measuring the expression level of many thousands of genes within a number of different experimental conditions simultaneously. Microarray data provide clues of a comprehensive view of underlying biology. Finding relationships and recognizing nonrandom patterns or structures in microarray data are essential for molecular characterization of a wide range of diverse phenomena, from disease states and gene functions, to the differences between cells of different types. As the scale of microarray data keeps growing, analysis of these large data sets poses numerous algorithmic challenges.

Microarray data, also called *gene expression data*, are the output of microarray experiments. A microarry experiment measures the volume of gene products from the probe–target hybridization. The volume of gene products represents gene expression levels. High volume means high expression levels. Expression levels of thousands of genes are measured simultaneously in one single microarray experiment. Then data from multiple experiments are combined to form a gene expression data matrix. Gene expressions across all the samples are called *gene expression* patterns.

Pattern Recognition in Computational Molecular Biology: Techniques and Approaches,
First Edition. Edited by Mourad Elloumi, Costas S. Iliopoulos, Jason T. L. Wang, and Albert Y. Zomaya.

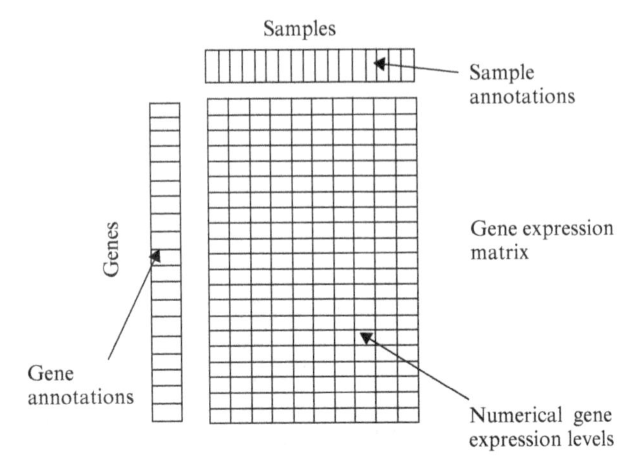

Figure 20.1 A conceptual view of a microarray data matrix. There are three parts in the data matrix: (i) numerical gene expression data, (ii) gene annotation, and (iii) sample annotation. The gene and sample annotations are important as the data only have meaning within the context of the underlying biology.

Figure 20.1 is a conceptual view of a gene expression data matrix. There are three parts in the data matrix, numerical expression data, gene annotation, and sample annotation. Normally, one column of the data matrix comes from one microarray experiment. Each element stands for the expression level of the corresponding gene in the corresponding sample. The rows of the matrix are gene expression patterns and the columns of the matrix are sample expression patterns.

Microarray data have properties of high noise, high variation, and high correlations. Increasing the experiment accuracy could reduce the noise to some extent, but still could not eliminate it. Microarray experiments have multiple sources of variation, for example, the measurement error (noise), the natural variability of measured attributes in biological system, and the known or unknown variability in experimental conditions. The high relevance in microarray data results in a large number of genes sharing very similar expression patterns in certain biological processes. In some research, duplicate patterns will be deleted as redundant. However, this may be inappropriate. One gene could be a trigger or a depressor to some other genes that have similar expression patterns. As a result, how to pick up interesting genes is a widely discussed topic.

Microarray data analysis often suffers from the inequality of a large number of genes versus a small number of samples. Microarray experiments typically involve measurements of expression levels of many thousands of genes in only dozens of biological samples. Because of the differences among microarray experiments both in technical and biological settings, there are limitations on combining microarray data sets from different experiments.

There are two straightforward ways that microarray data can be studied: studying genes with information provided by samples and studying samples with information provided by genes. In addition, both methods can be combined. When genes are under study, gene expression patterns provide information of the gene regulatory mechanism related to the experimental samples. From the sample analysis view, sample expression profile, or signature, can be thought of as a precise molecular definition of the cell in a specific state. Variations of gene/sample expression patterns provide rich underlying biological knowledge. The following are some of the topics that researchers have been working on through microarray data analysis:

- finding the functional role for a gene or group of genes and discovering the cellular processes that they participate in;
- finding the genes of which the expression levels differentially expressed across various diseases, compound treatments, or various cell types;
- finding the gene-regulatory mechanism, for example, the way genes and gene products interact and their interaction networks;
- finding the mechanism of cell-state transit.

The impact of microarray technology on biology depends heavily on data mining and statistical analysis. Sophisticated data-mining and analytical tools are needed to correlate all of the data obtained from multiple arrays in a meaningful way. The goal of microarray data analysis is to find relationships and recognize any nonrandom patterns or structures that require further exploration and ultimately achieve new insights into the underlying biology. To achieve this goal, the microarray data analysis problems should be efficiently formatted and solved using well-designed data-mining tools.

In this chapter, we will introduce the concepts and techniques for mining the informative patterns in microarray data that carry a biological meaning. Specifically, we will introduce several different types of nonrandom patterns and describe how they are mathematically formulated and measured, and then discuss some general methods used to find the patterns of interest. Since this field is still evolving, our emphasis has been on the methods that have a level of maturity.

20.2 PATTERNS WITH SIMILARITY

From similarity to closeness is one of the general ways of perception when there is little prior knowledge. Biologists have observed that genes involved in common processes are often coexpressed, for example, their expressions across samples are correlated. For the two straightforward ways to analyze microarray data, studying genes with information provided by samples and studying samples with information provided by genes, the general assumption is that genes or samples that share similar expression patterns may be coregulated and functionally related (for the genes) or coming from similar types of organisms (for the samples), respectively.

Both supervised and unsupervised learning methods based on various distance or similarity measurements are carried out to achieve this goal of finding genes/samples with similar expression patterns. So, first, let us take a look at the similarity measures that the algorithms rely on.

20.2.1 Similarity Measurement

Most of the microarray data analysis methods are based on pairwise comparisons among gene/sample expression patterns. Therefore, we first need a measure of similarity (or dissimilarity) to make comparisons among these objects, that is, gene/sample expression vectors. Among the many choices, Euclidean distance and Pearson correlation coefficient are the two commonly used measures. Other measures, such as Spearman's rank correlation, Minkowski distance, Manhattan distance, and mutual information, have also been applied in microarray data analysis.

The similarity matrix of microarray data G can be represented by an $n \times n$ matrix $A = (a_{ij})$, where n is the number of genes and a_{ij} stands for the similarity between two genes g_i and g_j. A is also called the *affinity matrix*. Sometimes, A is constructed as a dissimilarity matrix or a distance matrix.

Euclidean distance is one of the most intuitive ways to measure distance between two points in space. Suppose $G = \{g_{ij}\}$ is an $n \times m$ microarray data matrix. Each row of G stands for a gene and each column of G stands for a sample. Element g_{ik} stands for the expression of the ith gene g_i in the kth sample. The Euclidean distance between two genes g_i and g_j can be easily calculated using Equation 20.1. The larger the Euclidean distance, the less similarity these two gene expression patterns share. But it is not always the most appropriate measure for gene expression data. When density of the probe–target hybridization differs in each microarray experiment, the magnitude of gene expression levels from different samples will not be comparable. What is usually done to deal with this situation is to normalize the samples so that they have a similar background level, or use other similarity measurements.

$$d(g_i, g_j) = \sqrt{\sum_{k=1}^{m} (g_{ik} - g_{jk})^2} \tag{20.1}$$

Sometimes, the gene expression data represent comparative expression measurements, for example, ratio of two-color microarray or fold change. When the features have different scales, for example, one feature in the range of [0,100] and another feature in the range of $[-10, 10]$, a small variation in the feature with a large scale may cause a large variation in the Euclidean distance. Therefore, it is important to measure the similarity by comparing the "trends" of the expression patterns.

Pearson correlation coefficient (Eq. 20.2) measures the similarity between the trends of two gene expression vectors rather than absolute magnitude of the numerical values.

$$r(g_i, g_j) = \frac{\sum_{k=1}^{m}(g_{ik} - \overline{g}_i)(g_{jk} - \overline{g}_j)}{\sqrt{\sum_{k=1}^{m} (g_{ik} - \overline{g}_i)^2}\sqrt{\sum_{k=1}^{m} (g_{jk} - \overline{g}_j)^2}} \tag{20.2}$$

where $\overline{g}_i$ and $\overline{g}_j$ are the average expression levels of the two genes over all samples, respectively. Similarity between sample expression vectors is measured in a similar way. Pearson correlation coefficient is approximately 1 for a strong correlation, approximately 0 for a weak correlation, and approximately −1 for a strong negative correlation. The value gives us a clear impression of how similar (or not similar) two vectors are.

Spearman's rank correlation is a distance measure that is invariant to monotone changes, that is, ignores the magnitude of the changes if their relative ranks within the series are preserved. To be more precise, first we order the values in each of the series nondecreasingly and assign to each value a rank beginning with 1. For example, vectors $g_1 = (1, 4, -4, 8, 7)$ and $g_2 = (-2, 8, 2, 4, 6)$ are transformed into rank vectors $g_1' = (2, 3, 1, 5, 4)$ and $g_2' = (1, 5, 2, 3, 4)$, respectively. The rank correlation between g_1 and g_2 is the Pearson correlation between g_1' and g_2'. The advantage of rank correlation is that the effect from the scales of the features was ruled out and it is robust to outliers. The disadvantage is that some information may be lost during the ranking transform.

The above measures all rely on numerical comparison between two genes (or samples) across all measured features (samples or genes, respectively). These measures treat all features with equal importance. There are situations when some features are more reliable, for example, samples with more replicates are more reliable than those with one replicate. Weighted correlation coefficient was proposed to calculate the correlations between the features that have different degrees of importance.

$$\frac{\Sigma o_i \overline{x_i}\,\overline{y_i} - \Sigma o_i \overline{x_i} \Sigma o_i \overline{y_i} / \Sigma o_i}{\sqrt{(\Sigma o_i \overline{x_i}^2 - (\Sigma o_i \overline{x_i})^2 / \Sigma o_i)(\Sigma o_i \overline{y_i}^2 - (\Sigma o_i \overline{y_i})^2 / \Sigma o_i)}} \tag{20.3}$$

where x and y are two genes that are measured under k samples, and o_i is the confidence level of feature i. $\overline{x_i}$ and $\overline{y_i}$ represent, respectively, the mean values of x and y on the ith sample normalized by o_i. The samples with higher confidence level are more creditable, so they contribute more to the similarity measure. When all the o_is are equal, they cancel out, and Equation 20.3 becomes the same formula as that for a correlation coefficient (Eq. 20.2).

Weight correlation coefficient is also good for finding local similarities (see Section 20.2.3 for details). When similar patterns appear only under a subset of samples (or genes), they will be missed under measures such as Euclidean distance or Pearson correlation coefficient. Different weights are assigned to the genes and samples to differentiate the contributions in the similarity measure. Then weighted correlation coefficient is used to measure the local similarities in a global way. In addition, one can find local similarities, for example, biclusters, without breaking down the whole data structure.

There is no golden rule on how to choose the best similarity measurement in certain situations. One has to consider the nature of the data and the goals of the analysis, and then perhaps try one or a few different measures. Having said that, Pearson correlation coefficient is the most commonly used as it captures the similarity between trends, and is close to the intuitive goal and comparably robust.

20.2.2 Clustering

The goal of clustering is to group or segment a collection of objects into "clusters" (subsets of objects) such that objects within the same cluster are more similar (high intrasimilarity) compared to those assigned to different clusters (low intersimilarity). Central to the clustering analysis is the measurement of similarity (or dissimilarity) between two objects, between two clusters, or between an object and a cluster. Depending on the data properties, for example, continuous data or categorical data, one can choose an appropriate similarity measure from, but not limited to, those we have discussed in the previous section.

With large gene expression data set and little prior knowledge, it is intuitive to use the unsupervised learning method to explore the structure of the data set and find those nonrandom patterns in the data. Clustering is first and widely used in microarray data analysis as a tool to group together genes (or samples) with similar expression patterns. It can also be useful for discovering distinct patterns where each pattern represents objects with a substantially different activity across the features to reduce the scale for downstream analysis. Clustering can also be used for the detection of outliers.

Many of the well-known clustering algorithms have been successfully used in microarray data, including hierarchical clustering [9], k-means clustering [26], and self-organizing maps [23]. Principal components analysis [13] is also used to reduce the dimensionality of data sets. Further analysis is done after clustering, to determine whether the gene products share similar functional roles, through, for example, Gene Ontology (GO) enrichment analysis, pathway enrichment analysis. Besides, the resulting clusters can be used as prototypes for gene network construction.

Nowadays, clustering has become a fairly standard approach for microarray data analysis. There are many well-written tools and packages in software and programming environment, such as *Matlab*, R, and SAS. Figure 20.2 shows an example of a clustering result on iris data set [10] using the R function *heatmap. 2*. The iris data set has been widely used in the pattern-recognition literature. It has 150 instance/rows and 4 attributes. In the figure, both the rows and columns are clustered using hierarchical clustering. For details of the procedures of these algorithms, refer any textbook on pattern recognition or data mining. In this chapter, we will discuss more on how to use these algorithms with a proper setting.

20.2.2.1 Consensus Clustering. The fundamental issues of clustering include (i) how to determine the optimal number of clusters and (ii) how to assign confidence to the selected number of clusters, as well as to the induced cluster assignments. These are particularly important in microarray data analysis, for which little knowledge of a presumable number of clusters is available. In addition, the problem of a relatively small sample size is compounded by the very high dimensionality of the data available, making the clustering results especially sensitive to noise and susceptible to overfitting. A setting of the number of clusters is required for the clustering process as well as the analysis of the clustering result. Using k-means algorithm that groups the objects into k clusters, one has to assign the number of k before running the

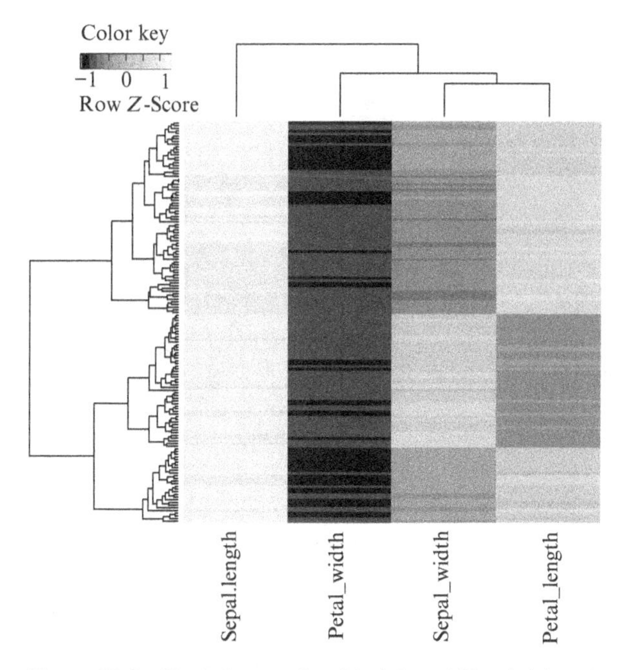

Figure 20.2 Clustering result on iris data, a 150 × 4 data set.

algorithm. For hierarchical clustering, one gets a tree structure representing the similarities among the objects. Then the tree may be cut at certain similarity threshold to get a set of clusters or the tree may be cut to get a specific number of clusters. Setting with a small number of clusters may cause the objects to become not well divided, while a large number of clusters may result in multiple clusters with similar patterns.

Finding an optimal number of clusters gives a clear separation of the objects and reduces redundancy. Methods have been proposed to find an optimal number of targeted clusters. Recent proposals include trying different settings and finding an optimal setting, postprocessing the clusters to merge the ones with high similarities, and using resampling and cross-validation techniques to simulate perturbations of the original data. Monti *et al.* [19] developed a general, model-independent resampling-based methodology of class discovery and clustering validation and visualization tailored to the task of analyzing microarray data. They termed it textitconsensus clustering. It has emerged as an important elaboration of the classical clustering problem.

Consensus clustering provides a method to represent the consensus across multiple runs of a clustering algorithm, to determine the number of clusters in the data, and to assess the stability of the discovered clusters. The method can also be used to represent the consensus over multiple runs of a clustering algorithm with random restart (such as k-means, hierarchical clustering, model-based Bayesian clustering, and SOM (Self-Organizing Map)) so as to account for its sensitivity to the initial conditions. Finally, it provides a visualization tool to inspect the cluster number, membership, and boundaries.

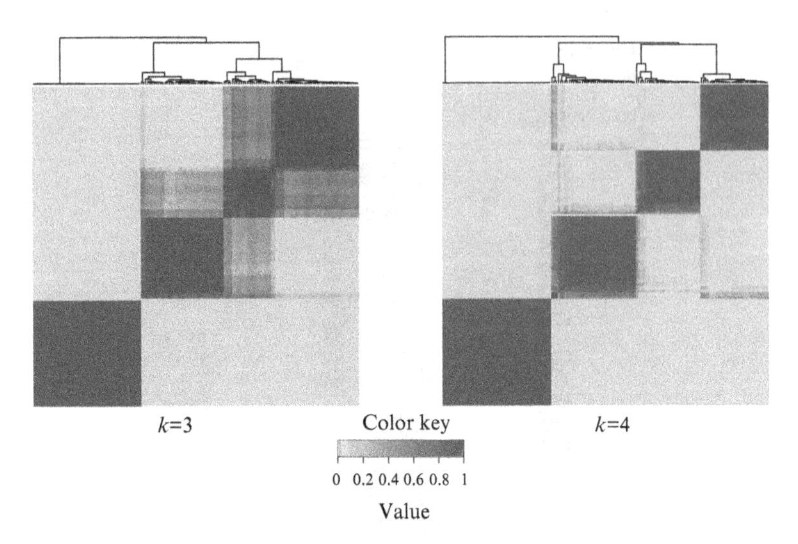

Figure 20.3 Consensus matrix for iris data with $k = 3$ and k=4.

Consensus of the clustering result is measured through a consensus matrix. The basic procedure is as follows: given a set of n genes $G = (g_i), i = 1, 2, \ldots, n$, the goal is to partition the genes into k sets of exhaustive and nonoverlapping clusters. Suppose that a resampling scheme and a clustering algorithm have been chosen. An $n \times n$ consensus matrix $C = \{c_{ij}\}, i, j = 1, 2, \ldots, n$, is constructed to represent and quantify the agreement among the repeated clustering runs. Initially, $C = 0$. In each run, a subset (e.g., 80% of the total genes from random sampling) of the whole data set is selected. The clustering algorithm is carried out on the sampled data to group the genes into k clusters. If g_i and g_j are grouped in the same cluster, the value of c_{ij} is increased by 1. The random sampling and clustering process are repeated for several runs. The final consensus matrix for cluster number = k is normalized by the total number of runs. So each element in C is a real number between 0 and 1.

Figure 20.3 (generated using R package *ConsensusClusterPlus* [31]) shows the consensus matrix of clustering iris data, 150 rows, with $k = 3$ and $k = 4$. Perfect consensus corresponds to a consensus matrix C with all 0s or 1s, which means that any two objects are either always been assigned in one cluster or never been assigned in one cluster in all runs (e.g., perfectly stable clusters). Furthermore, if the items in the matrix were arranged so that items belonging to the same cluster are adjacent to each other, as shown in Figure 20.3, perfect consensus would translate into a block-diagonal matrix with nonoverlapping blocks of 1s along the diagonal, each block corresponding to a different cluster surrounded by 0s.

20.2.2.2 Consensus Distribution. It is a statistical measure defined as the Area Under Curve (AUC) of the Cummulative Distribution Function (CDF) plot of the nondiagonal elements of the consensus matrix. There are two extreme cases corresponding to the minimum and the maximum value of the consensus distribution, respectively: k=1 and $k = n$ (n is the number of objects). When $k = 1$, all objects are always clustered together and the consensus matrix has only 1s, and when $k = n$, every object is a cluster and all the nondiagonal elements of the consensus matrix are 0s. So the consensus distribution provides a simplified and generalized measure of consensus matrix. Monti *et al.* [19] proposed a method to determine the optimal value of k by inspecting the increase in the consensus distribution when increasing the number k. A large increase in the consensus distribution suggests a significant increase in the clustering stability. For more details, refer to the original paper and manual of the R package *ConsensusClusterPlus*.

20.2.3 Biclustering

Clustering is able to find groups of similar gene (or sample) expression patterns. However, applying clustering algorithms to gene expression data has a significant difficulty when the patterns are not apparent in a global way. The clustering process builds on the assumption that related genes behave similarly across all measured samples. This assumption is reasonable when the data set contains only a few samples from a single, focused experiment, but does not hold for larger data sets containing hundreds of heterogeneous samples from many unrelated experiments.

Our general understanding of cellular processes leads us to expect subsets of genes to be coregulated and coexpressed only under certain experimental conditions, but to behave almost independently under other conditions. An example is shown in Figure 20.4. The gene expression patterns in the figure have no apparent similarity across the 10 measured samples (Figure 20.4a). However, if we pick up only a subset of the samples, the two genes show exactly the same trend (Figure 20.4b). Discovering such local expression patterns is the key to uncover many genetic pathways that are not apparent otherwise. Therefore, it is highly desirable to move beyond the clustering paradigm and to develop approaches capable of discovering local patterns in microarray data [10].

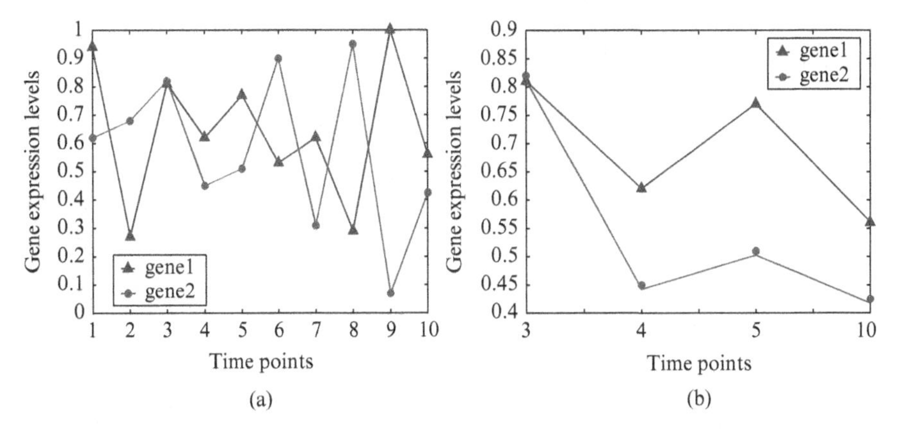

Figure 20.4 Two genes that coexpress only under a subset of the samples: (a) the original patterns and (b) under extracted subset of samples.

Biclustering or *coclustering* is clustering of the both dimensions of the data matrix simultaneously. Many biclustering algorithms have been proposed and used in the identification of coregulated genes, gene functional annotation, and sample classification. The specific problem addressed by biclustering can be defined as follows: given data matrix $G = \{g_{ij}\}$, with n rows and m columns. G can be defined by its set of rows, $X = \{x_1, \dots, x_n\}$, and its set of columns, $Y = \{y_1, \dots, y_m\}$. Conside that $I = \{i_1, \dots, i_k\} \subseteq X$ $(k \leq n)$ and $J = \{j_1, \dots, j_s\} \subseteq Y$ $(s \leq m)$ are subsets of the rows and columns, respectively. $B = (I, J)$ denotes the submatrix of G that contains only the elements g_{ij} belonging to the submatrix with the set of rows I and the set of columns J. Biclustering identifies a set of biclusters $B_k = (I_k, J_k)$ such that each bicluster B_k satisfies some specific characteristics of homogeneity.

20.2.4 Types of Biclusters

The first difference among various biclustering methods is in the way they measure the homogeneity of a bicluster. The criterion to evaluate a biclustering algorithm concerns the identification of the type of biclusters the algorithm is able to find. Figure 20.5 shows some examples of different kinds of biclusters. They can be divided into four major classes [18]:

1. Biclusters with constant values
2. Biclusters with constant values on rows or columns
3. Biclusters with coherent values
4. Biclusters with coherent evolutions.

The first three classes analyze directly the numeric values in the data matrix and try to find subsets of rows and/or subsets of columns with similar behaviors.

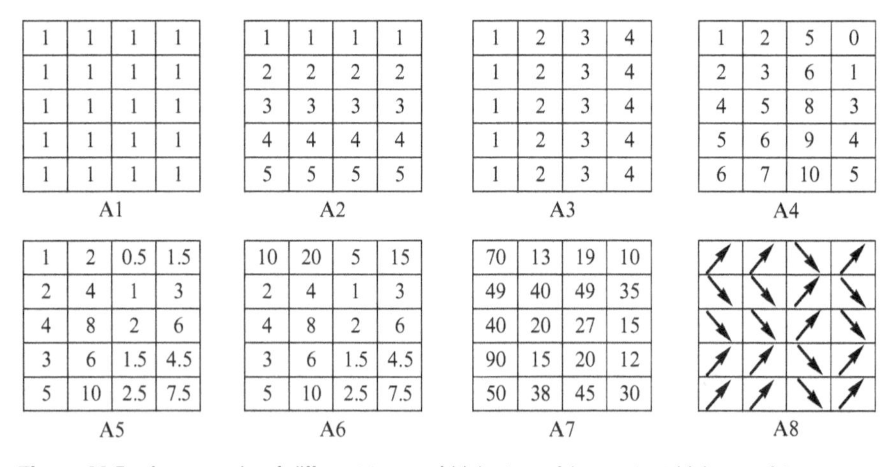

A1

1	1	1	1
1	1	1	1
1	1	1	1
1	1	1	1
1	1	1	1

A2

1	1	1	1
2	2	2	2
3	3	3	3
4	4	4	4
5	5	5	5

A3

1	2	3	4
1	2	3	4
1	2	3	4
1	2	3	4
1	2	3	4

A4

1	2	5	0
2	3	6	1
4	5	8	3
5	6	9	4
6	7	10	5

A5

1	2	0.5	1.5
2	4	1	3
4	8	2	6
3	6	1.5	4.5
5	10	2.5	7.5

A6

10	20	5	15
2	4	1	3
4	8	2	6
3	6	1.5	4.5
5	10	2.5	7.5

A7

70	13	19	10
49	40	49	35
40	20	27	15
90	15	20	12
50	38	45	30

Figure 20.5 An example of different types of biclusters. A1: constant bicluster, A2: constant rows, A3: constant columns, A4: coherent values (additive model), A5: coherent values (multiplicative model), A6: coherent values (multiplicative model), A7: coherent evolution on the columns, and A8: coherent sign changes on rows and columns.

These behaviors can be observed on the rows, on the columns, or in both dimensions of the data matrix, as in A1–A7. The fourth class aims to find coherent behaviors regardless of the exact numeric values in the data matrix. As such, biclusters with coherent evolutions view the elements in the data matrix as symbols. These symbols can be purely nominal or represent coherent positive and negative changes relative to a normal value, as in A8. The first two types of biclusters are in the category of bicluster with coherent values. A bicluster with coherent evolution is the most generalized form. Algorithms that could find biclusters in a more generalized form are regarded as more powerful in real applications.

20.2.4.1 Biclusters with Constant Values. In gene expression data, constant biclusters reveal subsets of genes with similar expression values within a subset of samples. A1 in Figure 20.5 is an example of this kind of bicluster. When the goal of a biclustering algorithm is to find constant biclusters, it is natural to consider ways of reordering the rows and columns of the matrix to arrange together similar rows and similar columns and discover biclusters with similar values. This approach produces good results only on data without noise. Hartigan [12] introduced a partition-based algorithm called *direct clustering*, which is known as *block clustering*. He used variance (Eq. 20.4) to evaluate the quality of each bicluster $B = (I, J)$:

$$variance(B) = \sum_{i \subseteq I, j \subseteq J} (g_{ij} - g_{IJ})^2 \tag{20.4}$$

where g_{IJ} is the mean of all elements in a bicluster. According to this criterion, a perfect bicluster is a submatrix with variance = 0 and this is a constant bicluster. Tibshirani et al. [29] improved Hartigan's block-splitting algorithm method [12] by adding a backward pruning method. They found more constant biclusters. Cho et al. [7] used the variance to find constant biclusters together with an alternative measure to enable the discovery of more complex biclusters.

20.2.4.2 Biclusters with Constant Values on Rows or Columns. A bicluster with constant values in the rows identifies the subset of genes with similar expression values across a subset of samples, allowing the expression levels to differ from gene to gene. Similarly, a bicluster with constant columns identifies the subset of samples within which a subset of genes exhibit similar expression values. There exists considerable practical interest in discovering biclusters that exhibit coherent variation on the rows and/or on the columns of the data matrix. A2 and A3 in Figure 20.5 are examples of biclusters with constant values on rows or columns, respectively.

This class of biclusters cannot be found simply by computing the variance of the values within the bicluster. The straightforward approach to identify these biclusters is to normalize the rows or the columns of the data matrix using the row mean and the column mean, respectively. So, these biclusters can be transformed into constant biclusters. Getz et al. [11] performed a relatively complex normalization step before their algorithm was implemented. They not only found biclusters with constant rows or constant columns but also managed to find more complex biclusters with coherent values. Moreover, as perfect biclusters with constant rows or columns are difficult to find in real data because of noise, there are approaches considering the possible existence of multiplicative noise or the values in the rows or columns belonging to a certain interval [5, 21, 22].

20.2.4.3 Biclusters with Coherent Values. Researchers are interested in identifying more general coherence among the genes and the samples. The coherence can be evaluated by high similarity scores or the additive/multiplicative relationships among the rows and columns. (Details about the additive and multiplicative models are discussed in Chapter 4.) A4, A5, and A6 in Figure 20.5 are examples of biclusters with coherent values. The rows and columns of A4 have additive relationship, while the rows and columns of A5 and A6 have multiplicative relationship. Their rows and columns can be turned into the same vector by adding or multiplying by a constant. Gene or sample expression patterns in biclusters with coherent values exhibit more universal coherence compared with biclusters with exactly the same values on rows or columns.

Several biclustering algorithms [6, 7, 11, 14, 16, 20, 25, 30, 32, 33] were proposed to find biclusters with coherent values. Cheng and Church [6] defined a bicluster as a subset of rows and subset of columns with a high mean-squared residue score. The mean-squared residue score (see Chapter 4 for details) was used to measure the homogeneity of the biclusters. It conforms to the additive model. Kluger et al. [15] also addressed the problem of discovering biclusters with coherent values and looked for checkerboard structures in the data matrix by integrating biclustering of rows and columns with normalization of the data matrix. They assumed that after appropriate normalization, biclusters can be accentuated if they exist.

Tang *et al.* [25] introduced the Interrelated Two-Way Clustering (ITWC) algorithm. After normalizing the rows of the data matrix, they computed the vector-angle cosine value between each row and a predefined stable pattern to test whether the row varies much among the columns and removed the ones with little variation. Then, they used a correlation coefficient to measure the strength of the linear relationship between two rows or two columns, to perform two-way clustering. As this similarity measurement depends only on the pattern and not on the absolute magnitude of the vector, ITWC also identifies biclusters with coherent values.

20.2.4.4 Biclusters with Coherent Evolutions. There are biclustering algorithms that address the problem of finding coherent evolutions across the rows and/or columns of the data matrix. A7 and A8 in Figure 20.5 are examples of biclusters with coherent evolutions. In A2, there are no apparent correlations among the rows and the columns. But in each row, the trend of expression value across two adjacent samples is the same. If we move up and down to express the change, as shown in A8, we see perfect correlation among the rows/columns. Identifying biclusters with coherent evolutions may be helpful if one is interested in finding a subset of genes that are upregulated or downregulated across a subset of samples without taking into account their actual expression values, or if one is interested in identifying a subset of samples that have always the same or opposite effects on a subset of genes.

The order-preserving cluster is a type of biclusters with coherent evolutions. Ben-Dor *et al.* [2] first defined the Order-Preserving SubMatirx (OPSM). They defined a bicluster as a group of rows whose values induce the same linear order across a subset of the columns. Their work focused on the relative orders of the columns in the bicluster rather than on the uniformity of the actual values in the data matrix. They developed a stochastic model to discover the best row-supported submatrix given a fixed size of samples.

Following the idea of Ben-Dor *et al.*, Liu and Wang [17] defined a bicluster as an OP-cluster if there is a permutation of its columns under which the sequences of values in every row are strictly increasing. They proposed an exhaustive bicluster enumeration algorithm to discover all significant OP-clusters. They used a compact tree structure, called the *OPC-tree*, to store the identities of the genes and the sequences sharing the same prefixes. The construction of the OPC-tree is very time-consuming and needs excessive memory resources. Pruning is needed to make the algorithm effective.

Bleuler and Zitzler [3] used a scoring function that combines the mean-squared residue score [6] with the OPSM concept. It allows to arbitrarilyscale the degree of orderliness required for a cluster. They proposed an evolutionary algorithm framework that allows simultaneous clustering over multiple time course experiments. A local search procedure based on a greedy heuristic was applied to each individual before evaluation. They found nonoverlapping clusters in one run. Teng and Chan [28] proposed two heuristic algorithms Growing Prefix and Suffix (GPS) and Growing Frequent Position (GFP) to find the common orders shared by a significant number of genes. Their methods, with comparatively lower space and computation cost, aimed to find a significantly larger OP-cluster more efficiently.

20.2.5 Measurement of the Homogeneity

The performance of a biclustering method is evaluated by the variety of resulting biclusters. As mentioned in Chapter 3, there are different types of biclusters, such as biclusters with constant values, biclusters with constant values on rows or columns, biclusters with coherent values, and biclusters with coherent evolutions. Some of them, such as the first three types, can be standardized using mathematical formulas, while, for the biclusters with coherent evolutions, there is no exact mathematical description.

Existing methods use different merit functions that conform to the formulation of a certain kind of biclusters and their results are evaluated in a relative way. On the other hand, the methods they used to evaluate the biclusters also reflect their usability in finding various kinds of biclusters. In the following, we first introduce some existing measures, then we propose the *Average Correlation Value* (ACV). ACV can evaluate more types of biclusters compared with the other alternatives.

20.2.5.1 The Additive and Multiplicative Models. The additive (Eq. 20.5) and multiplicative models (Eq. 20.6) are widely used to model biclusters with coherent values [18]. Several biclustering algorithms assume either additive or multiplicative models [5, 11, 12, 22]. Suppose $B = (b_{ij})$ is a perfect bicluster with coherent values, then all the values in B can be obtained using one of the following expressions.

$$b_{ij} = \mu + \alpha_i + \beta_j \tag{20.5}$$

$$b_{ij} = \mu' \times \alpha'_i \times \beta'_j \tag{20.6}$$

where b_{ij} is an element in the bicluster, μ and μ' are typical values within the bicluster, α_i and α'_i are the adjustments for row i, and β_j and β'_j are the adjustments for column j.

In an additive model, each element b_{ij} is seen as the summation of the typical value within the bicluster, μ, the adjustment for row i, α_i, and the adjustment for column j, β'_j. The bicluster A4 in Figure 20.5 is an example of a bicluster with coherent values on both rows and columns, of which the values can be described using an additive model. The biclusters A1, A2, and A3 in Figure 20.5 can be considered the special cases of this general additive model. For A2 and A3, the coherence of values can be observed on the rows and columns of the bicluster, respectively (when $\alpha_i = 0$ and $\beta_j = 0$).

The multiplicative model is equivalent to the additive model when $\mu = \log(\mu')$, $\alpha_i = \log(\alpha'_i)$, and $\beta_j = \log(\beta'_j)$. In this model, each element is seen as the product of the typical value within the bicluster, the adjustment for row, and the adjustment for column. The biclusters A5 and A6 in Figure 20.5 are examples of biclusters with coherent values on both rows and columns. They can be described using a multiplicative model. A5 can be changed to A6 after magnifying its first row.

These two models can be simplified to Equation 20.7 if we use σ_i to replace $\mu + \alpha_i$ and use σ'_i to replace $\sigma' \times \alpha'_i$.

$$\begin{aligned} b_{ij} &= \sigma_i + \beta_j \\ b_{ij} &= \sigma'_i \times \beta'_j \end{aligned} \tag{20.7}$$

The additive and multiplicative models are good to model the perfect biclusters, of which the value of each element can be easily computed by the value of other rows and columns. However, in real applications, perfect biclusters are rarely seen when noise is considered. Similarly to the A7 bicluster in Figure 20.5, it has no clear additive or multiplicative relationship among the rows and columns, or to the A8 bicluster in Figure 20.5, only general trends of the elements are involved. To discover these biclusters, we need more general evaluations.

20.2.5.2 *Mean-Squared Residue Score.* Cheng and Church [6] defined a bicluster as a subset of rows and a subset of columns with a high similarity score. The similarity score H, which was introduced and called the *mean-squared residue*, was used as a measure of the coherence of the rows and columns in the bicluster. The mean-squared residue score and some related criteria that are based on the evaluation of the residue have also been widely used in this field [7, 30, 33].

Suppose $B = \{b_{ij}\}$ is an $n \times m$ data matrix. The mean-squared residue score for B could be calculated by the following function:

$$H(B) = \frac{1}{nm} \sum_{i=1}^{n} \sum_{j=1}^{m} (b_{ij} - b_{iM} - b_{Nj} + b_{NM})^2 \tag{20.8}$$

where

$$b_{iM} = \frac{1}{m} \sum_{j=1}^{m} b_{ij}, b_{Nj} = \frac{1}{n} \sum_{i=1}^{n} b_{ij} \tag{20.9}$$

and

$$b_{NM} = \frac{1}{nm} \sum_{i=1}^{n} \sum_{j=1}^{m} b_{ij} = \frac{1}{n} \sum_{i=1}^{n} b_{iM} = \frac{1}{m} \sum_{j=1}^{m} b_{Nj} \tag{20.10}$$

A low mean-squared residue score, for example, $H(B) = 0$, means that each element b_{ij} can be uniquely defined by its row mean, b_{iM}, its column mean, b_{Nj}, and the bicluster mean, b_{NM}.

$$b_{ij} = b_{iM} + b_{Nj} - b_{NM} \tag{20.11}$$

Note that this is to consider $\mu = b_{NM}$, $\alpha_i = b_{iM} - b_{NM}$, and $\beta_j = b_{Nj} - b_{NM}$ in Equation 20.5.

Cheng and Church used the residue to quantify the difference between the actual value of an element b_{ij} and its expected value predicted from the corresponding row mean, column mean, and bicluster mean. They supposed that a low mean-squared residue score plus a large variation from the constant may be a good criterion for identifying a bicluster. They also defined the residues for one row and one column, respectively, as

$$residue\_row(i) = \frac{1}{m} \sum_{j=1}^{m} (b_{ij} - b_{iM} - b_{Nj} + b_{NM})^2 \tag{20.12}$$

$$residue\_column(j) = \frac{1}{n} \sum_{i=1}^{n} (b_{ij} - b_{iM} - b_{Nj} + b_{NM})^2 \tag{20.13}$$

They proposed several row/column removal methods to delete rows/columns that have large residues to find biclusters with mean-squared score lower than a threshold. Cheng and Church's method has the potential to be very fast. However, using their methods, appropriate normalization has to be taken out before biclusters other than those with additive models are discoverable. They may make wrong decisions, and by deleting rows/columns they could lose good biclusters in the early processing.

The mean-squared residue score can be used as a merit function to find biclusters of additive models. Since such biclusters have a minimum mean-squared residue score,

$$b_{iM} = \alpha_i + \frac{\Sigma\beta_j}{m}, b_{Nj} = \beta_j + \frac{\Sigma\alpha_i}{n}, b_{NM} = \frac{\Sigma\beta_j}{m} + \frac{\Sigma\alpha_i}{n} \tag{20.14}$$

So we have $H(B) = 0$. However, for biclusters, the multiplicative models are as follows:

$$b_{iM} = \alpha_i \frac{\Sigma\beta_j}{m}, a_{Nj} = \beta_j \frac{\Sigma\alpha_i}{n}, b_{NM} = \frac{\Sigma\beta_j \Sigma\alpha_i}{nm} \tag{20.15}$$

$$H(B) = \frac{1}{nm}\Sigma\Sigma\left[\left(\beta_j - \frac{\Sigma\beta_j}{m}\right)\left(\alpha_i - \frac{\Sigma\alpha_i}{n}\right)\right]^2 \neq 0 \tag{20.16}$$

The value of Equation 20.16 has no upper bound. A scaling (multiplying a row or column by a constant) can easily change the value. So the mean-squared residue score cannot be used as a criterion to evaluate the homogeneity of biclusters of the multiplicative model. Similarly to A5 and A6 in Figure 20.5, they are perfect biclusters in the multiplicative model, but they have a large difference in mean-square residue score.

20.2.5.3 *Average Correlation Value.* The Pearson correlation coefficient is widely used in microarray data analysis, as it conforms to the assumption of similarity in trend but not in magnitude among gene expressions. It is difficult to set a threshold between similar and nonsimilar expressions because of the differences in data background. However, in most studies, a Pearson correlation coefficient of 0.8 is considered high enough to declare that two genes/samples are coexpressed or coregulated.

There are positive as well as negative correlations and both of them are interesting for gene regulation analysis. So a bicluster in microarray data should be a subset of genes and samples in which the patterns are highly correlated either in a positive way or in a negative way. However, the existing criteria are not suitable for measuring negative correlations.

The *Average Correlation Value* (ACV) [27] of the genes (and samples) was proposed to evaluate the homogeneity of a bicluster. Matrix $B = \{b_{ij}\}$ has the ACV, which is defined by the following function:

$$\begin{aligned} \overline{R}(B) = \max\{ & \frac{\sum_{i=1}^{n}\sum_{j=1}^{n}|r\_row_{ij}| - n}{n^2 - n}, \\ & \frac{\sum_{k=1}^{m}\sum_{l=1}^{m}|r\_col_{kl}| - m}{m^2 - m}\} \\ \overline{R}(B) \subseteq [0, 1] & \end{aligned} \tag{20.17}$$

where r_row_{ij} is the correlation between the ith and jth rows of B and r_col_{kl} is the correlation between the kth and lth columns of B. So a high value of $\overline{R}(B)$ means that the rows or columns of B are highly correlated.

As we have mentioned in the previous section, mean-squared residue score [6] and some related criteria that are based on the evaluation of residue have been widely used in this field. However, mean-squared residue score is not feasible for biclusters

Table 20.1 Corresponding mean-squared residue score and average correlation value of A1–A7

	A1–A4	A5	A6	A7
$H(A_i)$	0	0.625	2.425	131.875
$\overline{R}(A_i)$	1	1	1	0.850

of multiplicative models. To compare the mean-squared residue score and ACV, we list the mean-squared residue score and ACV of A1–A7 of Figure 20.5 in Table 20.1, respectively.

ACV always gets the best value for both the additive model and the multiplicative model. So A1–A6 all have the best value. A7 has a fairly high ACV and this could be an evidence that A7 is a reasonable bicluster. ACV could be a criterion to measure whether it is necessary to do biclustering. A high average correlated value means that the samples or the genes in one data set are highly correlated in the whole scale. Then it is more reasonable and feasible to analyze the data by clustering methods considering the computation cost. In addition, a low average correlated value means neither the samples nor the genes are similar in the whole scale and it is more appropriate to use biclustering to find local similarities in the data set.

20.2.6 Biclustering Algorithms with Different Searching Schemes

Because of the NP-hardness of this problem, there is no method that guarantees a perfect, quick answer. Besides the exhaustive enumeration methods, most of the biclustering approaches are heuristic. They can be divided into five groups according to the searching scheme: iterative row and column clustering combination, divide and conquer, greedy iterative search, exhaustive bicluster enumeration, and distribution parameter identification (for a review see Reference [18]).

1. Iterative row and column clustering combination methods: Sometimes they are called two-way clustering (cf. References [4, 11, 25]). The results obtained by clustering each dimension of the data matrix, separately, are relied on; then in some way, the algorithms combine the results of clustering on two dimensions to form stable biclusters. However, when performing clustering on one dimension, the structure of the other dimension is unknown and potential biclusters can be broken.
2. Divide-and-conquer algorithms: For an example, the block clustering by Hartigan [12] is the first divide-and-conquer approach to perform biclustering. Block clustering is a top-down, row and column clustering method of the data matrix. Beginning with the entire data in one block (bicluster), at each iteration, it finds the row or column that produces the largest reduction in the total "within block" variance by splitting a given block into two pieces. In order to find the best split, the rows and columns of the data matrix are sorted by row and column means, respectively. Duffy and Quiroz [8] suggested the

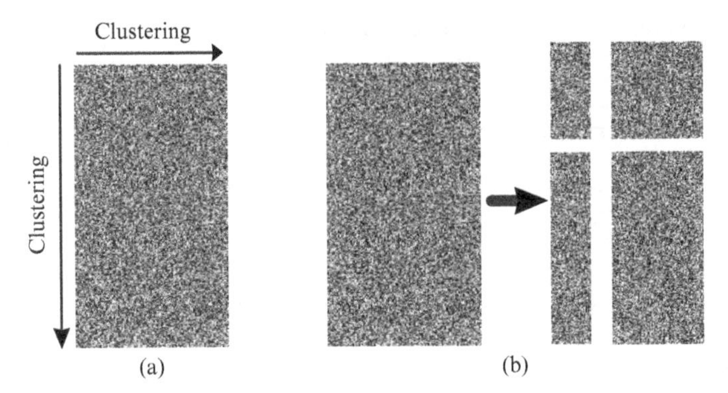

Figure 20.6 An illustration of the (a) iterative row and column clustering combination method, and (b) divide-and-conquer algorithm.

use of permutation tests to determine when a given block split is not significant. Tibshirani et al. [29] improved the algorithm of Hartigan by finding an optimal number of splicings. Those methods are potentially very fast. But the approaches have the significant drawback that good biclusters could be missed by splitting.

Figure 20.6 shows an illustration of the two biclustering methods we have just mentioned. By clustering on either dimension or splitting the data set, they break down the original data structures into smaller pieces. In this way, they first scale down the problem to a manageable scale, then solve it by finding biclusters in smaller pieces of the original data structure. However, as the breaking is heuristic, good biclusters can be split into different pieces and lost in the results.

3. Greedy iterative search methods: These approaches are based on the idea of creating biclusters by adding or removing rows/columns and they use certain criterion that maximizes a local gain. Cheng and Church proposed several greedy row/column removal/addition algorithms and combined them in an overall approach to find biclusters with low mean-squared residue scores. The mean-squared residue score (detailed in Chapter 4) and some related criteria, which are based on the residue, are widely used in gene expression biclustering [7, 30, 32, 33]. These algorithms [33] could be fast; however, they would make wrong decisions by deleting the rows/columns in the intermediate steps and miss good biclusters.
4. Exhaustive bicluster enumeration methods: Many researchers took the idea that the best biclusters can only be identified using an exhaustive enumeration of all possible biclusters existing in the matrix. Their algorithms (such as those defined in References [17, 24, 30]) certainly find the best biclusters, if they exist. Owing to the complexity, these methods have high computation cost and sometimes can only be executed under some restrictions, such as the size of the biclusters. Efficient algorithmic techniques were designed to make the enumeration feasible.

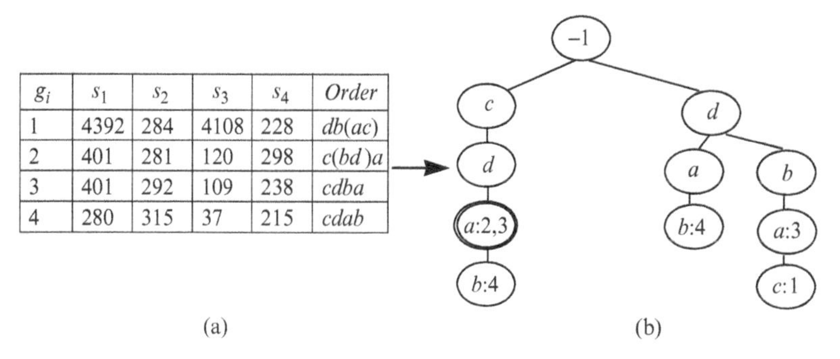

g_i	s_1	s_2	s_3	s_4	*Order*
1	4392	284	4108	228	*db(ac)*
2	401	281	120	298	*c(bd)a*
3	401	292	109	238	*cdba*
4	280	315	37	215	*cdab*

Figure 20.7 An example of the OPC-tree constructed by Liu and Wang [17]. (a)The sequences set and (b) the OPC-tree.

Here, we will elaborate on Liu and Wang's [17] work. They proposed an exhaustive bicluster enumeration algorithm to find OP-clusters with a minimum number of rows and columns. They implemented a compact tree structure, called the *OPC-tree*, to store the identities of the genes and the sequences sharing the same prefixes. Figure 20.7 shows an example in which the sequences set on the left is the original data and the OPC-tree they constructed is on the right.

As shown in the figure, the numerical data are first transformed to sequence data with each sequence representing the ranking of samples for each gene. Labels inside brackets indicate values with little difference. They can be considered interchangeable depending on the definition of the target biclusters.

Liu and Wang's algorithm takes the minimum number of genes and minimum number of samples as the threshold. Biclusters that satisfy these size limits will be suppressed. The construction of the OPC-tree is based on depth-first traversal. The major steps are as follows:

Step 1. A root is created and set as the current node. All the sequences are inserted in the tree such that all sequences sharing the same prefix fall on the same prefix of the tree.

Step 2. For each child of the current node, insert suffixes of its subtree to the node's child that has a matching label.

Step 3. Prune the current root's children to satisfy the minimum gene and the sample threshold.

Step 4. Take the first child of the current node and set it as the current node, repeat Steps 2 and 3.

Figure 20.7 shows the intermediate step of constructing the OPC-tree for four sequences. Current node "*a*:2,3" is shown in double circles. The sequence IDs are stored in the leaves. Sequence 2 "*c*(*bd*)*a*" and sequence 3 "*cdba*" are sharing an order of "*cda*" that is the path from the root to node "*a*:2,3.". The length of the subtree is the number of samples in the OP-cluster and the

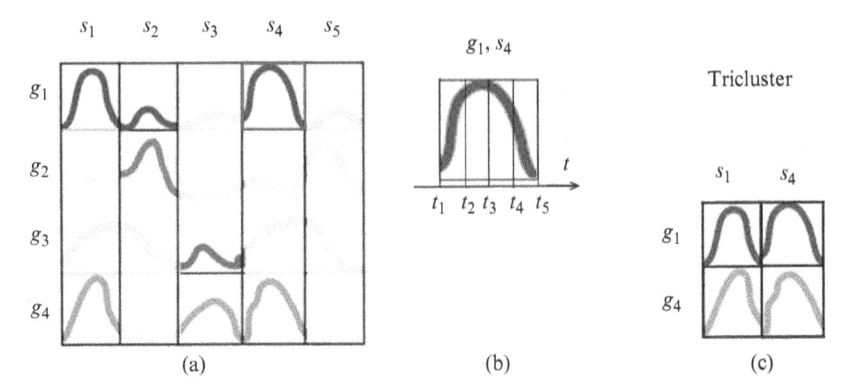

Figure 20.8 An illustration of triclustering. (a) The 3D gene–sample-time microarray data. (b) An enlargement of expression pattern of g_1 on sample s_4. (c) The tricluster that contains $\{g_1, g_4\}$ and $\{s_1, s_4\}$.

number of sequence IDs is the size of the genes. Subsequences "*ac*" and "*ab*" have been pruned for too few leaves. After this node has been processed, the current node will move to "*d*," the left sibling of the root. The procedure is time-consuming when the data set is large, but it discovers all OP-clusters with a user-specified threshold.

Jang *et al.* [14] proposed another tree data structure-based biclustering algorithm that finds coherent gene clusters. They have expanded the biclustering concept to 3D microarray data, a novel type of gene–sample-time microarray data. A coherent bicluster of the 3D data contains a subset of genes and a subset of samples such that the genes are coherent on the samples along the time series. Triclustering is another term for related proposals. Figure 20.8 is an illustration of the gene–sample-time microarray data and an example of the coherent gene clusters in which the genes have similar expressed patterns on the subset of samples. They proposed two efficient algorithms, the Sample–Gene Search and the Gene–Sample Search, to mine the complete set of coherent gene clusters. The basic idea is to enumerate all the possible combinations of the genes/samples and find the number of support samples/genes. With well-designed data structure and efficient pruning techniques, these approaches can effectively find all coherent gene clusters.

5. Distribution parameter identification methods: These methods assume a given statistical model and try to identify the distribution parameters used to generate the data by iteratively minimizing a certain criterion (such as the plaid model method by Lazzeroni and Owen [16] and [20–22]). However, these methods have excessive computation cost.

Many algorithms use variance or residue as the merit function. Those methods rely on normalization steps as preprocessing. Normalization can be carried out on the rows or the columns, or on both dimensions. There is no golden rule for normalization.

Many attempts will have to be made. Some could be complex to implement and the structures of the original biclusters could be changed.

Besides biclustering along the gene–sample dimensions, there has been a lot of interest in mining gene expression patterns across time points [1]. TriCluster [34] can mine arbitrarily positioned and overlapping clusters, and depending on different parameter values, it can mine different types of clusters, including those with constant or similar values along each dimension, as well as scaling and shifting expression patterns. Data complexity and volume increase with the development of experimental techniques. Researchers will continue thinking of and developing new methods to understand the data.

20.3 CONCLUSION

Microarray techniques measure the expression level of a large number of genes, perhaps all genes of an organism, within a number of different experimental samples. By finding relationships and recognition of any nonrandom patterns or structures that need further exploration in microarray data, we can ultimately achieve new insights into the underlying biology. In microarray data, the set of genes/samples of interest may take different forms. On the basis of the assumption that coexpressed genes (or samples) would share similar functional roles, we find genes (or samples) with similar expression patterns and predict unknown genes (or samples) with properties of known genes (or samples). Traditional analysis uses clustering to find groups of genes/samples that have similar patterns. Clustering works well when the similarity lies in a global way. Biclustering of microarray data is promising in finding local patterns, in which the subset of genes and the subset of samples show some homogeneity. This conforms to the general understanding that some genes may coexpress or coregulate only under a subset of experimental samples. That would be crucial for discovering the regulatory mechanism among genes.

In this chapter, we emphasized on the introduction of different types of informative patterns in microarray data including their mathematical formulation and their quality measurement, and briefly introduced the algorithms for finding these patterns. The approaches that we have discussed are also applicable in large-scale data-based research areas such as information retrieval, text mining, collaborative filtering, recommendation systems, target marketing, and market research. They are suitable for analyzing a large sparse data set in finding interesting patterns either on a global or a local scale.

REFERENCES

1. Bar-Joseph Z. Analyzing time series gene expression data. Bioinformatics 2004;20(16): 2493–2503.
2. Ben-Dor A, Chor B, Karp R, Yakhini Z. Discovering local structure in gene expression data: the order-preserving submatrix problem. Proceedings of RECOMB'02. Washington DC; ACM Press; 2002. p 49–57.

3. Bleuler S, Zitzler E. Order preserving clustering over multiple time course experiments. Proceedings of EvoBIO'05; Lausanne, Switzerland; 2005. p 33–43.
4. Busygin S, Jacobsen G, Kramer E. Double conjugated clustering applied to leukemia microarray data. Proceedings of the 2nd SIAM International Conference on Data Mining, Workshop Clustering High Dimensional Data; Arlington, VA; 2002.
5. Califano A, Stolovitzky G, Tu Y. Analysis of gene expression microarrays for phenotype classification. Proceedings of International Conference Computational Molecular Biology; Tokyo, Japan; 2000. p 75–82.
6. Cheng Y, Church G. Biclustering of expression data. Proceedings of ISMB'00. Menlo Park, CA; AAAI Press; 2000. p 93–103.
7. Cho H, Dhillon IS, Guan Y, Sra S. Minimum sum squared residue co-clustering of gene expression data. Proceedings of the 4th SIAM International Conference Data Mining; Lake Buena Vista, FL; 2004.
8. Duffy D, Quiroz A. A permutation based algorithm for block clustering. J Classif 1991;8: 65–91.
9. Eisen MB, Spellman PT, Brown PO, Bottstein D. Cluster analysis and display of genome-wide expression patterns. Proc Natl Acad Sci U S A 1998;95:14863–14868.
10. Fisher RA. The use of multiple measurements in taxonomic problems. Annu Eugen 1936; 7:179–188.
11. Getz G, Levine E, Domany E. Coupled two-way clustering analysis of gene microarray data. Proc Natl Acad Sci U S A 2000;97:12079–12084.
12. Hartigan JA. Direct clustering of a data matrix. J Am Stat Assoc 1972;67(337):123–129.
13. Hotelling H. Analysis of a complex of statistical variables into principal components. J Educ Psychol 1993;24:441.
14. Jiang D, Pei J, Ramananthan M, Tang C, Zhang A. Mining coherent gene clusters from gene-sample-time microarray data. Proceedings of the 19th ACM SIGKDD Conference on Knowledge Discovery and Data Minging; Chicago, IL; 2004. p 430–439.
15. Klugar Y, Basri R, Chang JT, Gerstein M. Spectral biclustering of microarray data: coclustering genes and conditions. Genome Res 2003;13:703–716.
16. Lazzeroni L, Owen A. Plaid models for gene expression data. Technical report. Stanford University; 2000.
17. Liu J, Wang W. OP-Cluster: clustering by tendency in high dimensional space. Proceedings of the 3rd IEEE International Conference Data Mining; Dallas, TX; 2003. p 187–194.
18. Madeira SC, Oliveira AL. Biclustering algorithms for biological data analysis: A survey. IEEE/ACM Trans Comput Biol Bioinform 2004;1:24–25.
19. Monti S, Tamayo P, Mesirov J, Golub T. Consensus clustering: a resampling-based method for class discovery and visualization of gene expression microarray data. Mach Learn 2003;52:91–118.
20. Segal E, Battle A, Koller D. Decomposing gene expression into cellular processes. Proc Pac Symp Biocomput 2003;8:89–100.
21. Segal E, Taskar B, Gasch A, Friedman N, Koller D. Rich probabilistic models for gene expression. Bioinformatics 2001;17:243–252.
22. Sheng Q, Moreau Y, De Moor B. Biclustering microarray data by Gibbs sampling. Bioinformatics 2003;19:196–205.
23. Tamayo P, Slonim D, Mesirov J, Zhu Q, Kitareewan S, Dmitrovsky E, Lander ES, Gloub TR. Interpreting patterns of gene expression with self-organizing maps:

methods and application to hematopoietic differentiation. Proc Natl Acad Sci U S A 1999;96:2907–2912.

24. Tanay A, Sharan R, Shamir R. Disvoering statistically significant biclusters in gene expression data. Bioinformatics 2002;18:136–144.
25. Tang C, Zhang L, Ahang I, Ramanathan M. Iterrelated two-way clustering: an unsupervised approach for gene expression data analysis. Proceedings of the 2nd IEEE International Symposium on Bioinformatics and Bioengineering; Bethesda, MD; 2001. p 41–48.
26. Tavazoie S, Hughes JD, Campbell MJ, Cho RJ, Church GM. Systematic determination of genetic network architecture. Nat Genet 1999;22:281–285.
27. Teng L, Chan L. Discovering biclusters by iteratively sorting with weighted correlation coefficient in gene expression data. J VLSI Signal Process Syst 2007;50(3):267–280.
28. Teng L, Chan L. Mining order preserving patterns in microarray data by finding frequent orders. Proceeding of the Seventh International Conference on Bioinformatics and Bioengineering; Boston, MA; 2007. p 1019–1026.
29. Tibshirani R, Hastie T, Eisen M, Ross D, Botstein D, Brown P. Clustering methods for the analysis of DNA microarray data. Technical report, Department of Health Research and Policy and Department of Genetics and Department of Biochemestry, Stanford University; 1999.
30. Wang H, Wang W, Yang J, Yu PS. Clustering by pattern similarity in large data sets. Proceedings of the 2002 ACM SIGMOD International Conference on Management of Data; Madison, WI; 2002. p 394–405.
31. Wilerson M, Waltman P. ConsensusClusterPlus: ConsensusClusterPlus. R package version 1.20.0; 2013.
32. Yang J, Wang W, Wang H, Yu P. δ-clusters: capturing subspace correlation in a large data set. Proceedings of the 18th IEEE International Conference on Data Engineering; Washington DC; 2002. p 517–528.
33. Yang J, Wang H, Wang W, Yu P. Enhanced biclustering on expression data. Proceedings of the 3rd IEEE Conference on Bioinformatics and Bioengineering; Bethesda, MD; 2003. p 321–327.
34. Zhao L, Zaki MJ. triCluster: an effectie algorithm for mining coherent clusters in 3D microarray data. Proceedings of SIGMOD'05; Chicago, IL; 2005. p 694–705.

CHAPTER 21

ARROW PLOT AND CORRESPONDENCE ANALYSIS MAPS FOR VISUALIZING THE EFFECTS OF BACKGROUND CORRECTION AND NORMALIZATION METHODS ON MICROARRAY DATA

Carina Silva[1,2], Adelaide Freitas[3,4], Sara Roque[3], and Lisete Sousa[2,5]

[1]*Lisbon School of Health Technology, Lisbon, Portugal*
[2]*CEAUL, Lisbon, Portugal*
[3]*Department of Mathematics, University of Aveiro, Aveiro, Portugal*
[4]*CIDMA, Aveiro, Portugal*
[5]*Department of Statistics and Operations Research, University of Lisbon, Lisbon, Portugal*

21.1 OVERVIEW

Microarray technology, developed in the mid-1990s, is aimed at simultaneously monitoring thousands of genes by quantifying the abundance of the transcripts under

Pattern Recognition in Computational Molecular Biology: Techniques and Approaches,
First Edition. Edited by Mourad Elloumi, Costas S. Iliopoulos, Jason T. L. Wang, and Albert Y. Zomaya.

the same experimental condition. The use of DNA arrays in comparative analysis of gene expression has provided significant advances in the identification of specific genes from certain tissues, with direct impact on the understanding of the processes that lead to cell differentiation. Among various available array technologies, double-channel cDNA microarray experiments provide numerous technical protocols associated with functional genomic studies and are also the focus of this chapter. In a two-color cDNA microarray experiment, two samples of mRNA (tissues), one labeled with Cy5 (red) and the other with Cy3 (green) fluorescent dyes, are simultaneously hybridized to a microscope slide (microarray) spotted with a large number of cDNA sequences (probes) according to some biological application. After the laboratory phase, for each fluorescent dye (red and green), an image of a two-dimensional array with the fluorescence intensity value detected in each pixel is produced by a scanning process. Each pixel value represents the level of hybridization at a specific location on the microarray and is classified as foreground intensity or background intensity, according to whether the pixel is inside the spot or in the surrounding area, respectively. Both mean and median intensity values of the pixel distribution of each spot are usually calculated to represent the foreground intensities (say, R_f and G_f for red and green fluorescent dyes, respectively) and background intensities (say, R_b and G_b for red and green fluorescent dyes, respectively) of each spot. A file with the raw data information is created for each microarray. While the foreground intensities, R_f and G_f, represent estimates of the specific binding of the labeled mRNA to the spotted probes, the background intensities, R_b and G_b, represent all nonspecific signals, such as chemicals and auto-fluorescence on the glass slides, because of infiltrating noise.

Microarray experiments involve many steps: spotting cDNA, extracting mRNA, labeling targets, hybridizing, scanning, analyzing images, and so on. Each step can introduce experimental bias and random fluctuations in the measured intensities that affect the quality of raw data. *Background Correction* (BC) and *Normalization* (NM) are preprocessing techniques aimed at correcting the raw data at these undesirable random and systematic fluctuations arising from technical factors and retaining as much as possible of the biological variations due to biological factors [26].

21.1.1 Background Correction Methods

For each spot, the intensity measured during the scanning process may be due not only to the hybridization of the target to the probe, but also to some unspecific hybridization. The BC, also referred to as the *adjusting signal* [5], aims (i) to remove unspecific background noise detected in each spot; (ii) to correct technical effects (e.g., the presence of other chemicals on the glass) in processing arrays; (iii) to adjust nonspecific or cross-hybridizations, and (iv) to adjust the estimates of the expressions to be linearly related to the concentration [16]. Assuming that any observed intensity value is affected by a background signal or noise, more accurate gene expression levels will be found if its elimination is provided. The application of BC strategies on the observed intensity values (R_f, G_f) has been suggested in order to obtain accurate measures of true intensities (R, G). A complete overview of BC methods is available in the R package `limma` [20]. We briefly describe six BC methods available through the function `backgroundCorrect` in this chapter (only formulas for the red fluorescence intensities are presented as the ones for the green intensities are equivalent [9]):

1. none (No BC is performed): $R = R_f$.
2. sub (Subtraction): $R = R_f - R_b$.
3. half (Half): $R = 0.5$, if $R_f - R_b < 0.5$; $R = R_f - R_b$ otherwise.
4. min(Minimum): $R = R_f - R_b$, if $R_f - R_b \geq 0$; $R = \min_{1 \leq i \leq g}(R_{fi} - R_{bi} : R_{fi} - R_{bi} \geq 0)$ otherwise, with g the total number of genes.
5. edw (Edwards): $R = R_f - R_b$, if $R_f - R_b \geq \delta$; $R = \delta e^{1-(R_b+\delta)/R_f}$ otherwise.
6. nexp (Normexp): This method is based on the convolution of a normal distribution with an exponential distribution. The true signals (R, G) are assumed to follow an exponential distribution and are independent of the background signals (R_b, G_b). In addition, the background signals are supposed to be normally distributed and the foreground intensities (R_f, G_f) are considered additive (i.e., $R_f = R + R_b$ and $G_f = G + G_b$). The corrected intensity values are then calculated as the conditional expected values of the true signal (R, G) given the observed foreground. Several variants can be considered depending on how the model parameters are estimated.

21.1.2 Normalization Methods

In addition to the inevitable random errors, there are also systematic errors due to the experimental protocols, the scanning parameters, and differences between chips, because they are produced in different periods. For the chips to be comparable, the expression levels of different arrays should have the same distribution. The main objective of NM methods is the removal (or minimization) of nonbiological variation (e.g., dye effect) in the measured signal intensity levels among arrays [5, 16]. This allows the detection of appropriate biological differences between gene expressions. Nevertheless, NM strategies inevitably alter the data and certainly introduce bias. There may be biological reasons for the distributions to be skewed, and every NM, in an artificial way, will naturally remove important biological differences. There is no well-known global method for standardization or calibration of the data, which means that it is prudent to apply different methods and compare results in subsequent analyses.

While BC is executed on the green and red intensity values (R_f, R_b, G_f, and G_b) separately, NM is applied on the combination of these intensities into a single measure called M-values, where $M = \log_2 R/G$. An overview of several location NM strategies that adjust spatial- and/or intensity-dependent dye biases, and scale NM strategies that adjust scale differences, are presented by Chiogna et al. [6] and Wu et al. [26]. In this chapter, besides nonnormalization, five NM methods are considered. All of them are available in the R package `marray`, provided by bioconductor, and are based on the loess regression function, also called *lowess* [7]. Graphically, the loess function corresponds to curves determined by locally weighted (linear or quadratic) polynomial regressions to smooth data. In the loess-based NM methods, the loess curve fits the data represented in scatterplots, called *MA*-plots, where *A*-values are displayed, $A = \log_2 \sqrt{RG}$, against the *M*-values, for one array or for all arrays. The formulas of the normalized *M*-values, say M', obtained by the six NM methods considered in our comparative study and also in Reference [9], are as follows:

1. NN (No NM is performed): $M' = M$.
2. lg (Igloess): $M' = M - c(A)$, where $c(A)$ is the loess curve fitted to the *MA*-plot.
3. ll (Illoess): $M' = M - c_i(A)$, where $c_i(A)$ is the loess curve fitted to the *MA*-plot for the *i*th print-tip group (i.e., group of genes spotted in the same grid on the microarray slide), $i = 1, \ldots, N$, with N the total number of print-tip groups.
4. Sl (Slloess): $M' = M - \text{loess}(l_i, c_i)$, where $\text{loess}(l_i, c_i)$ is a bidimensional loess function depending on the row location, l_i, and the column location, c_i, of the spots on the microarray for the *i*th print-tip group.
5. lg–Sl (Igloess–Slloess): This method is based on two steps: $M' = M - c(A)$ (first step) and $M'' = M' - loess(l_i, c_i)$ (second step).
6. ll–Sl (Illoess–Slloess): This method is based on two steps: $M' = M - c_i(A)$ (first step) and $M'' = M' - loess(l_i, c_i)$ (second step).

21.1.3 Literature Review

Studies involving gene expression microarray applications are essentially divided into three types: class comparison, class prediction, and class discovery [1, 17]. In the first, the goal is to correctly identify *Differentially Expressed* (DE) genes between two or more experimental conditions. In the second, the main objective is to predict the correct class provided by classifiers obtained from microarray data sets. In the last, the purpose is to detect similar patterns of genes (i.e., genes with similar expression profiles under the same subset of experimental conditions) in order to identify, for instance, regulatory properties of cellular processes [8].

The best microarray data preprocessing technique in at least one of those three contexts has been studied by several authors [1, 4, 6, 9, 11, 20, 21, 26]. Most of these works have focused on the effects of BC and NM, separately, and it is difficult to define a set of BC and NM common to several studies.

Ritchie et al. [20] evaluated eight BC methods and concluded that BC methods that stabilize the variance of the *M*-values across the whole processed intensity range present the best option for the identification of DE genes. Ambroise et al. [1] arrived at a similar conclusion when assessing the impact of four BC methods to recommend the use of the BC methods sub and edw for class prediction and the BC method edw for class comparison studies. The BC method nexp was recommended by Ritchie et al. [20] for class comparison. Sifakis et al. [21] evaluated seven BC methods on the basis of class comparison using four statistical descriptive measures. These measures suggested that two new BC methods proposed by Sifakis et al. [21] are globally better than the BC methods none and sub.

For the evaluation of NM methods in the context of class comparison, we can refer the following studies: (i) Fujita et al. [11] compared five NM methods in terms of the robustness of their normalization curve estimated in the *MA*-plot; for that, interquartile ranges and the existence of outliers in the sets of the normalized expression levels were analyzed; the authors observed that loess-based NM methods may not be sufficiently robust in the presence of outliers creating difficulties in the identification of DE genes; (ii) Chiogna et al. [6] studied 10 NM methods (including NN and lg) in terms of specificity and sensitivity highlighting that substantially different DE gene lists can be obtained using different NM techniques. Hence, these authors recommend a careful exploratory analysis before applying any NM method.

In the context of class discovery, Wu et al. [26] examined 41 NM methods on the predictive performance of *k*-nearest neighbor classifier. They concluded that two-step NM strategies, which remove both intensity effect at the global level and spatial effect at the local level (for instance, lg–Sl), tend to be more effective in reducing the classification error. Nevertheless, it is worth mentioning that, as highlighted by Freyhult et al. [10], until now, no NM strategy is generically known to have a good performance in all situations. Freyhult et al. [10] argued that NM methods seem to have a significant positive impact on the ability to cluster individuals correctly based on expression profiles; however, the evaluation of the performance of different NM methods needs more research.

On the basis of NM methods studied by Wu et al. [26], Freitas et al. [9] investigated the effects of six BC methods (none, sub, half, min, edw, and nexp), combined with six NM methods (NN, lg, ll, Sl, lg–Sl, and ll–Sl), on the predictive performance of two classifiers (*k-nearest neighbor* (kNN) and *support vector machine* with linear kernel (SVM)). The two classifiers were obtained from each of 36 preprocessing strategies, resulting from the combination of each of the six BC methods with each of the six NM techniques, and applied on three published cDNA microarray data sets: lymphoma (lym), lung (lun), and liver (liv) (all available in http://genome-www5.stanford.edu/), whose microarrays were already classified by cancer type. Freitas et al. [9] reported the observed estimates of the classification error rates using *Leave-One-Out Cross Validation* (LOO-CV). The results suggested that sub and nexp are the best methods in terms of the performance of classifiers.

In all above-mentioned works, there are no graphical representations aimed to investigate the effects of the preprocessing techniques on the studies involving gene expression microarray class. In this chapter, we provide a graphical representation of the impact of the six BC and the six NM methods on the cDNA microarray data sets lym, lun, and liv, in the context of class prediction and class comparison applications. For the former, we analyze the predictive performance of kNN and SVM classifiers using the table of the observed estimates of the classification error rates, which is reported in Freitas et al. [9]. For the latter, we analyze the DE genes resulting from the SAM [23] and the *Arrow plot* [22], after the application of each one of the 36 preprocessing strategies over each data set for SAM, and over data set liv for the *Arrow plot*. For graphical visualization, we organize each approach in a

6×6 matrix (6 BC $\times$ 6 NM) and obtain the graphical representation of each matrix using correspondence analysis. The main goal is to provide a visual comparison of both profiles of BC and NM methods simultaneously.

We start the chapter by detailing the *Arrow plot*, which is a recent graphical-based methodology to detect DE genes, and briefly mention the SAM procedure, which is, in contrast, quite well known. Next, we introduce the *Correspondence Analysis* (CA) and explain how the resultant graphic can be interpreted. Then, we execute CA in both class comparison and class prediction applications and over the data sets lym, lun, and liv. Finally, we present the conclusions.

21.2 ARROW PLOT

Numerous methods have been developed for selecting DE genes, but the *Fold-Change* (FC) method is generally used as the first approach. However, FC can lead to erroneous conclusions not only because of the uncertainty that is induced by the quotient between two levels of intensity, but also because it implicitly admits that the variance is constant for all genes. Several variants of the t-statistic have been explored, but the problem is that in these experiments, various hypotheses are tested for a very small number of replicates, leading to unstable statistics. For example, high values of the test statistic may occur owing to the presence of very small variances, even when the difference between the mean expression levels is very small. The disadvantages of the t-statistics and FC have been pointed out by several authors.

In many microarray experiments, the first goal is to sort the genes according to the value of a test statistic instead of calculating the corresponding p-values. This is because only a very small number of genes can then be monitored, even if more genes are considered significant. In this work, the approach for the selection of DE genes will be based on ranking and selecting the first genes in an orderly list of values obtained from two different statistics. However, as the goal is to select some genes from a very large list, thousands of hypotheses are being tested simultaneously and the issue of multiple testing should be addressed through the correction of the p-values. This issue is widely discussed in many publications. Selection of genes from the p-values, whether corrected or not, seems to be the preferred approach to select genes.

21.2.1 DE Genes Versus Special Genes

DE genes are usually detected using statistics based on means or medians. However, if there are genes expressed in different subclasses, the above-mentioned techniques will not select them because either mean or median values tend to be similar between the considered groups.

Genes with a bimodal or a multimodal distribution within a class (considering a binary study) may indicate the presence of unknown subclasses with different expression values [4, 5, 6], meaning that there are two separate peaks in the distribution: one peak due to a subclass clustered around a low expression level and a second peak due to a subclass clustered around a higher expression level. As a consequence, the identification of such subclasses may provide useful insights into the biological mechanisms underlying physiological or pathological conditions. In cancer research, a common approach for prioritizing cancer-related genes is to compare gene expression profiles between cancer and normal samples, selecting genes with consistently higher expression levels in cancer samples. Such an approach ignores tumor heterogeneity and is not suitable for finding cancer genes that are overexpressed in only a subgroup of a patient population. As a result, important genes differentially expressed in a subset of samples can be missed by gene selection criteria based on the difference between sample means. Silva-Fortes et al. [22] proposed a new tool, the *Arrow plot*, in order to address this issue.

The construction of the *Arrow plot* is based on the graphical representation of the estimated *Overlapping Coefficient* (OVL) between two densities and the area under the *Receiver Operating Characteristic* (ROC) curve of all the genes belonging to the experiment. We will then present the main characteristics of these statistics. Silva-Fortes et al. [22] derived this graphic on the basis of nonparametric estimates of the *Area Under the ROC Curve* (AUC) and OVL.

21.2.2 Definition and Properties of the ROC Curve

Consider that $X_1, X_2, X_3, \ldots, X_{n_0}$ and $Y_1, Y_2, \ldots, Y_{n_1}$ are two random and independent samples representing the expression levels of a gene in the control and the experimental populations, respectively. Let F_X be the distribution function associated with X_i and F_Y the distribution function associated with Y_j, and without loss of generality, for any cutoff point $c \in \mathbb{R}$, $F_X(c) > F_Y(c)$. The sensitivity and specificity are given by

$$q(c) = 1 - F_Y(c) \tag{21.1}$$

and

$$p(c) = F_X(c) \tag{21.2}$$

respectively. The ROC curve results from the graphical representation of all coordinate pairs $(1 - p(c), q(c))$ in the unit Cartesian plane by varying the cutoff point c along the support of the decision variable (Figure 21.1). This is a monotonically increasing function, and if for some cutoff point c, $1 - F_X(c) = 1 - F_Y(c)$, the classification model is uninformative, that is, the ROC curve coincides with the main diagonal (or reference line) of the unit plane. A perfect classification model completely separates the cases corresponding to the experimental population from the population control cases, that is, for a given cutoff point c, $1 - F_Y(c) = 1$ and $1 - F_X(c) = 0$, which fully corresponds to a curve that passes through vertex (0,1) of the unit plane.

In the context of the differential expression analysis, if a gene is called *differentially expressed*, it is equivalent to saying that the distribution of the expression levels of the control population is different from that of the experimental population. Thus, the ROC curve can be used to evaluate the separation of the distributions of the expression levels between control and experimental groups as $\forall\ c:\ F_X(c) \geq F_Y(c)$.

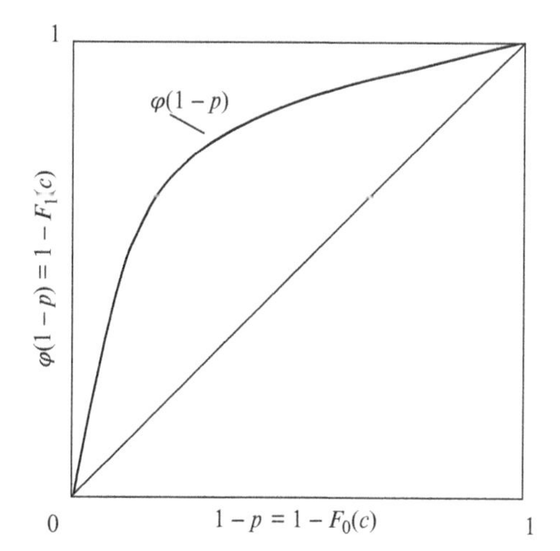

Figure 21.1 Example of a ROC curve.

21.2.3 AUC and Degenerate ROC Curves

The AUC is the index most commonly used in the ROC methodology to evaluate the discriminative power of a classification model. Bamber [2] showed that the AUC is obtained by integrating the curve in its domain, namely, $\int_0^1 \varphi(p)\, dp = P(X < Y)$. The AUC can accept values between 0.5 and 1, where AUC values close to 1 indicate that the classification model has a high discriminative power. Wolf and Hogg [25] recommended this index as a measure to evaluate differences between two distributions.

As already mentioned, the ROC curves are constructed on the basis of sensitivity and specificity of a system to classify the data into two mutually exclusive classes for all possible cutoff points of the decision variable. These probabilities depend on the classification rule, and in traditional ROC analysis, a high value of the decision variable corresponds to the presence of the artifact of interest, that is, the variable representing the expression levels of genes in the population control, X, is stochastically lower than that in the experimental population Y, or $F_X(c) > F_Y(c)$. However, if ROC curves are applied for all genes in a microarray experiment considering the same classification rule, these curves may be present below the positive diagonal of the unit plane. This is because the classification rule will not be the same for all genes in a microarray experiment, as some may be upregulated (positive regulation), $F_X(c) > F_Y(c)$, and others downregulated (negative regulation), $F_X(c) < F_Y(c)$.

This is not the only situation that leads to degenerate ROC curves. The presence of bimodal distributions in the experimental conditions with similar averages in the two groups leads to sigmoidal ROC curves that cross the reference line of the unit plane (Figure 21.2–a and b). In this study, genes with expression levels that present distributions as shown in Figure 21.2–a and b, shall be designated *special genes*. This designation is an alternative to what is usually the name of a DE gene, where, in particular, a gene is either upregulated or downregulated. Thus, the special genes, as the characteristics they present, will have a bimodal or a multimodal distribution.

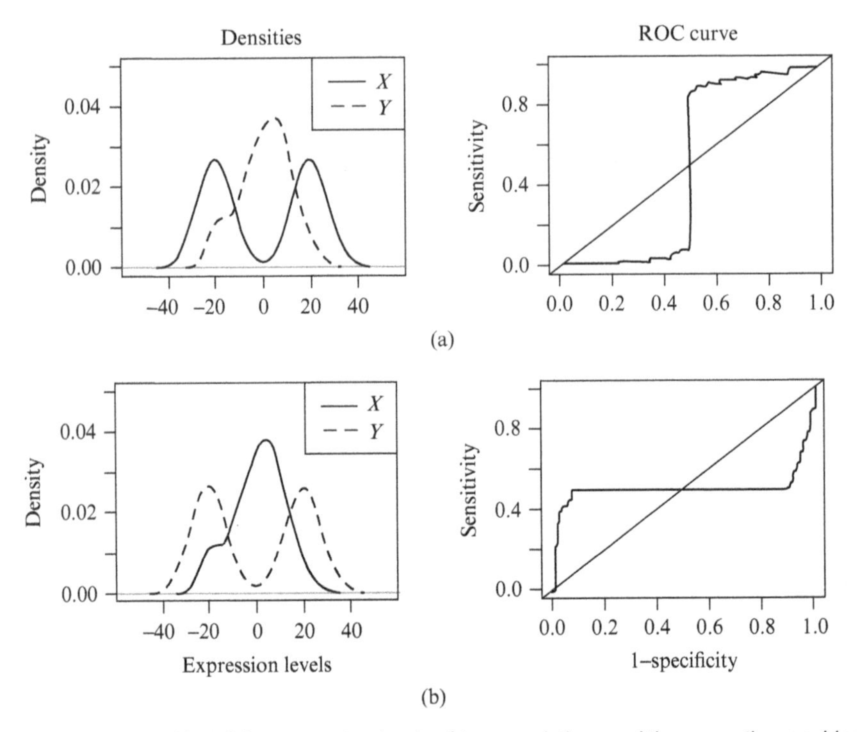

Figure 21.2 Densities of the expression levels of two populations and the respective empirical ROC curves. X represents the expression levels in the control group and Y represents the expression levels in the experimental group. The same classification rule was considered for estimating the ROC curves. The densities were estimated by the Kernel density estimator from two samples of size 100, simulated from two normal distributions. (a) $X \sim 0.5N(-20, 2) + 0.5N(20, 2)$, $Y \sim N(0, 11)$; (b) $X \sim N(0, 11)$, $Y \sim 0.5N(-20, 2) + 0.5N(20, 2)$.

The AUC is an index that evaluates the difference between the distributions of the groups. If the goal is to select DE genes, while keeping the same classification rule for all genes, those showing negative regulation have AUC lower than 0.5 and special genes present AUC values around 0.5.

In this work, we consider a nonparametric approach for estimating the AUC, as it does not require distributional assumptions, and in a microarray experiment, usually the number of replicates is small and the data are not drawn from a population with normal distribution [14]. The nonparametric approach has the disadvantage of the loss of efficiency, which is balanced with the risk reduction in interpreting results based on incorrect specifications. We consider the empirical and Kernel methods for estimating the ROC curve and the AUC [22].

21.2.4 Overlapping Coefficient

The *OVerLapping Coefficient* (OVL) of two densities is defined as the common area between both probability density functions. This coefficient is used as a measure of agreement between two distributions [15, 24].

The OVL expression can be represented in various forms. Weitzman [24] proposed the following expression:

$$\text{OVL} = \int_c \min\left[f_X(c), f_Y(c)\right] dc \tag{21.3}$$

where f_X and f_Y are the probability distribution functions of the random variables X and Y, respectively. The results are directly applicable to discrete distributions by replacing the integral by a summation. OVL is invariant with respect to scale transformations strictly increasing and differentiable in order to X and Y. This property turns out to be very important in the context of microarray data analysis, as in most cases the variables are subject to changes, particularly the base-2 logarithm. As for the AUC, the OVL has been estimated on the basis of a nonparametric approach developed by Silva-Fortes et al. [22].

21.2.5 Arrow Plot Construction

The *Arrow plot* is a graphical representation in a unit plane of the estimates of the AUC and the OVL, for all genes in a microarray experiment (Figure 21.3) [22]. From this graph, it is possible to select upregulated and downregulated genes, and genes that show the presence of subclasses mixtures (special genes). It is expected that

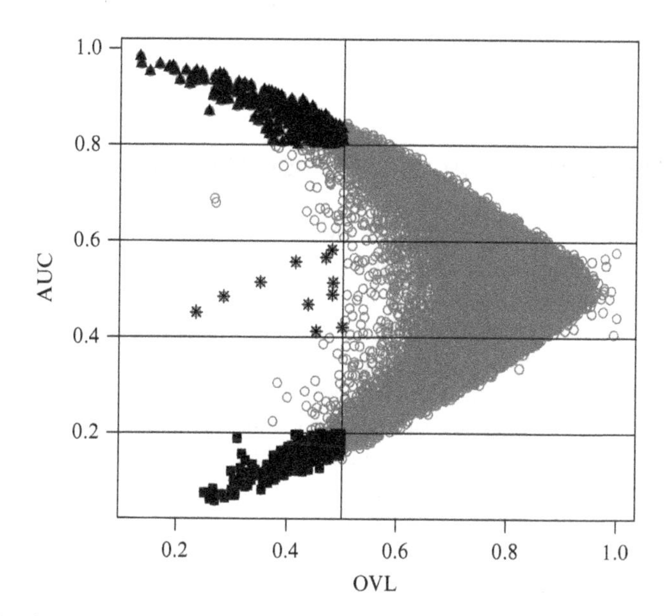

Figure 21.3 *Arrow plot*. To select upregulated genes, the conditions AUC ≥ 0.9 and OVL ≤ 0.5 were considered, corresponding to triangles on the plot. To select downregulated genes, the conditions AUC ≤ 0.2 and OVL ≤ 0.5 were considered, corresponding to squares on the plot. To select special genes, the conditions OVL ≤ 0.5 and 0.4 ≤AUC ≤ 0.6 were considered, corresponding to stars on the plot.

these genes exhibit near-zero values for the OVL and in what concerns the AUC, upregulated genes have values approximately 1, downregulated genes have values approximately 0, and special genes have AUC values approximately 0.5.

As already stated above, non-DE genes with the variance of the expression levels differing significantly in the two groups, with similar mean values and unimodal distributions, will be present in the same area of the graph as special genes. To overcome this situation, we developed an algorithm such that, among genes with low values for the OVL estimates and AUC estimates around 0.5, it would be possible to identify special genes.

In Reference [22], only one-channel microarrays were considered by Silva-Fortes et al., but the *Arrow plot* can also be used in two-channel microarrays as it will be illustrated below.

Graphic analysis is quite intuitive, as it allows us to obtain a global picture of the behavior of the genes and, based on its analysis, the user can choose the cutoff points for AUC and OVL, although this choice is arbitrary. A mixed selection of genes consists of two main steps. In the first step, all genes with AUC values approximately 0.5 are selected (e.g., $0.4 < \text{AUC} < 0.6$), as well as genes with low values for the OVL. These genes are candidates to be special genes. In the second step, the special genes are identified, among the selected genes in the first step, according to the presence of bimodality (or multimodality) in the distributions of the groups. For a gene to be considered special, it is sufficient to identify bimodality in at least one group.

The estimation of the AUC and the OVL for all genes are represented in a scatter diagram, where the abscissa axis represents the estimates of the OVL and the axis of ordinates represents the estimates of the AUC.

21.3 SIGNIFICANCE ANALYSIS OF MICROARRAYS

Significance Analysis of Microarrays (SAM) is a permutation-based statistical technique that considers gene-specific statistical tests and measures the strength of the relationship between gene expression and condition types in order to decide whether there are statistically significant differences in gene expression levels, controlling the *False Discovery Rate* (FDR). SAM was incorporated into the *SAM Program* as an extension to Microsoft Excel. It can be applied on many data types (e.g., one class, two unpaired and paired classes, survival, time course, and pattern discovery). More than one type of input data is allowed (oligo and cDNA arrays, SNP arrays and protein arrays). In the presence of missing values, automatic imputation via nearest-neighbor algorithm (kNN, with k=10 by default) is provided.

For each gene i, $i = 1, 2, \ldots, g$, SAM considers the following test statistic given by the relative difference:

$$d(i) = \frac{\bar{x}_{i(2)} - \bar{x}_{i(1)}}{\sqrt{\frac{\left(\frac{1}{n_1}+\frac{1}{n_2}\right)\left\{\sum_{j\in C_1}\left(x_{ij}-\bar{x}_{i(1)}\right)^2+\sum_{j\in C_2}\left(x_{ij}-\bar{x}_{i(2)}\right)^2\right\}}{n_1+n_2-2}} + s_0} \tag{21.4}$$

where x_{ij} denotes the observed expression level for gene i in the microarray j, $j = 1, 2, \ldots, n$, $\overline{x}_{i(l)}$ is the average of expression levels in the class C_l that contains n_l microarrays, $l = 1, 2$, and s_0 is an exchangeability factor chosen to minimize the coefficient of variation. The data are permuted and, for each permutation p, the relative differences $d_p(i)$, $i = 1, 2, \ldots, g$, are calculated. The expected relative differences are estimated by $d_E(i) = \sum_p \frac{d_p(i)}{P}$, with P the total number of permutations. The observed relative differences are plotted against the expected relative differences under the null hypothesis. The identification of DE genes is based on a chosen threshold Δ. Two lines at distance Δ from the diagonal are considered and all genes that lie above the upper Δ line and all genes that lie below the lower Δ line are considered significantly DE. The expected number of false positives is estimated and the FDR is finally calculated by

$$\text{FDR} = \frac{\text{Expected Number of False Positives}}{\text{Number of rejected hypotheses}} \tag{21.5}$$

21.4 CORRESPONDENCE ANALYSIS

Correspondence Analysis (CA) is an exploratory multivariate technique, developed by Jean-Paul Benzécri [3], which is generally aimed at transposing the rows and columns of any matrix of nonnegative data into a two-dimensional graphical display, where the rows and columns of the matrix are depicted as points [12, 13]. Primarily, CA has been applied for contingency tables but it may be applied to other types of data matrices whenever the interpretation of the relative values makes sense.

21.4.1 Basic Principles

Let $\mathbf{X} = (x_{ij})$ be an $r \times c$ nonnegative data matrix. Denoting x_{++} the total sum of the $r \times c$ elements of $\mathbf{X}$ and $\mathbf{t} = (1/x_{++}, \ldots, 1/x_{++})'$, then $\mathbf{r} = \mathbf{Xt}$ and $\mathbf{c} = \mathbf{t'X'}$ are the vectors of the marginal totals of $\mathbf{X}^* = (x_{ij}/x_{++})$. Let $\mathbf{D_r} = \text{diag}(\mathbf{r})$ and $\mathbf{D_c} = \text{diag}(\mathbf{c})$ be the diagonal matrices with the row and column marginal total of $\mathbf{X}^*$. Each row (column) of the matrix $\mathbf{D_r X^*}$ ($\mathbf{X^* D_c}$) is known as a *row (column) profile*.

To construct a plot for allowing a simultaneous graphical visualization of the row and column profiles as points in a low-dimensional space, CA uses the singular value decomposition as follows:

$$\mathbf{D_r}^{-1/2}(\mathbf{X}^* - \mathbf{rc'})\mathbf{D_r}^{-1/2} = \mathbf{UDV'} \tag{21.6}$$

where $\mathbf{D}$ represents the diagonal matrix of the singular values of the matrix $\mathbf{D_r}^{-1/2}(\mathbf{X}^* - \mathbf{rc'})\mathbf{D_r}^{-1/2}$ in descending order, and $\mathbf{U'U} = \mathbf{V'V} = \mathbf{I}$. Several variants for the coordinates of the r row profiles and the c column profiles with respect to principal axes can be taken for obtaining a representation of the $r + c$ profiles in a CA map: [18]:

- *Principal coordinates*:

  Row profile: $\mathbf{D_r}^{-1/2}\mathbf{UD}$

Column profile: $\mathbf{D_c}^{-1/2}\mathbf{VD}$

- *Standard coordinates*:

  Row profile: $\mathbf{D_r}^{-1/2}\mathbf{U}$

  Column profile: $\mathbf{D_c}^{-1/2}\mathbf{V}$

A graphical representation as points the first k columns of the (principal or standard) coordinates of the r row profiles jointly with that of the c column profiles determines the so-called *CA map* in the k-dimensional space (usually, in practice, $k = 2$). When the row and column profiles are represented with the same type of coordinates (i.e., both rows and columns are defined either in standard coordinates or in principal coordinates), the plot corresponds to a *symmetric CA map*, otherwise an *asymmetric CA map* is obtained. The quality of representation of the points onto the kth principal axis in the CA map is measured by the following expression [12]:

$$\frac{\lambda_k}{tr(\mathbf{D})} \times 100\% \tag{21.7}$$

where λ_k is the kth singular value in the diagonal of $\mathbf{D}$.

21.4.2 Interpretation of CA Maps

Let $\mathbf{a}_i$ denote the ith row of the matrix $\mathbf{A} = \mathbf{D_r}^{-1/2}\mathbf{UD}$ of the row profiles in principal coordinates. Since $\mathbf{V'V} = \mathbf{I}$ and from Equation 21.6, it follows that

$$\begin{aligned}\mathbf{AA'} &= \mathbf{D_r}^{-1/2}\mathbf{UD}\ \ \mathbf{V'V}\ \ \mathbf{DU'D_r}^{-1/2}\\ &= \mathbf{D_r}^{-1}(\mathbf{X}^* - \mathbf{rc'})\mathbf{D_c}^{-1/2}(\mathbf{D_r}^{-1}(\mathbf{X}^* - \mathbf{rc'})\mathbf{D_r}^{-1/2})'\end{aligned}$$

Hence, the distance between two row profiles in principal coordinates is given by the distance between the corresponding rows of the matrix

$$\mathbf{D_r}^{-1}(\mathbf{X}^* - \mathbf{rc'})\mathbf{D_c}^{-1/2} = \left(\frac{x_{++}}{x_{i+}}\left(\frac{x_{ij}}{x_{++}} - \frac{x_{i+}}{x_{++}}\frac{x_{+j}}{x_{++}}\right)\sqrt{\frac{x_{++}}{x_{+j}}}\right)$$

Concretely, the Euclidean distance between two row profiles in principal coordinates, say $\mathbf{a_i}$ and $\mathbf{a_l}$, is given by

$$d(\mathbf{a}_i, \mathbf{a}_l) = \sum_{j=1}^{c} \frac{\left(\frac{x_{ij}}{x_{i+}} - \frac{x_{lj}}{x_{l+}}\right)^2}{\frac{x_{+j}}{x_{++}}} \tag{21.8}$$

which coincides with the so-called *chi-squared distance* between the ith and the lth rows of the matrix $\mathbf{A}$.

Analogously, the distance between two column profiles in principal coordinates, say $\mathbf{b}_j$ and $\mathbf{b}_m$, the jth and the mth columns of the matrix $\mathbf{B} = \mathbf{D_c}^{-1/2}\mathbf{VD}$, is given by

$$d(\mathbf{b}_j, \mathbf{b}_m) = \sum_{i=1}^{r} \frac{\left(\frac{x_{ij}}{x_{+j}} - \frac{x_{im}}{x_{+m}}\right)^2}{\frac{x_{i+}}{x_{++}}} \tag{21.9}$$

Consequently, in a CA map, the distance between row points (column points) in principal coordinates is the chi-squared distance between the profiles of rows (columns) of the matrix **A** (**B**).

Doing so for the standard coordinates, comparisons between the positions of the row points (column points) cannot be achieved in terms of distance as there is no distance interpretation [13]. The standard coordinates in a CA map must be performed by the examination of their projections onto each principal axis, providing a possible interpretation of each axis. Concretely, the main goal is to look for groupings and contrasts in the configuration of the projected profiles on each axis in the CA map.

21.5 IMPACT OF THE PREPROCESSING METHODS

Now we fix on the six BC methods and the six NM methods described in Section 21.1. Each of the resultant 36 preprocessing strategies from the combination of one BC method with one NM method has been applied on three cDNA microarray data sets lym, lun, and liv, which are already classified by cancer type [9]. A brief description of these databases is summarized in Table 21.1. The CA is applied to all three databases in order to obtain graphical representations of BC and NM profiles in a two-dimensional reduced space. The main goal is allowing a visualization of features of the six BC methods and the six NM methods when they are applied to a cDNA microarray data set. Next, interpretations of CA maps constructed from 6×6 matrices are discussed to graphically highlight differences or similarities between BC and NM effects in both class prediction and class comparison contexts.

Table 21.1 Description of the three cDNA microarray data sets considered

Data set	lym	lun	liv
Number of classes	3	5	2
Number of samples per class	68/31/9	39/13/4/4/5	131/76
Number of genes	7079	22,646	21,901

21.5.1 Class Prediction Context

We start with analyzing the relationships among different BC and NM profiles in terms of predictive performance of classifiers.

For each data set, Freitas et al. [9] obtained the cancer-type classification error rates using LOO-CV for kNN and SVM classifiers, across different techniques of BC and NM. This procedure was implemented in order to compare the predictive ability of these classifiers for different preprocessing techniques on the microarray data. Now, a CA on these classification error rates is carried out for each data set, computing the first two dimensions ($k = 2$). Since standard coordinates of CA tended to appear near the origin referential in asymmetric maps, relationships between row points and column points may be poorly represented in these graphs. Therefore, symmetric maps for the three data sets lym, lun, and liv (Figure 21.4) are interpreted. The quality of representation of the results into each of the six CA maps, given by Equation 21.7, was observed as being high (at least 86%).

For the lym data set (graphs at the top in Figure 21.4), we conclude that (i) different effects are produced on the classification error rates when no NM (NN) is executed and one out of the methods IG, IL, IG–SL, and IL–SL is applied (the farthest uppercase marked points in the plan on the first principal axis, Dimension 1); and (ii) sub and min BC methods tend to produce similar profiles (the nearest lowercase marked points), and so the effect on the predictive ability of the classifiers when one of these two BC techniques is combined with any NM strategy is similar. For the lun data set, these two mentioned features, (i) and (ii), are not detected in its CA maps (graphs in the middle in Figure 21.4). Furthermore, no common effects are detected between the two classifiers in the CA maps. While, for kNN classifiers, no NM and local intensity-dependent NM (IL and IL–SL) present the most different profiles (some of the farthest uppercase marked points), for SVM classifiers, no NM and two-step NM methods (IG—SL and IL–SL) reflect higher differences (the farthest uppercase marked points). In addition, for kNN classifiers, both half and min BC techniques produce similar classification error profiles (the nearest lowercase marked points) and this is different from the results obtained when the edw BC method has been applied; for SVM classifiers, the profiles of the edw BC method and the non-application of BC methods (none) present the highest differences under the six NM methods fixed. Finally, for liv data set (graphs at the bottom in Figure 21.4), none and nexp BC methods tend to produce opposed profiles (the farthest lowercase marked points). It means that effects on the predictive performance of both kNN and SVM classifiers are expected to be different when these two BC techniques are considered for preprocessing the microarray data. Concerning applications of NM methods, no common effects between the two classifier types are observed in the two CA maps.

21.5.2 Class Comparison Context

Next, we investigate the difference between results in the detection of DE genes, when different BC and NM strategies are preprocessed on microarray data for removing technical noises and for calibration. For this, two different approaches are considered for the identification of DE genes: the popular algorithm SAM designed by

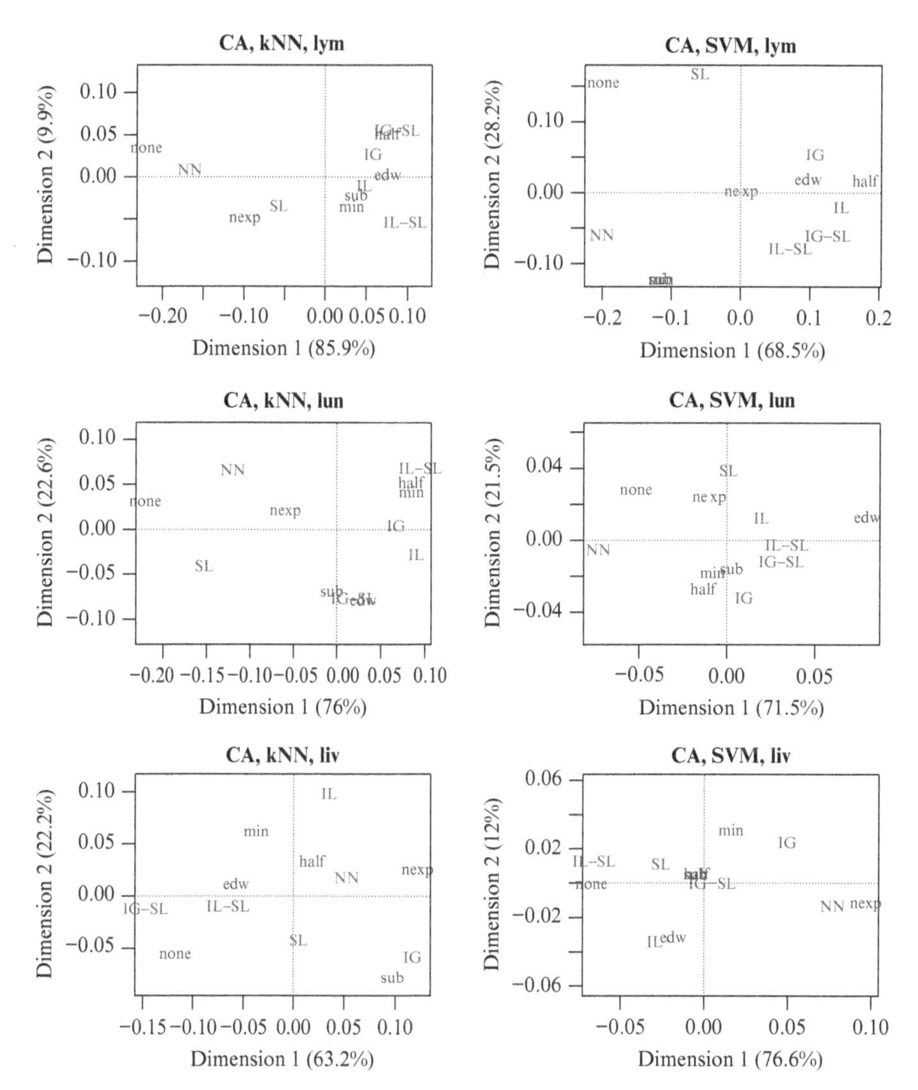

Figure 21.4 CA symmetric maps of the cancer classification leave-one-out cross-validation error rate estimates obtained from two classifiers: kNN (on the right side) and SVM (on the left side). The quality of representation of the points onto each axis/dimension in the CA map is indicated within brackets.

Tusher et al. [23] and the recent graphical methodology, the *Arrow plot*, developed by Silva-Fortes et al. [22].

First, SAM is addressed. For each data set, we investigate the number of genes selected as DE versus the FDR provided by SAM for each of the resultant 36 preprocessed microarray data matrices. For instance, fixing a value of the FDR equal to 0.005, we observe different numbers of DE genes emerging for each of the 36 preprocessing strategies per data set (data not shown). In order to compare the results of the SAM among preprocessing methods, we calculate the total sum of the FDR of the all genes with the FDR value lower than the lth lowest FDR value. To consider about 20% of the top-list genes with lower FDR, the following numbers of genes were obtained: $l_{\text{lym}} = 1400$, $l_{\text{lun}} = 5000$, $l_{\text{liv}} = 5000$. The total sum of the FDR of the 20% of the top-list gene with a lower FDR value, herein denoted by $S_{20\%top}$(FDR), was calculated for each BC–NM pair and for each data set (Table 21.2). Applying CA, three symmetric maps were constructed (Figure 21.5). The quality of representation of the results into each of the three CA maps was observed as being high (at least 93.9%).

Table 21.2 Values of $S_{20\%top}$(FDR) corresponding to 20% of the top DE genes with the lowest FDR induced by SAM for the 36 preprocessing strategies per data set: lym (at the top), lun (in the middle), and liv (at the bottom)

	none	sub	half	min	edw	nexp
NN	0.75	11.89	21.42	8.21	13.44	4.60
IG	0.21	0.03	2.07	0.14	0.80	0.15
IL	0.02	0.03	2.88	0.12	0.95	0.16
SL	0.35	0.16	3.19	0.13	1.08	0.50
IG–SL	0.11	0.05	3.44	0.11	0.74	0.45
IL–SL	0.14	0.19	2.17	0.16	0.88	0.29
NN	1.05	25.92	246.47	225.45	110.92	15.25
IG	30.08	30.75	108.97	111.51	88.83	109.49
IL	27.76	30.99	103.66	111.48	56.86	11.36
SL	10.23	21.28	125.58	137.02	83.36	85.61
IG–SL	24.25	17.34	123.67	136.86	63.84	105.75
IL–SL	26.00	32.97	128.89	109.59	62.90	72.94
NN	0.00	0.05	1.52	1.35	0.63	0.04
IG	0.05	0.02	0.50	0.51	0.15	0.03
IL	0.03	0.01	0.32	0.38	0.16	0.00
SL	0.14	0.03	0.36	0.36	0.17	0.03
IG–SL	0.00	0.04	0.14	0.14	0.15	0.02
IL–SL	0.02	0.00	0.31	0.28	0.13	0.00

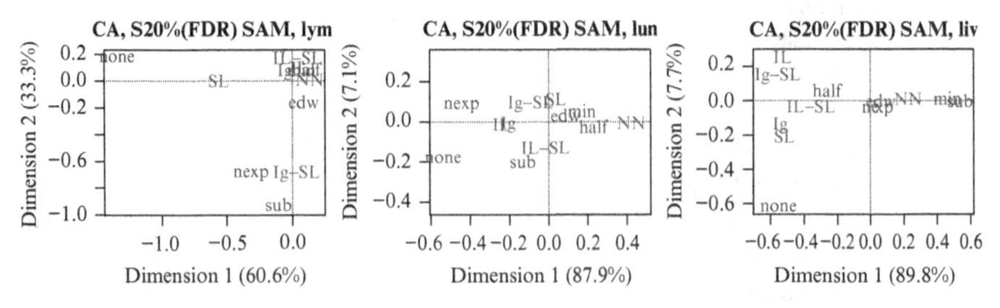

Figure 21.5 CA maps of the values of $S_{20\%top}$(FDR) corresponding to 20% of the top DE genes with the lowest FDR induced by SAM.

For the lym data set, we conclude that (i) different effects are produced on the detection of DE genes when no NM is executed and SL is applied; (ii) half and min BC methods tend to produce similar profiles, and so the effect on class comparison when one of these two BC techniques is combined with any NM strategy is similar; (iii) the effect of these two latter BC methods is in opposition to that when no BC is performed on the data. For lun data set, no NM strategy compared to both IL and IG techniques present the highest differences, however, the two latter NM methods show similar profiles. Regarding BC methods, half and none of the BC techniques tend to produce different profiles. Finally, for the liv data set, different effects are produced when no NM is executed and one out of the other five methods is applied. This means that the effects on the class comparison of NM techniques are expected to be different from that when no NM is implemented on the microarray data matrix liv. Also, the sub BC technique tends to produce a different profile from that when no BC method (none) is applied.

Now we analyze the results when the *Arrow plot* is applied. For this, besides separated quantification of the number of both up- and downregulated DE genes, the number of special genes is also provided. In order to illustrate the usage of the *Arrow plot*, we consider only one data set, liv. Following this, the results are depicted (Table 21.3) and analyzed. Applying CA, three symmetric maps were constructed (Figure 21.6). The quality of representation of the results into each three CA maps was observed as being high (at least 80.4%). No common effects for NM profiles and

Table 21.3 Number of DE genes as upregulated (at the top), special (in the middle), and downregulated (at the bottom) that were detected by the *Arrow plot* for the liv data set

	NN	IG	IL	SL	IG–IL	IG–SL
none	289	610	604	389	604	633
sub	371	654	671	489	603	620
half	374	598	621	378	560	586
min	374	604	530	378	534	563
edw	365	630	633	429	633	569
nexp	341	646	653	448	646	640
none	3	6	6	4	6	2
sub	2	9	12	9	5	14
half	5	15	19	11	17	12
min	5	13	16	11	19	20
edw	6	6	11	7	6	15
nexp	0	25	14	5	22	17
none	144	187	220	86	220	218
sub	165	245	276	250	259	282
half	144	228	252	271	239	214
min	144	218	228	271	250	266
edw	141	211	252	256	252	267
nexp	124	175	193	153	194	188

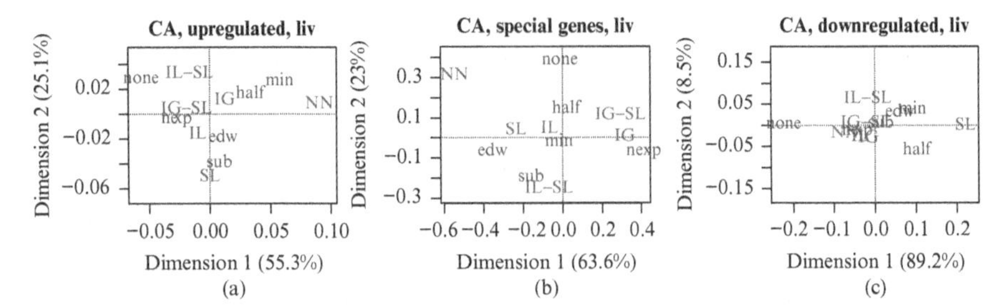

Figure 21.6 CA maps of the number of DE genes upregulated (a), number of DE genes downregulated (c), and number of special genes (b) detected by the *Arrow plot* for the liv data set.

for BC profiles were observed among the three CA maps. For upregulated genes, it was observed that no NM (NN) and two-step NM methods (IG–SL and IL–SL) reflect profiles with higher differences, whereas sub and none BC methods tend to produce more different profiles. In addition, for special genes, no NM (NN) and two-step NM methods (IG–SL and IL–SL) reflect profiles with higher differences, whereas edw and nexp BC methods tend to produce more different profiles. For the number of downregulated genes, the highest difference between conditional distributions is described when no NM (NN) technique is executed and the method SL is applied. Furthermore, both half and min BC techniques and no BC execution produce the highest difference between conditional distributions with regard to BC techniques.

21.6 CONCLUSIONS

In this chapter, the existence of different effects produced by the resultant 36 preprocessing methods of six BC methods combined with six NM methods were graphically investigated for both differential expression and predictive performance on supervised classification of cancer by using two-channel microarray data sets. For this, 6×6 data tables already obtained in a previous work [9] containing the error classification rates of classifiers induced from preprocessing microarray data sets, and data tables quantifying some specific features of DE genes detected using both the SAM and the *Arrow plot* from the same preprocessed microarray data sets, were studied via CA. CA is an exploratory multivariate technique that converts any matrix or table of nonnegative data into a particular type of two-dimensional graphical display in which the rows and columns of the matrix are depicted as points. The *Arrow plot* is a useful graphical methodology for the identification of up- and downregulated DE genes as well as special genes. In order to visualize relationships between BC methods and NM methods, the six column profiles and the six row profiles of each of those data tables were analyzed using a CA map.

CA maps showed to be a useful and easy graphical exploratory tool for visually investigating effect differences when results from applications of different preprocessing techniques are known. The interpretation of the CA maps showed differences in the impact of both BC and NM methods on cancer classification predictive performance of classifiers and identification of DE genes, resulting in different choices among competing preprocessing techniques. Furthermore, using CA maps, several and different effects of the BC and NM methods were revealed along the three data sets. Hence, it is is difficult to draw a general conclusion about the impact of different preprocessing methods in the context of class comparison and class prediction. This global idea is consistent with other studies regarding the evaluation of the performance of different NM methods, for instance [6, 10, 19]. Since there is no preprocessing strategy that outperforms others in all circumstances, it is highly recommended to carefully select the BC method and the NM method to apply on microarray data. Whenever possible, more than one preprocessing strategy on microarray data could be applied and results from preprocessed data should be compared before any conclusion and subsequent analysis.

All statistical analyses were performed with the R statistical package freely available on http://www.r-project.org/. Packages used: `ca`, `limma`, `marray`, `samr`. To construct *Arrow plots*, the R code available in Reference [22] was used.

ACKNOWLEDGMENTS

This work was partially supported by Portuguese funds through the CIDMA—Center for Research and Development in Mathematics and Applications, the CEAUL—Center of Statistics and Applications of Lisbon University, and the Portuguese Foundation for Science and Technology (FCT—Fundação para a Ciência e a Tecnologia), within projects UID/MAT/04106/2013 (A. Freitas), PEst-OE/MAT/UI0006/2014 (C. Silva and L. Sousa), and PTDC/MAT/118335/2010 (L. Sousa).

REFERENCES

1. Ambroise J, Bearzatto B, Robert A, Govaerts B, Macq B, Gala JL. Impact of the spotted microarray preprocessing method on fold-change compression and variance stability. BMC Bioinformatics 2011;12:413.
2. Bamber D. The area above the ordinal dominance graph and the area below the receiver operating graph. J Math Psychol 1975;12:387–415.
3. Benzécri JP. *L'analyses des donné. Tome 1: La taxinomie. Tome 2: L'analyses des correspondances*. Paris: Dunod; 1973.
4. Bolstad BM. *Low level analysis of high-density oligonucleotide array data: background, normalization and summarization* [PhD thesis]. Berkeley (CA): University of California; 2004.
5. Bolstad B, Irizarry RA, Astrand M, Speed T. A comparison of normalization methods for high density oligonucleotide array data based on variance and bias. Bioinformatics 2003;19:185–193.

6. Chiogna M, Massa MS, Risso D, Romualdi C. A comparison on effects of normalisations in the detection of differentially expressed genes. BMC Bioinformatics 2009;10:61.
7. Cleveland W, Devlin S. Locally weighted regression: an approach to regression analysis by local fitting. J Am Stat Assoc 1988;403:596–610.
8. Freitas A, Afreixo V, Pinheiro M, Oliveira JL, Moura G, Santos M. Improving the performance of the iterative signature algorithm for the identification of relevant patterns. Stat Anal Data Min 2011;4(1):71–83.
9. Freitas A, Castillo G, São Marcos A. Effect of background correction on cancer classification with gene expression data. *Proceedings of AIME 2009*, Lecture Notes in Artificial Intelligence in Medicine. Berlin; New York: Springer; 2009. p 416–420.
10. Freyhult E, Landfors M, Önskog J, Hvidsten T, Rydén P, Risso D, Romualdi C. Challenges in microarray class discovery: a comprehensive examination of normalization, gene selection and clustering. BMC Bioinformatics 2010;11:503.
11. Fujita A, Sato J, Rodrigues L, Ferreira C, Sogayar M. Evaluating different methods of microarray data normalization. BMC Bioinformatics 2006;7:469.
12. Greenacre MJ. *Theory and Applications of Correspondence Analysis*. London: Academic Press; 1984.
13. Greenacre M, Hastie T. The geometric interpretation of correspondence analysis. J Am Stat Assoc 1983;82:437–447.
14. Hardin J, Wilson J. Oligonucleotide microarray data are not normally distributed. Bioinformatics 2007; 10:1–6.
15. Inman HF, Bradley EL. The overlapping coefficient as a measure of agreement between two probability distributions and point estimation of the overlap of two normal densities. Commun Stat Theory Methods 1989;18(10):3851–3872.
16. Irizarry RA, Collin HB, Beazer-Barclay YD, Antonellis KJ, Scherf UU, Speed T. Exploration, normalization, and summaries of high density oligonucleotide array probe level. Biostatistics 2003;4:249–264.
17. Leung Y, Cavalieri D. Fundamentals of cDNA microarray data analysis. Trends Genet 2003;19:649–659.
18. Nenadić O, Greenacre M. Correspondence analysis in R, with two- and three-dimensional graphics: the `ca` package. Journal of Statistical Software 2007;20(3):1–13.
19. Önskog J, Freyhult E, Landfors M, Rydén P, Hvidsten T. Classification of microarrays; synergistic effects between normalization, gene selection and machine learning. BMC Bioinformatics 2011;12:390.
20. Ritchie ME, Silver J, Oshlack A, Holmes M, Diyagama D, Holloway A, Smyth GK. A comparison of background correction methods for two-colour microarrays. Bioinformatics 2007;23(20):2700-2707.
21. Sifakis E, Prentza A, Koutsouris D, Chatziioannou A. Evaluating the effect of various background correction methods regarding noise reduction, in two-channel microarray data. Comput Biol Med 2012;42:19–29.
22. Silva-Fortes C, Turkman MA, Sousa L. Arrow plot: a new graphical tool for selecting up and down regulated genes and genes differentially expressed on sample subgroups. BMC Bioinformatics 2012;13(147):15.
23. Tusher VG, Tibshirani R, Chu G. Significance analysis of microarrays applied to the ionizing radiation response. Proc Natl Acad Sci USA 2001;98(9):5116–5121.

24. Weitzman MS. Measure of the overlap of income distribution of white and negro families in the united states. Technical Report 22. Washington (DC): U.S. Department of Commerce, Bureau of the Census; 1970.
25. Wolfe DA, Hogg RV. On constructing statistics and reporting data. Am Stat 1971;25:27–30.
26. Wu W, Xing E, Myers C, Mian IS, Bissel MJ. Evaluation of normalization methods for cDNA microarray data by k-NN classification. BMC Bioinformatics 2005;6:191.

PART VI

PATTERN RECOGNITION IN PHYLOGENETIC TREES

CHAPTER 22

PATTERN RECOGNITION IN PHYLOGENETICS: TREES AND NETWORKS

David A. Morrison

Systematic Biology, Uppsala University, Norbyvägen, Uppsala, Sweden

22.1 INTRODUCTION

Complex networks are found in all areas of biology, graphically representing biological patterns and their causal processes [32]. These networks are graph-theoretical objects consisting of nodes (or vertices) representing the biological objects, which are connected by edges (or arcs, if they are directed) representing some form of biological association. They are currently used to model various aspects of biological systems, such as gene regulation, protein interactions, metabolic pathways, ecological interactions, and evolutionary histories.

Mostly, these are what might be called *observed networks*, in the sense that the nodes and edges represent empirical observations. For example, a food web consists of nodes representing animals with connecting arcs representing which animals eat which other animals. Similarly, in a gene regulation network the genes (nodes) are connected by arcs showing which genes affect the functioning of which other genes. In all cases, the presence of the nodes and edges in the graph is based on experimental data. These are collectively called *interaction networks* or *regulation networks*.

However, when studying historical patterns and processes, not all of the nodes and arcs can be observed. Instead, they are inferred as part of the data-analysis procedure. We can call these *inferred networks*. In this case, the empirical data consist solely of

Pattern Recognition in Computational Molecular Biology: Techniques and Approaches,
First Edition. Edited by Mourad Elloumi, Costas S. Iliopoulos, Jason T. L. Wang, and Albert Y. Zomaya.

the leaf nodes (i.e., those with incoming arcs but no outgoing arcs) and we infer the other nodes plus all of the arcs. For example, every person has two parents, and even if we do not observe those parents, we can infer their existence with confidence, as we also can for the grandparents, and so on back through time with a continuous series of ancestors. This can be represented as a *pedigree network*, when referring to individual organisms, or a *phylogenetic network* when referring to groups (populations, species, or larger taxonomic groups).

The data used to construct a phylogenetic network consist of measured characteristics of the leaf nodes. These characteristics may be phenotypic, such as morphological or anatomical features, or they may be genotypic, such as presence/absence of genes or even detailed gene sequences. For simplicity, this discussion is restricted to gene sequence data as an example, without affecting the generality of the discussion.

The objective of network construction is to quantitatively infer the characteristics of the ancestral nodes of the leaf nodes, and thus the pattern of arcs connecting these ancestral nodes with the leaf nodes and also with each other. Taking the human example again, we start with data concerning what the contemporary people look like, and we infer what their parents and grandparents, and so on, looked like, and we infer how many other descendants those ancestors have had. The resulting network will specify the evolutionary history of the group of people being studied.

In essence, we empirically observe a set of patterns among the leaf nodes, and we then infer the historical patterns (and processes) that led to these contemporary patterns. Evolution involves a series of unobservable historical events, each of which is unique, and we can neither make direct observations of them nor perform experiments to investigate them. This makes a phylogenetic study one of the hardest forms of data analysis known, as there is no mathematical algorithm for discovering unique historical accidents.

Nevertheless, methods have been, and are being, developed to tackle this problem. This chapter first discusses the characteristics of phylogenetic networks and how they are constructed. Then, the various biological processes that are known to determine the shape of a phylogenetic network are outlined. These historical processes create patterns in the contemporary organisms, and these patterns are discussed next. Finally, the concept of a multi-labeled tree, which is one of the more useful tools for constructing phylogenetic networks, is explained.

22.2 NETWORKS AND TREES

Mathematically, *rooted phylogenetic networks* are directed acyclic connected graphs (see Figure 22.1). These have labeled *leaf nodes* representing the contemporary organisms, *internal nodes* that are usually unlabeled (representing the inferred ancestors), and *arcs* connecting all of the nodes (representing the inferred history). The important characteristics of phylogenetic networks, if they are to represent evolutionary history, include [28] the following:

1. They are fully connected, so that there is at least one common ancestor for any given pair of leaf nodes, which indicates their historical relationship (i.e.,

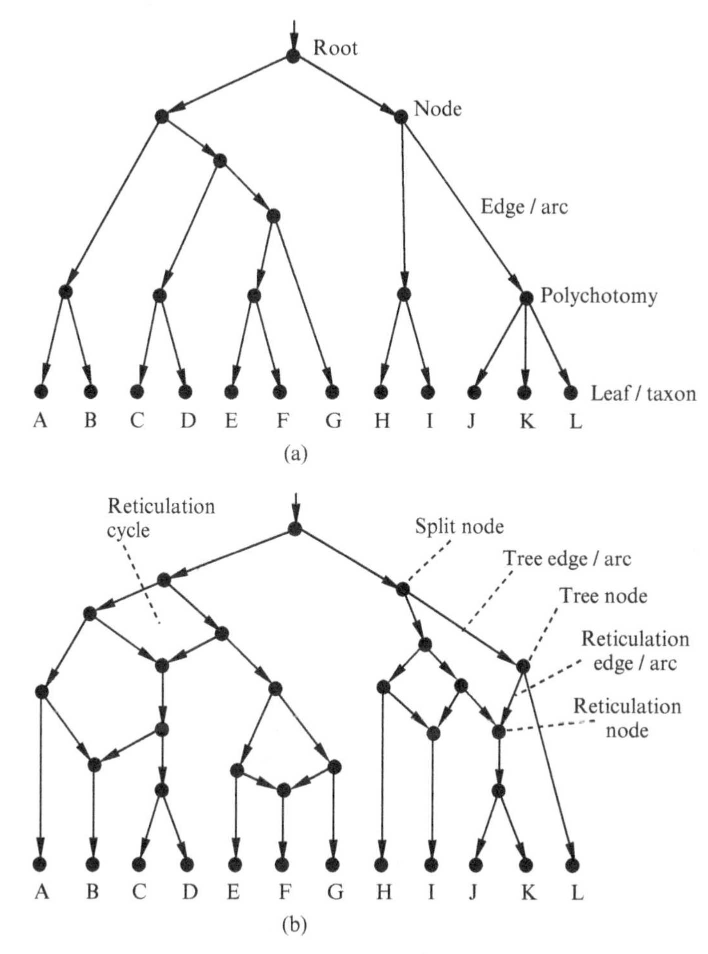

Figure 22.1 A rooted phylogenetic tree (a) and a rooted phylogenetic network (b) for 12 species (labeled A–L), illustrating the various terms used to describe the diagrams. The arrows indicate the direction of evolutionary history, away from the common ancestor at the root.

at least one undirected path can be traced between any given pair of leaf nodes).

2. They are directed, so that each edge is an arc, indicating the ancestor–descendant relationships among the internal nodes and between each leaf node and at least one internal node.
3. They have a single root, indicating the most recent common ancestor of the collection of leaf nodes, and this ultimately provides an unambiguous direction for each arc.
4. There are no directed cycles, as this would imply that an ancestor could be its own descendant (or a descendant could be its own ancestor).

5. Each edge (arc) has a single direction, away from the root (this follows from 1 and 4).
6. In species networks, the internal nodes are usually unlabeled, although in population networks some (or many) of them may be labeled.

Note that the main difference between a *population network* and a *species network* is that we might empirically sample some of the ancestors in the former, owing to the more recent time scale, so that some or all internal nodes are labeled.

In addition to these characteristics, we can also distinguish between tree nodes and reticulation nodes. Tree nodes have a single incoming arc and two or more outgoing arcs. (Note: leaf nodes have no outgoing arcs, and so tree nodes are always internal nodes.) A reticulation node, on the other hand, has more than one incoming arc. For algorithmic convenience, reticulation nodes are often restricted to having only two incoming arcs, but this is not a necessary biological restriction, although it will in general be true. Under these circumstances, multiple incoming arcs can be modeled as a series of consecutive reticulation nodes each with only two incoming arcs. For algorithmic convenience, reticulation nodes are usually restricted to having only one outgoing arc, but this is also not a necessary biological restriction. Once again, multiple outgoing arcs can be modeled as a series of consecutive nodes each with only one outgoing arc. Biologically, a leaf node could be a reticulation node with no outgoing arc, but this situation is rarely modeled algorithmically.

A network without reticulation nodes is usually called *tree*, as shown in Figure 22.1a; that is, a *phylogenetic tree* consists solely of tree nodes and leaf nodes, whereas a *phylogenetic network* has reticulation nodes as well as tree nodes (Figure 22.1b). For generality, this discussion focuses on networks, as a tree is simply a special case of a network (one that does not contain any reticulation nodes).

Reconstructing a simple tree-like phylogenetic history is conceptually straightforward. The objective is to infer the ancestors of the contemporary organisms, and the ancestors of those ancestors, and so on, all the way back to the Most Recent Common Ancestor (MRCA) or root of the group of organisms being studied. Ancestors can be inferred because the organisms share unique characteristics. That is, they have features that they hold in common and that are not possessed by any other organisms. The simplest biological explanation for this observation is that the features are shared because they were inherited from an ancestor. The ancestor acquired a set of heritable (i.e., genetically controlled) characteristics, and passed those characteristics on to its offspring. Empirically, we observe the offspring, note their shared characteristics, and thus infer the existence of the unobserved ancestor(s). If we collect a number of such observations, what we often find is that they form a set of nested groupings of the organisms. This can be represented as a tree-like network, with the leaf nodes representing the contemporary organisms, the internal nodes representing the ancestors, and the arcs representing the lines of descent.

Phylogenetic analysis thus attempts to arrange organisms on the basis of their common ancestry. Although there is no mathematical algorithm for discovering unique evolutionary events, the analysis can proceed via optimization algorithms. We specify an evolutionary model of some sort, involving specified patterns that can arise from known evolutionary events (described in the next section). We then seek the optimal graph based on that model. Many different mathematical methods have been developed, based on different optimality criteria: minimum distance, maximum parsimony, maximum likelihood, and Bayesian probability [10]. No particular criterion has been shown to be superior under all circumstances, but contemporary usage indicates that likelihood is the most popular criterion, with minimum distance being used when reduced analysis time is of importance. Severe computational intractability (NP-hardness or worse) is a recurring feature of phylogenetic analysis [10], and so heuristic methods are almost always used. The main practical issue is that the search space of rooted phylogenetic trees is vast, and locating the optimal graph cannot be guaranteed in polynomial time.

Generalizing these algorithmic approaches to a network with reticulations is also conceptually straightforward, but it is very difficult in practice. Essentially, the models are extended from their simple form, which includes only evolutionary processes such as nucleotide substitutions and insertions/deletions, to also include reticulation events such as hybridization, introgression, lateral gene transfer, and recombination (see the next section). (For completeness, note that in population studies, the parents–child relationship also involves a reticulation.) The optimality criteria are then expanded to include optimization across the expanded set of characteristics. We can conceptualize a reticulating network as a set of interlinked trees, and if we do so then the optimization procedure can be seen as optimizing one set of characteristics within each tree and optimizing another set of characteristics across the set of trees.

Not unexpectedly, tree space is tiny when compared with the conceptual space of rooted phylogenetic networks, and even navigating this space heuristically is a formidable computational challenge, let alone locating the optimal network (or networks). Heuristic techniques for network analysis therefore make simplifying assumptions that restrict the network space. This can be done on the basis of biological criteria, as specified by the analyst on a case by case basis [19], or it can be based on general mathematical criteria, such as the tree characteristics (e.g., specifying binary trees only) or the allowed interconnections between the trees (e.g., constraining the number of reticulations and/or their relative locations in the network).

In order to simplify the discussion in the next section, the trees embedded within a network are referred to as *gene trees*, whereas the network itself is viewed as a *species network*. This is based on the idea that (as the examples are restricted to gene sequence data) each non-recombining sequence block will have a single tree-like phylogenetic history, whereas the genome comprising these sequence blocks may have a more complex history [2, 37]. For simplicity, each sequence block is referred to as a *gene*, but they may actually be parts of genes rather than whole genes. The tree and network leaves are treated as species, more generally referred to as *taxa*.

Finally, it is worth emphasizing that the networks being discussed here have inferred ancestral nodes. There are other networks used in phylogenetics that do not have these nodes, and are therefore interaction networks, instead. These include the phylogenomic networks discussed by Dagan [7], including directed Lateral Gene

Transfer (LGT) networks and Minimal Lateral Networks. These networks connect observed taxa on the basis of their shared genes, and thus indicate possible LGT events, but they do not represent phylogenetic histories.

22.3 PATTERNS AND THEIR PROCESSES

Phylogenetic relationships are historical, often arising millions of years in the past, so that they cannot be directly observed by studying contemporary organisms. However, those organisms contain information about their historical relationships stored in their genes, and it is this information that is used to infer the evolutionary relationships. Genetic information relates to how an organism functions, not its history. Nevertheless, the functions arose in particular ways and at particular times in the past, and it is the pattern of sharing of these functions among organisms that can be used to reveal the past. Basically, if two organisms share part of their genome, then it is likely to be because they inherited that piece of genome from a shared common ancestor, and so we can infer the existence of that ancestor and its relationships from the patterns of sharing.

The key, then, is to understand the processes that create genotypic changes, and thus the genetic patterns that we can observe in the contemporary organisms. There are three types of processes: those that create evolutionary divergence (or branching), those that create evolutionary convergence (or reticulation), and those that create parallelism (multiple origins or chance similarity). For practical purposes, all of the rest of the biology can be viewed as stochastic variation, which causes variation in the length of the network arcs but does not lead to the creation of new nodes. This stochastic variation will also include estimation errors, which can be caused by issues such as incorrect data, inappropriate sampling, and model misspecification (e.g., Reference [20]).

If the convergence processes are not operating, then the resulting phylogeny will be a tree. Phylogenetics is founded on a widely held view of the mode of the evolutionary process: species are lineages undergoing divergent evolution with modification of their intrinsic attributes, the attributes being transformed through time from ancestral to derived states. The logic is as follows:

Definitions:

A *network* represents a series of overlapping groups.

A *tree* represents a set of nested groups.

Observation:

Each evolutionary event defines a group (consisting of all of the descendants of the ancestor in which the event occurred).

Conclusions:

Dichotomous speciation leads to a tree-like history, by definition.

Other processes will lead to a network, by definition.

Thus, we recognize that in biology there are both vertical (speciation) and horizontal (reticulation) evolutionary processes. Therefore, no biological data fit a tree perfectly (unless the data are carefully selected to do so). An important part of the use of genotypes for constructing phylogenies is thereby the distinction between vertical and horizontal flow of genetic information. The vertical components of descent are those from parent to offspring, while all other components are referred to as horizontal. The horizontal components arise due to phenomena such as hybridization and introgression, recombination, lateral gene transfer, and genome fusion.

This means that there are two types of data that can be analyzed in phylogenetics [29]:

1. Character-state changes (e.g., nucleotide substitution, nucleotide insertion/deletion) and
2. Character-block events (e.g., inversion, duplication/loss, transposition, recombination, hybridization, lateral gene transfer).

In the former, the evolutionary events occur at the level of individual nucleotides of the DNA, thus allowing each nucleotide position in a genome to have its own phylogenetic history. In the latter, groups of nucleotides are affected by the events simultaneously, so that a sequence block (or gene) has a shared phylogenetic history. It is the block events that are of particular interest (Table 22.1). Among these block events, some will lead to dichotomous speciation, such as inversion, duplication/loss, and transposition; others will lead to convergent histories, such as hybridization and introgression, recombination, and lateral gene transfer.

These processes are explained in more detail by Morrison [29], and only a brief summary is included here. *Hybridization* refers to the formation of a new species via sexual reproduction—the new (hybrid) species has a genome that consists of equal amounts of genomic material from each of the two parental species. Two distinct variants are usually recognized: in *homoploid hybridization* one copy of the genome is inherited from each parent species, while in *polyploid hybridization* multiple copies of the genome are inherited from each parent species. *Introgression* (or introgressive

Table 22.1 Summary of the processes involved in the evolution of chromosome blocks

Speciation processes	Reticulation processes
Inversion	Polyploid hybridization
Repeats	
Inverted repeats	Homoploid hybridization
Transposition	Introgression
Duplication/loss	Lateral gene transfer
	Intragenic recombination
	Intergenic recombination
	Reassortment
	Genome fusion

hybridization) is the transfer of genetic material from one species to another via sexual reproduction. *Lateral gene transfer*, on the other hand, is the transfer of genetic material from one species to another via any means that does not involve sex (e.g., transformation, transduction, conjugation, or gene transfer agents; see Reference [7]). No new species is created by either introgression or LGT, unlike in hybridization. *Homologous recombination* and *viral reassortment* are two processes that involve parts of a genome breaking apart and rearranging themselves. These processes usually occur within a species, or between closely related species. In eukaryotes, recombination usually occurs during sexual reproduction (via crossing-over during meiosis), so that the two genomes exchange material (called *meiotic recombination*).

It is important to note that reticulate evolution may function both at the species level, through homoploid and polyploid hybridization, and below the species level, through inter- and intragenic recombination; that is, at the population level reticulation is attributed to recombination, at the species level it is attributed to hybridization or introgression, and at higher levels it is attributed to LGT. Currently used methods to evaluate these processes and their patterns are reviewed by Morrison [29].

In addition to these processes, for the construction of networks, it is important to consider other processes that, if present, will confound the detection of reticulation patterns. The two most important of these are *Incomplete Lineage Sorting* (ILS; also called *Deep Coalescence*, or *Ancestral Polymorphism*) and *Gene Duplication–Loss* (D–L). These are both vertical evolutionary processes, but they make the construction of a phylogenetic tree difficult; that is, they can confound the construction of a network by creating spurious reticulations. ILS occurs when different populations within a species carry detectably different patterns of genetic diversity. This results in different gene fragments having different phylogenetic histories within a species (i.e., the gene tree is not the same as the species tree). D–L refers to duplication of genes followed by selective loss of some of the duplicates, resulting in incomplete data for some gene fragments.

It is important to note that, when reconstructing a phylogeny, any of the above processes may be capable of explaining the pattern of characteristics observed among the contemporary species. Indeed, ideally it is part of the network construction process to work out which process or processes have been involved. For example, Rheindt et al. [33] describe a situation where three closely related groups of birds have a mosaic pattern of genetic similarities: *group A* has a yellow belly and an inflected birdcall, *group B* has a gray belly and an inflected birdcall, and *group C* has a gray belly and a deflected birdcall. The authors suggest five possible phylogenetic hypotheses to explain the development of this situation:

Hypothesis 1: Incomplete lineage sorting—the ancestral population had a gray belly and an inflected birdcall (like group B, which has retained these characteristics), and groups A and C diverged from the ancestor by developing the yellow belly and deflected birdcall, respectively;

Hypothesis 2: Parallelism—group B is a recently diverged lineage from group A that has (independently of group C) acquired a gray belly similar to that of group C;

Hypothesis 3: Introgression—group B is a part of group A that has received genes from group C that give it the gray belly;

Hypothesis 4: Introgression—group B is a part of group C that has received genes from group A that give it the inflected birdcall;

Hypothesis 5: Hybridization—group B is a genomic hybrid derived from both groups A (where it got its birdcall) and C (where it got its gray belly).

Clearly, several other hypotheses could also be developed, but these were excluded *a priori*. In particular, LGT was excluded in favor of introgression, because the birds are closely related and therefore likely to be capable of exchanging genes via sexual reproduction. The authors' empirical data favor Hypothesis 3 but do not exclude Hypothesis 2.

Finally, it is also important to note that a phylogeny is a static picture of the patterns produced by an ongoing set of processes. Gene modification and elimination are continuing processes in species lineages, and we cannot know where in time during these processes we have taken our samples. So, in theory, we cannot treat all genes as being equal, because (i) some of them have been fixed for thousands or millions of years, (ii) others are in the process of being modified through evolutionary time, (iii) others have only relatively recently been acquired and are in the process of spreading throughout a species and becoming fixed, and (iv) yet others are in the process of being eliminated from that species. In practice, it may be difficult to tell these situations apart, and so we need to keep in mind that we are looking at dynamic processes through an image consisting of static patterns.

22.4 THE TYPES OF PATTERNS

The patterns formed by evolutionary processes can be of two types:

1. The gene tree topologies formed during evolutionary history; or
2. The spatial distributions of the different gene trees within the genome.

These are explained in this section.

Genomes are usually made up of independent chromosomes, each of which consists of many genes, along with inter- and intragenic spacers. There may be more than one copy of any gene within a genome. Furthermore, many organisms have multiple copies of their genome. In diploid organisms, for example, there are two copies of the nuclear genome plus one copy of the mitochondrial genome. Any one contiguous block of a chromosome will share the same evolutionary history, and that history will be tree-like due to evolutionary divergence. This shared history is referred to as a *gene tree*. (As noted above, this is a term of convenience, because not all sequence blocks will be genes.) However, the genome will consist of many different gene trees, and phylogenetic analysis consists of reconciling the different gene trees to work out the overall genomic history, called the *species phylogeny*. (This is also a term of

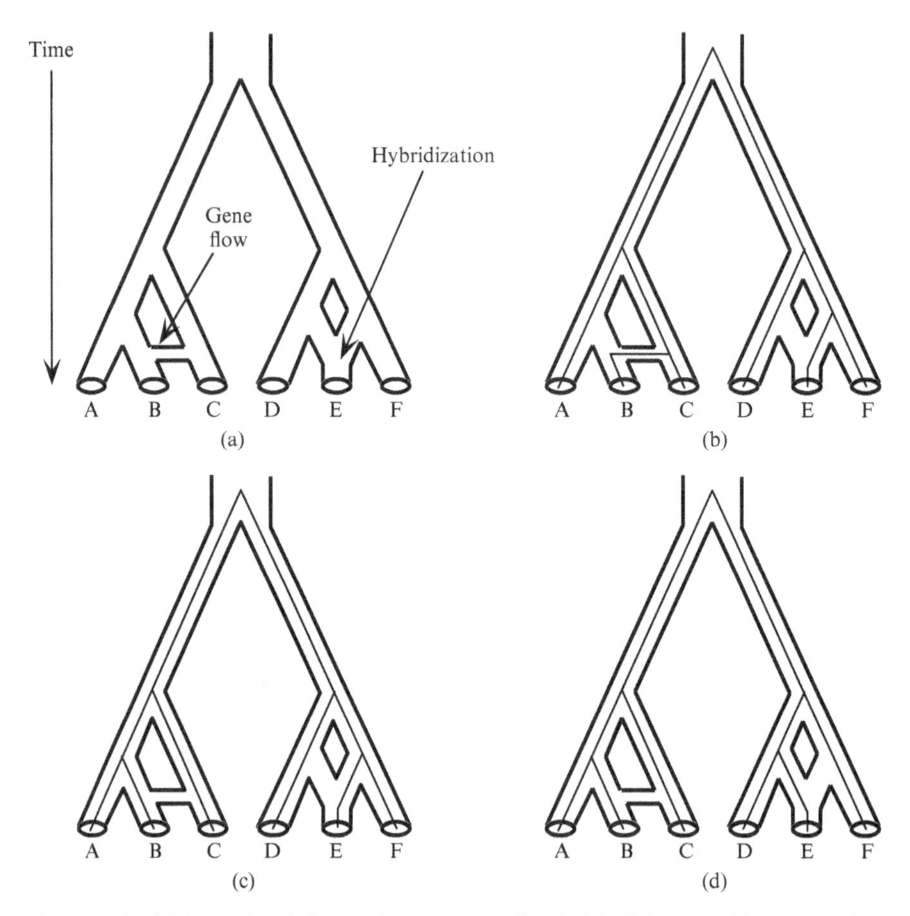

Figure 22.2 (a) A species phylogeny for six species (labeled A–F) in which there is gene flow between two species plus a hybridization event. (b–f) Possible gene trees that might arise from the species phylogeny. (Not all possible gene trees are shown.) The hybridization event involves species D and F, which hybridize to form species E. This would create gene trees that match either (c) or (d). The gene flow might involve introgression of genes from species C into species B. This would create gene trees that match either (b) or (c/d). (Similarly, the gene flow might involve lateral transfer of genes from species C to species B. This would also create gene trees that match either (b) or (c/d).) There might also be *Incomplete Lineage Sorting* (ILS), and in example (e) there is genic polymorphism in the immediate ancestor of species A and B. If the different forms are not all passed to each descendant, then the gene tree will not track the species phylogeny through time. There might also be gene *Duplication and Loss* (D–L), and in example (f) there is gene duplication in the most recent common ancestor of species A, B, and C. These gene copies then track the species phylogeny through the two subsequent divergence events. One of the gene copies is then lost in species A, and the other gene copy is lost in species B and C. Note that the gene trees shown in (b), (e), and (f) would be identical when sampled in the contemporary species, even though the biological mechanisms that formed them are quite different.

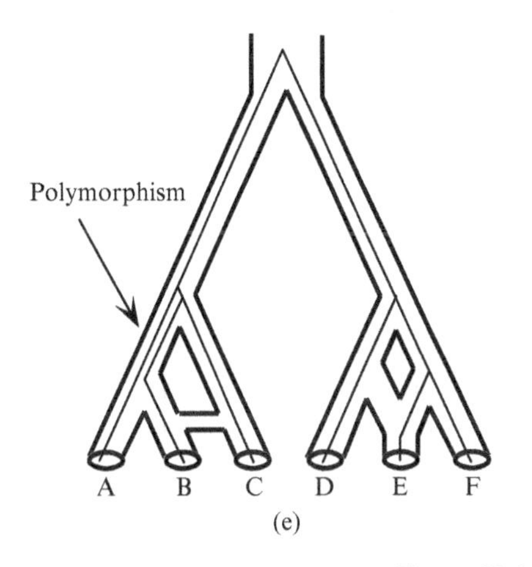

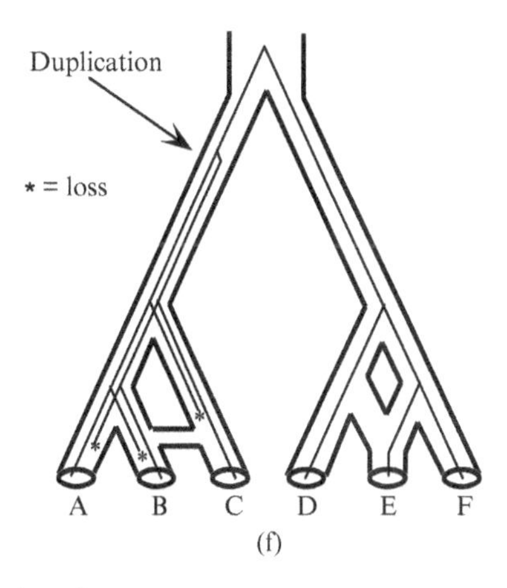

Figure 22.2 (*Continued*).

convenience, as not all taxa will be species.) Gene trees can thus be thought of as being "contained within" the species phylogeny.

The gene trees within any one genome will differ because of evolutionary parallelism and reversal (collectively called *homoplasy*), evolutionary convergence (reticulation due to horizontal evolutionary processes), and vertical processes such as ILS and D–L. This is illustrated in Figure 22.2. Figure 22.2a shows a species phylogeny as a set of tubes, within which is contained a set of gene trees. The phylogeny is for six species, showing four divergence events and two convergence events—one hybridization and one gene flow (which might be introgression, given that the species involved are closely related). Figure 22.2b–f shows five possible scenarios for any given gene tree contained within the species phylogeny. These are not the only possible scenarios, but there are three different gene tree topologies shown.

For example, as a result of the hybridization event involving species D and F (which hybridize to form species E), some chromosomal blocks will share the gene tree shown in Figure 22.2c while other blocks will share the gene tree shown in Figure 22.2d. The phylogenetic analysis will consist of trying to correctly reconstruct the species phylogeny given these two gene tree topologies. Alternatively, as a result of the introgressive gene flow involving species B and C, some chromosomal blocks will share the gene tree shown in Figure 22.2b while other blocks will share the gene trees shown in Figure 22.2c and d. Once again, the phylogenetic analysis will consist of trying to correctly reconstruct the species phylogeny given these gene tree topologies.

If all of the evolutionary patterns were created by divergence and convergence, then reconstructing the species phylogeny would be a conceptually straightforward optimization procedure. However, there is also the potential confounding effect of other vertical processes, notably ILS and D–L. These are illustrated by example in Figure 22.2e and f, respectively. Note that the two gene tree topologies illustrated are identical with that shown in Figure 22.2b, when the data are sampled in the six contemporary species. However, D–L involves multiple copies of the same gene within a single genome, whereas ILS involves different variants of a gene occurring in different genomes within the same species. So, the biological processes are different but the gene-tree topology observed when we sample only the contemporary species can be the same.

This highlights one of the essential problems of reconstructing a species phylogeny—gene-tree patterns can be the same even though the evolutionary mechanisms that formed them are quite different. In other words, it is difficult to reconstruct the processes from the gene-tree patterns alone. This fact has not been sufficiently emphasized in the literature. There is often an *a priori* assumption that certain evolutionary processes are of importance, and analytical methods have been developed to evaluate the resulting gene-tree patterns without any appreciation that these patterns are not necessarily unique to that particular process. Well-known examples involve the study of ILS, the study of D–L and the study of LGT—in each of these cases, the study process has been assumed to be the prime evolutionary mechanism, and the other processes have frequently been ignored [29].

There are a number of approaches to modeling the dependence between gene trees and a species phylogeny [30, 36], but the mathematical models that do integrate multiple processes have historically associated ILS with hybridization and D–L with LGT (see References [29, 36]). This pairwise association seems to reflect historical accident rather than any actual mathematical difference in procedure. Only Sjöstrand et al. [34, 35] and Chaudhary et al. [5] have tried to model LGT in the presence of both D–L and ILS.

The other important group of patterns formed by evolutionary processes concerns the spatial distributions of the different gene trees within the genome. Since a genome consists of many chromosomal blocks, each block can potentially have a different phylogenetic history compared to adjacent blocks. The block boundaries and their arrangements are determined by the process of *recombination*. This will be mainly meiotic recombination in eukaryotic organisms, homologous recombination in prokaryotes, and reassortment in viruses. Meiotic recombination is illustrated in Figure 22.3. Put simply, the blocks are rearranged among the chromosome copies, so that within any one organism, the genes inherited from each parent become mixed along each chromosome copy (e.g., instead of all of the genes from the father being on the same chromosome copy).

Recombination is closely associated with the other evolutionary processes. For example, recombination acts to integrate donor DNA into the recipient genome when horizontal gene flow occurs [11], such as hybridization, introgression and LGT. Also, recombination is associated with ILS, because ILS will be a more common occurrence in chromosomal regions with high recombination rates [31]. As noted above, recombination is basically a population-level process, introgression and hybridization are species-level processes, and LGT is a process at higher taxonomic levels.

The end result is that any one genome will consist of a mosaic of chromosomal blocks with different gene trees, as illustrated in Figure 22.4. The boundaries between the blocks can be the result of either recombination events or horizontal gene-flow

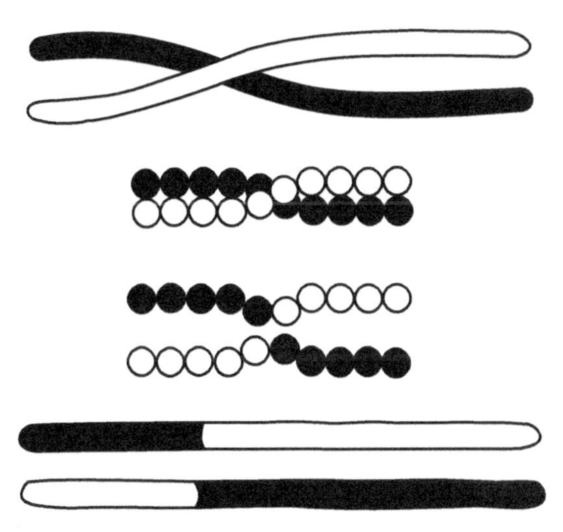

Figure 22.3 Recombination during meiosis. Chromosomes are represented as oblongs, and genes as circles. During meiosis, homologous chromosomes form pairs, and parts of the pairs potentially cross over each other, as shown in the top pair of pictures. If the touching parts break apart and rejoin, then the chromosomes will have exchanged some of their gene copies, as shown in the bottom pair of pictures. Modified from Reference [26].

events. Once again, these spatial patterns can be the same even though the biological mechanisms that formed them are quite different; that is, it is also difficult to reconstruct the processes from the spatial patterns in the genomes alone. For example, the earliest study of LGT [21] incorrectly interpreted the empirical patterns as resulting from sexual recombination rather than LGT.

The basic question for pattern recognition in phylogenetics is thus: do we need to model multiple reticulation processes in order to construct a useful network, or can we simply have a single generic "reticulation model"? The latter implies that the biologist will have to decide about the underlying processes *after* the network is constructed, whereas the former might imply deciding *before* constructing the network (e.g., perhaps deciding to model D–L or ILS but not both). Since the genome patterns and the associated set of gene-tree topologies turn out to be the same for most of the processes, in what way would different models produce different networks? It seems likely that they would not.

22.5 FINGERPRINTS

If phylogenetic processes cannot be reconstructed solely from the gene-tree topologies and/or spatial genomic patterns, then biologists will need to distinguish between them on the basis of other information. Indeed, it may be possible to distinguish different types of evolutionary events depending on other fingerprints or signatures that they leave in the data; that is, the genomic context within which the patterns occur might provide clues to the processes.

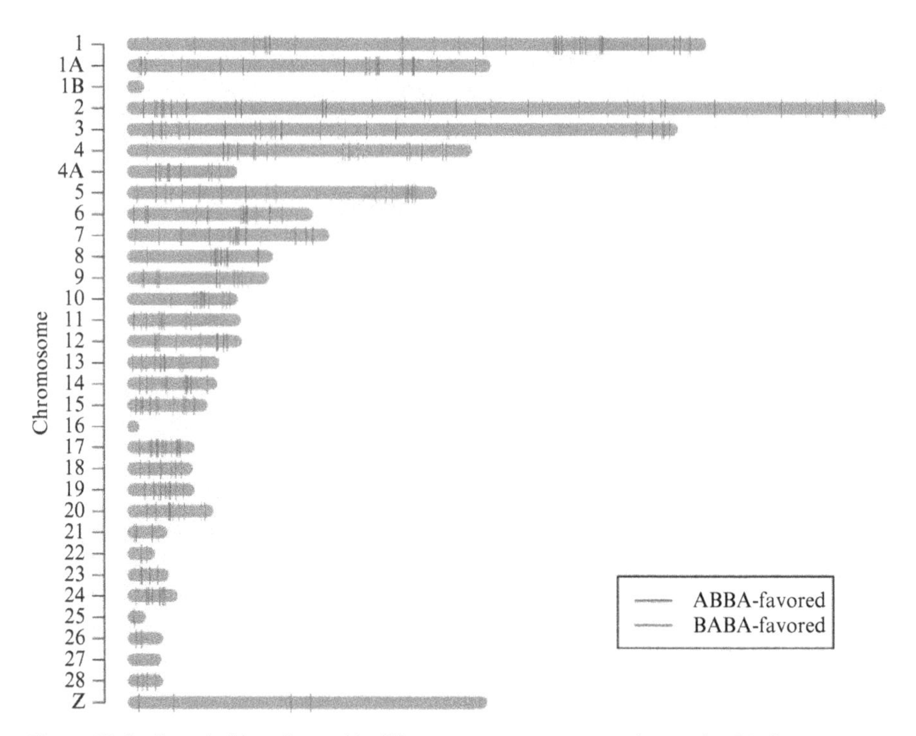

Figure 22.4 Sampled locations with different gene trees, mapped onto the 32 chromosomes of the nuclear genome of the zebra finch. Two different gene-tree patterns are shown, labeled ABBA (204 locations) and BABA (158 locations). Adjacent locations with the same gene tree may be part of the same chromosomal block, but adjacent locations with different trees are in different blocks. The authors attribute the different gene trees mainly to introgression. Reproduced with permission from Reference [33].

Some possible fingerprints of gene-flow events are listed in Table 22.2. For example, hybrid topologies can be topographically different from recombination topologies because in the former the parental lineages remain unchanged, with the intermixed DNA in a new evolutionary lineage. Consequently, there have been statistical tests suggested for detecting hybrids, which can be expected to have characteristics (phenotypic or genomic but not genic) intermediate between those of the parents. Indeed, for polyploid hybridization there is an increase in chromosome number that is easily detected, and which thus acts as a clear fingerprint [1].

If there is only a small number of genes involved in the apparent gene flow, then the process is likely to be introgression or LGT rather than hybridization (i.e., hybridization involves about 50% of the genome). Introgression and LGT can be distinguished because in the former the gene flow must be between closely related species (owing to the requirement for sex), whereas in the latter it can involve very distantly related species. Indeed, introgression is often tested by comparing the reconstructed nuclear gene trees with the mitochondrial gene tree (which is usually inherited from the

Table 22.2 Possible fingerprints for gene flow via different evolutionary processes

Process	Fingerprint
Introgression	Small number of genes from closely related species (possibly whole chromosomes)
Lateral gene transfer	Small number of genes from distantly related species (rarely whole chromosomes)
Homoploid hybridization	Equal number of genes from each parent (usually whole chromosomes)
Polyploid hybridization	Combined genomes from each parent (whole chromosomes)

mother alone) or the Y-chromosome (which is inherited from the father)—this allows the different ancestors to be distinguished.

Furthermore, horizontal gene flow (hybridization, introgression, LGT) requires that the species involved coexist, but D–L and ILS do not; that is, gene flow requires time consistency in the network [9]. This places a restriction on the possible network topologies, which can be used to constrain the network search.

However, in general, even with genome-scale data, incongruence fingerprints are often identical among phenomena, and it may be impossible to reliably distinguish homoploid hybridization, introgression, LGT, and intergenic recombination from each other by pattern analysis alone [12]. Statistically, detecting a particular reticulation process involves a null tree model, so that rejecting the null model infers the presence of the process. However, any mathematical test is a test of the model not the process [18], so that rejecting the null model has no direct biological interpretation.

Equally importantly, the fingerprints of reticulation events may diminish rapidly over time, owing to superimposed vertical evolutionary events (e.g., DNA mutations) or even other reticulation events, making the detection of earlier events extremely problematic. In particular, recombination can make it difficult to detect earlier evolutionary events. It is for this reason that the mitochondrion and the Y-chromosome are often targeted for DNA sequencing, as recombination is less likely in these two chromosomes.

Finally, reticulate evolution may be missed in studies using only a few genes, because the relevant patterns are likely to be missed during gene sampling [25].

22.6 CONSTRUCTING NETWORKS

Current algorithms for constructing phylogenetic trees involve a model in which character mutations are given relative costs; for example, nucleotide data are modeled on the basis of the probability of vertical processes such as substitutions and indels. Presence or absence of particular branches in the tree is then based on the probabilities of DNA mutations along those branches. This system can be generalized to a reticulating network merely by adding reticulation events to the model, which are also associated with some explicit cost. Under this model, a tree simply has an implicit assumption of infinite cost for reticulations [8], so that they cannot occur, whereas networks apply some smaller nonzero cost.

There are, unfortunately, a number of issues that make this generalization difficult in practice. First, when searching for optimal networks (instead of optimal trees), the space of rooted networks is vastly larger than the space of rooted trees. Searching this space even heuristically is a daunting computational challenge [19]. Second, adding reticulations to a tree or network will monotonically improve the value of any optimality criterion based on the character data, because a more complex network can never fit the data any worse than a simpler one [17, 24]. This overestimates the amount of reticulation in the data; and so there needs to be a separate optimality criterion for the number of reticulations, as well. Third, the usual strategy used for calculating a rooted phylogenetic tree, where we first produce an unrooted tree and then determine the root location (e.g., with an outgroup taxon), does not work for rooted networks [27]. When we add a root to any of the currently available methods for unrooted networks, in the resulting diagram (i) internal nodes will be present that do not all represent inferred ancestors, (ii) not all of the edges will have a unique direction, or (iii) both.

One long-standing approach to these issues has been to simply ignore the mutation part of the model, and to focus on the reticulation part alone. That is, in the absence of homoplasy, and with knowledge of the binary character states at the leaves and root, it is possible to reconstruct a unique reticulate network, provided that various quite strong assumptions are met [38]; and it is possible to reconstruct a unique reticulate network given the complete set of trees that it contains [39]. This is a combinatorial optimization problem in which the desired network is the one with the minimum number of reticulations. In practice, the optimization will be given some pragmatic constraints in order to reduce the search space, as all of the associated problems are NP-hard [4].

There are three basic ways to combine the phylogenetic signal contained in rooted trees into a network: via the tree topologies themselves, by combining the clusters they contain, or by combining their triplets [15]. These have often been referred to as *hybridization networks*, although (as noted above) the networks might actually involve any of the reticulation processes, not just hybridization. In general, the minimum number of reticulations required for a network representing the same set of trees (all with the same taxa) is triplet network $\leq$ cluster network $\leq$ tree-topology network [16]. One fundamental limitation with this approach is that, even given a complete collection of all of the subhistories (i.e., subtrees or subnetworks), it is not necessarily possible to reconstruct a unique overall network history [14]. This limitation does not apply to phylogenetic trees, which can be uniquely reconstructed from a collection of all of the possible subtrees.

If the optimal network is produced directly from the DNA character data, then it has often been referred to a *recombination network*, although (once again) the network might actually involve any of the reticulation processes, not just recombination. However, this model is based on the idea of minimizing the number of inferred crossover events necessary to explain the full set of incompatible character patterns [15].

The alternative strategy to combinatorial optimization is a statistical framework, which incorporates a stochastic model. The basic issue is to find a suitable mathematical model, one that captures the relationships across genomes as well as those within genomes, and which incorporates both vertical and horizontal evolutionary patterns.

A number of general statistical strategies based on likelihood have been suggested, especially in the context of analyzing genome-scale data. For example, Bloomquist and Suchard [3] have adopted a Bayesian approach based on Ancestral Recombination Graph*s* (ARGs). The biological model for an ARG is derived from population genetics—the coalescent with recombination; however, as noted above, this is likely to model all reticulation events. In this model, the likelihood is framed in terms of an ARG, but the prior distribution is based on the number of nonvertical nodes, and the posterior distribution is approximated via a reversible-jump Markov Chain Monte Carlo sampler. Alternately, Liu et al. [22] have adopted a Hidden Markov Model (HMM) approach. Their biological model is based on accounting for introgression and ILS, but it should successfully model other reticulation events. The HMM is intended to account for dependencies within genomes, while the network captures dependencies between genomes. Model training uses dynamic programming paired with a multivariate optimization heuristic, which then identifies parts of the genome with a reticulate history. As yet, both of these modeling approaches are too limited for general use beyond data sets with a very small number of taxa, but they provide the foundation for future work.

22.7 MULTI-LABELED TREES

Perhaps the most interesting consequence of the idea that all reticulation processes lead to the same set of gene-tree topologies is that these multiple gene trees can be represented by a single *Multi-Labeled Tree* (a MUL-tree), and that this tree can also be represented as a reticulating network. A MUL-tree has leaves that are not uniquely labeled by a set of species, as shown in Figure 22.5. So, multiple gene trees can be represented by a single MUL-tree, with different combinations of the leaf labels representing different gene trees. Furthermore, a phylogenetic network can be derived from the MUL-tree by combinatorial optimization, minimizing the number of reticulations in the network [13].

In a diploid organism (with two copies of each gene, one from the mother and one from the father), we usually need to sequence only one copy of each gene, because the other copy is very similar (at least for the genes that biologists usually work on). However, for a tetraploid hybrid, for example, there are four copies of each gene, and they come in two pairs (one pair from each of the two parent species). The gene copies are similar within each pair but can be quite different between pairs. So, sequence data are usually obtained for one copy from each pair. This means that the hybrid appears twice in the data set, once for each DNA sequence that we have. This results in a MUL-tree. Similarly, gene duplication also means that there are extra gene copies, but *within* each genome, instead. Once again, if we obtain data for each gene copy, then this creates exactly the same data pattern, and so it can be analyzed using a MUL-tree.

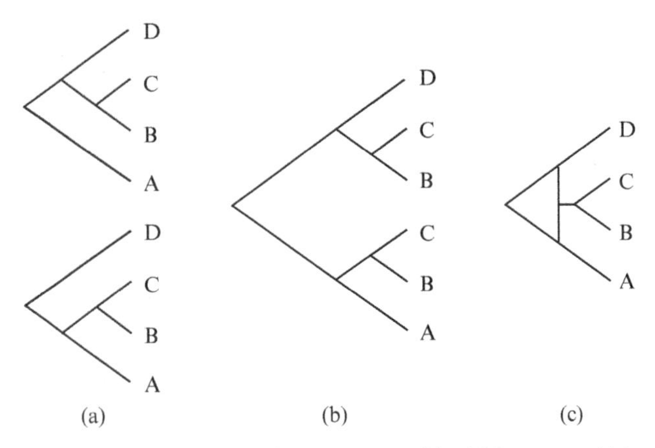

Figure 22.5 The relationships between (a) gene trees, (b) a MUL-tree, and (c) a phylogenetic network for four species (labeled A–D). Each gene tree can be derived from the MUL-tree by deleting duplicated labels, so that each label occurs once and once only in the tree. In this example, the two gene trees shown in (a) are not the complete set of trees that could be derived from the MUL-tree in (b). The network can be derived from the MUL-tree by minimizing the number of reticulations in the network. In this example, only one reticulation is needed.

Analogous arguments apply to any situation where there are multiple gene copies to be sampled, which can also pertain to introgression, LGT, and ILS.

The use of a multi-labeled tree thus provides a unified model for analyzing species that are subject to both vertical and horizontal evolution, with the different data samples being combined into a MUL-tree. Computer programs such as PADRE [23] can then be used to derive an optimal network from the MUL-tree. For completeness, note that heuristics have been developed to obtain an optimal species tree from a set of MUL-trees [5], using the MulRF program [6].

22.8 CONCLUSION

Phylogenetic analysis involves reconstructing inferred evolutionary patterns and processes from observed contemporary patterns. The processes involve evolutionary divergence (or branching), evolutionary convergence (or reticulation), and parallelism (multiple origins or chance similarity). Therefore, networks are replacing trees in phylogenetics for modeling the patterns, as they can accommodate reticulation in the species history derived from (nonreticulating) gene trees. Generalizing from a tree to network is conceptually straightforward but algorithmically hard in practice, owing to the increased search space to find the optimum. Unfortunately, almost all of the evolutionary processes can create the same set of gene trees, so that it is difficult to infer evolutionary processes directly from the network patterns. However, this does mean that we may need only one general reticulation mode, rather than a separate model for each process. Multi-labeled trees seem to be a practical tool for joint modeling of vertical and horizontal evolution.

REFERENCES

1. Adams KL, Wendel JF. Polyploidy and genome evolution in plants. Curr Opin Plant Biol 2005;8:135–141.
2. Blair C, Murphy RW. Recent trends in molecular phylogenetic analysis: where to next? J Heredity 2011;102:130–138.
3. Bloomquist EW, Suchard MA. Unifying vertical and nonvertical evolution: a stochastic ARG-based framework. Syst Biol 2010;59:27–41.
4. Bordewich M, Semple C. Computing the minimum number of hybridization events for a consistent evolutionary history. Disc Appl Math 2007;155:914–928.
5. Chaudhary R, Burleigh JG, Fernández-Baca D. Inferring species trees from incongruent multi-copy gene trees using the Robinson–Foulds distance. Algorithms Mol Biol 2013;8:28.
6. Chaudhary R, Fernández-Baca D, Burleigh JG. MulRF: a software package for phylogenetic analysis using multi-copy gene trees. Bioinformatics 2015;31:432–433.
7. Dagan T. Phylogenomic networks. Trends Microbiol 2011;19:483–491.
8. Dickerman AW. Generalizing phylogenetic parsimony from the tree to the forest. Syst Biol 1998;47:414–426.
9. Doyon JP, Ranwez V, Daubin V, Berry V. Models, algorithms and programs for phylogeny reconciliation. Brief Bioinform 2011;12:392–400.
10. Felsenstein J. *Inferring Phylogenies*. Sunderland MA: Sinauer Associates; 2004.
11. Hernández-López A, Chabrol O, Royer-Carenzi M, Merhej V, Pontarotti P, Raoult D. To tree or not to tree? Genome-wide quantification of recombination and reticulate evolution during the diversification of strict intracellular bacteria. Genome Biol Evol 2013;5:2305–2317.
12. Holder MT, Anderson JA, Holloway AK. Difficulties in detecting hybridization. Syst Biol 2001;50:978–982.
13. Huber KT, Oxelman B, Lott M, Moulton V. Reconstructing the evolutionary history of polyploids from multilabeled trees. Mol Biol Evol 2006;23:1784–1791.
14. Huber KT, van Iersel L, Moulton V, Wu T. How much information is needed to infer reticulate evolutionary histories? Syst Biol 2015;64:102–111.
15. Huson DH, Scornavacca C. A survey of combinatorial methods for phylogenetic networks. Genome Biol Evol 2011;3:23–35.
16. van Iersel L, Kelk S. When two trees go to war. J Theor Biol 2011;269:245–255.
17. Jin G, Nakhleh L, Snir S, Tuller T. Inferring phylogenetic networks by the maximum parsimony criterion: a case study. Mol Biol Evol 2007;24:324–337.
18. Joly S. JML: testing hybridization from species trees. Mol Ecol Resour 2012;12:179–184.
19. Kelk S., Linz S. Morrison D.A. Fighting network space: it is time for an SQL-type language to filter phylogenetic networks. 2013. http://arxiv.org/abs/1310.6844
20. Kück P, Misof B, Wägele J-W. Systematic errors in maximum-likelihood tree inference. In: Wägele J-W, Bartolomaeus T, editors. *Deep Metazoan Phylogeny: The Backbone of the Tree of Life*. Berlin: De Gruyter; 2014. p 563–583.
21. Lederberg J, Tatum EL. Gene recombination in *Escherichia coli*. Nature 1946;158:558.
22. Liu K.J., Dai J., Truong K., Song Y., Kohn M.H., Nakhleh L. An HMM-based comparative genomic framework for detecting introgression in eukaryotes. 2013. http://arxiv.org/abs/1310.7989

23. Lott M, Spillner A, Huber KT, Moulton V. PADRE: a package for analyzing and displaying reticulate evolution. Bioinformatics 2009;25:1199–1200.
24. Makarenkov V, Legendre P. Improving the additive tree representation of a given dissimilarity matrix using reticulation. In: Kiers HAL, Rasson JP, Groenen PJF, Schade M, editors. *Data Analysis, Classification, and Related Methods*. Berlin: Springer; 2000. p 35–46.
25. Moody ML, Rieseberg LH. Sorting through the chaff, nDNA gene trees for phylogenetic inference and hybrid identification of annual sunflowers (*Helianthus* sect. *Helianthus*). Mol Phylogenet Evol 2012;64:145–155.
26. Morgan TH. *A Critique of the Theory of Evolution*. Princeton NJ: Princeton University Press; 1916.
27. Morrison DA. Networks in phylogenetic analysis: new tools for population biology. Int J Parasitol 2005;35:567–582.
28. Morrison DA. Phylogenetic networks are fundamentally different from other kinds of biological networks. In: Zhang WJ, editor. *Network Biology: Theories, Methods and Applications*. New York: Nova Science Publishers; 2013. p 23–68.
29. Morrison DA. Phylogenetic networks: a review of methods to display evolutionary history. Annu Res Rev Biol 2014;4:1518–1543.
30. Nakhleh L. Computational approaches to species phylogeny inference and gene tree reconciliation. Trends Ecol Evol 2013;28:719–728.
31. Pease JB, Hahn MW. More accurate phylogenies inferred from low-recombination regions in the presence of incomplete lineage sorting. Evolution 2013;67:2376–2384.
32. Raval A, Animesh Ray A. *Introduction to Biological Networks*. Boca Raton FL: CRC Press; 2013.
33. Rheindt FE, Fujita MK, Wilton PR, Edwards SV. Introgression and phenotypic assimilation in *Zimmerius* flycatchers (Tyrannidae): population genetic and phylogenetic inferences from genome-wide SNPs. Syst Biol 2014;63:134–152.
34. Sjöstrand J, Sennblad B, Arvestad L, Lagergren J. DLRS: gene tree evolution in light of a species tree. Bioinformatics 2012;28:2994–2995.
35. Sjöstrand J, Tofigh A, Daubin V, Arvestad L, Sennblad B, Lagergren J. A Bayesian method for analyzing lateral gene transfer. Syst Biol 2014;63:409–420.
36. Szöllösi GJ, Tannier E, Daubin V, Boussau B. The inference of gene trees with species trees. Syst Biol 2015;64:e42–e62.
37. Whelan N. Species tree inference in the age of genomics. Trends Evol Biol 2011;3:e5.
38. Willson SJ. Reconstruction of certain phylogenetic networks from the genomes at their leaves. J Theor Biol 2008;252:338–349.
39. Willson SJ. Regular networks can be uniquely constructed from their trees. IEEE/ACM Trans Comput Biol Bioinform 2011;8:785–796.

CHAPTER 23

DIVERSE CONSIDERATIONS FOR SUCCESSFUL PHYLOGENETIC TREE RECONSTRUCTION: IMPACTS FROM MODEL MISSPECIFICATION, RECOMBINATION, HOMOPLASY, AND PATTERN RECOGNITION

Diego Mallo[1], Agustín Sánchez-Cobos[2], and Miguel Arenas[2,3]

[1]*Department of Biochemistry, Genetics and Immunology, University of Vigo, Vigo, Spain*
[2]*Bioinformatics Unit, Centre for Molecular Biology "Severo Ochoa" (CSIC), Madrid, Spain*
[3]*Institute of Molecular Pathology and Immunology, University of Porto (IPATIMUP), Porto, Portugal*

Pattern Recognition in Computational Molecular Biology: Techniques and Approaches,
First Edition. Edited by Mourad Elloumi, Costas S. Iliopoulos, Jason T. L. Wang, and Albert Y. Zomaya.

23.1 INTRODUCTION

Phylogenetic tree reconstruction provides an analysis of the evolutionary relationships among genetic sequences. In population genetics and molecular evolution, these relationships can be useful to understand processes such as species history and speciation [18, 62], demographic history of populations [20, 67], the evolution of genes and protein families [15, 57], the emergence of new protein functions [73, 102], coevolution [42], or comparative genomics [79]. Moreover, phylogenetic trees can be used to detect signatures of selection [49] or to perform ancestral sequence reconstruction [104]. These interesting applications resulted in an increasing number of studies on phylogenetic reconstruction (Figure 23.1).

Nowadays, phylogenetic tree reconstruction is straightforward for any researcher given the current variety of user-friendly software. Despite the fact that the first phylogenetic programs ran on the command line, current frameworks often implement a Graphical User Interface (GUI) where nonexperts can easily perform phylogenetic reconstructions [50, 95]. However, such attractive simplicity may sometimes generate incorrect phylogenetic reconstructions because of ignoring a variety of processes. For example, it is known that substitution model misspecification, that is, ignored recombination or homoplasy, can result in incorrect phylogenies [10, 56, 87, 91]. Other evolutionary processes such as Gene Duplication and Loss (GDL), Incomplete Lineage Sorting (ILS), Horizontal Gene Transfer (HGT), and gene flow between species can also bias species tree reconstruction by making discordant gene and species histories [32, 52, 62]. Briefly, GDL describes how a piece of genetic material (locus) is copied to a different place of the genome (duplication) or erased (loss). It is a well-known (from Ohno's seminal book [71] to recent reviews [112]) source of genetic variation and it is usually described as the main evolutionary force driving gene family evolution. However, ILS (also known as *deep coalescence*) describes a polymorphism present in an ancestral population that is retained along at least two speciation events, subsequently sorting in a way that is incongruent to the species phylogeny. Finally, HGT describes transferences of genetic material (one or a few loci) from one species to another. Since phylogenetic software always outputs a phylogeny, one could believe that such a phylogeny is correct, when maybe it is not. Therefore, it is important to remember the different evolutionary aspects that can affect phylogenetic tree reconstructions and how we can account for them.

In this chapter, we describe the influences of diverse evolutionary phenomena on phylogenetic tree reconstruction and, when possible, we provide strategies to consider such phenomena. From this perspective, we finally discuss the future of phylogenetic tree reconstruction frameworks.

23.2 OVERVIEW ON METHODS AND FRAMEWORKS FOR PHYLOGENETIC TREE RECONSTRUCTION

The phylogenetic reconstruction of a given genetic sample can be performed both at the gene and the species levels, hence studying two different biological histories. Note that gene and species trees may differ owing to evolutionary processes such as GDL,

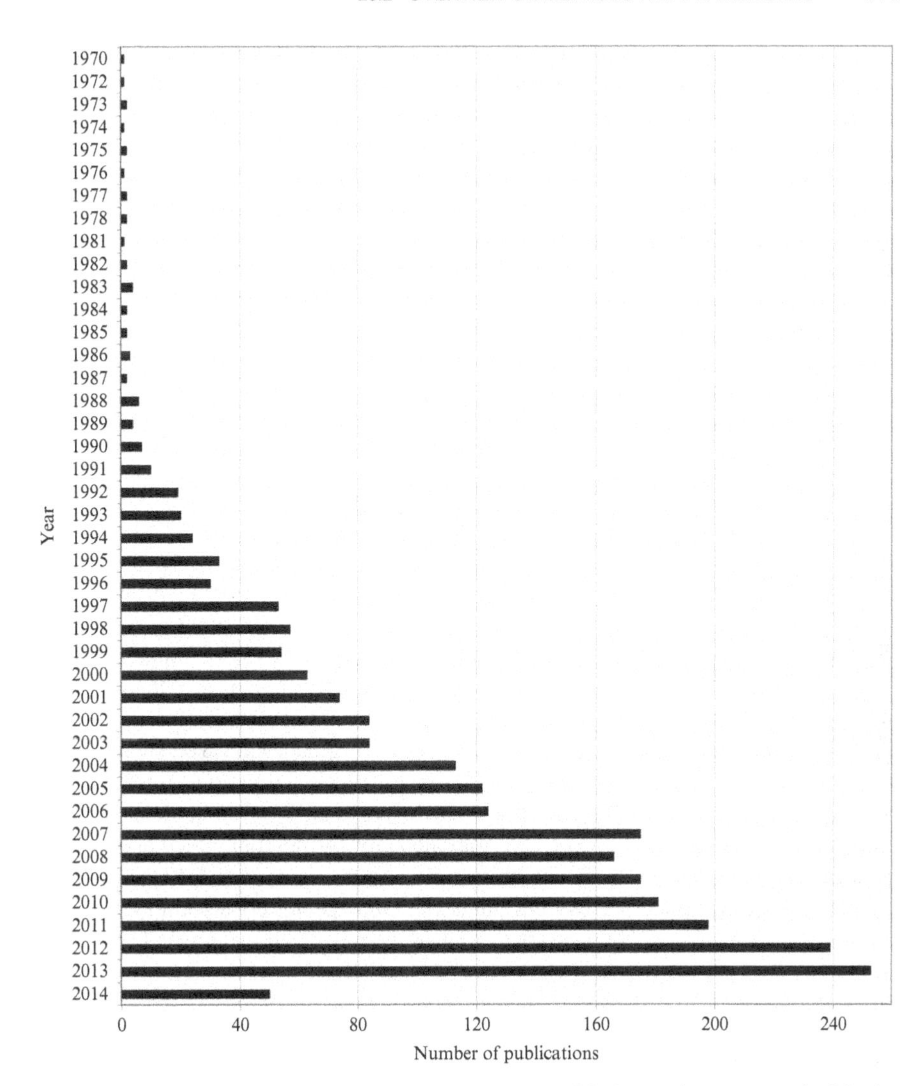

Figure 23.1 The number of articles per year with terms "phylogenetic reconstruction" in the title or the abstract, as measured by PubMed <http://www.ncbi.nlm.nih.gov/pubmed> (search conducted on 18 February 2014).

ILS, or HGT [62]. As a consequence, phylogenetic methods for gene tree reconstruction are based on approaches different to those applied for species tree reconstruction [58]. In this subsection, we briefly describe the commonly used methods and frameworks to perform phylogenetic tree reconstruction.

23.2.1 Inferring Gene Trees

Three main approaches are currently applied to reconstruct phylogenetic trees, namely, distance-based methods (Neighbor-Joining (NJ) [85]), Maximum Likelihood

(ML) methods [34], and Bayesian methods [45]. The goal of distance-based methods is a fast phylogenetic reconstruction from large amounts of data. ML methods are slower than distance-based methods but can generate much more accurate inferences owing to the consideration of a substitution model of evolution and ML optimizations. Finally, Bayesian methods differ from the previous methods on having an integrative point of view, estimating tree distributions instead of punctual estimates. Bayesian-based methods can also incorporate complex and flexible models such as demographics, relaxed molecular clocks, and longitudinal sampling [31]. Owing to their complexity, these methods usually require much more computational costs to yield reliable estimations (reaching convergence between different Markov Chain Monte Carlo (MCMC) runs). Thus, Bayesian methods may generate more information and sometimes provide more realistic reconstructions [31, 46]. For further details about these three approaches, we direct the reader to References [29, 36, 69, 103].

A number of programs have been developed to infer phylogenetic trees from nucleotide, codon, and amino acid sequences. Table 23.1 shows an updated list of the most currently used programs. Concerning distance-based phylogenetic inferences, we would recommend, from our practical experience, the programs *HyPhy* [50], *SplitsTree* [47], and *MEGA* [95], which support a variety of formats for the input Multiple Sequence Alignment (MSA) and are user-friendly. Concerning ML inferences, *RAxML* [92] is recommended for dealing with large amounts of data, *PhyML* [41] is one of the most famous programs, and both implement the interesting option of user-specified transition matrices. *MEGA* is also commonly used owing to its user-friendly GUI. Finally, Bayesian-based phylogenetic estimations are frequently performed with *MrBayes* [84] that allows for different substitution models along the sequences, and with *BEAST* [31], that may account for demographics and relaxed rates of evolution. Owing to the flexibility of Bayesian approaches, both programs can also be used to infer species trees under the multispecies coalescent model [82], see Section 23.2.2. Note that we only described the most well-established programs. Of course, there are many more phylogenetic tree reconstruction programs that continuously emerge with the implementation of new capabilities (see Reference [22]).

23.2.2 Inferring Species Trees

In spite of the fact that the explicit differences between species trees and gene trees had already been noted some decades ago [38, 72, 94], the species tree reconstruction paradigm has only recently blossomed, becoming one of the hottest topics in phylogenomics [24, 58, 108].

In Figure 23.2, the trees are composed by four species (A, B, C, and D) and four gene copies (A0, B0, C0, and D0). Figure 23.2 also shows the following:

(i) a gene duplication (square, GDL) and a gene loss (cross, GDL) in the species A;
(ii) a gene transfer (arrow, HGT) from C0 to replace the original gene copy of D0;
(iii) a deep coalescence (circle) showing the role of ILS. Dashed lines indicate lost lineages, either due to replacement by a transfer or to a loss.

Table 23.1 An updated list of the most commonly used phylogenetic tree reconstruction programs available up to date

Program	Approach	Substitution models	GUI	Source	Reference (last version)
SplitsTree	Distance matrix	–	Yes	http://www.splitstree.org/	[47]
HyPhy	Distance matrix	Nucleotide (All[a]), codon (NG[b]), amino acid (Dayhoff, JTT)	Yes	http://hyphy.org/w/index.php/Main_Page	[50]
MEGA	Distance matrix, ML	Nucleotide (All[a]), codon (NG[b]), amino acid (LG, Dayhoff, JTT, WAG, Mt, Rt)	Yes	http://www.megasoftware.net/	[95]
Mesquite	Distance matrix, ML	Nucleotide (JC, K2P, F81, F84)	Yes	http://mesquiteproject.org/mesquite/mesquite.html	[64]
Phylip	Distance matrix, ML	Nucleotide (JC, … ,HKY), amino acid (JTT, Dayhoff, PAM)	No	http://evolution.genetics.washington.edu/phylip.html	[35]
PhyML	ML	Nucleotide (All[a]), amino acid (All[b])	No	http://www.atgc-montpellier.fr/phyml/	[41]
CodonPhyML	ML	Nucleotide (All[a]), Codon (GY94, MG94, empirical models), amino acid (All[c])	No	http://sourceforge.net/projects/codonphyml/	[37]
RAxML	ML	Nucleotide (All[a]), amino acid (All[c])	Yes	http://sco.h-its.org/exelixis/web/software/raxml/index.html	[92]
MrBayes	Bayesian	Nucleotide (All[a]), amino acid (All[c])	No	http://mrbayes.sourceforge.net/	[84]
BEAST	Bayesian	Nucleotide (HKY, GTR), amino acid (Blosum62, Dayhoff, JTT, mtRev, cpRev, WAG)	Yes	http://beast.bio.ed.ac.uk/Main_Page	[31]

For each program, we indicate the underlying approach (distance matrix, ML, or Bayesian), kind of implemented substitution models (nucleotide, codon, or amino acid), development of a GUI, link, and reference.

Details about nucleotide, codon, and amino acid substitution models can be found in the References [3, 78, 105], respectively. Although many more software packages exist, here we have selected, from our point of view, those programs most commonly used and user-friendly.

[a] All: indicates all common nucleotide substitution models (included in *jModelTest2* [28]: JC, … ,GTR) are implemented.

[b] NG: Nei and Gojobori codon model [70].

[c] All: indicates all common amino acid substitution models (included in *ProtTest* [1]: Blosum62, … ,WAG) are implemented.

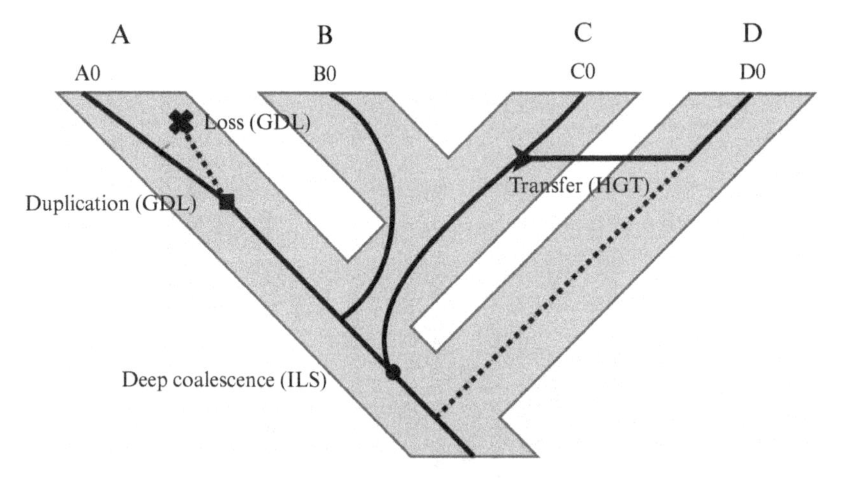

Figure 23.2 An illustrative example of a gene tree (thin lines) inside a species tree (gray tree in the background) and the effect of the three main evolutionary processes that can generate discordance between them.

Despite the complexity of the main approaches to infer species trees, we can roughly classify them into three categories:

(i) The *supermatrix approach* is based on the concatenation of different gene alignments followed by a global gene tree reconstruction to finally generate a "supergene" phylogeny. Thus, it assumes that the discordance between gene trees is the result of minor discordant phylogenetic signals, which would cancel out when using all the information at once. Therefore, the reconstructed supergene phylogeny would represent the species tree.

(ii) The *supertree approach* considers each gene alignment independently. Consequently, gene trees are independently reconstructed with any common phylogenetic tree reconstruction method and then subsequently used to reconstruct the species phylogeny. This last step can be performed just by trying to minimize the gene tree discordance or by considering a model that takes into account at least one evolutionary process. Examples of the process-oblivious approach are the consensus-based methods, Bayesian concordance (e.g., BUCKy [53]), MRP [21, 80], and *RF supertrees* [17, 26]. Model-based methods usually consider evolutionary processes leading to species tree/gene tree incongruence, that is, HGT, GDL, and ILS, in different combinations. Reconciliation-based methods (e.g., iGTP [25] and *SPRSupertrees* [98]), distance-based methods (e.g., STAR and STEAC [59]), pseudo-likelihood-based methods (e.g., STELLS [101] and MP-EST [60]) are examples of process-aware/model-based methods. Recently, our laboratory developed a novel Bayesian method that lies between these two categories (process-oblivious and process-aware/model-based), integrating error-based and reconciliation-based distances [66]. In general, these methods are much faster than the *supermatrix* or fully *probabilistic* approaches.

(iii) *Fully probabilistic models* are the most sophisticated and are usually implemented in Bayesian frameworks. They integrate both gene tree and species tree estimation. The computer programs *BEST* [58] and **BEAST* [43] consider the effect of ILS by implementing the multispecies coalescent model, while *PHYLDOG* [23] considers GDL using a birth–death model. In general, these fully probabilistic approaches are not only much more comprehensive but also more time consuming than the previous approaches. Another drawback is that they sometimes have convergence problems when tackling big data sets in terms of number of taxa.

The selection of a species tree reconstruction method to analyze a biological data set is a difficult task, mainly due to the variety of underlying assumptions considered in the different methods. In general, probabilistic models can be recommended beforehand, as long as the model can handle the evolutionary processes present in the data. Indeed, there is an important trade-off between computational costs, data set size, and accuracy. Species tree reconstruction is a very active research topic where the emergence of new methods and comprehensive benchmarks will probably lead to more realistic species tree estimations.

In the following sections, we describe the impact of different evolutionary phenomena on phylogenetic tree reconstruction and we provide alternative methodologies, when possible, to account for such phenomena.

23.3 INFLUENCE OF SUBSTITUTION MODEL MISSPECIFICATION ON PHYLOGENETIC TREE RECONSTRUCTION

It is clear that nucleotide, codon, or amino acid substitution models of evolution are fundamental in phylogenetic inferences because the substitution model may affect the estimates of topology, substitution rates, bootstrap values, posterior probabilities, or the molecular clock [56, 111]. Actually, a simulation study by Lemmon and Moriarty [56] pointed out that more simple models than the true model can increase phylogenetic reconstruction biases. As a consequence, researchers use to apply the most complex models, that is, the GTR substitution model at nucleotide level, but this overparameterization may increase the variance and noise of the estimates and also generate temporal inconsistencies [93].

Altogether, the researcher should first apply a statistical selection of best-fit substitution models using programs such as *jModelTest2* [28] (at nucleotide level) or *ProtTest* [1] (at amino acid level). Indeed, notice that different genomic regions can evolve under different substitution models [5, 8, 13]. Then, the selected model could be used to perform the phylogenetic reconstruction. Still with this consideration, it is known that commonly used substitution models can be unrealistic owing to assumptions such as site-independence evolution [99]. Site-dependent substitution models can better mimic the real evolutionary process [15, 39] but they have not been implemented yet in phylogenetic reconstruction methods because of their complexity, that is, the computation of the likelihood function from site-specific and branch-specific transition matrices is not straightforward, although significant advances have been recently made in this area [16].

23.4 INFLUENCE OF RECOMBINATION ON PHYLOGENETIC TREE RECONSTRUCTION

Genetic recombination, or other processes where genetic material is exchanged, is a fundamental evolutionary process in a variety of organisms, especially in viruses and bacteria [61, 76, 83]. Unfortunately, recombination can seriously bias phylogenetic tree reconstruction. As an example, Schierup and Hein [87] simulated DNA sequence evolution under the coalescent with recombination [44]. They found that ignoring recombination biases the inferred phylogenetic trees toward larger terminal branches, smaller times to the Most Recent Common Ancestor (MRCA), and incorrect topologies for both distance-based and ML methods [87]. In addition, it can lead to loss of molecular clock, apparent homoplasies, and overestimation of the substitution rate heterogeneity [77, 87, 88]. Real data also presented these phylogenetic incongruences [33, 100]. Analysis derived from this kind of incorrect phylogenies can be also seriously affected, for example, the Ancestral Sequence Reconstruction (ASR) [6, 11] or the estimation of positively selected sites [4, 9, 12].

The evolutionary history of a sample with ancestral recombination events can be represented as a phylogenetic network, usually called Ancestral Recombination Graph (ARG) [7, 40], rather than as a single phylogenetic tree because each recombinant fragment (or partition) can have a particular evolutionary history (Figure 23.3). As a consequence, there are two phylogenetic reconstruction methodologies accounting for recombination:

(i) *Phylogenetic network reconstruction* [68] by using programs such as *SplitsTree*. An example of this methodology applied to real data is described in Reference [14].

(ii) *Phylogenetic tree reconstruction* for each recombinant fragment. In the first step, recombination breakpoints are detected [65]. Then, a phylogenetic tree reconstruction can be performed for each recombinant fragment. Examples of this methodology applied to real data are described in References [74, 75].

Both the above-described methodologies are correct and the choice is just based on the evolutionary question to be answered. A phylogenetic network provides a complete visualization of clades and phylogenetic relationships (see Reference [48]), while methods such as ASR, or the detection of molecular adaptation, often require a tree for each recombinant fragment [4, 11].

In Figure 23.3, the ARG is based on two recombination events with breakpoints at positions 50 and 100, which result in three recombinant fragments (1–50, 51–100, and 101–300). Note that each recombinant fragment may be based on an evolutionary history that differs from the evolutionary history of other recombinant fragments. Dashed lines indicate branches derived from recombinant events.

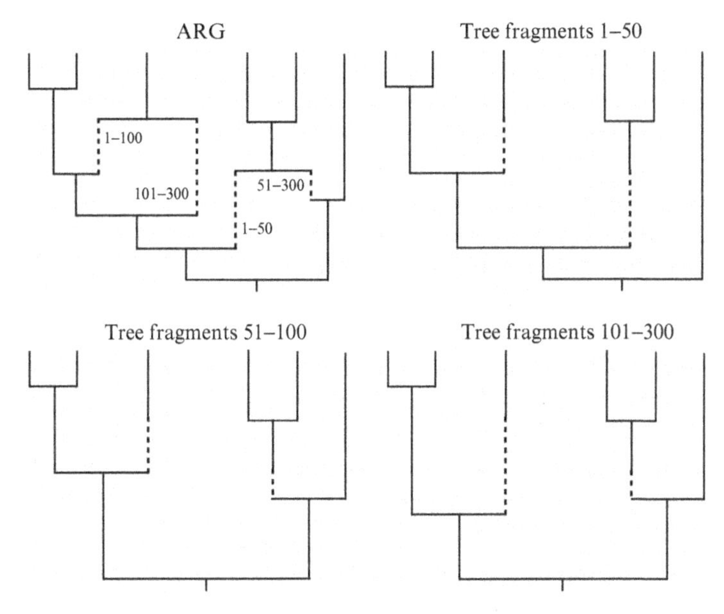

Figure 23.3 An illustrative example of an ARG and its corresponding embedded trees.

23.5 INFLUENCE OF DIVERSE EVOLUTIONARY PROCESSES ON SPECIES TREE RECONSTRUCTION

Since most of species tree reconstruction methods entail a gene tree reconstruction step, every evolutionary process that is able to mislead gene tree estimation can also, in a greater or lesser extent, affect the species tree reconstruction. Simulation studies by Bayzid and Warnow [19] have in fact shown that species tree reconstruction accuracy is highly dependent upon gene tree accuracy. Moreover, these authors also suggested that the most accurate species tree methods base part of their advantage on estimating gene trees much better than common gene tree estimation methods.

However, ILS-based distance methods are not strongly influenced by intralocus recombination as discussed by Lanier and Knowles [52]. In particular, these authors showed that tree lengths, and therefore the intensity of ILS, are the most important factors to determine the accuracy of the reconstruction when there are not additional, apart from ILS and intralocus recombination, evolutionary processes. In fact, recombination has to occur in unsorted lineages to affect species tree reconstruction under ILS-based models. Actually, the own effects of ILS are more misleading than recombination. Interestingly, sampling parameters, that is, number of loci and individuals, may properly control the species tree reconstruction accuracy and reduce the misleading effect of recombination.

Gene flow may generate important biases in species tree inference through different migration models and migration patterns. Attending to topological accuracy,

Eckert and Cartens [32] observed that supertree methods are robust to historical gene flow models in contraposition to the supermatrix approach. However, the stepping-stone, and more importantly the n-island, models extremely misled both species tree reconstruction approaches, especially under high migration rates. In addition, Leaché et al. [55] showed that both the migration model and the migration pattern could bias the species tree inference, by using both tree-based and fully probabilistic Bayesian methods. For example, these authors showed that gene flow between sister species increases both the probability of estimating the true species tree and the posterior probabilities of the clade where migration events take place. On the other hand, migration between nonsister species can strongly bias species tree reconstruction and even increase the posterior probabilities of wrongly estimated clades. Regarding other species tree reconstruction parameters, gene flow can generate overcompression, that is, underestimation of the branch lengths of the species tree, and dilatation, that is, overestimation of the population size, as a function of the gene flow intensity [55].

Hybridization can also lead to gene tree incongruence and, as a consequence, mislead species tree reconstruction methods. However, there are a bunch of species tree methods to detect hybridization [51, 107] and to estimate species trees in presence of both ILS and hybridization events, that is, species networks (see Reference [106]).

The most recognized processes able to generate gene tree discordance, that is, GDL, HGT, and ILS, have never been jointly modeled by any species tree reconstruction program, and therefore, the effect of unmodeled processes should also be explored for every kind of method. Unfortunately, there is a general lack of studies about the robustness to model misspecification of species tree methodologies, although there are some exceptions. For example *PHYLDOG*, a GDL-based probabilistic approach, can accurately reconstruct species trees under moderate levels of ILS, although it overestimates the number of duplications and losses [23]. Concerning the effect of HGT, ILS-based fully probabilistic methods showed a robust behavior when HGT was randomly distributed along the trees, although the accuracy drops when HGT is focalized into a specific branch of the species tree [27].

Apart from not having integrative models for those evolutionary processes, GDL, HGT, and ILS can bias species tree inference even when they are properly modeled. ILS not only generates gene tree incongruence, but also misleads species tree reconstruction, even when being explicitly modeled [19, 54, 63]. This pattern is not completely shared with GDL and HGT methods, which conversely obtain information from the modeled events, that is, duplications, losses, or transfers. Thus both high and low rates can mislead the estimation [89].

In summary, common evolutionary processes, such as GDL, HGT, and ILS, together with migration and hybridization, can bias species tree inferences although the performance of the methods differs. Nevertheless, we notice that additional simulation studies are still demanded to further analyze the effect of these processes on the broad plethora of species tree reconstruction methods.

23.6 INFLUENCE OF HOMOPLASY ON PHYLOGENETIC TREE RECONSTRUCTION: THE GOALS OF PATTERN RECOGNITION

The recurrence of similarity in evolution may generate homoplasy [86] where sequences are similar but are not derived from a common ancestor. Homoplasy can be generated by a variety of evolutionary processes such as recurrent mutation, gene flow, or recombination. As noted, phylogenetic tree reconstruction methods assume a common ancestor and, as a consequence, homoplasy can lead to phylogenetic uncertainty [91]. It is therefore fundamental to search for homology in the data before building a phylogenetic tree, for example, by using computer tools such as BLAST [2].

This consideration is related with the use of pattern recognition techniques [96] to reconstruct phylogenetic histories through cluster analysis [90]. In particular, a phylogeny can be reconstructed by following a particular pattern that can be recognized from the data [97, 109]. For example, one could directly apply a clustering algorithm and build a distance-based phylogenetic tree [97] or specify a particular pattern, that is, genes, domains, or sites, and then reconstruct a tree based on such a pattern [81, 113]. Pattern recognition can also be useful to study the influence of different phenomena on phylogenetic inferences. For example, this can be performed through the reconstruction of phylogenetic histories of different local patterns collected from the same raw data, that is, genome-wide data, and thus be able to identify a phylogeny for each pattern [110].

23.7 CONCLUDING REMARKS

In recent years, phylogenetic reconstruction frameworks have experienced a sharp increase in new methods and technical advances, for example, parallelization on multiprocessors, GUI environments, or specific tools for specific analysis. However, we notice that influences derived from the nature of the data on phylogenetic tree reconstruction are sometimes forgotten. For example, sometimes phylogenies are incorrectly reconstructed under the GTR substitution model without further statistical analysis (see the interesting article by Sumner et al. [93]) or recombination is not considered when inferring phylogenetic trees [30]. Here, we suggest that the consideration of evolutionary phenomena, that is, through the above-described alternative methodologies, is fundamental for successfully building phylogenies.

In our opinion, future phylogenetic frameworks should implement new models, for example, site-dependent substitution models, and technical advances but also should "internally" evaluate the data before performing a phylogenetic reconstruction, for example, as we suggest in the following. First, a homology search through pattern recognition methods would provide important information about the ancestral history, that is, homology, [91]. Note that a phylogenetic tree should not be inferred under lack of homology [36]. Second, a substitution model choice could be used to select the substitution model that best fits the data and then, such a fit to the selected substitution model could be evaluated according to a given threshold, that is, a likelihood score, under which the tree should not be computed, for example, data with lack of evolutionary information. Third, recombination breakpoints could be also internally detected [65] and in presence of recombination, a tree should be estimated for each recombinant fragment. In case of species tree reconstruction, an extra gene tree discordance analysis could be used to select a proper species tree method to perform the final estimation step.

Altogether, we believe that a tree should not be provided by a phylogenetic framework if the evolutionary history of the data cannot actually be explained or supported by such a tree from a biological perspective. This strategy (i.e., a framework that includes previous analyses of the data before computing a phylogeny) can probably increase computing times but it could help to avoid incorrect phylogenies. In summary, one should bear in mind that several evolutionary processes might impact the reconstructed phylogenies and we consequently recommend being cautious and contrasting the inferences.

ACKNOWLEDGMENTS

This study was supported by the Spanish Government through the "Juan de la Cierva" fellowship "JCI-2011-10452" to MA and through the "FPI" fellowship "BES-2010-031014" at the University of Vigo to DM. MA also wants to acknowledge the EMBO fellowship "ASTF 367-2013" and the Portuguese Government with the FCT Starting Grant IF/00955/2014. We thank Leonardo de Oliveira Martins for helpful comments about some species trees methodological aspects. We thank to an anonymous reviewer for detailed comments. We also want to thank the Editors for the invitation to contribute with this chapter.

REFERENCES

1. Abascal F, Zardoya R, Posada D. ProtTest: selection of best-fit models of protein evolution. Bioinformatics 2005;21:2104–2105.
2. Altschul SF, Madden TL, Schaffer AA, Zhang J, Zhang Z, Miller W, Lipman DJ. Gapped BLAST and PSI-BLAST: a new generation of protein database search programs. Nucleic Acids Res 1997;25:3389–3402.
3. Anisimova M, Kosiol C. Investigating protein-coding sequence evolution with probabilistic codon substitution models. Mol Biol Evol 2009;26:255–271.
4. Anisimova M, Nielsen R, Yang Z. Effect of recombination on the accuracy of the likelihood method for detecting positive selection at amino acid sites. Genetics 2003;164:1229–1236.
5. Arbiza L, Patricio M, Dopazo H, Posada D. Genome-wide heterogeneity of nucleotide substitution model fit. Genome Biol Evol 2011;3:896–908.
6. Arenas M. Computer programs and methodologies for the simulation of DNA sequence data with recombination. Front Genet 2013a;4:9.
7. Arenas M. The Importance and Application of the Ancestral Recombination Graph. Front Genet 2013b;4:206.

8. Arenas M. Advances in computer simulation of genome evolution: toward more realistic evolutionary genomics analysis by approximate bayesian computation. J Mol Evol 2015;80:189–192.
9. Arenas M, Posada D. Coalescent simulation of intracodon recombination. Genetics 2010;184:429–437.
10. Arenas M, Posada D. Computational design of centralized HIV-1 genes. Curr HIV Res 2010;8:613–621.
11. Arenas M, Posada D. The effect of recombination on the reconstruction of ancestral sequences. Genetics 2010;184:1133–1139.
12. Arenas M, Posada D. The influence of recombination on the estimation of selection from coding sequence alignments. In: Fares MA, editor. Natural Selection: Methods and Applications. Boca Raton: CRC Press/Taylor & Francis; 2014.
13. Arenas M, Posada D. Simulation of Genome-wide Evolution under Heterogeneous Substitution models and Complex Multispecies Coalescent Histories. Mol Biol Evol 2014;31:1295–1301.
14. Arenas M, Patricio M, Posada D, Valiente G. Characterization of phylogenetic networks with NetTest. BMC Bioinform 2010;11:268.
15. Arenas M, Dos Santos HG, Posada D, Bastolla U. Protein evolution along phylogenetic histories under structurally constrained substitution models. Bioinformatics 2013;29:3020–3028.
16. Arenas M, Sanchez-Cobos A, Bastolla U. Maximum likelihood phylogenetic inference with selection on protein folding stability. Mol Biol Evol 2015;32:2195–2207.
17. Bansal MS, Burleigh JG, Eulenstein O, Fernandez-Baca D. Robinson–Foulds supertrees. Algorithms Mol Biol 2010;5:18.
18. Barraclough TG, Nee S. Phylogenetics and speciation. Trends Ecol Evol 2001; 16:391–399.
19. Bayzid MS, Warnow T. Naive binning improves phylogenomic analyses. Bioinformatics 2013;29:2277–2284.
20. Benguigui M, Arenas M. Spatial and temporal simulation of human evolution. Methods, frameworks and applications. Curr Genom 2014;15:245–255.
21. Bininda-Emonds OR. The evolution of supertrees. Trends Ecol Evol 2004;19:315–322.
22. Blair C, Murphy RW. Recent trends in molecular phylogenetic analysis: where to next? J Hered 2011;102:130–138.
23. Boussau B, Szollosi GJ, Duret L, Gouy M, Tannier E, Daubin V. Genome-scale coestimation of species and gene trees. Genome Res 2012;23:323–330.
24. Capella-Gutierrez S, Kauff F, Gabaldon T. A phylogenomics approach for selecting robust sets of phylogenetic markers. Nucleic Acids Res 2014;42:e54.
25. Chaudhary R, Bansal MS, Wehe A, Fernandez-Baca D, Eulenstein O. iGTP: a software package for large-scale gene tree parsimony analysis. BMC Bioinform 2010;11:574.
26. Chaudhary R, Burleigh JG, Fernandez-Baca D. Inferring species trees from incongruent multi-copy gene trees using the Robinson-Foulds distance. Algorithms Mol Biol 2013;8:28.
27. Chung Y, Ane C. Comparing two Bayesian methods for gene tree/species tree reconstruction: simulations with incomplete lineage sorting and horizontal gene transfer. Syst Biol 2011;60:261–275.

28. Darriba D, Taboada GL, Doallo R, Posada D. jModelTest 2: more models, new heuristics and parallel computing. Nat Methods 2012;9:772.
29. Delsuc F, Brinkmann H, Philippe H. Phylogenomics and the reconstruction of the tree of life. Nat Rev Genet 2005;6:361–375.
30. Doria-Rose NA, Learn GH, Rodrigo AG, Nickle DC, Li F, Mahalanabis M, Hensel MT, Mclaughlin S, Edmonson PF, Montefiori D, Barnett SW, Haigwood NL, Mullins JI. Human immunodeficiency virus type 1 subtype B ancestral envelope protein is functional and elicits neutralizing antibodies in rabbits similar to those elicited by a circulating subtype B envelope. J Virol 2005;79:11214–11224.
31. Drummond AJ, Suchard MA, Xie D, Rambaut A. Bayesian phylogenetics with BEAUti and the BEAST 1.7. Mol Biol Evol 2012;29:1969–1973.
32. Eckert AJ, Carstens BC. Does gene flow destroy phylogenetic signal? The performance of three methods for estimating species phylogenies in the presence of gene flow. Mol Phylogenet Evol 2008;49:832–842.
33. Feil EJ, Holmes EC, Bessen DE, Chan M-S, Day NPJ, Enright MC, Goldstein R, Hood DW, Kalia A, Moore CE, Zhou J, Spratt BG. Recombination within natural populations of pathogenic bacteria: Short-term empirical estimates and long-term phylogenetic consequences. Proc Natl Acad Sci USA 2001;98:182–187.
34. Felsenstein J. Evolutionary trees from DNA sequences: A maximum likelihood approach. J Mol Evol 1981;17:368–376.
35. Felsenstein J. PHYLIP (Phylogeny Inference Package)3.5c ed.. Seattle: Department of Genetics, University of Washington; 1993.
36. Felsenstein J. *Inferring Phylogenies*. MA, Sunderland: Sinauer Associates; 2004.
37. Gil M, Zanetti MS, Zoller S, Anisimova M. CodonPhyML: fast maximum likelihood phylogeny estimation under codon substitution models. Mol Biol Evol 2013;30:1270–1280.
38. Goodman M, Czelusniak J, Moore G, Romero-Herrera A, Matsuda G. Fitting the gene lineage into its species lineage, a parsimony strategy illustrated by cladograms constructed from globin sequences. Syst Zool 1979;28:132–163.
39. Grahnen JA, Nandakumar P, Kubelka J, Liberles DA. Biophysical and structural considerations for protein sequence evolution. BMC Evol Biol 2011;11:361.
40. Griffiths RC, Marjoram P. An ancestral recombination graph. In: Donelly P, Tavaré S, editors. *Progress in Population Genetics and Human Evolution*. Berlin: Springer-Verlag; 1997.
41. Guindon S, Dufayard JF, Lefort V, Anisimova M, Hordijk W, Gascuel O. New algorithms and methods to estimate maximum-likelihood phylogenies: assessing the performance of PhyML 3.0. Syst Biol 2010;59:307–321.
42. Hafner MS, Nadler SA. Phylogenetic trees support the coevolution of parasites and their hosts. Nature 1988;332:258–259.
43. Heled J, Bryant D, Drummond AJ. Simulating gene trees under the multispecies coalescent and time-dependent migration. BMC Evol Biol 2013;13:44.
44. Hudson RR. Properties of a neutral allele model with intragenic recombination. Theor Popul Biol 1983;23:183–201.
45. Huelsenbeck JP, Ronquist F. MRBAYES: Bayesian inference of phylogeny. Bioinformatics 2001;17:754–755.
46. Huelsenbeck JP, Ronquist F, Nielsen R, Bollback JP. Bayesian inference of phylogeny and its impact on evolutionary biology. Science 2001;294:2310–2314.

47. Huson DH. SplitsTree: analyzing and visualizing evolutionary data. Bioinformatics 1998;14:68–73.

48. Huson DH, Bryant D. Application of phylogenetic networks in evolutionary studies. Mol Biol Evol 2006;23:254–267.

49. Kosakovsky Pond SL, Frost SD. Not so different after all: a comparison of methods for detecting amino Acid sites under selection. Mol Biol Evol 2005;22:1208–1222.

50. Kosakovsky Pond SL, Frost SD, Muse SV. HYPHY: hypothesis testing using phylogenies. Bioinformatics 2005;21:676–679.

51. Kubatko LS. Identifying hybridization events in the presence of coalescence via model selection. Syst Biol 2009;58:478–488.

52. Lanier HC, Knowles LL. Is recombination a problem for species-tree analyses? Syst Biol 2012;61:691–701.

53. Larget BR, Kotha SK, Dewey CN, Ane C. BUCKy: gene tree/species tree reconciliation with Bayesian concordance analysis. Bioinformatics 2010;26:2910–2911.

54. Leaché AD, Rannala B. The accuracy of species tree estimation under simulation: a comparison of methods. Syst Biol 2011;60:126–137.

55. Leaché AD, Harris RB, Rannala B, Yang Z. The influence of gene flow on species tree estimation: a simulation study. Syst Biol 2014;63:17–30.

56. Lemmon AR, Moriarty EC. The importance of proper model assumption in bayesian phylogenetics. Syst Biol 2004;53:265–277.

57. Lijavetzky D, Carbonero P, Vicente-Carbajosa J. Genome-wide comparative phylogenetic analysis of the rice and Arabidopsis Dof gene families. BMC Evol Biol 2003;3:17.

58. Liu L. BEST: Bayesian estimation of species trees under the coalescent model. Bioinformatics 2008;24:2542–2543.

59. Liu L, Yu L, Kubatko L, Pearl DK, Edwards SV. Coalescent methods for estimating phylogenetic trees. Mol Phylogenet Evol 2009;53:320–328.

60. Liu L, Yu L, Edwards SV. A maximum pseudo-likelihood approach for estimating species trees under the coalescent model. BMC Evol Biol 2010;10:302.

61. Lopes JS, Arenas M, Posada D, Beaumont MA. Coestimation of recombination, substitution and molecular adaptation rates by approximate Bayesian computation. Heredity 2014;112:255–264.

62. Maddison W. Gene trees in species trees. Syst Biol 1997;46:523–536.

63. Maddison WP, Knowles LL. Inferring phylogeny despite incomplete lineage sorting. Syst Biol 2006;55:21–30.

64. Maddison, W.P. and Maddison, D.R. (2010). Mesquite: a modular system for evolutionary analysis. 2.73, http://mesquiteproject.org.

65. Martin DP, Lemey P, Posada D. Analysing recombination in nucleotide sequences. Mol Ecol Resour 2011;11:943–955.

66. Martins LD, Mallo D, Posada D. A Bayesian Supertree Model for Genome-Wide Species Tree Reconstruction. Syst Biol 2014. doi: 10.1093/sysbio/syu082.

67. Mona S, Ray N, Arenas M, Excoffier L. Genetic consequences of habitat fragmentation during a range expansion. Heredity 2014;112:291–299.

68. Morrison DA. Networks in phylogenetic analysis: new tools for population biology. Int J Parasitol 2005;35:567–582.

69. Nakhleh L. Computational approaches to species phylogeny inference and gene tree reconciliation. Trends Ecol Evol 2013;28:719–728.
70. Nei M, Gojobori T. Simple method for estimating the numbers of synonymous and nonsynonymous nucleotide substitutions. Mol Biol Evol 1986;3:418–426.
71. Ohno S. *Evolution by Gene Duplication.* Berlin: Springer-Verlag; 1970.
72. Pamilo P, Nei M. Relationships between gene trees and species trees. Mol Biol Evol 1988;5:568–583.
73. Pellegrini M, Marcotte EM, Thompson MJ, Eisenberg D, Yeates TO. Assigning protein functions by comparative genome analysis: protein phylogenetic profiles. Proc Natl Acad Sci USA 1999;96:4285–4288.
74. Perez-Losada M, Posada D, Arenas M, Jobes DV, Sinangil F, Berman PW, Crandall KA. Ethnic differences in the adaptation rate of HIV gp120 from a vaccine trial. Retrovirology 2009;6:67.
75. Perez-Losada, M., Jobes, D.V., Sinangil, F., Crandall, K.A., Arenas, M., Posada, D. and Berman, P.W. (2011). Phylodynamics of HIV-1 from a phase III AIDS vaccine trial in Bangkok, Thailand. PLoS One, 6, e16902.
76. Perez-Losada M, Arenas M, Galan JC, Palero F, Gonzalez-Candelas F. Recombination in viruses: Mechanisms, methods of study, and evolutionary consequences. Infect Genet Evol 2015;30C:296–307.
77. Posada D. Unveiling the molecular clock in the presence of recombination. Mol Biol Evol 2001;18:1976–1978.
78. Posada D. Selecting models of evolution. In: Vandemme A, Salemi M, editors. *The Phylogenetic Handbook.* Cambridge, UK: Cambridge University Press; 2003.
79. Postlethwait JH, Woods IG, Ngo-Hazelett P, Yan YL, Kelly PD, Chu F, Huang H, Hill-Force A, Talbot WS. Zebrafish comparative genomics and the origins of vertebrate chromosomes. Genome Res 2000;10:1890–1902.
80. Ragan MA. Phylogenetic inference based on matrix representation of trees. Mol Phylogenet Evol 1992;1:53–58.
81. Rajendran KV, Zhang J, Liu S, Peatman E, Kucuktas H, Wang X, Liu H, Wood T, Terhune J, Liu Z. Pathogen recognition receptors in channel catfish: II. Identification, phylogeny and expression of retinoic acid-inducible gene I (RIG-I)-like receptors (RLRs). Dev Comp Immunol 2012;37:381–389.
82. Rannala B, Yang Z. Bayes estimation of species divergence times and ancestral population sizes using DNA sequences from multiple loci. Genetics 2003;164:1645–1656.
83. Robertson DL, Sharp PM, Mccutchan FE, Hahn BH. Recombination in HIV-1. Nature 1995;374:124–126.
84. Ronquist F, Teslenko M, Van Der Mark P, Ayres DL, Darling A, Hohna S, Larget B, Liu L, Suchard MA, Huelsenbeck JP. MrBayes 3.2: efficient Bayesian phylogenetic inference and model choice across a large model space. Syst Biol 2012;61:539–542.
85. Saitou N, Nei M. The neighbor-joining method: a new method for reconstructing phylogenetic trees. Mol Biol Evol 1987;4:406–425.
86. Sanderson M, Hufford L. *Homoplasy: The Recurrence of Similarity in Evolution.* New York: Academic Press; 1996.
87. Schierup MH, Hein J. Consequences of recombination on traditional phylogenetic analysis. Genetics 2000;156:879–891.

88. Schierup MH, Hein J. Recombination and the molecular clock. Mol Biol Evol 2000;17:1578–1579.
89. Sennblad B, Lagergren J. Probabilistic orthology analysis. Syst Biol 2009;58:411–424.
90. Sharaf MA, Kowalski BR, Weinstein B. Construction of phylogenetic trees by pattern recognition procedures. Z Naturforsch C 1980;35:508–513.
91. Smouse P. To tree or not to tree. Mol Ecol 1998;7:399–412.
92. Stamatakis A, Aberer AJ, Goll C, Smith SA, Berger SA, Izquierdo-Carrasco F. RAxML-Light: a tool for computing terabyte phylogenies. Bioinformatics 2012;28: 2064–2066.
93. Sumner JG, Jarvis PD, Fernandez-Sanchez J, Kaine BT, Woodhams MD, Holland BR. Is the general time-reversible model bad for molecular phylogenetics? Syst Biol 2012;61:1069–1074.
94. Takahata N. Gene geneology in three related populations: consistency probability between gene and population trees. Genetics 1989;122:957–966.
95. Tamura K, Stecher G, Peterson D, Filipski A, Kumar S. MEGA6: molecular evolutionary genetics analysis version 6.0. Mol Biol Evol 2013;30:2725–2729.
96. Vogt NB, Knutsen H. SIMCA pattern recognition classification of five infauna taxonomic groups using non-polar compounds analysed by high resolution gas chromatography. Marine Ecol 1985;26:145–156.
97. Wei K. Stratophenetic tracing of phylogeny using SIMCA pattern recognition technique: a case study of the late Neogene planktic foraminifera Globoconella clade. Paleobiology 1994;20:52–65.
98. Whidden C, Zeh N, Beiko RG. Supertrees Based on the Subtree Prune-and-Regraft Distance. Syst Biol 2014;63:566–581.
99. Wilke, C.O. (2012). Bringing molecules back into molecular evolution. PLoS Comput Biol, 8, e1002572.
100. Worobey M, Holmes EC. Evolutionary aspects of recombination in RNA viruses. J Gen Virol 1999;80:2535–2543.
101. Wu Y. Coalescent-based species tree inference from gene tree topologies under incomplete lineage sorting by maximum likelihood. Evolution 2012;66:763–775.
102. Yang Z. The power of phylogenetic comparison in revealing protein function. Proc Natl Acad Sci USA 2005;102:3179–3180.
103. Yang Z. *Computational Molecular Evolution*. Oxford, England: Oxford University Press; 2006.
104. Yang Z. PAML 4: phylogenetic analysis by maximum likelihood. Mol Biol Evol 2007;24:1586–1591.
105. Yang Z, Nielsen R, Masami H. Models of amino acid substitution and applications to mitochondrial protein evolution. Mol Biol Evol 1998;15:1600–1611.
106. Yu, Y. and Nakhleh, L. (2012). Fast algorithms for reconciliation under hybridization and incomplete lineage sorting. arXiv:1212.1909.
107. Yu Y, Than C, Degnan JH, Nakhleh L. Coalescent histories on phylogenetic networks and detection of hybridization despite incomplete lineage sorting. Syst Biol 2011;60:138–149.
108. Yu Y, Ristic N, Nakhleh L. Fast algorithms and heuristics for phylogenomics under ILS and hybridization. BMC Bioinform 2013;14(Suppl 15):S6.

109. Zahid MaH, Mittal A, Joshi RC. A pattern recognition-based approach for phylogenetic network construction with constrained recombination. Pattern Recogn 2006;39: 2312–2322.
110. Zamani N, Russell P, Lantz H, Hoeppner MP, Meadows JR, Vijay N, Mauceli E, Di Palma F, Lindblad-Toh K, Jern P, Grabherr MG. Unsupervised genome-wide recognition of local relationship patterns. BMC Genomics 2013;14:347.
111. Zhang J. Performance of likelihood ratio tests of evolutionary hypotheses under inadequate substitution models. Mol Biol Evol 1999;16:868–875.
112. Zhang J. Evolution by gene duplication: an update. Trends Ecol Evol 2003;18:292–298.
113. Zhang J, Liu S, Rajendran KV, Sun L, Zhang Y, Sun F, Kucuktas H, Liu H, Liu Z. Pathogen recognition receptors in channel catfish: III phylogeny and expression analysis of Toll-like receptors. Dev Comp Immunol 2013;40:185–194.

CHAPTER 24

AUTOMATED PLAUSIBILITY ANALYSIS OF LARGE PHYLOGENIES

David Dao[1], Tomáš Flouri[2], and Alexandros Stamatakis[1,2]

[1]*Karlsruhe Institute of Technology, Institute for Theoretical Informatics, Postfach 6980, 76128, Karlsruhe, Germany*
[2]*Heidelberg Institute for Theoretical Studies, 69118 Heidelberg, Germany*

24.1 INTRODUCTION

Disentangling the evolutionary history and diversity of species has preoccupied mankind for centuries. Ever since Darwin's work on evolutionary theory [7], evolutionary trees or phylogenies are typically used to represent evolutionary relationships among species.

Although phylogenies have been used for almost 150 years, statistical, computational, and algorithmic work on phylogenies—often referred to as computational phylogenetics—is only 50 years old. The analysis of phylogenetic trees does not only serve human curiosity but also has practical applications in different fields of science. Phylogenies help to address biological problems such as drug design [5], multiple sequence alignment [8], protein structure [27], gene function prediction [9], or studying the evolution of infectious diseases [13].

In the past decade, the molecular revolution [24] has led to an unprecedented accumulation of molecular data for inferring phylogenies. Public databases such

Pattern Recognition in Computational Molecular Biology: Techniques and Approaches,
First Edition. Edited by Mourad Elloumi, Costas S. Iliopoulos, Jason T. L. Wang, and Albert Y. Zomaya.

as GenBank [3] grow exponentially, which, in conjunction with scalable software, allows for computing extremely large phylogenies that contain thousands or even tens of thousands of species (see References [14, 28], for instance). In practice, however, our ability to infer such comprehensive trees resurrects old problems and gives rise to novel challenges in computational phylogenetics.

First, reconstructing the phylogeny that best fits the data is a combinatorial optimization problem. The number of phylogenetic trees increases super-exponentially with the number of taxa [10]. For example, there are three distinct unrooted trees for four species. However, only for 23 species, there already exist 1.32×10^{25} possible unrooted trees, which is 10 times the number of estimated stars in the universe.

Owing to the continuous progress in processor technology, for instance, in transistor size [4] and parallel processing techniques [17], as well as phylogenetic analysis software such as RAxML [29] and MrBayes [25], biologists can now routinely reconstruct trees with approximately 100–1000 species for their studies. However, for even larger phylogenies with up to tens of thousands of species, we are not sure whether we are obtaining a plausible, let alone correct answer, given the literally astronomical size of the tree search space. Reconstructing large phylogenies is particularly difficult when the alignment length remains constant. In other words, accuracy decreases as we add taxa while keeping the number of sites in the alignment fixed [16, 23].

A further problem is the nonuniform sampling of data. Although genomic data are becoming available at an increasing rate, many species are undersampled because it is difficult to obtain or collect the sequencing samples. For example, prokaryotic organisms that have small genomes, or model organisms and model genes are sampled more frequently than other species.

For that reason, increasing the number of taxa in a data set typically also increases the amount of missing data [26], which leads to potentially biasing the results and, therefore, to decreased phylogenetic accuracy [30]. So given a large phylogeny, how can we assess the biological plausibility of such a large tree?

Visual inspection to assess the plausibility of trees with more than 50,000 taxa is not an option because (i) it is impossible to do so for humans and (ii) there are only but a few tools available for visualizing such large phylogenies.

One can follow two avenues to address the plausibility assessment problem: either devise novel visual tools for phylogenetic data exploration or design algorithms for automated plausibility analysis. Here, we focus on the latter idea.

We introduce a new approach to assess the plausibility of large phylogenies by computing all pairwise topological *Robinson–Foulds* (RF) distances [22] of a 55,000 taxon tree of plants [28], for instance, and a set containing a large number of substantially smaller reference trees. These small reference trees are available in curated databases, such as STBase [15], and comprise a subset of taxa of the large tree. The underlying assumption is that small trees are substantially more accurate. Hence, the lower the average RF distance, or any other reasonable topological distance, between all small reference trees and the large tree, the more plausible will be the large tree. While we use a reference database containing a billion of small reference trees, one could also use reference trees from the literature or from the treebase database [21].

Our main contribution is the design and production-level implementation in RAxML [1] of an effective algorithm for extracting induced subtrees from the large tree with respect to the taxon set of the small reference tree. This step is necessary to compute the RF distance.

The rest of this chapter is organized as follows: initially, we discuss the preliminaries and present the formal problem description. Subsequently, we first present a naïve and then an effective algorithm for inducing subtrees. Next, we provide an experimental evaluation of both algorithms using simulated and real data from STBase [15] and studies by Smith et al. [28]. Finally, we present a brief summary and conclusion.

24.2 PRELIMINARIES

A *rooted tree* is a connected, directed, and acyclic graph with an *internal node* r, (i.e., $deg(r) > 2$ designated as the *root node*. We say a node v has *height* $h(v)$ if the length of the longest path from v to some leaf is $h(v)$.

We denote a node v of a tree T as *central* if there is no other node u such that the length of its longest undirected path to a leaf is shorter than that of v. Trivially, a tree can have at most two central nodes, otherwise there would exist a cycle.

We say that node w is the *Lowest Common Ancestor* (LCA) of nodes u and v *if and only if* (iff) u and v are both descendants of w and there is no node w' with descendants u and v such that $h(w') \leq h(w)$. We denote the LCA of nodes u and v by $lca(u, v)$. We further denote the path from node u to v in a tree T by $u \rightsquigarrow v$.

An *unrooted tree* is a connected, undirected, and acyclic graph with no root node. The notion of node height (and hence LCA) is not defined in unrooted trees. Therefore, we henceforth imply that we are dealing with rooted trees when we use the terms node height and LCA. We further only consider rooted binary trees and unrooted trees that only consist of inner nodes of degree 3 and leaves (degree 1), that is, strictly bifurcating trees.

Definition 24.1 (Euler tour of a rooted tree) *The Euler tour of a rooted tree is a sequence of nodes formed by executing* Step(v) *with the root node* r *as input parameter.* Step(v)*: List/print node* v*. If* v *is not a leaf: call* Step(v) *with the left child as parameter, then call* Step *with the right child as parameter and finally list/print* v*.*

The length of the Euler tour for a rooted binary tree of n nodes is exactly $2(n - 1)$. The number of leaves in such a tree is $\frac{n+1}{2}$. We list each leaf only once, each inner node with the exception of the root node three times, and the root node exactly twice. Hence this sums to exactly $2(n - 1)$ elements. See Figure 24.1 for a graphical illustration of the Euler tour on a tree.

Definition 24.2 (Inorder traversal of a rooted tree) *The inorder traversal of a rooted tree is a sequence of nodes formed by executing* Step(v) *with the root node*

[1] https://github.com/stamatak/standard-RAxML

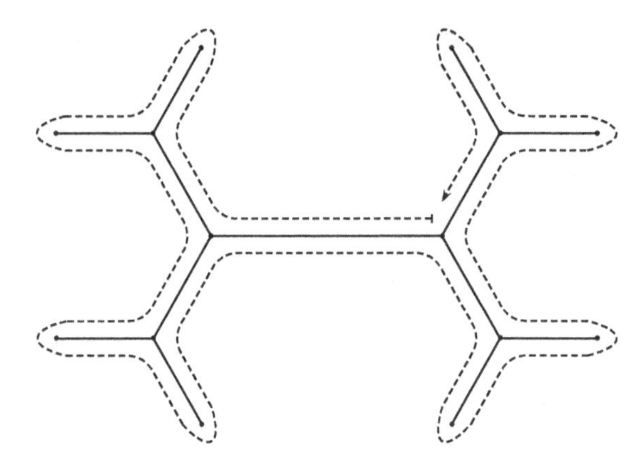

Figure 24.1 Euler traversal of a tree.

r as input parameter. Step*(v): Call* Step*(v) with left child as parameter, list v and call* Step*(v) with right child as parameter.*

Analogously to the Euler tour, the length of the inorder traversal sequence for a tree with n nodes is exactly n.

Lemma 24.1 *Let $v_1, v_2, \dots, v_n$ be the inorder notation of a tree T. For every $k = 1, 2, \dots, \lfloor n/2 \rfloor$ it holds that $v_{2k} = lca(v_{2k-1}, v_{2k+1})$.*

Proof. We prove the lemma by induction on the size of T. We first show that the lemma holds for our base case, which is a binary tree with three nodes. Then, under the assumption that our claim holds for all trees with up to m nodes, we prove that the claim also holds for trees with $m + 1$ nodes.

- Let T be a tree with root node u and two child nodes v and w. The inorder traversal of T yields the sequence $v, u,$ *and* w and hence the base case holds.
- Assuming that the claim holds for trees with up to m nodes, we now prove that the claim holds for any tree T with $m + 1$ nodes. Let u be the root node of T and $u_1, u_2, \dots, u_k$ its direct descendants. The inorder traversal of T yields the sequence $I(T(u_1)), u, I(T(u_2)), u, \dots, u, I(T(u_k))$ where $I(T(u_i))$ is the inorder notation of the subtree rooted at the ith direct descendant of u. Trivially, it holds that $|T(u_i)| < m$, for $1 \leq i \leq k$, and based on our assumption, the claim holds for any node in $I(T(u_i))$. Now, consider the ith occurrence of u in the sequence. Trivially, we observe that a node from $T(u_i)$ (specifically its rightmost leaf) appears immediately before u and a node from $T(u_{i+1})$ (specifically its leftmost leaf) immediately after u. Since the LCA of any pair (p, q) such that $p \in T(u_i)$ and $q \in T(u_{i+1})$ is u, the lemma holds.

For instance, let node u be the parent of two nodes v and w where v is the root node of subtree $T(v)$ and w is the root node of subtree $T(w)$. Let p be the last node in the inorder traversal of $T(v)$ and q the first node in the inorder traversal of $T(w)$. By definition of inorder traversal, p is the rightmost leaf of $T(v)$ and q is the leftmost leaf of $T(w)$. Hence, the LCA of p and q is u.

Definition 24.3 (Preorder traversal of a rooted tree) *The preorder traversal of a rooted tree is a sequence of nodes formed by executing* Step(v) *with the root node r as parameter.* Step(v): *List node v. Call* Step(v) *with the left child as parameter. Call* Step *with the right child as parameter.*

As for the inorder traversal, the generated sequence is n elements long.

Next, we define the binary relation $< \subset V^2$ on a tree with nodes drawn from set V, such that $v < u$ *iff* the preorder id of v is smaller than the preorder id of u.

Definition 24.4 (Induced subgraph) *Let T be a tree such that L is the set of its leaves and $L' \subsetneq L$ a proper subset of its leaves. We call induced subgraph the minimal subgraph that connects the elements of L'.*

We now give the formal definition of an *induced subtree.*

Definition 24.5 (Induced subtree) *Let $G(T, L')$ be the induced subgraph of a tree T on some leaf set L'. Remove nodes $v_2, \ldots, v_{q-1}$ if there exist paths of the form $v_1, v_2, \ldots, v_q$ and $r, v_2, v_3, \ldots, v_q$ such that r is the root, $deg(v_1) > 2$, $deg(v_q) \neq 2$, and $deg(v_2), \ldots, deg(v_{q-1}) = 2$, and replace the corresponding edges with a single edge (v_1, v_q) and (r, v_q).*

Definition 24.6 (Bipartition) *Let T be a tree. Removing an edge from T disconnects the tree into two smaller trees, which we call T_a and T_b. Cutting T also induces a bipartition of the set S of taxa of T into two disjoint sets A of taxa of T_a and B of taxa of T_b. We call a bipartition trivial, when the size of either A or B is 1. These bipartitions are called trivial because, as opposed to nontrivial bipartitions, they do not contain any information about the tree structure; a trivial bipartition A of size 1 occurs in every possible tree topology for S. We also denote by $B(T)$ the set of all trivial bipartitions of tree T.*

Definition 24.7 (Robinson–Foulds distance) *Given a set S of taxa, two phylogenetic trees T_1 and T_2 on S, and their respective sets of nontrivial bipartitions $B(T_1)$ and $B(T_2)$, the RF distance between T_1 and T_2 is $\mathrm{RF}(T_1, T_2) = \frac{1}{2}((B(T_1) \setminus B(T_2)) \cup (B(T_2) \setminus B(T_1)))$. In other words, the RF distance is the number of bipartitions that occur in one of the two trees, but not in both. This measure of dissimilarity is easily seen to be a metric [22] and we can compute it in linear time [19].*

24.3 A NAÏVE APPROACH

In the following, we introduce the PLAUSIBILITY-CHECK algorithm.

The algorithm assesses whether a comprehensive phylogenetic tree T is plausible or not by comparing it with a set of smaller reference trees that contain a proper subset of taxa of T.

We denote an induced tree as $T|t_i$ and read it as the tree induced by the taxon set of t_i in T.

Algorithm 24.1: PLAUSIBILITY-CHECK

Input : Tree T of n nodes, set $F = \{t_1, t_2, \dots, t_m\}$ of small reference trees that contain a proper subset of the taxa in T

Output: m pairwise RF distances $\text{RF}(T, F) = \{r_1, r_2, \dots, r_m\}$ between induced tree $T|t_i$ and t_i

1 $B(T) \leftarrow$ EXTRACT-NONTRIVIAL-BIPARTITIONS(T)
2 **for** $i \leftarrow 1$ **to** m **do**
3   ▷ Extract leaf-set L_i' from t_i
4     $B(T|t_i) \leftarrow$ COMPUTE-INDUCED-SUBTREE-NBP(T, L'_i)
5     $B(t_i) \leftarrow$ EXTRACT-NONTRIVIAL-BIPARTITIONS(t_i)
6     $r_i \leftarrow$ RF-DISTANCE($B(T|t_i), B(t_i)$)
7 **end**
8 Let $\text{RF}(T, F) = r_1, r_2, \dots, r_m$

PLAUSIBILITY-CHECK (Algorithm 24.1) takes as input parameters a large tree T and a set F of small reference trees. It is important to ensure that trees in F only contain proper subsets of the taxa in T. In a preprocessing phase, the algorithm extracts all bipartitions of T, which we denote as $B(T)$, and stores them in a hash table. Then, the algorithm iterates over all small trees t_i in F and, for every small tree t_i, it extracts the corresponding leaf set L_i'. After obtaining L_i', PLAUSIBILITY-CHECK computes the induced subtree $T|t_i$, its bipartitions $B(T|t_i)$, and hence the Robinson–Foulds distance for $T|t_i$ and t_i. The algorithm finishes when all small trees have been processed and returns a list of m pairwise RF distances.

Algorithm 24.2: EXTRACT-NONTRIVIAL-BIPARTITIONS

Input : Tree T of n nodes

Output: List of all nontrivial bipartitions $B(T)$ of T

1 ▷ Extract all bipartitions from T
2 Let $B(T) = b_1, b_2, \dots, b_{n-3}$ be the list of nontrivial bipartitions of T

EXTRACT-NONTRIVIAL-BIPARTITIONS (Algorithm 24.2) calculates all nontrivial bipartitions of T. The implementation requires $\mathcal{O}(n^2)$ time for traversing the tree and

Algorithm 24.3: COMPUTE-INDUCED-SUBTREE-NBP

Input : List of all nontrivial bipartitions $B(T) = b_1, b_2, \dots, b_{n-3}$ of T leaf-set L'

Output: List of all nontrivial bipartitions $B(T|t_i)$ of induced tree $T|t_i$

```
1 ▷ Iterate through all bipartitions B(T)
2 for i ← 1 to |B(T)| do
3 |   ▷ Filter bipartition b with L'
4 |   forall the taxa in b_i do
5 |   |   if taxon is in L' then
6 |   |   |   b'_i ← taxon
7 |   |   end
8 |   end
9 end
```

storing all bipartitions in a suitable data structure, typically, a hash table. For further implementation details, see Chapter 4.1 in Reference [18].

COMPUTE-INDUCED-SUBTREE-NBP is a naïve approach to extract the induced subtree $T|t_i$ and return its nontrivial bipartitions. These steps are described in Algorithm 24.3. It iterates through all bipartitions b_i of T stored in the hash table and generates the induced nontrivial bipartitions b_i' for $T|t_i$ by deleting the taxa that are not present in the leaf set L'. The resulting induced bipartitions are then rehashed. Therefore, COMPUTE-INDUCED-SUBTREE-NBP has a complexity of $\mathcal{O}(n^2)$, where n is the number of leaves of T. Note that we can reduce the complexity to $\mathcal{O}(\frac{n^2}{w})$ using a bit-parallel implementation, where w is the vector width of the target architecture (e.g., 128 bits for architectures with SSE3 support).

In summary, PLAUSIBILITY-CHECK has an overall complexity of $\mathcal{O}(\frac{n^2mk}{w})$, where n is the size of the large tree, k is the number of small trees in the set F, and m is the average size of small trees.

24.4 TOWARD A FASTER METHOD

In this section, we present a novel method for speeding up the computation of induced subtrees from a given leaf set. The key idea is to root the large tree at an inner node and compute the LCA of each and every pair of leaves in the leaf set. We can then build the induced subtree from the leaves and the LCAs. Although at first glance this may seem computationally expensive, with a series of lemmas we show that it is sufficient to compute the LCAs of only specific pairs of leaves. In fact, we only have to compute the LCAs of $m - 1$ particular pairs, and with a special type of preprocessing, we can compute the LCA of each pair in $\mathcal{O}(1)$ time.

The induced subgraph for a leaf set of size two is a path of nodes and edges from one taxon to the other, and hence the induced tree is a simple line segment with the two taxa as end points. Therefore, in the rest of this chapter, we only consider leaf sets of three taxa and more. For an arbitrary leaf set L', we partition the set V of vertices of the induced subgraph into three disjoint sets

$$V = V_1 \uplus V_2 \uplus V_3$$

such that $V_i = \{ v \mid v \in V, deg(v) = i \}$. From the properties of unrooted binary trees, we obtain the size of V_3 as exactly $|L'| - 2$.

In the next lemma, we show that all nodes in V_3 are LCAs of some pairs of leaves in L' when rooting the tree at an arbitrary inner node (Lemma 24.2). In fact, V_3 consists of all LCAs of all possible pairs of leaves in L' except possibly at most one LCA, which will be in V_2. We prove this last claim in Lemma 24.3.

Lemma 24.2 *Let $G(T)$ be the induced subgraph of an unrooted tree T. Rooting T at an arbitrary inner node r allows us to compute V_3 from the LCAs of pairs of leaves in L'.*

Proof. By contradiction.

- Let us assume that there exists a node v in V_3 that is not the lowest common ancestor of any two nodes from L'. Because $deg(v) = 3$, in the rooted tree there exist exactly two paths $v \rightsquigarrow u$ and $v \rightsquigarrow w$, where $u, w \in L'$. Now let $p = lca(u, w)$ be the least common ancestor of u and w, $r \rightsquigarrow u$ and $r \rightsquigarrow w$ the two paths leading from root r to u and w, and $r \rightsquigarrow p$ their common path. However, $p \neq v$ implies a cycle in the subgraph.

Therefore, any node in V_3 is the LCA of two leaves in L'.

Figure 24.2 depicts any node v from set V_3, that is, the root node of a rooted tree with three subtrees t_1, t_2, and t_3. Since v is in V_3, we know that all three subtrees t_1, t_2, and t_3 contain at least one leaf from L'.

The next lemma proves that V_3 is the set of all LCAs for all pairs of leaves from L', except possibly at most one LCA v. This node is part of V_2 (of degree 2) and results from rooting the tree at a node that is in V_2 (in which case v is the root) or at a node that does not appear in V.

Lemma 24.3 *There may exist at most one node in V_2 that is the LCA of two leaves from L'. That node appears* if and only if *the root is not part of the induced subgraph or if it is the root itself.*

Proof. First, we show a necessary condition that must hold for a node to be in V_2 and to simultaneously be an LCA of two leaves. Then, we show that only one such node exists.

- An internal node v (as depicted in Figure 24.2), which is the LCA of two leaves after rooting the tree at an arbitrary point, ends in V_2 only if one of the subtrees,

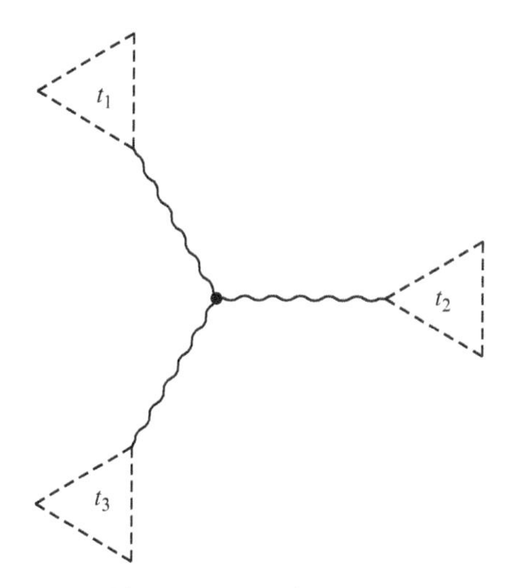

Figure 24.2 Node from V_3.

for instance, t_1, does not contain leaves from L'. Moreover, the root must either be at node v or be in t_1.

- Now, assume that there exists a node v' in t_3 or t_2 that is an LCA and belongs to V_2. This node must have degree 3 in the rooted tree, and hence connect three subtrees t'_1, t'_2, and t'_3. By definition, two of the subtrees must contain leaves from L' and the third subtree must not (and must contain the root), such that node v' is in V_2. However, this is a contradiction as the third subtree is either the subtree that contains t_1, t_2, and v (in case v' is in t_3) or t_1, t_3, and v (in case v' is in t_2).

To generate the induced tree from the induced subgraph $G(T)$, we remove all nodes from V_2 and replace all edges $(v_1, v_2), (v_2, v_3), \ldots, (v_{n-1}, v_n)$ formed by paths $v_1, v_2, \ldots, v_n$ such that $v_1, v_n \notin V_2$ and $v_i \in V_2$, for all $1 < i < n$, which is represented by a single edge (v_1, v_n). We have already shown that it is sufficient to compute set V_3, which, together with L', can generate the induced tree. Therefore, computing $|L'| - 2$ LCAs is sufficient to induce the tree for leaf set L'. However, there exist $\frac{|L'|(|L'|-1)}{2}$ pairs of leaves. The main question now is how to choose the pairs for computing the unique $|L'| - 2$ LCAs.

Let C denote the set of all LCAs of all pairs of a leaf set L', that is,

$$C = \{ p \mid p = lca(u, v), u \neq v, u, v \in L' \}$$

The following lemma proves a fundamental property of LCA computation using the preorder notation. For three nodes u, v, and w, where u appears before v and v before w in the preorder traversal of a rooted tree T, we show a transitive property that dictates that knowing the LCAs p of (u, v) and q of (v, w) is sufficient to determine the

LCA of (u, w). In fact, the LCA of (u, w) is the node that appears first in the preorder traversal of T between node p and node q.

This property allows us to formulate the main theorem that proves that there exist at most $L' - 1$ unique LCAs for the leaf set L' (and a rooted tree), and states which pairs of leaves to use for obtaining the required LCAs. In the following, we denote the preorder identifier of a leaf v as $pid(v)$.

Lemma 24.4 *Let $u, v, w \in L'$ such that $u < v < w$. It holds that $lca(u, w) = v'$, such that $pid(v') = \min(pid(p), pid(q))$, where $p = lca(u, v)$ and $q = lca(v, w)$.*

Proof. Proof by contradiction. Let r be the root of the tree. Let us assume that $p = lca(u, v)$ and $q = lca(v, w)$. By definition of the LCA, q may appear only along the path $r \rightsquigarrow p$ or $p \rightsquigarrow v$, that is, along the path $r \rightsquigarrow v$. We split the proof into two cases: node q appears on either $r \rightsquigarrow p$ or $p \rightsquigarrow v$.

- In case q appears along $p \rightsquigarrow v$. Let us assume that $v' = lca(u, w)$ is not p. It can then appear along the path $r \rightsquigarrow u$. If it appears anywhere except p, we have a cycle. Therefore, $lca(u, w) = p$ and it holds that $pid(p) = \min(pid(p), pid(q))$.
- In case q appears along $r \rightsquigarrow p$. Let us assume that $v' = lca(u, w)$ is not q. It can appear along the path $r \rightsquigarrow u$. If it appears anywhere except p, we have a cycle. Therefore, $lca(u, w) = q$ and it holds that $pid(q) = \min(pid(p), pid(q))$.

The lemma holds.

Specifically, with the next theorem, we show that computing the set

$$C' = \{\ p \mid p = lca(u, v), u < v, u, v \in L', \nexists w\ :\ u < w < v\ \}$$

is not only sufficient, but also that $C' = C$.

Theorem 24.1 *Given leaves $v_1, v_2, \ldots, v_n$ such that $v_i < v_{i+1}$ for $1 \leq i < n$, it holds that $lca(v_j, v_k) = u$ and $pid(u) = \min(pid(u_j), pid(u_{j+1}), \ldots, pid(u_{k-1}))$ for $1 \leq j < k \leq n$ and $u_i = lca(v_i, v_{i+1})$ for $j \leq i < k$.*

Proof. By strong induction on the range.

- Let $k - j = 2$. The claim holds as shown by Lemma 24.4 and this forms our base case.
- Let m be a positive integer greater than 2, and let us assume that the claim holds for $k - j \leq m$. This forms our induction hypothesis and we must now prove that the claim holds for $m + 1$.

- Let $k - j = m + 1$. From this interval, let us consider nodes v_j, v_{k-1}, and v_k. From the induction hypothesis, we obtain that $u_\ell = lca(v_j, u_{k-1})$ such that $pid(u_\ell) = \min \bigcup_{i=j}^{k-1}(pid(u_i))$. We also have that $lca(v_{k-1}, v_k) = u_{k-1}$. From Lemma 24.4, we can easily obtain the desired proof that $lca(v_j, v_k)$ is the node that has the smallest preorder identifier between u_ℓ and u_{k-1} and hence our claim holds.

Theorem 24.1 implies that it is sufficient to sort the leaf set in ascending order according to the preorder identifiers of the corresponding leaves in the rooted tree. Then, for the sequence $u_1, u_2, \ldots, u_{|L'|}$ of leaves, one can compute the LCA of $|L'| - 1$ pairs (u_i, u_{i+1}), for $i \leq 1 < |L'|$, to obtain the desired $|L'| - 1$ unique LCAs. Note that, in case the selected root node is in V_3, it will appear twice: once as the LCA of two leaves from t_1 and t_2 (see Figure 24.2), and once as the LCA of two leaves from t_2 and t_3. On the other hand, if the root node is not in V_3, then we obtain one additional LCA from V_2, which will not appear in the induced tree.

So far, we have proved that, given a leaf set L', we can compute all common ancestors of all pairs of leaves in L' by simply computing the LCA of only $|L'| - 1$ pairs. Each pair consists of two vertices u_i and u_j so that there is no vertex u_k such that $u_i < u_k < u_j$. Moreover, we have shown that there are exactly $|L'| - 2$ common ancestors of degree 3 that will be present in the induced tree. In case the root node (when rooting the unrooted tree) is not part of these nodes, we will obtain one additional common ancestor that will have degree 2 in the induced subgraph, but that will not appear in the induced tree.

24.5 IMPROVED ALGORITHM

We are now in a position to present and explain the actual algorithm for inducing a tree given a leaf set L' and an unrooted tree T. We divide the algorithm into two steps—preprocessing and inducing the trees. In the preprocessing step, we use a dedicated data structure created from tree T that we can query for LCAs in constant time. In the second step, we show how to query for LCAs and build the resulting induced tree.

Moreover, if we are given several leaf sets at once, we can implement the algorithm with one of the following two variants, each of which has different asymptotic space and time complexity. The first requires loading all leaf sets in memory prior to computing the induced subtrees, and hence runs in $\Theta(n + km)$ time and space. The second variant does not require preloading, uses only $\Theta(n)$ space, and runs in $\mathcal{O}(n + m \log m)$ time.

24.5.1 Preprocessing

For fast LCA queries, it is necessary to root T at an arbitrary point and then create a data structure that allows for $\mathcal{O}(1)$ queries. To achieve this, we consider the close relationship between LCA computation and *Range Minimum Queries* (RMQs). As shown in Reference [2], we construct a sequence of node identifiers that correspond to an Euler tour of the rooted tree. Identifiers are assigned to nodes upon the first visit and in increasing order. We then preprocess this sequence for RMQs. The size of the succinct preprocessed data structure for a tree of n nodes is at most $2n + o(n)$ bits [12]. Algorithm 24.4 lists the preprocessing phase. We omit the details of constructing an RMQ structure and rather point the interested reader to the available literature [11, 12].

Algorithm 24.4: PREPROCESS-ROOTED-TREE

Input : Tree T of n nodes
Output: Preprocessed data structure RMQ(T)

1 ▷ Root tree T
2 **if** T *is unrooted* **then**
3 | Root T at an arbitrary node
4 **end**
5 ▷ Build Euler tour of T
6 Let $E(T) = s_1, s_2, \dots, s_{2n-1}$ be the Euler tour of T
7 ▷ Prepare a RMQ data structure
8 Let $P(T) = pid(s_1), pid(s_2), \dots, pid(s_{2n-1})$ be the list of preorder identifiers of $E(T)$
9 Let RMQ(T) = RANGE-MINIMUM-QUERY-PREPROCESS($P(T)$)

24.5.2 Computing Lowest Common Ancestors

Once we have constructed the preprocessed data structure via the Euler tour, it is possible to compute the LCA of two leaf nodes in time $\mathcal{O}(1)$. To do so, we need one additional data structure. Let L be the set of leaves of the rooted tree T of n nodes that we preprocessed. The new data structure represents the mapping

$$f : L \rightarrow \langle 1, 2(n-1) \rangle$$

of the position where each leaf appears for the first time in the Euler tour. We can now compute the LCA of two leaves $u, v \in L$ by finding the node with the lowest preorder identifier in the Euler tour in the range $\langle i,j \rangle$, where $i = \min(f(u),f(v))$ and $j = \max(f(u),f(v))$. With the preprocessed Euler tour, we can compute this minimum value, and hence the ancestor of u and v, in time $\mathcal{O}(1)$ using RMQs.

24.5.3 Constructing the Induced Tree

Given a list L' sorted (in ascending order) by the preorder identifiers of the nodes in tree T, we determine the common ancestor of every pair of adjacent nodes. Let $v_1, v_2, \dots, v_{|L'|}$ be the list of nodes in L' such that $v_i < v_{i+1}$ for $1 \leq i < |L'|$. We compute the LCA of every pair (v_i, v_{i+1}) and construct a new sequence $I =$

$v_1, lca(v_1, v_2), v_2, lca(v_2, v_3), v_3, \ldots, v_{|L'|-1}, lca(v_{|L'|-1}, v_{|L|}), v_{|L'|}$ of size $2|L'| - 1$ by interleaving the LCA node between each pair of adjacent nodes. These steps are described in Algorithm 24.5.

Algorithm 24.5: COMPUTE-INDUCED-TREE

Input : Preprocessed RMQ structure RMQ(T)
leaf-set L'
Mapping $f : L' \rightarrow \langle 1, \ldots, n \rangle$

Output: Induced unrooted tree $T|t_i$

1 ▷ Sort leaf-set according to preorder traversal identifiers of T
2 Let $L_S{}' = (u_1, u_2, \ldots, u_{|L'|})$ such that $u_{i-1} < u_i < u_{i+1}$, for $1 < i < |L'|$
3 ▷ Compute common ancestors
4 **for** $i \leftarrow 2$ **to** $|L'|$ **do**
5 | $c_i \leftarrow lca(u_{i-1}, u_i)$
6 **end**
7 ▷ Sort the resulting nodes and construct the induced tree
8 Let $V' = L_S{}' \cup \bigcup_{i=2}^{|L'|}(c_i)$ and name the nodes as $u_{1,2}, \ldots, u_{|V'|}$
9 Let $V_S{}' = (u_1, u_2, \ldots, u_{|V'|})$ such that $u_{i-1} < u_i < u_{i+1}$, for $1 < i < |L'|$
10 $T|t_i \leftarrow$ BUILD-INDUCED-TREE($V_S{}'$)

The resulting sequence corresponds to the inorder notation of the induced rooted tree (see Lemma 24.1). While we can construct the induced tree in $\mathcal{O}(|L'|)$ time directly from the inorder sequence, we will show a different approach that requires an additional sorting step and is substantially simpler to explain.

By sorting the inorder sequence in ascending order, we obtain the preorder notation of the induced rooted tree. The first node in the sequence is the root node, and we can build the induced tree by applying Algorithm 24.6 on the preorder sequence, which we include for completeness. The algorithm is simple and builds the induced rooted tree in a depth-first order. After building the tree and in case the root node is of degree 2 (see Lemma 24.3), we remove the root and connect the two children by an edge.

Note that, as an alternative, it is possible to extract the required bipartitions after sorting the sequence without building the actual tree. Using Algorithm 24.7, we can calculate the nontrivial bipartitions based on the preorder sequence. The algorithm determines all subtrees of the induced tree $T|t_i$ and extracts the corresponding bipartitions by separating these subtrees from $T|t_i$.

■ **EXAMPLE 24.1** Compute the induced tree, given the query tree in Figure 24.3 and a leaf set L' that consists of the leaves marked in gray.

First, we transform the unrooted tree into a rooted one by designating one node as the root. We then assign a preorder traversal identifier to each node as shown in

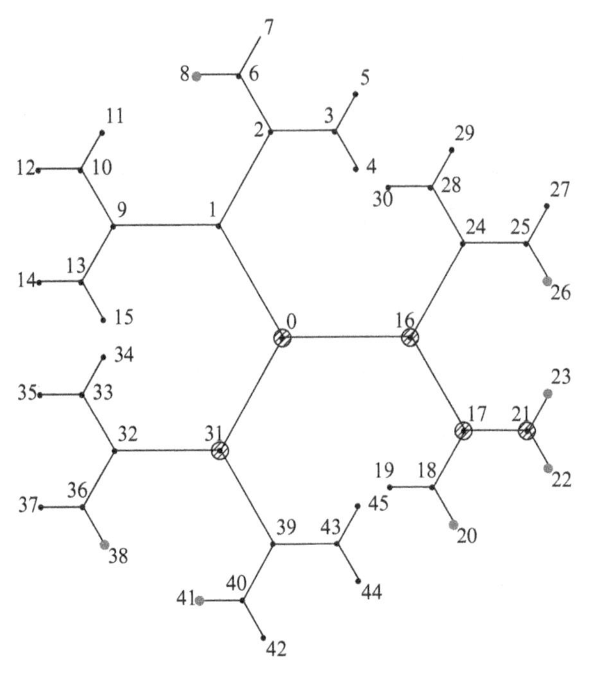

Figure 24.3 Unrooted phylogeny of 24 taxa (leaves). The taxa for which we induce a subtree are marked in gray. The lines are inner nodes that represent the common ancestors and hence the minimum amount of nodes needed to maintain the evolutionary relationships among the selected taxa. The numbers denote the order of each node in the preorder traversal of the tree assuming that we root it at node 0.

Figure 24.4, starting from 0. The numbers at each node in the figure indicate the preorder traversal identifiers assigned to that particular node. For this example, the Euler traversal is the sequence

0 1 2 3 4 3 5 3 2 6 7 6 8 6 2 1
9 10 11 10 12 10 9 13 14 13 15 13 9 1 0 16
17 18 19 18 20 18 17 21 22 21 23 21 17 16 24 25
26 25 27 25 24 28 29 28 30 28 24 16 0 31 32 33
34 33 35 33 32 36 37 36 38 36 32 31 39 40 41 40
42 40 39 43 44 43 45 43 39 31

which is preprocessed for RMQ queries. We then sort the sequence of leaves of L' in ascending order, that is,

$$8, 20, 22, 23, 26, 38, 41$$

Then, we compute the LCAs of node pairs

$$(8, 20), (20, 22), (22, 23), (23, 26), (26, 38), (38, 41)$$

Algorithm 24.6: BUILD-INDUCED-TREE

Input : Sorted list of nodes $(u_1, u_2, \dots, u_n)$
Output: Induced unrooted tree

```
1  ▷ Check whether the root is of degree 2 or 3
2  r ← NEW-NODE
3  push r; push r
4  if u_1 = u_2 then
5      push r
6      start ← 3
7      deg(r) ← 3
8  else
9      start ← 2
10     deg(r) ← 2
11 for i ← start to n do
12     pop p
13     q ← NEW-NODE
14     APPEND-CHILD(p, q)
15     if u_i is a leaf then
16         push q; push q
17         deg(q) ← 3
18     else
19         deg(q) ← 1
20 if deg(r) = 2 then
21     Connect the two children of r with an edge and remove r
```

and obtain the sequence

$$8, 0, 20, 17, 22, 21, 23, 16, 26, 0, 38, 31, 41,$$

which represents the inorder notation of the induced tree. We can now build the induced tree directly from this inorder notation, or sort the sequence and build the tree using Algorithm 24.6. Figure 24.4 depicts the induced tree.

24.5.4 Final Remarks

As stated earlier, it is possible to implement the algorithm in two different ways, depending on the amount of available memory. The difference between the two variants is in the way how the initial sorting of each query leaf set is done.

Let T be a large tree of n nodes and let $L'_1, L'_2, \dots, L'_k$ be k leaf sets with an average size of m. One can now sort each leaf set, compute the LCAs from the sorted sequence (and the already preprocessed Euler tour of the query tree) using Algorithm 24.5, and then apply Algorithm 24.6 to construct the induced tree. The asymptotic time and

Algorithm 24.7: EXTRACT-BIPARTITIONS-PREFIX

Input : Sorted list of nodes $(u_1, u_2, \dots, u_n)$
Output: Nontrivial bipartitions from induced subtree $B(T|t_i) = b_1, b_2, \dots, b_{n-3}$

```
1  k = 0
2  for i ← n to 1 do
3      if k < number of splits then
4          V[i] ← 1
5          if u_i is not a leaf then
6              for j ← 1 to deg(u_i) do
7                  V[i] ← V[i] + V[i + V[i]]
8              for j ← 1 to V[i] do
9                  if u_{i+j} is a leaf then
10                     Add u_{i+j} into b_i
11                 else
12                     Add all nodes from b_{i+j} into b_i
13                     j ← j + V[i + j]
14         k = k + 1
```

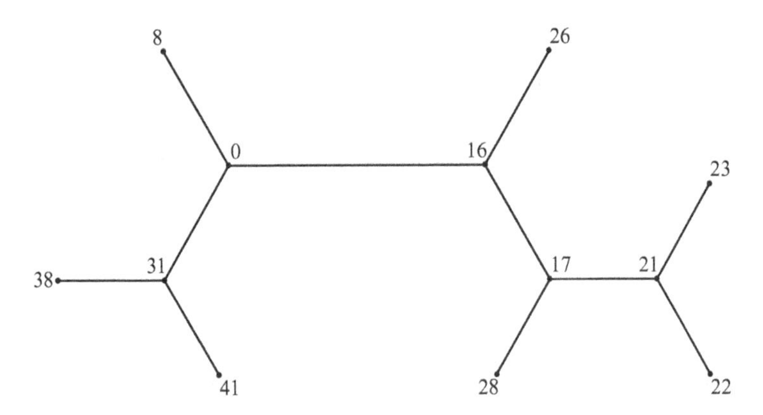

Figure 24.4 Induced tree for Example 24.1.

space complexity for this variant is $\mathcal{O}(n)$ time and space for preprocessing T and $\mathcal{O}(km \log\ m)$ time for inducing k trees.

The alternative variant is to avoid sorting each of the k leaf sets individually. Instead, one can store all of them in memory at the same time and sort them using a bucket sort method. Since the range of values in the k leaf sets is $\langle 1, n \rangle$, we can sort them all in a single pass in conjunction with the preprocessing step in $\mathcal{O}(\max(n, km))$ time and space. Thereafter, we can build the k induced trees in $\mathcal{O}(km)$ time, assuming that we construct the induced tree directly from the inorder notation.

24.6 IMPLEMENTATION

In the following, we present a straightforward implementation of the PLAUSIBILITY-CHECK algorithm. We have implemented the algorithm in C as part of RAxML. Furthermore, we address how to efficiently implement the fast method from Section 24.5 for bifurcating unrooted trees.

24.6.1 Preprocessing

First, we need to preprocess the large phylogenetic tree by assigning preorder identifiers to every node. Therefore, we root the tree at an arbitrary inner node and traverse it to assign preorder identifiers and store them in an array. We will use this array in the following steps to efficiently look up preorder identifiers for every node.

We now traverse our tree for a second time via an Euler traversal. We can also avoid this second tree traversal by assigning preorder identifiers on the fly during the Euler traversal. However, this method requires additional memory for marking already visited nodes. Note that the resulting array consists of $4|L| - 5$ elements because the Euler traversal visits $|L| - 3$ inner nodes (all inner nodes except for the root) three times, all other $|L|$ nodes once, and the root four times. To further optimize the induced tree reconstruction phase, we use an additional array, which we denote by FASTLOOKUP, that stores the index of the first appearance of each taxon during the Euler tour. This information allows us to speed up RMQ queries in the reconstruction phase and also compute it on-the-fly during the Euler traversal.

While we choose to use arrays for storing node information such as preorder identifiers or Euler labels, one could also use hash tables to reduce memory storage or list data structures, for instance.

Based on the Euler tour, we can now construct an RMQ data structure. For this, we use a source code developed by Fischer and Heun [12], which we modify and adapt to our purposes.

24.6.2 Reconstruction

Initially, we extract the leaf set from our small reference tree by traversing the small tree and storing its taxon set in an auxiliary array called SMALLTREETAXA. As before, we denote the number of taxa in the small reference tree by m. In the following, we use SMALLTREETAXA each time we need to iterate through the leaf set of the small tree.

Now, for every taxon in the reference tree, we look up at which index position it first appeared in the Euler tour using the FASTLOOKUP array. Because of the auxiliary FASTLOOKUP array, this procedure has a time complexity of $\mathcal{O}(m)$. Without this additional array, we would have to search through the entire Euler tour to find the corresponding indices, which would require $\mathcal{O}(nm)$ time. Thereafter, we sort all resulting indices in ascending order using quicksort. Note that this is analogous to sorting the preorder identifiers, which is necessary for computing the induced tree as outlined in Section 24.5. By querying the RMQ data structure, we can now find the least common ancestor of two taxa in constant time and reconstruct the induced tree using Algorithm 24.6.

24.6.3 Extracting Bipartitions

To finally compute the RF distance, we extract all nontrivial bipartitions by traversing the small reference tree and the induced tree using the bipartition hash function, which has been thoroughly discussed by Pattengale et al. [18].

To reduce memory consumption and to improve running times, we store bipartitions in bit vectors with m instead of n bits. We achieve this by consistently using the taxon indices from SMALLTREETAXA instead of the original taxon index in the large tree. Bit vectors are well suited for storing sets with a predefined number of m elements such as bipartitions. They only need $\Theta(m)$ bits of space and can be copied efficiently with C functions such as ~`memcpy ()`~. These bit vectors are then hashed to a hash table and can be looked up efficiently.

As stated earlier in Section 24.5, it is possible to extract all nontrivial bipartitions directly from the preorder sequence without relying on an instance of a tree structure, as outlined in Algorithm 24.7. We deploy this approach because it does not require building the induced tree at all.

However, for both implementation options, we need a mechanism to avoid storing (and thus checking for) complementary bipartitions. To avoid distinct, yet identical representations of one and the same bipartition (the bipartition and its bit-wise complement), we hash bipartitions in a canonical way. We only hash a bipartition if it contains a specific taxon (in our case the first taxon in SMALLTREETAXA). If our bit vector does not contain the specific taxon, we compute and then hash its complement instead.

24.7 EVALUATION

In the following, we describe the experimental setup and the results.

24.7.1 Test Data Sets

24.7.1.1 Real-World Data Sets. For real-world datasets, we used the mega-phylogeny of 55,473 plant species by Smith et al. [28]. To obtain a reference tree set, we queried all trees in STBase [15], which are proper subsets of the large tree. Our reference tree set consists of 175,830 trees containing 4 up to 2065 taxa and is available for download at http://www.exelixis-lab.org/material/plausibilityChecker.tar.bz2.

24.7.1.2 Simulated Data Sets. As large trees, we used 15 trees with 150 up to 2554 taxa from Reference [20], which are available for download at http://lcbb.epfl.ch/BS.tar.bz2. For each large tree, we generated 30,000 corresponding reference trees containing 64 taxa. We used the following procedure to simulate and build the reference trees: first we extract the taxon labels of the large tree. Thereafter, we randomly select a proper subset of these taxa and construct the trees using an algorithm that is similar to Algorithm 24.6.

Moreover, we also want to assess how long it will take our algorithm to run on a very large number of reference trees for a mega-phylogeny. To this end, we extracted 1 million reference trees with 128 taxa each from the empirical mega-phylogeny with 55,000 taxa.

24.7.2 Experimental Results

All experiments were conducted on a 2.2-GHz AMD Opteron 6174 CPU running 64-bit Linux Ubuntu. We invoked the plausibility check algorithm as implemented in standard RAxML with the following command:

```
raxmlHPC-AVX -f R -m GTRCAT -t largetree
  -zreferencetrees -n T1
```

In all experiments, we verified that both algorithms yield exactly identical results.

24.7.2.1 Mega-Phylogeny. For the mega-phylogeny, we obtain an average relative RF distance of 0.318 (see Table 24.1) between the large tree and the reference trees from STBase. We consider this average topological distance of approximately 32% to be rather low because of the substantially larger tree search space for the 55 K taxon tree. For a tree with 2000 taxa, there are approximately 3.00×10^{6328} possible unrooted tree topologies, whereas for 55,000 taxa there exist approximately $2.94 \times 10^{253,380}$ possible unrooted tree topologies. In other words, the tree search space of the 55K taxon tree is approximately 10^{247052} times larger than that for the 2000 taxon tree. Taking into account that different procedures were used to automatically construct the corresponding alignments and that the trees have also partially been constructed from different genes, an average error of approximately 30% appears to be low. However, the interpretation of these results is subject to an in-depth empirical analysis, which is beyond the scope of this paper. Figure 24.5 illustrates the overall distribution of RF distances, whereas Figure 24.6 shows the corresponding distribution for the 20,000 largest reference trees. Using our improved algorithm, we can process the 175,830 small reference trees by five orders of magnitude faster than with the naïve algorithm. In total, the naïve algorithm required 67,644s for all reference trees, while the effective algorithm required less than 7.14 s, after a preprocessing time of 0.042 s. If we only consider the inducing steps and ignore the time for parsing every single tree, the naïve algorithm needs 67,640 s for reconstructing the induced tree whereas the effective approach only takes 3.11 s. Hence, the effective algorithm is five orders of magnitude faster than the naïve version.

24.7.2.2 Simulated Data. The naïve algorithm needs more time for larger phylogenies as discussed in Section 24.3 because it iterates over all taxa of the large tree for each small tree. In contrast to this, our new approach only preprocesses the large tree once. As we show in Figure 24.7, the runtime of the effective algorithm is independent of the size of input tree. It induces the subtree in time that is proportional to

Table 24.1 Test results for a mega-phylogeny of 55,473 taxa

Average Robinson–Foulds distance	0.318 [-]
Total time for inducing (naïve)	67,640.00 s
Total time for inducing (improved)	3.11 s
Total execution time (naïve)	67,643.00 s
Total execution time (improved)	7.14 s

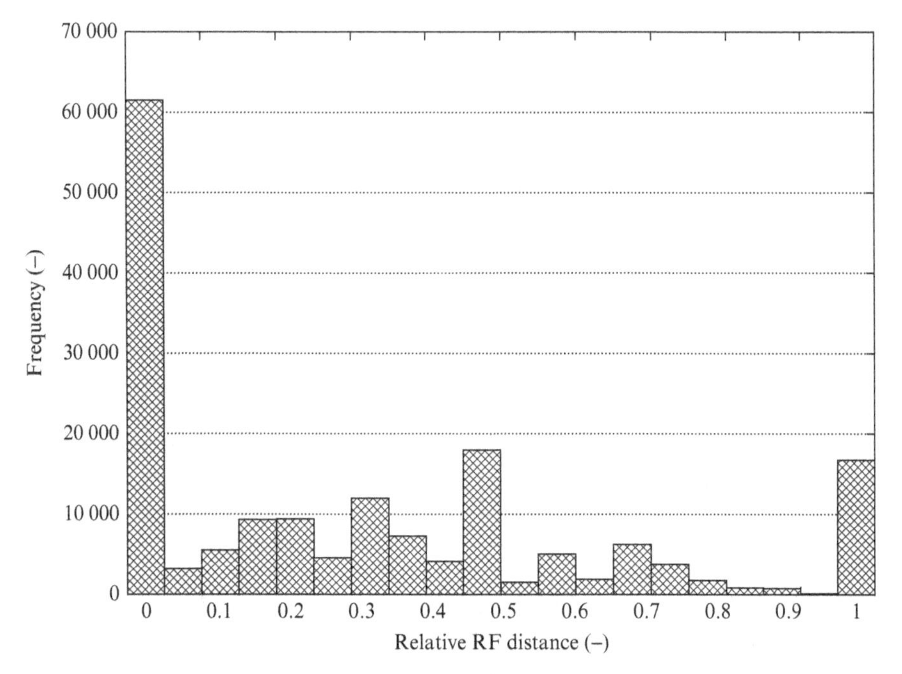

Figure 24.5 Distribution of all relative RF distances between the large mega-phylogeny and the reference trees from STBase.

the size of each small tree. This yields a significant runtime improvement for our new algorithm (see Table 24.2). In the following, we calculated the speedup by comparing the run times for the inducing step in both algorithms. Figure 24.8 shows the speedup for the optimized, induced subtree version of PLAUSIBILITY-CHECK compared with the naïve approach. As theoretically expected, the speedup improves with an increase in the size of the input phylogeny T. For example, on the large tree with 2458 tips, the effective approach is approximately 19 times faster than the naïve algorithm, which is consistent with our theory. In each run, the naïve algorithm has to traverse the large tree, which is approximately 40 times the size of the small tree (64 tips), whereas the efficient method only traverses the small reference tree. However, due to additional sorting and traversing of the small tree, we suffer a loss in runtime performance that explains the resulting speedup. If the difference between the size of the large tree and

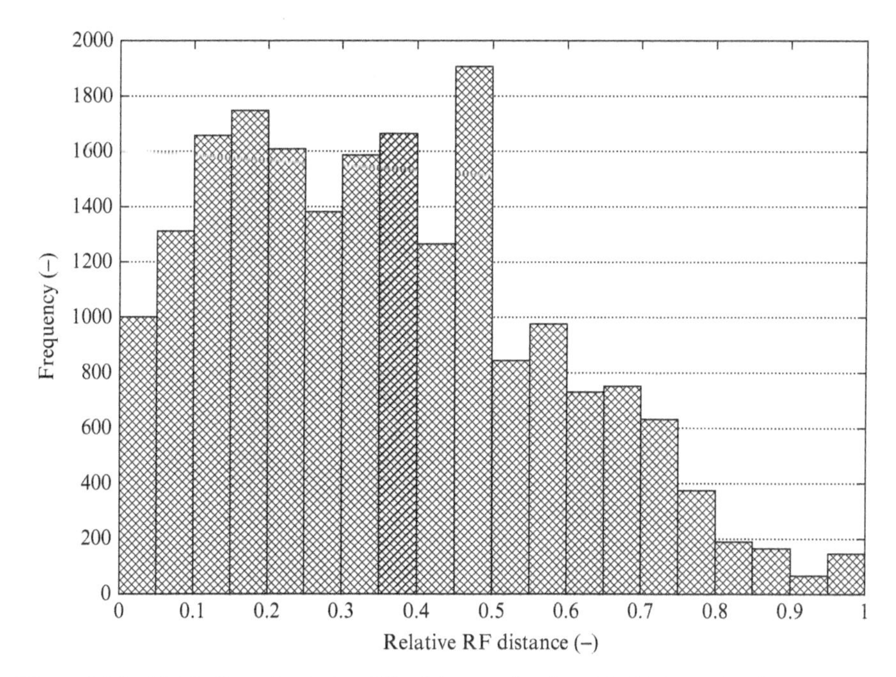

Figure 24.6 Distribution of relative RF distances for the 20,000 largest reference trees (30–2065 taxa).

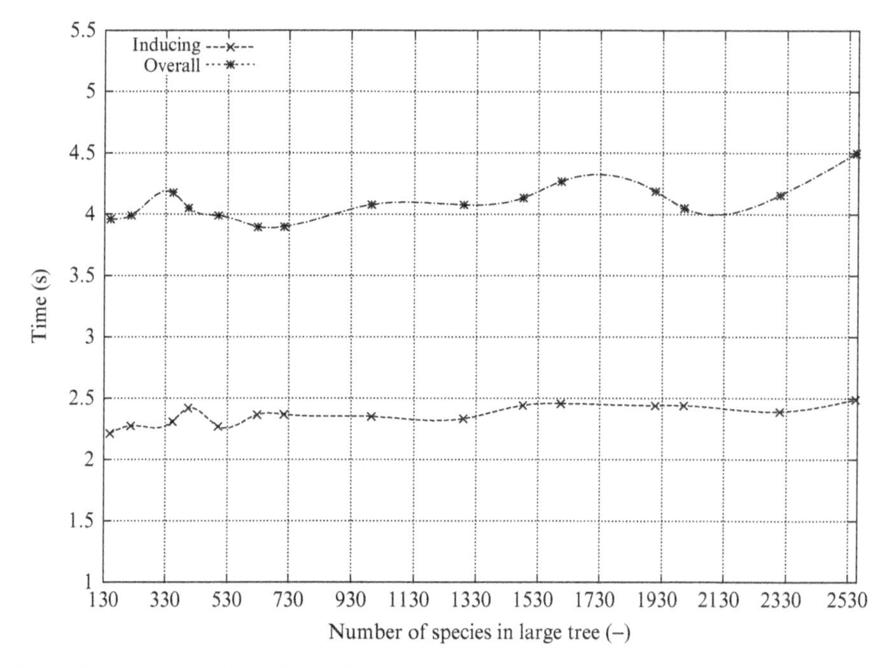

Figure 24.7 Running time of the effective inducing step (dashed) compared to the overall execution time of the effective algorithm (dotted).

Table 24.2 Test results for different input tree sizes (150–2554 taxa). The algorithm is executed on 30,000 small trees for each run. Each small tree contains exactly 64 taxa

Number of Taxa in Large Tree	Inducing Time (Naïve)	Inducing Time (Improved)	Preprocessing Time	Overall Execution Time (Naïve)	Overall Execution Time (Effective)
150	1.805363	2.21387	0.00030	3.200785	3.959420
218	2.173332	2.27510	0.00031	3.614717	3.989973
354	3.318583	2.30837	0.00036	4.935320	4.178407
404	3.683192	2.42039	0.00037	4.904781	4.053480
500	4.318119	2.26976	0.00038	5.648583	3.990615
628	6.077749	2.36570	0.00046	7.312694	3.895842
714	7.063149	2.36753	0.00048	8.399326	3.897443
994	10.290771	2.35056	0.00056	11.840957	4.079138
1288	16.531953	2.33238	0.00077	18.346817	4.078463
1481	20.654801	2.44133	0.00080	22.444981	4.134798
1604	23.317732	2.45706	0.00086	25.385845	4.269186
1908	29.793863	2.44010	0.00100	31.903671	4.188301
2000	30.726621	2.43945	0.00106	32.648712	4.050954
2308	39.535349	2.39014	0.00119	41.739811	4.157518
2554	46.642499	2.48903	0.00125	48.698793	4.498240

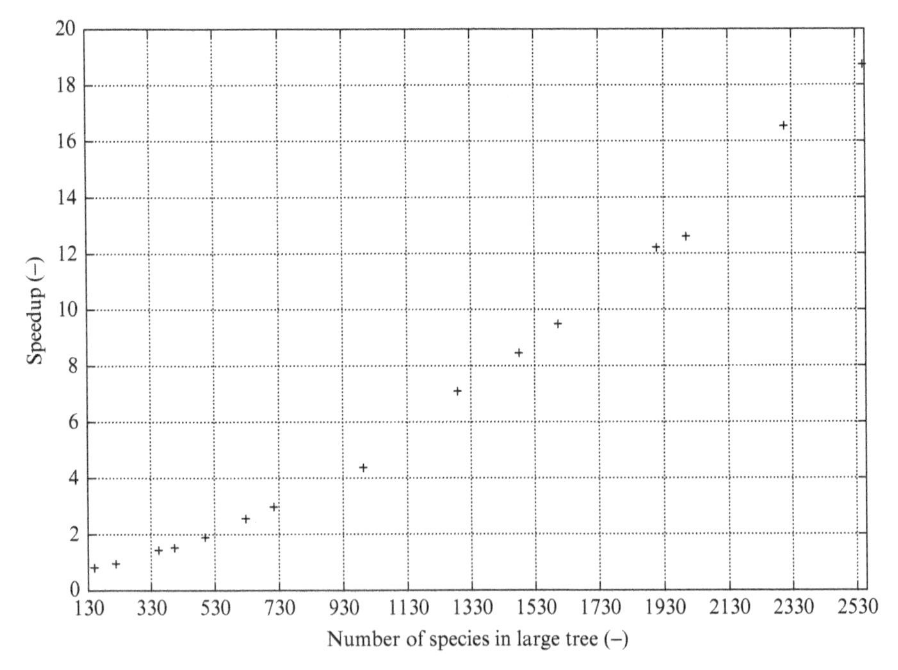

Figure 24.8 Speedup of the effective inducing tree approach. The speedup is calculated by dividing the overall naïve inducing time with the effective inducing time.

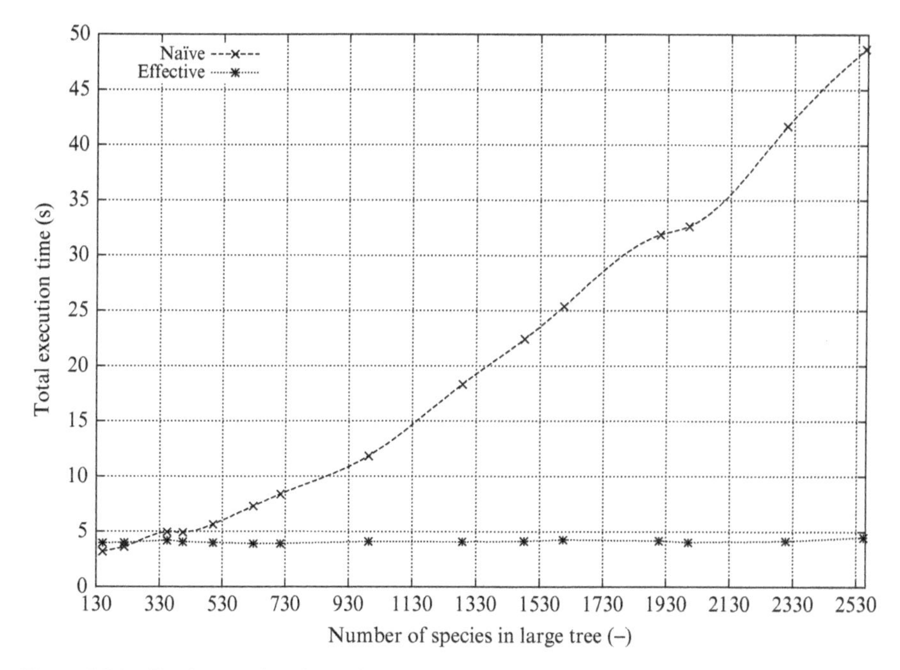

Figure 24.9 Total execution time of the naïve algorithm (dashed) compared to the effective approach (dotted).

the small reference tree is small, both algorithms will have approximately the same runtime. However, this is not the standard use case for our algorithm. Figure 24.9 shows the overall execution times for both algorithms, while Figure 24.10 shows the preprocessing time for the effective algorithm that depends on the size of T. The preprocessing time is negligible compared to the overall execution time. Table 24.3 illustrates the huge differences between the effective and the naïve algorithms on extremely large data sets. For 1 million reference trees, our naïve algorithm required 113 (approximately 5 days) whereas the effective algorithm required less than 8 min.

24.8 CONCLUSION

In view of the increasing popularity of mega-phylogeny approaches in biological studies [14, 28], one of the main challenges is to assess the plausibility of such large phylogenies. Because of the availability of a large number of curated smaller phylogenies, the methods and software we introduce here allow to automatically assess and quantify the plausibility of such large trees. Moreover, they can be used to compare such large phylogenies with each other by means of their respective average RF distances. Here, we use the RF distance metric, but any other, potentially more advanced, topological distance metric, such as the quartet distance [6], can be used.

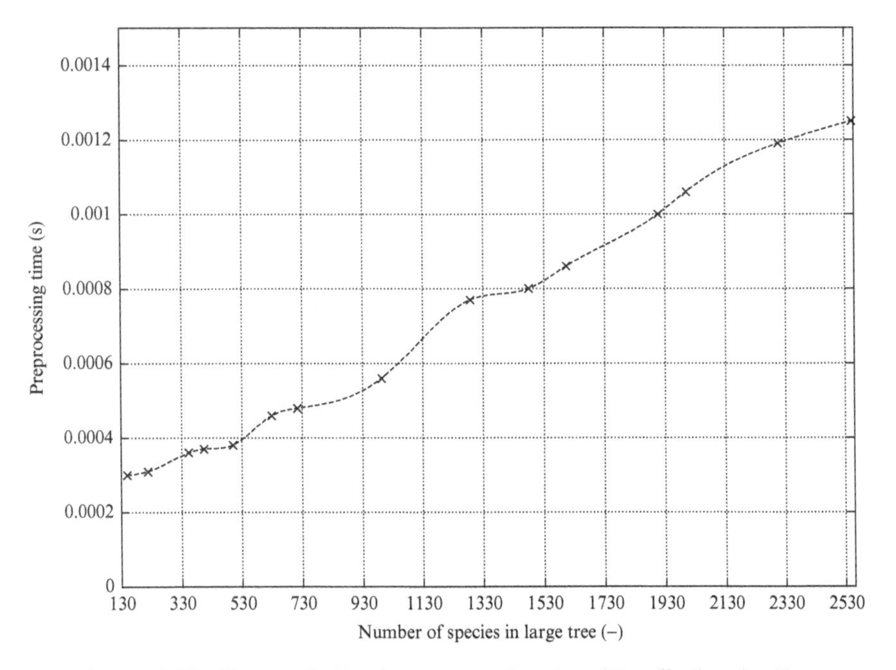

Figure 24.10 Time needed for the preprocessing step of the effective algorithm.

Table 24.3 Test results for 1 million simulated reference trees (each containing 128 taxa)

Number of taxa in large tree	55,473
Number of small trees	1,000,000
Total time for inducing (naïve)	406,159.00 s
Preprocessing time	0.045 s
Total time for inducing (improved)	238.37 s
Total execution time (naïve)	405,902.00 s
Total execution time (improved)	448.40 s

We consider the average RF distance of 32% that we obtained using empirical reference trees for the 55 K taxon tree to be surprisingly small with respect to the size of the tree search space. The histograms with the distribution of the RF distances can be used to identify problematic clades in mega-phylogenies. One could also establish an iterative procedure that removes taxa from the mega-phylogeny in such a way that the average RF distance drops below a specific threshold. This may also give rise to novel optimization problems. For instance, one may consider the problem of finding the smallest set of taxa to prune, whose removal yields a 10% improvement of the average RF distance. This kind of optimization problems might also be connected to recent algorithms for rogue taxon identification [1]. Apart from these practical considerations, we showed that our method runs in $\mathcal{O}(km)$ or $\mathcal{O}(km \log m)$ time. This is an important finding because the time complexity, except for the preprocessing phase, is independent of the size of the mostly very large input phylogeny. Our experimental findings are in line with our theoretical results and the implementation exhibits a substantial speedup over the naïve algorithm. Nevertheless, there are still several open problems that need to be addressed. Is it possible to design an algorithm for our method that runs in linear time as a function of the leaf set of the small reference tree? Furthermore, our method examines the extent to which a large phylogeny corresponds to existing, smaller phylogenies. At present, the small trees have to contain a proper taxon subset of the large phylogeny. An open problem is how to handle small trees that contain taxa that do not form part of the large tree.

Finally, we simply do not know whether large trees that attain high plausibility scores (low average RF distance) do indeed better represent the evolutionary history of the organisms at hand.

ACKNOWLEDGMENT

The authors thank Mike Sanderson, from the University of Arizona at Tucson, for sharing the problem statement with them and implementing an option to extract small reference trees in STBase.

REFERENCES

1. Aberer AJ, Krompass D, Stamatakis A. Pruning rogue taxa improves phylogenetic accuracy: an efficient algorithm and webservice. Syst Biol 2013;62(1):162–166.
2. Bender MA, Farach-Colton M. The LCA problem revisited. In: Gonnet GH, Panario D, Viola A, editors. *LATIN*. Volume 1776, Lecture Notes in Computer Science. Springer Berlin Heidelberg; 2000. p 88–94.
3. Benson DA, Karsch-Mizrachi I, Lipman DJ, Ostell J, Sayers EW. GenBank. Nucleic Acids Res 2010;38 Suppl 1:D46–D51.
4. Borkar S, Chien AA. The future of microprocessors. Commun ACM 2011;54(5):67–77.
5. Brown JR, Warren PV. Antibiotic discovery: is it all in the genes? Drug Discov Today 1998;3(12):564–566.
6. Bryant D, Tsang J, Kearney P, Li M. Computing the quartet distance between evolutionary trees. Proceedings of the 11th Annual ACM-SIAM Symposium on Discrete Algorithms, SODA '00. Philadelphia (PA): Society for Industrial and Applied Mathematics; 2000. p 285–286.
7. Darwin C. *The Origin of Species*. Everyman's Library Dent; London; 1936.
8. Edgar RC. MUSCLE: multiple sequence alignment with high accuracy and high throughput. Nucleic Acids Res 2004;32(5):1792–1797.
9. Eisen JA, Wu M. Phylogenetic analysis and gene functional predictions: phylogenomics in action. Theor Popul Biol 2002;61(4):481–487.
10. Felsenstein J. The number of evolutionary trees. Syst Biol 1978;27(1):27–33.
11. Fischer J, Heun V. Theoretical and practical improvements on the RMQ-problem, with applications to LCA and LCE. In: Lewenstein M, Valiente G, editors. *CPM*. Volume 4009, Lecture Notes in Computer Science. Springer Berlin Heidelberg; 2006. p 36–48.

12. Fischer J, Heun V. A new succinct representation of RMQ-information and improvements in the enhanced suffix array. In: Chen B, Paterson M, Zhang G, editors. *ESCAPE*. Volume 4614, Lecture Notes in Computer Science. Springer Berlin Heidelberg; 2007. p 459–470.
13. Fitch WM, Bush RM, Bender CA, Subbarao K, Cox NJ. The Wilhelmine E Key 1999 invitational lecture. Predicting the evolution of human influenza A. J Hered 2000;91(3):183–185.
14. Goloboff PA, Catalano SA, Marcos Mirande J, Szumik CA, Salvador Arias J, Källersjö M, Farris JS. Phylogenetic analysis of 73 060 taxa corroborates major eukaryotic groups. Cladistics 2009;25:211–230.
15. McMahon MM, Deepak A, Fernández-Baca D, Boss D, Sanderson MJ. STBase: one million species trees for comparative biology. PLoS ONE 2015; 10:e0117987.
16. Moret BME, Roshan U, Warnow T. Sequence-length requirements for phylogenetic methods. In: Guigó R, Gusfield D, editors. *Algorithms in Bioinformatics*. Volume 2452, Lecture Notes in Computer Science. Berlin, Heidelberg: Springer-Verlag; 2002. p 343–356.
17. Owens JD, Houston M, Luebke D, Green S, Stone JE, Phillips JC. GPU computing. Proc IEEE 2008;96(5):879–899.
18. Pattengale ND, Alipour M, Bininda-Emonds OR, Moret BM, Stamatakis A. How many bootstrap replicates are necessary? Proceedings of the 13th Annual International Conference on Research in Computational Molecular Biology, RECOMB 2009; Berlin, Heidelberg: Springer-Verlag; 2009. p 184–200.
19. Pattengale ND, Gottlieb EJ, Moret BME. Efficiently computing the Robinson–Foulds metric. J Comput Biol 2007;14(6):724–735.
20. Pattengale ND, Swenson KM, Moret BME. Uncovering hidden phylogenetic consensus. In: Borodovsky M, Gogarten JP, Przytycka TM, Rajasekaran S, editors. *ISBRA*. Volume 6053, Lecture Notes in Computer Science. Springer Berlin Heidelberg; 2010. p 128–139.
21. Piel WH, Chan L, Dominus MJ, Ruan J, Vos RA, Tannen V. TreeBASE v. 2: a database of phylogenetic knowledge. e-BioSphere 2009; Springer Berlin Heidelberg; London, UK. 2009.
22. Robinson DF, Foulds LR. Comparison of phylogenetic trees. Math Biosci 1981;53:131–147.
23. Rokas A, Carroll SB. More genes or more taxa? The relative contribution of gene number and taxon number to phylogenetic accuracy. Mol Biol Evol 2005;22(5):1337–1344.
24. Ronaghi M, Uhlén M, Nyrén Paal. A sequencing method based on real-time pyrophosphate. Science 1998;281(5375):363–365.
25. Ronquist F, Teslenko M, van der Mark P, Ayres DL, Darling A, Höhna S, Larget B, Liu L, Suchard MA, Huelsenbeck JP. MrBayes 3.2: efficient Bayesian phylogenetic inference and model choice across a large model space. Systematic Biology 2012;61(3):539–542.
26. Roure B, Baurain D, Philippe H. Impact of missing data on phylogenies inferred from empirical phylogenomic datasets. Mol Biol Evol 2013;30(1):197–214.
27. Shindyalov IN, Kolchanov NA, Sander C. Can three-dimensional contacts in protein structures be predicted by analysis of correlated mutations? Protein Eng 1994;7(3):349–358.
28. Smith SA, Beaulieu JM, Stamatakis A, Donoghue MJ. Understanding angiosperm diversification using small and large phylogenetic trees. Am J Bot 2011;98(3):404–414.
29. Stamatakis A. RAxML-VI-HPC: maximum likelihood-based phylogenetic analyses with thousands of taxa and mixed models. Bioinformatics 2006;22(21):2688–2690.
30. Wiens JJ. Missing data, incomplete taxa, and phylogenetic accuracy. Syst Biol 2003;52(4):528–538.

CHAPTER 25

A NEW FAST METHOD FOR DETECTING AND VALIDATING HORIZONTAL GENE TRANSFER EVENTS USING PHYLOGENETIC TREES AND AGGREGATION FUNCTIONS

Dunarel Badescu, Nadia Tahiri, and Vladimir Makarenkov

Département d'informatique, Université du Québecà Montréal, Succ. Centre–Ville, Montréal Québec, Canada

25.1 INTRODUCTION

Until recently, the traditional view of bacterial evolution has been based on divergence and periodic selection. Mutation has been assumed to be the main diversifying force and selection was the main unifying one, until the accumulation of mutations led to a speciation event. A new bacterial evolutionary model slowly emerges, in which *Horizontal Gene Transfer* (HGT) [12, 18] is the main diversifying force and recombination is the main unifying one, speciation being an ecological adaptation. The mechanisms by which bacteria and viruses adapt to changing environmental conditions are well known. These mechanisms include homologous recombination [17],

Pattern Recognition in Computational Molecular Biology: Techniques and Approaches,
First Edition. Edited by Mourad Elloumi, Costas S. Iliopoulos, Jason T. L. Wang, and Albert Y. Zomaya.

nucleotide substitutions, insertions–deletions [13], and HGT [6]. The variation of the DNA composition is spread throughout bacterial genomes leading to the formation of different polymorphic strands of the same group of organisms. The survival of these strands depends on their ability to overcome environmental changes [15]. Multiple mechanisms can overlap and limits between groups are sometimes "fuzzy" [11]. The classical Linnaean paradigm of biological classification is the most widely used framework for interpreting Darwinian evolution. According to it, the most narrowly defined biological group is the species, and the formation of a new lineage is a speciation, which entails the diversification of one species into two different species. Inside the species, a free exchange of genetic information is allowed, but outside the species boundaries, genetic information is passed solely to descendant individuals.

This model of evolution is challenged on the bacterial level, where there exists experimental evidence of massive transfer of genetic material between species [10]. Such a transfer can occur by two distinct routes: homologous recombination and HGT [21]. Homologous recombination is often limited to closely related organisms, having sufficient sequence similarity to allow for efficient integration [16]. HGT can occur between both closely related and genetically distinct organisms.

In this chapter, we present a new algorithm, called *HGT-QFUNC*, for detecting genomic regions that may be associated with complete HGT events. A complete HGT involves a transfer of an entire gene (or a group of genes) from one species to another. The transferred gene can then replace a homologous gene of the host to ensure a better species adaptation to changing conditions (e.g., new environmental conditions). The aggregation functions described in Reference [1], which yielded good results in detecting genomic regions associated with a disease, will be tested in the context of HGT identification. Moreover, novel aggregation functions, which perform better in presence of HGT and recombination, will be also introduced.

The rest of the chapter is organized as follows. In Section 25.2, we present the details of the method for inferring complete HGT events. In Section 25.3, we validate the obtained results with p-values calculated using a Monte Carlo approach. The main advantage of the proposed algorithm is its quadratic time complexity on the number of considered species. In Section 25.4, we estimate the rates of complete HGT among prokaryotes comparing our results to the highly accurate, but much slower, HGT-Detection algorithm [6] based on the calculation of bootstrap support of considered gene trees. We also compare the results provided by HGT-QFUNC and HGT-Detection using simulated data, which will be representative of the prokaryotic landscape. We show that the proposed new functions and algorithm are capable of providing good detection rates for most of the highly probable HGT events. This makes them applicable to the study of large genomic data sets. Note that the proposed HGT-QFUNC algorithm yields better performances than a simple conservation approach, running at the same quadratic asymptotic time. The obtained results confirm the prime importance of HGT in the light of prokaryotic evolution. In Section 25.5, we provide a conclusion for the chapter.

25.2 METHODS

In this section, we present the details of the method for inferring complete HGT events.

25.2.1 Clustering Using Variability Functions

Considering a collection of different prokaryotes, classified as belonging to different taxonomic groups, we can model the simplest case of HGT as the transfer of one single gene sequence between two different species (e.g., x_0 and y_0) coming from two different monophyletic groups (e.g., X and Y). If there was a genetic transfer from source strain y_0 to destination species x_0, then species x_0 would have the same or very similar genetic sequence as the source species y_0. This would lead to an inverse direction shift in phylogenetic classification. HGT can involve either heterologous (e.g., site-specific) or homologous recombination, or direct host gene replacement followed by intragenic recombination. The end result at the phylogenetic level is the integration of species x_0 into the group Y closely to source species y_0. This situation is depicted in Figure 25.1. Genetic transfer direction and resulting phylogenetic neighborhood are dependent on the relative fractions of species x_0 and y_0 involved in the process of recombination. In this chapter, we consider the case of complete HGT when source species is integrated into the host genome without intragenic recombination (i.e., without formation of a mosaic gene).

Here, we describe the HGT detection problem mathematically. To perform the clustering of our data, we first define the following sets, involving the HGT-related

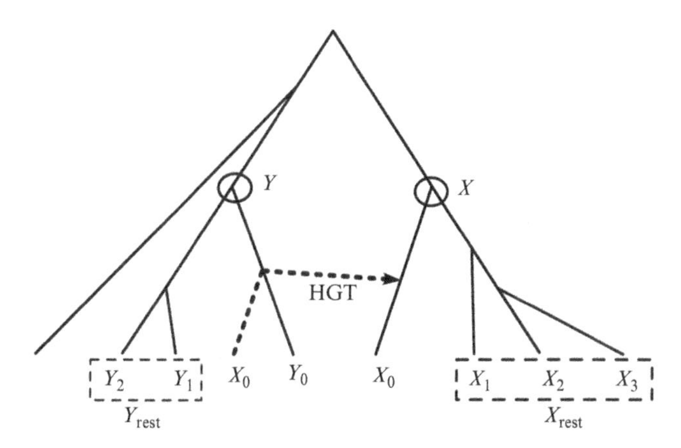

Figure 25.1 Intra- and intergroup phylogenetic relationships following an HGT. A horizontal gene transfer from species y_0 of the group Y to species x_0 of the group X is shown by an arrow; the dotted line shows the position of species x_0 in the tree after the transfer; X_{rest} denotes the rest of the species of the group X and Y_{rest} denotes the rest of the species of the group Y. Each species is represented by a unique nucleotide sequence.

species x_0 and y_0 (Eqs 25.1–25.3):

$$R = \{x_0 \cup y_0\} \tag{25.1}$$

$$X_{\text{rest}} = X \setminus x_0 \tag{25.2}$$

$$Y_{\text{rest}} = Y \setminus y_0 \tag{25.3}$$

Note that in a general case, x_0 and y_0 can be clusters (i.e., subtrees) including several species. We now define inter- and intragroup variability. Consider two groups of species A and B not having common members. The measures in question are calculated as the means of the Hamming distances (any other evolutionary distance can be used instead), $dist_h$, among the sequences from the same group A (or B) only, and among the sequences from the distinct groups A and B.

First, the intragroup variability of the group A, denoted by $V(A)$, is defined by Equation 25.4:

$$V(A) = \sum_{\{a_1, a_2 \in A | a_1 \neq a_2\}} dist_h(a_1, a_2) \tag{25.4}$$

We then normalize $V(A)$ by the number of possible different pairs of elements in A (Eq. 25.5):

$$V_{norm}(A) = \frac{V(A)}{N(A) \times (N(A) - 1)/2} \tag{25.5}$$

where $N(A)$ is the number of elements in the group A.

The intergroup variability of the groups A and B, denoted by $D(A, B)$, is defined as follows:

$$D(A, B) = \sum_{a \in A, b \in B} dist_h(a, b) \tag{25.6}$$

We then normalize $D(A, B)$ by the number of possible pairs of species:

$$D_{norm}(A, B) = \frac{D(A, B)}{N(A) \times N(B)} \tag{25.7}$$

Using previously described groups and functions, we introduce a new aggregation function Q_7 as follows (see References [1, 2] for definition of the functions Q_1–Q_6; these functions are not adapted for detecting HGT events):

$$Q_7(R) = Max(D_{norm}(R, X_{\text{rest}}); D_{norm}(R, Y_{\text{rest}})) - V_{norm}(R) \tag{25.8}$$

where R is defined by Equation 25.1.

When a complete HGT occurs (Figure 25.1), the transferred gene is assumed to replace a homologous gene in the host genomes. As a result of this event, destination species x_0 migrates close to source species y_0 into the phylogenetic network representing the evolution of the given gene (Figure 25.1). Thus, in the obtained gene tree, destination species x_0 will be a part of the group Y to which source species y_0 belongs.

Equation 25.8 reflects such a principle. Also $V_{norm}(R)$, in this particular case, defines the distance between species x_0 and y_0.

We also introduce the aggregation function, Q_8, similar to the function that provided good results in detecting lineage-specific selection in Reference [2], (i.e., $Q_6 = | V(A)/V(B) |$).

$$Q_8(R) = \frac{D_{norm}(R, XY_{\text{rest}})}{V_{norm}(R)} \tag{25.9}$$

Because this function uses the division instead of the summation, such a function underlines the asymmetry between the two groups. Note that both HGT and lineage-specific selection exhibit asymmetrical properties.

Let us now define the function $Q_9(R)$:

$$Q_9(R) = -V_{norm}(R) \tag{25.10}$$

Another clustering option would be to consider both interacting species as a destination group and merge the rest of the sequences into a source group.

$$XY_{\text{rest}} = \{X \cup Y \setminus R\} \tag{25.11}$$

25.2.2 Other Variants of Clustering Functions Implemented in the Algorithm

Here we describe particular cases of formulas used in our implementation when the HGT occurs between the tree leaves (i.e., individual species of X and Y; Eqs 25.12–25.18). Let us define the following:

$$D(x_0, y_0) = dist_h(x_0, y_0) \tag{25.12}$$

$$D(x_0, X_{\text{rest}}) = \sum_{\{x_i \in X;\ x_i \neq x_0\}} dist_h(x_0, x_i) = D(x_0, X) \tag{25.13}$$

$$D(y_0, Y_{\text{rest}}) = \sum_{\{y_i \in Y;\ y_i \neq y_0\}} dist_h(y_0, y_i) = D(y_0, Y) \tag{25.14}$$

$$D(x_0, Y) = \sum_{\{y_i \in Y\}} dist_h(x_0, y_i) \tag{25.15}$$

$$D(y_0, X) = \sum_{\{x_i \in X\}} dist_h(y_0, x_i) \tag{25.16}$$

$$D(x_0, Y_{\text{rest}}) = \sum_{\{y_i \in Y;\ y_i \neq y_0\}} dist_h(x_0, y_i) = D(x_0, Y) - D(x_0, y_0) \tag{25.17}$$

$$D(y_0, X_{\text{rest}}) = \sum_{\{x_i \in X;\ x_i \neq x_0\}} dist_h(y_0, x_i) = D(y_0, X) - D(x_0, y_0) \tag{25.18}$$

$$n = N(X) \tag{25.19}$$

$$m = N(Y) \tag{25.20}$$

We also introduce an epsilon (ϵ) value (i.e., in our implementation we set the value of ϵ equal to 0.00001) to avoid division by zero. Some other constants are also considered in Equations 25.9–25.22 in order to normalize the results and to obtain the same minimum value.

$$Q_7(x_0, y_0) = Max\left(\frac{D(x_0, X_{\text{rest}}) + D(y_0, X_{\text{rest}})}{2(n-1)}, \frac{D(x_0, Y_{\text{rest}}) + D(y_0, Y_{\text{rest}})}{2(m-1)}\right) - D(x_0, y_0) + 2 \quad (25.21)$$

$$Q_{8a}(x_0, y_0) = \frac{D(x_0, X_{\text{rest}}) + D(y_0, X_{\text{rest}}) + D(x_0, Y_{\text{rest}}) + D(y_0, Y_{\text{rest}}) + \varepsilon}{2(xy + \varepsilon)(n + m - 2)} \quad (25.22)$$

In our implementation, we used the following variant of the functions Q_8 and Q_9 as well:

$$Q_{8b}(x_0, y_0) = \frac{\frac{D(x_0,X)+D(y_0,X)}{n-1} + \frac{D(x_0,Y)+D(y_0,Y)}{m-1} + \varepsilon}{2(D(x_0, y_0) + \varepsilon)} \text{ and} \quad (25.23)$$

$$Q_9(x_0, y_0) = -D(x_0, y_0) + 2 \quad (25.24)$$

For each pair of species (x, y) belonging to two different groups, we maximize Q_7, Q_{8a}, and Q_{8b}, over all possible sets of R in order to identify the best HGT candidates. At the same time, maximizing Q_9 is equivalent to minimizing the distance between x_0 and y_0.

25.2.3 Description of the New Algorithm

Here, we present a new algorithm allowing one to estimate the values of functions Q_7, Q_{8a}, Q_{8b}, and Q_9, and to validate the results using the p-value estimation procedure. Our algorithm takes as input a *Multiple Sequence Alignment* (MSA) of n species, a set of groups (e.g., species families) and a unique association of each specie to one of these groups. The algorithm's output consists of pairs of clusters that could be involved in HGTs. The detailed algorithmic scheme is presented in Algorithm 25.1. The p-value estimation is done by carrying out a Monte Carlo procedure with a fixed p-value threshold. For a constant number of steps, this procedure simulates permuted MSAs. A constant number of nucleotides are permuted within each of the original sequences. Then, we compare the obtained values of the selected function Q to the reference value obtained with the original data. The detailed p-value estimation scheme is presented in Algorithm 25.2. The main algorithm consists of three major steps. The first step calculates the pairwise distance matrix between all given species. The second calculates the distance between each species and all other species belonging to the other groups. The third estimates the inter- and intragroup distances and aggregation solutions by using Equations 25.12–25.24. There is one more step needed to

Algorithm 25.1 HGT-QFUNC algorithm for detecting species related to each other by the way of complete HGT

Require:
MSA: Multiple sequence alignment,
FI: Aggregation function to be optimized Q_7, Q_{8a}, Q_{8b}, or Q_9,
GR: Groups,
SG: Unique association of each sequence in the MSA to one group (G),
Ensure:
QVAL: Matrix of Q_{FI} values for each pair of sequences $\in$ MSA

1: MSA_N $\leftarrow$ Number of sequences in MSA
2: GR_N $\leftarrow$ Number of groups
3: N_SEQS [GR_N] $\leftarrow$ Number of sequences in each group
4: D_SEQ_SEQ $\leftarrow$ Matrix[MSA_N][MSA_N] //sequence to sequence distance matrix
5: **for all** *seq_i* $\in$ MSA **do**
6:     **for all** *seq_j* $\in$ MSA **do**
7:         D_SEQ_SEQ[*seq_i*][*seq_j*] $= D(seq\_i, seq\_j) = dist_h(seq\_i, seq\_j)$
8:     **end for**
9: **end for**
10: D_SEQ_GR $\leftarrow$ Matrix[MSA_N][GR_N] //sequence to group distance matrix
11: **for all** *seq_i* $\in$ MSA **do**
12:     **for all** *seq_j* $\in$ MSA **do**
13:         $gr\_j \leftarrow$ SG[seq_j]
14:         D_SEQ_GR [*seq_i*][*gr_j*]$+= D\_SEQ\_SEQ[seq\_i][seq\_j]$
15:     **end for**
16: **end for**
17: QVAL *gets*Matrix[MSA_N][MSA_N]
18: **for all** *seq_i* $\in$ MSA **do**
19:     **for all** *seq_j* $\in$ MSA **do**
20:         $gr\_i \leftarrow$ SG[*seq_i*]
21:         $gr\_j \leftarrow$ SG[*seq_j*]
22:         $n \leftarrow$ N_SEQS[*gr_i*]
23:         $m \leftarrow$ N_SEQS[*gr_j*]
24:         **go to 37 if** $(seq\_i \geqslant seq\_j)$ **or** $(n < 2)$**or**$(m < 2)$
25:         $D(x_0, y_0) =$ D_SEQ_SEQ[*seq_i*][*seq_j*]
26:         $D(x_0, X_{\text{rest}}) = D(x_0, X) =$ D_SEQ_GR[*seq_i*][*gr_i*]
27:         $D(y_0, Y_{\text{rest}}) = D(y_0, Y) =$ D_SEQ_GR[*seq_j*][*gr_j*]
28:         $D(x_0, Y) =$ D_SEQ_GR[*seq_i*][*gr_j*]
29:         $D(x_0, Y_{\text{rest}}) = D(x_0, Y) - D(x_0, y_0)$
30:         $D(y_0, X) =$ D_SEQ_GR[*seq_j*][*gr_i*]
31:         $D(y_0, X_{\text{rest}}) = D(y_0, X) - D(x_0, y_0)$
32:         $Q_7(x_0, y_0) = Max\left(\frac{D(x_0,X_{\text{rest}})+D(y_0,X_{\text{rest}})}{2(n-1)}, \frac{D(x_0,Y_{\text{rest}})+D(y_0,Y_{\text{rest}})}{2(m-1)}\right) - D(x_0, y_0)+2$
33:         $Q_{8a}(x_0, y_0) = \frac{D(x_0,X_{\text{rest}})+D(y_0,X_{\text{rest}})+D(x_0,Y_{\text{rest}})+D(y_0,Y_{\text{rest}})+\varepsilon}{2(xy+\varepsilon)(n+m-2)}$
34:         $Q_{8b}(x_0, y_0) = \frac{\frac{D(x_0,X)+D(y_0,X)}{n-1} + \frac{D(x_0,Y)+D(y_0,Y)}{m-1} + \varepsilon}{2(D(x_0,y_0)+\varepsilon)}$

Algorithm 25.1 *(continued)*

```
35:         Q_9(x_0, y_0) = −D(x_0, y_0)+2
36:         QVAL[seq_i][seq_j] ← Q_FI(x_0, y_0)
37:     end for
38: end for
39: return QVAL
```

Algorithm 25.2 *p*-Values validation for HGT-QFUNC algorithm (see Algorithm 25.1) using Monte Carlo estimation

Require:
MSA: Multiple sequence alignment,
FI: Aggregation function to be optimized Q_7, Q_{8a}, Q_{8b}, or Q_9,
GR: Groups,
SG: Unique association of each sequence in the MSA, to one group (G),
PVST : Constant number of p-value steps,
PERM: Nucleotide permutation percentage.
Ensure:
QVAL: Matrix of Q_{FI} p-values for each pair of sequences ∈ MSA

```
 1: PQVAL ← Matrix[MSA_N][MSA_N]
 2: QVAL ← call algorithm 25.1 (MSA and FI, GR, SG)
 3: for i ∈ (1, … , PVST) do
 4:     MSA_PERM ← MSA
 5:     //introduce a level of uncertainty
 6: end for
 7: //calculate regular values with permuted MSA
 8: QVAL_PERM ← call algorithm 25.1 (MSA_PERM and FI, GR, SG)
 9: //test if the obtained values are at least as good as those obtained without permutation
10: for all seq_i ∈ MSA do
11:     for all seq_j ∈ MSA do
12:         if QVAL_PERM[seq_i][seq_j] ⩾ QVAL[seq_i][seq_j] then
13:             PQVAL [seq_i][seq_j]++
14:         end if
15:     end for
16: end for
17: //end permutations
18: //update p-value
19: for all seq_i ∈ MSA do
20:     for all seq_j ∈ MSA do
21:         PQVAL [seq_i][seq_j]++
22:         PQVAL [seq_i][seq_j]/ = (PVST + 1)++
23:     end for
24: end for
25: return QVAL, PQVAL
```

complete the detection of HGT. The obtained potential HGTs are ranked according to the value of the corresponding Q-function, first, then p-value, second. Those HGTs whose Q-function values are greater than a fixed threshold are considered valid. This threshold can be set on the basis of the p-values or a fixed percentage of the total number of species (called here *percentage of positive values*). We also consider an alternative results ranking: by p-value, first, and by the Q-function value, second. Such a ranking strategy allowed us to better emphasize the strength of a statistical signal. All the tests of the new HGT detection algorithm are carried out in parallel with both ranking strategies.

25.2.4 Time Complexity

The time complexity of the new algorithm in a general case, carried out over an MSA of n species and DNA or amino acid sequences of size l, is $\mathcal{O}\ (ln^2 + n^4)$. When we consider HGT between individual species only (i.e., tree leaves), the most common case in HGT analysis, the time complexity is $\mathcal{O}\ (ln^2)$ only (see Algorithm 25.1). The p-value estimation procedure (see Algorithm 25.2) adds a constant overhead to the algorithm's running time in order to maintain the desired p-value precision. This constant is usually 100, 1000, or 10,000, for a precision of 0.01, 0.001, or 0.0001, respectively.

25.3 EXPERIMENTAL STUDY

In this section, we discuss the details of our implementation and describe the data sets we examine in this study.

25.3.1 Implementation

The presented algorithm can be parallelized to improve its performances. At least three different parallelization schemes exist. The first one uses fine-grained parallelism with global atomic reductions that would be better suited for graphic cards. The second one involves the parallelization of higher granularity, implying the p-value estimation steps. It would be better suited to multicore processors. The third scheme, which we implemented in our program, proceeds by mapping each group into a CPU core. Although this is not the most efficient scheme, it has the advantage to accelerate calculations even in the absence of the p-value estimation step. We developed a C++ code for this algorithm for multicore CPUs, parallelizing using OpenMP, and SIMD vectorizing using SSE3 instructions.

25.3.2 Synthetic Data

In this section, we first describe synthetic data we use, followed by a simulation study that includes comparison with the HGT-Detection algorithm.

25.3.2.1 Description of Synthetic Data. For a simulation study conducted with synthetic data, we used the real prokaryotic data set as a basis, in order to maintain the same real-world relationships between sequences, and the same limitations for our detection algorithm as it would be in a real-world situation. To simulate our data, we chose as benchmark the gene *hisH*, which is the gene with the highest number of different prokaryotic strains (i.e., 99) in which the HGT-Detection algorithm did not find any HGT at the bootstrap level of 50%. This threshold was considered as a minimum quality requirement in our study.

25.3.2.2 Simulation with Synthetic Data and Comparison to HGT-Detection. After we have explored the real-life detection performances of the HGT-QFUNC algorithm, we tested its ability to recover correct HGT events by simulating different HGT rates in artificially generated multiple sequence alignments. We performed a series of tests involving random nonreciprocal sequence transfers between the species of different prokaryotic groups. Both complete (involving only gene replacement) and partial (involving intragenic recombination and creation of mosaic genes) HGT cases were considered in this simulation.

All simulated transfers were supposed to occur between single species of the considered MSA (a single species always corresponds to a tree leaf in a phylogenetic tree). From an evolutionary standpoint, such transfers are the most recent, and also the most recoverable ones [6]. Therefore, they are also the most reported ones. We considered the cases with 1–128 simulated transfers, following the logarithmic scale (i.e., 1, 2, 4, 8, 16, 32, 64, and 128 transfers). One of our goals in this simulation was to discriminate between the functions Q_7, Q_{8a}, Q_{8b}, and Q_9 when detecting different numbers of complete HGT. We set a maximum allowed number of positive values as the double of the number of transfers (i.e., 2, 4, 8, 16, 32, 64, 128, and 256 transfers, respectively). Note that in the above-described real-life experiments, we allowed 100, 200, and 300 transfers, depending on the bootstrap support fixed for the HGT-Detection algorithm. It must be mentioned that for the artificial data, the algorithm was carried out under more restrictive conditions (lower number of positive values) than those imposed in the experiments with real-life prokaryotic data.

We first simulated gene transfers without recombination (i.e., complete HGT), as a simple replacement of the source sequence by the destination sequence. Second, we added an average percentage of recombination of 25% to the data (i.e., partial HGT). This process was simulated as a random recombination between the source and destination sequences. The new resulting sequence (i.e., mosaic gene) contained 75% of the source sequence and 25% of the destination sequence. We also considered the case of a maximum recombination rate of 50%, where the resulting mosaic sequence was a hybrid of 50% source and 50% destination sequence. Every combination of the simulation parameters was tested with 50 replicates. The distribution of the obtained average results based on the Q functions ordering is shown in Figures 25.2 and 25.3. The additional results based on the p-value ordering, with a maximum threshold of 0.05, are shown in Figures 25.4–25.6.

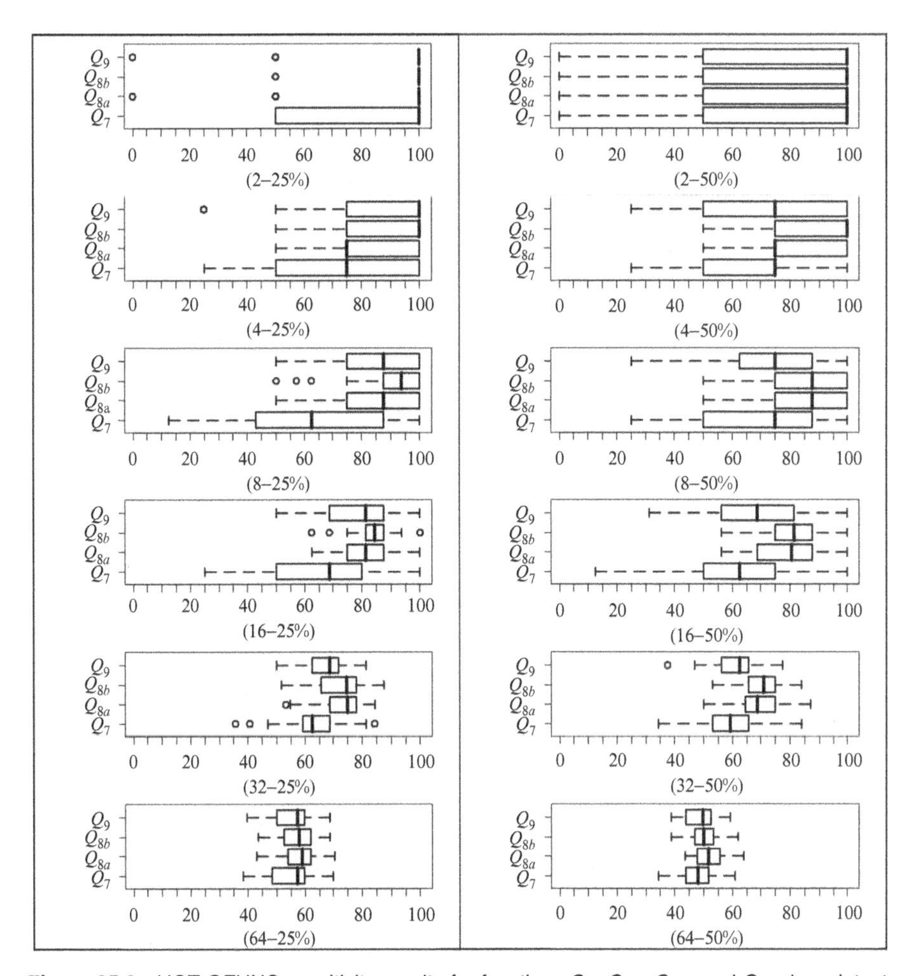

Figure 25.2 HGT-QFUNC sensitivity results for functions Q_7, Q_{8a}, Q_{8b}, and Q_9 when detecting partial HGT in a synthetic data set based on Q-value ordering—boxplot representation. The abscissa represents the sensitivity percentage and the ordinate the tested function. The median value is shown by a vertical black line within each box. Simulations for 2, 4, 8, 16, 32, and 64 random nonreciprocal sequence transfers between prokaryotic species (first value between parentheses) were carried out. Average simulation results under the medium degree of recombination (when 25% of the resulting sequence belong to one of the parent sequences) are depicted in the left panel. Average simulation results under the highest level of recombination (when 50% of the resulting sequence belong to the source sequence and 50% to the destination sequence) is depicted in the right panel. For each data set, the maximum allowed number of positive values was the double of the number of transfers (i.e., 4, 8, 16, 32, 64, and 128, respectively). Calculations were done over 50 replicates for each combination of parameters.

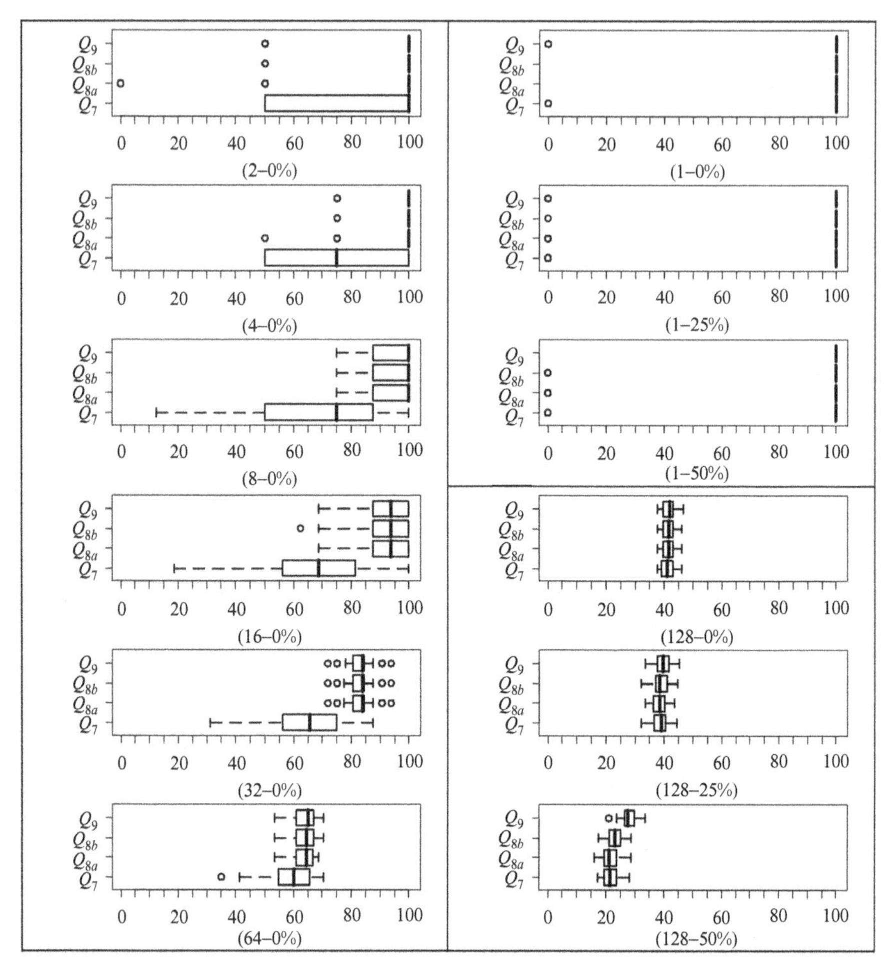

Figure 25.3 Remaining HGT-QFUNC sensitivity results for functions Q_7, Q_{8a}, Q_{8b}, and Q_9 when detecting complete and partial HGT in a synthetic data set based on Q-value ordering—boxplot representation. The abscissa represents the sensitivity percentage and the ordinate the tested function. The median value is shown by a vertical black line within each box. Simulations for 2, 4, 8, 16, 32, and 64 random nonreciprocal sequence transfers between prokaryotic species (first value between parentheses) were carried out. Average simulation results for data without recombination are depicted in the left panel. The right panel depicts the results of the same simulations, for the cases of 1 and 128 transfers, with recombination levels of 0% (no recombination), 25%, and 50%. Average simulation results under the highest level of recombination (when 50% of the resulting sequence belong to the source sequence and 50% to the destination sequence) is depicted in the right panel. For each data set, the maximum allowed number of positive values was the double of the number of transfers (i.e., 4, 8, 16, 32, 64, and 128, respectively). Calculations were done over 50 replicates for each combination of parameters.

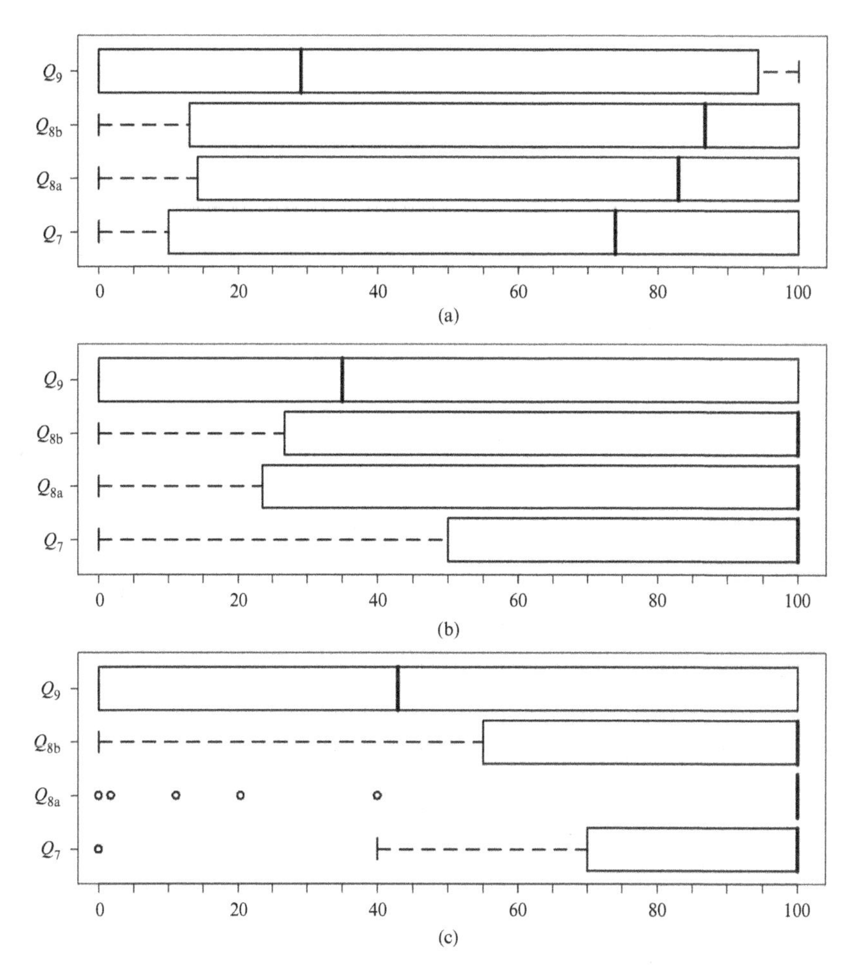

Figure 25.4 HGT-QFUNC sensitivity results for functions Q_7, Q_{8a}, Q_{8b}, and Q_9 when detecting complete HGT in a prokaryotic data set based on p-value ordering (maximum p-value of 0.05)—boxplot representation. The abscissa represents the sensitivity percentage and the ordinate the tested function. The median value is shown by a vertical black line within each box. The HGT-QFUNC algorithm was limited to the following maximum numbers of positive values: (a) 300 HGT (corresponds to 50% bootstrap support in the HGT-Detection algorithm); (b) 200 HGT (corresponds to 75% bootstrap support in the HGT-Detection algorithm); (c) 100 HGT (corresponds to 90% bootstrap support in the HGT-Detection algorithm).

25.3.3 Real Prokaryotic (Genomic) Data

In this section, we describe the real prokaryotic (genomic) data, followed by a simulation study which includes comparison with the HGT-Detection algorithm.

25.3.3.1 Description of Real Prokaryotic (Genomic) Data. We assembled a real-world data set, representative of the prokaryotic genomic landscape, to serve as

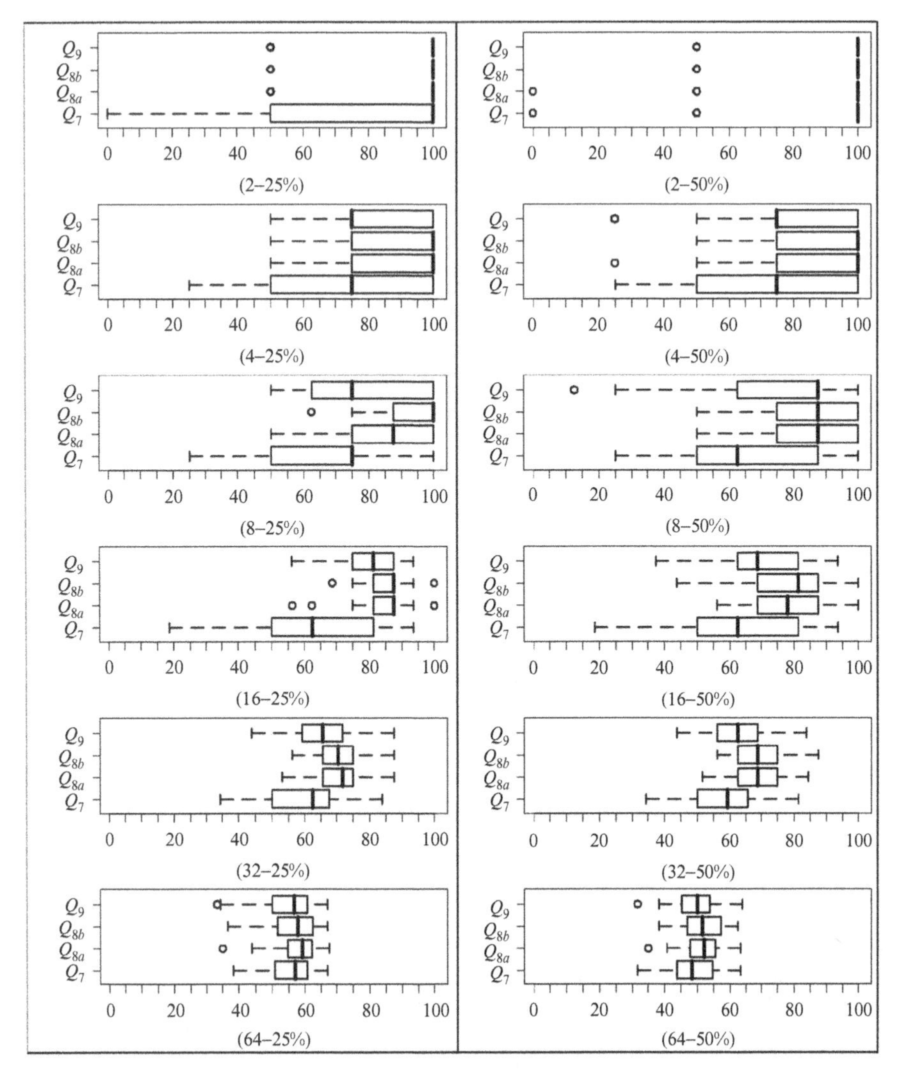

Figure 25.5 HGT-QFUNC sensitivity results for functions Q_7, Q_{8a}, Q_{8b}, and Q_9 when detecting partial HGT in a synthetic data set based on p-value ordering (maximum p-value of 0.05)—boxplot representation. The abscissa represents the sensitivity percentage and the ordinate the tested function. The median value is shown by a vertical black line within each box. Simulations for 2, 4, 8, 16, 32, and 64 random nonreciprocal sequence transfers between prokaryotic species (first value between parentheses) were carried out. Average simulation results under medium degree of recombination (when 25% of the resulting sequence belong to one of the parent sequences) are depicted in the left panel. Average simulation results under the highest level of recombination (when 50% of the resulting sequence belong to the source sequence and 50% to the destination sequence) is depicted in the right panel. For each data set, the maximum allowed number of positive values was the double of the number of transfers (i.e., 4, 8, 16, 32, 64, and 128, respectively). Calculations were done over 50 replicates for each combination of parameters.

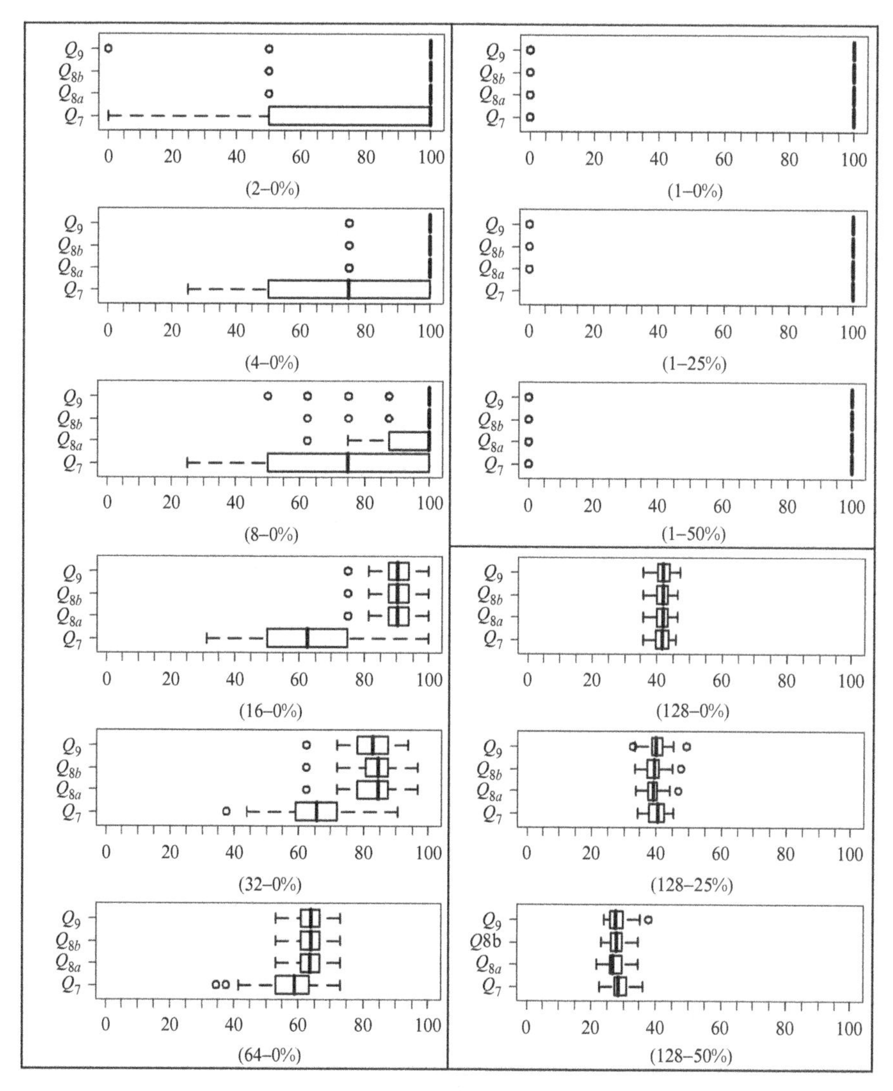

Figure 25.6 Remaining HGT-QFUNC sensitivity results for functions Q_7, Q_{8a}, Q_{8b}, and Q_9 when detecting complete and partial HGT in a synthetic data set based on p-value ordering (maximum p-value of 0.05)—boxplot representation. The abscissa represents the sensitivity percentage and the ordinate the tested function. The median value is shown by a vertical black line within each box. Simulations for 2, 4, 8, 16, 32, and 64 random nonreciprocal sequence transfers between prokaryotic species (first value between parentheses) were carried out. Average simulation results for data without recombination are depicted in the left panel. The right panel depicts the results of the same simulations, for the cases of 1 and 128 transfers, with recombination levels of 0% (no recombination), 25%, and 50%. Average simulation results under the highest level of recombination (when 50% of the resulting sequence belong to the source sequence and 50% to the destination sequence) is depicted in the right panel. For each data set, the maximum allowed number of positive values was the double of the number of transfers (i.e., 4, 8, 16, 32, 64, and 128, respectively). Calculations were done over 50 replicates for each combination of parameters.

a basis for testing our algorithm against an accurate HGT-Detection v.3.4 algorithm [6] available on the T-Rex website [5]. Here we outline the most important details of the data being examined.

All of the completely sequenced prokaryotic genomes available at the NCBI Genomes ftp site (1465 as of November 2011) were considered. Among them, we first selected 100 of the most complete genomes in terms of the number of genes. Then, we added to them 11 additional species to ensure that our data set included at least one representative from each of the 23 available prokaryotic families. This yielded us a total number of 111 species. Detailed information on the considered species can be found in Table ST1 available at http://www.info2.uqam.ca/~makarenkov_v/SM_genes_Dunarel.pdf. We also identified 110 of the most complete genes (see Table ST2) from the selected set of 111 species (see below). It must be mentioned that multiple sequence alignments we considered often contained multiple versions of the same gene (i.e., multiple alleles) belonging to the same species.

Id	Taxon-id	Scientific name	Id	Taxon-id	Scientific name
1	228908	*Nanoarchaeum equitans* Kin4-M	57	545693	*Bacillus megaterium* QM B1551
2	374847	*Candidatus Korarchaeum cryptofilum* OPF8	58	281309	*Bacillus thuringiensis* serovar konkukian str. 97-27
3	272557	*Aeropyrum pernix* K1	59	288681	*Bacillus cereus* E33L
4	273057	*Sulfolobus solfataricus* P2	60	637380	*Bacillus cereus* biovar anthracis str. CI
5	768679	*Thermoproteus tenax* Kra 1	61	361100	*Bacillus cereus* Q1
6	188937	*Methanosarcina acetivorans* C2A	62	279010	*Bacillus licheniformis* ATCC 14580
7	362976	*Haloquadratum walsbyi* DSM 16790	63	224308	*Bacillus subtilis* subsp. subtilis str. 168
8	309800	*Haloferax volcanii* DS2	64	655816	*Bacillus subtilis* subsp. spizizenii str. W23
9	348780	*Natronomonas pharaonis* DSM 2160	65	326423	*Bacillus amyloliquefaciens* FZB42
10	634497	*Haloarcula hispanica* ATCC 33960	66	692420	*Bacillus amyloliquefaciens* DSM 7
11	272569	*Haloarcula marismortui* ATCC 43049	67	177437	*Desulfobacterium autotrophicum* HRM2
12	243090	*Rhodopirellula baltica* SH 1	68	448385	*Sorangium cellulosum* 'So ce 56'
13	190304	*Fusobacterium nucleatum* subsp. nucleatum ATCC 25586	69	404380	*Geobacter bemidjiensis* Bem
14	240015	*Acidobacterium capsulatum* ATCC 51196	70	273121	*Wolinella succinogenes* DSM 1740
15	743525	*Thermus scotoductus* SA-01	71	382638	*Helicobacter acinonychis* str. Sheeba
16	224324	*Aquifex aeolicus* VF5	72	693745	*Helicobacter pylori* B8
17	484019	*Thermosipho africanus* TCF52B	73	62928	*Azoarcus sp.* BH72
18	255470	*Dehalococcoides sp.* CBDB1	74	266264	*Cupriavidus metallidurans* CH34
19	311424	*Dehalococcoides sp.* VS	75	1042878	*Cupriavidus necator* N-1
20	330214	*Candidatus Nitrospira defluvii*	76	381666	*Ralstonia eutropha* H16
21	379066	*Gemmatimonas aurantiaca* T-27	77	375286	*Janthinobacterium sp* . Marseille
22	267671	*Leptospira interrogans* serovar Copenhageni str. Fiocruz L1-130	78	757424	*Herbaspirillum seropedicae* SmR1
23	759914	*Brachyspira pilosicoli* 95/1000	79	1005048	*Collimonas fungivorans* Ter331
24	565034	*Brachyspira hyodysenteriae* WA1	80	452662	*Sphingobium japonicum* UT26S
25	167539	*Prochlorococcus marinus* subsp. marinus str. CCMP1375	81	272942	*Rhodobacter capsulatus* SB 1003
26	1148	*Synechocystis sp.* PCC 6803	82	375451	*Roseobacter denitrificans* OCh 114
27	43989	*Cyanothece sp.* ATCC 51142	83	272568	*Gluconacetobacter diazotrophicus* PAI 5
28	481448	*Methylacidiphilum infernorum* V4	84	414684	*Rhodospirillum centenum* SW
29	716544	*Waddlia chondrophila* WSU 86-1044	85	258594	*Rhodopseudomonas palustris* CGA009
30	765952	*Parachlamydia acanthamoebae* UV7	86	224911	*Bradyrhizobium japonicum* USDA 110
31	194439	*Chlorobium tepidum* TLS	87	288000	*Bradyrhizobium sp* . BTAi1
32	269798	*Cytophaga hutchinsonii* ATCC 33406	88	311403	*Agrobacterium radiobacter* K84
33	411154	*Gramella forsetii* KT0803	89	347834	*Rhizobium etli* CFN 42
34	402612	*Flavobacterium psychrophilum* JIP02/86	90	491916	*Rhizobium etli* CIAT 652
35	1034807	*Flavobacterium branchiophilum* FL-15	91	706191	*Pantoea ananatis* LMG 20103
36	405948	*Saccharopolyspora erythraea* NRRL 2338	92	321314	*Salmonella enterica* subsp. enterica serovar Choleraesuis str. SC-B67
37	227882	*Streptomyces avermitilis* MA-4680	93	585035	*Escherichia coli* S88
38	216594	*Mycobacterium marinum* M	94	585395	*Escherichia coli* O103:H2 str. 12009
39	1048245	*Mycobacterium canettii* CIPT 140010059	95	585055	*Escherichia coli* 55989
40	572418	*Mycobacterium africanum* GM041182	96	573235	*Escherichia coli* O26:H11 str. 11368
41	419947	*Mycobacterium tuberculosis* H37Ra	97	585397	*Escherichia coli* ED1a
42	83332	*Mycobacterium tuberculosis* H37Rv	98	585034	*Escherichia coli* IAI1
43	233413	*Mycobacterium bovis* AF2122/97	99	405955	*Escherichia coli* APEC O1
44	410289	*Mycobacterium bovis* BCG str. Pasteur 1173P2	100	364106	*Escherichia coli* UTI89
45	561275	*Mycobacterium bovis* BCG str. Tokyo 172	101	585056	*Escherichia coli* UMN026
46	991791	*Clostridium acetobutylicum* DSM 1731	102	544404	*Escherichia coli* O157:H7 str. TW14359
47	208596	*Carnobacterium sp.* 17-4	103	585396	*Escherichia coli* O111:H- str. 11128
48	684738	*Lactococcus lactis* subsp. lactis KF147	104	199310	*Escherichia coli* CFT073
49	272623	*Lactococcus lactis* subsp. lactis Il1403	105	701177	*Escherichia coli* O55:H7 str. CB9615
50	543734	*Lactobacillus casei* BL23	106	413997	*Escherichia coli* B str. REL606
51	568704	*Lactobacillus rhamnosus* Lc 705	107	585057	*Escherichia coli* IAI39
52	568703	*Lactobacillus rhamnosus* GG	108	574521	*Escherichia coli* O127:H6 str. E2348/69
53	358681	*Brevibacillus brevis* NBRC 100599	109	511145	*Escherichia coli* str. K-12 substr. MG1655
54	398511	*Bacillus pseudofirmus* OF4	110	316385	*Escherichia coli* str. K-12 substr. DH10B
55	315750	*Bacillus pumilus* SAFR-032	111	595496	*Escherichia coli* BW2952
56	592022	*Bacillus megaterium* DSM 319			

Afterward, we constructed 110 multiple sequence alignments (one MSA per selected gene) from which we excluded misclassified paralogs using TribeMCL [9]. The latter tool, which uses a *Markov Chain Clustering* (MCL) algorithm [22] on all-to-all BLASTP hits, is known to be conservative in the number of groups

[14]. We carried out the TribeMCL version bundled with "mcl" v11.294, with default parameters ($I = 2.0$). To obtain more accurate results of BLASTP, we set a Smith–Waterman backend and an E-value threshold of 10^{-4}. Using this procedure, 1% of initial alleles were identified as paralogs and excluded from the original MSA. Nucleotide sequences were retrieved from protein sequences identified above. They were aligned using MUSCLE v3.8.31 [8] with default parameters and trimmed with Gblocks v0.91b [7]. In our analysis, we were less restrictive than the default option of Gblocks, allowing 50% of the number of sequences for flank positions ($-b2$ parameter), a maximum of 10 contiguous nonconserved positions ($-b3$ parameter), minimum block length of 5 ($-b4$ parameter) and half-gap positions ($-b5$ parameter).

The obtained MSA were then used as a basis for the detection of complete HGT. Species taxonomy (i.e., species tree in the HGT context) was retrieved from the NCBI Taxonomy website [4]. Taxonomic groups were those assigned by the NCBI Genome Project. Each species was then assigned to one established prokaryotic family. We constructed the gene trees using the RAxML method [20]. We used the RAxML v.7.2.8—multithreaded implementation, GTR Gamma model, 20 starting random trees, and 100 bootstrap trees as RAxML input options.

25.3.3.2 *Simulation with the Real Prokaryotic Data Set and Comparison to HGT-Detection.* We tested the ability of the described HGT-QFUNC algorithm to detect complete HGT events by comparing it to a highly accurate but much slower HGT-Detection algorithm [6]. We used the presented functions Q_7, Q_{8a}, Q_{8b}, and Q_9 side by side, in order to identify their strengths and limitations in general use-case scenarios on the real prokaryotic data. We used the *Sensitivity* measure as a comparative statistic. *Sensitivity*, which reflects the ability to detect true positives is defined as follows:

$$Sensitivity = \frac{true\_positives}{true\_positives + false\_negatives} \tag{25.25}$$

The true positive and false negative HGTs were determined by comparing the obtained results to those provided by the HGT-Detection algorithm (i.e., the transfers found by HGT-Detection were considered as true positives). We excluded from the analysis the alignments where HGT-Detection did not return any HGT to avoid division by zero in the *Sensitivity* formula. We assured comparability of the output formats between the two compared algorithms. HGT-Detection provides its results as a list of pairs of HGT-related source and destination branches defined by the corresponding nodes of the species tree. We decomposed HGT-Detection transfer scenarios into a list of all affected leaf pairs between respective source and destination subtrees. As the species tree was not always completely resolved in our case, some trivial transfers (i.e., transfers among branches of the same multifurcating node) could occur. For quality reasons, we discarded such trivial transfers in this simulation study. HGT-Detection first applies sophisticated phylogenetic tree-based manipulations and then filters results by HGT bootstrap values. The output of the HGT-Detection program usually contains a very small number of transfers due to the applied evolutionary constraints and imposed bootstrap threshold. We tried to mimic this behavior by limiting the HGT-QFUNC algorithm to a restrictive p-value threshold of 0.001, but still had too many (compared to HGT-Detection) HGT events identified in the end. Therefore, we limited the number of detected HGT events by imposing a fixed threshold (as described below). We carried out the HGT-Detection algorithm over our prokaryotic data set with minimum bootstrap supports of 50%, 75%, and 90%, respectively. The obtained results are shown in Figure 25.7, while the corresponding results of

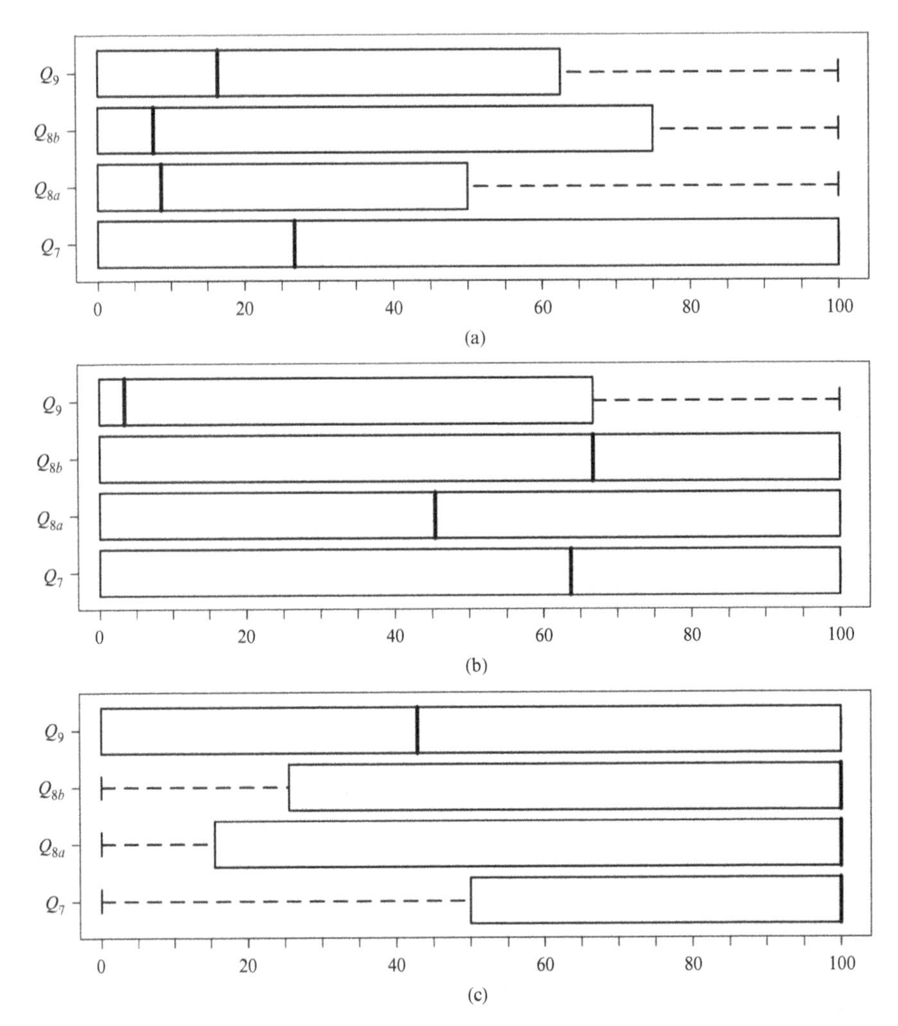

Figure 25.7 HGT-QFUNC sensitivity results for functions Q_7, Q_{8a}, Q_{8b}, and Q_9 when detecting complete HGT in prokaryotic data set based on Q-value ordering—boxplot representation. The abscissa represents the sensitivity percentage and ordinate the tested function. The median value is shown by a vertical black line within each box. The HGT-QFUNC algorithm was limited to the following maximum numbers of positive values: (a) 300 HGT (corresponds to 50% bootstrap support in the HGT-Detection algorithm); (b) 200 HGT (corresponds to 75% bootstrap support in the HGT-Detection algorithm); (c) 100 HGT (corresponds to 90% bootstrap support in the HGT-Detection algorithm).

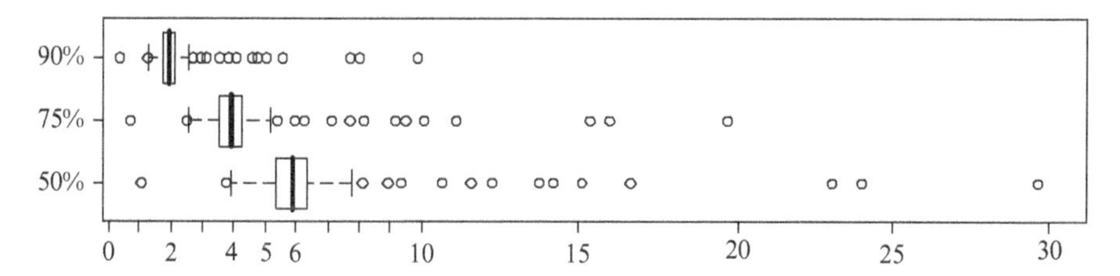

Figure 25.8 Distribution of the HGT-QFUNC maximum percentages of positive values chosen for prokaryotic data. The abscissa represents the percentage of the maximum possible number of HGTs between individual species. The ordinate represents the corresponding HGT-Detection bootstrap confidence level. Average values correspond to less than 6%, 4%, and 2% of the maximum possible number of HGTs for the 50%, 75%, and 90% bootstrap confidence levels, respectively.

HGT-QFUNC, based on the p-value ordering, are shown in Figure 25.4. Corresponding runs of HGT-QFUNC had a maximum allowed number of HGTs per alignment of 300, 200, and 100 events, respectively. How restrictive these thresholds are, in terms of possible number of detected HGT events for each considered gene, is shown in Figure 25.8.

25.4 RESULTS AND DISCUSSION

In this chapter, we have described a new algorithm for determining genomic regions that may be related to HGT and recombination, and introduced three new clustering functions Q_7, Q_{8a}, and Q_{8b}. We compared the performances of these functions to those yielded by a simple distance measure Q_9. All of the considered aggregation functions were tested on the real-life genomic data (see Figures 25.7 and 25.8) as well as on the synthetic data (see Figures 25.2 and 25.3).

25.4.1 Analysis of Synthetic Data

We also tested the detection sensitivity of our method for randomly generated HGTs between terminal tree branches using synthetically generated data and different levels of recombination. In the case of artificial data, the functions Q_7, Q_{8a}, and Q_{8b} provided better performances than the function Q_9 only when recombination was considered. The results obtained for the Q-function ordering are shown in Figure 25.2 (for the 25% recombination level in the left column and for the 50% recombination level in the right one). Figure 25.3 presents in the left panel the results for HGT with no recombination (i.e., 0%) and in the right panel, the limits of our simulation, where there is no difference between the functions involved.

The p-value-based ordering, with the threshold of 0.05, shows only minor improvements, especially when the number of simulated transfers is low (i.e., 2, 4, and 8) (see Figures 25.5 and 25.6). Specifically, for 1 simulated transfer we obtained an almost perfect detection rate, while for 128 transfers we obtained the worst HGT recovery rates that were around 40%. Thus, the following general trend can be observed: the higher number of transfers we have the lower detection rates are. Higher degrees of recombination also lead to a lower detection rate for all the functions, but favor the functions Q_{8a} and Q_{8b}, as their performance degrades less, especially in the middle range. The function Q_7, which showed very good performances for the real-life prokaryotic data, does not outperform the function Q_9 in this particular testing framework. It is important to note that even without recombination, the functions Q_{8a} and Q_{8b} can be also used as they yield almost the same detection rates as the function Q_9.

25.4.2 Analysis of Prokaryotic Data

For all the functions we introduced in this study, that is, Q_7, Q_{8a}, Q_{8b}, and Q_9, we can observe the following trend: better detection sensitivity corresponds to higher HGT bootstrap confidence thresholds. The function Q_7 always provided better results than Q_9, while Q_{8a} and Q_{8b} were better than Q_9 only for 75% and 90% bootstrap thresholds (based on the median values shown by a vertical black line on each of the boxes in Figure 25.7).

The p-value-based ordering, established with the threshold of 0.05, yields very good detection results for all of the tested functions Q_7, Q_{8a}, Q_{8b}, and Q_9 (see Figure 25.4). The functions Q_{8a} and Q_{8b} provided better results than Q_9 for the HGT detection threshold of 50% bootstrap support. Moreover, the presented results suggest that the function Q_{8a} is able to detect almost all of high confidence HGT (90% bootstrap support). The main differences can be observed in the tail of the distribution, for the lower 25% quartile, as the median and high quartile are already at the same maximum value (of 100%). It should be noted that for the 75% HGT detection threshold, which was our benchmark threshold throughout this study, the best average results were provided by the function Q_7.

One of the limitations of the HGT-QFUNC algorithm, compared to the HGT-Detection algorithm [6], is that our new algorithm imposes a fixed number of positive values (100, 200, or 300 in our case) regardless of the number of species in the given multiple sequence alignment. These constant values were selected in order to find on average less than 2%, 4%, and 6% of the maximum possible number of transfers between individual species for, respectively, 90%, 75%, and 50% bootstrap support levels adopted by the HGT-Detection algorithm (see Figure 25.8).

25.5 CONCLUSION

HGT is a well-structured evolutionary paradigm as the recent studies show higher levels of transfers between certain prokaryotic groups [3] or certain ecological habitats [19]. The impact of horizontal transfers on the evolution of many bacteria and viruses, as well as the cumulative effect of recombination over multiple generations, remains to be investigated in greater detail.

Despite the general availability of quality-controlling HGT detection methods based on complex phylogenetic analyses, simple distance measures can still be useful for recovering HGT events. The computational complexity of more precise HGT detection methods and the high volume of considered genomic data are the main motivations behind the development of fast and effective HGT detection algorithms.

In this chapter, we described a new fast HGT detection algorithm, HGT-QFUNC, which runs in quadratic time when HGTs between terminal branches are considered. It allows for an efficient parallel implementation. The discussed method also benefits from a Monte Carlo p-value validation procedure, obviously at the cost of the associated validation constant needed for maintaining precision. Because of its low time complexity, the new algorithm can be used in complex phylogenetic and genomic studies involving thousands of species.

Even though the presented method is designed to identify complete HGT, we investigated how it copes with partial HGT as well (i.e., HGT followed by the intragenic sequence recombination and formation of mosaic genes) and showed that in many cases it can be useful for detecting both complete and partial HGT. The new variability clustering functions Q_7, Q_{8a}, Q_{8b}, and Q_9 were introduced and tested in our simulations. Overall, the functions Q_7 and Q_{8a} provided the best HGT detection performances.

REFERENCES

1. Badescu D, Boc A, Diallo AB, Makarenkov V. Detecting genomic regions associated with a disease using variability functions and adjusted rand index. BMC Bioinformatics 2011;12 Suppl 9:S9.
2. Badescu D, Diallo AB, Makarenkov V. Identification of specific genomic regions responsible for the invasivity of Neisseria Meningitidis. *Classification as a Tool for Research.* Springer-Verlag Berlin Heidelberg; 2010. p 491–499.
3. Beiko RG, Harlow TJ, Ragan MA. Highways of gene sharing in prokaryotes. Proc Natl Acad Sci U S A 2005;102(40):14332–14337.
4. Benson DA, Karsch-Mizrachi I, Lipman DJ, Ostell J, Sayers EW. Genbank. Nucleic Acids Res 2009;37 Suppl 1:D26–D31.
5. Boc A, Diallo AB, Makarenkov V. T-REX: a web server for inferring, validating and visualizing phylogenetic trees and networks. Nucleic Acids Res 2012;40(W1):W573–W579.
6. Boc A, Philippe H, Makarenkov V. Inferring and validating horizontal gene transfer events using bipartition dissimilarity. Syst Biol 2010;59(2):195–211.
7. Castresana J. Selection of conserved blocks from multiple alignments for their use in phylogenetic analysis. Mol Biol Evol 2000;17(4):540–552.
8. Edgar RC. MUSCLE: multiple sequence alignment with high accuracy and high throughput. Nucleic Acids Res 2004;32(5):1792–1797.
9. Enright AJ, Van Dongen S, Ouzounis CA. An efficient algorithm for large-scale detection of protein families. Nucleic Acids Res 2002;30(7):1575–1584.
10. Fraser C, Hanage WP, Spratt BG. Recombination and the nature of bacterial speciation. Science 2007;315(5811):476–480.

11. Hanage WP, Fraser C, Spratt BG. Fuzzy species among recombinogenic bacteria. BMC Biol 2005;3(1):6.
12. Nikoh N, Koga R, Ross L, Duncan RP, Fujie M, Tanaka M, Satoh N, Bachtrog D, Wilson ACC, von Dohlen CD, Fukatsu T, Husnik F, McCutcheon JP. Horizontal gene transfer from diverse bacteria to an insect genome enables a tripartite nested mealybug symbiosis. Cell 2013;153(7):1567–1578.
13. Kimura M. *The Neutral Theory of Molecular Evolution*. Cambridge University Press; Cambridge; 1984.
14. Li J, Dai X, Liu T, Zhao PX. LegumeIP: an integrative database for comparative genomics and transcriptomics of model legumes. Nucleic Acids Res 2012;40(D1):D1221–D1229.
15. Moran PAP. *The Statistical Processes of Evolutionary Theory*. Oxford: Clarendon Press; 1962.
16. Ochman H, Lawrence JG, Groisman EA. Lateral gene transfer and the nature of bacterial innovation. Nature 2000;405(6784):299–304.
17. Posada D, Crandall KA. Evaluation of methods for detecting recombination from DNA sequences: computer simulations. Proc Natl Acad Sci U S A 2001;98(24):13757–13762.
18. Chen WH, Ternes CM, Barbier GG, Shrestha RP, Stanke M, Bräutigam A, Baker BJ, Banfield JF, Garavito RM, Carr K, Wilkerson C, Rensing SA, Gagneul D, Dickenson NE, Oesterhelt C, Lercher MJ, Schönknecht G, Weber AP. Gene transfer from bacteria and archaea facilitated evolution of an extremophilic eukaryote. Science 2013;339(6124):1207–1210.
19. Smillie CS, Smith MB, Friedman J, Cordero OX, David LA, Alm EJ. Ecology drives a global network of gene exchange connecting the human microbiome. Nature 2011;480(7376):241–244.
20. Stamatakis A. RAxML-VI-HPC: maximum likelihood-based phylogenetic analyses with thousands of taxa and mixed models. Bioinformatics 2006;22(21):2688–2690.
21. Thomas CM, Nielsen KM. Mechanisms of, and barriers to, horizontal gene transfer between bacteria. Nat Rev Microbiol 2005;3(9):711–721.
22. Van Dongen SM. Graph clustering via a discrete uncoupling process. SIAM J Matrix Anal Appl 2008;30(1):121–141.

PART VII

PATTERN RECOGNITION IN BIOLOGICAL NETWORKS

CHAPTER 26

COMPUTATIONAL METHODS FOR MODELING BIOLOGICAL INTERACTION NETWORKS

Christos Makris[1] and Evangelos Theodoridis[2]

[1]*Department of Computer Engineering and Informatics, University of Patras, Patras, Greece*
[2]*Computer Technology Institute and Press "Diophantus", University of Patras, Patras, Greece*

26.1 INTRODUCTION

The chapter focuses on the representation of computational methods for partitioning and modeling biological networks, giving special emphasis to applications on *Protein–Protein Interaction* (PPI) networks. We smoothly introduce the reader to various notions such as different types of biological interaction networks and fundamental measures and metrics used in most methods of biological interaction networks. Finally, we focus on basic concepts of reconstructing and partitioning biological networks along with developments in PPI databases and PPI prediction algorithms.

Graphs appear in a large number of applications in Bioinformatics and Computation Molecular Biology as they can be used to model interactions between the various macromolecules and, in particular, they can be used to model PPIs and metabolic pathways or for forming transcriptional regulatory complexes. Networks are basically represented by the mathematical notion of the underlying graph. There are different

Pattern Recognition in Computational Molecular Biology: Techniques and Approaches,
First Edition. Edited by Mourad Elloumi, Costas S. Iliopoulos, Jason T. L. Wang, and Albert Y. Zomaya.

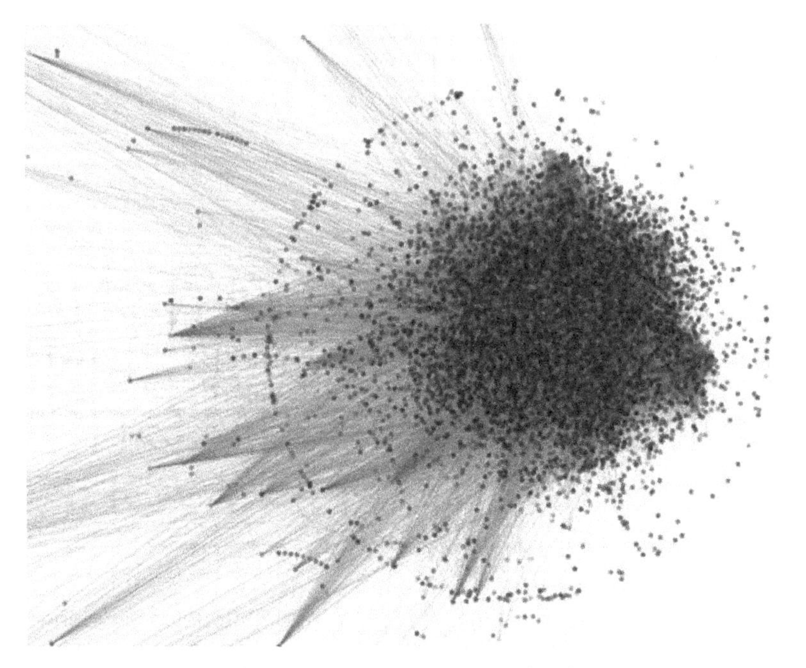

Figure 26.1 A PPI network. *Source:* https://en.wikipedia.org/wiki/Protein%E2%80%93protein_interaction#/media/File:Human_interactome.jpg. CC BY-SA 3.0.

kinds of networks/graphs such as directed, undirected, weighted, and unweighted. The basic instantiations of biological networks are [22] PPI networks, metabolic networks, transcriptional regulatory networks, and, in general, networks modeling genetic information (e.g., Figures 26.1 and 26.2).

The rest of the chapter is organized as follows. In Section 26.2, we introduce a set of measures and metrics that can structurally characterize a network. In Section 26.3, we describe the basic models of biological networks, while in Section 26.4, the basic techniques for reconstructing and partitioning biological networks are presented. In Section 26.5, we refer to techniques for efficiently handling PPI networks, and in Section 26.6, we refer to algorithms for mining PPI networks; finally, in Section 26.7 we present our conclusions.

26.2 MEASURES/METRICS

There are a set of measures and metrics that can structurally characterize a network [9], Chapter 2, the most significant being the node degree distribution, the node degree correlation, the shortest path length, the clustering coefficient, the assortativity, reciprocity, centrality measures (betweenness centrality, closeness centrality, degree centrality, eigenvector centrality), and modularity.

The *node degree* is defined as the number of edges a node has. The distribution of degrees across all nodes usually is used in order to model the number of edges the

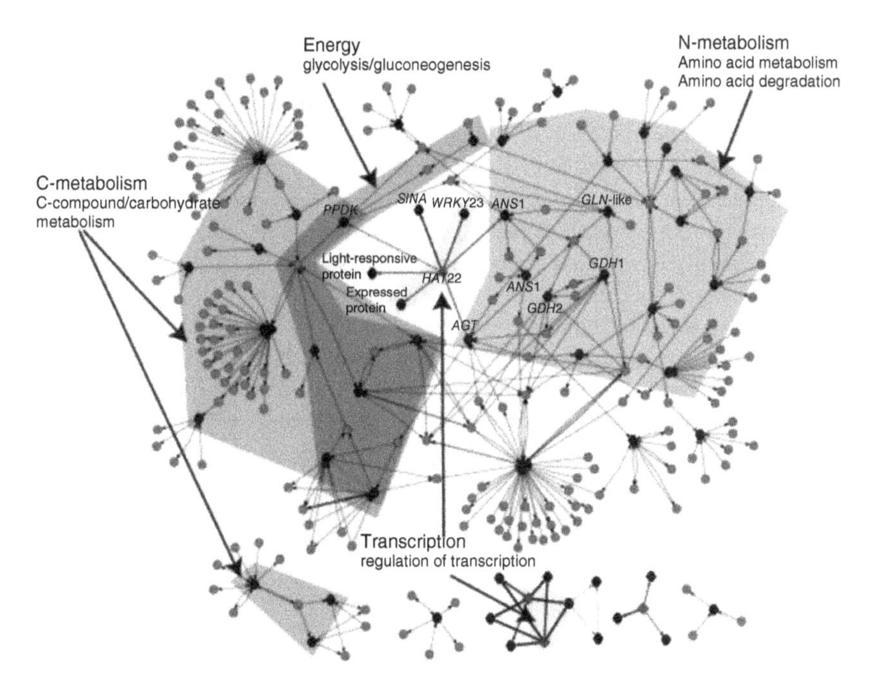

Figure 26.2 A metabolic network [90].

various nodes in the network have. On the basis of this distribution, a network can be distinguished as (i) a *scale-free* network (where degree distribution follows a power law distribution), (ii) a *broad-scale* network (where the degree distribution follows a power law distribution that has a sharp cut off at its tail), and (iii) *single-scale* network (where the degree distribution is decaying fast). The degree correlation measures the expected degree of the neighbors of a node and it is connected with the notion of *assortativity*.

As regards shortest paths, these networks usually use the simple geodesic distance (the length of the shortest path), the average shortest path, the diameter of the network, which is defined as the maximum distance between two nodes. The clustering coefficient measures the density of the edges in the neighbors of a node and it is defined as the number of edges among the neighbors divided by the total number of edges among the neighbors. The network average clustering coefficient is the average of the local clustering coefficients. The clustering coefficient can be used to characterize a network as small if the average clustering coefficient is significantly higher than a random graph constructed on the same vertex set, and if the graph has approximately the same mean-shortest path length as the corresponding random graph. Related notions are the global clustering coefficient that is based on triplets of nodes, and the transitivity ratios that mainly give higher weights higher degree nodes.

It is possible to characterize the relationship between the degrees of nodes by employing the assortative coefficient. Assortativity depicts the tendency of nodes to attach to others that are similar in some way, and is defined as the Pearson correlation coefficient [64] of degree between pairs of linked nodes. Positive values of the metric indicate that the network depicts correlation between nodes of high degree, while negative values indicate a higher dependencies between nodes of different degrees. Biological networks generally depict negative values, while social and technological networks generally have high positive values of assortativity.

Reciprocity is defined as the ratio of the number of links pointing in both directions in a network to the total number of links; when its value is equal to 1 we have a completely bidirectional network, while for a completely unidirectional network, it is equal to 0. Moreover, in Reference [26], a new measure of reciprocity is proposed that defines it as the correlation coefficient between the entries of the adjacency matrix of a real network. As the authors of the publication point out, their findings show that, using this metric, real networks are either negatively or positively correlated, while networks of the same type (ecological, social, etc.) display similar values of the reciprocity.

Another very popular metric is centrality and its various variants. The various notions of centrality are of particular interest:

- The *degree centrality* characterizes the compactness of link presence in various nodes and is essentially equal to the degree of the node (in-degree and out-degree).
- The *closeness centrality* is based on the shortest path length between two nodes. It is essentially the inverse of the farness, which is defined as sum of distances of a node to the other nodes.
- The *betweenness centrality*, as in the case of closeness centrality, is based on the shortest path between nodes, and, in particular, it quantifies the number of times a node acts as a bridge along the shortest path between two other nodes.
- The *eigenvector centrality* is a more abstract version of degree centrality. The degree centrality is only based on the number of neighbors; however, the eigenvector centrality can consider the centralities of neighbors.
- The *modularity* characterizes how well a partition affects the characteristic of a network. In particular, modularity can be used in many techniques that optimize modularity directly including greedy techniques, simulated annealing, external optimization, and spectral optimization. The modularity of a partition is a scalar value between -1 and 1 that measures the density of links inside communities as compared to links between communities (see also Reference [12]).

The measures defined above cannot be used as metrics because they do not uniquely identify the structure of a graph. In order to fill that gap, a new metric was proposed [9], Chapter 6 that could be used to uniquely identify graph structures. The new metric, the so called *Weighted Spectral Distribution* (WSD) compares graphs based on the distribution of a decomposition of their structure. It employs spectral properties of the normalized Laplacian matrix and it is defined as the normalized sum of N-cycles. Since the computation is expensive, the metric is computed using pivoting and *Sylvester's Law of Inertia* [89]. Moreover, it is tightly connected to the clustering coefficient.

26.3 MODELS OF BIOLOGICAL NETWORKS

There are a number of models that have been proposed to model biological networks; however, these models fail to efficiently capture the performance of networks. The major models that have been used are *Random Networks* and the *Barabasi Albers Network* (BA), whichis useful in modeling scale-free networks. They represent a model that was proposed by Erdos and Renyi, which is quite popular as it is the main theoretical vehicle in the study of the practical and theoretical properties of real graphs. According to the random graph model, a network is modeled as a set of n vertices with edges appearing between each pair of them with probability equal to p. As remarked in Reference [32], the random graph model basically studies ensembles of graphs; an interesting aspect of its properties is the existence of a giant component.

Despite their popularity, random networks fail to capture the behavior of scale-free networks that are basically governed by power law distributions. These graphs appear in bioinformatics and related applications and they can be handled by a variety of models, the most popular being the Barabasi–Albert model. This model is based on the assumption that a network evolves with the addition of new vertices and that the new vertices are connected to the previous vertices according to their degree. Despite the fact that the BA model explains scale-free properties, there exist variations that try to suit better biological explanations. In Ref. [9], Chapter 3, a set of new models is presented that better capture the evolvement of biological networks. The authors initially present a simple unified model that evolves by using simple rules: (i) it begins with small-sized complete graph. (ii) Every time, the network evolves in size by merging a module of the same size to the graph. (iii) When merging the modules, the rule is to select nodes according to their degrees along with the so-called *fitness of nodes*. The fitness is related to studies based on competition dynamics and weighted networks. The authors also present two specific models, a metabolic network model and bipartite network model that is useful in modeling real bipartite relationships.

26.4 RECONSTRUCTING AND PARTITIONING BIOLOGICAL NETWORKS

In the following, we will present the main algorithmic concepts for reconstructing biological networks based mainly on the descriptions provided in References [87], Chapter 27 and [9], Chapter 1. In Ref. [9], a set of approaches for reconstructing directed and undirected networks from biological data is presented, while in Reference [87] four different popular modeling frameworks are presented.

Biological data can appear in various forms and from various data sources, but they can easily be discretized. They basically appear in the form of gene microarrays and gene sets. Gene microarrays contain the expression of thousands of genes, while gene sets contain data describing common characteristics for a set of genes. The types of networks that can be inferred [9], Chapter 1 are Gaussian graphical models, Boolean networks, probabilistic boolean networks, differential equation-based networks, mutual information networks, collaborative graph model, frequency method, network inference, and so on. In Ref. [87], these were separated into the following categories, graph models, boolean networks, Bayesian networks, and differential equation models.

In graph theoretical models, the networks are represented by a *graph structure* where the nodes represent the various biological elements and the edges represent the relationships between them. When reconstructing graph models, the process entails the identification of genes and their relationships. Special cases of networks that are worth the effort to deal with are co-expression networks and the collaborative graph model. In co-expression networks, edges designate the co-expression of various genes, while in the collaborative graph model [44] a weight matrix is employed that uses weighted counts in order to estimate and qualitatively designate the connections between the genes.

A *boolean network* consists of a set of variables (that are mapped to genes) and a set of boolean functions on these variables that are mapped to the variables. Each gene can be on (1) or off (0), and the boolean network evolves with time, as the values of the variables at time t designates the values at time $t + 1$. Boolean networks were firstly introduced by Kaufman [39, 40] and have found a maximum applications in molecular biology. Probabilistic boolean networks are an extension of boolean networks; they are equipped with probabilities and are able to model uncertainties by utilizing the probabilistic setting of Markov chains. For a nice introduction to this area, see Reference [80].

Another graph model that is used when formulating biological problems is Bayesian networks, a formalism that has been fruitfully exploited [37] in Computer Science. Bayesian networks are probabilistic graphical models that can model evidence and with the chain rule calculate probabilities for the handling of various events. More typically, a Bayesian network consists of a set of variables, a set of directed edges that form a *Directed Acyclic Graph*, and to each variable a table of conditional probabilities obtained from its parents. Bayesian networks have been employed in a variety of biological applications, the most prominent being the application presented in References [1, 10, 38, 54, 57, 63, 71, 94, 97].

The last model (differential equation models) is based in the employment of a set of differential equations that are elegantly defined in order to model complex dynamics between gene expressions. In particular, these equations express the rates of changes of gene expressions as a function of expressions of other genes and they can incorporate many other parameters. The main representative of these models are *Linear Additive Models* where the corresponding functions are linear; details about these models can be found in Reference [87], Chapter 27.

There is a lot of work from different areas for creating communities from graphs. The problem is related to social algorithms, web searching algorithms, and bibliometrics; for a thorough review of this area, one should go through References [16, 24, 25, 46, 59, 74].

In particular, with regard to community detection, various algorithms have been proposed in the literature. A breakthrough in the area is the algorithm proposed in Reference [27], for identifying the edges lying between communities and their successive removal, a procedure that after some iterations leads to the isolation of the communities [27]. The majority of the algorithms proposed in the area are based on spectral partitioning techniques, that is, techniques that partition objects by using the eigenvectors of matrices that are form in the specific set [41, 60, 77, 78]. One should also mention techniques that are using modularity, a metric that designates the density of links inside communities against the density outside communities [24, 58] with the most popular being the algorithm proposed in Reference [12].

The algorithm is based on a bottom-up approach where initially all nodes belong to a distinct community and at each stage we examine all the neighbors j of a node i and change the i community of nodes by examining the gain of modularity we could attain by putting i in the j community. From all the neighbors, we choose the one that depicts the greater gain in modularity and if the modularity is positive, we perform the exchange. This stage of the algorithm stops when no gain of modularity can be further achieved, that is, a local maxima has been attained. In the second phase, a new network is formed with nodes from the aforementioned communities, with edge weights the sum of the weight of the links between nodes in the corresponding two communities. The process is reapplied, producing a hierarchy and the height of the hierarchy that is constructed is determined by the number of passes and is generally a small number.

Besides finding emerging communities, estimating authorities has also attracted attention. When a node (considered as valuable and informative) is usually pointed to by a large number of hyperlinks (so it has large in-degree), it is called an *authority*. A node that points to many authority nodes is itself a useful resource and is called a *hub*. The field is related to link analysis in the web with the PageRank citation metric [62], the HITS algorithm proposed by Kleinberg [43], and their variants proposed [45] as cornerstones.

PageRank is a simple metric based on the importance of the incoming links, while HITS uses two metrics, emphasizing the dual role of a web page as a hub and as an authority for information. Both metrics have been improved in various forms and a review of this can be found in Reference [45]. Note that if non-principal eigenvectors are being explored, HITS can be used in order to compute communities.

26.5 PPI NETWORKS

The application of various biological experiments has led to the discovery of many biochemical interactions, such as interactions among proteins. The PPIs are responsible for the signal transduction among cells. In this way, cells manage subcellular processes. Some proteins interact for a long period of time, forming units called *protein complexes*, whereas other proteins interact for a short time, when they spatially contact each other. There is a vast amount of existing proteins that vary in different species, so the number of possible interactions is significantly big. Many verified protein interactions are stored in repositories and are used as knowledge base for predicting protein interactions. Thus, there are two large groups of recorded PPIs, the verified ones and *in silico* predicted ones, using predicting algorithms [14].

Owing to great interest of protein interactions, a large amount of PPI data have been accumulated and stored in databases (usually implemented with *Relational Database Management Systems*—RDBMS). Examples of active PPI databases are the *Biomolecular Interaction Network Database* (BIND) [7], the *Database of Interacting Proteins* (DIP) [73], the *Munich Information Center for Protein Sequence* (MIPS) [53], the *Molecular Interaction Database* (MINT) [17], the *Human Protein Reference Database* (HPRD) [42], and others. Each one of these databases focuses on a different kind of interactions either by targeting small-scale or large-scale detection experiments, by targeting different organisms, or humans only, and so on. As result, there exist various PPI databases, categorized on the basis of the kind of interactions they have or the methodology used to collect and curate information [14]. All of them usually offer downloading of their data sets or custom web graphical user interfaces for retrieving information. Building a list of all of an organism's PPIs can be useful for the molecular-level understanding of the core of complex traits. Therefore, the collection of all PPIs can be important for understanding the underlying mechanisms of diseases, facilitating the process of drug design, elucidating the functions of newly identified proteins, predicting their subcellular location, and gaining insight into the evolution of some interaction or metabolic pathways, among other biological aspects of a cell or organism [93].

The most common representation of PPI data sets is by large interaction networks, known as *Protein Interaction Networks* (PINs) (Figure 26.3). The common representation of these networks is by using graphs where at each node and edge there is a large number of attributes expressing properties of proteins (nodes) and of interactions (edges), respectively. Managing and mining large graphs introduce significant computational and storage challenges [75]. As a result, issues arise about the management and processing of PPI data. Moreover, most of the databases are maintained and updated in different sites on web. Concerning these distributed data sources, Semantic Web technologies, such as RDF,[1] propose an attractive solution for the integration of the remote data sets and for the interconnection and alignment of entities among these different data sets. In this direction, linked data[2] is a paradigm of exposing, sharing, and connecting data via dereferenceable URIs on the Web. This is done by exposing *Resource Description Framework* (RDF) files. RDF is a data conceptual model which is based on the idea of making statements about resources in the form of subject–predicate–object expressions, forming a directed, labeled graph. This implies, that RDF could be an interesting, alternative solution for the management of

[1] http://www.w3.org/RDF/.
[2] http://linkeddata.org/.

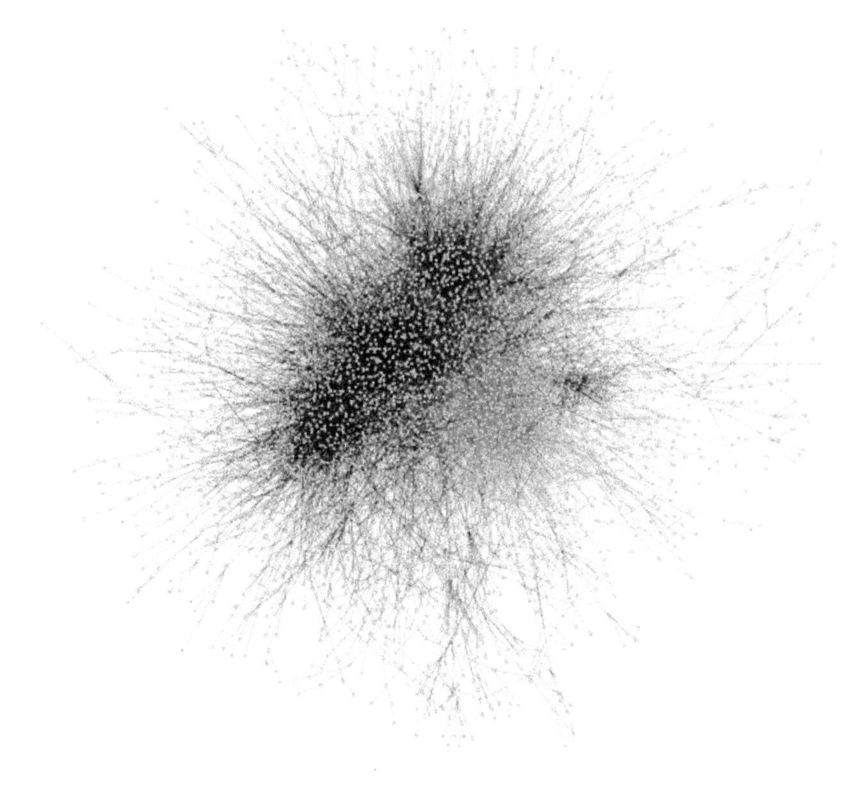

Figure 26.3 Human PPI network. Proteins are shown as graph nodes. [35].

PPI data compared to RDBMS. Furthermore, RDF modeling facilitates keeping data in more flexible schema formats, allowing extensions of the data schema, annotations and alignment of data to other ontologies and data sets.

In the case of developing a PPI data management system, either a website for public access, a desktop visualization application, or a PPI graph mining application can be usen; the crucial question that arises is which database management system (RDBMS or RDF storage) is the most appropriate for data storage. On one hand RDF storages seem to provide flexibility, enhanced functionalities, easy interconnection with WWW, and on the other hand, RDBMS, although they are more strict, are more mature as software frameworks and more optimized to certain query types.

The majority of PPI databases harvest PPIs from the scientific literature, mainly with manual curation or with manual curation supported by text-mining tools. Most of them contain interactions experimentally detected in lab, while other databases contain predicted protein interactions. Curators of the databases review the published articles and according to various criteria (e.g., type of the experiment and detection method, scale of the experiment.), they acknowledge a PPI and make an entry in the data set. The process of curating protein interactions is very time-consuming and demands a great amount of research. DIP database [73], contains experimentally verified interactions in different organisms. Data are entered manually and it offers the user the ability to make queries through a web-based interface or download a subset of data in various formats. BIND [7], contains protein interactions with molecular function annotations, according to literature. It offers various modes of search and the extracted information can be depicted using the BIND interaction viewer, where networks are represented as graphs, labeled with ontological information. MINT [17], contains physical interactions among proteins and other molecules, which are curated by experts. The user can make queries through a web-based interface and results are depicted in an interactive table. Moreover, MIPS [53] consists of various databases of genomic data. It supports the PSI-MI Standard[3] both for downloading and uploading data, and queries can be made through a web-based interface. Finally, the HPRD [42] contains manually curated proteomic information, mostly human. Protein information annotations are curated by expert biologists using published literature. The data are freely available to academic users and information can be retrieved through a web user interface.

Recently, some systems have been developed for the management of PPI data. In Ref. [15], authors have presented an application server for managing PPI data using RDF, based on the D2RQ[4] framework. This framework merges existing PPI data and annotations extracted from Gene Ontology[5] and stores the data in a relational database. The resulting PPI database, contains for each protein, the protein identifier and the annotations from Gene Ontology including biological process, molecular function, and cellular component. Using the D2RQ platform, the users gain access to the database through RDF graphs. The system supports querying based on a protein identifier or an annotation (Key-Based Querying), through a semantic-based interface. Moreover, in References [91, 92], the authors proposed a PPI unification system in the field of human proteome. The system mines PPI data from databases such as HPRD, INTACT, MINT, DIP, and BioGRID and after appropriate filtering, the PPIs are normalized to UniprotKB protein identifier level, determining the extent to which human proteome is covered by the interactome. Advanced data transformation methods Extract Tranform Load (ETL) were developed for this purpose. Finally, in Reference [67] a practical evaluation was presented assesing the performance of RDF and RDMS systems for storing and quering PPI data sets.

The aggregation of remote web data sources would be a challenging task. Producing links and connections between different databases is crucial for enriching and extending knowledge. This is a common problem in Semantic Web area. Graph databases, such as RDF storages, could offer this type of functionalities. In graph databases, schema and data are represented as graphs and data are managed using graph-based operations [5]. Consequently, representing a PPI network as graph could enable the use of graph-based algorithms for the detection of biological properties. PPI data querying, could also benefit from this approach, as users are facilitated to

[3]http://www.psidev.info/.
[4]http://d2rq.org/.
[5]http://www.geneontology.org/.

make graph queries easily. Nevertheless, there is no standard for the physical organization of large graph data. Many graph algorithms are not applicable in very large graphs because they rely on linear indexes [75]. This implies that the selection of a graph storage system should be made carefully according to the size and requirements of the corresponding graph.

26.6 MINING PPI NETWORKS—INTERACTION PREDICTION

Interactions present and captured in the aforementioned public databases, record only a small fraction of the real PPI networks [30, 31]. Physical PPIs can be identified using experimental methods, such as the yeast two-hybrid assay, mass spectrometry, protein microarrays, phage display, X-ray crystallography, fluorescence resonance energy transfer, surface plasmon resonance, atomic force microscopy, and electron microscopy [81]. However, these experiments, in addition to being expensive and time demanding, are not suitable for all proteins and do not report all interactions that can occur in cells or organisms [30, 81, 83].

Therefore, computational approaches capable of reliably predicting PPIs can identify the potential interactions to be further interrogated using experimental approaches. Graph or network theory has been used to model complex interaction systems either modeling social or biological phenomena [20, 47, 50] providing good abstractions and holistic views of the phenomena. In biological networks, the nodes represent biological entities, such as proteins and edges represent physical or functional interactions among these biological entities. These networks have shown to be useful for describing the behavior of systems and their construction can significantly reduce the costs of experimental identification of novel interactions. In most of the cases, the predictions are based on similarity-based algorithms, maximum likelihood methods, or probabilistic methods. Typically, applications of these algorithms concern the reconstruction of networks, evaluation of network-evolving mechanism, and classification of partially labeled networks [50]. In the molecular networks, the links can be predicted on the basis of the node features (e.g., protein sequences or domains), by the similarities of edge neighborhood, by comparisons to an appropriated model, and by network topology. Proteins can interact with many other biomolecules in cells, such as DNA, RNA, metabolites, and other proteins.

There are quite a few computational methods that provide complementary information for experimental approaches and try to predict interactions between proteins. Some of these methods use not only sequence information but also localization data [65], structural data [2, 3, 61], expression data [51, 86], and interactions from orthologs [23, 33] and phylogenetic profile analysis. See Reference [82] for an extensive review of these methods. Furthermore, gene co-expression profiles, sequence co-evolution, synthetic lethality data, *in silico* two-hybrid systems, gene cluster and gene neighbor analysis, protein domain information [82, 83], protein interface analysis, protein docking methods [6, 84], and orthology and ontology information [21] are being used.

The main limitation of these methods is that they rely on previously recorded knowledge that might be difficult to access. There are quite a few approaches in the literature that try to detect potential interactions of proteins using no secondary sources of information [11, 13, 19, 29, 49, 52, 56, 76, 88, 95]. Most of these methods transform protein sequences into feature vectors and adopt *Machine Learning* (ML) techniques to analyze these feature vectors. In Reference [55], a method based on codon usage has been proposed, which utilizes DNA sequence for feature extraction, utilizing *Open Reading Frame* (ORF) information. The machine learning methods commonly used consist of random decision forests [19] and *Support Vector Machine*s (SVMs) [11, 29, 52, 76, 95]. These methods usually perform well when the input data sets consist of equal number of interacting and noninteracting protein pairs. However, this ratio is not balanced in nature, and these methods were not comprehensively evaluated with respect to the effect of the large number of noninteracting pairs in naturally unbalanced data sets [8, 85]. Recently, in Reference [95], the authors have tried to deal with unbalanced data also using the primary structure of proteins. Moreover, several data sources can be used as examples for training several ML algorithms in methods of classification [19, 34, 82, 95, 96, 99]. In addition, it is possible to combine more than one method and data set to predict protein interaction as presented in References [11, 36, 70, 82, 98].

The aforementioned methods use data sets that are not available for all species or all proteins. For example, some methods use information from complete genome sequencing, gene-expression data, or protein information that is not available for all sequences. Moreover, some methods have some limitation as in the case where gene expression cannot determine physical PPIs; molecular evolutionary events can disturb the prediction of PPIs in phylogenetic profile methods; functional relationships based on genomic organization are only suitable for prokaryotes; some methods are based on sequence homology, which is more difficult to determine than finding genes; and other computational and biological aspects may render the prediction very difficult or do not provide concrete results in many cases [18, 66, 83]. See also Reference [93]. The primary protein structure, not considering domains, motifs, ontology, phylogenetic trees, and orthology, has sufficient information to estimate the propensity for PPIs [4, 20]. Moreover, amino acid sequences are the most universal protein feature and thus appear to be the most appropriate information for methods of predicting PPIs that are applicable to all proteins [76]. Many methods, as mentioned before, using primary sequences have been developed lately using machine learning approaches [28, 29, 48, 49, 52, 56, 57, 68, 69, 72, 76, 79, 95, 96, 99]. However, most of the methods based on machine learning and primary sequences have weaknesses, such as the building of negative data sets, the small number of examples, and the large number of attributes (vectors) that code protein pairs.

In Reference [93], the authors present a predictor of PPIs. This method is also a machine learning based approach that uses features associated with amino acid sequences for building models for predicting PPIs. It is a probabilistic model trying to deal with negative training examples used to construct the training data sets that are all experimentally supported no-PPIs; the amount and diversity of PPIs and no-PPIs used as training examples are much higher than some of those used by other methods; and with increased computational cost. Finally, the produced models are decision trees that can be easily analyzed and the most important features capable of distinguishing PPIs from no-PPIs can be promptly identified.

26.7 CONCLUSIONS

Graphs and Networks have been widely adopted in a large number of research problems in Bioinformatics and Computation Molecular Biology, as they efficiently model the interactions between various macromolecules; in particular, they can be used to model PPIs, metabolic pathways, or to form transcriptional regulatory complexes. Networks are strongly connected with the mathematical notion of the underlying graph, but in some applications one can also use this notion for the set of objects in the underlying network. The basic instantiations of biological networks are PPI networks, metabolic networks, transcriptional regulatory networks, and genetic networks. There are quite a few data set and databases that represent biological networks and record either actual interactions between biological entities or predicted ones.

In this chapter, we have briefly reviewed the various types of biological interaction networks and the fundamental measures and metrics used in most data mining methods of the biological interaction networks. Further, we have presented the basic algorithmic concepts of reconstructing and partitioning biological networks along with current developments in protein–protein interaction databases and PPI prediction algorithms.

REFERENCES

1. Agichtein E, Castillo C, Donato D, Gionis A, Mishne G. Finding high-quality content in social media. Proceedings of the 2008 International Conference on Web Search and Data Mining, WSDM '08. New York: ACM; 2008. p 183–194.
2. Aloy P, Russell RB. Interrogating protein interaction networks through structural biology. Proc Natl Acad Sci U S A 2002;99(9):5896–5901.
3. Aloy P, Russell RB. InterPreTS: protein interaction prediction through tertiary structure. Bioinformatics 2003;19(1):161–162.
4. Anfinsen CB. Principles that govern the folding of protein chains. Science 1973;181(4096):223–230.
5. Angles R, Gutierrez C. Survey of graph database models. ACM Comput Surv 2008;40(1):1–39.
6. Aytuna AS, Gürsoy A, Keskin O. Prediction of protein-protein interactions by combining structure and sequence conservation in protein interfaces. Bioinformatics 2005;21(12):2850–2855.
7. Bader GD, Donaldson I, Wolting C, Ouellette BF, Pawson T, Hogue CW. BIND–the biomolecular interaction network database. Nucleic Acids Res 2001;29(1):242–245.
8. Bader GD, Hogue CW. Analyzing yeast protein-protein interaction data obtained from different sources. Nat Biotechnol 2002;20(10):991–997.

9. Basak S, Dehmer M. *Statistical and Machine Lerning Approaches for Network Analyses.* New Jersey: John Wiley and Sons; 2012.
10. Beal MJ, Falciani F, Ghahramani Z, Rangel C, Wild DL. A Bayesian approach to reconstructing genetic regulatory networks with hidden factors. Bioinformatics 2005;21(3):349–356.
11. Ben-Hur A, Noble WS. Kernel methods for predicting protein-protein interactions. ISMB (Supplement of Bioinformatics); Detroit, MI 2005. p 38–46.
12. Blondel VD, Guillaume J-L, Lambiotte R, Lefebvre E. Fast unfolding of communities in large networks. J Stat Mech Theory Exp 2008;2008(10):P10008.
13. Cai YD, Chou KC. Predicting protein-protein interactions from sequences in a hybridization space. J Proteome Res 2006;5(2):316–322.
14. Cannataro M, Guzzi PH, Veltri P. Protein-to-protein interactions: technologies, databases, and algorithms. ACM Comput Surv 2010;43(1):1:1–1:36.
15. Cannataro M, Guzzi PH, Veltri P. Using RDF for managing protein-protein interaction data. Proceedings of the 1st ACM International Conference on Bioinformatics and Computational Biology, BCB '10. New York: ACM; 2010. p 664–670.
16. Carrington PJ, Scott J, Wasserman S, editors. *Models and Methods in Social Network Analysis.* Cambridge: Cambridge University Press; 2005.
17. Chatr-Aryamontri A, Ceol A, Montecchi-Palazzi L, Nardelli G, Schneider MV, Castagnoli L, Cesareni G. MINT: the molecular interaction database. Nucleic Acids Res 2007;35(Database-Issue):572–574.
18. Chen X-W, Jeong JC. Sequence-based prediction of protein interaction sites with an integrative method. Bioinformatics 2009;25(5):585–591.
19. Chen X-W, Liu M. Prediction of protein-protein interactions using random decision forest framework. Bioinformatics 2005;21(24):4394–4400.
20. Csermely P, Korcsmáros T, Kiss HJM, London G, Nussinov R. Structure and dynamics of molecular networks: a novel paradigm of drug discovery. A comprehensive review; 2012. CoRR, abs/1210.0330.
21. De Bodt S, Proost S, Vandepoele K, Rouzé P, de Peer YV. Predicting protein-protein interactions in Arabidopsis thaliana through integration of orthology, gene ontology and co-expression. BMC Genomics 2009;10(1):288.
22. Elloumi M, Zomaya A. *Algorithms in Computational Molecular Biology, Techniques, Approaches and Applications.* New Jersey: John Wiley and Sons, Inc.; 2011.
23. Espadaler J, Romero-Isart O, Jackson RM, Oliva B. Prediction of protein-protein interactions using distant conservation of sequence patterns and structure relationships. Bioinformatics 2005;21(16):3360–3368.
24. Fortunato S. Community detection in graphs. Phys Rep 2010;486(3-5):75–174.
25. Fortunato S, Lancichinetti A. Community detection algorithms: a comparative analysis: invited presentation, extended abstract. Proceedings of the 4th International ICST Conference on Performance Evaluation Methodologies and Tools, VALUETOOLS '09, ICST. Brussels, Belgium: Institute for Computer Sciences, Social-Informatics and Telecommunications Engineering; 2009. p 27:1–27:2.
26. Garlaschelli D, Loffredo MI. Patterns of link reciprocity in directed networks. Phys Rev Lett 2004;93:268701.
27. Girvan M, Newman MEJ. Community structure in social and biological networks. Proc Natl Acad Sci U S A 2002;99(12):7821–7826.

28. Guo Y, Li M, Pu X, Li G, Guang X, Xiong W, Li J. PRED_PPI: a server for predicting protein-protein interactions based on sequence data with probability assignment. BMC Res Notes 2010;3(1):145.
29. Guo Y, Yu L, Wen Z, Li M. Using support vector machine combined with auto covariance to predict protein-protein interactions from protein sequences. Nucleic Acids Res 2008;36(9):3025–3030.
30. Han J-D, Dupuy D, Bertin N, Cusick ME, Vidal M. Effect of sampling on topology predictions of protein-protein interaction networks. Nat Biotechnol 2005;23(7):839–844.
31. Hart GT, Ramani AK, Marcotte EM. How complete are current yeast and human protein-interaction networks? Genome Biol 2006;7(11):120.
32. Havlin S, Cohen R. *Complex Networks Structure, robustness and Function.* Cambridge UK: Cambridge University Press; 2010.
33. Huang T-W, Tien A-C, Huang W-S, Lee Y-C, Peng C-L, Tseng H-H, Kao C-Y, Huang C-Y. POINT: a database for the prediction of protein-protein interactions based on the orthologous interactome. Bioinformatics 2004;20(17):3273–3276.
34. Hue M, Riffle M, Vert JP, Noble WS. Large-scale prediction of protein-protein interactions from structures. BMC Bioinformatics 2010;11:144.
35. James FE Jr. Question & Answer Q& A: systems biology. J Biol 2009;8:Article 2.
36. Jansen R, Yu H, Greenbaum D, Kluger Y, Krogan NJ, Chung S, Emili A, Snyder M, Greenblatt JF, Gerstein M. A Bayesian networks approach for predicting protein-protein interactions from genomic data. Science 2003;302(5644):449–453. Evaluation Studies.
37. Jensen FV, Nielsen TD. *Bayesian Networks and Decision Graphs.* 2nd ed. New York, NY: Springer Publishing Company, Inc.; 2007.
38. Jurczyk P, Agichtein E. Discovering authorities in question answer communities by using link analysis. Proceedings of the 16th ACM Conference on Information and Knowledge Management, CIKM '07. New York: ACM; 2007. p 919–922.
39. Kauffman S. Metabolic stability and epigenesis in randomly constructed genetic nets. J Theor Biol 1969;22(3):437–467.
40. Kauffman SA. *The Origins of Order: Self-Organization and Selection in Evolution.* New York: OUP; 1993.
41. Kernighan BW, Lin S. An efficient heuristic procedure for partitioning graphs. Bell Syst Tech J 1970;49(2):291–307.
42. Keshava Prasad TS, Goel R, Kandasamy K, Keerthikumar S, Kumar S, Mathivanan S, Telikicherla D, Raju R, Shafreen B, Venugopal A, Balakrishnan L, Marimuthu A, Banerjee S, Somanathan DS, Sebastian A, Rani S, Ray S, Harrys Kishore CJ, Kanth S, Ahmed M, Kashyap MK, Mohmood R, Ramachandra YL, Krishna V, Rahiman BA, Mohan S, Ranganathan P, Ramabadran S, Chaerkady R, Pandey A. Human protein reference database - 2009 update. Nucleic Acids Res 2009;37(Database-Issue):767–772.
43. Kleinberg JM. Authoritative sources in a hyperlinked environment. J ACM 1999;46(5):604–632.
44. Kubica J, Moore A, Cohn D, Schneider J. cGraph: A fast graph-based method for link analysis and queries. In: Grobelnik M, Milic-Frayling N, Mladenic D, editors. *Proceedings of the 2003 IJCAI Text-Mining & Link-Analysis Workshop*; Acapulco, Mexico; 2003. p 22–31.
45. Langville AN, Meyer CD. *Google's PageRank and Beyond: The Science of Search Engine Rankings.* Princeton (NJ): Princeton University Press; 2006.

46. Leskovec J, Lang KJ, Mahoney M. Empirical comparison of algorithms for network community detection. Proceedings of the 19th International Conference on World Wide Web, WWW '10. New York: ACM; 2010. p 631–640.

47. Liben-Nowell D, Kleinberg J. The link-prediction problem for social networks. J Am Soc Inf Sci Technol 2007;58(7):1019–1031.

48. Liu X, Liu B, Huang Z, Shi T, Chen Y, Zhang J. SPPS: a sequence-based method for predicting probability of protein-protein interaction partners. PLoS ONE 2012;7(1):e30938.

49. Lo SL, Cai CZ, Chen YZ, Chung MCM. Effect of training datasets on support vector machine prediction of protein-protein interactions. Proteomics 2005;5(4):876–884.

50. Lü L, Zhou T. Link prediction in complex networks: A survey. Physica A 2011;390(6):1150–1170.

51. Marcotte EM, Pellegrini M, Ng HL, Rice DW, Yeates TO, Eisenberg D. Detecting protein function and protein-protein interactions from genome sequences. Science (New York) 1999;285(5428):751–753.

52. Martin S, Roe DC, Faulon J-L. Predicting protein-protein interactions using signature products. Bioinformatics 2005;21(2):218–226.

53. Mewes H-W, Frishman D, Güldener U, Mannhaupt G, Mayer KFX, Mokrejs M, Morgenstern B, Münsterkötter M, Rudd S, Weil B. MIPS: a database for genomes and protein sequences. Nucleic Acids Res 2002;30(1):31–34.

54. Murphy K, Mian S. Modelling gene expression data using dynamic bayesian networks. Technical report. Berkeley (CA): Department of Computer Science, University of California; 1999. Available at http://www.cs.ubc.ca/~murphyk/Papers/ismb99.pdf

55. Najafabadi HS, Salavati R. Sequence-based prediction of protein-protein interactions by means of codon usage. Genome Biol 2008;9(5):R87.

56. Nanni L, Lumini A. An ensemble of K-local hyperplanes for predicting protein-protein interactions. Bioinformatics 2006;22(10):1207–1210.

57. Needham CJ, Bradford JR, Bulpitt AJ, Westhead DR. A primer on learning in bayesian networks for computational biology. PLoS Comput Biol 2007;3(8):e129.

58. Newman ME. Fast algorithm for detecting community structure in networks. Phys Rev E Stat Nonlin Soft Matter Phys 2004;69(6 Pt 2):066133.

59. Newman M. *Networks: An Introduction.* New York: Oxford University Press; 2010.

60. Ng AY, Jordan MI, Weiss Y. On spectral clustering: analysis and an algorithm. In: *Advances in Neural Information Processing Systems 14.* Cambridge, MA: MIT Press; 2001. p 849–856.

61. Ogmen U, Keskin O, Aytuna AS, Nussinov R, Gürsoy A. PRISM: protein interactions by structural matching. Nucleic Acids Res 2005;33(Web-Server-Issue):331–336.

62. Page L, Brin S, Motwani R, Winograd T. *The pagerank citation ranking: bringing order to the web.* Technical Report 1999-66. Stanford InfoLab; 1999. Previous number = SIDL-WP-1999-0120.

63. Pal A, Counts S. Identifying topical authorities in microblogs. Proceedings of the 4th ACM international conference on Web search and data mining, WSDM '11. New York: ACM; 2011. p 45–54.

64. Pearson K. Notes on regression and inheritance in the case of two parents. Proc R Soc London 1895;58:240–242.

65. Pellegrini M, Marcotte EM, Thompson MJ, Eisenberg D, Yeates TO. Assigning protein functions by comparative genome analysis: protein phylogenetic profiles. Proc Natl Acad Sci U S A 1999;96(8):4285–4288.
66. Pitre S, Alamgir M, Green JR, Dumontier M, Dehne F, Golshani A. Computational methods for predicting protein-protein interactions. Adv Biochem Eng Biotechnol 2008;110:247–267.
67. Rapti A, Theodoridis E, Tsakalidis AK. Evaluation of protein-protein interaction management systems. 24th International Workshop on Database and Expert Systems Applications, DEXA 2013; Prague, Czech Republic; August 26-29, 2013; 2013. p 100–104.
68. Ren X, Wang Y-C, Wang Y, Zhang X-S, Deng N-Y. Improving accuracy of protein-protein interaction prediction by considering the converse problem for sequence representation. BMC Bioinformatics 2011;12:409.
69. Res I, Mihalek I, Lichtarge O. An evolution based classifier for prediction of protein interfaces without using protein structures. Bioinformatics 2005;21(10):2496–2501.
70. Rhodes DR, Tomlins SA, Varambally S, Mahavisno V, Barrette T, Kalyana-Sundaram S, Ghosh D, Pandey A, Chinnaiyan AM. Probabilistic model of the human protein-protein interaction network. Nat Biotechnol 2005;23(8):951–959.
71. Roberts S, Husmeier D, Dybowski R. *Probabilistic Modelling in Bioinformatics and Medical Informatics*. 2005 ed. Springer-Verlag; 2005.
72. Roy S, Martinez D, Platero H, Lane T, Werner-Washburne M. Exploiting amino acid composition for predicting protein-protein interactions. PloS ONE 2009;4:e781.
73. Salwínski L, Miller CS, Smith J, Pettit K, Bowie JU, Eisenberg D. The database of interacting proteins: 2004 update. Nucleic Acids Res 2004;32(Database-Issue):449–451.
74. Scott J. *Social Network Analysis: A Handbook*. 2nd ed. London: Sage Publications; 2000.
75. Shao B, Wang H, Xiao Y. Managing and mining large graphs: systems and implementations. Proceedings of the 2012 ACM SIGMOD International Conference on Management of Data, SIGMOD '12. New York: ACM; 2012. p 589–592.
76. Shen JW, Zhang J, Luo XM, Zhu WL, Yu KQ, Chen KX, Li YX, Jiang HL. Predicting protein-protein interactions based only on sequences information. Proc Natl Acad Sci U S A 2007;104(11):4337–4341.
77. Shi J, Malik J. Normalized cuts and image segmentation. Proceedings of the 1997 Conference on Computer Vision and Pattern Recognition (CVPR '97), CVPR '97. Washington (DC): IEEE Computer Society; 1997. p 731–737.
78. Shi J, Malik J. Normalized cuts and image segmentation. IEEE Trans Pattern Anal Mach Intell 2000;22(8):888–905.
79. Shi M-G, Xia J-F, Li X-L, Huang D-S. Predicting protein-protein interactions from sequence using correlation coefficient and high-quality interaction dataset. Amino Acids 2010;38(3):891–899.
80. Shmulevich I, Dougherty ER, Zhang W. From Boolean to probabilistic Boolean networks as models of genetic regulatory networks. Proc IEEE 2002;90(11):1778–1792.
81. Shoemaker BA, Panchenko AR. Deciphering protein-protein interactions. Part I. Experimental techniques and databases. PLoS Comput Biol 2007;3(3):e42.
82. Shoemaker BA, Panchenko AR. Deciphering protein-protein interactions. Part II. Computational methods to predict protein and domain interaction partners. PLoS Comput Biol 2007;3(4):e43.

83. Skrabanek L, Saini HK, Bader GD, Enright AJ. Computational prediction of protein-protein interactions. Mol Biotechnol 2008;38:1–17.

84. Smith GR, Sternberg MJ. Prediction of protein-protein interactions by docking methods. Curr Opin Struct Biol 2002;12(1):28–35.

85. Snyder M, Kumar A. Protein complexes take the bait. Nature 2002;415(6868):123–124.

86. Soong T-T, Wrzeszczynski KO, Rost B. Physical protein–protein interactions predicted from microarrays. Bioinformatics 2008;24(22):2608–2614.

87. Srinivas A. *Handbook of Computational Molecular Biology*. Chapman & All/Crc Computer and Information Science Series. Florida: Chapman & Hall/CRC; 2006.

88. Sylvain P. PIPE: a protein-protein interaction prediction engine based on the re-occurring short polypeptide sequences between known interacting protein pairs [PhD thesis]. Ottawa, Ontario, Canada: Carleton University; 2010. AAINR63858.

89. Sylvester JJ. A demonstration of the theorem that every homogeneous quadratic polynomial is reducible by real orthogonal substitutions to the form of a sum of positive and negative squares. Philos Mag (Ser 4) 1852;4(23):138–142.

90. Thum KE, Shin MJ, Gutiérrez RA, Mukherjee I, Katari MS, Nero D, Shasha D, Coruzzi GM. An integrated genetic, genomic and systems approach defines gene networks regulated by the interaction of light and carbon signaling pathways in Arabidopsis. BMC Syst Biol 2008;2:31.

91. Tsafou K, Theodoridis E, Klapa M, Tsakalidis A, Moschonas NK. Reconstruction of the known human protein-protein interaction network from five major literature-curated databases. 6th Conference of the Hellenic Society for Computational Biology and Bioinformatics (HSCBB11); Patras, Greece; 2011.

92. Tsafou K, Theodoridis E, Klapa M, Tsakalidis A, Moschonas NK. Reconstruction of the experimentally supported human protein interactome: what can we learn? BMC Syst Biol 2013;7:96. DOI: 10.1186/1752-0509-7-96.

93. Valente GT, Acencio ML, Martins C, Lemke N. The development of a universal in silico predictor of protein-protein interactions. PLoS ONE 2013;8(5):e65587. Published online 2013 May 31. DOI: 10.1371/journal.pone.0065587.

94. Weng J, Lim E-P, Jiang J, He Q. TwitterRank: finding topic-sensitive influential twitterers. Proceedings of the 3rd ACM International Conference on Web Search and Data Mining, WSDM '10. New York: ACM; 2010. p 261–270.

95. Yu C-Y, Chou L-C, Chang DT. Predicting protein-protein interactions in unbalanced data using the primary structure of proteins. BMC Bioinformatics 2010;11:167.

96. Zaki N, Lazarova-Molnar S, El-Hajj W, Campbell P. Protein-protein interaction based on pairwise similarity. BMC Bioinformatics 2009;10:150.

97. Zhang J, Ackerman MS, Adamic L. Expertise networks in online communities: structure and algorithms. Proceedings of the 16th International Conference on World Wide Web, WWW '07. New York: ACM; 2007. p 221–230.

98. Zhang LV, Wong SL, King OD, Roth FP. Predicting co-complexed protein pairs using genomic and proteomic data integration. BMC Bioinformatics 2004;5:38.

99. Zhou Y, Zhou YS, He F, Song J, Zhang Z. Can simple codon pair usage predict protein-protein interaction? Mol Biosyst 2012;8:1396–1404.

CHAPTER 27

BIOLOGICAL NETWORK INFERENCE AT MULTIPLE SCALES: FROM GENE REGULATION TO SPECIES INTERACTIONS

Andrej Aderhold[1], V Anne Smith[1], and Dirk Husmeier[2]

[1]*School of Biology, University of St Andrews, St Andrews, UK*
[2]*School of Mathematics and Statistics, University of Glasgow, Glasgow, UK*

27.1 INTRODUCTION

Mathematics is transforming biology in the same way it shaped physics in the previous centuries [11]. The underlying paradigm shift that distinguishes modern quantitative systems biology from more traditional nonquantitative approaches is based on a conceptualization of molecular reactions in the cell, the elementary building block of life, as a complex network of interactions. The approach is intrinsically holistic, aiming to understand the properties of cells, tissues, and organisms functioning as a complex system. Besides aiming for a deeper theoretical understanding of molecular processes and their emergent properties, modern systems biology sees a huge range of potential applications, ranging from the targeted genetic modifications of plants for improved resistance, yield, and a variety of agronomically desired traits [54], to unraveling the causes of neurodegenerative diseases, cancer, and ultimately aging [40, 58].

Pattern Recognition in Computational Molecular Biology: Techniques and Approaches,
First Edition. Edited by Mourad Elloumi, Costas S. Iliopoulos, Jason T. L. Wang, and Albert Y. Zomaya.

The new challenge for systems biology is to encompass all the facets of biology. For instance, from the molecular scale of gene regulatory networks to organ-level fluid pressure and velocity fields, or from chemotaxis of cancer cells during invasion and metastasis, to mass movement and migration patterns in locusts and wildebeest. The ultimate quest is the elucidation of the common principles spanning several spatial and temporal scales. The following are important questions to be addressed: Which common mechanisms determine both aberrant behavior in groups of migrating animals and the movement of cancer cells in the human body? Which organizational principles are common to the response of eukaryotic gene regulatory networks to environmental stress and the response of trophic species interaction networks to climate change? How can mathematical models of biochemical signal transduction pathways in plants link cell-level regulation to metabolism, biomass production, and properties at whole organism level?

The emphasis of the present chapter is to focus on the modeling commonalities between systems biology at the molecular and species levels by investigating the application of computational methods from molecular systems biology to whole ecosystems. As with molecular regulatory networks, the complexity of ecological networks is staggering, with hundreds or thousands of species interacting in multiple ways, from competition and predation to commensalism (whereby one species profits from the presence of another) and mutualism (where two species exist in a relationship in which each benefits from the other). Understanding the networks that form ecosystems is of growing importance, for example, to understand how a change in one population can lead to dramatic effects on others [12], drive them to alternative stable states [6], or even cause catastrophic failure [51]. Ecologists need to understand how populations and ecosystems as a whole respond to these changes. This is of enormous importance during a period of rapid climate change [53] that affects not only land use and agriculture [39], but can also cause a reduction in biodiversity in terrestrial and aquatic ecosystems [47].

Inferring the interactions in complex ecosystems is not a straightforward task to accomplish. Direct observation requires minute observations and detailed fieldwork, and is capable of measuring only certain types of species interactions, such as those between predators and their prey, or between pollinators and their host plants. The majority of interactions are not directly observable. This includes competition for resources, commensalism, or mutualism. This restriction calls for the development of novel computational inference techniques to learn networks of species interactions directly from observed species concentrations.

Network inference can be generally defined as the process of "reverse engineering," or learning, the interactions between components of a system given its observed data. Interactions play such an important role because a system's behavior and that of its parts is defined not only by external factors but also by internal influences represented by the components of the system itself. In the context of gene regulation networks, this involves the control of gene expression through regulation factors, such as proteins or microRNAs [31]. However, these underlying processes are often hidden from direct observation, causing poor understanding or complete lack of knowledge of how the components interact. The challenge arises with the fact that observable quantities of a system have to be sufficient to be used as a guide to identify the driving architecture, that is, the interaction networks that cause its behavior and observed characteristics. Previous knowledge about the system can help to improve network inference. However, in many instances little or no knowledge is available and patterns have to be inferred directly from the observed data. One such pattern is the gene regulation network, which is responsible for the control of the majority of molecular processes essential for growth and survival of any organism on earth. This can involve the control of organism development [5], response of the immune system to pathogens [46], or the adaptation to changing environmental conditions through stress responses [50].

In the last decade, molecular biology has been the driving force for the development of inference methods that help to identify such patterns. This becomes more important in light of the growing amount of biomolecular data from high-throughput techniques, such as DNA microarray experiments [49]. Ecological modeling could benefit from methods used in systems biology. Several commonalities between molecular and ecological systems exist: Genes are the basic building blocks of gene regulation networks and can be compared to organisms as principal participants in the formation of ecological networks; gene profile measurements from DNA or mRNA assays can be compared to population data gathered in field surveys; expression profiles of genes or proteins can be matched with population densities or species coverage; gene regulation compares to species interactions, and different conditions compare to different environments. It seems natural to apply the same methodology with certain modification to both the molecular and ecological data.

In this chapter, we show how established methods from systems biology (Section 27.2) can be modified to infer ecological networks (Section 27.3). The structure is in the following form: Section 27.4 introduces the mathematical notation (Section 27.4.1), state-of-the art regression methods (Sections 27.4.2, and 27.4.3), and a scoring metric to measure reconstruction accuracy (Section 27.4.4). Section 27.5 demonstrates an application to a circadian clock regulation system in the plant *Arabidopsis thaliana* involving simulated data and *Real-Time Polymerase Chain Reaction* (rtPCR) gene expression data (Sections 27.5.3 and 27.5.4), and the modification of a basic regression method to enable the handling of dynamic gene regulatory changes over time. Section 27.6 describes the method modifications that allow us to apply the previously defined methods to an ecological problem setting. This is realized with the expansion of the data domain from one-dimensional time to two dimensions of space. In addition, methods that can learn the spatial segmentation on a global scale (Section 27.6.2) and local scale using the *Mondrian process* (Section 27.6.3) are described. The learning performance of these methods is compared on synthetic data and realistic ecological simulated data (Sections 27.6.4 and 27.6.6). Finally, network reconstruction is demonstrated on a real-world case using plant coverage and environmental data (Section 27.6.6).

27.2 MOLECULAR SYSTEMS

The inference of molecular regulatory networks from postgenomic data has been a central topic in computational systems biology for over a decade. Following up on the seminal paper by Friedman et al. [21], a variety of methods have been proposed [59], and several procedures have been pursued to objectively assess the network reconstruction accuracy [32, 59, 60], for example, of the RAF kinase pathway, which is a cellular signaling network in human immune system cells [46]. It has been demonstrated that machine learning techniques can not only serve to broaden our biological knowledge [14, 19] but also handle the increasing amount of data from high-throughput measurements in a more efficient way than was previously attempted [29]. We describe four state-of-the art methods that are widely used for the inference of gene regulation networks. The accuracy of inference is assessed with a benchmark of a realistic gene and protein regulation system in the context of circadian regulation. The data are simulated from a recently published regulatory network of the circadian clock in *A. thaliana*, in which protein and gene interactions are described by a Markov jump process based on Michaelis–Menten kinetics. We closely follow recent experimental protocols, including the entrainment of seedlings to different *Light–Dark* (LD) cycles and the knockout of various key regulatory genes. Our study provides relative assessment scores for the comparison of the presented methods and investigates the influence of systematically missing values related to unknown protein concentrations and mRNA transcription rates.

27.3 ECOLOGICAL SYSTEMS

While interaction networks at the molecular level have been the forefront of modern biology, due to the ever increasing amount of available postgenomic data, interaction networks at other scales are drawing attention. This concerns, in particular, ecological networks, owing to their connection with climate change and biodiversity, which poses new challenges and opportunities for machine learning and computational statistics. Similar to molecular systems, ecological systems are complex dynamical systems with interconnected networks of interactions among species and abiotic factors of the environment. This interconnectedness can lead to seemingly unpredictable behavior: changing numbers of one species can influence unexpected changes in others [30]; the whole system can transit between different stable states [6]. Perturbations from features such as climate change and invasive species can affect both biodiversity and the stability of ecosystems [48]. Being able to make predictions on these scales requires an understanding of the ecological networks underlying the system [15].

The challenges for computational inference specific to ecological systems are that, first, the interactions take place in a spatially explicit environment which must be taken into account, and second, the interactions can vary across this environment depending on the makeup of the elements (species and abiotic factors) present. Here, we meet these challenges by showing the necessary modifications to an inference method from systems biology [35] for temporally explicit (one-dimensional) gene expression data to infer ecological interactions from spatially explicit species abundance data on a two-dimensional grid. We describe a nonhomogeneous Bayesian regression model based on the Bayesian hierarchical regression model of Andrieu and Doucet [4], using a multiple global *change-point* process as proposed in Reference [1]. We modify the latter method with a *Mondrian process* that implements a spatial partitioning at different levels of resolution [2]. We make further use of the spatially explicit nature of ecological data by correcting for spatial autocorrelation with a regulator node (in Bayesian network terminology) that explicitly represents the spatial neighborhood of a node. The performance of these methods is demonstrated on synthetic and realistic simulated data and we infer a network from a real world data set. The results show that ecological modeling could benefit from these types of methods, and that the required modifications do not conflict with, but extend the basic methodology used in systems biology.

27.4 MODELS AND EVALUATION

This section introduces the notations used throughout this chapter, two well-established sparse regression methods, and the basic framework of a homogeneous Bayesian regression model that is applied both to infer gene regulation networks and ecological networks. The naming *homogeneous* implies that the data are treated as a single monolithic block. This simplifies inference but insufficiently reflects the actual nature of underlying biological processes, which can change over time and space. For instance, morphological changes of an insect with the distinct phases of an embryo, larva, pupa, and adult [5] are matched by changes in gene expression profiles and interconnectedness, for example, regulation through transcription factors that are proteins. A heterogeneous model allows the partitioning of data to account for different phases and associated changes in the interaction networks and network parameters. Section 27.5.1 describes this model with the possibility to set fixed phase boundaries (so-called *change-points*) and learn model parameters for each of the phases. This technique is extended to a segmentation in two dimensions with an additional inference scheme for the *change-points* in the case that the segmentation of the data is not known (Section 27.6).

27.4.1 Notations

The following notations are used throughout this chapter for both gene regulation and species interaction. For the regression models, which we use to infer the network interactions, we have target variables y_n $(n = 1, \dots, N)$ that can each represent a temporal mRNA concentration gradient of a particular gene n or the abundance of a particular species also denoted as n. The realizations of each target variable y_n can then be written as a vector $\mathbf{y}_n = (y_{n,1}, \dots, y_{n,T})^T$, where $y_{n,t}$ is the realization of y_n in observation t. The potential regulators are either gene or protein concentrations in the case of gene regulation networks or species abundances for species interaction networks. The task is to infer a set of regulators π_n for each response variable y_n. The collective set of regulators $\{\pi_1, \dots, \pi_N\}$ defines a regulatory interaction network, $\mathcal{G}$. In $\mathcal{G}$, the regulators and the target variables represent the nodes, and from each regulator in π_n, a directed interaction (or *edge*) points to the target node n, attributed with the interaction strength vector, or regression coefficients, $\mathbf{w}_n = (w_n^p)_{p \in \pi_n}$. The complete set of regulatory observations is contained in the design matrix $\mathbf{X}$. Realizations of the regulators in the set π_n are collected in $\mathbf{X}_{\pi_n}$, where the columns of $\mathbf{X}_{\pi_n}$ are the realizations of the regulators π_n. Design matrix $\mathbf{X}$ and $\mathbf{X}_{\pi_n}$ are extended by a constant element equal to 1 for the intercept.

27.4.2 Sparse Regression and the LASSO

A widely applied linear regression method that encourages network sparsity is the *Least Absolute Shrinkage and Selection Operator* (LASSO) introduced in Reference [55]. The LASSO optimizes the parameters of a linear model based on the residual sum of squares subject to an $L1$-norm penalty constraint on the regression parameters, $\|\mathbf{w}_n\|_1$, which excludes the intercept [20]:

$$\hat{\mathbf{w}}_n = \text{argmin} \left\{ \|\mathbf{y}_n - \mathbf{X}^T \mathbf{w}_n\|_2^2 + \lambda_1 \|\mathbf{w}_n\|_1 \right\} \tag{27.1}$$

where λ_1 is a regularization parameter controlling the strength of shrinkage. Equation 27.1 constitutes a convex optimization problem, with a solution that tends to be sparse. Two disadvantages of the LASSO are arbitrary selection of single predictors from a group of highly correlated variables and saturation at T predictor variables. To avoid these problems, the *Elastic Net* method was proposed in Reference [63], which combines the LASSO penalty with a ridge regression penalty of the standard squared $L2$-norm $\|\mathbf{w}_n\|_2^2$ excluding the intercept:

$$\hat{\mathbf{w}}_n = \text{argmin} \left\{ \|\mathbf{y}_n - \mathbf{X}^T \mathbf{w}_n\|_2^2 + \lambda_1 \|\mathbf{w}_n\|_1 + \lambda_2 \|\mathbf{w}_n\|_2^2 \right\} \tag{27.2}$$

Similar to Equation 27.1, Equation 27.2 constitutes a convex optimization problem, which we solve with cyclical coordinate descent [20] implemented in the R software package *glmnet*. The regularization parameters λ_1 and λ_2 were optimized by 10-fold cross-validation.

27.4.3 Bayesian Regression

We follow Reference [25] in the definition of the Bayesian regression model and will further refer to it as "homogBR". It is assumed to be a linear regression model for the targets:

$$\mathbf{y}_n | (\mathbf{w}_n, \sigma_n, \pi_n) \sim \mathcal{N}(\mathbf{X}_{\pi_n}^{\mathrm{T}} \mathbf{w}_n, \sigma_n^2 \mathbf{I}) \tag{27.3}$$

where σ_n^2 is the noise variance, and $\mathbf{w}_n$ is the vector of regression parameters, for which we impose a Gaussian prior:

$$\mathbf{w}_n | (\sigma_n, \delta_n, \pi_n) \sim \mathcal{N}(\mathbf{0}, \delta_n \sigma_n^2 \mathbf{I}) \tag{27.4}$$

δ_n can be interpreted as the *Signal-to-Noise Ratios* (SNR) [25]. For the posterior distribution, we get

$$\mathbf{w}_n|(\sigma_n, \delta_n, \pi_n, \mathbf{y}_n) \sim \mathcal{N}(\mathbf{\Sigma}_n \mathbf{X}_{\pi_n} \mathbf{y}_n, \sigma_n^2 \mathbf{\Sigma}_n) \tag{27.5}$$

where $\mathbf{\Sigma}_n^{-1} = \delta_n^{-1}\mathbf{I} + \mathbf{X}_{\pi_n}\mathbf{X}_{\pi_n}^{\mathrm{T}}$, and the marginal likelihood can be obtained by application of standard results for Gaussian integrals [7]:

$$\mathbf{y}_n|(\sigma_n, \delta_n, \pi_n) \sim \mathcal{N}(\mathbf{0}, \sigma_n^2(\mathbf{I} + \delta_n \mathbf{X}_{\pi_n}^{\mathrm{T}} \mathbf{X}_{\pi_n})) \tag{27.6}$$

For σ_n^{-2} and δ_n^{-2} we impose conjugate gamma priors, $\sigma_n^{-2} \sim \mathrm{Gam}(\nu/2, \nu/2)$, and $\delta_n^{-1} \sim \mathrm{Gam}(\alpha_\delta, \beta_\delta)$.[1] The integral resulting from the marginalization over σ_n^{-2},

$$P(\mathbf{y}_n|\pi_n, \delta_n) = \int_0^\infty P(\mathbf{y}_n|\sigma_n, \delta_n, \pi_n) P(\sigma_n^{-2}|\nu) d\sigma_n^{-2}$$

is then a multivariate Student t-distribution with a closed-from solution [7, 25]. Given the data for the potential regulators of y_n, symbolically y, the objective is to infer the set of regulators π_n from the marginal posterior distribution:

$$P(\pi_n|y, \mathbf{y}_n, \delta_n) = \frac{P(\pi_n)P(\mathbf{y}_n|\pi_n, \delta_n)}{\sum_{\pi_n^\star} P(\pi_n^\star)P(\mathbf{y}_n|\pi_n^\star, \delta_n)} \tag{27.7}$$

where the sum is over all valid regulator sets $\pi_n^\star$, $P(\pi_n)$ is a uniform distribution over all regulator sets subject to a maximal cardinality, $|\pi_n| \leq 3$, and δ_n is a nuisance parameter, which can be marginalized over. We sample sets of regulators π_n, signal-to-noise hyperparameter δ_n, and noise variances σ_n^2 from the joint posterior distribution with *Markov Chain Monte Carlo* (MCMC), following a Metropolis–Hastings within a partially collapsed Gibbs scheme [25].

27.4.4 Evaluation Metric

The previously described methods provide a means by which interactions can be ranked in terms of their significance or influence. If the true network is known, this ranking defines the *Receiver Operating Characteristic* (ROC) curve as shown in Figure 27.1, where for all possible threshold values, the *sensitivity* or recall is plotted against the complementary *specificity*.[2] By numerical integration we then obtain the *Area Under the ROC* curve (AUROC) as a global measure of network reconstruction accuracy, where larger values indicate a better performance, starting from AUROC = 0.5 to indicate random expectation, to AUROC = 1 for perfect network reconstruction. There have been suggestions that the *Area Under Precision-RECall*

[1] We set: $\nu = 0.01$, $\alpha_\delta = 2$, and $\beta_\delta = 0.2$, as in Reference [25].

[2] The *sensitivity* is the proportion of true interactions that have been detected, the *specificity* is the proportion of noninteractions that have been avoided.

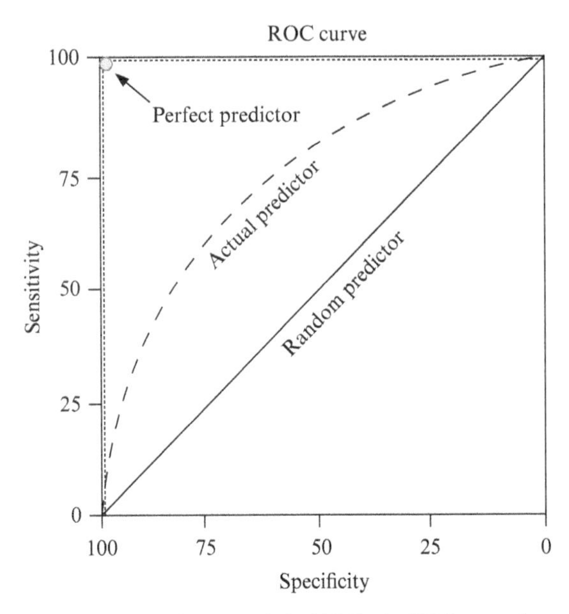

Figure 27.1 *Receiver Operating Characteristic* (ROC). An ROC curve for a perfect predictor, random expectation, and a typical predictor between these two extremes is shown. The *Area Under the ROC* curve (AUROC) is used as scoring metric.

curves (AUPREC) indicate differences in network reconstruction performance more clearly than AUROC curves [13]. While this is true for large, genome-wide networks, a study in Reference [26] has indicated that in networks with a low number of nodes, as with the studied networks in Figure 27.3, the difference between the two scoring schemes should be negligible. We therefore evaluate the performance of all methods with AUROC scores, due to their more straightforward statistical interpretation [28].

27.5 LEARNING GENE REGULATION NETWORKS

A typical feature of molecular regulation networks is the variability of interactions among genes and proteins that can occur over time, caused by changing internal and external conditions. The homogeneous Bayesian regression model described in Section 27.4.3 is unable to pick up such time-varying relationships because it treats all observations as coming from the same underlying regulatory process. This section describes a nonhomogeneous Bayesian regression model with a fixed *change-point* model that is able to approximate such nonhomogeneity. In the species interaction study (Section 27.6) we show that this model can be adapted to a spatial *change-point* process in two dimensions. In addition, protein–gene interactions affect transcription rates, but both these rates as well as protein concentrations might not be available from the wetlab assays. In such situations, mRNA concentrations have to be taken as proxy for protein concentrations, and rates have to be approximated by finite difference quotient or analytic derivation (Section 27.5.2). We compare these setups on a simulated data set (Section 27.5.3) and select the most realistic setup with the best performing method for a real-world data application (Section 27.5.4). The content in this chapter is in part based on a recent study [3] on circadian regulation.

27.5.1 Nonhomogeneous Bayesian Regression

Underlying regulatory relationships can vary over time, as they are dependent on external and internal influences, for example, changing environmental conditions such as light and dark (e.g., circadian regulation in plants) or different stages of development (e.g., from embryo to adult). This implies nonhomogeneous dynamics over time that can be modeled by introducing so-called *change-point*s into the regression model. *Change-points* partition the observations on the timeline into individual segments that are combined to a piecewise linear model that approximates nonhomogeneity. We follow Reference [25] and extend the Bayesian regression model from Section 27.4.3 with a multiple *change-point* process and refer to it with the name "nonhomogBR". The *change-point* process imposes a set of $H_n - 1$ *change-points*, $\{\tau_{n,h}\}_{1\leq h\leq(H_n-1)}$ with $\tau_{n,h} < \tau_{n,h+1}$, to divide the temporal observations of a variable into H_n disjunct segments. With the two *pseudo-change-points* $\tau_{n,0} := 1$ and $\tau_{n,H_n} := T$ each segment $h \in \{1, \dots, H_n\}$ is defined by two demarcating *change-points*, $\tau_{n,h}$ and $\tau_{n,h+1}$. The vector of the target variable realizations, $\mathbf{y}_n = (y_{n,1}, \dots, y_{n,T})^{\mathrm{T}}$, is thus divided into H_n subvectors, $\{\mathbf{y}_{n,h}\}_{h=1,\dots,H_n}$, where each subvector corresponds to a temporal segment: $\mathbf{y}_{n,h} = (y_{n,(\tau_{n,h}+1)}, \dots, y_{n,\tau_{n,h+1}})^{\mathrm{T}}$. Following Reference [25], the distances between two successive *change-points*, $T_{n,h} = \tau_{n,h+1} - \tau_{n,h}$, are assumed to have a negative binomial distribution, symbolically $T_{n,h} \sim NBIN(p, k)$.

We keep the regulator set, π_n, fixed among the H_n segments[3] and we apply the linear Gaussian regression model, defined in Equation 27.3, to each segment h:

$$\mathbf{y}_{n,h}|(\mathbf{X}_{\pi_n,h}, \mathbf{w}_{n,h}, \sigma_n^2) \sim \mathcal{N}(\mathbf{X}_{\pi_n,h}^T \mathbf{w}_{n,h}, \sigma_n^2 \mathbf{I}) \tag{27.8}$$

where $\mathbf{X}_{\pi_n,h}$ is the segment-specific design matrix, which can be built from the realizations of the regulator set π_n in segment h, and $\mathbf{w}_{n,h}$ is the vector of the segment-specific regression parameters for segment h. As in Section 27.4.3, we impose an inverse Gamma prior on σ_n^2, symbolically $\sigma_n^{-2} \sim \mathrm{Gam}(\nu/2, \nu/2)$. For the segment-specific regression parameters, $\mathbf{w}_{n,h}$ $(h = 1, \dots, H_n)$, we assume Gaussian priors:

$$\mathbf{w}_{n,h}|(\sigma_n, \delta_n, \mathbf{X}_{\pi_n,h}) \sim \mathcal{N}(0, \delta_n \sigma_n^2 \mathbf{I}) \tag{27.9}$$

with the hyperprior $\delta_n^{-1} \sim \mathrm{Gam}(A_\delta, B_\delta)$.

[3]The regulator set, that is, network structure, can differ between segments as shown in Reference [35]. However, here we limit the model assumption to a single network to decrease model complexity and based on the fact that interactions are more likely to change than to disappear.

As with the previously defined homogeneous Bayesian regression model (Section 27.4.3), posterior inference is again carried out with the Metropolis–Hastings within partially collapsed Gibbs sampling scheme [25]. The marginal likelihood in Equation 27.7 has to be replaced by

$$P(\pi_n|\mathbf{X}, \delta_n, \{\tau_{n,h}\}_{1\leq h\leq(H_n-1)}) \propto P(\pi_n)\prod_{h=1}^{H_n} P(\mathbf{y}_{n,h}|\mathbf{X}_{\pi_n,h}, \delta_n) \tag{27.10}$$

where $P(\mathbf{y}_{n,h}|\mathbf{X}_{\pi_n,h}, \delta_n)$ $(h = 1, \ldots, H_n)$ can be computed in closed-form; see Reference [25] for a mathematical derivation. The full conditional distribution of $\mathbf{w}_{n,h}$ is now given by [25]:

$$\mathbf{w}_{n,h}|(\mathbf{y}_{n,h}, \mathbf{X}_{\pi_n,h}, \sigma_n^2, \delta_n) \sim \mathcal{N}(\tilde{\mathbf{m}}_{n,h}, \sigma_n^2\mathbf{\Sigma}_{n,h}) \tag{27.11}$$

with $\mathbf{\Sigma}_{n,h}^{-1} = \delta_n^{-1}\mathbf{I} + \mathbf{X}_{\pi_n,h}\mathbf{X}_{\pi_n,h}^T$, and estimated mean $\tilde{\mathbf{m}}_{n,h} = \mathbf{\Sigma}_{n,h}\mathbf{X}_{\pi_n,h}\mathbf{y}_{n,h}$.

27.5.2 Gradient Estimation

The machine learning and statistical models applied in our study predict the rate of gene transcription from the concentrations of the putative regulators. With *de novo* mRNA profiling assays, the rate of transcription could in principle be measured, but these data are often not available. We therefore applied a numerical and analytical procedure to obtain the transcription rates. Appreciating that the transcription rate is just the time derivative of the mRNA concentration $c(t)$, the first approach is to approximate it by a difference quotient:

$$\frac{dc}{dt} \approx \frac{c(t+\delta t) - c(t-\delta t)}{2\delta t} \tag{27.12}$$

Although this is a straightforward procedure, it is well known that differencing noisy time series leads to noise amplification. As an additional procedure, an approach based on smooth interpolation with Gaussian processes is used. We follow Reference [52] and exploit the fact that the derivative of a Gaussian process is a Gaussian process again; hence analytic expressions for the mean and the standard deviation of the derivative are available (see Reference [52]). For the covariance of the Gaussian process, we used the squared exponential kernel, which is the standard default setting in the R package *gptk* [33].

27.5.3 Simulated Bio-PEPA Data

We generated data from the central circadian gene regulatory network in *A. thaliana*, as proposed in Reference [27] and depicted in Figure 27.3(a). Following Reference [61], the regulatory processes of transcriptional regulation and posttranslational protein modification were described with a Markov jump process based on Michaelis–Menten kinetics, which defines how mRNA and protein concentrations change in dependence on the concentrations of other interacting components in the system (see Appendix of Reference [27] for detailed equations). We simulated mRNA and protein concentration time courses with the Gillespie algorithm [23], using the Bio-PEPA modeling framework [10]. We created 11 interventions in consistency with standard biological protocols [16]. These include knockouts of proteins GI, LHY, PRR7/PRR9, TOC1, and varying photo-periods of 4, 6, 8, 12, or 18 h of light in a 24-h LD cycle. For each intervention, we simulated protein and mRNA concentration time courses over 6 days. The first 5 days served as entrainment to the indicated LD cycles. This was followed by a day of persistent *Darkness* (DD) or *light* (LL), during which concentrations of mRNAs and proteins were measured in 2-h intervals. Combining 13 observations for each intervention yielded 143 observations in total for each network type. All concentrations were standardized to unit standard deviation. The temporal mRNA concentration gradient was approximated by the two alternative schemes described in Section 27.5.2: a difference quotient (Eq. 27.12) of mRNA concentrations with −2 and +2 h[4] (termed *coarse gradient*), and smooth interpolation with a Gaussian process yielding a derivative (gradient) at each observation time point (termed *interp gradient*), followed by z-score standardization.

When trying to reconstruct the regulatory network from the simulated data, we ruled out self-loops, such as from LHY (modified) protein to LHY mRNA, and adjusted for mRNA degradation by enforcing mRNA self-loops, such as from the LHY mRNA back to itself. Protein ZTL was included in the stochastic simulations, but excluded from structure learning because it has no direct effect on transcription. We carried out two different network reconstruction tasks. The first was based on complete observation, including both protein and mRNA concentration time series. The second was based on incomplete observation, where only mRNA concentrations were available, but protein concentrations were systematically missing. All network reconstructions were repeated on five independent data sets.

27.5.4 Real mRNA Expression Profile Data

In addition to the realistic data simulated from a faithful mathematical description of the molecular interaction processes, as described earlier, we used real transcription profiles for the key circadian regulatory genes in the model plant *A. thaliana*. The objective is to infer putative gene regulatory networks with the statistical methods described in Section 27.4, and then to compare these predictions with network models of the circadian clock from the biological literature [38, 41–43], of which two are displayed in Figure 27.3(a) and (b). It is important to note that, as opposed to the realistic data described in the previous subsection, we do not have a proper biological understanding. Besides the fact that the models in References [38, 41–43] show noticeable variations, they were not obtained on the basis of proper statistical

[4]Two-hour intervals are a realistic sampling frequency for plants in the wetlab, although, smaller intervals would be favorable.

model selection, as described, for example, in Reference [57]. Nevertheless, a qualitative comparison will reveal to what extent the postulated interaction features and structural network characteristics from the literature are consistent with those inferred from the data.

The data used in our study come from the EU project TiMet [56], whose objective is the elucidation of the interaction between circadian regulation and metabolism in plants. The data consist of transcription profiles for the core clock genes from the leaves of various genetic variants of *A. thaliana*, measured with rtPCR. The study encompasses two wildtypes of the strains Columbia (Col-0) and Wasilewski (WS) and four clock mutants, namely, a double knockout LHY/CCA1 in the WS strain, a single knockout of GI and TOC1 in the strain Col-0, and a double knockout PRR7/PRR9 in strain Col-0. The plants were grown in the following three light conditions: a diurnal cycle with 12 h light and 12 h darkness (12L/12D), an extended night with full darkness for 24 h (DD), and an extended light with constant light (LL) for 48 h. Samples were taken every 2 h to measure mRNA concentrations. Further information on the data and the experimental protocols is available from Reference [18]. The mRNA profiles for the genes LHY, CCA1, PRR5, PRR7, PRR9, TOC1, ELF3, ELF4, LUX, and GI were extracted from the TiMet data [18], yielding a total of 288 samples per gene. We used the log mean copy number of mRNA per cell and applied a gene-wise z-score transformation for data standardization. An additional binary light indicator variable with 0 for darkness and 1 for light was included to indicate the status of the experimentally controlled light condition.

27.5.5 Method Evaluation and Learned Networks

We evaluate the reconstruction accuracy with AUROC scores (Section 27.4.4) of LASSO, *Elastic Net* (Section 27.4.2), the homogeneous Bayesian regression model (homogBR, Section 27.4.3), and the nonhomogeneous Bayesian regression model (nonhomogBR, Section 27.5.1). The method with the highest score was applied to the real-world data from the previous section. Since light may have a substantial effect on the regulatory relationships of the circadian clock, we set the *change-points* of the nonhomogBR method according to the day/night transitions yielding two segments, $h = 1$ (light) and $h = 2$ (darkness). This reflects the nature of the laboratory experiments, where *A. thaliana* seedlings are grown in an artificial light chamber whose light is switched on or off. It would be straightforward to generalize this approach to more than two segments to allow for extended dawn and dusk periods in natural light. Given that the light phase is known, we consider the segmentation as fixed.

27.5.5.1 *Simulated Data.* The marginal posterior probabilities of all potential interactions are computed for the Bayesian regression methods. For LASSO and *Elastic Net*, we record the absolute values of nonzero regression parameters. Both measures provide a means by which interactions between genes and proteins can be ranked in terms of their significance or influence. The AUROC scores given these values are calculated with the reference network *P2010* shown in Figure 27.3(a).

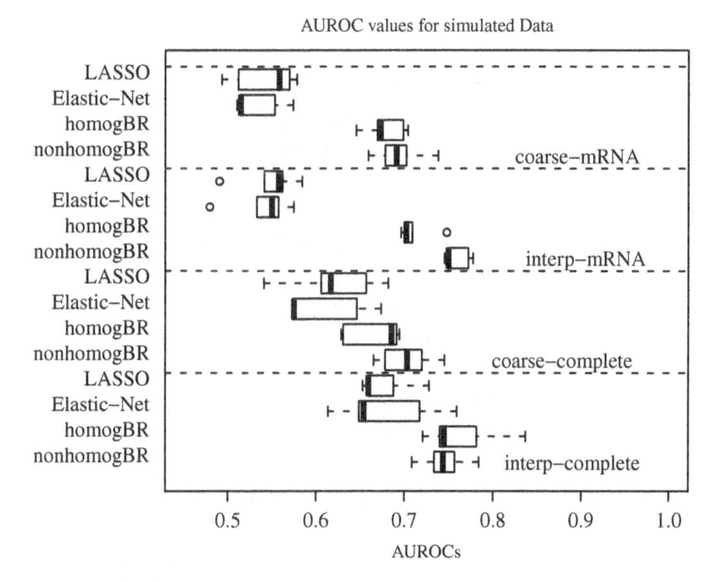

Figure 27.2 AUROC scores obtained for different reconstruction methods, and different experimental settings. Boxplots of AUROC scores obtained from LASSO, *Elastic Net* (both in Section 27.4.2), homogBR (homogeneous Bayesian regression, Section 27.4.3), and nonhomogBR (nonhomogeneous Bayesian regression, Section 27.5.1). The latter utilizes light- induced partitioning of the observations. The subpanels are *coarse-mRNA*: incomplete data, only with mRNA concentrations and coarse gradient, *interp-mRNA*: incomplete data with interpolated gradient, *coarse-complete*: complete data with protein and mRNA concentrations, and *interp-complete*: complete data with interpolated gradient. The *coarse* gradients are computed from Equation 27.12 from 4-h intervals, and the interpolated gradients (*interp*) are derived from a Gaussian process as described in Section 27.5.2.

They can assume values from around AUROC = 0.5 (random expectation) to AUROC = 1 for perfect learning accuracy (see Figure 27.1). The result of this study is shown in Figure 27.2 and includes four different experimental settings for the choice of predictor variables (mRNA or proteins) and gradient calculation (coarse or interpolated gradient). The *coarse-mRNA* setting is the most realistic experimental condition that is ordinarily met in wetlabs working with plants because it only involves mRNA sequencing and samples taken in rather coarse time intervals of 2 h.[5] However, we can use the mRNA profiles as proxies for missing protein data to predict mRNA target gradients. Protein profiles (as used in the *coarse-complete* and *interp-complete* setup) are in contrast less abundant but more suitable to predict target mRNA activities because they typically act as transcription factors. The *interp-mRNA* and *interp-complete* setup derive the mRNA gradient for the response from the derivative of a Gaussian process that interpolates the mRNA concentration (Section 27.5.2). A comparison of the AUROC scores in Figure 27.2 reveals that interpolated gradients (*interp*) improve inference over *coarse* gradients. Among

[5]The time interval can vary and depends on the time needed to extract the samples.

the compared inference methods, both Bayesian models, homogeneous Bayesian regression (homogBR) and nonhomogeneous Bayesian regression (nonhomogBR), display a significant improvement over the sparse regression models LASSO and *Elastic Net*. The nonhomogBR method shows slightly better AUROC scores for the *interp-mRNA* and *coarse-complete* setups and will be used in the following real data study.

27.5.5.2 *Real Data.* The nonhomogeneous Bayesian regression method (nonhomogBR) was applied to the real mRNA profile data set from the TiMet project (Section 27.5.4). Since this data lacks protein profiles and we previously discovered that an interpolated response gradient is superior to a coarse interval gradient, we use the *interp-mRNA* setup to learn the network. The observations were divided into day and night segments to allow for separate linear regression models for these two different conditions. MCMC simulations were run for 50,000 iterations (keeping every 10th sampled parameter configuration), with a burn-in of 40,000, after which standard convergence diagnostics based on the Rubin–Gelman scale reduction factors [22] were met.

Figure 27.3(c) shows the gene regulatory network learned from the TiMet data and Figures 27.3(a) and (b) show two hypothetical networks published in References [42] and [43], respectively. The networks contain two groups of genes: the morning genes LHY/CCA1, PRR5, PRR7, and PRR9 (which, as the name suggests, are expressed in the morning) and the evening genes GI, TOC1, LUX, ELF3, and ELF4 (which are expressed in the evening). A node in the graph represents both the gene as well as the corresponding protein. The subscript "mod" indicates a modified protein isoform. Solid lines represent transcriptional regulation and dashed lines represent protein complex formation. The latter cannot be learned from transcriptional data and are thus systematically missing. This explains, for instance, why the protein complexes EC and ZTL are detached from the remaining network. Various features of the inferred network are consistent with the published networks, such as the activation of PRR9 by LHY/CCA1, the inhibition of the evening genes by the morning genes, and the activation of PRR7 by PRR9. Similar to the network from Reference [43], LHY/CCA1 is inhibited by the evening genes, although the details of the inhibition are slightly different: the inhibition is caused by GI rather than TOC1. Consistent with both published network structures, PRR9 is the target of an inhibition. The regulating node is different from the two publications, however, which is indicative of the current level of uncertainty. Note that the two publications do not agree on this regulation pattern: according to Reference [42], PRR9 is inhibited by TOC1, whereas according to Reference [43], PRR9 is inhibited by EC, and according to the inferred network, PRR9 is inhibited by PRR5. Some interactions are correctly learned, but with opposite signs. For instance, the inferred network predicts that LHY/CCA1 is regulated by PRR9, which is consistent with References [42] and [43]. In these publications, PRR9 is an inhibitor of LHY/CCA1; however, in the inferred network, it is an activator. This ambiguity points to the intrinsic difficulty of the inference task due to the flexibility of the model, as a result of which similar gene expression profiles can be obtained with different interaction and sign configurations. Striking evidence

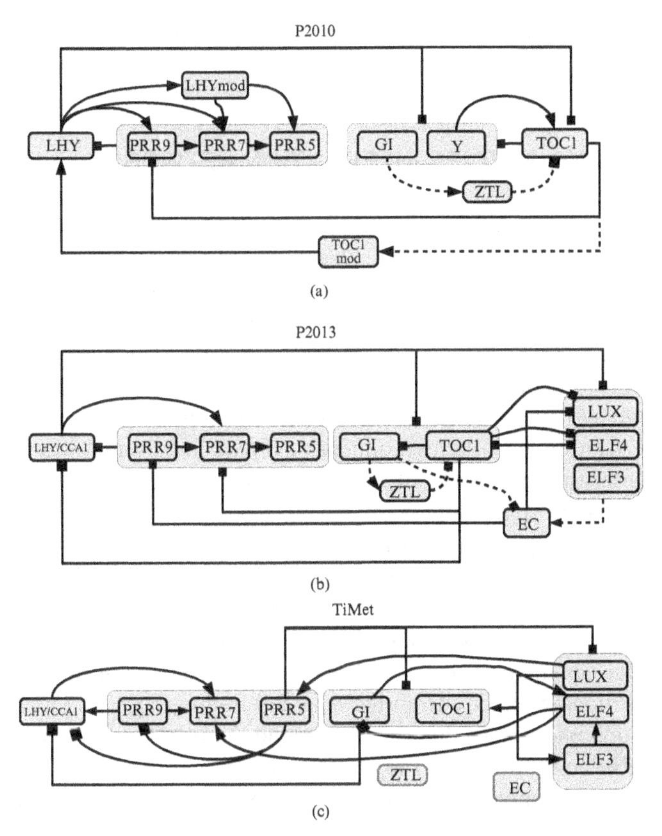

Figure 27.3 Hypothetical circadian clock networks from the literature and that inferred from the TiMet gene expression data. The panels *P2010* (a) and *P2013* (b) constitute hypothetical networks from the literature [42, 43]. The *TiMet* network (c) displays the reconstructed network from the TiMet data, described in Section 27.5.4, using the hierarchical Bayesian regression model from Section 27.5.1. Gene interactions are shown by black lines with arrowhead; protein interactions are shown by dashed lines. The interactions in the reconstructed network were obtained from their estimated posterior probabilities. Those above a selected threshold were included in the interaction network; those below were discarded. The choice of the cutoff threshold is, in principle, arbitrary. For optimal comparability, we selected the cutoff such that the average number of interactions from the published networks was matched (0.6 for molecular interactions).

can be found in the regulatory influence of TOC1 on LHY/CCA1: according to Reference [42], TOC1 is an activator, while according to Reference [43], TOC1 is an inhibitor. This inconsistency between the published networks underlines the fact that the true gene interactions of the circadian clock in *A. thaliana* are still unknown, and that different models focus on different aspects of the data. It is, thus, encouraging to note that several features of the published networks have been inferred with our statistical/machine learning model, using the data alone without any assumption of biological prior knowledge. This finding suggests that the statistical/machine learning models discussed in the present chapter, while too limited to uncover the whole ground truth, still provide powerful tools for the generation of biological hypothesis.

27.6 LEARNING SPECIES INTERACTION NETWORKS

Ecological systems share several commonalities with molecular systems as was already pointed out in Section 27.1. One distinctive feature of species networks, however, is the so-called autocorrelation effect and dispersion of species, which can impact population densities in neighboring regions. We account for this effect by introducing a virtual node into the regression model as described in Section 27.6.1. Another major difference rests on the fact that observations in ecological field surveys spread over a two-dimensional space. Although it would be straightforward to process the two-dimensional data in the same way as a series of molecular observations, one has to consider that ecological processes can be highly nonhomogeneous in space. The main driving forces in this respect are changes in the environment and population densities that produce direct or indirect effects on species interaction dynamics, for example, leading to the formation of ecological niches. Hence, a challenge in ecological modeling is the partitioning of space into local neighborhoods with similar population dynamics. Prediction methods using this knowledge can improve their model accuracy. In addition, it can be beneficial to learn the partitioning directly from the data and in this way gain knowledge about potential neighborhoods. To this end, we extend the one-dimensional *change-point* model described in Section 27.5.1 to two dimensions, and extend it with an inference mechanism that attempts to learn the local neighborhoods from the observed population densities. The multiple *change-point* model introduced in Reference [1], uses a global partitioning scheme and is described in Section 27.6.2. The method is complemented by an improved version that uses a local partitioning scheme based on a *Mondrian process* (Section 27.6.3) introduced in Reference [2]. These methods are evaluated together with the previously defined homogeneous Bayesian regression and sparse regression methods on synthetic data (Section 27.6.4) and simulated data based on a realistic ecological niche model (Section 27.6.5.3). The best performing method is applied to a real data set involving plant coverage and soil attribute measurements (Section 27.6.6).

27.6.1 Regression Model of Species interactions

For all species n, the random variable $y_n(x_1, x_2)$ refers to the abundance of species n at location (x_1, x_2). Within any partition h, this abundance depends on the abundance levels of the species in the regulator set of species n, π_n. The regulator set π_n is defined to be the same in all partitions $h \in \{1, \ldots, H\}$ to rule out fundamental changes to the network, as network changes among partitions are less likely to occur than a change in interaction strength. We model the partition-specific linear regression model with the set of parameters $\{(w^p_{n,h})_{p\in\pi_n}, \sigma_{n,h}\}$, where $w^p_{n,h} \in \mathbb{R}$ is a regression coefficient and $\sigma^2_{n,h} > 0$ is the noise variance for each segment h and target species n. For all species

n and all locations (x_1, x_2) in segment h, the response species $y_n(x_1, x_2)$ depends on the abundance variable of the predictor species $\{y_p(x_1, x_2)\}_{p\in\pi_n}$ according to

$$y_n(x_1, x_2) = w^0_{n,h} + \sum_{p\in\pi_n} w^p_{n,h} y_p(x_1, x_2) + \epsilon_n(x_1, x_2) + w^A_{n,h} A_n(x_1, x_2) \qquad (27.13)$$

where $\epsilon_n(x_1, x_2)$ is assumed to be white Gaussian noise with mean 0 and variance $\sigma^2_{n,h}$, $\epsilon_n(x_1, x_2) \sim N(0, \sigma^2_{n,h})$. We define $\mathbf{w}_{n,h} = (w^0_{n,h}, \{w^p_{n,h}\}_{p\in\pi_n}, w^A_{n,h})$ to denote the vector of all regression parameters of species n in partition h. This includes the parameters defining the strength of interactions with other species p, $w^p_{n,h}$, as well as a species-specific offset term, that is, the intercept or bias $w^0_{n,h}$. Spatial autocorrelation effects are represented with $A_n(x_1, x_2)$ weighted by an additional edge $w^A_{n,h}$. They reflect the influence of neighboring cells that can have a strong effect on model performance [36]. $A_n(x_1, x_2)$ denotes the average densities in the vicinity of (x_1, x_2), weighted inversely proportional to the distance of the neighbors:

$$A_n(x_1, x_2) = \frac{\sum_{(\tilde{x}_1,\tilde{x}_2)\in\mathcal{N}(x_1,x_2)} d^{-1}[(x_1, x_2), (\tilde{x}_1, \tilde{x}_2)] Y_n(\tilde{x}_1, \tilde{x}_2)}{\sum_{(\tilde{x}_1,\tilde{x}_2)\in\mathcal{N}(x_1,x_2)} d^{-1}[(x_1, x_2), (\tilde{x}_1, \tilde{x}_2)]} \qquad (27.14)$$

where $\mathcal{N}(x_1, x_2)$ is the spatial neighborhood of location (x_1, x_2) (e.g., the four nearest neighbors), and $d[(x_1, x_2), (\tilde{x}_1, \tilde{x}_2)]$ is the Euclidean distance between (x_1, x_2) and $(\tilde{x}_1, \tilde{x}_2)$.

27.6.2 Multiple Global Change-Points

We assume a two-dimensional grid with observations sampled at the locations (x_1, x_2) with $x_1 \in \{1, \dots, T^1\}$ and $x_2 \in \{1, \dots, T^2\}$, where T^i specifies the number of observations in the horizontal direction with $i = 1$ and vertical direction with $i = 2$. The regulatory relationships among the species may be influenced by latent variables, which are represented by spatial *change-points*. We assume that latent effects in close spatial proximity are likely to be similar, but locations where spatially close areas are not similar are separated by *change-points*. They are modeled with two *a priori* independent multiple *change-point* processes for each target species n along the two orthogonal spatial directions: $\boldsymbol{\tau}^i_n = (\tau^i_{n,1}, \dots, \tau^i_{n,k^i})$, with $i \in \{1, 2\}$ and k^i_n defining the number of *change-points* along the horizontal ($i = 1$) and vertical ($i = 2$) direction. The *pseudo-change-points* $\tau^i_{n,0} := 1$ and $\tau^i_{n,k^i+1} := T^i$ define the boundaries of the two-dimensional grid (see Figure 27.4). The *change-point* vectors $\boldsymbol{\tau}^1_n$ and $\boldsymbol{\tau}^2_n$ can be set to fixed values but we attempt to learn the partitioning and, thus, $\boldsymbol{\tau}^i_n$ contains an *a priori* unknown number of k^i_n *change-points*. The vectors $\boldsymbol{\tau}^i_n$ and *pseudo-change-points* $\tau^i_{n,0} := 1, \tau^i_{n,k^i+1} := T^i$ partition the space into $H_n = (k^1_n + 1)(k^2_n + 1)$ nonoverlapping segments. We denote the latent variable associated with a segment by $h_n \in \{1, \dots, H_n\}$. Figure 27.4 illustrates an example partitioning of a two-dimensional grid.

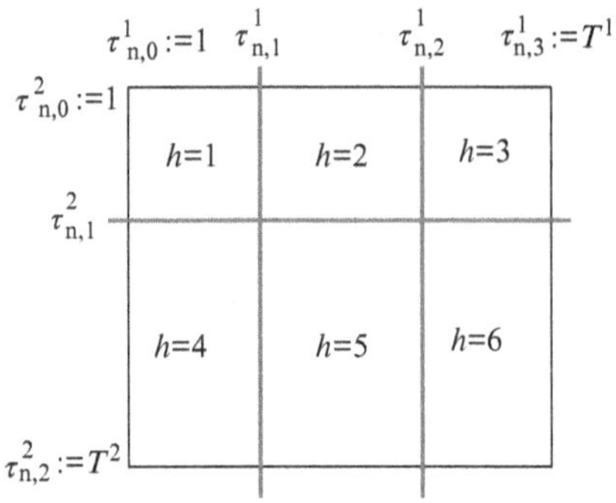

$\tau^1_{n,0} := 1$ $\tau^1_{n,1}$ $\tau^1_{n,2}$ $\tau^1_{n,3} := T^1$

$\tau^2_{n,0} := 1$

$h=1$ $h=2$ $h=3$

$\tau^2_{n,1}$

$h=4$ $h=5$ $h=6$

$\tau^2_{n,2} := T^2$

Figure 27.4 Multiple global *change-point* example. Partitioning with a horizontal *change-point* vector $\tau_n^{i=1} = (\tau^1_{n,1}, \tau^1_{n,2})$ and vertical vector $\tau_n^{i=2} = (\tau^2_{n,1})$. The *pseudo-change-points* $\tau^1_{n,0} = \tau^2_{n,0} := 1$ define the left and upper boundaries; $\tau^1_3 = T^1$ and $\tau^2_2 = T^2$ define the lower and right boundaries, where T^1 and T^2 are the number of locations along the horizontal and vertical directions, respectively. The number of *change-points* is $k_n^1 = 2$, $k_n^2 = 1$ and the number of segments $H_n = 6$.

The priors for the parameter $\mathbf{w}_{n,h}$, regulator set π_n, variance σ_n^2, and signal-to-noise hyperparameter δ_n^2, as well as the likelihood, are defined in a similar way as in the nonhomogeneous Bayesian regression model in Section 27.5.1 with the modification that considers the different partitioning scheme. In addition, the design matrix $\mathbf{X}_{n,h}$ is extended by the spatial autocorrelation variable A_n, and $\mathbf{w}_{n,h}$ includes an additional fixed edge $w^A_{n,h}$. The number of *change-points*, k_n^1, k_n^2, and locations, $\boldsymbol{\tau}_n^1$, $\boldsymbol{\tau}_n^2$, for a species n is sampled from the marginal posterior distribution

$$P(\boldsymbol{\tau}_n^i, k_n^i|\mathbf{X}, \delta_n) \propto \prod_{i=1}^{2} P(\boldsymbol{\tau}_n^i|k_n^i)P(k_n^i|\lambda_n) \prod_{h=1}^{H_n} P(\mathbf{y}_{n,h}|\mathbf{X}_{\pi_n,h}, \delta_n) \tag{27.15}$$

using an RJMCMC scheme [24], conditional on the following *change-point* priors: for both spatial directions $i \in \{1, 2\}$, the $k_n^i + 1$ intervals are delimited by k_n^i *change-points* and boundary *change-points* $\boldsymbol{\tau}^i_{n,0}$ and $\boldsymbol{\tau}^i_{n,k_n^i+1}$, where k_n^i is distributed *a priori* as a truncated Poisson random variable with mean λ_n and maximum $\overline{k^i} = T^i - 1$: $P(k_n^i|\lambda_n) \propto \frac{\lambda_n^{k_n^i}}{k_n^i!} 1\!\!1_{\{k_n^i \leq \overline{k^i}\}}$. Conditional on k_n^i *change-points*, the *change-point* position vector $\boldsymbol{\tau}_n^i = (\boldsymbol{\tau}^i_{n,1}, \ldots, \boldsymbol{\tau}^i_{n,k^i})$ takes nonoverlapping integer values, which we take to be uniformly distributed *a priori*. There are $(T^i - 1)$ possible positions for the k_n^i *change-points*, thus vector $\boldsymbol{\tau}_n^i$ has the prior density $P(\boldsymbol{\tau}_n^i|k_n^i) = 1/\binom{T^i - 1}{k_n^i}$. Reference [1] provides a more detailed description of the model.

27.6.3 Mondrian Process Change-Points

The global *change-point* process described in the previous section lacks the capability to create segmentations with spatially varying length scales and different local fineness and coarseness characteristics. In fact, introducing global *change-points* that might improve segmentation in one region can introduce artifacts in the form of undesired partitioning in other regions. A local approach to partitioning was proposed in Reference [2] using the so-called *Mondrian process* [45]. It is a generative recursive process for self-consistently partitioning the two-dimensional domain in the following way. A hyperparameter λ (the so-called *budget*) determines the average number of cuts in the partition. At each stage of the recursion, a Mondrian sample can either define a trivial partition $\Theta_1 \times \Theta_2$, that is, a segment, or a cut that creates two subprocesses $m_<$ and $m_>$: $m = \langle d, \chi, \lambda', m_<, m_> \rangle$, where $d \in \{0, 1\}$ is a binary indicator for the horizontal and vertical directions and χ the position of the cut. The direction d and position χ are drawn from a binomial and a uniform distribution, respectively, both depending on Θ_1 and Θ_2. The process of cutting a segment is limited by the budget λ associated with each segment and the cost E of a cut. Conditional on the half-perimeter $\tau = |\Theta_1| + |\Theta_2|$, a cut is introduced yielding $m_<$ and $m_>$ if the cost $E \sim \exp(\tau)$ does not exceed the budget λ, that is, satisfies $\lambda' = \lambda - E > 0$ with λ' defining the budget that is assigned to the subprocesses $m_<$ and $m_>$. The process is recursively repeated on $m_<$ and $m_>$ until the budgets are exhausted. This creates a binary tree with the initial Mondrian sample $m_{k=1}$ as the root node spanning the unit square $[0; 1]^2$ and subnodes representing Mondrian samples $m_{1<k\leq K}$, $k \in \{1, \dots, K\}$ where K is the total number of nodes in the tree, for example, $K = 15$ in Figure 27.5. The leaf nodes present nonoverlapping segments and are associated each with a latent variable $h(k)$ labeled with $m^{h(k)}$ (Figure 27.5). These latent variables determine the interactions among species, as described in Section 27.6.1. We denote by H the number of uncut segments, for example, $H = 8$ in Figure 27.5(a), and $h(k) \in \{1, \dots, H\}$.

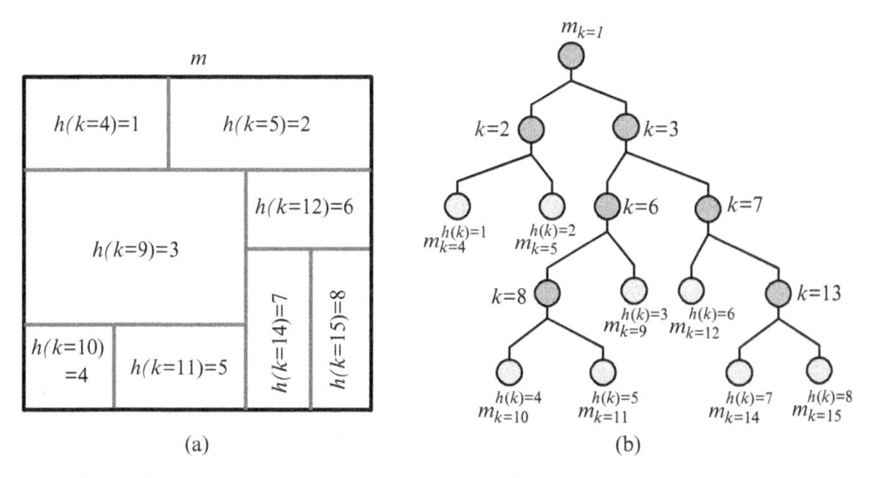

Figure 27.5 *Mondrian process* example. (a) An example partitioning with a *Mondrian process*. (b) The associated tree with labels of the latent variable $h(k)$ identifying each nonoverlapping segment with leaf nodes (light gray) designated as $m_k^{h(k)}$, where k indexes all tree nodes.

An essential step of the inference procedure is to sample a new *Mondrian process* segment m or remove existing ones. This is realized with an RJMCMC scheme described in detail in Section 2.8 of Reference [2]. The inference of model parameters $\mathbf{w}_n$ and structure π are the same as with the previously defined nonhomogeneous Bayesian regression model.

27.6.4 Synthetic Data

For an objective evaluation of network inference, we test the ability of the previously described methods to recover the true network structure from test data generated from a piecewise linear regression model following Equation 27.13. The data was partitioned by two-dimensional *change-points* that resemble a *Mondrian process* following Reference [2], that is, the data grid is iteratively subdivided into local segments (e.g., as shown in Figure 27.5a). The number of observations was selected to be 15 in each direction. The number of nodes n was set to 10 and the number of regulators for each node was sampled from a Poisson distribution. The regression coefficients $\mathbf{w}_{n,h}$ together with the intercept $w^0_{n,h}$ of each segment h were sampled from a uniform distribution in the interval of $[-1;-0.5]$ and $[0.5, 1.0]$. The noise ϵ_n was sampled from a normal distribution. Nodes without an incoming edge were initialized to a Gaussian random number. The values of the remaining nodes were calculated at each grid cell following Equation 27.13.

27.6.5 Simulated Population Dynamics

For a realistic evaluation, we followed Reference [17] and generated data from an ecological simulation that combines a niche model [62] with a stochastic population model [34] in a two-dimensional lattice.

27.6.5.1 *Niche Model and Species Interactions.* The niche model defines the structure of the trophic network and has two parameters: the number of species N and the connectivity (or network density) defined as L/N^2 where L is the number of interactions (edges) in the network. Each species n is assigned a niche value x_n, drawn uniformly from $[0, 1]$. This gives an ordering of the species, where higher values mean that species are higher up in the food chain. For each species, a niche range R_n is drawn from a beta distribution with expected value $2C$ (where C is the desired connectivity), and species n consumes all species falling in a range R_n that is placed by uniformly drawing the centre c_n of the range from $[R_n/2, x_n]$ as illustrated in Figure 27.6 and introduced in Reference [62]. Despite its simplicity, it was shown there that the resulting networks share many characteristics with real food webs.

27.6.5.2 *Stochastic Population Dynamics.* The population model is defined by a stochastic differential equation where the dynamics of the log abundance $X_n(t)$ of species n at time t can be expressed as

$$\frac{dX_n(t)}{dt} = r_n + \frac{\sigma_d}{\sqrt{e^{X_n(t)}}}\frac{dA_n(t)}{dt} + \sigma_e\frac{dB_n(t)}{dt} - \gamma X_n(t) - \Omega(X) + \sigma_E\frac{dE(t)}{dt} \quad (27.16)$$

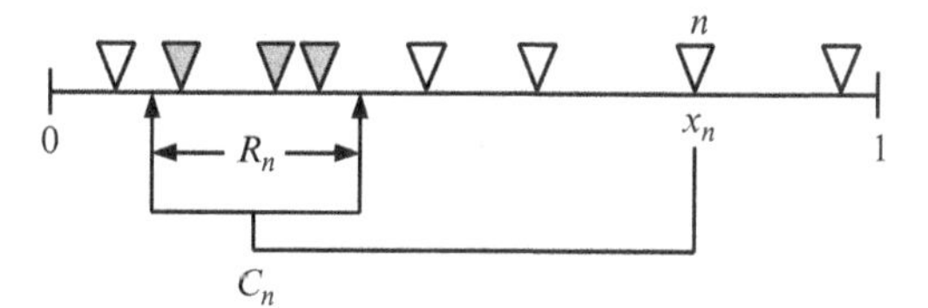

Figure 27.6 Diagram of the niche model. Species are indicated by triangles. A species n is placed with a niche value x_n into the interval $[0, 1]$. A value c_n is uniformly drawn that defines the centre of the range R_n. All species with a value x inside this interval, that is, $c_n - \frac{R_n}{2} \leq x \leq c_n + \frac{R_n}{2}$, as indicated by the gray triangles, are consumed ("eaten") by species n. Diagram adapted from Reference [62].

where X is the set of all $X_N(t)$, r_n is the growth rate of species n, σ_d is the standard deviation of the demographic effect, $A_n(t)$ is the species-specific demographic effect, σ_e is the standard deviation of the species-specific environmental effect, $B_n(t)$ is the species-specific environmental effect, γ is the intraspecific density dependence, Ω is the effect of competition for common resources, σ_E is the standard deviation of the general environmental effect, and $E(t)$ is the general community environment. The growth rates r_n are location dependent (depending on the cell of a rectangular grid), with a spatial pattern that is generated by noise with spectral density f^β (with $\beta < 0$, and f denoting the spatial frequency at which the noise is measured). An illustration is given in Figure 27.7. To model species migration, we included an exponential dispersal model, where the probability of a species moving from one location to another is determined by the Euclidean distance between the locations. To incorporate the niche model, we modified the term Ω in (27.16) to include predator–prey interactions in the Lotka–Volterra form. A detailed description is available in Reference [17].

27.6.5.3 *Simulation.* We applied this model to 10 species living in a 25×25 rectangular grid. We simulated the dynamics of this model for 3000 steps and then recorded species abundance levels in all grid cells at the final step; this corresponds to an ecological survey carried out at a fixed moment in time. For each grid cell, we

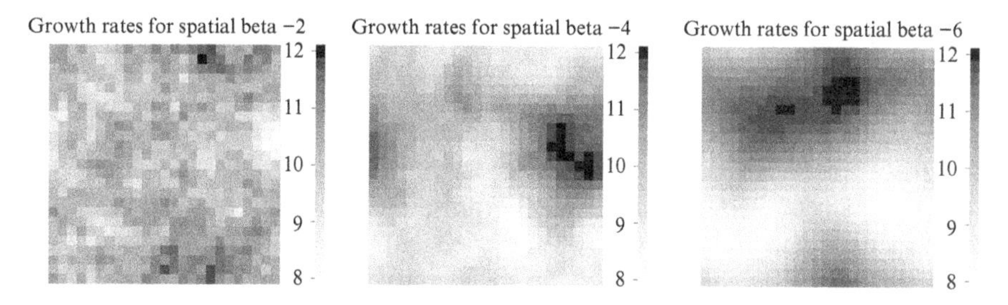

Figure 27.7 Spatial distribution. Shown are the spatial distributions of growth rates r_n entering Equation 27.16 as the spatial β parameter (Section 27.6.1) decreases from −2 to −6. A value of 0 corresponds to uniformly random noise, and −2 is Brownian noise.

counted the number of species that went extinct. These counts were added up over all cells, yielding the total number of extinctions. A simulation was rejected if these extinctions exceeded the value 50. For each of the spatial β parameters displayed in Figure 27.9, 30 surveys were collected by running the simulation repeatedly with different networks and parameter initializations.

27.6.6 Real World Plant Data

We have applied the method to real-world data from Reference [37], including 106 vascular plants and 12 environmental variables collected from a 200 m × 2162 m Machair vegetation land stripe at the western shore of the Outer Hebrides. Samples were taken at 217 locations, each 1 m ×1 m in size, equally distributed with a 50 m spacing. Plant samples were measured as ground coverage in percentage and physical samples as absolute values (such as moisture, pH value, organic matter, and slope). The data was log-normal transformed after observing substantial skewness in the distributions. Each sample point was mapped into a two-dimensional grid ignoring locations with no sample data available. The spatial autocorrelation value for each plant and location was calculated from neighbors inside a radius of 70 m. Since we are interested only in plant interactions, we defined each plant to have all 12 physical soil variables as fixed input, that is, permanent predictor variables.

27.6.7 Method Evaluation and Learned Networks

27.6.7.1 Synthetic Data. To provide a fair comparison between the methods, we disabled the spatial autocorrelation variables for all methods because no dispersion effect was simulated in the synthetic data model (Section 27.6.4). Figure 27.8 shows the AUROC scores for BRAMP (Bayesian regression with *Mondrian process change-points*, Section 27.6.3), BRAM (Bayesian regression with global *change-points*, Section 27.6.2), homogBR (homogeneous Bayesian regression without *change-points*, Section 27.4.3), LASSO and *Elastic Net* (Section 27.4.2). BRAMP outperforms all other schemes, which is not surprising, in that the data have been generated from a process that is consistent with the modeling assumptions of BRAMP. Notable is also the improvement of BRAMP over BRAM, which indicates that BRAMP is more flexible in terms of identifying local segments. It is reassuring that both nonhomogeneous Bayesian regression schemes (BRAMP and BRAM) can handle the increased model complexity, and improve network reconstruction accuracy compared to the competing methods homogBR, LASSO, and *Elastic Net*.

27.6.7.2 Simulated Population Dynamics. For a fair comparison, additional spatial autocorrelation variables (Section 27.6.1) were added to all methods, enabling them to account for dispersion effects intrinsic to this data set. The *Elastic Net* method was excluded from comparison because of the very similar results to LASSO. Figure 27.9 shows the AUROC scores for four different settings of the spatial β parameter that controls the heterogeneity of species distributions as

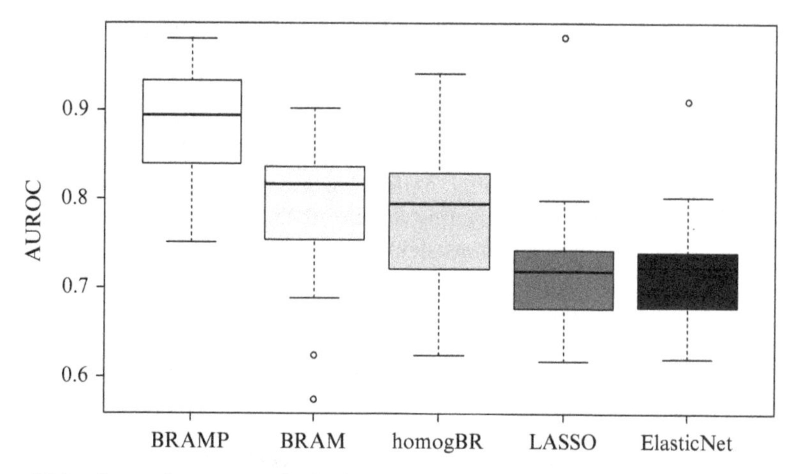

Figure 27.8 Comparison on synthetic data. Boxplots of AUROC scores obtained with five methods on the synthetic data described in Section 27.5: A *Bayesian regression model with* Mondrian process *change-points* (BRAMP, Section 27.6.4), a *Bayesian regression model with global change-points* (BRAM, Section 27.6.2), a Bayesian linear regression model without *change-points* (homogBR, Section 27.4.3), L1-penalized sparse regression (LASSO, Section 27.4.2), and the sparse regression *Elastic Net* method (Section 27.4.2). The boxplots show the distributions of the scores for 30 independent data sets with higher scores indicating better learning performance.

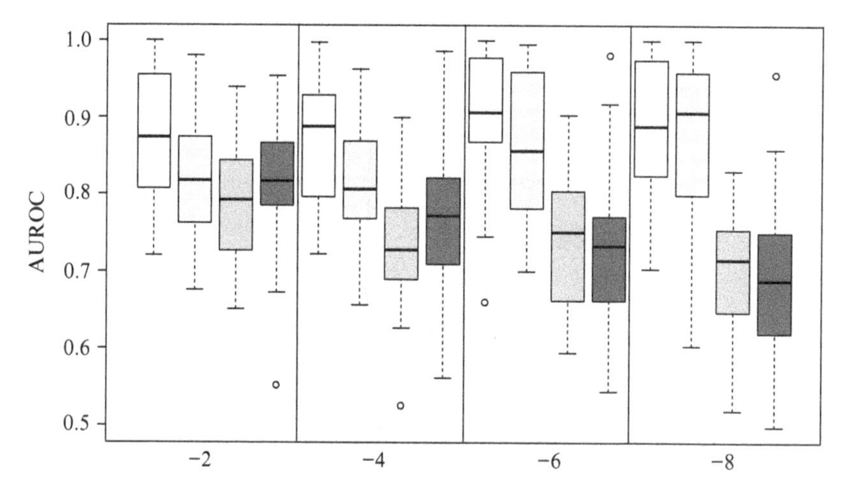

Figure 27.9 Comparative evaluation of four network reconstruction methods for the stochastic population dynamics data. Boxplots of AUROC scores obtained on the realistic simulated data described in Section 27.6.5.3 for different settings of the spatial β parameter, with lower values causing stronger heterogeneity in the data. Box color scheme: BRAMP (white), BRAM (light gray), homogBR (gray), and LASSO (dark gray).

illustrated in Figure 27.7. Lower values of β lead to the formation of clusters or "neighborhoods" of similar species concentrations, making it more difficult to learn underlying structures when averaging over the whole data domain. By treating these neighborhoods separately with a partitioning scheme, we would expect improved inference. In fact, Figure 27.9 shows that methods with a *change-point* process, BRAMP and BRAM, produce better AUROC scores than the competing methods homogBR and LASSO, which lack this feature. BRAMP, the Bayesian regression model with local partitioning, consistently outperforms the other methods, as displayed in Figure 27.9 with the single exception of BRAM and a spatial $\beta = -8$. Table 27.1 summarizes the corresponding p-values of paired Wilcoxon tests for the AUROC scores comparing BRAMP against BRAM, homogBR, and LASSO. The low p-values indicate a significant performance gain of BRAMP and suggest

Table 27.1 Improvement of the Bayesian regression model with *Mondrian process change-points* (BRAMP) on the stochastic population dynamics data

Spatial β:	−2	−4	−6	−8
BRAM	2.2e−04	1.9e−04	6.4e−03	0.14
homogBR	1.2e−06	2.9e−07	1.0e−07	1.9e−09
LASSO	6.1e−04	7.2e−04	1.3e−08	9.3e−10

p-Values for paired one-sided Wilcoxon tests for the difference of AUROC scores between BRAMP and the competing methods (BRAM, homogBR, LASSO) for several spatial β values. The alternative hypothesis states that BRAMP scores are greater than the competing methods with low p-values < 0.05 indicating significant performance gain of BRAMP.

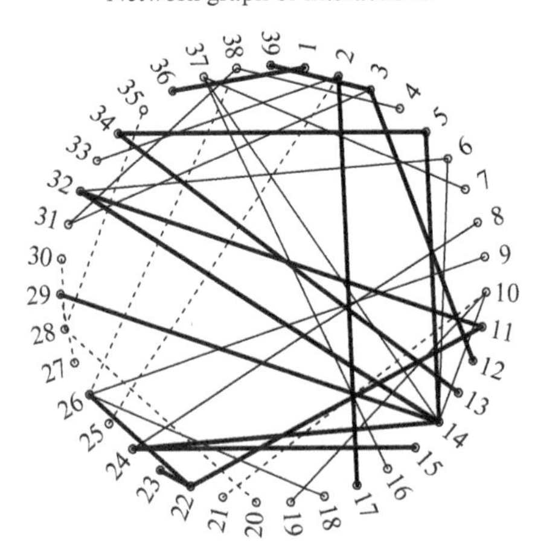

Figure 27.10 Species interaction network. Species interactions as inferred with BRAMP (Section 27.6.3), with an inferred marginal posterior probability of 0.5 (thick lines) and 0.1 (thin lines). Solid lines are positive interactions (e.g., mutualism, commensalism) and dashed are negative interactions (e.g., resource competition). Species are represented by numbers and have been ordered phylogenetically as displayed in Table 27.2.

Table 27.2 Indices with full scientific names as appearing in Figure 27.10

ID	Name	ID	Name
1	*Anagallis tenella*	21	*Aira praecox*
2	*Calluna vulgaris*	22	*Anthoxanthum odoratum*
3	*Drosera rotundifolia*	23	*Cynosurus cristatus*
4	*Epilobium palustre*	24	*Festuca rubra*
5	*Galium verum*	25	*Festuca vivipara*
6	*Hypochaeris radicata*	26	*Holcus lanatus*
7	*Leontodon autumnalis*	27	*Koeleria macrantha*
8	*Lychnis flos-cuculi*	28	*Molinia caerulea*
9	*Odontites verna*	29	*Poa pratensis*
10	*Plantago lanceolata*	30	*Juncus effusus*
11	*Potentilla erecta*	31	*Juncus kochii*
12	*Potentilla palustris*	32	*Luzula campestris*
13	*Prunella vulgaris*	33	*Luzula pilosa*
14	*Ranunculus bulbosus*	34	*Carex arenaria*
15	*Ranunculus repens*	35	*Carex demissa*
16	*Sagina procumbens*	36	*Carex dioica*
17	*Succisa pratensis*	37	*Carex flacca*
18	*Trifolium repens*	38	*Carex nigra*
19	*Viola riviniana*	39	*Eriophorum angustifolium*
20	*Agrostis capillaris*		

These plants can be assigned to four taxonomies of forbs (1–19), grasses (20–29), rushes (30–33), and sedges (34–39).

that the *Mondrian process* better captures spatial heterogeneity. In fact, both nonhomogeneous Bayesian regression models, BRAMP and BRAM, achieve high AUROC scores for the data simulated with low spatial β values, that is, high data heterogeneity. In contrast, the performance of homogBR and LASSO deteriorates as expected with higher data heterogeneity (i.e., lower spatial β).

27.6.7.3 *Real-World Plant Data.* We have applied BRAMP to the plant abundance data from the ecological survey described in Section 27.6.6. We sampled interaction network structures from the posterior distribution with MCMC and computed the marginal posterior probabilities of the individual potential species interactions. We kept all species interactions with a marginal posterior probability above a certain threshold, as shown in Figure 27.10, resulting in 39 out of 106 species with relevant interaction in the reconstructed network shown in Figure 27.10. Since we had defined the 12 soil attributes as fixed predictors to each plant, the interactions in this network represent plant–plant interactions not mediated by similar soil preferences. This network can lead to the formation of new ecological hypotheses. For instance, *Ranunculus bulbosus* (species 14) is densely connected with five interspecific links above the threshold. Can that be related to its tolerance for nutrient-poor soil and its preferred occurrence in species-rich patches? There is a noticeable imbalance between positive and negative interactions. An initial consultation with ecologists indicates the fact that our analysis tends to find more positive than negative links is interesting as it points to a dominance of commensalism over competition. The importance of commensalism was emphasized in Reference [9]. Ecologists also suggest that positive interactions may be more characteristic for harsh environments (e.g., in Reference [8]) as is found in the Machair vegetation. These results demonstrate that the proposed method provides a useful tool for exploratory data analysis in ecology with respect to both species interactions and spatial heterogeneity.

27.7 CONCLUSION

We have addressed the problem of reconstructing gene regulation networks from molecular expression data and modified the corresponding methods to reconstruct species interaction networks from species abundance data. To this end, we have described and applied two sparse regression methods, LASSO and *Elastic Net*, and a Bayesian regression method. The latter was extended from a homogeneous model (homogBR) to a nonhomogeneous model that can approximate nonhomogeneity with multiple *change-points*. This nonhomogeneous Bayesian regression method (nonhomogBR) was successfully applied to the inference of a circadian regulation network and benefited in terms of better performance from light-induced day and night phases. We have seen that several commonalities exist between molecular and ecological systems. They can be exploited to adapt established inference methods from systems biology to ecological applications. This was demonstrated on the nonhomogeneous Bayesian regression method in two ways: the partitioning of time was modified to segment a two-dimensional spatial domain, and an additional variable that reflects species dispersion effects was introduced. As a result, a global *change-point* model (BRAM) and a spatially varying, local partitioning scheme (BRAMP) were applied to realistic simulated and real data. We found that both approaches greatly benefit from learning the spatial segmentation in terms of the quality of inferring species interactions. They showed superior performance compared to LASSO and homogBR.

These results are encouraging and suggest that ecological modeling could greatly benefit from the methodologies established in systems biology during the last decade. This is of particular relevance given the great complexity of ecological systems and the increasing amount of data that has to be analyzed. A substantial improvement of the methodology could help to better understand the ecological processes, such as the impact of global warming on biodiversity [47], land use and agriculture [39], or the stability of ecological systems that are threatened by a radical extinction of plant and animal life, as occurring in the Amazonian rain forest [44].

REFERENCES

1. Aderhold A, Husmeier D, Lennon J, Beale C, Smith V. Hierarchical Bayesian models in ecology: reconstructing species interaction networks from non-homogeneous species abundance data. Ecol Inform 2012;11:55–64.

2. Aderhold A, Husmeier D, Smith V. Reconstructing ecological networks with hierarchical Bayesian regression and Mondrian processes. Proceedings of the 16th International Conference on Artificial Intelligence and Statistics; Phoenix, AZ; 2013. p 75–84.
3. Aderhold A, Husmeier D, Smith V, Millar A, Grzegorczyk M. Assessment of regression methods for inference of regulatory networks involved in circadian regulation. 10th International Workshop on Computational Systems Biology (WCSB); Tampere, Finland; 2013. p 30–42.
4. Andrieu C, Doucet A. Joint Bayesian model selection and estimation of noisy sinusoids via reversible jump MCMC. IEEE Trans Signal Process 1999.;47(10):2667–2676.
5. Arbeitman M, Furlong E, Imam F, Johnson E, Null B, Baker B, Krasnow M, Scott M, Davis R, White K. Gene expression during the life cycle of *Drosophila melanogaster*. Science 2002;297(5590):2270–2275.
6. Beisner B, Haydon D, Cuddington K. Alternative stable states in ecology. Front Ecol Environ 2003;1(7):376–382.
7. Bishop C. *Pattern Recognition and Machine Learning*. Singapore: Springer-Verlag; 2006.
8. Brooker R, Callaghan T. The balance between positive and negative plant interactions and its relationship to environmental gradients: a model. Oikos 1998;81(1):196–207.
9. Bruno J, Stachowicz J, Bertness M. Inclusion of facilitation into ecological theory. Evolution 2003;18(3):119–125.
10. Ciocchetta F, Hillston J. Bio-PEPA: A framework for the modelling and analysis of biological systems. Theor Comput Sci 2009;410(33):3065–3084.
11. Cohen J. Mathematics is biology's next microscope, only better; biology is mathematics' next physics, only better. PLoS Biol 2004;2(12):e439.
12. Cohen J, Schoenly K, Heong K, Justo H, Arida G, Barrion A, Litsinger J. A food web approach to evaluating the effect of insecticide spraying on insect pest population dynamics in a Philippine irrigated rice ecosystem. J Appl Ecol 1994;31:747–763.
13. Davies J, Goadrich M. The relationship between Precision-Recall and ROC curves. Proceedings of the 23rd International Conference on Machine Learning; Pittsburgh, PA; 2006. p 233–240.
14. D'haeseleer P, Liang S, Somogyi R. Genetic network inference: from co-expression clustering to reverse engineering. Bioinformatics 2000;16(8):707–726.
15. Dunne J, Williams R, Martinez N. Network structure and biodiversity loss in food webs: robustness increases with connectance. Ecol Lett 2002;5:558–567.
16. Edwards KD, Akman OE, Knox K, Lumsden PJ, Thomson AW, Brown PE, Pokhilko A, Kozma-Bognar L, Nagy F, Rand DA, Millar AJ. Quantitative analysis of regulatory flexibility under changing environmental conditions. Mol Syst Biol 2010;6:424.
17. Faisal A, Dondelinger F, Husmeier D, Beale C. Inferring species interaction networks from species abundance data: a comparative evaluation of various statistical and machine learning methods. Ecol Inform 2010;5(6):451–464.
18. Flis A, Fernandez P, Zielinski T, Sulpice R, Pokhilko A, McWatters H, Millar A, Stitt M, Halliday K. Biological regulation identified by sharing timeseries data outside the 'omics; 2013. Submitted.
19. Fogelberg C, Palade V. Machine learning and genetic regulatory networks: a review and a roadmap. In: *Foundations of Computational Intelligence*. Volume 1. Berlin, Germany: Springer-Verlag; 2009. p 3–34.

20. Friedman J, Hastie T, Tibshirani R. Regularization paths for generalized linear models via coordinate descent. J Stat Softw 2010;33(1):1–22. [Online]. Available at http://www.jstatsoft.org/v33/i01/.
21. Friedman N, Linial M, Nachman I, Pe'er D. Using Bayesian networks to analyze expression data. J Comput Biol 2000;7:601–620.
22. Gelman A, Rubin D. Inference from iterative simulation using multiple sequences. Stat Sci 1992;7:457–511.
23. Gillespie D. Exact stochastic simulation of coupled chemical reactions. J Phys Chem 1977;81(25):2340–2361.
24. Green P. Reversible jump Markov chain Monte Carlo computation and Bayesian model determination. Biometrika 1995;82:711–732.
25. Grzegorczyk M, Husmeier D. A non-homogeneous dynamic Bayesian network with sequentially coupled interaction parameters for applications in systems and synthetic biology. Stat Appl Genet Mol Biol (SAGMB) 2012;11(4):Article 7.
26. Grzegorczyk M, Husmeier D. Regularization of non-homogeneous dynamic Bayesian networks with global information-coupling based on hierarchical Bayesian models. Mach Learn 2013;91(1):105–154.
27. Guerriero M, Pokhilko A, Fernández A, Halliday K, Millar A, Hillston J. Stochastic properties of the plant circadian clock. J R Soc Interface 2012;9(69):744–756.
28. Hanely J, McNeil B. The meaning and use of the area under a receiver operating characteristic (ROC) curve. Radiology 1982;143(1):29–36.
29. Hayete B, Gardner T, Collins J. Size matters: network inference tackles the genome scale. Mol Syst Biol 2007;3:77.
30. Henneman M, Memmott J. Infiltration of a Hawaiian community by introduced biological control agents. Science 2001;293(5533):1314–1316.
31. Hertel J, Lindemeyer M, Missal K, Fried C, Tanzer A, Flamm C, Hofacker IL, Stadler PF, Students of Bioinformatics Computer Labs 2004 and 2005. The expansion of the metazoan microRNA repertoire. BMC Genomics 2006;7(1):25.
32. Husmeier D. Sensitivity and specificity of inferring genetic regulatory interactions from microarray experiments with dynamic Bayesian networks. Bioinformatics 2003;19:2271–2282.
33. Kalaitzis A, Honkela A, Gao P, Lawrence N. gptk: Gaussian Processes Tool-Kit; 2013, r package version 1.06. [Online]. Available at http://CRAN.R-project.org/package=gptk. Accessed 2015 Aug 8.
34. Lande R, Engen S, Saether B. *Stochastic Population Dynamics in Ecology and Conservation*. New York: Oxford University Press; 2003.
35. Lèbre S, Becq J, Devaux F, Stumpf M, Lelandais G. Statistical inference of the time-varying structure of gene-regulation networks. BMC Syst Biol 2010;4:130.
36. Lennon J. Red-shifts and red herrings in geographical ecology. Ecography 2000;23:101–113.
37. Lennon J, Beale C, Reid C, Kent M, Pakeman R. Are richness patterns of common and rare species equally well explained by environmental variables. Ecography 2011;34:529–539.
38. Locke JC, Southern MM, Kozma-Bognar L, Hibberd V, Brown PE, Turner MS, Millar AJ. Extension of a genetic network model by iterative experimentation and mathematical analysis. Mol Syst Biol 2005;1:2005.0013.

39. Mendelsohn R, Nordhaus W, Shaw D. The impact of global warming on agriculture: a ricardian analysis. Am Econ Rev 1994;84:753–771.
40. Passos J, Nelson G, Wang C, Richter T, Simillion C, Proctor CJ, Miwa S, Olijslagers S, Hallinan J, Wipat A, Saretzki G, Rudolph KL, Kirkwood TB, von Zglinicki T. Feedback between p21 and reactive oxygen production is necessary for cell senescence. Mol Syst Biol 2010;6:347.
41. Pokhilko A, Fernández A, Edwards K, Southern M, Halliday K, Millar A. The clock gene circuit in *Arabidopsis* includes a repressilator with additional feedback loops. Mol Syst Biol 2012;8:574.
42. Pokhilko A, Hodge SK, Stratford K, Knox K, Edwards KD, Thomson AW, Mizuno T, Millar AJ. Data assimilation constrains new connections and components in a complex, eukaryotic circadian clock model. Mol Syst Biol 2010;6:416.
43. Pokhilko A, Mas P, Millar A. Modelling the widespread effects of TOC1 signalling on the plant circadian clock and its outputs. BMC Syst Biol 2013;7(1):1–12.
44. Rammig A, Thonicke K, Jupp T, Ostberg S, Heinke J, Lucht W, Cramer W, Cox P. Estimating Amazonian rainforest stability and the likelihood for large-scale forest dieback. EGU General Assembly Conference Abstracts, Volume 12; 2010. p 14289.
45. Roy DM. Computability, inference and modeling in probabilistic programming [PhD dissertation]. Massachusetts Institute of Technology; 2011.
46. Sachs K, Perez O, Pe'er D, Lauffenburger D, Nolan G. Causal protein-signaling networks derived from multiparameter single-cell data. Science 2005;308(5721):523–529.
47. Sala OE, Chapin FS, Armesto JJ, Berlow E, Bloomfield J, Dirzo R, Huber-Sanwald E, Huenneke LF, Jackson RB, Kinzig A, Leemans R, Lodge DM, Mooney HA, Oesterheld M, Poff NL, Sykes MT, Walker BH, Walker M, Wall DH. Global biodiversity scenarios for the year 2100. Science 2000;287(5459):1770–1774.
48. Scheffer M, Carpenter S, Foley JA, Folke C, Walker B. Catastrophic shifts in ecosystems. Nature 2001;413(6856):591–596.
49. Schena M, Shalon D, Davis R, Brown P. Quantitative monitoring of gene expression patterns with a complementary DNA microarray. Science 1995;270(5235):467–470.
50. Shinozaki K, Yamaguchi-Shinozaki K, Seki M. Regulatory network of gene expression in the drought and cold stress responses. Curr Opin Plant Biol 2003;6(5):410–417.
51. Sinclair A, Byrom A. Understanding ecosystem dynamics for conservation of biota. J Anim Ecol 2006;75(1):64–79.
52. Solak E, Murray-Smith R, Leithead W, Leith D, Rasmussen C. Derivative observations in Gaussian process models of dynamic systems. In: *Proceedings of Neural Information Processing Systems*. Vancouver, Canada: MIT Press; 2002.
53. Solomon S. *Climate Change 2007 - The Physical Science Basis: Working Group I Contribution to the Fourth Assessment Report of the IPCC*. Volume 4. Cambridge, United Kingdom: Cambridge University Press; 2007.
54. Tait R. The application of molecular biology. Curr Issues Mol Biol 1999;1(1):1–12.
55. Tibshirani R. Regression shrinkage and selection via the Lasso. J R Stat Soc Ser B (Methodological) 1995;58(1):267–288. [Online]. Available at http://www.jstor.org/stable/2346178.
56. TiMet-Consortium. The TiMet Project - Linking the clock to metabolism; 2012. [Online] Available at http://timing-metabolism.eu. Accessed 2015 Aug 8.

57. Vyshemirsky V, Girolami M. Bayesian ranking of biochemical system models. Bioinformatics 2008;24:833–839.
58. Wang G, Zhu X, Hood L, Ao P. From Phage lambda to human cancer: endogenous molecular-cellular network hypothesis. Quant Biol 2013;1(1):1–18.
59. Weirauch MT, Cote A, Norel R, Annala M, Zhao Y, Riley TR, Saez-Rodriguez J, Cokelaer T, Vedenko A, Talukder S, DREAM5 Consortium, Bussemaker HJ, Morris QD, Bulyk ML, Stolovitzky G, Hughes TR. Evaluation of methods for modeling transcription factor sequence specificity. Nat Biotechnol 2013;31(2):126–134.
60. Werhli AV, Grzegorczyk M, Husmeier D. Comparative evaluation of reverse engineering gene regulatory networks with relevance networks, graphical Gaussian models and Bayesian networks. Bioinformatics 2006;22:2523–2531.
61. Wilkinson D. *Stochastic Modelling for Systems Biology*. Volume 44. Boca Raton, FL: CRC Press; 2011.
62. Williams R, Martinez N. Simple rules yield complex food webs. Nature 2000;404(6774): 180–183.
63. Zou H, Hastie T. Regularization and variable selection via the Elastic net. J R Stat Soc Ser B (Stat Methodol) 2005;67(2):301–320. [Online]. DOI: 10.1111/j.1467-9868.2005.00503.x.

CHAPTER 28

DISCOVERING CAUSAL PATTERNS WITH STRUCTURAL EQUATION MODELING: APPLICATION TO TOLL-LIKE RECEPTOR SIGNALING PATHWAY IN CHRONIC LYMPHOCYTIC LEUKEMIA

Athina Tsanousa[1], Stavroula Ntoufa[2,3], Nikos Papakonstantinou[2,3], Kostas Stamatopoulos[2,3], and Lefteris Angelis[1]

[1]*Department of Informatics, Aristotle University of Thessaloniki, Thessaloniki, Greece*

[2]*Hematology Department and HCT Unit, G. Papanikolaou Hospital, Thessaloniki, Greece*

[3]*Institute of Applied Biosciences, Centre for research and technology Hellas, Thessaloniki, Greece*

28.1 INTRODUCTION

Statistical methods are applied in many scientific fields where complex systems have to be studied and there is the need to discover relations and patterns among data or

Pattern Recognition in Computational Molecular Biology: Techniques and Approaches,
First Edition. Edited by Mourad Elloumi, Costas S. Iliopoulos, Jason T. L. Wang, and Albert Y. Zomaya.

verify already found ones. Structural Equation Modeling (SEM) is an appropriate method that is able to conduct such an analysis. In SEM, models represent causal relationships, while the covariance as well as the direction of causality among variables can be estimated [23]. The early "form" of SEM is path analysis. Path analysis was introduced by the geneticist S. Wright in the early 1920s in order to verify already existing relations among genes [34].

Path discovery and estimation is not the only utility of SEM and path analysis. The property that makes them distinguished among other multivariate data analysis methods is the ability to concurrently analyze multiple relations. A structural equation model consists of several equations, each one describing causal relations, which are built based on data and theoretical assumptions [47]. This ability made SEM a popular method in various scientific fields, such as economics, biology, and social sciences among others.

In the behavioral and financial sciences, the actual causal relations are rarely known and the models are built on hypothetical assumptions [34]. In such cases, there are constructs that cannot be directly measured but are estimated through others. These constructs are defined as latent variables. SEM has the additional advantage of allowing the study of such variables [34].

As mentioned earlier, the first "version" of SEM was path analysis introduced in Reference [58]. Wright needed to "study the direct influence of one variable on another in experiments where all other possible causes of variation were eliminated." This property is rarely found in biological sciences, because of the existence of complex interactions. Statistical methods established until then could estimate the correlations among variables but could not define the direction of these relationships or whether the relations were direct or indirect [58]. Wright applied path analysis to biological data with already established relationships and tried to estimate the magnitude of the effects. That was the first application of this method in the field of biology, but its usefulness and the scientific fields of application were expanded afterward [34].

In Reference [51], the concept of structure was explicitly discussed, while a formal presentation of Structural Equation Models was given by Pearl [47]. SEM is an extension of path analysis, providing the ability to perform analysis for latent variables. Both methods have a wide area of applications. Owing to their nature, path analysis is mostly found in biological studies and SEM mostly in sociology and psychology.

In biology and medicine, SEM has been applied to study causal relations in pathways in numerous papers [11, 12, 37]. The biology-related applications of SEM may usually involve only observed variables; however, there are a number of cases where unmeasured quantities must be estimated [43, 56]. Psychology and social sciences are fields in which SEM has been applied extensively. In fact, a number of social scientists have contributed toward the enrichment of SEM theory [10, 34]. In studies of these fields, the most useful property of SEM is the estimation of latent variables as there are numerous constructs that are measured through observed quantities [26]. Applications of SEM can also be found in econometric journals [21, 39].

Lately, there has been a growing interest in gene pathway identification, a scientific field that is closely related to biological data knowledge discovery. Hence, SEM, in general and path analysis in particular are essentially methods for discovering causal patterns among biological structures such as proteins and genes which are expressed as numerical variables measured from samples either in clinical trials or experiments. These patterns and their variations among groups of individuals can prove especially useful in studying and treating certain diseases. In this chapter, SEM is applied in such measurements in order to statistically investigate the Toll-Like Receptor (TLR) signaling pathway in *Chronic Lymphocytic Leukemia* (CLL).

CLL is the most common type of leukemia in the western hemisphere [18] and presents great interest due to the subgroups of the patients. The patients are categorized as Mutated *CLL* (M-CLL) and Unmutated *CLL* (U-CLL), according to the mutational status of specific genes. Extended study of the expression profiles of genes participating in the TLR signaling pathway from samples obtained from CLL patients have shown that the TLR pathway is active in CLL [6, 22, 42, 44]. Moreover, the existence of significant differences between M-CLL and U-CLL [44] may imply that these differences will be observed in the signaling pathway too. The research conducted so far was peripheral to the TLR signaling pathway of the CLL patients. Therefore, a method to study the exact path of signals needed to be applied in order to explore the pathway itself and not just the properties of the genes participating in it.

The TLR pathway is statistically investigated through SEMs separately for the M-CLL and the U-CLL patients, as these two subgroups perform differently. The signal transferred from a gene to another is statistically interpreted as a causal effect. The order of the signaling pathway goes along with the assumption of causal diagrams that states that the variable causing the effect must occur before the variable receiving it [34]. A part of our recent research efforts have been already presented in the BIOKDD workshop of the DEXA conference [54]. Here, we extend the previous research by building more models and by reporting additional results. It should be noted that as there are no latent variables in the models under study, it can be considered that we applied path analysis. However, we refer to the method principally as SEM, as it is the most general methodology and the models can potentially be expanded by considering latent variables whenever there is a rationale to do so.

This chapter is organized as follows: In Section 28.2, we provide the biological background of our research regarding CLL and TLR pathways. In Section 28.3, we describe the theory and methodology regarding SEM, in Section 28.4 we present the application of the method to a data set by building models and estimating the causal relationships of the variables, and finally in Section 28.5, we discuss the results.

28.2 TOLL-LIKE RECEPTORS

This section aims at introducing the reader to some basic biological terminology needed for this chapter, which includes definition of the proteins and description of the interactions among those that play a crucial part in CLL.

28.2.1 Basics

TLRs are a class of proteins that play critical roles in host defense against infection; they recognize conserved motifs called Pathogen Associated Molecular Patterns

(PAMPs) predominantly found in microbial components but not in vertebrates. Stimulation of TLRs induces an immediate signaling cascade resulting in the production of inflammatory cytokines and the expression of costimulatory molecules [3, 4].

28.2.2 Structure and Signaling of TLRs

TLRs are type 1 transmembrane receptors. Their extracellular domain includes Leucine-Rich Repeat (LRR) motifs that recognize the ligands. In general, when TLRs recognize a ligand, they form dimers in order to induce a signal. These dimers are either homodimers, as in the case of TLR9, or heterodimers, as in the case of TLR2/TLR1 or TLR2/TLR6 [46]. Ligand binding induces symmetrical dimerization of TLR ectodomains. As a result, the cytoplasmic parts of two TLR molecules come into close proximity through conformational changes. The TLR cytoplasmic region is known as the Toll/IL-1R (TIR) domain because it shares a conserved section of 200 amino acids with the *Interleukin-1 Receptor* (IL-1R) superfamily. The homology in the TIR motif is restrained to three conserved boxes that contain amino acids crucial for signaling. It is generally thought that TIR–TIR interaction at the receptor level provides the location for interaction with the TIR domain-containing adapters that mediate signaling by linking the receptors to downstream intracellular proteins [2].

The first TIR domain-containing adapter found to be essential for the TLR signaling cascade was Myeloid Differentiation primary response gene (88) (MyD88). MyD88 is a member of the Toll/IL-1 receptor family and consequently, its function is crucial for signaling induced by IL-1, as well as for all TLRs except TLR3 [40]. Later on, more adapter molecules homologous to MyD88 were found to be required for TLR signal transduction, including Mal/TIRAP [19]. TIRAP (toll-interleukin 1 receptor (TIR) domain containing adaptor protein) protein is required for TLR2 and TLR4 signaling; in fact, TIRAP is the bridging adapter that recruits MyD88 to the receptor and MyD88 is the signaling adapter leading to the downstream signaling. In contrast to MyD88, TIRAP can associate with TRAF6, a TNF (Tumor Necrosis Factor) receptor-associated factor. It seems that TIRAP is responsible for the recruitment of TRAF6 in the TLR2 and TLR4 signaling complexes [38].

TRIF (TIR-domain-containing adapter-inducing interferon-b) is another adapter utilized by TLR3 and also required for TLR4 signaling [45]. TRIF has been shown to be important for the induction of the apoptotic pathway in response to TLR stimulation. Along with TRIF, the apoptotic pathway also involves proteins Fas-Associated Protein with Death Domain (FADD) and caspase-8. TLR3 interacts with TRIF directly. On the other hand, *Translocating Chain-Associating Membrane* (TRAM) protein is required for TLR4 signaling in order to engage TRIF. Thus, TRAM is also a bridging adapter, similar to TIRAP, and is different from all other adapters because it is utilized only by TLR4 [48].

TLR signaling pathways lead to the activation of NF-κB and MAP kinases. All the functional TLRs, apart from TLR3, use the adapter MyD88 in order to transduce the signal. After TLR stimulation, MyD88 is recruited to the receptor complex through TIR–TIR interactions [40]. MyD88 recruits IRAK-4 and IRAK-1 (interleukin-1 receptor-associated kinase 4 and 1, respectively) in the complex. For an illustration of the biological interactions, we refer the reader to the following site (http://www.kegg.jp/kegg/pathway.html).

IRAK-1 associates with the Toll-interacting protein (Tollip, a negative regulator of IRAK-1), MyD88, and TRAF6. Phosphorylation of IRAK-1 by IRAK-4 triggers IRAK-1 hyperphosphorylation and dissociation from the complex along with TRAF6 [41]. In this way, the IRAKs link the receptor complexes with TRAF6 and induce TRAF6 activation. The IRAK-1/TRAF6 complex interacts with another preformed complex consisting of MAP3K7 (Mitogen-activated protein kinase kinase kinase 7), also known as TAK1, and two adaptor proteins, MAP3K7IP1 (TAB1) and TAB2.

TAB1 is an activator of TAK1. TAB2 functions as an adaptor molecule, linking TAK1 to TRAF6, thereby facilitating TAK1 activation [1]. The interaction of the two complexes induces phosphorylation of TAB2 and TAK1, which then translocate together with TRAF6 and TAB1 to the cytosol. TAK1 is then activated, leading to the activation of the NF-κB pathway [1]. TAK1 activation is dependent on ubiquitination. TRAF6 interacts with the Ubc13 in the ubiquitin-conjugating enzyme complex Ubc13/Uev1A. This Ubc/TRAF6 complex triggers TAK1 activation through a unique proteasome-independent mechanism. TAK1 seems to be the main kinase that phosphorylates and activates the NF-κB pathway. TAK1 is also linked to the MAPK cascades [1, 2, 41]. Both pathways are very commonly used in B-cells in response to different receptors and signals and, depending on the signal, they lead to production of proinflammatory cytokines, proliferation differentiation, or apoptosis.

28.2.3 TLR Signaling in Chronic Lymphocytic Leukemia

CLL is a malignancy of mature B cells and the most frequent leukemia in the western hemisphere [18]. It is now unequivocally accepted that CLL development probably involves aberrant activation and proliferation of antigen-experienced B-cells stimulated through the BCR (Breakpoint Cluster Region Protein) by both self and exogenous antigens [17].

CLL patients can be assigned to two main groups, namely, *Mutated* (M-CLL) and *Unmutated* (U-CLL), defined by the mutational status of the immunoglobulin genes expressed by the antigen receptors. U-CLL is associated with a more aggressive course, while M-CLL is usually more indolent [18]. This categorization is also in accordance with BCR signaling, highlighting the important role of BCR signaling in CLL. Indeed, U-CLL cases can signal through their BCR, while M-CLL cases are thought to have attained a functional status of anergy [24].

Recent studies have shown that CLL cells recognize common pathogens, that is, bacteria or apoptotic remnants [35]. Such structures are also recognized by TLRs which are known to be expressed on B-cells. Authors participating in this study and others have shown that TLRs also play a crucial role in CLL B-cells [6, 22, 42, 44]. By analyzing a cohort of 192 CLL cases in more detail, we recorded high expression for TLR7 and CD180, intermediate for TLR1, TLR6, and TLR10 and low expression for TLR2 and TLR9. TLR4 and TLR8 were characterized by low-to-undetectable expression, with significant variation between patients, while TLR3 and TLR5 were not expressed in any case.

The vast majority of the adaptors (e.g., MyD88, TICAM1, TRAF6), effectors (UBE2N), and members of the NFKB, JNK/p38, NF/IL6, and IRF signaling pathways downstream of TLR signaling were intermediately-to-highly expressed, while the inhibitors of TLR activity were generally low-to-undetectable (TOLLIP, SIGIRR/TIR8), suggesting that the TLRs signaling pathway is active in CLL. Further comparison of M-CLL versus U-CLL (124 and 67 cases, respectively) revealed upregulation of CD80, CD86, IL6, IFNG, and TLR4 and downregulation of TLR8 and NFKBIL1 in the former [6]. In a subsequent study, we found differential profiles of functionality for specific TLRs within M-CLL and U-CLL [44]. These data suggest that CLL clones with distinct antigen reactivity are collaborating with different TLRs, alluding to specific recognition of and selection by the respective ligands.

28.3 STRUCTURAL EQUATION MODELING

SEM is a technique used for analyzing multivariate data. Structural equation models can generally be interpreted "as carriers of substantive causal or probabilistic information" [47]. Multivariate analysis techniques are useful for studying and analyzing relationships among data, in case there are more than one outcome variables. One common issue of these techniques is that they are able to access one single relationship at a time. SEM solves this problem by offering the researcher the advantage of examining multiple interdependence relationships at the same time [23].

There are three main aspects of SEM:

1. The role of variables: In the relationships being studied, a variable that is characterized as independent in one equation may be treated as dependent in another equation.
2. The discovery and evaluation of causal relationships among these variables: Causality is depicted through path diagrams, which are being evaluated for their fit and their statistical significance. New relationships are revealed and others are removed. There is a consecutive process of reconstruction of the model (or path diagram) and reevaluation of its fit.
3. The study of relationships between latent and observed variables: Latent variables (factors or constructs) cannot be directly measured, but can be approached through variables that can be measured (observed or manifest variables).These are called *factor loadings* or *indicators* of the latent variable [23].

28.3.1 Methodology of SEM Modeling

In SEM theory, there are two categorizations of variables: (i) exogenous and endogenous and (ii) latent and observed variables. Independent variables are characterized as exogenous and dependent ones as endogenous. Each endogenous variable is accompanied by an error term [13]. As mentioned earlier, a variable that is endogenous in one equation can be treated as exogenous in another equation.

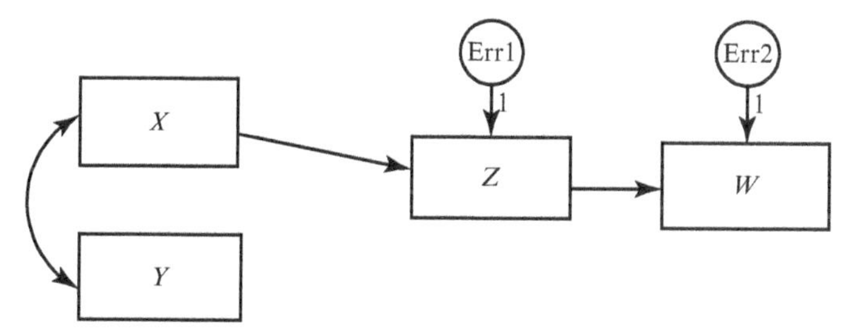

Figure 28.1 Path analysis: X and Y are the exogenous variables. The bidirectional arrow means that they are correlated. Z and W are endogenous and accompanied by an error term. Z can also be considered as exogenous, as it originates an effect that goes to W. Z receives a direct effect from X and W receives a direct effect from Z. W also receives an indirect effect of X through Z.

The relationships between these variables are graphically represented through path diagrams. The causal relations are depicted by arrows, which originate from an exogenous variable and result in an endogenous one, showing the derivation and the destination of the effect, respectively. Bidirectional arrows can be used to symbolize correlation but they do not imply causality [23]. A graphical depiction of a path diagram can be seen in Figure 28.1. Path diagrams undergo some assumptions that are mentioned explicitly later.

Latent variables, that is, variables that cannot be measured directly, are mostly used in the behavioral and financial sciences. The model that defines relations among the unobserved (latent) variables only is called a *structural model* [13]. The model that describes and studies the relationships between observed and unobserved variables is called a *measurement model*. The measurement model is somewhat similar to factor analysis. The difference is that in SEM the researcher determines the indicators of each latent variable, while in factor analysis, we have the ability to determine only the number of factors and not of their loadings [23]. When the models consist of observed variables only, path analysis is actually applied.

Special statistical packages have been developed in order to process SEM and path analysis. The first developed package for this kind of analysis was LISREL, which was developed by Joereskog in the 1970s [31]. Other popular software packages are MPLUS, Spss AMOS, SEM package in R, and CALIS package in SAS. In most of the packages, input data can be both in the form of a covariance matrix or individual observations. In case the input matrix consists of individual observations, the program converts it into covariance or correlation matrices [13], as the main goal of SEM is to analyze patterns of relationships, not individual cases [23].

28.3.2 Assumptions

Kline [34] stated that in most SEM-relevant studies, the assumptions are not given appropriate attention. This might result in invalid results. In summary, there are three assumptions that should be met in SEM. These assumptions are also found in other multivariate techniques: the observations should be independent, the respondents should be randomly chosen, and the relations should be linear.

There are special assumptions regarding path diagrams. In general, path diagrams have the following properties: (i) all causal relationships are indicated by theory, that is, there should be theoretical justification for the existence or the absence of a causal relationship. (ii) Causal relationships are assumed to be linear. Nonlinear relationships cannot be directly estimated in SEM [23]. (iii) There is no other exogenous cause that provokes the effect. More assumptions regarding directionality of the paths state that: the cause must occur before the effect and there is an observed covariance between them. The correct specification of the direction of causality is the most critical assumption according to Kline. Assumptions regarding the distribution and the form of the data vary depending on the estimation method used [34].

28.3.3 Estimation Methods

The parameters are estimated through the comparison of the actual covariance matrix with the estimated one of the best-fitting model and the regression weights or the path coefficients are tested for being statistically significant. Nonsignificant paths and correlations should be removed from the model. Other indications of "offensive" estimates include negative or nonsignificant error variances, standardized coefficients exceeding or being very close to 1, and very large standard errors [23]. The estimation of the regression weights is of great importance for SEM as they express the expected change in the endogenous variable when the exogenous variable changes by one unit.

Various estimation methods are applied in SEM, some of which are maximum likelihood, generalized least squares, unweighted least squares, and distribution-free methods [5]. The most frequently used are the maximum likelihood and generalized least squares methods. Both require continuous data following a multivariate normal distribution [13]. Maximum likelihood estimation is the method used in this chapter, as the most frequent and suitable choice.

28.3.4 Missing Data

Multivariate methods, among them SEM, require complete data in order to estimate the parameters. In real data, we rarely meet fully completed data sets. In SEM software packages, there are several options for dealing with this issue. We briefly refer to some methods dealing with missing data: (i) Listwise Deletion (LD), (ii) Pairwise Deletion (PD), and (iii) imputation.

In LD, there is the assumption that the data are Missing Completely at Random (MCAR), which means that there is no dependence between the probability of finding a specific pattern in missing data and the observed data values. Listwise deletion removes incomplete cases. Even if there is a missing value in one variable, the whole case is removed. In case of a small sample size, this can result in a great reduction in the sample [14].

PD also assumes that data are MCAR [14]. PD removes only the observation with the missing value, if it is used in a calculation. In calculations containing more than one variable, a pair of observations is removed, if there is a missing value in one of the variables or in both of them [5].

In case of imputation (iii), estimated values are entered in the place of missing ones. There are various ways of calculating the values used for the replacement. The missing values can be substituted with the overall mean value of the particular variable, by a predicted value calculated by multiple regression of the non-missing values, and by values calculated by the Estimation–Maximization algorithm (EM). EM is an iterative algorithm, in which an estimation step is followed by a maximization step [14]. In the current application, EM was used in order to impute missing cases. EM seemed more preferable to the other available methods, as it uses information not only from the variable with the missing values but from all variables in order to impute missing cases. This procedure is more suitable for the data, because as some genes are correlated, the value of a variable that represents the expression level of the respective gene is not independent of the other observations.

28.3.5 Goodness-of-Fit Indices

In this section, the fit indices used for a particular application are described. As in any modeling procedure in statistics, the creation of a model is followed by testing its fit to the data. In SEM, there is a variety of fit indices available, evaluating different aspects of the proposed model. In order for them to be more comprehensible, they are represented here in categories that are formed according to the comparisons made for their calculation.

1. Absolute measures of fit
   - GFI-AGFI: The Goodness-of-Fit Index (GFI) and Adjusted Goodness-of-Fit Index (AGFI) can be classified in this category, because they basically do not compare the hypothesized model with any model [13]. The GFI was proposed by Joereskog and Sorbom [29] as best suited for models that are estimated with maximum likelihood estimation or with unweighted least squares. Tanaka and Huba [53] generalized this to other estimation methods as well [5]. GFI compares the covariance structure of the model to that of the sample data [16]. AGFI is an adjustment of GFI that takes into consideration the number of degrees of freedom of the specified model [13]. It is widely known that the values of GFI and AGFI range between 0 and 1; however, there are some references in the literature that this interval is not absolute [20]. AGFI is bounded above by 1; however, it is not bounded below by 0, as is the GFI [5]. Joereskog and Sorbom [30] noted that there is a theoretical possibility for them to be negative [13]. A general rule is that the closer these indices are to 1, the better the fit of the model. Perfect fit is achieved when the indices are equal to 1, which is not usually met, but if their values are equal to or greater than 0.9, this is a suggestion of meaningful realistic models [7]. There are researchers that question the practical importance of these indices [20]. GFI and AGFI are sensitive to sample size [13].

- RMSEA: The Root Mean Square Error of Approximation (RMSEA) is a measure that represents the goodness of fit that could be expected if the model was estimated in the population, not in the sample [23]. Values less than 0.05 indicate good fit and values close to 0.08 represent reasonable errors of approximation. Values varying from 0.08 to 0.10 represent mediocre fit and values greater than 0.10, poor fit. The ideal fit is achieved when RMSEA = 0, which is not a realistic value [13]. RMSEA is sensitive to models with misspecified factor loadings [28]. It is calculated by (Equation 28.1)

$$\text{RMSEA} = \frac{\sqrt{X^2 - \text{df}}}{\sqrt{\text{df}(N-1)}} \tag{28.1}$$

  where *df* stands for the degrees of freedom and *N* for the sample size.

2. Relative measures of fit
   - NFI: This was introduced by Bentler and Bonnet [9]. Normed Fit Index (NFI) is a relative measure of the discrepancy between a baseline model, which is usually the independence (null) model where all variables are uncorrelated, and a hypothesized model, as can be seen mathematically in (Eq. 28.2) [23]. The discrepancy function measures the residuals between the sample and the estimated covariance matrix [13]. This index is sensitive to sample size and tends to underestimate fit in small samples. Bollen found that the means of the sampling distributions of NFI, GFI, and AGFI tended to increase with sample size. There is also a reference that the mean of NFI is positively correlated to sample size [8] and that NFI values tend to be less than 1 when sample size is small [5].

$$\text{NFI} = \frac{X^2_{\text{null}} - X^2_{\text{hypothesized}}}{X^2_{\text{null}}} \tag{28.2}$$

   - CFI: In order to encounter the issues of NFI's performance in small samples, Bentler revised it so that it could take sample size into account, by introducing the Comparative Fit Index (CFI)[13]. When using AMOS for the analysis, values larger than 1 are reported as 1, while values smaller than 0 are reported as 0. CFI is moderately sensitive to simple model misspecification but is very sensitive to complex model misspecification. It is usually not influenced by the estimation method used. This fit index is less sensitive to distribution and sample size [28].

   Values for both NFI and CFI indices range between 0 and 1. The cutoff value that indicates a well-fitted model can be 0.9 or 0.95. There is literature supporting both cases [13].

3. Parsimonious fit measures

   Parsimonious fit measures relate the goodness of fit of the model to the number of estimated coefficients required to achieve this level of fit. They focus on diagnosing if this fit has been achieved by overfitting the data. The procedure is similar to the adjustment of R^2 in multiple regression [23].

- AIC: This is a measure used to compare models with differing relations among variables. The index obtained for the estimated model is compared with those of the independence and the saturated models. The independence model is the one in which all correlations among variables are considered to be zero and the saturated model is a model where the number of estimated parameters equals the number of data points [13]. The model with the smallest Akaike's Information Criterion (AIC) value is considered the best one. The computational formula can be found in Equation 28.3 [23].

$$\text{AIC} = X^2 + 2(\text{Number of estimated parameters}) \tag{28.3}$$

- BIC: Bayes Information Criterion (BIC), Similar to AIC, this is a measure used to compare models. BIC imposes greater penalties than AIC in order to avoid overfitting of the model [50]. In SEM, BIC is calculated for the independence and the saturated models. The model with the smallest value is considered the best [13].

AIC and BIC are criteria that are used in various methods besides SEM, in order to assess the fit of a model.

4. Hoelter's critical N: This [25] is a measure that focuses directly on the adequacy of sample size rather than on model fit. It estimates a sample size sufficient to obtain an adequate model fit for a Chi-square test [27]. Hoelter proposed that a value above 200 is indicative of a model that sufficiently represents the sample data. Amos provides the index for both 0.05 and 0.01 significance levels (denoted by HOELTER.05 and HOELTER.01) [5].

General comments on fit indices: As stated earlier, there has been a conflict over which goodness-of-fit indices are most appropriate given some characteristics of the model. Reference 53 have suggested that GFI is more appropriate than NFI in finite samples and across different estimation methods. Using a large empirical data set, absolute-fit indices (i.e., GFI and RMSEA) tend to behave more consistently across estimation methods than do incremental fit indices (i.e., NFI). This is mostly observed when there is a good fit between the hypothesized model and the observed data. As the degree of fit between hypothesized models and observed data decreases, GFI and RMSEA behave less consistently [28].

28.3.6 Other Indications of a Misspecified Model

Besides the study of the goodness-of-fit indices and the examination for offending estimates, there are some other points that indicate a misspecified model.

- *Residuals*: Residuals are defined as the difference between observed and fitted covariances. Their pattern can indicate a potential problem with the model. The existence of extreme outliers, for example, is an indication of a misspecified model [36].

- *Modification indices*: Modification Indices (MIs) capture evidence of misfit [30]. They are of great importance for the respecification of the model as they guide the user to add potentially significant regression weights that could improve the fit [13]. The best approach for using MIs is to proceed sequentially. Add the suggested path, rerun the model, test the estimates of the coefficients and the goodness-of-fit indices, and then recalculate the MIs. It is important to note that the respecification should not be based only on the statistical results but it should go along with the theoretical findings concerning the model. From a mathematical point of view, a modification index is actually the first derivative of the objective function with respect to a model parameter that has been fixed. The first derivative is the slope of the function; therefore, a modification index reveals how much improvement one might expect by adding the suggested regression weight. The MI values mentioned in the application actually reveal how much the discrepancy between the sample and estimated covariance matrix will fall if the suggested path coefficient is added to the causal diagram [5].

28.4 APPLICATION

In this section, we examine the relationships among genes included in the TLR signaling pathway, from a statistical point of view. The goal is to track interactions between genes, which translates into finding causality relations using a statistical method, in particular, SEM. The data set consists of 84 genes relevant to the TLR signaling pathway, obtained from 191 patients, 124 of them being characterized as M-CLL and 67 as U-CLL. Genes are represented by variables whose values show the relative expression of each gene, measured using real-time quantitative PCR (Polymerase Chain reaction) and the 2(−Delta Delta $C(T)$) Method [49]. For this chapter, 11 genes were combined to create different pathways. Two of these models, Model B (depicting the TLR2/6 pathway) and Model D (depicting the TLR9 pathway), were included in the paper presented in the BIOKDD 2013 workshop [54]. Previous work on the same data set [6] revealed differences in gene expression profiles among M-CLL and U-CLL subgroups. The flowchart describing the procedure can be found in Figure 28.2.

The models are created according to the TLR pathway provided by the KEGG pathway database [32]. The KEGG pathway used as a template was derived from healthy genes. Since the available sample size does not allow the investigation of the total pathway, the examination of separate pathways initiating from different genes was considered as the most optimal solution for obtaining accurate results. Each part of the TLR pathway is depicted with a path diagram and inserted in the statistical software in order to be assessed. Each model is tested against two data sets, the M-CLL patients and the U-CLL patients. The statistical package used for the analysis is SPSS AMOS [5]. The models (or path diagrams) are reconstructed according to the results provided by the method and, of course, without violating important aspects of the

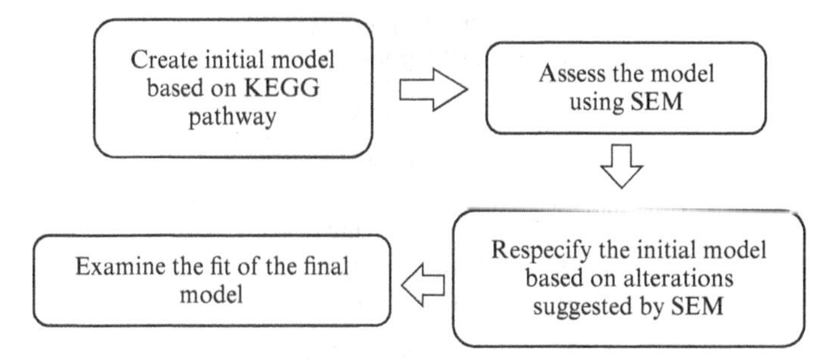

Figure 28.2 Description of the steps of analysis.

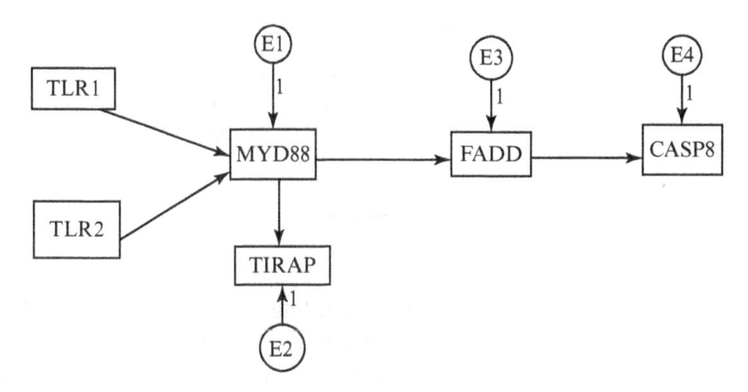

Figure 28.3 Initial model for the TLR1/2 pathway.

underlying biological theory. Since variables represent genes, the effect of one gene on another is depicted by an arrow. Arrows are statistically interpreted as regression weights or path coefficients. Both terms are used in this section. When two genes are interacting in order to produce an effect on another gene, this interaction is expressed statistically by a correlation, that is, a bidirectional arrow. For each pathway, the initial and the final respecified models are provided, along with the goodness-of-fit indices and the table of direct and indirect effects. All examined models are recursive, which means that they do not allow for any reciprocal effects [13].

1. Model A: TLR1/2 pathway The initial model for this pathway is shown in Figure 28.3. In the KEGG model, TLR1 and TLR2 interact; however, we started modeling without depicting this interaction, in order to test whether the method could reveal this correlation.
   - M-CLL dataset

     The program produced a significant correlation between TLR1 and TLR2 (MI = 13.958), which actually exists according to theory. The regression weight from TLR2 to MyD88 was found insignificant (sig. = 0.421), therefore the arrow was removed. However, another arrow was

Table 28.1 Effects of the TLR1/2 pathway on M-CLL patients

	TLR1	MyD88	FADD
		Total effects	
MyD88	0.387	0	0
FADD	0.171	0.442	0
CASP8	0.270	0.213	0.481
TIRAP	0.520	0.159	0
		Direct effects	
MyD88	0.387	0	0
FADD	0	0.442	0
CASP8	0.188	0	0.481
TIRAP	0.459	0.159	0
		Indirect effects	
MyD88	0	0	0
FADD	0.171	0	0
CASP8	0.082	0.213	0
TIRAP	0.062	0	0

added from TLR1 to TIRAP, according to the respective modification index (MI = 19.663). It seems that the magnitude of the effect of TLR1 to TIRAP was so significant that TIRAP is influenced by TLR1 both directly and indirectly (through MyD88). The final suggestion of MIs was the addition of a coefficient from TLR1 to CASP8 (MI = 5.821). Although, to the best of our knowledge, these relations are not supported by the biological literature, the significant regression weights show remarkable covariance between the aforementioned genes, which may practically indicate high expression profiles for the M-CLL patients.

Table 28.1 shows the total, direct, and indirect effects among genes. The total effect between two variables is calculated as the sum of the direct and indirect effects between these variables. The direct effect of one variable on the other equals the respective regression weight and shows how much the endogenous variable will go up when the exogenous goes up by 1.

The respecification of the initial model improved its fit. The final model (Figure 28.4) has a Chi-square value equal to 14.351(sig. = 0.073). The Chi-square value, represents the minimum value of the discrepancy function, which follows a Chi-square distribution, when using the maximum likelihood estimation method [13]. The significance value (>0.05) suggests that the departure of the data from the model is not significant [5]. The goodness-of-fit indices presented in Table 28.3, reveal very good fit to the data. The first four indices are above 0.9, which indicates very good fit. The RMSEA value corresponds to a model of mediocre fit. Hoelter's index for both significance levels is below the suggested

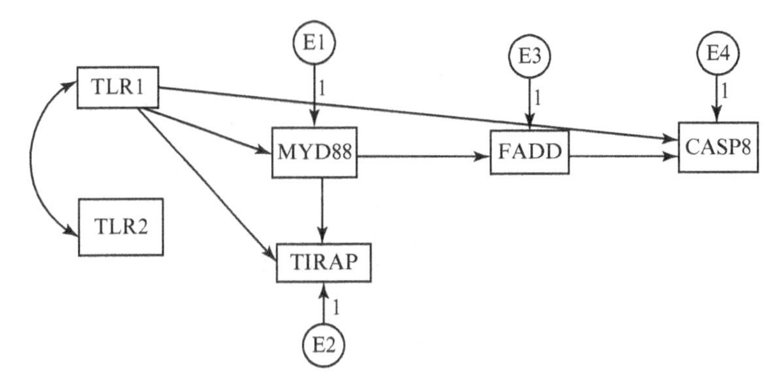

Figure 28.4 Final model of the TLR1/2 pathway for M-CLL patients.

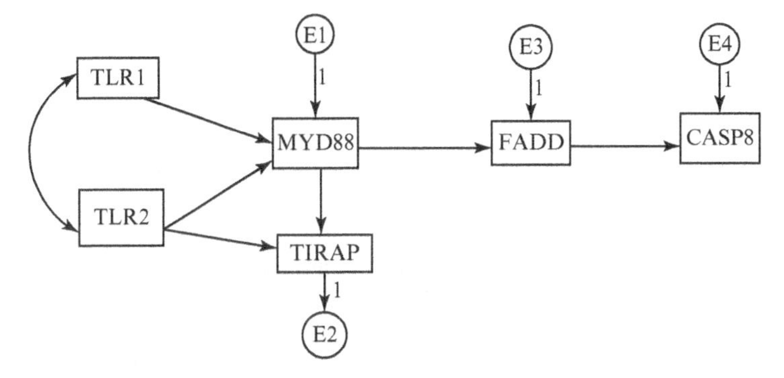

Figure 28.5 Final model of the TLR1/2 pathway for U-CLL patients.

limit, which indicates that the current sample may not be adequate for the examination of the fitted model. This may be caused by the number of variables participating in the model, although the rule of at least 10 observations for one variable [13] is not violated (124 M-CLL patients).

The study of the TLR1/2 pathway using SEM showed important covariances between some variables representing genes. These covariances could reveal interesting biological relations. The main indication of the current pathway is the importance of the role of TLR1 in it.

- U-CLL dataset

  The correlation between TLR1 and TLR2 was again proposed by the MIs (MI=17.863). The other single difference from the initial path diagram, is the addition of an arrow originating from TLR2 and resulting in TIRAP (MI=8.187). The modification indices and the estimates of the regression weights led us to this path diagram (Figure 28.5). Table 28.2 shows the total, direct, and indirect effects among variables.

  The respecified model produced a satisfactory but not quite good fit (Table 28.3) and a Chi-square value of 23.569 with sig. =0.003, which suggests that the model does not fit the data well. However, that should

Table 28.2 Effects of the TLR1/2 pathway on U-CLL patients

	TLR2	TLR1	MyD88	FADD
		Total Effects		
MyD88	0.217	0.471	0	0
FADD	0.120	0.261	0.554	0
TIRAP	0.337	0.185	0.393	0
CASP8	0.125	0.271	0.574	1.037
		Direct Effects		
MyD88	0.217	0.471	0	0
FADD	0	0	0.554	0
TIRAP	0.252	0	0.393	0
CASP8	0	0	0	1.037
		Indirect effects		
MyD88	0	0	0	0
FADD	0.120	0.261	0	0
TIRAP	0.085	0.185	0	0
CASP8	0.125	0.271	0.574	0

Table 28.3 Goodness-of-fit indices for TLR 1/2 pathways

	M-CLL pathway	U-CLL pathway
GFI	0.963	0.898
AGFI	0.903	0.731
NFI	0.914	0.889
CFI	0.958	0.921
RMSEA	0.08	0.172
AIC	40.351 (sat=42, ind=178.181)	49.569 (sat=42, ind=224.820)
BIC	77.015 (sat=101.226, ind=195.103)	78.230 (sat=88.299, ind=238.049)
HOELTER.05	133	44
HOELTER.01	173	57

not lead to the rejection of the model, as Chi-square is not used as a determinant factor for examining the fit in SEM, but just for a quick overview of the model [13]. For the first four indices the cutoff value that indicates good fit is about 0.9 and only CFI exceeds this cutoff. The values of GFI and NFI are less than 0.9 but close to it and AGFI is close to 0.7. These indices reveal satisfactory or mediocre fit. The value of RMSEA indicates poor fit, since it is bigger than 0.1 [13]. The value of AIC for the proposed model is smaller than that of the independence model but not of the saturated one, which implies that the proposed model is not the best-fitted one. However, the value of BIC for the proposed model is smaller than

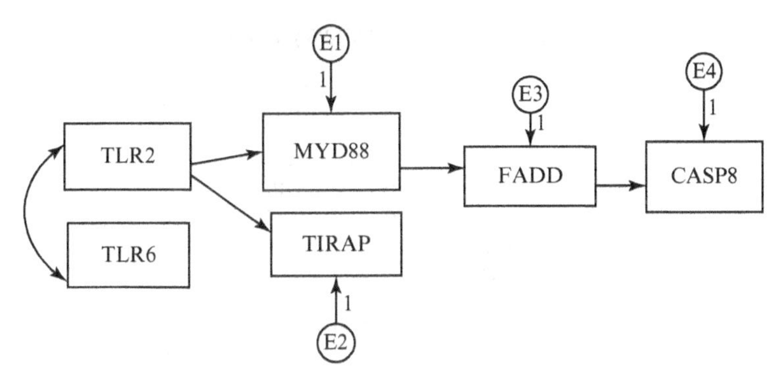

Figure 28.6 Initial model for TLR2/6 pathway.

those of the independence and the saturated models, which indicates that the proposed is the best-fitted model compared to the others. The values of Hoelter for both significance levels are much less than 200, which is the proposed acceptable value of that index, indicating that the data are not adequate for the testing of this model. A probable reason for this is the sample size of the U-CLL file, which contains 67 cases and six variables involved in this model.

We may have not achieved a well-fitted model for the U-CLL patients' pathway, however we obtained the important indication that TLR1 plays a crucial role in M-CLL patients and seems to diminish the effect of TLR2 with which is correlated. This result is not observed in U-CLL patients.

2. MODEL B: TLR2/6 PATHWAY

The initial model was that of Figure 28.6, created according to KEGG pathway model [32]. TLR2 and TLR6, act together in order to produce an effect for the genes, MyD88 and TIRAP. The bidirectional arrow on TLR2 and TLR6 depicts the fact that these genes are "co-operating" in order to produce the effect. On the basis of biological knowledge, the signal derives from TLR2, which is why the arrows originate from this gene. The model is examined against the two data sets of M-CLL and U-CLL patients.

- M-CLL dataset

  According to modification indices, four regression weights, some of which reveal existing and some nonexisting relationships, were added to the path diagram. MIs suggested adding more path coefficients; however, this suggestion was rejected, as the proposed relations violated important aspects of biological theory. The newly produced relations were the following: From TLR6 to MyD88 (MI=8.946), from TLR6 to FADD (MI=19.108), from TLR6 to TIRAP (MI=14.541), and from MyD88 to TIRAP (MI=9.386). The last proposed relation exists biologically. The interactions between TLR6 and FADD and between TLR6 and TIRAP, do not exist biologically; therefore, this result could encourage further

Table 28.4 Effects of the TLR2/6 pathway on M-CLL patients

	TLR6	TLR2	MyD88	FADD
		Total effects		
MyD88	0.328	0	0	0
FADD	0.495	0	0.352	0
CASP8	0.261	0	0.186	0.528
TIRAP	0.378	0.160	0.190	0
		Direct effects		
MyD88	0.328	0	0	0
FADD	0.380	0	0.352	0
CASP8	0	0	0	0.528
TIRAP	0.315	0.160	0.190	0
		MyD88 Indirect effects		
	0	0	0	0
FADD	0.115	0	0	0
CASP8	0.261	0	0.186	0
TIRAP	0.062	0	0	0

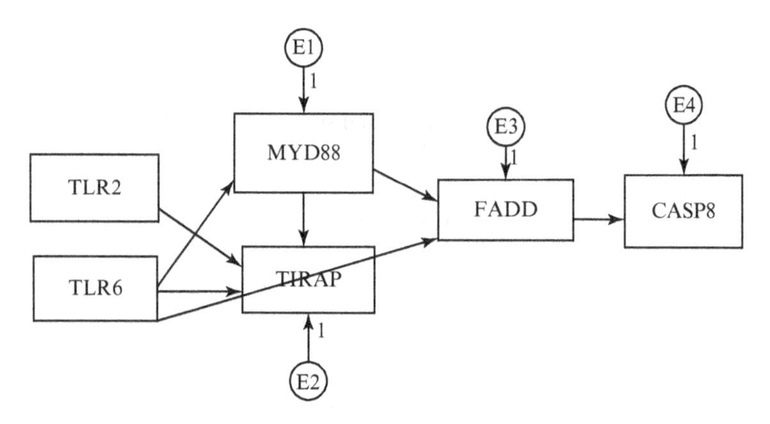

Figure 28.7 Final model of the TLR2/6 pathway for M-CLL patients.

research regarding these genes. The correlation between TLR2 and TLR6 was found insignificant (sig.=0.376), so the bidirectional arrow was removed. The effects among variables are shown in Table 28.4. The direct effects represent the regression weights.

The fit of the final model (Figure 28.7) is very good (Table 28.6) [54]. The first four indices are above 0.9, the value of RMSEA also indicates good fit, and the values of AIC and BIC are smaller than the respective values of the saturated and independence models, suggesting that the proposed model fits the data best. Hoelter is above 200 for both significance

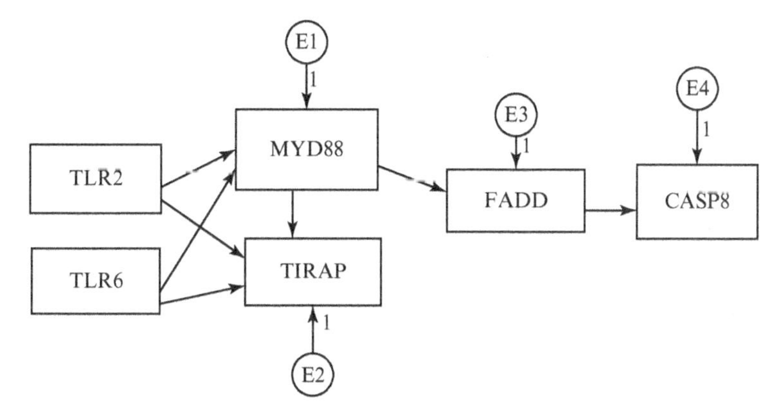

Figure 28.8 Final model of the TLR2/6 pathway for U-CLL patients.

levels, implying that the sample size is adequate for the study of this path diagram. The Chi-square of the final model is 9.569 (sig.=0.297). TLR6 seems to have a significant impact in this pathway.

- U-CLL dataset

  The correlation of TLR2 with TLR6 was again found insignificant (sig. = 0.073). Three relations were added according to modification indices: from TLR6 to MyD88 (MI = 8.799), from TLR6 to TIRAP (MI = 13.867), and from MyD88 to TIRAP (MI = 8.746). The Chi-square value is 17.724 (sig. = 0.023). The model for the U-CLL patients (Figure 28.8) is quite similar to that for M-CLL patients. In both cases, there is no statistical proof for the existence of a correlation between TLR2 and TLR6. There is a direct effect from MyD88 to TIRAP. The differences are that in the M-CLL pathway, exists a direct effect from TLR6 to FADD and in U-CLL patients, the regression weight of TLR2 to MyD88 is not found insignificant [54].

  Table 28.5 shows the total, direct, and indirect effects among variables. The fit of the respecified model is good (Table 28.6); however, we obtained better goodness-of-fit indices for the final model of the M-CLL data set. In the U-CLL data set, AGFI is less than 0.9, which is the cutoff value that represents well-fitted models. The RMSEA value (>0.1) indicates poor fit. AIC for the proposed model is smaller than that of the independence one, but not of the saturated one. The BICs value for the proposed model is smaller than those of the other two mentioned models. Hoelter's indices are much less than 200, which means that the data are not sufficient for the examination of this causal diagram. This might be a reason why we could not receive satisfying values for all the goodness-of-fit indices.

3. MODEL C: TLR7 PATHWAY

   The initial model for the pathway initiating from TLR7 is shown in Figure 28.9. This pathway consists of few variables, therefore the results are expected to be quite accurate. In the KEGG pathway database [32], there

Table 28.5 Effects of the TLR2/6 pathway on U-CLL patients

	TLR6	TLR2	MyD88	FADD
		Total effects		
MyD88	0.449	0.305	0	0
FADD	0.249	0.169	0.554	0
CASP8	0.258	0.175	0.574	1.037
TIRAP	0.516	0.334	0.270	0
		Direct effects		
MyD88	0.449	0.305	0	0
FADD	0	0	0.554	0
CASP8	0	0	0	1.037
TIRAP	0.395	0.252	0.270	0
		Indirect effects		
MyD88	0	0	0	0
FADD	0.249	0.169	0	0
CASP8	0.258	0.175	0.574	0
TIRAP	0.121	0.082	0	0

Table 28.6 Goodness-of-fit indices for TLR 2/6 pathways

	M-CLL pathway	U-CLL pathway
GFI	0.977	0.920
AGFI	0.938	0.791
NFI	0.937	0.909
CFI	0.989	0.946
RMSEA	0.040	0.136
AIC	35.569 (sat = 42, ind = 163.844)	43.724 (sat = 42, ind = 206.387)
BIC	72.232(sat = 101.226, ind = 180.765)	72.385 (sat = 88.299, ind = 219.615)
HOELTER.05	200	58
HOELTER.01	259	75

is one more gene interfering with MyD88 and IRAK1, but this was not included in our data set.

- M-CLL dataset

  The difference between the initial model and the respecified one (Figure 28.10) is the addition of the effect of MYD88 to TRAF6 (MI = 14.888) and the absence of the effect of IRAK1 to TRAF6, which was found insignificant (sig. = 0.140). In the initial model, this effect was indirect, while in the resulting model this is direct. The magnitude of the effects among variables can be seen in Table 28.7. The respecified model fits the data very well (Table 28.9) and has a Chi-square = 0.900

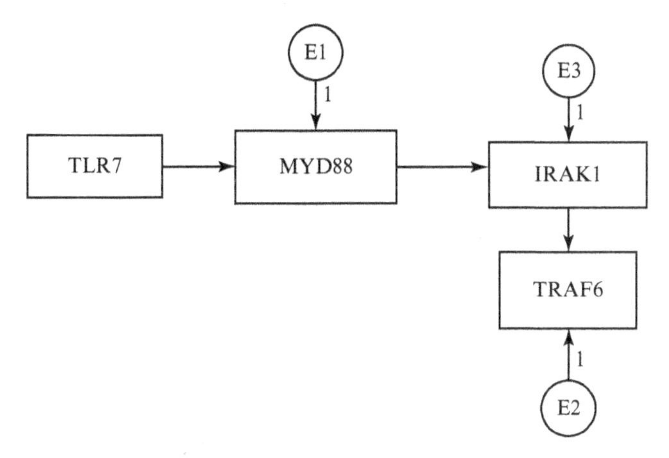

Figure 28.9 Initial model for the TLR7 pathway.

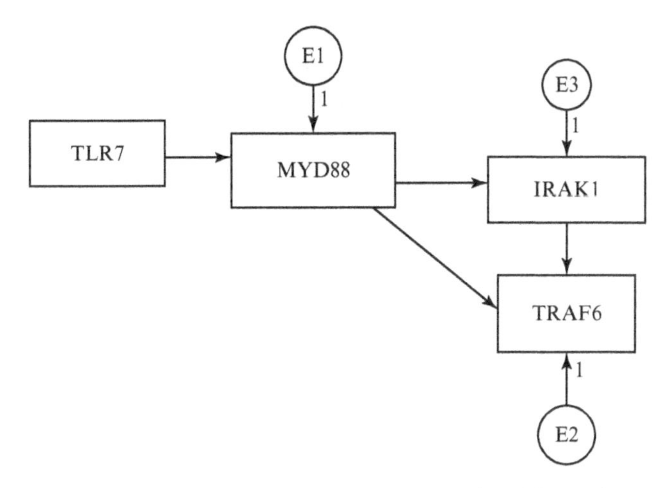

Figure 28.10 Final model of the TLR7 pathway for M-CLL patients.

(sig. = 0.825). The first four indices are above 0.9, actually CFI equals 1, which is an unrealistic value, however there was no indication that the model is misspecified, therefore we cannot conclude that this value was obtained due to an error. RMSEA shows absolute fit. AIC and BIC of the hypothesized model are smaller than those of the independence and saturated models. Hoelter's indices are much higher than 200, indicating that this sample size was more than sufficient for the study of this model. The sample size compared to the variables composing the model might be the reason of the exceptionally good values of fit indices.

- U-CLL dataset

Table 28.7 Effects of the TLR7 pathway on M-CLL patients

	TLR7	MyD88
	Total effects	
MyD88	0.351	0
TRAF6	0.117	0.334
IRAK1	0.236	0.672
	Direct effects	
MyD88	0.351	0
TRAF6	0	0.334
IRAK1	0	0.672
	Indirect effects	
MyD88	0	0
TRAF6	0.117	0
IRAK1	0.236	0

Table 28.8 Effects of the TLR7 pathway on U-CLL patients

	TLR7	MyD88
	Total effects	
MyD88	0.409	0
TRAF6	0.175	0.427
IRAK1	0.303	0.742
	Direct effects	
MyD88	0.409	0
TRAF6	0	0.427
IRAK1	0	0.742
	Indirect effects	
MyD88	0	0
TRAF6	0.175	0
IRAK1	0.303	0

The same result was also obtained for the U-CLL file. A regression weight from MyD88 to TRAF6 was suggested by the MIs (MI = 12.578). After this addition, the IRAK1 to TRAF6 path coefficient was found insignificant (sig. = 0.948). The regression coefficients represented by the direct effects along with the indirect effects, are shown in Table 28.8. The final model (Figure 28.11) fits the data very well (Table 28.9) with a Chi-square = 1.290 (sig. = 0.732). All indices show very good fit. CFI and RMSEA reveal perfect fit, a result also obtained for the M-CLL model. The Hoelter's indices are very good, but lower than the respective values

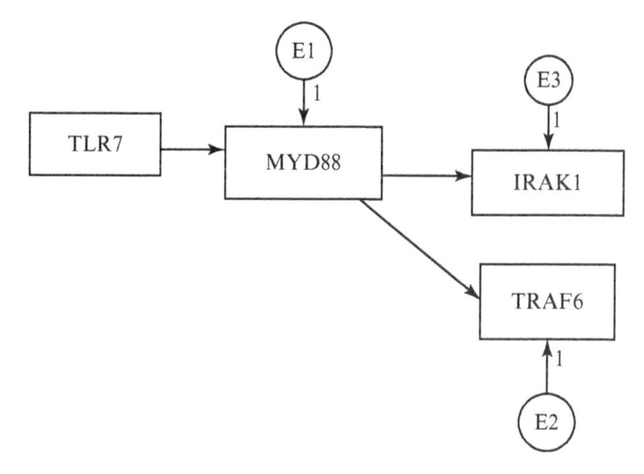

Figure 28.11 Final model of the TLR7 pathway for U-CLL patients.

Table 28.9 Goodness-of-fit indices for TLR7 pathway

	M-CLL pathway	U-CLL pathway
GFI	0.996	0.990
AGFI	0.988	0.968
NFI	0.983	0.979
CFI	1	1
RMSEA	0	0
AIC	14.900 (sat = 20, ind = 60.268)	15.290 (sat = 20, ind = 70.695)
BIC	34,642 (sat = 48.203, ind = 71.549)	30.723 (sat = 42.047-ind = 79.514)
HOELTER.05	1068	400
HOELTER.01	1551	581

of the M-CLL model, which is expected, as the U-CLL file consists of fewer patients.

4. MODEL D: TLR9 PATHWAY

   The hypothetical model inserted in the program is shown in Figure 28.12. This model was also included in the paper presented in the BIOKDD workshop [54]. Besides the first gene, the rest of the causal diagram is the same as that of the TLR7 pathway (Figure 28.9). The test of the fit of the model against M-CLL and U-CLL patients is presented below.

   • M-CLL dataset

     In the M-CLL path, the effect of MyD88 to TRAF6 (MI = 14.888) has proved to be direct and not with IRAK1 mediating between the mentioned genes. There is also an effect produced from TLR9 and resulting in TRAF6 (MI = 14.100), suggesting a causal relation that does not exist in the KEGG pathway. For the M-CLL patients, the regression weight of

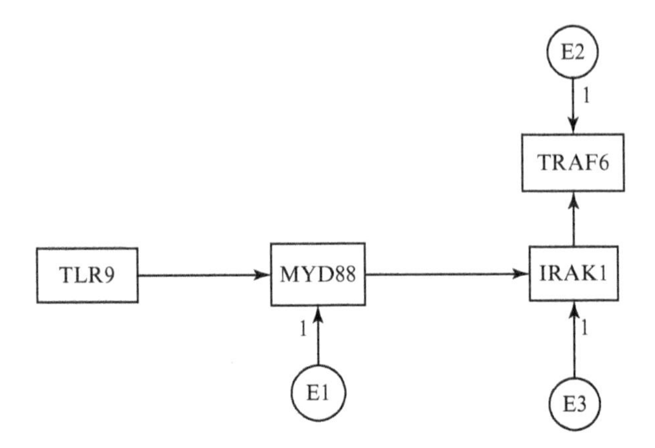

Figure 28.12 Initial model for TLR9 pathway.

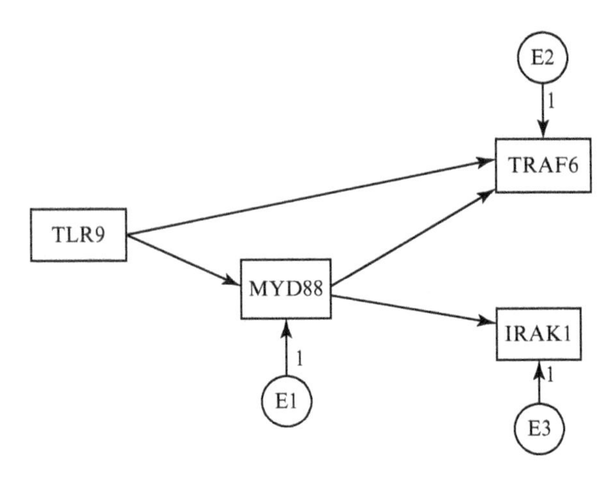

Figure 28.13 Final model of the TLR9 pathway for M-CLL patients.

IRAK1 to TRAF6 was found insignificant from the initial evaluation of the model (sig. = 0.140). The final model (Figure 28.13) has a Chi-square = 2.623 (sig. = 0.269). The arrow from TLR9 to TRAF6 indicates a direct effect that does not exist in the KEGG pathway or an indirect effect. The effects among genes are shown in Table 28.10.

The fit of the model (Table 28.12) was very good. The first four indices are equal or above 0.95. The RMSEA value also indicates very good fit. The values of AIC and BIC are smaller for the hypothesized model and Hoelter's indices indicate that the sample size is adequate.

Table 28.10 Effects of the TLR9 pathway on M-CLL patients

	TLR9	MyD88
	Total effects	
MyD88	0.244	0
TRAF6	0.255	0.266
IRAK1	0.164	0.672
	Direct effects	
MyD88	0.244	0
TRAF6	0.190	0.266
IRAK1	0	0.672
	Indirect effects	
MYD88	0	0
TRAF6	0.065	0
IRAK1	0.164	0

Table 28.11 Effects of the TLR9 pathway on U-CLL patients

	TLR9	MyD88
	Total effects	
MyD88	0.451	0
TRAF6	0.193	0.427
IRAK1	0.334	0.742
	Direct effects	
MyD88	0.451	0
TRAF6	0	0.427
IRAK1	0	0.742
	Indirect effects	
MyD88	0	0
TRAF6	0.193	0
IRAK1	0.334	0

- U-CLL dataset

  In the U-CLL model, the direct effect of MyD88 to TRAF6 (MI = 12.578) was produced again. According to MIs, there was the suggestion to also add a regression weight from TLR9 to TRAF6 (MI = 8.622), which was found insignificant after rerunning the model, and therefore removed. So was the IRAK1 to TRAF6 weight (sig. = 0.877). The fit of the respecified model (Figure 28.14) was very good (Table 28.12). All indices represent very good fit. The values of AIC and BIC suggest that

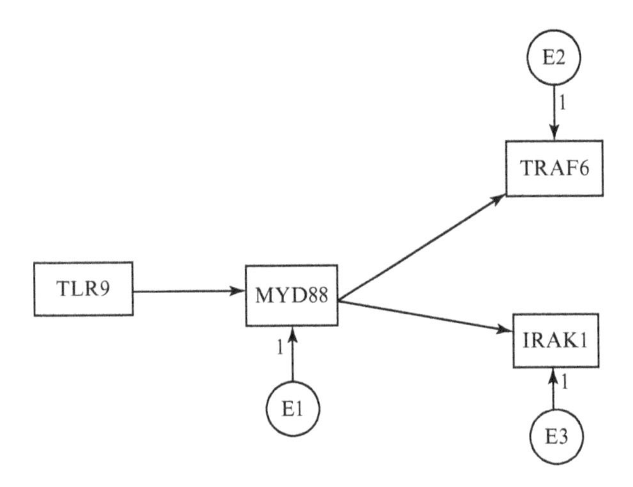

Figure 28.14 Final model of the TLR9 pathway for U-CLL patients.

Table 28.12 Goodness-of-fit indices for TLR 9 pathway

	M-CLL pathway	U-CLL pathway
GFI	0.990	0.975
AGFI	0.948	0.918
NFI	0.959	0.955
CFI	0.989	0.994
RMSEA	0.050	0.046
AIC	18.623(sat; = 20, ind = 71.845)	17.428 (sat = 20, ind = 84.297)
BIC	41.185 (sat = 48.203, ind = 83.126)	32.861 (sat = 42.047, ind = 93.115)
HOELTER.05	281	151
HOELTER.01	432	219

the hypothesized model is the best fitted to the data, compared to the independence and saturated models and Hoelter's indices for both significance levels show adequacy of the sample size.

28.5 CONCLUSION

Structural equation models applied for the justification of the TLR pathway in patients with chronic lymphocytic leukemia confirm some already existing assumptions regarding relations and reveal some new associations. Consistency with the reference pathway was observed mostly for the M-CLL subgroup, and much less for the U-CLL subgroup. These results agree with the already established differences in immune signaling between these subgroups [6, 44, 52] and also with recently published single cell network profiling studies [15].

Some of the causal relations produced by the method may not exist in the KEGG pathway model; however, there is available literature to justify the existence of association between some genes. In model A, an effect originating from TLR2 and resulting in TIRAP was found. According to recent papers, there is association between these genes [55]. In model C, a regression weight from MyD88 to TRAF6 was suggested by the method. On the basis of References [33, 60], this association exists. Finally, in Model D, TLR9 causes an effect that results in TRAF6, genes that have been found to interact with each other [59]. Since SEM is a statistical technique that reveals significant correlations and directionality of causation among data, it can be used as a reference tool for the prediction of unexpected connections between genes that should be further studied [57].

Although the signaling pathways acting downstream of the different TLRs were initially thought to be identical, it is now apparent that individual TLRs activate different sets of signaling pathways and exert distinct biological effects. This diversity of signaling is most likely the result of the combinatorial action of adaptor molecules.

The entire procedure of building causal models from data is difficult and laborious. In general, the models are built after several steps and several rejections of inappropriate models, which either do not fit well to the data or are not meaningful in a biological sense. This is why the building of these models requires continuous collaboration and interaction between researchers from both statistical and biomedical disciplines. In general, causal models can stimulate further research by opening new directions of research which can be further explored by obtaining more data through laboratory measurements or experiments.

REFERENCES

1. Akira S. Toll-like receptor signaling. J Biol Chem 2003;278(40):38105–38108.
2. Akira S, Takeda K. Toll-like receptor signalling. Nat Rev Immunol 2004;4(7):499–511.
3. Akira S, Takeda K, Kaisho T. Toll-like receptors: critical proteins linking innate and acquired immunity. Nat Immunol 2001;2(8):675–680.
4. Akira S, Uematsu S, Takeuchi O. Pathogen recognition and innate immunity. Cell 2006;124(4):783–801.
5. Arbuckle JL. *IBM SPSS® Amos™ 19 User's Guide*. Crawfordville (FL): Amos Development Corporation; 2010.
6. Arvaniti E, Ntoufa S, Papakonstantinou N, Touloumenidou T, Laoutaris N, Anagnostopoulos A, Lamnissou K, Caligaris-Cappio F, Stamatopoulos K, Ghia P, Muzio M, Belessi C. Toll-like receptor signaling pathway in chronic lymphocytic leukemia: distinct gene expression profiles of potential pathogenic significance in specific subsets of patients. Haematologica 2011;96(11):1644–1652.
7. Bagozzi RP, Yi Y. On the evaluation of structural equation models. J Acad Mark Sci 1988;16(1):74–94.
8. Bearden WO, Sharma S, Teel JE. Sample size effects on chi square and other statistics used in evaluating causal models. J Mark Res 1982;19(4):425–430.
9. Bentler PM, Bonett DG. Significance tests and goodness of fit in the analysis of covariance structures. Psychol Bull 1980;88(3):588.

10. Bollen KA, Scott Long J. *Testing Structural Equation Models*, Volume 154. USA: Sage Publications; 1993.
11. Bottomly D, Ferris MT, Aicher LD, Rosenzweig E, Whitmore A, Aylor DL, Haagmans BL, Gralinski LE, Bradel-Tretheway BG, Bryan JT, Threadgill DW, de Villena FP, Baric RS, Katze MG, Heise M, McWeeney SK. Expression quantitative trait Loci for extreme host response to influenza a in pre-collaborative cross mice. G3: Genes- Genomes-Genetics 2012;2(2):213–221.
12. Büchel C, Friston KJ. Modulation of connectivity in visual pathways by attention: cortical interactions evaluated with structural equation modelling and fMRI. Cereb Cortex 1997;7(8):768–778.
13. Byrne BM. *Structural Equation Modeling with AMOS: Basic Concepts, Applications, and Programming*. USA: CRC Press; 2009.
14. Carter RL. Solutions for missing data in structural equation modeling. Res Pract Assess 2006;1(1):1–6.
15. Cesano A, Perbellini O, Evensen E, Chu CC, Cioffi F, Ptacek J, Damle RN, Chignola R, Cordeiro J, Yan X-J, Hawtin RE, Nichele I, Ware JR, Cavallini C, Lovato O, Zanotti R, Rai KR, Chiorazzi N, Pizzolo G, Scupoli MT. Association between B-cell receptor responsiveness and disease progression in B-cell chronic lymphocytic leukemia: results from single cell network profiling studies. Haematologica 2013;98(4):626–634.
16. Cheung GW, Rensvold RB. Evaluating goodness-of-fit indexes for testing measurement invariance. Struct Equ Model 2002;9(2):233–255.
17. Chiorazzi N, Ferrarini M. Cellular origin(s) of chronic lymphocytic leukemia: cautionary notes and additional considerations and possibilities. Blood 2011;117(6):1781–1791.
18. Chiorazzi N, Rai KR, Ferrarini M. Chronic lymphocytic leukemia. N Engl J Med 2005;352(8):804–815.
19. Fitzgerald KA, Palsson-McDermott EM, Bowie AG, Jefferies CA, Mansell AS, Brady G, Brint E, Dunne A, Gray P, Harte MT, McMurray D, Smith DE, Sims JE, Bird TA, O'Neill LA. Mal (myd88-adapter-like) is required for toll-like receptor-4 signal transduction. Nature 2001;413(6851):78–83.
20. Fox J. Bootstrapping regression models. *An R and S-PLUS Companion to Applied Regression: A Web Appendix to the Book*. Thousand Oaks (CA): Sage Publications; 2002. Available at http://cran. r-project.org/doc/contrib/Fox-Companion/appendix-bootstrapping.pdf.
21. Goldberger AS. Structural equation methods in the social sciences. Econometrica 1972;40(6):979–1001.
22. Grandjenette C, Kennel A, Faure GC, Béné MC, Feugier P. Expression of functional toll-like receptors by B-chronic lymphocytic leukemia cells. Haematologica 2007;92(9):1279–1281.
23. Hair JF Jr., Anderson RE, Tatham RL, Black WC. *Multivariate Data Analysis, Structural Equation Modeling*. 5th ed. Chapter 11; Pearson Prentice Hall; 1998.
24. Hamblin TJ, Davis Z, Gardiner A, Oscier DG, Stevenson FK. Unmutated Ig V(H) genes are associated with a more aggressive form of chronic lymphocytic leukemia. Blood 1999;94(6):1848–1854.
25. Hoelter JW. The analysis of covariance structures goodness-of-fit indices. Sociol Methods Res 1983;11(3):325–344.
26. Hoyle RH, Smith GT. Formulating clinical research hypotheses as structural equation models: a conceptual overview. J Consult Clin Psychol 1994;62(3):429.

27. Hu L-T, Bentler PM. Evaluating model fit. In: Hoyle RH, editor. *Structural Equation Modeling: Concepts, Issues, and Applications*. Thousand Oaks (CA): Sage Publications; 1995. p 76–99.
28. Hu L-T, Bentler PM. Fit indices in covariance structure modeling: sensitivity to underparameterized model misspecification. Psychol Methods 1998;3(4):424.
29. Joreskog KG, Sorbom D. *Lisrel vi User's Guide*. Mooresville (IN): Scientific Software; 1984.
30. Jöreskog KG, Sörbom D. *LISREL 8: Structured Equation Modeling with the Simplis Command Language*. Scientific Software International; 1993.
31. Joreskog KG, Van Thillo M. *LISREL: A General Computer Program for Estimating a Linear Structural Equation System Involving Multiple Indicators of Unmeasured Variables*, ETS Research Bulletin Series. Princeton, NJ: John Wiley and Sons, Inc.; 1972.
32. Kanehisa M, Goto S. KEGG: kyoto encyclopedia of genes and genomes. Nucleic Acids Res 2000;28(1):27–30.
33. Kawai T, Sato S, Ishii KJ, Coban C, Hemmi H, Yamamoto M, Terai K, Matsuda M, Inoue J-I, Uematsu S, Takeuchi O, Akira S. Interferon-α induction through toll-like receptors involves a direct interaction of IRF7 with MyD88 and TRAF6. Nat Immunol 2004;5(10):1061–1068.
34. Kline RB, Hoyle R. Assumptions in structural equation modeling. In: Hoyle R, editor. *Handbook of Structural Equation Modeling*. New York: Guilford Press; 2011.
35. Kostareli E, Gounari M, Agathangelidis A, Stamatopoulos K. Immunoglobulin gene repertoire in chronic lymphocytic leukemia: insight into antigen selection and microenvironmental interactions. Mediterr J Hematol Infect Dis 2012;4(1):e2012052.
36. Lattin JM, Carroll JD, Green PE. *Analyzing Multivariate Data*. Pacific Grove (CA): Thomson Brooks/Cole; 2003.
37. Liu B, de la Fuente A, Hoeschele I. Gene network inference via structural equation modeling in genetical genomics experiments. Genetics 2008;178(3):1763–1776.
38. Mansell A, Brint E, Gould JA, O'Neill LA, Hertzog PJ. Mal interacts with tumor necrosis factor receptor-associated factor (TRAF)-6 to mediate NF-κB activation by toll-like receptor (TLR)-2 and TLR4. J Biol Chem 2004;279(36):37227–37230.
39. Manski CF, McFadden D. *Structural Analysis of Discrete Data with Econometric Applications*. Cambridge (MA): MIT Press; 1981.
40. Medzhitov R, Preston-Hurlburt P, Kopp E, Stadlen A, Chen C, Ghosh S, Janeway CA Jr. MyD88 is an adaptor protein in the hToll/iL-1 receptor family signaling pathways. Mol Cell 1998;2(2):253–258.
41. Miggin SM, O'Neill LAJ. New insights into the regulation of TLR signaling. J Leukoc Biol 2006;80(2):220–226.
42. Muzio M, Scielzo C, Bertilaccio MTS, Frenquelli M, Ghia P, Caligaris-Cappio F. Expression and function of toll like receptors in chronic lymphocytic leukaemia cells. Br J Haematol 2009;144(4):507–516.
43. Neale MCCL, Cardon LR. *Methodology for Genetic Studies of Twins and Families*. Volume 67. USA: Springer-Verlag; 1992.
44. Ntoufa S, Vardi A, Papakonstantinou N, Anagnostopoulos A, Aleporou-Marinou V, Belessi C, Ghia P, Caligaris-Cappio F, Muzio M, Stamatopoulos K. Distinct innate immunity pathways to activation and tolerance in subgroups of chronic lymphocytic leukemia with distinct immunoglobulin receptors. Mol Med 2012;18(1):1281.

45. O'Neill LAJ, Fitzgerald KA, Bowie AG. The Toll–IL-1 receptor adaptor family grows to five members. Trends Immunol 2003;24(6):286–289.
46. Ozinsky A, Underhill DM, Fontenot JD, Hajjar AM, Smith KD, Wilson CB, Schroeder L, Aderem A. The repertoire for pattern recognition of pathogens by the innate immune system is defined by cooperation between toll-like receptors. Proc Natl Acad Sci U S A 2000;97(25):13766–13771.
47. Pearl J. *Causality: Models, Reasoning and Inference*. Volume 29. Cambridge University Press; 2000.
48. Rowe DC, McGettrick AF, Latz E, Monks BG, Gay NJ, Yamamoto M, Akira S, O'Neill LA, Fitzgerald KA, Golenbock DT. The myristoylation of trif-related adaptor molecule is essential for Toll-like receptor 4 signal transduction. Proc Natl Acad Sci U S A 2006;103(16):6299–6304.
49. Schmittgen TD, Livak KJ. Analyzing real-time PCR data by the comparative C(T) method. Nat Protoc 2008;3(6):1101–1108.
50. Schwarz G. Estimating the dimension of a model. Ann Stat 1978;6(2):461–464.
51. Simon HA. Causal ordering and identifiability. *Models of Discovery*. Springer-Verlag; 1977. p 53–80.
52. Stevenson FK, Krysov S, Davies AJ, Steele AJ, Packham G. B-cell receptor signaling in chronic lymphocytic leukemia. Blood 2011;118(16):4313–4320.
53. Tanaka JS, Huba GJ. A fit index for covariance structure models under arbitrary GLS estimation. Br J Math Stat Psychol 1985;38(2):197–201.
54. Tsanousa A, Angelis L, Ntoufa S, Papakwnstantinou N, Stamatopoulos K. A structural equation modeling approach of the Toll-like receptor signaling pathway in chronic lymphocytic leukemia. BIOKDD 2013: Proceedings of the 24th International DEXA conference on Database and Expert Systems Applications. Prague: IEEE; 2013.
55. Ulrichts P, Tavernier J. MAPPIT analysis of early Toll-like receptor signalling events. Immunol Lett 2008;116(2):141–148.
56. Wahlgren CM, Magnusson PK. Genetic influences on peripheral arterial disease in a twin population. Arterioscler Thromb Vasc Biol 2011;31(3):678–682.
57. Wiestner A. Emerging role of kinase-targeted strategies in chronic lymphocytic leukemia. Blood 2012;120(24):4684–4691.
58. Wright S. Correlation and causation. J Agric Res 1921;20(7):557–585.
59. Yang M, Wang C, Zhu X, Tang S, Shi L, Cao X, Chen T. E3 ubiquitin ligase CHIP facilitates Toll-like receptor signaling by recruiting and polyubiquitinating Src and atypical PKCζ. J Exp Med 2011;208(10):2099–2112.
60. Zhao J, Kong HJ, Li H, Huang B, Yang M, Zhu C, Bogunovic M, Zheng F, Mayer L, Ozato K, Unkeless J, Xiong H. IRF-8/interferon (IFN) consensus sequence-binding protein is involved in Toll-like receptor (TLR) signaling and contributes to the cross-talk between TLR and IFN-{gamma} signaling pathways. Sci Signal 2006;281(15):10073.

CHAPTER 29

ANNOTATING PROTEINS WITH INCOMPLETE LABEL INFORMATION

Guoxian Yu[1], Huzefa Rangwala[2], and Carlotta Domeniconi[2]

[1]*College of Computer and Information Science, Southwest University, Chongqing, China*
[2]*Department of Computer Science, George Mason University, Fairfax, VA, USA*

29.1 INTRODUCTION

Proteins are the essential and versatile macromolecules of life. The knowledge of protein functions can promote the development of new drugs, better crops, and even synthetic biochemicals [21]. High-throughput biological techniques produced various proteomic data (i.e., *Protein–Protein Interaction* (PPI) networks, protein sequences), but the functions of these proteomic data cannot be annotated at the same pace. The availability of vast amount of proteomic data enables researchers to computationally annotate proteins. Therefore, automated protein function prediction is one of the grand challenges in computational biology, and many computational models have been developed to reduce the cost associated with experimentally annotating proteins in the wet lab [6, 7, 27, 35].

Various computational models (including classification, clustering, association rule mining) from the data mining domain have been developed for protein function prediction [21, 25]. Classification-based methods often take proteins (i.e., protein sequences, PPI, and gene expressions) and annotations associated with these proteins

Pattern Recognition in Computational Molecular Biology: Techniques and Approaches,
First Edition. Edited by Mourad Elloumi, Costas S. Iliopoulos, Jason T. L. Wang, and Albert Y. Zomaya.

as input and then train predictive classifiers on these inputs [11, 18, 30, 33, 34]. In contrast, clustering-based models exploit the similarity or structures among proteins to explore the function modules among them [19, 26], as densely connected function modules often perform similar functions [21, 25]. Approaches based on association rules have been developed for protein function prediction [14, 31]. In addition, text mining approaches are also proposed to infer protein functions [9]. For more details on the various models in protein functions, one can refer to a comprehensive literature survey [21].

Proteins often involve multiple biological processes and thus have multiple functions. Each function can be viewed as a label and these function labels are often correlated with each other. Early protein function prediction models often divided the prediction problem into multiple binary classification problems [18, 30] and ignored the correlation between labels. However, it has been observed that function correlation can be utilized to boost the prediction accuracy [22, 28]. Multi-label learning, an emerging machine learning paradigm, can take advantage of label correlations and assign a set of functions to a protein [37]. Therefore, multi-label learning approaches are widely applied in recent protein function prediction [16, 36]. In addition, another kind of approaches firstly learn a binary classifier for each function label, and then organize these classifiers in a hierarchical (tree or direct acyclic graph) structure according to the *Function Catalog* (FunCat) [23][1] or Gene Ontology [2][2] database [3].

Traditional techniques often assume that the available annotations of the training proteins are complete, without any missing labels. Actually, owing to various reasons (i.e., the updating Gene Ontology, experimental limitations), we may just have a subset of the functions of a protein, while it is unknown whether other functions are missing. In other words, proteins are incompletely (or partially) annotated [6, 27]. Learning from partially labeled multi-label samples is called *multi-label weak-label learning*, a much less studied problem in multi-label learning [27] and protein function prediction [35].

In this chapter, we study protein function prediction using partially annotated proteins. This learning scenario is often associated with two prediction tasks, which are illustrated in Figure 29.1. In the first task, we have partially labeled proteins: given a protein, some of its functions are specified, and some may be missing. The task we address is, how can we use these incomplete annotations to replenish the missing functions (Figure 29.1(b)? In the second task, we address the following issue: how do we utilize the incomplete annotated proteins to annotate proteins that are completely unlabeled (Figure 29.1(c))?

The rest of the chapter is organized as follows: In Section 29.2, we review related work on multi-label learning algorithms for network-based protein function prediction and weak-label learning approaches. We introduce *Protein function prediction using Dependency Maximization* (ProDM) in Section 29.3 and detail the

[1]http://mips.helmholtz-muenchen.de/proj/funcatDB/.
[2]http://www.geneontology.org/.

	f1	f2	f3
p1	1	1	0
p2	0	1	1
p3	1	1	0
p4	0	1	1
p5	1	0	0

(a) Original

	f1	f2	f3
p1	1	1	0
p2	0	1	1
p3	1	1	0
p4	0	1	?
p5	?	0	0

(b) Task1

	f1	f2	f3
p1	1	1	0
p2	0	1	1
p3	1	1	0
p4	?	?	?
p5	?	?	?

(c) Task2

Figure 29.1 The two tasks studied in this chapter: "**?**" denotes the missing functions; "1" means the protein has the corresponding function; "0" in (a), (b), and (c) means the protein does not have the corresponding function. Task1 replenishes the missing functions and Task2 predicts the function of proteins **p4** and **p5**, which are completely unlabeled. Some figures are from Reference [35].

experimental setup in Section 29.4. In Section 29.5, we analyze the experimental results. In Section 29.6, we provide the conclusions and some future directions.

29.2 RELATED WORK

A protein often interacts with its partners to accomplish certain functions and the function of the protein can be defined precisely by the topological features of the network it belongs to [12]. In this network, each protein corresponds to a node and the edge's weight reflects the reliability of the interaction (or similarity between proteins). Therefore, various network-based methods have been developed for protein function prediction [25]. Schwikowski et al. [24] observed that the functions of a protein can be determined on the basis of annotations of its direct interacting partners. This observation is recognized as the "guilt by association" rule. Indirectly interacting proteins also share few functions. Chua et al. [8] exploited this knowledge and extended the PPI network by integrating the level-1 (direct) and level-2 (indirect) neighbors using different weights for protein function prediction. To assign more than one function to a protein, these methods use thresholding on the predicted likelihood vectors. However, these methods do not explicitly utilize the correlation among functions.

More recently, multi-label learning approaches [37] have been introduced for protein function prediction. Pandey et al. [22] incorporated function correlations within a weighted k-nearest neighbor classifier and observed that incorporating function correlations can boost the prediction accuracy. Jiang and McQuay [16] applied the learning with local and global consistency model [38] on a tensor graph to predict protein functions. Zhang and Dai [36] included a function correlation term within the manifold regularization framework [4] to annotate proteins. Jiang [15] conducted label propagation on a bi-relation graph to infer protein functions. To avoid the risk of overwriting functions during label propagation, Yu et al. [33] introduced a *Transductive Multi-label Classifier* (TMC) on a directed bi-relation graph to annotate proteins. Chi and Hou [7] considered the fact that the proteins' functions can influence the similarity between pairs of proteins and proposed an iterative model called the *Cosine Iterative Algorithm* (CIA). In each iteration of CIA, the most confidently predicted function of an unlabeled protein is appended to the function set of this protein. Next, the pairwise similarity between training proteins and testing proteins is updated on the basis of the similar functions within the two sets for each protein. CIA uses the updated similarity, function correlations, and PPI network structures to predict the functions on the unlabeled proteins in the following iteration.

All the above multi-label learning approaches focus on utilizing function correlation in various ways and assume that the function annotations on the training proteins are complete and accurate (without missing functions). However, owing to various reasons (e.g., the evolving Gene Ontology scheme or limitations of experimental methods), we may be aware of some of the functions of a protein, but we may do not know whether other functions are associated with the same protein. In other words, proteins are partially annotated. Learning from partially (or incomplete) labeled data is different from learning from partial labels [10]. In the latter case, one learns from a set of candidate labels of an instance and assumes that only one label in this set is the ground-truth label. Learning from partially labeled data is also different from semi-supervised and supervised learning, as they both assume complete labels. In this chapter, we present how to leverage partially annotated proteins, a less studied but yet realistic scenario in protein function prediction and multi-label learning literature [6, 27, 35].

Several multi-label, weak-label learning approaches have been proposed [27, 32]. Sun et al. [27, 32] introduced a method called *WEak Label Learning* (WELL). WELL is based on three assumptions: (i) the decision boundary for each label should go across low-density regions; (ii) any given label should not be associated to the majority of samples; and (iii) there exists a group of low rank-based similarities, and the approximate similarity between samples with different labels can be computed on the basis of these similarities. WELL uses convex optimization and quadratic programming to replenish the missing labels of a partially labeled sample. As such, WELL is computationally expensive. Buncak et al. [6] annotated unlabeled images using partially labeled images and proposed a method called *MLR-GL*. MLR-GL optimizes the ranking loss and group Lasso in a convex optimization form. Yu et al. [35] proposed a method called *Protein function prediction using Weak-label Learning* (ProWL). ProWL can replenish the missing functions of partially annotated proteins and can predict the functions of completely unlabeled proteins using the partially annotated ones. However, ProWL depends heavily on function correlations and performs the prediction for one function label at a time.

Here, we introduce a new protein function prediction approach called ProDM. ProDM can alleviate the drawbacks associated with ProWL. In particular, ProDM uses function correlations, the "guilt by association" rule [24] and maximizes the dependency between the features and function labels of proteins to complete the prediction for all the function labels at one time. In the empirical study, we observe that ProDM performs better than the other competitive methods in replenishing the missing functions and in predicting functions for completely unlabeled proteins.

29.3 PROBLEM FORMULATION

For the task of *replenishing* missing functions, we have available n partially annotated proteins. The goal is to replenish the missing functions using such partially annotated proteins. For the task of *predicting* the functions of completely unlabeled proteins, we have a total of $n = l + u$ proteins, where the first l proteins are partially annotated and the last u proteins are completely unlabeled. The goal here is to use the l partially annotated proteins to annotate the u unlabeled ones.

Let $\mathbf{Y} = [\mathbf{y}_1; \mathbf{y}_2; \ldots ; \mathbf{y}_n] \in \mathbb{R}^{n \times C}$ be the currently available function set of n proteins, with $y_{ic} = 1$ if protein i $(i = 1, 2, \ldots , n)$ has the cth $(c = 1, 2, \ldots , C)$ function, and $y_{ic} = 0$ otherwise. At first, we can define a function correlation matrix $\mathbf{M}' \in \mathbb{R}^{C \times C}$ based on cosine similarity as follows:

$$\mathbf{M}'_{st} = \frac{\mathbf{Y}(\cdot, s)^T \mathbf{Y}(\cdot, t)}{\|\mathbf{Y}(\cdot, s)\| \|\mathbf{Y}(\cdot, t)\|} \tag{29.1}$$

where $\mathbf{M}'_{st}$ is the correlation between functions s and t $(s, t = 1, 2, \ldots , C)$ and $\mathbf{Y}(\cdot, s)$ represents the sth column of $\mathbf{Y}$. There exists a number of ways (e.g., Jaccard coefficient [36] and Lin's similarity [22]) to define function correlation. Here, we use the cosine similarity for its simplicity and wide application [7, 29, 35]. If $\mathbf{Y}$ is represented in a probabilistic function assignment form (e.g., $y_{ic} \in [0, 1]$), Equation 29.1 can also be applied to compute the cosine similarity between functions.

From Equation 29.1, we can see that $\mathbf{M}'_{st}$ measures the fraction of times function s and t coexist in a protein. We normalize $\mathbf{M}'$ as follows:

$$\mathbf{M}_{st} = \frac{\mathbf{M}'_{st}}{\sum_{c=1}^{C} \mathbf{M}'_{sc}} \tag{29.2}$$

$\mathbf{M}_{st}$ can be viewed as the probability that a protein has function t given that it is annotated with function s.

Now, let us consider the scenario with incomplete annotations and extend the observed function set $\mathbf{Y}$ to $\tilde{\mathbf{Y}} = \mathbf{YM}$. Our motivation in using $\tilde{\mathbf{Y}}$ is to append the missing functions using the currently known functions and their correlations. More specifically, suppose the currently confirmed functions $\mathbf{y}_i$ for the ith protein have a large correlation with the cth function (which may be missing), then it is likely that this protein will also have function c. On the basis of this assumption, we define the first part of our objective function as follows:

$$\Psi_1(\mathbf{F}) = \frac{1}{2} \sum_{i=1}^{n} \sum_{c=1}^{C} (f_{ic} - \tilde{y}_{ic})^2 = \frac{1}{2} \sum_{i=1}^{n} \|\mathbf{F} - \tilde{\mathbf{Y}}\|_2^2 \tag{29.3}$$

where f_{ic} is the predicted likelihood of protein i with respect to the cth function, $\tilde{y}_{ic}$ is the extended function annotation of protein i with respect to the cth function, and $\mathbf{F} = [\mathbf{f}_1; \mathbf{f}_2; \ldots ; \mathbf{f}_n] \in \mathbb{R}^{n\times C}$ is the prediction for the n proteins.

Since a protein has multiple functions, and the overlap between the function sets of two proteins can be used to measure their similarity, the larger the number of shared functions, the more similar the proteins are. This function-induced similarity between proteins was used successfully by Chi and Hou [7] and Wang et al. [29]. The function annotations of a protein can be used to enrich its feature representation. Thus, we define the function-based similarity matrix $\mathbf{W}^f \in \mathbb{R}^{n\times n}$ between n proteins as follows:

$$\mathbf{W}^f_{ij} = \frac{\mathbf{y}_i^T \mathbf{y}_j}{\|\mathbf{y}_i\| \|\mathbf{y}_j\|} \tag{29.4}$$

Note that $\mathbf{W}^f_{ij}$ measures the pairwise similarity (induced by the function sets of two proteins) between proteins i and j, whereas $\mathbf{M}_{st}$ in Equation 29.2 describes the pairwise function correlations.

We now define a composite similarity $\mathbf{W}$ between pairwise proteins as

$$\mathbf{W} = \mathbf{W}^p + \eta \mathbf{W}^f \tag{29.5}$$

where $\mathbf{W}^p \in \mathbb{R}^{n\times n}$ describes the feature-induced similarity between pairs of proteins. Here, $\mathbf{W}^p$ can be set on the basis of the amino acid sequence similarity of a protein pair (i.e., string kernel [18] for protein sequence data), or by using the frequency of interactions found in multiple PPI studies (i.e., PPI networks in BioGrid[3]), or the weighted pairwise similarity based on reliability scores from all protein identifications by mass spectrometry (e.g., Krogan et al. [17][4]). η is a predefined parameter to balance the trade-off between $\mathbf{W}^p$ and $\mathbf{W}^f$, it is specified as

$$\eta = \frac{\sum_{i=1,j=1}^{n,n} \mathbf{W}^p_{ij}}{\sum_{i=1,j=1}^{n,n} \mathbf{W}^f_{ij}} \tag{29.6}$$

The second part of our objective function leverages the knowledge that proteins with similar amino acid sequences are likely to have similar functions. In other words, we capture the "guilt by association" rule [24], which states that interacting proteins are more likely to share similar functions. This rule is widely used in network-based protein function prediction approaches [7, 24, 25, 35]. As in learning with local and global consistency [38], we include a smoothness term as the second part of our objective function:

$$\Psi_2(\mathbf{f}) = \frac{1}{2} \sum_{i,j=1}^{n} \left\| \frac{\mathbf{f}_i}{\sqrt{\mathbf{D}_{ii}}} - \frac{\mathbf{f}_j}{\sqrt{\mathbf{D}_{jj}}} \right\|^2 \mathbf{W}_{ij}$$

[3]http://thebiogrid.org/.

[4]http://www.nature.com/nature/journal/v440/n7084/suppinfo/nature04670.html.

$$= \text{tr}\,(\mathbf{F}^T(\mathbf{I} - \mathbf{D}^{-(1/2)}\mathbf{W}\mathbf{D}^{-(1/2)})\mathbf{F})$$
$$= \text{tr}\,(\mathbf{F}^T\mathbf{L}\mathbf{F}) \tag{29.7}$$

where $\mathbf{D}$ is a diagonal matrix with $\mathbf{D}_{ii} = \sum_{j=1}^{n} \mathbf{W}_{ij}$. $\mathbf{I}$ is an $n \times n$ identity matrix, $\mathbf{L} = \mathbf{I} - \mathbf{D}^{-(1/2)}\mathbf{W}\mathbf{D}^{-(1/2)}$, and $\text{tr}(\cdot)$ is the matrix trace operation.

Here, we assume the function labels of a protein depend on the feature representation of this protein. We encode this assumption as the third part of our objective function. This assumption is also widely used in most pattern recognition and data mining algorithms, which presume that the label of a sample depends on its features [5]. Each protein is composed of a linear chain of amino acids. Different combinations and organizations of these amino acids result in different proteins and hence different functions. To capture the dependency between the function labels and the features of proteins, we take advantage of the *Hilbert–Schmidt Independence Criterion* (HSIC) [13]. HSIC computes the squared norm of the cross-covariance operator over the feature and label domains in Hilbert Kernel Reproduced Space to estimate the dependency. We choose HSIC because of its computational efficiency, simplicity, and solid theoretical foundation. The empirical estimation of HSIC is given by

$$HSIC\,(\mathbf{F}, \mathbf{Y}) = \frac{\text{tr}(\mathbf{KHSH})}{(\text{n}-1)^2} = \frac{\text{tr}(\mathbf{HKHS})}{(\text{n}-1)^2} \tag{29.8}$$

where $\mathbf{H}, \mathbf{K}, \mathbf{S} \in \mathbb{R}^{n \times n}$, $\mathbf{K}_{ij}$ is used to measure the kernel-induced similarity between two samples i and j, $\mathbf{S}_{ij}$ is the label-induced similarity between them, $\mathbf{H}_{ij} = \delta_{ij} - \frac{1}{n}$, $\delta_{ij} = 1$ if $i = j$, otherwise $\delta_{ij} = 0$. HSIC makes use of kernel matrices to estimate the dependency between labels and features of samples, thus it can also be applied in the case that there is no explicit feature representation for the n samples, as in the case of PPI network data. Although there are many other ways to initialize $\mathbf{K}$ and $\mathbf{S}$, here we just set $\mathbf{K} = \mathbf{W}$ and $\mathbf{S}_{ij} = \mathbf{y}_i^T\mathbf{y}_j$ for its simplicity and its strong empirical performance. Alternative initializations of $\mathbf{K}$ and $\mathbf{S}$ will be investigated in future study.

29.3.1 The Algorithm

By integrating the three objective functions introduced above, we obtain the overall objective function of ProDM:

$$\Psi(\mathbf{F}) = \text{tr}(\mathbf{F}^T\mathbf{L}\mathbf{F}) + \alpha\|\mathbf{F} - \tilde{\mathbf{Y}}\|_2^2 - \beta\text{tr}(\mathbf{HKHF}\mathbf{F}^T) + \gamma\text{tr}(\mathbf{F}^T\mathbf{F}) \tag{29.9}$$

where $\alpha > 0$ and $\beta > 0$ are used to balance the trade-off between the three terms. Our motivation to minimize $\Psi(\mathbf{F})$ is threefold: (i) two proteins with similar sequences (or frequently interacting) should have similar functions, which corresponds to the smoothness assumption in label propagation [38]; (ii) predictions in $\mathbf{F}$ should not change too much from the extended function labels $\tilde{\mathbf{Y}}$; and (iii) the dependency between the function labels and the features of a protein should be maximized. In Equation 29.9, we also add a term $\text{tr}(\mathbf{F}^T\mathbf{F})$ (weighted by $\gamma > 0$) to enforce the sparsity of $\mathbf{F}$, as each function is often associated with a relatively small number of proteins.

ProWL [35] makes use of function correlations and the "guilt by association" rule to replenish the missing functions of partially annotated proteins. In addition, ProDM incorporates the assumption of dependency maximization. ProWL relies on the function correlation matrix **M** to extend the observed function annotations and to define the weight of each function label of a protein (see Eq. (3) in Reference [35]). In contrast, ProDM exploits the function correlations to expand the incomplete function sets. As the number of missing functions increases, the function correlation matrix **M** becomes less reliable [35]. Therefore, when the number of missing functions is large, ProDM outperforms ProWL. In addition, ProWL predicts each function label separately and computes the inverse of a matrix for each label. ProDM, instead, predicts all C labels at once, and computes the inverse of a matrix only once. As a result, ProDM is faster than ProWL. These advantages of ProDM with respect to ProWL are corroborated in the following experiments.

Equation 29.9 can be solved by taking the derivative of $\Psi(\mathbf{F})$ with respect to $\mathbf{F}$:

$$\frac{\partial \Psi(\mathbf{F})}{\partial \mathbf{F}} = 2(\mathbf{LF} + \alpha(\mathbf{F} - \tilde{\mathbf{Y}}) - \beta\mathbf{HKHF} + \gamma\mathbf{F}) \tag{29.10}$$

By setting $\frac{\partial \Phi(\mathbf{F})}{\partial \mathbf{F}} = 0$, we obtain

$$\mathbf{F} = \alpha(\mathbf{L} + \alpha\mathbf{I} - \beta\mathbf{HKH} + \gamma\mathbf{I})^{-1}\tilde{Y} \tag{29.11}$$

In Equation 29.11, the complexity of the matrix multiplication **HKH** is $O(n^3)$ and the complexity of the matrix inverse operation is $O(n^3)$. Thus, the time complexity of ProDM is $O(n^3)$. In practice, however, **L**, **H**, and **K** are all sparse matrices, and Equation 29.11 can be computed more efficiently. In particular, the complexity of sparse matrix multiplication is $O(nm_1)$, where m_1 is the number of nonzero elements in **K**. In addition, instead of computing the inverse of $(\mathbf{L} + \alpha\mathbf{I} - \beta\mathbf{HKH} + \gamma\mathbf{I})$ in Equation 29.11, we can use iterative solvers (i.e., *Conjugate Gradient* (CG)). CG is guaranteed to terminate in n iterations. In each iteration, the most time-consuming operation is the product between an $n \times n$ sparse matrix and a label vector (one column of $\tilde{\mathbf{Y}}$). Thus, in practice, the time complexity of ProDM is $O(m_1 n + tm_2 nC)$, where C is the number of function labels, m_2 is the number of nonzero elements in $(\mathbf{L} + \alpha\mathbf{I} - \beta\mathbf{HKH} + \gamma\mathbf{I})$, and t is the number of CG iterations. CG often terminates in no more than 20 iterations. The ProDM algorithm is described in **Algorithm 1**.

29.4 EXPERIMENTAL SETUP

29.4.1 Data sets

We investigate the performance of ProDM on replenishing missing functions and predicting protein functions on three different PPI benchmarks. The first data set, Saccharomyces cerevisiae *PPI*s (ScPPI), is extracted from BioGrid.[5] We annotate

[5]http://thebiogrid.org/.

Algorithm 29.1 ProDM: Protein Function Prediction using Dependency Maximization

---

Input:$^{\approx}$
 Weight matrix $\mathbf{W}^p$
 Incomplete annotations $\mathbf{Y} = [\mathbf{y}_1, \mathbf{y}_2, \cdots, \mathbf{y}_n]$
 α, β, γ
Output:$^{\approx}$
 1: Predicted likelihood matrix $\mathbf{F}$
 2: Compute $\mathbf{M}$ using Equation 29.2 and $\mathbf{W}^f$ using Equation 29.4.
 3: Calculate $\mathbf{W}$ using $\mathbf{W}^f$ and $\mathbf{W}^p$, and $\mathbf{L} = \mathbf{I} - \mathbf{D}^{-\frac{1}{2}}\mathbf{W}\mathbf{D}^{-\frac{1}{2}}$.
 4: Let $\tilde{\mathbf{Y}} = \mathbf{Y}\mathbf{C}$, initialize $\mathbf{H}$, $\mathbf{K}$ in Equation 29.8.
 Compute $\mathbf{F}$ using Equation 29.11.

---

these proteins according to the FunCat [23] database and use the largest connected component of ScPPI for experiments, which includes 3041 proteins. FunCat organizes function labels in a tree structure. We filtered the function labels and used the 86 informative functions. Informative functions [16, 36] are the ones that have at least 30 proteins as members and within the tree structure these functions do not have a particular descendent node with more than 30 proteins. The weight matrix $\mathbf{W}^p$ of ScPPI is specified by the number of PubMed IDs, where 0 means no interaction between two proteins and $q > 0$ implies the interaction is supported by q distinct PubMed IDs. The second data set, KroganPPI, is obtained from the study of Krogan et al. [17].[6] We use its largest connected component for the experiments and annotate these proteins according to FunCat. After the preprocessing, KroganPPI contains 3642 proteins annotated with 90 informative functions. The weight matrix $\mathbf{W}^p$ of KroganPPI is specified by the provider. The third data set, HumanPPI is obtained from the study of Mostafavi and Morris [20].[7] HumanPPI is extracted from the multiple data types of Human Proteomic data. The proteins in HumanPPI are annotated according to the Gene Ontology [2]. Similarly to References [16, 20], we use the largest connected components of HumanPPI and the functions that have at least 30 annotated proteins. The weight matrix $\mathbf{W}^p$ of HumanPPI is specified by the provider. The characteristics of these processed data sets are listed in Table 29.1.

29.4.2 Comparative Methods

We compare the proposed method with (i) ProWL [35], (ii) WELL [27],[8] (iii) MLR-GL [6],[9] (iv) TMC [33], and (v) CIA [7]. The first three approaches are multi-label learning models with partially labeled data and the last two methods

[6] http://www.nature.com/nature/journal/v440/n7084/suppinfo/nature04670.html.
[7] http://morrislab.med.utoronto.ca/ sara/SW/.
[8] http://lamda.nju.edu.cn/code_WELL.ashx.
[9] http://www.cse.msu.edu/ bucakser/MLR_GL.rar.

Table 29.1 Data set statistics (Avg ± Std means average number of functions for each protein and its standard deviation)

Data Set	#Proteins	#Functions	Avg ± Std
ScPPI	3041	86	2.49 ± 1.70
KroganPPI	3642	90	2.20 ± 1.60
HumanPPI	2950	200	3.80 ± 3.77

are recently proposed protein function prediction algorithms based on multi-label learning and PPI networks. WELL and MLR-GL need an input kernel matrix. We substitute the kernel matrix with $\mathbf{W}^p$, which is semidefinite positive and can be viewed as a Mercer kernel [1]. WELL was proposed to replenish the missing functions of partially annotated proteins. We adopt it here to predict the functions of completely unlabeled proteins by including the unlabeled proteins in the input kernel matrix. MLR-GL is targeted at predicting the functions of completely unlabeled proteins using partially annotated proteins. We adapt it to replenish the missing functions of partially annotated proteins by using all the proteins as training and testing set. Similarly, we also adapt TMC to replenish the missing functions. Owing to the iterative procedure of CIA, it cannot be easily adapted to replenish missing functions. The parameters of WELL, MLR-GL, ProWL, TMC, and CIA are set as the authors specified in their codes, or reported in the papers. For ProDM, we search for optimal α values in the range [0.5, 1] with step size 0.05 and β values in the range [0.01, 0.1] with step size 0.01. In our experiments, we set α and β to 0.99 and 0.01, respectively, because we observed that the performance with respect to the various metrics does not change as we vary α and β around the fixed values. Similarly to ProWL, we set γ to 0.001.

29.4.3 Experimental Protocol

In order to simulate the incomplete annotation scenario, we assume the annotations on the currently labeled proteins are complete and mask some of the ground truth functions. The masked functions are considered missing. For presentation, we define a term called *Incomplete Function* (IF) ratio, which measures the ratio between the number of missing functions and the number of ground truth functions. For example, if a protein has five functions (labels) and two of them are masked (two 1s are changed to two 0s), then the IF ratio is $2/5 = 40\%$.

29.4.4 Evaluation Criteria

Protein function prediction can be viewed as a multi-label learning problem and evaluated using multi-label learning metrics [16, 35]. Various evaluation metrics have been developed for evaluating multi-label learning methods [37]. Here we use five metrics: *MicroF1*, *MacroF1*, *HammingLoss*, *RankingLoss*, and adapted *AUC* [6]. These metrics were also used to evaluate WELL [27], MLR-GL [6], and ProWL

[35]. In addition, we design *Replenishing Accuracy* (RAccuracy) to evaluate the performance of replenishing missing functions. Suppose the predicted function set of n proteins is $\mathcal{F}_p$, the initial incomplete annotated function set is $\mathcal{F}_q$, and the ground truth function set is $\mathcal{Y}$. *RAccuracy* is defined as follows:

$$RAccuracy = \frac{|(\mathcal{Y} - \mathcal{F}_q) \cap \mathcal{F}_p|}{|(\mathcal{Y} - \mathcal{F}_q)|} \tag{29.12}$$

where $|(\mathcal{Y} - \mathcal{F}_q)|$ measures how many functions are missing among n proteins and $|(\mathcal{Y} - \mathcal{F}_q) \cap \mathcal{F}_p|$ counts how many missing functions are correctly replenished.

MicroF1 calculates the $F1$ measure on the predictions of different functions as a whole:

$$MicroF1 = \frac{1}{C} \frac{\sum_{c=1}^{C} 2p_c r_c}{\sum_{c=1}^{C} p_c + r_c} \tag{29.13}$$

where p_c and r_c are the precision and recall of the cth function.

MacroF1 is the average $F1$ scores of different functions:

$$MacroF1 = \frac{1}{C} \sum_{c=1}^{C} \frac{2p_c r_c}{p_c + r_c} \tag{29.14}$$

Hamming loss evaluates how many times an instance–label pair is misclassified:

$$HammLoss = \frac{1}{n} \sum_{i=1}^{n} \frac{\|\mathbf{f}_i \oplus \mathbf{y}_i\|}{C} \tag{29.15}$$

where $\oplus$ stands for the XOR operation and $\mathbf{y}_i$ are the ground-truth functions of protein i. When $HammLoss = 0$, the performance is perfect.

Ranking loss evaluates the average fraction of function label pairs that are not correctly ordered.

$$RankLoss = \frac{1}{n} \sum_{i=1}^{n} \frac{1}{|\mathcal{Y}_i||\overline{\mathcal{Y}}_i|} |\{(c_1, c_2) \in \mathcal{Y}_i \times \overline{\mathcal{Y}}_i | \mathbf{F}(i, c_1) \leq \mathbf{F}(i, c_2)\}| \tag{29.16}$$

where $\mathcal{Y}_i$ is the function set of the ith protein and $\overline{\mathcal{Y}}_i$ is the complement set of $\mathcal{Y}_i$. The performance is perfect when $RankLoss = 0$.

The adapted *Area Under the Curve* (AUC) for multi-label learning was introduced in Reference [6]. AUC first ranks all the functions for each test protein in descending order of their scores; it then varies the number of predicted functions from 1 to the total number of functions, and computes the receiver operating curve score by calculating true and false positive rates for each number of predicted functions.

It finally computes the AUC of all functions to evaluate the multi-label learning methods.

To maintain consistency with other evaluation metrics, we report *1-HammLoss* and *1-RankLoss*. Thus, similarly to other metrics, the higher the values of *1-HammLoss* and *1-RankLoss*, the better the performance.

29.5 EXPERIMENTAL ANALYSIS

In this section, we perform experiments to investigate the performance of ProDM on replenishing the missing functions of n partially labeled proteins, and on predicting the functions of completely unlabeled proteins on the three data sets.

29.5.1 Replenishing Missing Functions

To this end, we consider all the proteins in each data set as training and testing data. To perform comparisons against the other methods, we vary the IF ratio from 30% to 70%, with an interval of 20%. A few proteins in the PPI networks do not have any functions. To make use of the "guilt by association" rule and keep the PPI network connected, we do not remove them and test the performance of replenishing missing functions on the proteins with annotations. We repeat the experiments 20 times with respect to each IF ratio. In each run, the missing functions are randomly masked for each protein according to the IF ratio. $\mathbf{F} \in \mathbb{R}^{n \times C}$ in Equation 29.10 is a predicted likelihood matrix. *MicroF1*, *MacroF1*, *1-HammLoss*, and *RAccuracy* require $\mathbf{F}$ to be a binary indicator matrix. Here, we consider the functions corresponding to the r largest values of $\mathbf{f}_i$ as the functions of the ith protein, where r is determined by the number of ground-truth functions of this protein. To simulate the incomplete annotation scenario, we assume the given functions of the ith protein in a data set are ground-truth functions, and mask some of them to generate the missing functions. The experimental results are reported in Tables 29.2–29.4. In these tables, best and comparable results are in boldface (statistical significance is examined via pairwise t-test at 95% significance level).

From Tables 29.2–29.4, we can observe that ProDM performs much better than the competitive methods in replenishing the missing functions of proteins across all the metrics. Both ProDM and ProWL take advantage of function correlations and of the "guilt by association" rule, but ProDM significantly outperforms ProWL. The difference in performance between ProDM and ProWL confirms our intuition that maximizing the dependency between functions and features of proteins is effective. The performance of WELL is not comparable to that of ProDM. The possible reason is that the assumptions used in WELL may be not suitable for the PPI network data sets. The performance of MLR-GL varies because it is targeted at predicting functions of unlabeled proteins using partially annotated proteins, whereas here it is adapted for replenishing missing functions. TMC is introduced to predict functions for completely unlabeled proteins using completely labeled ones; TMC sometimes outperforms ProWL and WELL. This happens because the missing functions can be appended in the bi-relation graph. In fact, TMC also makes use of function correlations and

Table 29.2 Results of replenishing missing functions on ScPPI

Metric	IF Ratio (%)	ProDM	ProWL	WELL	MLR-GL	TMC
RAccuracy	30	**49.24 ± 1.28**	38.05 ± 1.07	23.94 ± 1.55	46.18 ± 1.04	46.01 ± 1.52
	50	**46.57 ± 0.71**	32.14 ± 0.92	18.83 ± 1.01	35.59 ± 0.91	42.46 ± 0.76
	70	**44.18 ± 1.03**	31.41 ± 1.03	17.12 ± 0.12	33.89 ± 0.74	41.42 ± 0.82
MicroF1	30	**93.88 ± 0.12**	86.28 ± 0.14	60.49 ± 0.54	23.67 ± 0.50	91.80 ± 0.20
	50	**79.09 ± 0.28**	68.36 ± 0.36	47.42 ± 0.74	26.98 ± 0.49	77.09 ± 0.28
	70	**71.67 ± 0.51**	60.09 ± 0.51	42.06 ± 0.04	27.15 ± 0.59	69.79 ± 0.44
MacroF1	30	**94.05 ± 0.18**	86.28 ± 0.18	55.35 ± 0.52	24.06 ± 0.79	90.98 ± 0.24
	50	**78.39 ± 0.33**	67.81 ± 0.36	43.80 ± 0.55	27.45 ± 0.72	74.72 ± 0.35
	70	**70.05 ± 0.45**	59.45 ± 0.62	38.25 ± 0.87	27.98 ± 0.72	67.34 ± 0.52
1-HammLoss	30	**99.65 ± 0.01**	99.20 ± 0.01	97.71 ± 0.03	95.58 ± 0.03	99.52 ± 0.01
	50	**98.79 ± 0.02**	98.17 ± 0.02	96.95 ± 0.04	95.77 ± 0.03	98.67 ± 0.02
	70	**98.36 ± 0.03**	97.69 ± 0.03	96.64 ± 0.00	95.78 ± 0.03	98.25 ± 0.03
1-RankLoss	30	**99.67 ± 0.02**	95.16 ± 0.02	94.78 ± 0.07	44.38 ± 0.39	99.65 ± 0.02
	50	96.80 ± 0.12	91.95 ± 0.24	90.41 ± 0.24	41.43 ± 0.66	**97.06 ± 0.10**
	70	**94.92 ± 0.17**	88.03 ± 0.24	89.01 ± 0.26	38.06 ± 0.77	94.52 ± 0.29
AUC	30	**98.79 ± 0.05**	94.92 ± 0.04	93.09 ± 0.04	55.63 ± 0.38	**98.77 ± 0.04**
	50	95.63 ± 0.14	92.07 ± 0.16	88.24 ± 0.24	54.01 ± 0.66	**95.97 ± 0.10**
	70	**93.09 ± 0.22**	88.85 ± 0.20	86.08 ± 0.35	52.60 ± 0.46	**93.04 ± 0.29**

Table 29.3 Results of replenishing missing functions on KroganPPI

Metric	IF Ratio (%)	ProDM	ProWL	WELL	MLR-GL	TMC
RAccuracy	30	**44.97 ± 1.63**	14.90 ± 0.98	9.24 ± 0.66	30.90 ± 1.48	23.89 ± 1.30
	50	**42.20 ± 0.63**	11.04 ± 0.77	7.03 ± 0.22	23.83 ± 0.71	27.89 ± 0.61
	70	**36.25 ± 0.75**	14.89 ± 0.61	7.68 ± 0.44	21.69 ± 0.80	27.06 ± 0.65
MicroF1	30	**95.51 ± 0.13**	93.05 ± 0.08	61.04 ± 0.27	14.78 ± 0.23	88.67 ± 0.12
	50	**79.46 ± 0.22**	68.39 ± 0.27	48.54 ± 0.67	16.18 ± 0.29	70.93 ± 0.22
	70	**70.23 ± 0.35**	60.25 ± 0.29	43.72 ± 0.19	16.09 ± 0.34	61.82 ± 0.31
MacroF1	30	**95.70 ± 0.18**	94.57 ± 0.15	58.24 ± 0.20	13.71 ± 0.28	88.41 ± 0.12
	50	**78.92 ± 0.25**	71.51 ± 0.32	52.09 ± 1.08	15.12 ± 0.34	69.20 ± 0.33
	70	**69.01 ± 0.40**	62.30 ± 0.46	48.79 ± 0.52	14.92 ± 0.35	60.20 ± 0.44
1-HammLoss	30	**99.78 ± 0.01**	99.66 ± 0.00	98.08 ± 0.01	95.81 ± 0.01	99.44 ± 0.01
	50	**98.99 ± 0.01**	98.44 ± 0.01	97.47 ± 0.03	95.87 ± 0.01	98.57 ± 0.01
	70	**98.53 ± 0.02**	98.04 ± 0.01	97.23 ± 0.01	95.87 ± 0.02	98.12 ± 0.02
1-RankLoss	30	**99.75 ± 0.02**	99.61 ± 0.02	96.50 ± 0.03	39.88 ± 0.37	99.52 ± 0.02
	50	**96.87 ± 0.12**	94.55 ± 0.12	91.60 ± 0.09	39.99 ± 0.27	96.20 ± 0.16
	70	**94.37 ± 0.14**	91.02 ± 0.25	89.89 ± 0.06	38.48 ± 0.39	93.28 ± 0.19
AUC	30	**98.87 ± 0.04**	98.58 ± 0.04	94.90 ± 0.05	45.49 ± 0.28	98.59 ± 0.05
	50	**95.47 ± 0.12**	92.55 ± 0.15	88.88 ± 0.14	46.65 ± 0.32	94.63 ± 0.18
	70	**91.91 ± 0.16**	86.90 ± 0.35	85.87 ± 0.10	46.45 ± 0.37	90.58 ± 0.24

Table 29.4 Results of replenishing missing functions on HumanPPI

Metric	IF Ratio (%)	ProDM	ProWL	WELL	MLR-GL	TMC
RAccuracy	30	**80.39 ± 0.80**	71.86 ± 0.79	20.50 ± 0.59	30.92 ± 1.09	53.35 ± 0.83
	50	**73.14 ± 0.96**	46.78 ± 0.55	18.23 ± 0.62	23.92 ± 0.56	48.66 ± 0.63
	70	**63.28 ± 0.97**	32.76 ± 0.53	15.09 ± 0.82	21.41 ± 0.45	45.36 ± 0.55
MicroF1	30	**96.60 ± 0.14**	95.12 ± 0.14	86.21 ± 0.10	15.76 ± 0.30	91.90 ± 0.15
	50	**88.48 ± 0.41**	77.18 ± 0.24	64.93 ± 0.26	16.36 ± 0.21	77.98 ± 0.27
	70	**79.20 ± 0.55**	61.91 ± 0.30	51.91 ± 0.46	16.10 ± 0.29	69.05 ± 0.31
MacroF1	30	**96.21 ± 0.16**	94.76 ± 0.16	87.95 ± 0.03	15.79 ± 0.27	91.43 ± 0.15
	50	**87.49 ± 0.46**	76.86 ± 0.30	70.43 ± 0.18	16.00 ± 0.26	77.05 ± 0.31
	70	**77.58 ± 0.53**	62.19 ± 0.30	59.05 ± 0.37	15.45 ± 0.26	67.67 ± 0.35
1-HammLoss	30	**99.87 ± 0.01**	99.81 ± 0.01	99.48 ± 0.00	96.80 ± 0.01	99.69 ± 0.01
	50	**99.56 ± 0.02**	99.13 ± 0.01	98.67 ± 0.01	96.82 ± 0.01	99.16 ± 0.01
	70	**99.21 ± 0.02**	98.55 ± 0.01	98.17 ± 0.02	96.82 ± 0.01	98.83 ± 0.01
1-RankLoss	30	**99.81 ± 0.02**	99.74 ± 0.03	97.19 ± 0.03	54.78 ± 0.32	99.73 ± 0.02
	50	**98.73 ± 0.07**	96.90 ± 0.21	87.55 ± 0.44	58.09 ± 0.29	98.31 ± 0.12
	70	**97.50 ± 0.15**	93.56 ± 0.41	83.97 ± 0.08	58.35 ± 0.36	96.76 ± 0.21
AUC	30	**98.65 ± 0.04**	98.52 ± 0.05	93.51 ± 0.13	54.32 ± 0.22	98.44 ± 0.04
	50	**97.37 ± 0.09**	95.86 ± 0.15	83.05 ± 0.20	55.90 ± 0.21	96.82 ± 0.10
	70	**95.48 ± 0.14**	91.31 ± 0.28	76.12 ± 0.48	55.69 ± 0.26	94.64 ± 0.18

the "guilt by association" rule, but it still loses to ProDM. The reason is that ProDM maximizes the dependency between proteins' functions and features. The margin in performance achieved by ProDM with respect to ProWL and TMC demonstrates the effectiveness of using *dependency maximization* in replenishing the missing functions of proteins.

We also observe that, as more functions are masked, ProWL downgrades much more rapidly than ProDM. As the IF ratio increases, the function correlation matrix **M** becomes less reliable. ProWL uses **M** to estimate the likelihood of missing functions and to weigh the loss function. ProDM only utilizes **M** to estimate the probability of missing functions and makes additional use of dependency maximization. Thus ProDM is less dependent on **M**. Taking *RAccuracy* on ScPPI as an example, ProDM on average is 33.55% better than ProWL, 49.60% better than WELL, 19.31% better than MLR-GL, and 8.21% better than TMC. These results confirm the effectiveness of ProDM in replenishing the missing functions. Overall, these experimental results confirm the advantages of combining the "guilt by association" rule, function correlations, and dependency maximization.

29.5.2 Predicting Unlabeled Proteins

We conduct here another set of experiments to study the performance of ProDM in predicting the function of completely unlabeled proteins using partially labeled ones. In this scenario, $l < n$ proteins are partially annotated and $n - l$ proteins are completely unlabeled. At first, we partition each data set into a *training* set (accounting for 80% of all the proteins) with partial annotations and into a *testing* set (accounting for the remaining 20% of all the proteins) with no annotations. We run the experiments 20 times for each data set. In each round, the data set is randomly divided into training and testing data sets. We simulate the setting of missing functions (IF ratio = 50%) in the training set as in the experiments in Section 29.5.1, but r is determined as the average number of functions (round to the next integer) of all proteins. From Table 29.1: r is set to 3 for ScPPI and KroganPPI, and to 4 for HumanPPI. The results (average of 20 independent runs) are listed in Tables 29.5–29.7. Since *RAccuracy* is not suitable for the settings of predicting completely unlabeled proteins, the results for this metric are not reported.

From Tables 29.5–29.7, we can observe that ProDM achieves the best (or comparable to the best) performance among all the comparing methods on various evaluation metrics. ProDM and ProWL have similar performance in the task of predicting the functions of completely unlabeled proteins. One possible reason is that **F** is initially set to $\tilde{\mathbf{Y}}$ and $\{\tilde{\mathbf{y}}_j\}_{j=l+1}^{n}$ are zero vectors. WELL works better than MLR-GL in replenishing the missing functions and it loses to MLR-GL in predicting the functions of unlabeled proteins. One possible cause is that WELL is targeted at replenishing missing functions and here it is adjusted to predict functions on completely unlabeled proteins. MLR-GL predicts protein functions under the assumption of partially annotated proteins and it is outperformed by ProDM. MLR-GL optimizes the ranking loss and the group Lasso loss, whereas ProDM optimizes an objective function based on the function correlations, the "guilt by association" rule, and the dependency between

Table 29.5 Prediction results on completely unlabeled proteins of ScPPI

Metric	ProDM	ProWL	WELL	MLR-GL	TMC	CIA
MicroF1	**32.78 ± 1.37**	30.06 ± 1.15	16.75 ± 2.03	24.15 ± 1.40	3.67 ± 0.38	20.78 ± 0.38
MacroF1	**31.91 ± 1.48**	**31.33 ± 1.74**	5.19 ± 0.71	26.25 ± 1.50	2.00 ± 0.39	26.27 ± 0.39
1-HammLoss	**95.73 ± 0.10**	95.56 ± 0.09	94.69 ± 0.16	95.19 ± 0.09	93.89 ± 0.05	94.96 ± 0.05
1-RankLoss	**73.13 ± 2.72**	60.37 ± 1.64	73.57 ± 0.05	41.56 ± 1.06	28.29 ± 0.70	21.82 ± 0.70
AUC	**78.40 ± 1.57**	**78.63 ± 0.74**	77.00 ± 0.53	61.47 ± 1.26	55.72 ± 0.84	63.38 ± 0.84

Table 29.6 Prediction results on completely unlabeled proteins of KroganPPI

Metric	ProDM	ProWL	WELL	MLR-GL	TMC	CIA
MicroF1	**22.55 ± 1.35**	**22.40 ± 0.97**	14.35 ± 1.25	13.58 ± 0.86	3.32 ± 0.52	13.78 ± 0.52
MacroF1	**18.26 ± 1.53**	**17.68 ± 1.11**	1.47 ± 0.30	12.80 ± 0.92	2.05 ± 0.41	13.85 ± 0.41
1-HammLoss	**96.40 ± 0.08**	**96.40 ± 0.08**	96.04 ± 0.03	95.99 ± 0.07	95.52 ± 0.06	95.99 ± 0.06
1-RankLoss	66.69 ± 1.19	**75.41 ± 0.88**	**75.43 ± 0.22**	48.40 ± 1.13	61.26 ± 0.89	18.43 ± 0.89
AUC	72.26 ± 0.73	**74.78 ± 0.73**	**74.16 ± 0.12**	58.80 ± 1.10	61.35 ± 0.68	59.45 ± 0.68

Table 29.7 Prediction results on completely unlabeled proteins of HumanPPI

Metric	ProDM	ProWL	WELL	MLR-GL	TMC	CIA
MicroF1	**24.57 ± 1.03**	23.18 ± 1.24	16.43 ± 1.78	12.87 ± 0.76	1.91 ± 0.28	12.86 ± 0.28
MacroF1	**20.58 ± 1.18**	19.32 ± 0.90	15.55 ± 1.30	11.95 ± 0.78	1.61 ± 0.26	9.90 ± 0.26
1-HammLoss	**97.17 ± 0.05**	97.11 ± 0.09	96.85 ± 0.10	96.73 ± 0.05	96.33 ± 0.07	96.72 ± 0.07
1-RankLoss	**76.70 ± 1.07**	**76.64 ± 2.01**	62.98 ± 1.82	67.89 ± 1.44	50.93 ± 0.77	33.87 ± 0.77
AUC	**78.82 ± 1.19**	77.41 ± 0.92	62.30 ± 1.38	66.23 ± 0.85	51.78 ± 1.21	67.08 ± 1.21

the function labels and the features of proteins. We can claim that ProDM is more faithful to the characteristics of proteomic data than MLR-GL. For the same reasons, ProDM often outperforms WELL, which takes advantage of low density separation and low-rank-based similarity to capture function correlations and data distribution.

TMC sometimes performs similarly to ProDM in the task of replenishing the missing functions. However, TMC is outperformed by other methods when making predictions for completely unlabeled proteins. A possible reason is that TMC assumes the training proteins are fully annotated, and the estimated function correlation matrix $\mathbf{M}$ may be unreliable when IF ratio is set to 50%. CIA also exploits function-based similarity and PPI networks to predict protein functions but it is always outperformed by ProDM and by ProWL. There are two possible reasons for this. First, CIA does not account for the weights of interaction between two proteins. Second, CIA mainly relies on the function-induced similarity $\mathbf{W}^f$, and when training proteins are partially annotated, $\mathbf{W}^f$ becomes less reliable. CIA performs better than TMC. One reason might be that CIA exploits a neighborhood count algorithm [24] to initialize the functions on unlabeled proteins in the kick-off step of CIA, whereas TMC does not. All these results show the effectiveness of ProDM in predicting unlabeled proteins by considering the partial annotations on proteins.

29.5.3 Component Analysis

To investigate the benefit of using the "guilt by association" rule and of exploiting function correlations, we introduce two variants of ProDM, namely, ProDM_nGBA and ProDM_nFC. ProDM_nGBA corresponds to ProDM without using the "guilt by association" rule. ProDM_nGBA is based on Equation 29.8 without the first term; that is, ProDM_nGBA uses only the partial annotations and function correlations to replenish the missing functions. ProDM_nFC corresponds to Protein function prediction using Dependency Maximization with no Function Correlation. In ProDM_nFC, $\mathbf{Y}$ is used in Equation 29.8 instead of $\tilde{\mathbf{Y}}$. We increase the IF ratio from 10% to 90% at intervals of 10%, and record the results of ProDM, ProDM_nGBA, and ProDM_nFC with respect to each IF ratio. For brevity, in Figure 29.2 we just report the results with respect to *MicroF1* and *AUC* on HumanPPI.

From Figure 29.2, we can observe that ProDM, ProDM_nGBA, and ProDM_nFC have similar performance when few functions are missing. This indicates that both the "guilt by association" rule rule and function correlations can be utilized to replenish the missing functions. However, as the number of missing function increases, ProDM generally outperforms ProDM_nGBA and ProDM_nFC. The reason is that ProDM, unlike ProDM_nGBA and ProDM_nFC, makes use of *both* the "guilt by association" rule rule and function correlations. This fact shows that it is important and reasonable to integrate these two components in replenishing missing functions.

29.5.4 Run Time Analysis

In Table 29.8, we record the average run time of each of the methods on the three data sets. The experiments were conducted on Windows 7 platform with Intel E31245

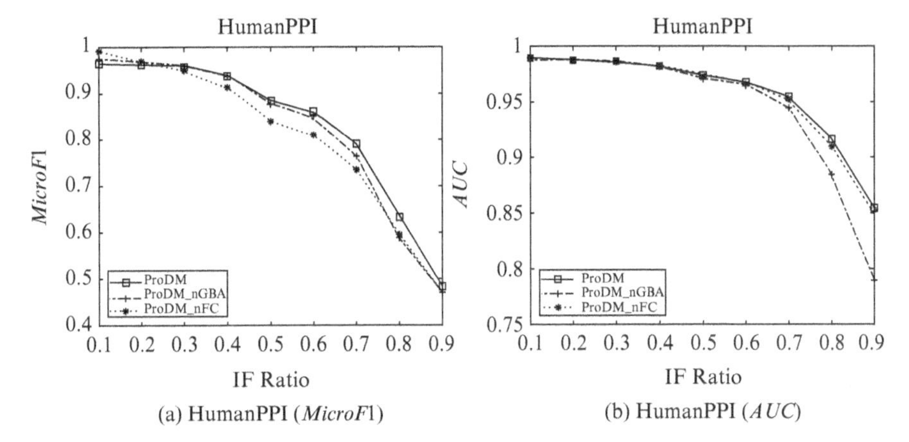

Figure 29.2 The benefit of using both the "guilt by association" rule and function correlations (ProDM_nFC is ProDM with no function correlation, and ProDM_nGBA is ProDM with no "guilt by association" rule).

Table 29.8 Runtime analysis (s)

Data Set	ProDM	ProWL	WELL	MLR-GL	TMC
ScPPI	60.77	83.09	1687.09	22.66	2.29
KroganPPI	80.60	134.94	3780.24	32.40	3.62
HumanPPI	64.02	194.62	5445.97	50.68	3.49
Total	178.37	412.65	10,913.30	105.74	9.40

processor and 16 GB memory. TMC assumes the training proteins are accurately annotated, and it takes much less time than the other methods. MLR-GL relaxes the convex–concave optimization problem into a *Second-Order Cone Programming* (SOCP) [6] problem, and it ranks second (from fast to slow). ProDM takes less time than ProWL, as ProDM infers the functions of a protein in one step, whereas ProWL divides the prediction into C subproblems. WELL uses eigen-decomposition and convex optimization and it costs much more than the other methods. As such, it is desirable to use ProDM for protein function prediction.

29.6 CONCLUSIONS

In this chapter, protein function prediction using partially annotated proteins is investigated and the ProDM method is introduced. ProDM integrates the maximization of dependency between features and function labels of proteins, the "guilt by association" rule rule, and function correlations to replenish the missing functions of partially annotated proteins, and to predict the functions of completely unlabeled proteins. The empirical study on three PPI networks data sets shows that the introduced ProDM performs significantly better than the competitive methods. In addition, the experimental results demonstrate the benefit of integrating the "guilt by association" rule rule, function correlations, and dependency maximization in protein function prediction. In future work, we plan to incorporate more background information of proteomic data and hierarchical structure among function labels for protein function prediction.

ACKNOWLEDGMENTS

We thank the reviewers for helpful comments on improving this chapter. This work is partially supported by NSF IIS-0905117 and NSF Career Award IIS-1252318, NSFC (Nos 61101234, 61402378), Natural Science Foundation of CQ CSTC (No. cstc2014jcyjA40031), Fundamental Research Funds for the Central Universities of China (Nos XDJK2014C044 and 2362015XK07), Doctoral Fund of Southwest University (No. SWU113034), and *China Scholarship Council* (CSC).

REFERENCES

1. Aizerman A, Braverman EM, Rozoner LI. Theoretical foundations of the potential function method in pattern recognition learning. Autom Remote Control 1964;25:821–837.
2. Ashburner M, Ball CA, Blake JA, Botstein D, Butler H, Cherry JM, Davis AP, Dolinski K, Dwight SS, Eppig JT, Harris MA, Hill DP, Issel-Tarver L, Kasarskis A, Lewis S, Matese JC, Richardson JE, Ringwald M, Rubin GM, Sherlock G. Gene ontology: tool for the unification of biology. Nat Genet 2000;25(1):25–29.
3. Barutcuoglu Z, Schapire RE, Troyanskaya OG. Hierarchical multi-label prediction of gene function. Bioinformatics 2006;22(7):830–836.
4. Belkin M, Niyogi P, Sindhwani V. Manifold regularization: a geometric framework for learning from labeled and unlabeled examples. J Mach Learn Res 2006;7:2399–2434.
5. Bishop CM. *Pattern Recognition and Machine Learning*. Berlin, German: Springer-Verlag; 2006.
6. Bucak SS, Jin R, Jain AK. Multi-label learning with incomplete class assignments. Proceedings of 24th IEEE Conference on Computer Vision and Pattern Recognition; Colorado Springs, CO: 2011. p 2801–2808.
7. Chi X, Hou J. An iterative approach of protein function prediction. BMC Bioinformatics 2011;12(1):437.
8. Chua HN, Sung WK, Wong L. Exploiting indirect neighbours and topological weight to predict protein function from protein-protein interactions. Bioinformatics 2006;22(13):1623–1630.
9. Cohen AM, Hersh WR. A survey of current work in biomedical text mining. Brief Bioinform 2005;6(1):57–71.
10. Cour T, Sapp B, Taskar B. Learning from partial labels. J Mach Learn Res 2011;12:1501–1536.

11. Elisseeff A, Weston J. A Kernel Method for Multi-Labelled Classification In: *Proceedings of Advances in Neural Information Processing Systems*, Volume 14; 2001. Vancouver, Canada: MIT Press. p. 681–687.
12. Fraser AG, Marcotte EM. A probabilistic view of gene function. Nat Genet 2004; 36(6):559–564.
13. Gretton A, Bousquet O, Smola A, Schölkopf B. Measuring statistical dependence with Hilbert-Schmidt norms. Algorithmic Learning Theory; Berlin, German: Springer-Verlag; 2005. p 63–77.
14. Hu H, Yan X, Huang Y, Han J, Zhou X. Mining coherent dense subgraphs across massive biological networks for functional discovery. Bioinformatics 2005;21 Suppl 1:i213–i221.
15. Jiang J. Learning protein functions from Bi-relational graph of proteins and function annotations. Algorithms in Bioinformatics; Berlin, German: Springer-Verlag; 2011. p 128–138.
16. Jiang JQ, McQuay LJ. Predicting protein function by multi-label correlated semi-supervised learning. IEEE/ACM Trans Comput Biol Bioinform 2012;9(4): 1059–1069.
17. Krogan NJ, Cagney G, Yu H, Zhong G, Guo X, Ignatchenko A, Li J, Pu S, Datta N, Tikuisis AP, Punna T, Peregrín-Alvarez JM, Shales M, Zhang X, Davey M, Robinson MD, Paccanaro A, Bray JE, Sheung A, Beattie B, Richards DP, Canadien V, Lalev A, Mena F, Wong P, Starostine A, Canete MM, Vlasblom J, Wu S, Orsi C, Collins SR, Chandran S, Haw R, Rilstone JJ, Gandi K, Thompson NJ, Musso G, St. Onge P, Ghanny S, Lam MH, Butland G, Altaf-Ul AM, Kanaya S, Shilatifard A, O'Shea E, Weissman JS, Ingles CJ, Hughes TR, Parkinson J, Gerstein M, Wodak SJ, Emili A, Greenblatt JF. Global landscape of protein complexes in the yeast Saccharomyces cerevisiae. Nature 2006;440(7084):637–643.
18. Leslie CS, Eskin E, Cohen A, Weston J, Noble WS. Mismatch string kernels for discriminative protein classification. Bioinformatics 2004;20(4):467–476.
19. Lin C, Cho Y, Hwang W, Pei P, Zhang A. Clustering methods in protein-protein interaction network. In: *Knowledge Discovery in Bioinformatics: Techniques, Methods and Application*. Hoboken, NJ: Wiley; 2007. p 1–35.
20. Mostafavi S, Morris Q. Fast integration of heterogeneous data sources for predicting gene function with limited annotation. Bioinformatics 2010;26(14):1759–1765.
21. Pandey G, Kumar V, Steinbach M. *Computational approaches for protein function prediction*. Technical Report 06-028. Twin Cities (MN): Department of Computer Science and Engineering, University of Minnesota; 2006.
22. Pandey G, Myers CL, Kumar V. Incorporating functional inter-relationships into protein function prediction algorithms. BMC Bioinformatics 2009;10(1):142.
23. Ruepp A, Zollner A, Maier D, Albermann K, Hani J, Mokrejs M, Tetko I, Güldener U, Mannhaupt G, Münsterkötter M, Mewes HW. The FunCat, a functional annotation scheme for systematic classification of proteins from whole genomes. Nucleic Acids Res 2004;32(18):5539–5545.
24. Schwikowski B, Uetz P, Fields S. A network of protein-protein interactions in yeast. Nat Biotechnol 2000;18(12):1257–1261.
25. Sharan R, Ulitsky I, Shamir R. Network-based prediction of protein function. Mol Syst Biol 2007;3(88):1–13.
26. Shih YK, Parthasarathy S. Identifying functional modules in interaction networks through overlapping markov clustering. Bioinformatics 2012;28(18):i473–i479.

27. Sun Y, Zhang Y, Zhou Z. Multi-label learning with weak label. Proceedings of 24th AAAI Conference on Artificial Intelligence; San Francisco, CA; 2010. p 593–598.
28. Valentini G. *Hierarchical Ensemble Methods for Protein Function Prediction*. New York, NY; ISRN Bioinformatics; 2014.
29. Wang H, Huang H, Ding C. Function-function correlated multi-label protein function prediction over interaction networks. Proceedings of the 16th International Conference on Research in Computational Molecular Biology; Berlin, German: Springer-Verlag; 2012. p 302–313.
30. Weston J, Leslie C, Ie E, Zhou D, Elisseeff A, Noble WS. Semi-supervised protein classification using cluster kernels. Bioinformatics 2005;21(15):3241–3247.
31. Xiong H, He XF, Ding C, Zhang Y, Kumar V, Holbrook SR. Identification of functional modules in protein complexes via hyperclique pattern discovery. Proceedings of Pacific Symposium on Biocomputing; Big Island of Hawaii, USA 2005. p 221–232.
32. Yang S, Jiang Y, Zhou Z. Multi-instance multi-label learning with weak label. Proceedings of the 23rd International Joint Conference on Artificial Intelligence; Beijing, China 2013. p 1869–1875.
33. Yu G, Domeniconi C, Rangwala H, Zhang G, Yu Z. Transductive multi-label ensemble classification for protein function prediction. Proceedings of the 18th ACM SIGKDD International Conference on Knowledge Discovery and Data Mining; Beijing, China 2012. p 1077–1085.
34. Yu G, Rangwala H, Domeniconi C, Zhang G, Zhang Z. Protein function prediction by integrating multiple kernels. Proceedings of the 23rd International Joint Conference on Artificial Intelligence; Beijing, China 2013. p 1869–1875.
35. Yu G, Zhang G, Rangwala H, Domeniconi C, Yu Z. Protein function prediction using weak-label learning. Proceedings of the ACM Conference on Bioinformatics, Computational Biology and Biomedicine; Orlando, FL 2012. p 202–209.
36. Zhang X, Dai D. A framework for incorporating functional interrelationships into protein function prediction algorithms. IEEE/ACM Trans Comput Biol Bioinform 2012;9(3):740–753.
37. Zhang M, Zhou Z. A review on multi-label learning algorithms. IEEE Trans Knowl Data Eng 2014;26(8):1819–1837.
38. Zhou D, Bousquet O, Lal TN, Weston J, Schölkopf B. Learning with local and global consistency In: *Proceedings of Advances in Neural Information Processing Systems*; Vancouver, Canada 2003. p 321–328.

INDEX

Note: Page numbers in italics denote figures and bolded page numbers denote tables

Pattern Recognition in Computational Molecular Biology: Techniques and Approaches,
First Edition. Edited by Mourad Elloumi, Costas S. Iliopoulos, Jason T. L. Wang, and Albert Y. Zomaya.

Wiley Series on

Bioinformatics: Computational Techniques and Engineering

Bioinformatics and computational biology involve the comprehensive application of mathematics, statistics, science, and computer science to the understanding of living systems. Research and development in these areas require cooperation among specialists from the fields of biology, computer science, mathematics, statistics, physics, and related sciences. The objective of this book series is to provide timely treatments of the different aspects of bioinformatics spanning theory, new and established techniques, technologies and tools, and application domains. This series emphasizes algorithmic, mathematical, statistical, and computational methods that are central in bioinformatics and computational biology.

Series Editors: **Professor Yi Pan** and **Professor Albert Y. Zomaya**
pan@cs.gsu.edu albert.zomaya@sydney.edu.au

Knowledge Discovery in Bioinformatics: Techniques, Methods, and Applications
Xiaohua Hu and Yi Pan

Grid Computing for Bioinformatics and Computational Biology
Edited by El-Ghazali Talbi and Albert Y. Zomaya

Bioinformatics Algorithms: Techniques and Applications
Ion Mandiou and Alexander Zelikovsky

Machine Learning in Bioinformatics
Yanqing Zhang and Jagath C. Rajapakse

Biomolecular Networks: Methods and Applications in Systems Biology
Luonan Chen, Rui-Sheng Wang, and Xiang-Sun Zhang

Computational Systems Biology
Huma Lodhi

Analysis of Biological Networks
Edited by Björn H. Junker and Falk Schreiber

Computational Intelligence and Pattern Analysis in Biological Informatics
Edited by Ujjwal Maulik, Sanghamitra Bandyopadhyay, and Jason T. L. Wang

Mathematics of Bioinformatics: Theory, Practice, and Applications
Matthew He and Sergey Petoukhov

Introduction to Protein Structure Prediction: Methods and Algorithms
Huzefa Rangwala and George Karypis

Algorithms in Computational Molecular Biology: Techniques, Approaches and Applications
Edited by Mourad Elloumi and Albert Y. Zomaya

Mathematical and Computational Methods in Biomechanics of Human Skeletal Systems: An Introduction
Jiří Nedoma, Jiří Stehlík, Ivan Hlaváček, Josef Daněk, Taťjana Dostálová, and Petra Přečková

Rough-Fuzzy Pattern Recognition: Applications in Bioinformatics and Medical Imaging
Pradipta Maji and Sankar K. Pal

Data Management of Protein Interaction Networks
Mario Cannataro and Pietro Hiram Guzzi

Algorithmic and Artificial Intelligence Methods for Protein Bioinformatics
Yi Pan, Jianxin Wang, and Min Li

Classification Analysis of DNA Microarrays
Leif E. Petersen

Biological Knowledge Discovery Handbook: Processing, Mining, and Postprocessing of Biological Data
Edited by Mourad Elloumi and Albert Y. Zomaya

Pattern Recognition in Computational Molecular Biology: Techniques and Approaches
Mourad Elloumi, Costas S Iliopoulos, Jason T.L Wang, and Albert Y. Zomaya